The Pesticide Guide 2005

Editor: R. Whitehead BA, MSc

CABI Publishing (a division of CAB International) is one of the world's foremost publishers of databases, books and journals in agriculture, and applied life sciences. It has a worldwide reputation for producing high quality, value-added information, drawing on its links with the scientific community. From its Headquarters in Wallingford, UK, CABI Publishing runs a worldwide operation, distributing books, journals and electronic products to customers in over 150 countries, and selling its products through a network of international agents. For further information, please contact CABI Publishing, CAB International, Wallingford, Oxon OX10 8DE, UK.

Telephone: (01491) 832111
Fax: (01491) 833508
e-mail: publishing@cabi.org
Web: www.cabi-publishing.org

BCPC (British Crop Production Council) is a self-supporting limited company with charity status, which was formed in 1968 to promote the knowledge and understanding of the science and practice of crop protection. The corporate members include government departments and research councils; advisory services; associations concerned with the farming industry; agrochemical manufacturers; agricultural engineering, agricultural contracting and distribution services; universities; scientific societies; organisations concerned with the environment; and some experienced independent members. Further details available from BCPC, 7 Omni Business Centre, Omega Park, Alton, Hampshire GU34 2QD, UK.

Telephone: (01420) 593200
Fax: (01420) 593209
e-mail: md@bcpc.org
Web: www.bcpc.org

ISBN 0 85199 024 X

Typeset and printed by Page Bros, Norwich, UK

Contents

Disclaimer v
The Dangerous Preparations Directive v
Editor's Note vi
The Voluntary Initiative vii
Changes Since 2004 edition viii
About *The UK Pesticide Guide* xv
How To Use *The UK Pesticide Guide* xvii

SECTION 1 CROP/PEST GUIDE 1
Crop/Pest Guide Index 3
Crop/Pest Guide 5

SECTION 2 PESTICIDE PROFILES 65

SECTION 3 PRODUCTS ALSO REGISTERED 449

SECTION 4 ADJUVANTS 511

SECTION 5 USEFUL INFORMATION 533
Pesticide Legislation 535
The Food and Environment Protection Act 1985 (FEPA) and Control of Pesticides Regulations 1986 (COPR) 535
The Plant Protection Products Directive 535
The Review Programme 536
Derogations for 'Essential Uses' Permitted until 31 December 2007 536
The Control of Substances Hazardous to Health Regulations 1988 (COSHH) 541
Certificates of Competence – the roles of BASIS and NPTC 541
Maximum Residue Levels 542
Approval (On-label and Off-label) 543
Statutory Conditions of Use 543
Types of Approval 543
Withdrawal of Approval 544
Approval of Commodity Substances 544
Off-label Extension of Use 544
Long-Term Arrangements for Extension of Use 545
Specific restrictions for extension of use 545
Using Crop Protection Chemicals 551
Use of Herbicides In or Near Water 551
Use of Pesticides in Forestry 552
Pesticides Used as Seed Treatments 554
Aerial Application of Pesticides 557
Resistance Management 560
Preparation in advance 560
Using crop protection products 560
International action committees 561
Poisons and Poisoning 562
Chemicals Subject to the Poisons Law 562
Part I Poisons 562
Part II Poisons 562
Occupational Exposure Limits 562
First Aid Measures 563
General measures 563
Specific measures 563
Reporting pesticide poisoning 564
Additional information 564

Environmental Protection 565
Protection of Bees 565
Honey bees 565
Wild bees 565
The Campaign Against Illegal Poisoning of Wildlife 565
Water Quality 566
Protecting surface waters 566
Local Environmental Risk Assessments for Pesticides 567
Groundwater regulations 568

SECTION 6 APPENDICES **569**

Appendix 1 Suppliers of Pesticides and Adjuvants 571
Appendix 2 Useful Contacts 579
Appendix 3 Keys to Crop and Weed Growth Stages 582
Appendix 4 Key to Hazard Classifications and Safety Precautions 590
Appendix 5 Key to Abbreviations and Acronyms 596
Appendix 6 Definitions 599
Appendix 7 References 601

INDEX OF PROPRIETARY NAMES OF PRODUCTS **603**

Disclaimer

Every effort has been made to ensure that the information in this book is correct at the time of going to press, but the Editor and the publishers do not accept liability for any error or omission in the content, or for any loss, damage or other accident arising from the use of the products listed herein. Omission of a product does not necessarily mean that it is not approved or that it is unavailable.

It is essential to follow the instructions on the approved label when handling, storing or using any crop protection product. Approved 'off-label' extensions of use are undertaken entirely at the risk of the user.

The Dangerous Preparations Directive

The Dangerous Preparations Directive (1999/45/EEC) came into force in UK for pesticide and biocidal products on 30 July 2004. Its aim is to achieve a uniform approach to the classification and labelling of most dangerous preparations, including crop protection products. The Directive is implemented in UK under the Chemicals (Hazard Information and Packaging Supply) Regulations 2002, often referred to under the shorthand acronym 'CHIP3'. In most cases the Regulations have led to additional hazard symbols, and associated risk and safety phrases, relating to environmental and health hazards appearing on the label. All products affected by the Directive entering the supply chain from the implementation date above must be so labelled.

Most of the products in this edition are shown with their new environmental hazard classification, and risk phrases detailed in the pesticide profiles in Section 2. Unfortunately some suppliers were unable to supply the Editor with amended labels of their products in time for this edition. No advantage should be inferred for those products that do not show the additional hazards and risks.

Editor's Note

Correct and responsible use of pesticides, so as to optimise their performance and minimise their environmental impact, is a vital element of modern farming and the major focus of the Voluntary Initiative. Pesticides are a powerful tool in the fight against the depredations of weeds, pests and diseases, but they are not the only answer, nor should they be regarded as such. This Guide provides farmers, their advisors, spray operators and others involved with the use of pesticides, with much of the information needed to make a correct choice of products.

The UK Pesticide Guide is the only printed reference catalogue of pesticides registered for use in agriculture, forestry, horticulture and amenity in the UK. For this 18th edition the vast majority of product entries have been updated to include their new environmental classification under the Dangerous Preparations Directive, which became applicable to pesticides in July 2004. In addition, we have taken the opportunity to change the order of the contents to bring the key product listings nearer to the front of the volume, while the supporting background information, known to be valuable to many users, has been moved to the back. The information contained in the Guide has not changed but, as usual, the entire contents have been reviewed and updated.

Users of the Guide are reminded that, in addition to the products notified to the Editor for inclusion in Section 2 (and therefore confirmed by the supplier as available for purchase), many more have extant approval but, for a number of reasons, may not be available. These products are shown in Section 3 to enable comparisons to be made with the main list.

There are about 100 new product entries in this edition, but a substantially greater number of deletions reflecting the general loss of products to the industry, largely as a result of the EU Pesticide Review Programme, but also due to range rationalisation by the manufacturers. Three new active substances are listed, indicating a general slow-down in the introduction of new molecules.

Any important new developments or new products notified to the Editor during the year will again be shown on our website at www.ukpesticideguide.co.uk

As always, criticisms and suggestions are welcome and, in particular, notification of any errors or omissions.

R. Whitehead
Editor

The information in this publication has been selected from official sources and from suppliers' product labels and product manuals of pesticides approved for use in the UK under the Control of Pesticides Regulations 1986 or the Plant Protection Products Regulations 1995.

The content is based on information received by the Editor up to October 2004.

For developments since publication of this edition, and new products notified to the Editor during the year, see:

www.ukpesticideguide.co.uk

The Voluntary Initiative

The Voluntary Initiative (VI) is a five-year programme of measures, agreed by the crop protection industry and related organisations with Government, to minimise the environmental impacts of crop protection products. By protecting water quality and enhancing farmland biodiversity the programme aims to reduce the risk of a pesticide tax. Anyone involved with the products in this *Guide*, whether as a user or advisor, is likely to be affected directly or indirectly by the projects in the programme.

This edition of *The UK Pesticide Guide* can help in two specific ways.

- It is illegal to use, or even store, products whose approval has expired. While mere storage is unlikely to have any environmental impact, their incorrect or inappropriate disposal could undoubtedly do so. All the products listed in Sections 2 and 3 of this *Guide* have extant approval for all or part of 2005. A date is shown for any whose approval will expire before the end of the year. Therefore any products not listed here are likely to be obsolete, although a check should be made with the manufacturer or supplier before disposal.
- Product labels contain much information that is often too crowded and complicated. Another project in the VI, undertaken by the Crop Protection Association (CPA) together with the Pesticides Safety Directorate and the Health and Safety Executive, is a review of product labels and the production of Best Practice Labelling Guidance. CPA member companies have adapted their labels to be compliant with these guidelines at the same time as making the mandatory changes required by the Dangerous Preparations Directive. The presentation of information in this *Guide* reflects this improved clarity as far as possible. In the sister publication, *The e-UK Pesticide Guide*, we have gone further by using the new symbols for environmental hazard, product activity and LERAP category, as well as providing links to company websites to enable access to product Environmental Information Sheets (also being produced as a VI project).

Further information on the Voluntary Initiative is available at www.voluntaryinitiative.org.uk

Changes Since 2004 Edition

Pesticides and adjuvants added or deleted since the 2004 edition are listed below. Every effort has been made to ensure the accuracy of this list, but late notifications to the Editor are not included.

Products Added

Products new to this edition are shown here. In addition, products that were listed in the previous edition whose PSD/HSE registration number *and* supplier have changed are included. Where only the PSD/HSE registration number has changed, the product is not listed.

Product	Reg. No.	Supplier
AC 650	11102	United Phosphorus
Acanto Prima	11864	Syngenta
Agate	12016	Makhteshim
Ag-Chem Metribuzin	12007	Ag-Chem
Ag-Chem Prefect	11985	Ag-Chem
Agri-50E	–	Fargro
Agricola Lens Plus	11587	Interfarm
Agriguard Metribuzin	09853	AgriGuard
AgriGuard Pro-Turf	11837	AgriGuard
Aligran	11761	Nufarm UK
Alistell	11053	United Phosphorus
Alpha Metamitron	11081	Makhteshim
Alpha Prochloraz 40 EC	11002	Makhteshim
Alphaguard 100 EC	10772	Interfarm
Alto Xtra	11839	Syngenta
Amistar Opti	11863	Syngenta
Appeal	12022	Chiltern
Asana	11934	BASF
Attract	12023	Chiltern
Autumn Kite	11549	Nufarm UK
BAS 493F	11748	BASF
Bema	11756	Makhteshim
Bettix 70 WG	11154	United Phosphorus
Bettix Flo	11959	United Phosphorus
Biplay PX	10990	DuPont
Blaster	10571	Headland Amenity
Brits	11792	Doff Portland
Bulldog	11723	Makhteshim
Cabaret	11915	Bayer CropScience
Capitan 40	10914	DuPont
Cerround	A0510	Helena
Citation 70	11929	United Phosphorus
Cleancrop GYR	10646	Makhteshim
Cleancrop Phenmedipham	11586	United Phosphorus
Consento	11889	Bayer CropScience
Conserve	12058	Fargro
Consul	11523	Headland
Covershield	10900	BASF
Crawler	11840	Makhteshim
Dow Agrosciences Glyphosate 360	11552	Dow
DP 911 PX	10992	DuPont
DUK 118	11923	Headland
Duplosan KV	12073	Nufarm UK
ethylene	–	various
Fernpath Dart	11970	AgriGuard
Fernpath Haptol	11951	AgriGuard
Fernpath Lenzo Flo	11919	AgriGuard
Fernpath Pronto	11988	AgriGuard
Finish PX	10989	DuPont
Flash	11989	AgriGuard
Flexity	11775	BASF
Fox	11981	Makhteshim
Frupica SC	12067	Certis
Galera	11961	Dow
GF 184	10878	Dow
Glyphosate 360	11726	Monsanto
Goldbeet	11538	Makhteshim
Green Gold	A0250	Intracrop
Greencrop Amaize	11973	Greencrop
Greencrop Biscay	12132	Greencrop
Greencrop Doonbeg WG	12199	Greencrop
Greencrop Estuary	12095	Greencrop
Greencrop Frond	11912	Greencrop
Greencrop Monogram	12048	Greencrop

Product	Reg. No.	Supplier
Greencrop Tabloid	11969	Greencrop
Guilder	11894	Nufarm UK
High Load Mircam	11930	Nufarm UK
Hiker	11451	Dow
I T Fosetyl-AL	11717	I T Agro
Insignia	11865	Vitax
Intracrop Neotex	A0460	Intracrop
Intracrop Predict	A0503	Intracrop
Intracrop Questor	A0495	Intracrop
Intracrop Retainer	A0508	Intracrop
Intracrop Rigger	A0479	Intracrop
Intracrop Saturn	A0494	Intracrop
Intracrop Status	A0506	Intracrop
Intracrop Super Rapeze MSO	A0491	Intracrop
Inzacur	12014	Nufarm UK
Jenton	11898	BASF
Karmex	09475	Headland
Kinetic	A0252	Helena
Kinto	12038	BASF
KN 540	12009	Monsanto
Landgold Deputy	11975	Landgold
Landgold Tepraloxydim	11808	Landgold
Leaf-Koat	A0511	Helena
Liberator	12032	Bayer CropScience
Luxan 9363 Red	11480	Luxan
Marnoch Chlorothalonil	09763	Marnoch
Me2 Booty 2	12010	Me2
Me2 Puddy	12164	Me2
Me2 Succotash	11815	Me2
Me2 Sylvester	12021	Me2
Medley	10933	DuPont
New 5C Quintacel	12074	Nufarm UK
Nomix Diuron 80	12079	Nomix Enviro
Nomix Diuron Flowable	12078	Nomix Enviro
Nomix Garlon 4	12081	Nomix Enviro
Nomix Mosskiller	12080	Nomix Enviro
Novall	12031	BASF
Nufarm Adjuvant Oil	A0474	Nufarm UK
Oberon	11819	Bayer CropScience
Option	11834	DuPont
Pan Ethephon	11020	Pan Agriculture
Pan Magician	11992	Pan Agriculture
Panoctine Plus	11757	Makhteshim
Pasturol Plus	10278	FCC
Phantom	11954	Syngenta

Product	Reg. No.	Supplier
Pico Stomp	11987	BASF
Pierce	11924	Nufarm UK
Pirlid	11946	AgriGuard
Pluton	10957	DuPont
Poncho Beta	11999	Bayer CropScience
Profit Oil	A0294	Microcide
Prospero	11931	Makhteshim
Prostore 157 UL	12017	Nickerson
Prostore 420 EC	12036	Nickerson
Raptor	11092	Makhteshim
Ravine Plus	11832	Makhteshim
Reward Oil	A0295	Microcide
Roundup Pro-Green	11907	Monsanto
Roundup ProVide	11721	Monsanto
Roundup Rail	11874	Monsanto
Rowent	12024	Chiltern
Rubigan	11069	Gowan
Safeguard	12020	AgriGuard
Sahara	11905	Bayer CropScience
Sakarat Bromabait	H7902	Killgerm
Samson	10433	Syngenta
Saracen	A0368	Interagro
Shogun	10584	Makhteshim
Shotput	11960	Makhteshim
SmartFresh	11799	Landseer
Snare	11873	Headland Amenity
Solace	11936	Nufarm UK
Solan 40	11897	Nufarm UK
Solfa WG	11602	Nufarm UK
Spitfire	12002	AgriGuard
Stalwart	11877	United Phosphorus
Standon Metazachlor 500	12012	Standon
Standon Mimas WG	12075	Standon
Starane 2	12018	Dow
Stika	A0452	Loveland
Sting ECO	10337	Monsanto
Swipe P	11955	Nufarm UK
T2 Green	11925	Nufarm UK
Tangent	11872	Headland Amenity
Tarot	11896	Makhteshim
Thunder	11186	Makhteshim
Titus	11895	Makhteshim
Tolurex 90 WDG	11403	Makhteshim
Tordon 101	05816	Dow
Transcend	A0333	Helena

Product	Reg. No.	Supplier
Triflurex 48 EC	07947	Makhteshim
Tripart Culmus	06619	Tripart
Trustee Elite	11556	Barclay
Try-Flex	A0489	Greenaway
Tumbleweed Pro-Active	11963	Scotts
Typhoon 360	11175	Makhteshim
Validate	A0500	Loveland
Venus	11856	Headland

Products Deleted

The appearance of a product name in the following list may not necessarily mean that it is no longer available. In cases of doubt, refer to the supplier.

Product	Reg. No.	Supplier
Admire	07481	Bayer
Agate	08826	Bayer
Agate	11235	Bayer CropScience
Agriguard 5C Chlormequat 460	09851	AgriGuard
Agriguard Aldicarb	10481	AgriGuard
Agriguard Alpha-Cyper	09920	AgriGuard
Agriguard Chlormequat 700	09782	AgriGuard
Agriguard Chlormequat 760	10290	AgriGuard
Agriguard Chlorpyrifos	10626	AgriGuard
Agriguard Clopyralid	10625	AgriGuard
Agriguard Cypermethrin	12134	AgriGuard
Agriguard Deltamethrin	10770	AgriGuard
Agriguard Duo	10453	AgriGuard
Agriguard Isoproturon	09769	AgriGuard
Agriguard Isoxaben	11652	AgriGuard
Agriguard Metamitron	09859	AgriGuard
Agriguard Metsulfuron	09569	AgriGuard
Agriguard Phenmedipham	10712	AgriGuard
Agriguard Triclopyr	10679	AgriGuard
Agropen	A0143	Intracrop
Alfacron 10 WP	09439	Novartis A H
Alfadex	H7221	Novartis A H
Aliette 80 WG	09156	Certis
Ally WSB	06588	DuPont
Amazon	10266	Aventis
Apex 5E	08739	Novartis A H
Applaud	10930	Certis
Artist	10658	Bayer
AS Elite	A0406	Monsanto
Atlas 5C Quintacel	11130	Nufarm UK
Atlas Adjuvant Oil	A0021	Nufarm UK
Atlas Cropguard	09123	Nufarm UK
Atlas Solan 40	07726	Nufarm UK
Attribut	10671	Bayer
Aurora 50 WG	09807	FMC
Bacara	09976	Bayer CropScience
Barclay Goalpost	11382	Barclay
Barclay Keeper	11375	Barclay
Barclay Keeper 500 FL	11376	Barclay
Base 50 W	10202	Interfarm
Base 50 W	10765	Dow
BASF MCPA Amine 50	00209	BASF
Bavistin DF	03848	BASF
Bavistin FL	00218	BASF
Baytan Flowable	02593	Bayer
Baytan Secur	09510	Bayer
Beam	10411	Bayer
Beam	11255	Bayer CropScience
Beetup Flo 160	10382	United Phosphorus
Benlate Fungicide	00229	DuPont
Beret Gold	08390	Syngenta
Besiege WSB	08075	DuPont
Bettix 70 WG	11019	United Phosphorus
Biothene	A0360	Intracrop
Brits	10116	Doff Portland
Caddy 240 EC	10631	Bayer
Calypso	10342	Bayer
Casoron G4	09215	Crompton
CDA Vanquish	09927	Bayer Environ.
Centium 360 CS	10720	FMC
Chum	A0420	Greenhill
Compass	10041	Aventis
Croptex Bronze	04087	Certis

Product	Reg. No.	Supplier
Curzate M68 WSB	08073	DuPont
Cutinol	A0281	Greenhill
Cutinol Plus	A0395	Greenhill
Dacthal W-75	10623	Certis
Decoy Wetex	09707	Bayer
Deosan Rataway	08803	Johnson Diversey
Deosan Rataway Bait Bags	08805	Johnson Diversey
Discovery	10416	United Phosphorus
Doff Horticultural Slug Killer Blue Mini Pellets	09666	Doff Portland
Doxstar	06050	Dow
Druid	08714	Aventis
Duplosan KV	09431	BASF
Easel	11146	Nufarm UK
Eclipse	07361	BASF
Elan	A0392	Intracrop
Electis 75 WG	10565	Interfarm
Elvaron Multi	10080	Bayer
Embark Lite	08749	Intracrop
Emerald	A0031	Intracrop
Enhance	A0147	Greenhill
Enhance Low Foam	A0148	Greenhill
Ethokem	A0353	Greenhill
Ethos	10736	Bayer
Euroagkem Pace	A0305	Euroagkem
Evade	08071	SumiAgro Amenity
Evict	09150	Bayer
Evidence	06934	Aventis
Exit Wetex	10149	Bayer
Fernpath Banjo	10182	AgriGuard
Fernpath Haptol	10446	AgriGuard
Fernpath Tangent	10341	AgriGuard
Fieldgard	11541	Nufarm UK
Fiesta T	10260	BASF
Flamenco	09913	Aventis
Flexidor 125	05104	Landseer
Foil	10905	Aventis
Folicur	08691	Bayer
Force ST	11085	Bayer
Fort	11606	Bayer CropScience
Frupica	10944	Certis
Fumite Dicloran Smoke	09291	Certis
Fumyl-O-Gas	04833	Brian Jones
Fungaflor	11326	Certis

Product	Reg. No.	Supplier
Fury 10 EW	10718	FMC
Fusilade 250 EW	10525	Syngenta
Gaucho	06590	Bayer
Genesis	09993	Sipcam
Glydate	10999	Nufarm UK
Glyper	11095	Interfarm
Glyphosate 360	09151	Dow
Glyphosate 360	11579	Cardel
Goltix 90	08654	Bayer
Goltix Flowable	08986	Bayer
Goltix WG	02430	Bayer
Granit	09995	Aventis
Greencrop Biscay	12132	Greencrop
Greencrop Cajole	11402	Greencrop
Greencrop Coolfin	09449	Greencrop
Greenhill Spreader	A0196	Greenhill
Greenmaster Autumn	07508	Scotts
Greenmaster Mosskiller	07509	Scotts
Greenor	07848	Rigby Taylor
Grovex Zinc Phosphide	H6800	Killgerm
GS 800	A0354	Greenhill
Headland Neptune	10230	Headland
Herbasan	10734	Nufarm UK
Heritage	11383	Scotts
Hive	10953	Nufarm UK
Huron	10148	Bayer
HY-D	06278	Agrichem
Intracrop Green Oil	A0262	Intracrop
Intracrop Non-Ionic Wetter	A0009	Intracrop
Intracrop Rapeze	A0390	Intracrop
Intracrop Rapide Beta	A0175	Intracrop
Iona	A0232	Intracrop
Iona Low Foam	A0255	Intracrop
Isoguard	10829	Nufarm UK
Isomec	09881	Nufarm UK
Javelin	09998	Bayer CropScience
Jockey	10076	Aventis
Jockey F	10074	Aventis
Jogral	A0226	Intracrop
Jubilee (WSB)	06082	DuPont
Judge	10000	Sipcam
Karan	09637	Bayer
Kaspar	10615	Certis
Katamaran	09049	BASF
Kerb 50 W	10200	Interfarm
Kerb Flo	10160	Interfarm

Product	Reg. No.	Supplier
Kerb Pro Flo	10158	SumiAgro Amenity
Killgerm Fenitrothion 40 WP	H4858	Killgerm
Killgerm Fenitrothion 50 EC	H4722	Killgerm
Knot Out	05163	Vitax
Konker	03988	BASF
Kruga 5EC	09863	Interfarm
Kubist 2	11126	Nufarm UK
Laminator DG	11073	Interfarm
Laminator FL	11072	Interfarm
Laminator WP	11071	Interfarm
Landgold CCC 720	08527	Landgold
Landgold Difenoconazole	09964	Landgold
Landgold Epoxiconazole FM	08806	Landgold
Landgold Ethofumesate 200	08980	Landgold
Landgold Fenpropidin 750	08973	Landgold
Landgold Flusilazole MBC	08528	Landgold
Landgold Isoproturon	06012	Landgold
Landgold Rimsulfuron	08959	Landgold
Landgold Strobilurin KE	09908	Landgold
Landgold Strobilurin KF	09196	Landgold
Landgold TFS 50	08941	Landgold
Landgold Vinclozolin SC	06459	Landgold
Lexus Class WSB	08637	DuPont
Lexus XPE WSB	08542	DuPont
Leyclene	07263	Bayer CropScience
Lo Dose G	A0349	Reabrook
Lotus	09231	BASF
Lupus	09638	Bayer
Luxan 2,4-D	09379	Luxan
Luxan Chlorotoluron 500 Flowable	09165	Luxan
Luxan Dichlorvos 600	08297	Luxan
Luxan Dichlorvos Aerosol 15	08298	Luxan
Luxan Gro-Stop Basis	08601	Luxan
Luxan Isoproturon 500 Flowable	07437	Luxan
Luxan MCPA 500	07470	Luxan
Maneb 80	01276	Rohm & Haas
Marshal 10G	11682	Belchim
Mascot Mosskiller	02439	Rigby Taylor

Product	Reg. No.	Supplier
Mascot Systemic	08776	Rigby Taylor
Matador 200 SC	11058	Margarita
Me2 Notin	11655	Me2
Me2 Snap!	10476	Me2
Merit	10384	BASF
Metham Sodium 400	08051	United Phosphorus
Micromite	H4480	Killgerm
Minuet EW	10716	FMC
Mircam Plus	11004	Nufarm UK
Mithras 80 EDF	10164	Interfarm
Monceren DS	04160	Bayer
Monceren IM	06259	Bayer
n2n Diuron 80	11821	Nomix Enviro
n2n Diuron Flowable	11822	Nomix Enviro
n2n Garlon 4	11806	Nomix Enviro
n2n Mosskiller	11823	Nomix Enviro
Nemathorin 10G	08915	Syngenta
Nemolt	07012	Fargro
Neon	08337	Bayer
Neon	11226	Bayer CropScience
Neporex 2SG	H7222	Novartis A H
New Draza	11082	Bayer
Novak	08020	Aventis
Novak	11693	BASF
Nu Film 17	A0055	Intracrop
Nuvan 500 EC	08590	Novartis A H
Opus Team	07362	BASF
Pal	A0396	Greenhill
Panther	10008	Aventis
Panther WDG	10650	Aventis
Partna	A0428	Syngenta
Pastor	07440	Dow
Patriot EC	08990	Aventis
Perm-E8	A0304	Intracrop
Pin-o-Film	A0209	Intracrop
Platform S	09706	FMC
Plover	08429	Syngenta
Poraz	10474	Aventis
Posse 10G	10725	Bayer
Premiere Granules	07987	Dow
Premis	10177	Aventis
Profile	08134	Aventis
Propose	11195	Interfarm
Quantum	06270	DuPont
Quaver Flo	10162	Interfarm
Racumin Contact Powder	H6755	Bayer Environ.
Ragox	11145	Nufarm UK

Product	Reg. No.	Supplier
Raxil S	06974	Bayer
Raxil Secur	08966	Bayer
RCR Zinc Phosphide	H6801	Killgerm
Redeem Flo	10236	Interfarm
Redeem Flo	10764	Dow
Repulse	07641	Certis
Reward 5EC	09862	Interfarm
Rhythm	09636	Interfarm
Rivet	09512	Bayer
Rizolex Flowable	09358	Certis
Robust	10176	Aventis
Ronstar Liquid	08974	Certis
Rovral Flo	10013	Aventis
Rovral Liquid FS	10551	Aventis
Rovral WP	10015	Aventis
Roxam 75 WG	10566	Interfarm
Sage	10413	Bayer
Salute	07660	United Phosphorus
Scala	07806	Aventis
Sceptre	08043	Bayer CropScience
Sectacide 50 EC	H4939	Killgerm
Sencorex WG	03755	Bayer
Seradix 1	10422	Certis
Seradix 2	10423	Certis
Seradix 3	10424	Certis
Shark	11054	FMC
Shark	11616	Belchim
sHYlin	10030	Agrichem
Sibutol	07238	Bayer
Sibutol Secur	09131	Bayer
Sierraron G	09675	Scotts
Silvacur	06387	Bayer
Sinbar	01956	DuPont
SL 567A	08811	Syngenta
Slaymor	H6729	Novartis A H
Slaymor Bait Bags	H6730	Novartis A H
Snooker	10018	Aventis
Sobrom BM 100	04381	Brian Jones
Sobrom BM 98	04189	Brian Jones
Solfa	06959	Nufarm UK
Speedway 2	09273	Scotts
Sportak 45 EW	07996	Aventis
Sportak Delta 460 HF	07431	Aventis
Spotlight 24 EC	10702	FMC
Standon Aldicarb 10G	11024	Standon
Standon Chlorothalonil 500	08597	Standon
Standon CIPC 300 HN	09187	Standon

Product	Reg. No.	Supplier
Standon Cymoxanil Extra	09442	Standon
Standon Cyprodinil	09345	Standon
Standon Dichlobenil 6G	08874	Standon
Standon Ethofumesate 200	09360	Standon
Standon Fluazinam 500	08670	Standon
Standon Flupyrsulfuron MM	09098	Standon
Standon Flusilazole Plus	07403	Standon
Standon Glyphosate 360	05582	Standon
Standon Metamitron	07885	Standon
Standon Metsulfuron	05670	Standon
Standon Pendimethalin 400 SC	10729	Standon
Standon Pyrimethanil	10576	Standon
Standon Triflusulfuron	09487	Standon
Starion	09795	FMC
Sterilite Tar Oil Winter Wash 60% Stock Emulsion	05061	Coventry Chemicals
Sterilite Tar Oil Winter Wash 80% Miscible Quality	05062	Coventry Chemicals
Stomp Pico	10738	BASF
Storite Clear Liquid	08982	Banks Cargill
Super Selective Plus	11928	Rigby Taylor
Superflor 6% Metaldehyde Slug Killer Mini Pellets	09773	CMI
Surpass 5EC	09861	Interfarm
suSCon Indigo	09902	Fargro
Swift SC	10884	Bayer
Systhane 6 Flo	07334	Landseer
Systol M	08085	DuPont
T 25	A0352	Reabrook
Takron	06237	BASF
Talstar	10835	Certis
Talzene	A0233	Intracrop
Tech E	A0346	Greenhill
Tech G	A0345	Greenhill
Teldor	08955	Bayer
Terbine	11407	Nufarm UK
Titus	07908	DuPont
Tolkan Liquid	10023	Aventis
Tolkan Liquid	10895	Aventis
Topas C 50 WP	08459	Novartis
Torch	08336	Bayer

Product	Reg. No.	Supplier
Trimanzone	09584	Intracrop
Tripart Accendo	06110	Tripart
Tripart Acer	A0097	Tripart
Tripart Brevis	03754	Tripart
Tripart Faber	04549	Tripart
Tripart Lentus	A0117	Tripart
Tripart Minax	A0108	Tripart
Tripart Orbis	A0204	Tripart
Tripart Sentinel	03250	AgriGuard
Tripart Trifluralin 48 EC	02215	Tripart
Triumph	08740	Aventis
Tuberon	10613	Unicrop
Tumbleweed Pro	11083	Scotts
Turf Systemic Fungicide	09349	SumiAgro Amenity
Twist	10587	Bayer

Product	Reg. No.	Supplier
Ultrafaber	05627	Tripart
Unix	08764	Syngenta
UPL Diuron 80	07619	United Phosphorus
Uplift	07527	United Phosphorus
Veto F	08057	Bayer
Vindex	05470	Dow
Vizzler	11409	Interfarm
Wayfarer	A0045	Certis
Weedazol-TL	02979	Bayer
Weedazol-TL	11430	Nufarm UK
X-Spor SC	08077	United Phosphorus
Zapper	09944	Bayer Environ.
Zenon	09193	Bayer

About *The UK Pesticide Guide*

Purpose

The primary aim of this book is to provide a practical handbook of pesticides, plant growth regulators and adjuvants that the farmer or grower can realistically and legally obtain in the UK, and to indicate the purposes for which they may be used. It is designed to help in the identification of products appropriate for a particular problem, and a Crop/Pest Guide (Section 1) is included to facilitate this. In addition to uses recommended on product labels, details are provided of those uses that do not appear on labels but which have been granted specific off-label approval (SOLA). Such uses are not endorsed by the manufacturer, and are undertaken entirely at the risk of the user.

As well as identifying the products available, the book provides guidance on how to use them safely and effectively, but without giving details of doses, volumes, spray schedules or approved tank mixtures. Sections 5 and 6 provide essential background information on a wide range of pesticide-related issues including legislation, codes of practice, poisons and treatment of poisoning, products for use in special situations, and weed and crop growth stage keys.

While we have tried to cover all other important factors, this book does **not** provide a full statement of product recommendations. **Before using any pesticide product it is essential that the user should read the label carefully and comply strictly with the instructions it contains**.

Scope

Individual profiles are provided of some 350 individual active ingredients, and a further 150 profiles of mixtures containing these active ingredients. Each profile has a list of available approved products for use in arable agriculture, horticulture (including amenity horticulture), forestry and areas near water, together with the name of the supplier from whom each may be obtained. Within these fields of use, all types of pesticide covered by the Control of Pesticides Regulations 1986 or the Plant Protection Products Regulations are included. This embraces acaricides, algicides, fungicides, herbicides, insecticides, lumbricides, molluscicides, nematicides and rodenticides, together with plant growth regulators used as straw shorteners, sprout inhibitors and for various horticultural purposes. The total number of pesticides covered in the *Guide* is about 1150.

Products are included in the *Guide* only if requested by the supplier and supported by evidence of approval. This is intended to ensure that only marketed products are listed.

In addition, the *Guide* gives information on more than 100 authorised adjuvants which, although not themselves pesticides, may be added to pesticides to improve their effectiveness in use.

The *Guide* does **not** include products approved solely for amateur, home garden, domestic, food storage, public health or animal health uses.

Sources of Information

The information for this edition has been drawn from these authoritative sources:

- approved labels and product manuals received from suppliers of pesticides up to October 2004
- websites of the Pesticides Safety Directorate (PSD, www.pesticides.gov.uk) and the Health and Safety Executive (HSE, www.hse.gov.uk)
- *The Pesticides Monitor* (incorporating *The Pesticides Register*), published monthly by PSD and HSE, listing new UK approvals (including off-label approvals), up to and including the final issue for June/July 2004
- PSD lists of approval expiries.

Criteria for Inclusion

To be included in the *Guide*, a product must meet the following conditions:

- it must have extant approval under UK pesticides legislation
- information on the approved uses must have been provided by the supplier
- it must be expected to be on the UK market during the currency of this edition.

When a company changes its name, whether by merger, take-over or joint venture, it is obliged to re-register its products in the name of the new company, and new Ministry registration numbers are normally assigned. Where stocks of the previously registered products remain legally available, both old and new identities are included in the *Guide* and remain until approval for the former lapses, or stocks are notified as exhausted.

Products that have been withdrawn from the market and whose approval will finally lapse during 2005 are identified in each profile. After the indicated date, sale, storage or use of the product bearing that approval number becomes illegal. Where there is a direct replacement product, this is indicated.

How To Use *The UK Pesticide Guide*

The book consists of six main sections:

1 Crop/Pest Guide
2 Pesticide Profiles
3 Products Also Registered
4 Adjuvants
5 Useful Information
6 Appendices

1 Crop/Pest Guide

This section enables the user to identify which active ingredients are approved for a particular crop/pest combination. The crops are grouped as shown in the Crop Index. For convenience, some crops and targets have been grouped into generic units (for example, 'cereals' are handled as a group, not individually). Therefore indications from the Crop/Pest Guide must always be followed up by reference to the specific entry in the pesticide profiles section because a product may not be approved on all crops in the group. Chemicals indicated as having uses in cereals, for example, may be approved for use only on winter wheat and barley, not for other cereals. Because of differences in the wording of product labels it may sometimes be necessary to refer to broader categories of organism in addition to specific organisms (e.g. annual grasses and annual weeds in addition to annual meadow grass).

2 Pesticide Profiles

Each active ingredient has a separate, numbered profile entry, as does each mixture of active ingredients. The entries are arranged in alphabetical order using the common names approved by the British Standards Institution, and used in *The Pesticide Manual* (BCPC, 2003). Where an active ingredient is available only in mixtures, this is stated. The ingredients of the mixtures are themselves ordered alphabetically, and the entries appear at the correct point for the first named ingredient.

Within each profile entry, a table lists the products approved and available on the market, in the following style:

Product name	Main supplier	Active ingredient content	Formulation type	Registration number
1 Broadsword	United Phosphorus	200:85:65 g/l	EC	09140
2 Greengard	SumiAgro Amenity	200:85:65 g/l	EC	11715
3 Nu-Shot	Nufarm UK	200:85:65 g/l	EC	11148

Many of the **product names** are registered trademarks, but no special indication of this is given. Individual products are indexed under their entry number in the Index of Proprietary Names at the back of the book. The **main supplier** indicates the marketing outlet through which the product may be purchased. Full addresses and telephone/fax numbers of suppliers are listed in Appendix 1. Some website and e-mail addresses are also given. For mixtures, the **active ingredient contents** are given in the same order as they appear in the profile heading. The **formulation types** are listed in full in the Key to Abbreviations and Acronyms (Appendix 5). The **registration number** normally refers to registration with PSD. In cases where products registered with the Health and Safety Executive are included, HSE numbers are quoted, e.g. H0462.

Below the product table, a **Uses** section lists all approved uses notified by suppliers to the Editor by October 2004, giving both the principal target organisms and the recommended crops or situations (identified in ***bold italics***). Where there is an important condition of use, or the approval is off-label, this is shown in parentheses, e.g. (off-label). Numbers in square brackets refer to the numbered products in the table above. Thus, in the example shown above, a use approved for Nu-Shot (product 3) but not for Broadsword or Greengard (products 1 and 2) appears as:

Annual dicotyledons in ***rotational grassland*** [3]

Any **Specific Off-Label Approvals (SOLAs)** for products in the profile are detailed below the list of approved uses. Each SOLA has a separate entry and shows the crops to which it applies, the Notice of Approval number, the expiry date (if any) and the product reference number in square brackets.

Below the SOLA paragraph, **Notes** are listed under the following headings. Unless otherwise stated, any reference to dose is made in terms of product rather than active ingredient. Where notes refer to particular products, rather than to the entry generally, this is indicated by numbers in square brackets as described above.

Approval information	Information of a general nature about the approval status of the active ingredient or products in the profile is given here. Notes on approval for aerial or ULV application are given. Where a product approval will finally expire in 2005, the expiry date is shown here together with the registration number of its direct replacement, if any.
Efficacy	Factors important in making the most effective use of the treatment. Certain factors, such as the need to apply chemicals uniformly, the need to spray at appropriate volume rates, and the need to prevent settling out of active ingredient in spray tanks, are assumed to be important for all treatments and are not emphasised in individual profiles.
Restrictions	Notes are included in this section where products are subject to the Poisons Law, and where the label warns about organophosphorus and/or anticholinesterase compounds. Factors important in optimising performance and minimising the risk of crop damage, including statutory conditions relating to the maximum permitted number of applications, are listed. Any restrictions on crop varieties that may not be sprayed are mentioned here.
Environmental safety	Where the label specifies an environmental hazard classification, it is noted, together with the associated risk phrases. Any other special operator precautions, and any conditions concerning withholding livestock from treated areas, are specified. Where any of the products in the profile are subject to Category A, Category B or broadcast air-assisted buffer zone restrictions under the LERAP scheme, the relevant classification is shown. Other environmental hazards are also noted here, including potential dangers to livestock, game, wildlife, bees and fish. The need to avoid drift onto neighbouring crops and to wash out equipment thoroughly after use are important with all pesticide treatments, but may receive special mention here if of particular significance.
Crop-specific information	Instructions about timing of application or cultivations that are specific to a particular crop, rather than generally applicable, are listed here. The latest permitted use and harvest intervals (if any) are shown for individual crops.
Following crops guidance	Any specific label instructions about what may be grown after a treated crop, whether harvested normally or after failure, are shown here.

Hazard classification and safety precautions

The label hazard classification(s) and precautions are shown using a series of letter or number codes which are explained in Appendix 4. The codes are listed under a number of sub-headings as follows:

- Hazard
- Risk phrases
- Operator protection
- Environmental protection
- Consumer protection
- Storage and disposal
- Treated seed
- Vertebrate/rodent control products
- Medical advice

This section is given for information only and should not be used for the purpose of making a COSHH assessment without reference to the actual label of the product to be used.

3 Products also Registered

Products with extant approval for all or part of the year of the edition in which they appear are listed in this section if they have not been requested by their supplier or manufacturer for inclusion in Section 2. Details shown are the same (apart from formulation) as those in the product tables in Section 2 and, where relevant, an approval expiry date is shown. Not all the products listed here will be available in the market, but this list, together with the products in Section 2, comprises a comprehensive listing of all approved products for uses within the scope of the *Guide* for the year in question.

4 Adjuvants

Adjuvants are listed in a table ordered alphabetically by product name. For each product, details are shown of the main supplier, the authorisation number and the mode of action (e.g. wetter, sticker, etc.) as shown on the label. A brief statement of the uses of the adjuvant is given. Protective clothing requirements and label precautions are listed using the codes from Appendix 4.

5 Useful Information

This section summarises legislation covering approval, storage, sale and use of pesticides in the UK. There are brief articles on the broader issues of using crop protection chemicals, including resistance management. Lists are provided of products approved for use in or near water, in forestry, as seed treatments, and for aerial application. Chemicals subject to the Poisons Laws are listed, and there is a summary of first aid measures if pesticide poisoning should be suspected. Finally, this section provides guidance on environmental protection issues and covers the protection of bees and the use of pesticides in or near water.

6 Appendices

Appendix 1

Gives names, addresses, telephone and fax numbers of all companies listed as main suppliers in the pesticide profiles. E-mail and website addresses are also listed where available.

Where a supplier is no longer trading under that name (usually following a merger or takeover) but products in that name are still listed in the *Guide* (because they are still available in the supply chain), a cross-reference indicates the new 'parent' company from which technical or commercial information can be obtained.

Appendix 2

Gives names, addresses, telephone and fax numbers of useful contacts, including the National Poisons Information Service. Website addresses are included where available.

Appendix 3	Gives details of the keys to crop and weed growth stages, including the publication reference for each. The numeric codes are used in the descriptive sections of the pesticide profiles (Section 2).
Appendix 4	Shows the full text for code letters and numbers used in the pesticide profiles (Section 2) to indicate personal protective equipment requirements and label precautions.
Appendix 5	Shows the full text for the formulation abbreviations used in the pesticide profiles (Section 2). Other abbreviations and acronyms used in the *Guide* are explained here.
Appendix 6	Provides full definitions of officially agreed descriptive phrases for crops or situations used in the pesticide profiles (Section 2) where misunderstandings can occur.
Appendix 7	Shows a list of useful reference publications which amplify the summarised information provided in this section.

SECTION 1
CROP/PEST GUIDE

Crop/Pest Guide Index

Important Note: The Crop/Pest Guide Index refers to pages on which the subject can be located.

Arable and vegetable crops

Brassicas
- Brassica seed crops 5
- Brassicas, general 5
- Fodder brassicas 5
- Leaf and flowerhead brassicas 6
- Mustard 7
- Root brassicas 7
- Salad brassicas 8

Cereals
- Barley 9
- Cereals, general 13
- Maize/sweetcorn 13
- Oats 13
- Rye/triticale 15
- Undersown cereals 16
- Wheat 16

Edible fungi
- Mushrooms 21

Fruiting vegetables
- Aubergines 21
- Cucurbits 21
- Peppers 22
- Tomatoes 22

Herb crops
- Herbs 22

Leafy vegetables
- Endives 23
- Lettuce 23
- Spinach 24
- Watercress 24

Legumes
- Beans (Phaseolus) 25
- Beans (Vicia) 25
- Clovers 26
- Forage legumes, general 26
- Lupins 26
- Peas 27

Miscellaneous arable
- Miscellaneous arable crops 28
- Miscellaneous arable situations 28

Miscellaneous field vegetables
- All vegetables 28

Oilseed crops
- Linseed/flax 29
- Miscellaneous oilseeds 29
- Oilseed rape 30
- Soya 31
- Sunflowers 31

Root and tuber crops
- Beet crops 31
- Carrots/parsnips/celeriac 33
- Potatoes 34

Stem and bulb vegetables
- Asparagus 35
- Celery/chicory 35
- Globe artichokes/cardoons 36
- Onions/leeks/garlic 36

Flowers and ornamentals

Flowers
- Bedding plants, general 37
- Bulbs/corms 38
- Miscellaneous flowers 39
- Pot plants 39
- Protected bulbs/corms 40
- Protected flowers 40

Miscellaneous flowers and ornamentals
- Miscellaneous uses 40
- Soils 40

Ornamentals
- Nursery stock 40
- Trees and shrubs 42
- Woody ornamentals 42

Forestry

Forest nurseries, general
- Forest nurseries 43

Forestry plantations, general
- Forestry plantations 43

Miscellaneous forestry situations
- Cut logs/timber 43

Woodland on farms
- Woodland 44

Fruit and hops

Bush fruit
- Bilberries/blueberries/cranberries 44
- Currants 44
- Gooseberries 45
- Miscellaneous bush fruit 45
- Protected bush fruit 46

Cane fruit
- Outdoor cane fruit 46
- Protected cane fruit 47

Hops, general
- Hops 47

Miscellaneous fruit situations
- Fruit crops, general 47
- Fruit nursery stock 48
- Orchards 48
- Protected miscellaneous fruit 48

Miscellaneous nuts
- Nuts 48

Other fruit
- Rhubarb 48
- Vines 48

Soft fruit
Miscellaneous soft fruit 49
Strawberries 49
Tree fruit
Miscellaneous tree fruit 50
Pome fruit 50
Stone fruit 51

Grain/crop store uses

Stored produce
Food/produce storage 52
Stored seed
Stored grain/rape/linseed 52

Grass

Grass seed
Grass seed crops 52
Grassland
Leys 52
Permanent pasture 53
Turf/amenity grass
Amenity grassland 54
Managed amenity turf 54

Non-crop pest control

Farm buildings/yards
Farm buildings 55
Farmyards 55
Farmland pest control
Farmland situations 55
Miscellaneous non-crop pest control
Manure/rubbish 55
Miscellaneous pest control situations 56

Protected salad and vegetable crops

Protected brassicas
Protected brassica vegetables 56
Protected salad brassicas 56
Protected crops, general
All protected crops 56
Glasshouses 57
Protected vegetables, general 57
Soils and compost 57
Protected fruiting vegetables
Protected aubergines 57
Protected cucurbits 57
Protected tomatoes 58
Protected herb crops
Protected herbs 59
Protected leafy vegetables
Mustard and cress 59
Protected leafy vegetables 60
Protected spinach 60
Protected legumes
Protected peas and beans 60
Protected root and tuber vegetables
Protected carrots/parsnips/celeriac .. 60
Protected potatoes 61
Protected root brassicas 61
Protected stem and bulb vegetables
Protected asparagus 61
Protected celery/chicory 61
Protected onions/leeks/garlic 61

Total vegetation control

Aquatic situations, general
Aquatic situations 62
Non-crop areas, general
Farm buildings/yards 62
Miscellaneous non-crop situations ... 62
Non-crop farm areas 62
Paths/roads etc 62

Crop/Pest Guide

Important note: For convenience, some crops and pests or targets have been brought together into generic groups in this guide, e.g. 'cereals', 'annual grasses'. It is essential to check the profile entry in Section 2 *and* the label to ensure that a product is approved for a specific crop/pest combination, e.g. winter wheat/blackgrass.

Arable and vegetable crops

Brassicas - Brassica seed crops

Diseases	Alternaria	iprodione
	Botrytis	iprodione
Weeds	Broad-leaved weeds	carbetamide, propyzamide, trifluralin (*off-label*)
	Crops as weeds	carbetamide
	Grass weeds	carbetamide, propyzamide, trifluralin (*off-label*)

Brassicas - Brassicas, general

Diseases	Alternaria	iprodione (*seed treatment*)
	Downy mildew	chlorothalonil
	Ring spot	chlorothalonil
Pests	Aphids	chlorpyrifos
	Caterpillars	chlorpyrifos
	Cutworms	chlorpyrifos
	Flies	chlorpyrifos
	Leatherjackets	chlorpyrifos

Brassicas - Fodder brassicas

Diseases	Alternaria	iprodione, iprodione (*seed treatment*)
	Black rot	copper oxychloride (*off-label*)
	Damping off	fosetyl-aluminium (*off-label*)
	Downy mildew	fosetyl-aluminium (*off-label*), fosetyl-aluminium (*off-label - seedlings*)
	Ring spot	tebuconazole (*off-label*)
	Spear rot	copper oxychloride (*off-label*)
	White blister	boscalid + pyraclostrobin (*off-label*)
Pests	Aphids	carbosulfan, imidacloprid (*off-label - seed treatment*), nicotine, pirimicarb, pymetrozine (*off-label*)
	Beetles	alpha-cypermethrin, carbosulfan, deltamethrin
	Caterpillars	alpha-cypermethrin, Bacillus thuringiensis (*off-label*), cypermethrin, deltamethrin, nicotine
	Flies	carbosulfan, chlorpyrifos (*off-label*)
	Leaf miners	nicotine
	Pests, miscellaneous	dimethoate (*off-label*)
	Weevils	carbosulfan

Weeds	Broad-leaved weeds	chlorthal-dimethyl, chlorthal-dimethyl + propachlor, clopyralid, cyanazine (*off-label*), propachlor, sodium monochloroacetate, trifluralin
	Crops as weeds	fluazifop-P-butyl (*stockfeed only*)
	Grass weeds	chlorthal-dimethyl + propachlor, fluazifop-P-butyl (*stockfeed only*), propachlor, trifluralin
	Mayweeds	clopyralid

Brassicas - Leaf and flowerhead brassicas

Diseases	Alternaria	boscalid + pyraclostrobin, chlorothalonil, chlorothalonil + metalaxyl-M (*moderate control*), difenoconazole, iprodione, tebuconazole
	Black rot	azoxystrobin (*off-label*), copper oxychloride (*off-label*)
	Botrytis	chlorothalonil
	Damping off	fosetyl-aluminium (*off-label*), thiram (*seed treatment*)
	Damping off and foot rot	etridiazole, etridiazole (*seeds and seedlings*)
	Damping off and wirestem	tolclofos-methyl
	Downy mildew	chlorothalonil, chlorothalonil + metalaxyl-M, fosetyl-aluminium (*off-label*), fosetyl-aluminium (*off-label - seedlings*), propamocarb hydrochloride
	Light leaf spot	tebuconazole
	Phytophthora	etridiazole, etridiazole (*transplants*), propamocarb hydrochloride
	Powdery mildew	tebuconazole
	Pythium	propamocarb hydrochloride
	Ring spot	boscalid + pyraclostrobin, chlorothalonil, chlorothalonil + metalaxyl-M (*reduction*), difenoconazole, tebuconazole, tebuconazole (*off-label*)
	Spear rot	copper oxychloride (*off-label*)
	Storage rots	metalaxyl-M (*off-label*)
	White blister	boscalid + pyraclostrobin, boscalid + pyraclostrobin (*off-label*), boscalid + pyraclostrobin (*qualified minor use*), chlorothalonil + metalaxyl-M, mancozeb + metalaxyl-M (*off-label*)
Pests	Aphids	bifenthrin, carbosulfan, chlorpyrifos, cypermethrin, dimethoate, fatty acids, imidacloprid (*off-label - seed treatment*), lambda-cyhalothrin + pirimicarb, nicotine, pirimicarb, pymetrozine (*off-label*), triazamate
	Beetles	alpha-cypermethrin, carbosulfan, deltamethrin, lambda-cyhalothrin + pirimicarb
	Birds/mammals	aluminium ammonium sulphate
	Caterpillars	alpha-cypermethrin, Bacillus thuringiensis, Bacillus thuringiensis (*off-label*), bifenthrin, chlorpyrifos, cypermethrin, deltamethrin, diflubenzuron, lambda-cyhalothrin, lambda-cyhalothrin + pirimicarb, nicotine
	Cutworms	chlorpyrifos
	Flies	carbosulfan, chlorpyrifos, chlorpyrifos (*off-label*)
	Leaf miners	nicotine
	Leatherjackets	chlorpyrifos
	Mealybugs	fatty acids

	Pests, miscellaneous	dimethoate (*off-label*)
	Scale insects	fatty acids
	Slugs/snails	methiocarb
	Spider mites	fatty acids
	Weevils	carbosulfan
	Whiteflies	bifenthrin, chlorpyrifos, cypermethrin, fatty acids, lambda-cyhalothrin, lambda-cyhalothrin + pirimicarb
Weeds	Broad-leaved weeds	carbetamide, chlorthal-dimethyl, chlorthal-dimethyl + propachlor, clopyralid, cyanazine (*off-label*), metazachlor, pendimethalin, propachlor, pyridate, sodium monochloroacetate, sodium monochloroacetate (*off-label*), trifluralin
	Crops as weeds	carbetamide, cycloxydim, pendimethalin, tepraloxydim
	Grass weeds	carbetamide, chlorthal-dimethyl + propachlor, cycloxydim, fluazifop-P-butyl (*off-label*), metazachlor, pendimethalin, propachlor, tepraloxydim, trifluralin
	Mayweeds	clopyralid

Brassicas - Mustard

Crop control	Pre-harvest desiccation	diquat (*off-label*), glyphosate
Diseases	Alternaria	iprodione, iprodione (*seed treatment*)
	Botrytis	iprodione
Pests	Beetles	deltamethrin
	Midges	deltamethrin
	Weevils	deltamethrin
Weeds	Broad-leaved weeds	chlorthal-dimethyl + propachlor, diquat (*off-label*), glyphosate, propachlor, trifluralin
	Crops as weeds	glyphosate
	Grass weeds	chlorthal-dimethyl + propachlor, glyphosate, propachlor, trifluralin
	Weeds, miscellaneous	glyphosate

Brassicas - Root brassicas

Diseases	Alternaria	fenpropimorph (*off-label*), iprodione (*seed treatment*), iprodione + thiophanate-methyl (*off-label*), tebuconazole
	Canker	iprodione (*off-label*)
	Crown rot	fenpropimorph (*off-label*), iprodione + thiophanate-methyl (*off-label*)
	Damping off	thiram (*seed treatment*)
	Downy mildew	propamocarb hydrochloride (*off-label*)
	Fungus diseases	tebuconazole (*off-label*)
	Powdery mildew	fenpropimorph (*off-label*), sulphur, tebuconazole
	Rhizoctonia	azoxystrobin (*off-label*), tolclofos-methyl (*off-label - with fleece covers*), tolclofos-methyl (*off-label - without covers*)
	White blister	mancozeb + metalaxyl-M (*off-label*), metalaxyl-M (*off-label*), propamocarb hydrochloride (*off-label*)
Pests	Aphids	nicotine, pirimicarb, pirimicarb (*off-label*)

	Beetles	deltamethrin
	Caterpillars	deltamethrin, nicotine
	Flies	chlorpyrifos (*off-label*)
	Leaf miners	nicotine
	Pests, miscellaneous	dimethoate (*off-label*)
	Weevils	lambda-cyhalothrin (*off-label*)
Plant growth regulation	Quality/yield control	sulphur
Weeds	Broad-leaved weeds	chlorthal-dimethyl, chlorthal-dimethyl + propachlor, clopyralid, ethofumesate (*off-label*), glyphosate, metazachlor, propachlor, propachlor (*off-label*), trifluralin, trifluralin (*off-label*)
	Crops as weeds	cycloxydim, fluazifop-P-butyl (*stockfeed only*), glyphosate
	Grass weeds	chlorthal-dimethyl + propachlor, cycloxydim, ethofumesate (*off-label*), fluazifop-P-butyl (*stockfeed only*), glyphosate, metazachlor, propachlor, propachlor (*off-label*), propaquizafop, trifluralin, trifluralin (*off-label*)
	Mayweeds	clopyralid
	Weeds, miscellaneous	ethofumesate (*off-label*), glyphosate

Brassicas - Salad brassicas

Diseases	Alternaria	azoxystrobin (*off-label - for baby leaf production*), iprodione (*seed treatment*)
	Black rot	copper oxychloride (*off-label*)
	Damping off	fosetyl-aluminium (*off-label*)
	Damping off and wirestem	tolclofos-methyl
	Downy mildew	azoxystrobin (*off-label - for baby leaf production*), fosetyl-aluminium (*off-label*), fosetyl-aluminium (*off-label - seedlings*), propamocarb hydrochloride
	Phytophthora	propamocarb hydrochloride
	Pythium	propamocarb hydrochloride
	Ring spot	tebuconazole (*off-label*)
	Spear rot	copper oxychloride (*off-label*)
	White blister	boscalid + pyraclostrobin (*off-label*), boscalid + pyraclostrobin (*off-label - baby leaf production*)
Pests	Aphids	chlorpyrifos, cypermethrin (*off-label*), deltamethrin (*off-label*), nicotine, pirimicarb, pirimicarb (*off-label - for baby leaf production*)
	Beetles	deltamethrin (*off-label*), deltamethrin (*off-label - baby leaf production*), deltamethrin (*off-label - for baby leaf production*)
	Birds/mammals	aluminium ammonium sulphate
	Caterpillars	alpha-cypermethrin (*off-label - for baby leaf production*), Bacillus thuringiensis (*off-label*), chlorpyrifos, cypermethrin (*off-label*), deltamethrin (*off-label*), deltamethrin (*off-label - baby leaf production*), deltamethrin (*off-label - for baby leaf production*), nicotine
	Cutworms	chlorpyrifos

	Flies	chlorpyrifos
	Leaf miners	nicotine
	Leatherjackets	chlorpyrifos
	Pests, miscellaneous	deltamethrin (*off-label*), dimethoate (*off-label*)
	Whiteflies	chlorpyrifos
Weeds	Broad-leaved weeds	propachlor (*off-label*), propachlor (*off-label - for baby leaf production*)
	Grass weeds	fluazifop-P-butyl (*off-label*), propachlor (*off-label*)

Cereals - Barley

Crop control	Pre-harvest desiccation	diquat (*stockfeed only*), glyphosate
Diseases	Brown foot rot and ear blight	carboxin + thiram (*seed treatment*), fludioxonil (*seed treatment*), fluquinconazole + prochloraz (*seed treatment*), fuberidazole + imidacloprid + triadimenol (*seed treatment*), fuberidazole + imidacloprid + triadimenol (*seed treatment - reduction*), fuberidazole + triadimenol (*seed treatment*), guazatine (*seed treatment - reduction*), guazatine + imazalil (*seed treatment*), guazatine + imazalil (*seed treatment - moderate control*), imazalil + triticonazole (*seed treatment - moderate control*), prochloraz + triticonazole (*seed treatment*)
	Covered smut	carboxin + thiram (*seed treatment*), fludioxonil (*seed treatment*), fluquinconazole (*seed treatment*), fluquinconazole + prochloraz (*seed treatment*), fuberidazole + imidacloprid + triadimenol (*seed treatment*), fuberidazole + triadimenol (*seed treatment*), guazatine + imazalil (*seed treatment*), imazalil + triticonazole (*seed treatment*), prochloraz + triticonazole (*seed treatment*)
	Eyespot	azoxystrobin + cyproconazole (*reduction*), carbendazim, carbendazim + flusilazole, chlorothalonil + cyproconazole, cyproconazole, cyproconazole + cyprodinil, cyprodinil, cyprodinil + picoxystrobin, epoxiconazole (*reduction*), epoxiconazole + fenpropimorph (*reduction*), epoxiconazole + fenpropimorph + kresoxim-methyl (*reduction*), epoxiconazole + kresoxim-methyl (*reduction*), flusilazole, prochloraz, prochloraz + propiconazole, prochloraz + tebuconazole
	Foot rot	guazatine (*seed treatment*)
	Leaf stripe	carboxin + thiram (*seed treatment*), carboxin + thiram (*seed treatment - reduction*), fludioxonil (*seed treatment - reduction*), fuberidazole + triadimenol (*seed treatment*), guazatine (*seed treatment*), guazatine + imazalil (*seed treatment*), imazalil + triticonazole (*seed treatment*), imidacloprid + tebuconazole + triazoxide (*seed treatment*), prochloraz + triticonazole (*seed treatment*), tebuconazole + triazoxide (*seed treatment*)
	Loose smut	carboxin + thiram (*seed treatment - reduction*), fluquinconazole (*seed treatment*), fluquinconazole + prochloraz (*seed treatment*), fuberidazole + imidacloprid + triadimenol (*seed treatment*), fuberidazole + triadimenol (*seed treatment*), imazalil + triticonazole (*seed treatment*), imidacloprid + tebuconazole + triazoxide (*seed treatment*), prochloraz + triticonazole

(*seed treatment*), tebuconazole + triazoxide (*seed treatment*)

Net blotch	azoxystrobin, azoxystrobin + chlorothalonil, azoxystrobin + cyproconazole, azoxystrobin + fenpropimorph, carboxin + thiram (*seed treatment*), chlorothalonil + cyproconazole, chlorothalonil + tetraconazole, chlorothalonil + tetraconazole (*autumn sown*), cyproconazole, cyproconazole + cyprodinil, cyproconazole + propiconazole, cyproconazole + trifloxystrobin, cyprodinil, cyprodinil + picoxystrobin, epoxiconazole, epoxiconazole + fenpropimorph, epoxiconazole + fenpropimorph + kresoxim-methyl, epoxiconazole + fenpropimorph + pyraclostrobin, epoxiconazole + kresoxim-methyl, epoxiconazole + kresoxim-methyl + pyraclostrobin, epoxiconazole + pyraclostrobin, famoxadone + flusilazole, fenpropimorph + flusilazole, fenpropimorph + pyraclostrobin, flusilazole, fuberidazole + imidacloprid + triadimenol (*seed treatment - seed-borne only*), guazatine (*seed treatment*), guazatine + imazalil (*seed treatment - moderate control*), imazalil + triticonazole (*seed treatment - moderate control*), imidacloprid + tebuconazole + triazoxide (*seed treatment*), iprodione, mancozeb, metconazole (*reduction*), picoxystrobin, prochloraz, prochloraz + tebuconazole, propiconazole + trifloxystrobin, pyraclostrobin, spiroxamine + tebuconazole, tebuconazole, tebuconazole + triadimenol, tebuconazole + triazoxide (*seed treatment*), trifloxystrobin
Powdery mildew	azoxystrobin, azoxystrobin + cyproconazole, azoxystrobin + fenpropimorph, carbendazim + flusilazole, chlorothalonil + cyproconazole, chlorothalonil + tetraconazole, chlorothalonil + tetraconazole (*autumn sown*), cyproconazole, cyproconazole + cyprodinil, cyproconazole + propiconazole, cyproconazole + trifloxystrobin, cyprodinil, cyprodinil + picoxystrobin, epoxiconazole, epoxiconazole + fenpropimorph, epoxiconazole + fenpropimorph + kresoxim-methyl, epoxiconazole + fenpropimorph + pyraclostrobin, epoxiconazole + kresoxim-methyl, fenpropidin, fenpropimorph, fenpropimorph + flusilazole, fenpropimorph + kresoxim-methyl, fenpropimorph + pyraclostrobin, fenpropimorph + quinoxyfen, fluquinconazole + prochloraz, flusilazole, flutriafol, fuberidazole + imidacloprid + triadimenol (*seed treatment*), fuberidazole + triadimenol (*seed treatment*), metconazole, metconazole (*moderate control*), metrafenone, picoxystrobin, prochloraz, prochloraz + tebuconazole, propiconazole, propiconazole + trifloxystrobin, quinoxyfen, spiroxamine, spiroxamine + tebuconazole, sulphur, tebuconazole, tebuconazole + triadimenol, tetraconazole, trifloxystrobin
Rhynchosporium	azoxystrobin, azoxystrobin + chlorothalonil, azoxystrobin + cyproconazole (*moderate control*), azoxystrobin + fenpropimorph, carbendazim, chlorothalonil, chlorothalonil + cyproconazole, chlorothalonil + tetraconazole, chlorothalonil + tetraconazole (*autumn sown*), cyproconazole, cyproconazole + cyprodinil, cyproconazole + propiconazole, cyproconazole + trifloxystrobin, cyprodinil, cyprodinil + picoxystrobin, epoxiconazole,

		epoxiconazole + fenpropimorph, epoxiconazole + fenpropimorph + kresoxim-methyl, epoxiconazole + fenpropimorph + pyraclostrobin, epoxiconazole + kresoxim-methyl, epoxiconazole + kresoxim-methyl + pyraclostrobin, epoxiconazole + pyraclostrobin, famoxadone + flusilazole, fenpropidin, fenpropimorph, fenpropimorph + flusilazole, fenpropimorph + kresoxim-methyl, fenpropimorph + pyraclostrobin, fluquinconazole + prochloraz, flusilazole, flutriafol, mancozeb, metconazole, picoxystrobin, prochloraz, prochloraz + propiconazole, prochloraz + tebuconazole, propiconazole, propiconazole + trifloxystrobin, pyraclostrobin (*moderate*), spiroxamine (*reduction*), spiroxamine + tebuconazole, tebuconazole, tebuconazole + triadimenol, trifloxystrobin
	Rust	azoxystrobin, azoxystrobin + chlorothalonil, azoxystrobin + cyproconazole, azoxystrobin + fenpropimorph, carbendazim + flusilazole, chlorothalonil + cyproconazole, chlorothalonil + tetraconazole, chlorothalonil + tetraconazole (*autumn sown*), cyproconazole, cyproconazole + propiconazole, cyproconazole + trifloxystrobin, cyprodinil + picoxystrobin, epoxiconazole, epoxiconazole + fenpropimorph, epoxiconazole + fenpropimorph + kresoxim-methyl, epoxiconazole + fenpropimorph + pyraclostrobin, epoxiconazole + kresoxim-methyl, epoxiconazole + kresoxim-methyl + pyraclostrobin, epoxiconazole + pyraclostrobin, famoxadone + flusilazole, fenpropidin, fenpropimorph, fenpropimorph + flusilazole, fenpropimorph + pyraclostrobin, fluquinconazole (*seed treatment*), fluquinconazole + prochloraz, fluquinconazole + prochloraz (*seed treatment*), flusilazole, flutriafol, fuberidazole + imidacloprid + triadimenol (*seed treatment*), fuberidazole + triadimenol (*seed treatment*), mancozeb, metconazole, picoxystrobin, prochloraz + tebuconazole, propiconazole, propiconazole + trifloxystrobin, pyraclostrobin, spiroxamine, spiroxamine + tebuconazole, tebuconazole, tebuconazole + triadimenol, tetraconazole, trifloxystrobin
	Septoria diseases	flusilazole
	Snow mould	fludioxonil (*seed treatment*), guazatine + imazalil (*seed treatment*)
	Snow rot	fuberidazole + imidacloprid + triadimenol (*seed treatment - reduction*)
	Sooty moulds	mancozeb
	Take-all	fluquinconazole (*seed treatment - reduction*), fluquinconazole + prochloraz (*seed treatment - reduction*), silthiofam (*seed treatment*)
Pests	Aphids	alpha-cypermethrin, bifenthrin, chlorpyrifos, cypermethrin, cypermethrin (*autumn sown*), deltamethrin, esfenvalerate, fuberidazole + imidacloprid + triadimenol (*seed treatment*), imidacloprid + tebuconazole + triazoxide (*seed treatment*), lambda-cyhalothrin, lambda-cyhalothrin + pirimicarb, pirimicarb, tau-fluvalinate, zeta-cypermethrin
	Birds/mammals	aluminium ammonium sulphate

	Flies	alpha-cypermethrin, chlorpyrifos, cypermethrin, cypermethrin (*autumn sown*), deltamethrin, tefluthrin (*seed treatment*)
	Leatherjackets	chlorpyrifos
	Slugs/snails	thiodicarb
	Thrips	chlorpyrifos
	Wireworms	fuberidazole + imidacloprid + triadimenol (*reduction of damage*), imidacloprid + tebuconazole + triazoxide (*reduction of damage*), tefluthrin (*seed treatment*)
Plant growth regulation	Growth control	2-chloroethylphosphonic acid, 2-chloroethylphosphonic acid + mepiquat chloride, chlormequat, chlormequat + 2-chloroethylphosphonic acid, chlormequat + 2-chloroethylphosphonic acid + mepiquat chloride, chlormequat with choline chloride, trinexapac-ethyl
	Quality/yield control	2-chloroethylphosphonic acid + mepiquat chloride (*low lodging situations*), chlormequat, chlormequat with choline chloride, sulphur
Weeds	Broad-leaved weeds	2,4-D, 2,4-D + MCPA, 2,4-DB + MCPA, amidosulfuron, amidosulfuron + iodosulfuron-methyl-sodium, bifenox, bromoxynil, bromoxynil + diflufenican + ioxynil, bromoxynil + ioxynil, bromoxynil + ioxynil + mecoprop-P, carfentrazone-ethyl, carfentrazone-ethyl + mecoprop-P, carfentrazone-ethyl + metsulfuron-methyl, chlorotoluron, chlorotoluron + isoproturon, clopyralid, dicamba + MCPA + mecoprop-P, dicamba + mecoprop-P, dichlorprop-P, dichlorprop-P + MCPA + mecoprop-P, diflufenican + flufenacet, diflufenican + flurtamone, diflufenican + flurtamone + isoproturon, diflufenican + isoproturon, diflufenican + trifluralin, florasulam, florasulam + fluroxypyr, flufenacet + pendimethalin, fluroxypyr, glyphosate, imazamethabenz-methyl, isoproturon, isoproturon (*off-label*), isoproturon + pendimethalin, isoproturon + simazine, isoproturon + trifluralin, linuron, linuron + trifluralin, MCPA, MCPA + MCPB, MCPB, mecoprop-P, metsulfuron-methyl, metsulfuron-methyl + thifensulfuron-methyl, metsulfuron-methyl + tribenuron-methyl, pendimethalin, pendimethalin + picolinafen, thifensulfuron-methyl, thifensulfuron-methyl + tribenuron-methyl, tribenuron-methyl, trifluralin
	Crops as weeds	amidosulfuron + iodosulfuron-methyl-sodium, diclofop-methyl + fenoxaprop-P-ethyl, diflufenican + flurtamone, diflufenican + flurtamone + isoproturon, florasulam, fluroxypyr, glyphosate, imazamethabenz-methyl, pendimethalin, tralkoxydim, tralkoxydim (*autumn sown*)
	General weed control	diquat (*stockfeed only*)
	Grass weeds	chlorotoluron, chlorotoluron + isoproturon, diclofop-methyl + fenoxaprop-P-ethyl, diflufenican + flufenacet, diflufenican + flurtamone, diflufenican + flurtamone + isoproturon, diflufenican + isoproturon, diflufenican + trifluralin, flufenacet + pendimethalin, glyphosate, imazamethabenz-methyl, isoproturon, isoproturon (*off-label*), isoproturon + pendimethalin, isoproturon + simazine, isoproturon + trifluralin, linuron, linuron + trifluralin, pendimethalin, pendimethalin + picolinafen, tralkoxydim, tralkoxydim (*autumn sown*), tralkoxydim (*autumn sown - qualified minor use*), tralkoxydim (*qualified minor use*), tri-allate, trifluralin

	Mayweeds	amidosulfuron + iodosulfuron-methyl-sodium, bromoxynil + diflufenican + ioxynil, clopyralid, dicamba + MCPA + mecoprop-P, dicamba + mecoprop-P, dichlorprop-P + MCPA + mecoprop-P, diflufenican + flufenacet, florasulam + fluroxypyr, metsulfuron-methyl, metsulfuron-methyl + thifensulfuron-methyl, metsulfuron-methyl + tribenuron-methyl, thifensulfuron-methyl, thifensulfuron-methyl + tribenuron-methyl, tribenuron-methyl
	Polygonums	dicamba + MCPA + mecoprop-P, dicamba + mecoprop-P, metsulfuron-methyl + thifensulfuron-methyl
	Speedwells	bifenox, bromoxynil + diflufenican + ioxynil, carfentrazone-ethyl + mecoprop-P, diflufenican + flufenacet, flufenacet + pendimethalin, metsulfuron-methyl + thifensulfuron-methyl, pendimethalin + picolinafen
	Weeds, miscellaneous	glyphosate

Cereals - Cereals, general

Pests	Leatherjackets	methiocarb (*reduction*)
	Slugs/snails	methiocarb
Weeds	Broad-leaved weeds	2,4-DB + MCPA, MCPA + MCPB

Cereals - Maize/sweetcorn

Diseases	Damping off	thiram (*seed treatment*)
Pests	Aphids	nicotine, pirimicarb, pirimicarb (*off-label*), pymetrozine (*off-label*)
	Caterpillars	nicotine
	Flies	chlorpyrifos, lambda-cyhalothrin (*off-label*)
	Leaf miners	nicotine
	Slugs/snails	methiocarb
Weeds	Broad-leaved weeds	atrazine, bromoxynil, bromoxynil (*off-label*), bromoxynil + prosulfuron, clopyralid, fluroxypyr, nicosulfuron, pendimethalin, pendimethalin (*off-label*), pyridate, rimsulfuron
	Crops as weeds	fluroxypyr, nicosulfuron, pendimethalin, rimsulfuron
	Grass weeds	atrazine, nicosulfuron, nicosulfuron (*off-label*), pendimethalin
	Mayweeds	bromoxynil + prosulfuron, clopyralid, nicosulfuron

Cereals - Oats

Crop control	Pre-harvest desiccation	diquat (*stockfeed only*), glyphosate
Diseases	Brown foot rot and ear blight	carboxin + thiram (*seed treatment*), fludioxonil (*seed treatment*), fuberidazole + imidacloprid + triadimenol (*seed treatment - reduction*), fuberidazole + triadimenol (*seed treatment*), guazatine (*seed treatment - reduction*)
	Covered smut	carboxin + thiram (*seed treatment*)
	Crown rust	azoxystrobin, azoxystrobin + cyproconazole, chlorothalonil + tetraconazole, cyproconazole,

		epoxiconazole + fenpropimorph + pyraclostrobin, epoxiconazole + kresoxim-methyl + pyraclostrobin, epoxiconazole + pyraclostrobin, fuberidazole + triadimenol (*seed treatment*), picoxystrobin, propiconazole, pyraclostrobin, tebuconazole (*reduction*), tebuconazole + triadimenol, tetraconazole
	Eyespot	epoxiconazole + fenpropimorph + kresoxim-methyl (*reduction*), epoxiconazole + kresoxim-methyl (*reduction*)
	Fusarium root rot	bitertanol + fuberidazole (*seed treatment*), bitertanol + fuberidazole + imidacloprid (*seed treatment*)
	Loose smut	carboxin + thiram (*seed treatment - reduction*), fuberidazole + imidacloprid + triadimenol (*seed treatment*), fuberidazole + triadimenol (*seed treatment*)
	Powdery mildew	azoxystrobin, azoxystrobin + cyproconazole, cyproconazole, epoxiconazole, epoxiconazole + fenpropimorph, epoxiconazole + fenpropimorph + kresoxim-methyl, epoxiconazole + fenpropimorph + pyraclostrobin, epoxiconazole + kresoxim-methyl, fenpropimorph, fenpropimorph + kresoxim-methyl, fenpropimorph + quinoxyfen, fuberidazole + imidacloprid + triadimenol (*seed treatment*), picoxystrobin, propiconazole, quinoxyfen, sulphur, tebuconazole, tebuconazole + triadimenol
	Pyrenophora leaf spot	fludioxonil (*seed treatment*), fuberidazole + imidacloprid + triadimenol (*seed treatment*), fuberidazole + triadimenol (*seed treatment*), guazatine (*seed treatment*), guazatine + imazalil (*seed treatment*)
Pests	Aphids	bifenthrin, bitertanol + fuberidazole + imidacloprid (*seed treatment*), chlorpyrifos, cypermethrin, deltamethrin, fuberidazole + imidacloprid + triadimenol (*seed treatment*), lambda-cyhalothrin, lambda-cyhalothrin + pirimicarb, pirimicarb, zeta-cypermethrin
	Birds/mammals	aluminium ammonium sulphate
	Flies	chlorpyrifos, cypermethrin, deltamethrin, tefluthrin (*seed treatment*)
	Leatherjackets	chlorpyrifos
	Midges	chlorpyrifos
	Slugs/snails	thiodicarb
	Thrips	chlorpyrifos
	Wireworms	bitertanol + fuberidazole + imidacloprid (*seed treatment - reduction of damage*), fuberidazole + imidacloprid + triadimenol (*reduction of damage*), tefluthrin (*seed treatment*)
Plant growth regulation	Growth control	chlormequat, chlormequat with choline chloride, trinexapac-ethyl
Weeds	Broad-leaved weeds	2,4-D, 2,4-D + MCPA, 2,4-DB + MCPA, amidosulfuron, bromoxynil, bromoxynil + ioxynil, bromoxynil + ioxynil + mecoprop-P, carfentrazone-ethyl, carfentrazone-ethyl + mecoprop-P, carfentrazone-ethyl + metsulfuron-methyl, clopyralid, dicamba + MCPA + mecoprop-P, dicamba + mecoprop-P, dichlorprop-P, dichlorprop-P + MCPA + mecoprop-P, florasulam, florasulam + fluroxypyr, fluroxypyr, glyphosate, linuron, MCPA, MCPA + MCPB, MCPB, mecoprop-P, metsulfuron-methyl, metsulfuron-methyl + thifensulfuron-methyl, metsulfuron-methyl + tribenuron-methyl, tribenuron-methyl
	Crops as weeds	florasulam, fluroxypyr, glyphosate

	General weed control	diquat (*stockfeed only*)
	Grass weeds	glyphosate, linuron
	Mayweeds	clopyralid, dicamba + MCPA + mecoprop-P, dicamba + mecoprop-P, dichlorprop-P + MCPA + mecoprop-P, florasulam + fluroxypyr, metsulfuron-methyl, metsulfuron-methyl + thifensulfuron-methyl, metsulfuron-methyl + tribenuron-methyl, tribenuron-methyl
	Polygonums	dicamba + MCPA + mecoprop-P, dicamba + mecoprop-P
	Speedwells	carfentrazone-ethyl + mecoprop-P, metsulfuron-methyl + thifensulfuron-methyl
	Weeds, miscellaneous	glyphosate

Cereals - Rye/triticale

Diseases	Blue mould	fuberidazole + triadimenol (*seed treatment*)
	Brown foot rot and ear blight	carboxin + thiram (*seed treatment*), fuberidazole + triadimenol (*seed treatment*)
	Eyespot	epoxiconazole + fenpropimorph + kresoxim-methyl (*reduction*), epoxiconazole + kresoxim-methyl (*reduction*), prochloraz, prochloraz + tebuconazole
	Fusarium root rot	bitertanol + fuberidazole (*seed treatment*)
	Powdery mildew	azoxystrobin, azoxystrobin + cyproconazole, cyproconazole, epoxiconazole, epoxiconazole + fenpropimorph, epoxiconazole + fenpropimorph + kresoxim-methyl, epoxiconazole + kresoxim-methyl, fenpropimorph, fenpropimorph + kresoxim-methyl, fenpropimorph + quinoxyfen, prochloraz, prochloraz + tebuconazole, propiconazole, quinoxyfen, spiroxamine, spiroxamine + tebuconazole, sulphur, tebuconazole, tebuconazole + triadimenol
	Rhynchosporium	azoxystrobin, azoxystrobin (*off-label*), azoxystrobin + cyproconazole (*moderate control*), epoxiconazole, epoxiconazole + fenpropimorph, epoxiconazole + fenpropimorph + kresoxim-methyl, epoxiconazole + kresoxim-methyl, fenpropimorph + kresoxim-methyl, prochloraz, prochloraz + tebuconazole, propiconazole, spiroxamine (*reduction*), tebuconazole
	Rust	azoxystrobin, azoxystrobin (*off-label*), azoxystrobin + cyproconazole, cyproconazole, epoxiconazole, epoxiconazole + fenpropimorph, epoxiconazole + fenpropimorph + kresoxim-methyl, epoxiconazole + kresoxim-methyl, fenpropimorph, fuberidazole + triadimenol (*seed treatment*), prochloraz + tebuconazole, propiconazole, spiroxamine, spiroxamine + tebuconazole, tebuconazole, tebuconazole + triadimenol
	Septoria diseases	epoxiconazole, epoxiconazole + fenpropimorph, epoxiconazole + fenpropimorph + kresoxim-methyl, epoxiconazole + kresoxim-methyl, fenpropimorph + kresoxim-methyl (*reduction*), prochloraz, propiconazole
Pests	Aphids	cypermethrin, dimethoate, lambda-cyhalothrin + pirimicarb, pirimicarb

	Flies	cypermethrin, dimethoate
	Slugs/snails	thiodicarb
Plant growth regulation	Growth control	2-chloroethylphosphonic acid, 2-chloroethylphosphonic acid + mepiquat chloride, chlormequat, chlormequat with choline chloride, trinexapac-ethyl
Weeds	Broad-leaved weeds	2,4-D, amidosulfuron, amidosulfuron + iodosulfuron-methyl-sodium, bifenox, bromoxynil + diflufenican + ioxynil, bromoxynil + ioxynil, bromoxynil + ioxynil + mecoprop-P, carfentrazone-ethyl, carfentrazone-ethyl + metsulfuron-methyl, chlorotoluron, dicamba + MCPA + mecoprop-P, dicamba + mecoprop-P, diflufenican + isoproturon, diflufenican + trifluralin, fluroxypyr, iodosulfuron-methyl-sodium, isoproturon, isoproturon + pendimethalin, linuron + trifluralin, MCPA, metsulfuron-methyl, metsulfuron-methyl + tribenuron-methyl, pendimethalin, tribenuron-methyl
	Crops as weeds	amidosulfuron + iodosulfuron-methyl-sodium, fluroxypyr, pendimethalin, tralkoxydim
	Grass weeds	chlorotoluron, clodinafop-propargyl, diflufenican + isoproturon, iodosulfuron-methyl-sodium (*from seed*), isoproturon, isoproturon + pendimethalin, linuron + trifluralin, pendimethalin, tralkoxydim, tralkoxydim (*qualified minor use*), tri-allate
	Mayweeds	amidosulfuron + iodosulfuron-methyl-sodium, bromoxynil + diflufenican + ioxynil, dicamba + MCPA + mecoprop-P, dicamba + mecoprop-P, iodosulfuron-methyl-sodium, metsulfuron-methyl, metsulfuron-methyl + tribenuron-methyl, tribenuron-methyl
	Polygonums	dicamba + MCPA + mecoprop-P, dicamba + mecoprop-P
	Speedwells	bifenox, bromoxynil + diflufenican + ioxynil, iodosulfuron-methyl-sodium

Cereals - Undersown cereals

Weeds	Broad-leaved weeds	2,4-DB (*undersown with lucerne*), 2,4-DB + linuron + MCPA (*undersown with clover*), 2,4-DB + MCPA, 2,4-DB + MCPA (*red or white clover*), bentazone + MCPA + MCPB, bromoxynil + ioxynil, dicamba + MCPA + mecoprop-P, dicamba + MCPA + mecoprop-P (*grass only*), dicamba + mecoprop-P, MCPA, MCPA (*red clover or grass*), MCPA + MCPB, MCPB, mecoprop-P
	Mayweeds	dicamba + MCPA + mecoprop-P, dicamba + MCPA + mecoprop-P (*grass only*), dicamba + mecoprop-P
	Polygonums	2,4-DB + MCPA, dicamba + MCPA + mecoprop-P, dicamba + MCPA + mecoprop-P (*grass only*), dicamba + mecoprop-P

Cereals - Wheat

Crop control	Pre-harvest desiccation	glyphosate
Diseases	Alternaria	iprodione
	Blue mould	fuberidazole + imidacloprid + triadimenol (*seed treatment*), fuberidazole + triadimenol (*seed treatment*)
	Botrytis	iprodione

Brown foot rot and ear blight	bitertanol + fuberidazole + imidacloprid (*seed treatment - reduction*), carboxin + thiram (*seed treatment*), dimoxystrobin + epoxiconazole, epoxiconazole, epoxiconazole (*reduction*), epoxiconazole + fenpropimorph, epoxiconazole + fenpropimorph (*reduction*), epoxiconazole + fenpropimorph + kresoxim-methyl (*reduction*), epoxiconazole + fenpropimorph + pyraclostrobin (*reduction*), epoxiconazole + kresoxim-methyl (*reduction*), fludioxonil (*seed treatment*), fuberidazole + imidacloprid + triadimenol (*seed treatment - reduction*), fuberidazole + triadimenol (*seed treatment*), guazatine (*seed treatment - reduction*), guazatine + triticonazole (*seed treatment*), metconazole (*reduction*), prochloraz + triticonazole (*seed treatment*), spiroxamine + tebuconazole, tebuconazole, tebuconazole + triadimenol
Drechslera leaf spot	dimoxystrobin + epoxiconazole
Eyespot	azoxystrobin + cyproconazole (*reduction*), carbendazim, carbendazim + flusilazole, chlorothalonil + cyproconazole, cyproconazole, cyproconazole + cyprodinil, cyprodinil, cyprodinil + picoxystrobin (*moderate control*), epoxiconazole (*reduction*), epoxiconazole + fenpropimorph (*reduction*), epoxiconazole + fenpropimorph + kresoxim-methyl (*reduction*), epoxiconazole + kresoxim-methyl (*reduction*), flusilazole, metrafenone (*reduction*), picoxystrobin (*reduction*), prochloraz, prochloraz + propiconazole, prochloraz + tebuconazole
Fusarium root rot	bitertanol + fuberidazole (*seed treatment*), bitertanol + fuberidazole + imidacloprid (*seed treatment*), fluquinconazole + prochloraz (*seed treatment*)
Late ear diseases	azoxystrobin, azoxystrobin + fenpropimorph, chlorothalonil, chlorothalonil + flutriafol, cyproconazole, cyproconazole + propiconazole, picoxystrobin, prochloraz + tebuconazole
Loose smut	bitertanol + fuberidazole (*seed treatment - reduction*), bitertanol + fuberidazole + imidacloprid (*seed treatment - reduction*), fuberidazole + imidacloprid + triadimenol (*seed treatment*), fuberidazole + triadimenol (*seed treatment*), prochloraz + triticonazole (*seed treatment*)
Powdery mildew	azoxystrobin, azoxystrobin + cyproconazole, carbendazim + flusilazole, chlorothalonil + cyproconazole, chlorothalonil + flutriafol, chlorothalonil + tetraconazole, chlorothalonil + tetraconazole (*autumn sown*), cyproconazole, cyproconazole + cyprodinil, cyproconazole + propiconazole, cyproconazole + trifloxystrobin, cyprodinil, epoxiconazole, epoxiconazole + fenpropimorph, fenpropidin, fenpropimorph, fenpropimorph + flusilazole, fenpropimorph + quinoxyfen, fluquinconazole, fluquinconazole + prochloraz (*moderate control*), flusilazole, flutriafol, fuberidazole + imidacloprid + triadimenol (*seed treatment*), fuberidazole + triadimenol (*seed treatment*), metconazole (*moderate control*), metrafenone, prochloraz, prochloraz + tebuconazole, propiconazole, quinoxyfen, spiroxamine, spiroxamine + tebuconazole, sulphur, tebuconazole, tebuconazole + triadimenol, tetraconazole

Rust	azoxystrobin, azoxystrobin + chlorothalonil, azoxystrobin + cyproconazole, azoxystrobin + fenpropimorph, carbendazim + flusilazole, chlorothalonil + cyproconazole, chlorothalonil + flutriafol, chlorothalonil + tetraconazole, chlorothalonil + tetraconazole (*autumn sown*), cyproconazole, cyproconazole + cyprodinil, cyproconazole + propiconazole, cyproconazole + trifloxystrobin, cyprodinil + picoxystrobin, difenoconazole, dimoxystrobin + epoxiconazole, epoxiconazole, epoxiconazole + fenpropimorph, epoxiconazole + fenpropimorph + kresoxim-methyl, epoxiconazole + fenpropimorph + pyraclostrobin, epoxiconazole + kresoxim-methyl, epoxiconazole + kresoxim-methyl + pyraclostrobin, epoxiconazole + pyraclostrobin, famoxadone + flusilazole, fenpropidin, fenpropimorph, fenpropimorph + flusilazole, fluquinconazole, fluquinconazole (*seed treatment*), fluquinconazole + prochloraz, fluquinconazole + prochloraz (*seed treatment*), flusilazole, flutriafol, fuberidazole + imidacloprid + triadimenol (*seed treatment*), fuberidazole + triadimenol (*seed treatment*), mancozeb, metconazole, picoxystrobin, prochloraz + tebuconazole, propiconazole, propiconazole + trifloxystrobin, pyraclostrobin, spiroxamine, spiroxamine + tebuconazole, tebuconazole, tebuconazole + triadimenol, tetraconazole, trifloxystrobin
Septoria diseases	azoxystrobin, azoxystrobin + chlorothalonil, azoxystrobin + cyproconazole, azoxystrobin + fenpropimorph, bitertanol + fuberidazole (*seed treatment - reduction*), carbendazim + flusilazole, carboxin + thiram (*seed treatment*), chlorothalonil, chlorothalonil + cyproconazole, chlorothalonil + flutriafol, chlorothalonil + mancozeb, chlorothalonil + tetraconazole, chlorothalonil + tetraconazole (*autumn sown*), cyproconazole, cyproconazole + cyprodinil, cyproconazole + propiconazole, cyproconazole + trifloxystrobin, cyprodinil + picoxystrobin, difenoconazole, dimoxystrobin + epoxiconazole, epoxiconazole, epoxiconazole + fenpropimorph, epoxiconazole + fenpropimorph + kresoxim-methyl, epoxiconazole + fenpropimorph + pyraclostrobin, epoxiconazole + kresoxim-methyl, epoxiconazole + kresoxim-methyl + pyraclostrobin, epoxiconazole + pyraclostrobin, famoxadone + flusilazole, fenpropidin, fenpropimorph + flusilazole, fenpropimorph + kresoxim-methyl (*reduction*), fluquinconazole, fluquinconazole (*seed treatment*), fluquinconazole + prochloraz, fluquinconazole + prochloraz (*seed treatment*), flusilazole, flutriafol, fuberidazole + imidacloprid + triadimenol (*seed treatment*), fuberidazole + imidacloprid + triadimenol (*seed treatment - reduction*), fuberidazole + triadimenol (*seed treatment*), guazatine (*seed treatment*), iprodione, mancozeb, metconazole, picoxystrobin, prochloraz, prochloraz + propiconazole, prochloraz + tebuconazole, prochloraz + triticonazole (*seed treatment*), propiconazole, propiconazole + trifloxystrobin, pyraclostrobin, spiroxamine + tebuconazole, tebuconazole, tebuconazole + triadimenol, tetraconazole, trifloxystrobin

	Snow mould	fludioxonil (*seed treatment*)
	Sooty moulds	chlorothalonil + tetraconazole, epoxiconazole (*reduction*), epoxiconazole + fenpropimorph (*reduction*), epoxiconazole + fenpropimorph + kresoxim-methyl (*reduction*), epoxiconazole + kresoxim-methyl (*reduction*), mancozeb, propiconazole, spiroxamine + tebuconazole, tebuconazole, tebuconazole + triadimenol, tetraconazole
	Stinking smut	bitertanol + fuberidazole (*seed treatment*), bitertanol + fuberidazole + imidacloprid (*seed treatment*), carboxin + thiram (*seed treatment*), fludioxonil (*seed treatment*), fluquinconazole (*seed treatment*), fluquinconazole + prochloraz (*seed treatment*), fuberidazole + imidacloprid + triadimenol (*seed treatment*), fuberidazole + triadimenol (*seed treatment*), guazatine + triticonazole (*seed treatment*), prochloraz + triticonazole (*seed treatment*)
	Take-all	fluquinconazole (*seed treatment - reduction*), fluquinconazole + prochloraz (*seed treatment - reduction*), silthiofam (*seed treatment*)
Pests	Aphids	alpha-cypermethrin, bifenthrin, bitertanol + fuberidazole + imidacloprid (*seed treatment*), chlorpyrifos, cypermethrin, cypermethrin (*autumn sown*), deltamethrin, dimethoate, esfenvalerate, fuberidazole + imidacloprid + triadimenol (*seed treatment*), lambda-cyhalothrin, lambda-cyhalothrin + pirimicarb, pirimicarb, tau-fluvalinate, zeta-cypermethrin
	Birds/mammals	aluminium ammonium sulphate
	Flies	alpha-cypermethrin, chlorpyrifos, cypermethrin, cypermethrin (*autumn sown*), deltamethrin, dimethoate, lambda-cyhalothrin, tefluthrin (*seed treatment*)
	Leatherjackets	chlorpyrifos
	Midges	chlorpyrifos
	Slugs/snails	thiodicarb
	Thrips	chlorpyrifos
	Wireworms	bitertanol + fuberidazole + imidacloprid (*seed treatment - reduction of damage*), fuberidazole + imidacloprid + triadimenol (*reduction of damage*), tefluthrin (*seed treatment*)
Plant growth regulation	Growth control	2-chloroethylphosphonic acid, 2-chloroethylphosphonic acid + mepiquat chloride, chlormequat, chlormequat + 2-chloroethylphosphonic acid, chlormequat + 2-chloroethylphosphonic acid + mepiquat chloride, chlormequat + mepiquat chloride, chlormequat with choline chloride, chlormequat with choline chloride + imazaquin, fuberidazole + imidacloprid + triadimenol (*seed treatment - reduction*), trinexapac-ethyl
	Quality/yield control	chlormequat with choline chloride + imazaquin, sulphur
Weeds	Broad-leaved weeds	2,4-D, 2,4-D + MCPA, 2,4-DB + MCPA, amidosulfuron, amidosulfuron + iodosulfuron-methyl-sodium, bifenox, bromoxynil, bromoxynil + diflufenican + ioxynil, bromoxynil + ioxynil, bromoxynil + ioxynil + mecoprop-P, carfentrazone-ethyl, carfentrazone-ethyl + mecoprop-P, carfentrazone-ethyl + metsulfuron-methyl, chlorotoluron, chlorotoluron + isoproturon, clodinafop-propargyl + trifluralin, clopyralid, dicamba + MCPA +

	mecoprop-P, dicamba + mecoprop-P, dichlorprop-P, dichlorprop-P + MCPA + mecoprop-P, diflufenican + flufenacet, diflufenican + flurtamone, diflufenican + flurtamone + isoproturon, diflufenican + isoproturon, diflufenican + trifluralin, florasulam, florasulam + fluroxypyr, flufenacet + pendimethalin, flupyrsulfuron-methyl + thifensulfuron-methyl, fluroxypyr, glyphosate, imazamethabenz-methyl, iodosulfuron-methyl-sodium, iodosulfuron-methyl-sodium + mesosulfuron-methyl, isoproturon, isoproturon (*autumn sown*), isoproturon + pendimethalin, isoproturon + pendimethalin (*autumn sown*), isoproturon + simazine, isoproturon + trifluralin, linuron, linuron + trifluralin, MCPA, MCPA + MCPB, MCPB, mecoprop-P, metsulfuron-methyl, metsulfuron-methyl + thifensulfuron-methyl, metsulfuron-methyl + tribenuron-methyl, pendimethalin, pendimethalin + picolinafen, sulfosulfuron, thifensulfuron-methyl, thifensulfuron-methyl + tribenuron-methyl, tribenuron-methyl, trifluralin
Crops as weeds	amidosulfuron + iodosulfuron-methyl-sodium, diclofop-methyl + fenoxaprop-P-ethyl, diflufenican + flurtamone, diflufenican + flurtamone + isoproturon, florasulam, fluroxypyr, glyphosate, imazamethabenz-methyl, pendimethalin, tralkoxydim, tralkoxydim (*autumn sown*)
Grass weeds	chlorotoluron, chlorotoluron + isoproturon, clodinafop-propargyl, clodinafop-propargyl + trifluralin, diclofop-methyl + fenoxaprop-P-ethyl, diflufenican + flufenacet, diflufenican + flurtamone, diflufenican + flurtamone + isoproturon, diflufenican + isoproturon, diflufenican + trifluralin, fenoxaprop-P-ethyl, flufenacet + pendimethalin, flupyrsulfuron-methyl + thifensulfuron-methyl, glyphosate, imazamethabenz-methyl, iodosulfuron-methyl-sodium (*from seed*), iodosulfuron-methyl-sodium + mesosulfuron-methyl, isoproturon, isoproturon (*autumn sown*), isoproturon + pendimethalin, isoproturon + pendimethalin (*autumn sown*), isoproturon + simazine, isoproturon + trifluralin, linuron, linuron + trifluralin, pendimethalin, pendimethalin + picolinafen, propoxycarbazone-sodium, sulfosulfuron, tralkoxydim, tralkoxydim (*autumn sown*), tralkoxydim (*autumn sown - qualified minor use*), tralkoxydim (*qualified minor use*), tri-allate, trifluralin
Mayweeds	amidosulfuron + iodosulfuron-methyl-sodium, bromoxynil + diflufenican + ioxynil, clopyralid, dicamba + MCPA + mecoprop-P, dicamba + mecoprop-P, dichlorprop-P + MCPA + mecoprop-P, diflufenican + flufenacet, florasulam + fluroxypyr, iodosulfuron-methyl-sodium, iodosulfuron-methyl-sodium + mesosulfuron-methyl, metsulfuron-methyl, metsulfuron-methyl + thifensulfuron-methyl, metsulfuron-methyl + tribenuron-methyl, sulfosulfuron, thifensulfuron-methyl, thifensulfuron-methyl + tribenuron-methyl, tribenuron-methyl
Polygonums	2,4-DB + MCPA, dicamba + MCPA + mecoprop-P, dicamba + mecoprop-P, metsulfuron-methyl + thifensulfuron-methyl
Speedwells	bifenox, bromoxynil + diflufenican + ioxynil, carfentrazone-ethyl + mecoprop-P, diflufenican + flufenacet, flufenacet + pendimethalin, iodosulfuron-

		methyl-sodium, metsulfuron-methyl + thifensulfuron-methyl, pendimethalin + picolinafen
	Weeds, miscellaneous	glyphosate

Edible fungi - Mushrooms

Diseases	Bacterial blotch	sodium hypochlorite (commodity substance)
	Fungus diseases	formaldehyde (commodity substance) (*spray or fumigant*)
	Trichoderma	carbendazim (*off-label - spawn treatment*)
Pests	Flies	deltamethrin (*off-label*), diflubenzuron

Fruiting vegetables - Aubergines

Diseases	Blight	azoxystrobin (*off-label*)
	Botrytis	azoxystrobin (*off-label*), pyrimethanil (*off-label*)
	Didymella stem rot	azoxystrobin (*off-label*)
	Phytophthora	propamocarb hydrochloride
	Powdery mildew	azoxystrobin (*off-label*), sulphur (*off-label*)
	Pythium	propamocarb hydrochloride
Pests	Aphids	nicotine, Verticillium lecanii
	Leaf miners	deltamethrin (*off-label*), oxamyl (*off-label*)
	Leafhoppers	nicotine
	Thrips	deltamethrin (*off-label*), nicotine
	Whiteflies	buprofezin, nicotine, Verticillium lecanii
Weeds	Broad-leaved weeds	paraquat (*off-label - inter-row directed treatment*)
	Grass weeds	paraquat (*off-label - inter-row directed treatment*)

Fruiting vegetables - Cucurbits

Diseases	Botrytis	iprodione
	Damping off and foot rot	etridiazole (*seeds and seedlings*)
	Phytophthora	etridiazole (*transplants*), propamocarb hydrochloride
	Powdery mildew	bupirimate (*off-label*), bupirimate (*outdoor only*), copper ammonium carbonate, fenarimol (*off-label*)
	Pythium	propamocarb hydrochloride
Pests	Aphids	fatty acids, nicotine, pirimicarb, pirimicarb (*off-label*)
	Caterpillars	nicotine
	Leaf miners	nicotine
	Mealybugs	fatty acids
	Scale insects	fatty acids
	Spider mites	fatty acids
	Whiteflies	fatty acids
Weeds	Grass weeds	propyzamide (*off-label*)
	Weeds, miscellaneous	chlorthal-dimethyl (*off-label*)

Fruiting vegetables - Peppers

Diseases	Damping off	propamocarb hydrochloride (*off-label*)
	Phytophthora	propamocarb hydrochloride
	Pythium	propamocarb hydrochloride
	Root rot	propamocarb hydrochloride (*off-label*)
Pests	Aphids	fatty acids, nicotine, pirimicarb
	Caterpillars	nicotine
	Leaf miners	nicotine
	Mealybugs	fatty acids
	Scale insects	fatty acids
	Spider mites	fatty acids
	Whiteflies	fatty acids

Fruiting vegetables - Tomatoes

Diseases	Blight	Bordeaux mixture, copper ammonium carbonate, copper oxychloride, propamocarb hydrochloride
	Botrytis	carbendazim, iprodione, pyrimethanil (*off-label*), thiram
	Damping off	copper oxychloride
	Damping off and foot rot	etridiazole (*seeds and seedlings*)
	Foot rot	copper oxychloride
	Phytophthora	copper oxychloride, etridiazole (*transplants*), propamocarb hydrochloride
	Pythium	propamocarb hydrochloride
Pests	Aphids	fatty acids, nicotine, pirimicarb
	Caterpillars	nicotine
	Leaf miners	nicotine
	Mealybugs	fatty acids
	Scale insects	fatty acids
	Spider mites	fatty acids
	Whiteflies	fatty acids
Plant growth regulation	Fruiting control	(2-naphthyloxy)acetic acid, 2-chloroethylphosphonic acid

Herb crops - Herbs

Crop control	Pre-harvest desiccation	diquat (*off-label*)
Diseases	Alternaria	fenpropimorph (*off-label*)
	Botrytis	iprodione (*off-label*), propamocarb hydrochloride (*off-label*)
	Crown rot	fenpropimorph (*off-label*)
	Downy mildew	fosetyl-aluminium (*off-label*), propamocarb hydrochloride (*off-label*)
	Fungus diseases	mancozeb + metalaxyl-M (*off-label*)
	Powdery mildew	azoxystrobin (*off-label*), fenpropimorph (*off-label*), tebuconazole (*off-label*)
	Ring spot	azoxystrobin (*off-label*), prochloraz (*off-label*)
	Rust	azoxystrobin (*off-label*), tebuconazole (*off-label*)

	Seed-borne diseases	thiram (*seed soak*)
Pests	Aphids	nicotine, pirimicarb (*off-label*)
	Beetles	deltamethrin (*off-label*)
	Caterpillars	Bacillus thuringiensis (*off-label*), deltamethrin (*off-label*), nicotine
	Cutworms	lambda-cyhalothrin (*off-label*)
	Flies	lambda-cyhalothrin (*off-label*)
	Leaf miners	nicotine
	Leafhoppers	deltamethrin (*off-label*)
	Thrips	pirimicarb (*off-label*)
Weeds	Broad-leaved weeds	asulam (*off-label*), chlorpropham, chlorpropham + pentanochlor, chlorthal-dimethyl, chlorthal-dimethyl (*off-label*), clopyralid (*off-label*), diquat (*off-label*), linuron, metamitron (*off-label*), paraquat (*off-label*), pendimethalin, pendimethalin (*off-label*), pentanochlor, prometryn, prometryn (*off-label*), propachlor, propachlor (*off-label*), propyzamide (*off-label*), trifluralin, trifluralin (*off-label*)
	Crops as weeds	pendimethalin, propaquizafop (*off-label*)
	Grass weeds	chlorpropham, chlorpropham + pentanochlor, linuron, paraquat (*off-label*), pendimethalin, pentanochlor, prometryn, prometryn (*off-label*), propachlor, propachlor (*off-label*), propaquizafop (*off-label*), propyzamide (*off-label*), trifluralin, trifluralin (*off-label*)
	Mayweeds	clopyralid (*off-label*)
	Polygonums	chlorpropham

Leafy vegetables - Endives

Diseases	Botrytis	iprodione (*off-label*)
Pests	Aphids	nicotine, pirimicarb (*off-label*)
	Caterpillars	nicotine
	Leaf miners	cypermethrin (*off-label*), nicotine
	Thrips	pirimicarb (*off-label*)
Weeds	Broad-leaved weeds	propachlor (*off-label - under crop covers*)
	Grass weeds	propachlor (*off-label - under crop covers*)

Leafy vegetables - Lettuce

Diseases	Big vein	carbendazim (*off-label*)
	Botrytis	iprodione (*outdoor crops*), propamocarb hydrochloride (*off-label*), thiram (*outdoor crops*)
	Damping off	thiram (*seed treatment*)
	Downy mildew	fosetyl-aluminium (*off-label*), mancozeb, propamocarb hydrochloride (*off-label*)
	Fungus diseases	mancozeb + metalaxyl-M (*off-label*)
	Ring spot	prochloraz (*off-label*)
	Sclerotinia	azoxystrobin (*off-label*)
Pests	Aphids	cypermethrin, cypermethrin (*outdoor crops*), deltamethrin (*off-label*), dimethoate, fatty acids, imidacloprid (*off-label*), lambda-cyhalothrin + pirimicarb,

		nicotine, pirimicarb (*outdoor crops*), pymetrozine (*off-label*)
	Caterpillars	Bacillus thuringiensis (*off-label*), cypermethrin, cypermethrin (*outdoor crops*), deltamethrin (*off-label*), nicotine
	Cutworms	cypermethrin, cypermethrin (*outdoor crops*), deltamethrin, lambda-cyhalothrin, lambda-cyhalothrin + pirimicarb
	Leaf miners	cypermethrin (*off-label*), nicotine
	Mealybugs	fatty acids
	Scale insects	fatty acids
	Slugs/snails	methiocarb
	Spider mites	fatty acids
	Whiteflies	fatty acids
Weeds	Broad-leaved weeds	chlorpropham, paraquat (*off-label - inter-row directed treatment*), propachlor (*off-label*), propachlor (*off-label - under crop covers*), propyzamide, propyzamide (*outdoor crops*), trifluralin
	Grass weeds	chlorpropham, paraquat (*off-label - inter-row directed treatment*), propachlor (*off-label*), propachlor (*off-label - under crop covers*), propyzamide, propyzamide (*outdoor crops*), trifluralin
	Polygonums	chlorpropham

Leafy vegetables - Spinach

Diseases	Downy mildew	fosetyl-aluminium (*off-label*), metalaxyl-M (*off-label*)
Pests	Aphids	cypermethrin (*off-label*), nicotine, pirimicarb (*off-label*)
	Caterpillars	cypermethrin (*off-label*), nicotine
	Leaf miners	nicotine
	Slugs/snails	methiocarb
Weeds	Broad-leaved weeds	chlorpropham + fenuron, clopyralid (*off-label*)
	Crops as weeds	propaquizafop (*off-label*)
	Grass weeds	chlorpropham + fenuron, propaquizafop (*off-label*)
	Mayweeds	clopyralid (*off-label*)

Leafy vegetables - Watercress

Diseases	Downy mildew	fosetyl-aluminium (*off-label - during propagation*), propamocarb hydrochloride (*off-label - during propagation*), propamocarb hydrochloride (*off-label - under protection*)
	Phytophthora	etridiazole (*off-label*), fosetyl-aluminium (*off-label - during propagation*), propamocarb hydrochloride (*off-label - during propagation*), propamocarb hydrochloride (*off-label - under protection*)
	Pythium	copper oxychloride (*off-label*), copper oxychloride (*off-label - during propagation*), etridiazole (*off-label*), fosetyl-aluminium (*off-label - during propagation*), metalaxyl-M (*off-label*), propamocarb hydrochloride (*off-label - during propagation*), propamocarb hydrochloride (*off-label - under protection*)
	Rhizoctonia	copper oxychloride (*off-label*), copper oxychloride (*off-label - during propagation*)

Pests	Beetles	deltamethrin (*off-label*)
	Caterpillars	Bacillus thuringiensis (*off-label*)
	Leafhoppers	deltamethrin (*off-label*)

Legumes - Beans (Phaseolus)

Crop control	Pre-harvest desiccation	diquat (*off-label*)
Diseases	Botrytis	azoxystrobin (*off-label*), iprodione (*off-label*), vinclozolin
	Damping off	thiram (*seed treatment*)
	Rust	tebuconazole (*off-label*)
	Sclerotinia	iprodione (*off-label*)
Pests	Aphids	fatty acids, nicotine, pirimicarb
	Caterpillars	lambda-cyhalothrin (*off-label*), nicotine
	Flies	chlorpyrifos (*off-label - harvested as a dry pulse*)
	Leaf miners	nicotine
	Mealybugs	fatty acids
	Pests, miscellaneous	nicotine (*off-label*)
	Scale insects	fatty acids
	Spider mites	fatty acids
	Whiteflies	fatty acids
Weeds	Broad-leaved weeds	bentazone, chlorpropham + fenuron (*off-label*), chlorthal-dimethyl, chlorthal-dimethyl (*off-label*), fomesafen, pendimethalin (*off-label*), simazine (*off-label*), trifluralin
	Crops as weeds	chlorthal-dimethyl (*off-label*), cycloxydim, fomesafen
	Grass weeds	cycloxydim, fluazifop-P-butyl (*off-label*), simazine (*off-label*), trifluralin

Legumes - Beans (Vicia)

Crop control	Pre-harvest desiccation	diquat (*stock or pigeon feed only*), glufosinate-ammonium, glyphosate
Diseases	Ascochyta	azoxystrobin (*off-label*), thiabendazole + thiram (*seed treatment*)
	Botrytis	azoxystrobin (*off-label*)
	Chocolate spot	carbendazim, chlorothalonil, chlorothalonil + cyproconazole, cyproconazole (*with chlorothalonil*), iprodione, iprodione + thiophanate-methyl, tebuconazole, vinclozolin
	Damping off	thiabendazole + thiram (*seed treatment*), thiram (*seed treatment*)
	Downy mildew	chlorothalonil + metalaxyl-M, cymoxanil + fludioxonil + metalaxyl-M (*off-label - seed treatment*), fosetyl-aluminium
	Rust	azoxystrobin, chlorothalonil + cyproconazole, cyproconazole, fenpropimorph, tebuconazole, tebuconazole (*off-label*)
Pests	Aphids	fatty acids, nicotine, pirimicarb
	Birds/mammals	aluminium ammonium sulphate
	Caterpillars	nicotine
	Leaf miners	nicotine

	Mealybugs	fatty acids
	Scale insects	fatty acids
	Spider mites	fatty acids
	Weevils	alpha-cypermethrin, cypermethrin, deltamethrin, lambda-cyhalothrin, lambda-cyhalothrin + pirimicarb, zeta-cypermethrin
	Whiteflies	fatty acids
Plant growth regulation	Quality/yield control	chlormequat with di-1-p-menthene
Weeds	Broad-leaved weeds	bentazone, carbetamide, clomazone, cyanazine, cyanazine (*Scotland only*), cyanazine + pendimethalin, fomesafen + terbutryn, fomesafen + terbutryn (*spring sown*), glyphosate, pendimethalin (*off-label*), propyzamide, simazine, terbuthylazine + terbutryn, trifluralin
	Crops as weeds	carbetamide, cycloxydim, fluazifop-P-butyl, glyphosate, propyzamide, quizalofop-P-ethyl, tepraloxydim
	Grass weeds	carbetamide, cyanazine, cyanazine (*Scotland only*), cyanazine + pendimethalin, cycloxydim, fluazifop-P-butyl, fluazifop-P-butyl (*off-label*), glyphosate, propaquizafop, propyzamide, quizalofop-P-ethyl, simazine, tepraloxydim, terbuthylazine + terbutryn, tri-allate, trifluralin
	Weeds, miscellaneous	glyphosate

Legumes - Clovers

Crop control	Pre-harvest desiccation	diquat
Weeds	Broad-leaved weeds	asulam (*off-label*), carbetamide, MCPA + MCPB, MCPB, MCPB (*seed crops*), propyzamide
	Crops as weeds	carbetamide
	Grass weeds	carbetamide, propyzamide, tri-allate

Legumes - Forage legumes, general

Weeds	Broad-leaved weeds	2,4-DB, carbetamide, chlorpropham, MCPA + MCPB, paraquat (*off-label*), propyzamide
	Crops as weeds	carbetamide, fluazifop-P-butyl (*off-label*)
	Grass weeds	carbetamide, chlorpropham, fluazifop-P-butyl (*off-label*), paraquat (*off-label*), propyzamide, tri-allate
	Polygonums	chlorpropham

Legumes - Lupins

Weeds	Broad-leaved weeds	terbuthylazine + terbutryn
	Grass weeds	terbuthylazine + terbutryn

Legumes - Peas

Crop control	Pre-harvest desiccation	diquat, glufosinate-ammonium, glufosinate-ammonium (*not for seed*), glyphosate, sulphuric acid (commodity substance)
Diseases	Ascochyta	azoxystrobin, chlorothalonil, cymoxanil + fludioxonil + metalaxyl-M (*seed treatment*), thiabendazole + thiram (*seed treatment*), vinclozolin
	Botrytis	chlorothalonil, chlorothalonil + cyproconazole, iprodione (*off-label*), vinclozolin
	Damping off	cymoxanil + fludioxonil + metalaxyl-M (*seed treatment*), thiabendazole + thiram (*seed treatment*), thiram (*seed treatment*)
	Downy mildew	cymoxanil + fludioxonil + metalaxyl-M (*seed treatment*), fosetyl-aluminium (*off-label*), fosetyl-aluminium (*off-label - seed treatment*)
	Mycosphaerella	chlorothalonil, vinclozolin
	Sclerotinia	iprodione (*off-label*)
Pests	Aphids	alpha-cypermethrin, alpha-cypermethrin (*reduction*), cypermethrin, deltamethrin, fatty acids, lambda-cyhalothrin, lambda-cyhalothrin + pirimicarb, nicotine, pirimicarb, triazamate, zeta-cypermethrin
	Birds/mammals	aluminium ammonium sulphate
	Caterpillars	alpha-cypermethrin, cypermethrin, deltamethrin, lambda-cyhalothrin, lambda-cyhalothrin + pirimicarb, nicotine, zeta-cypermethrin
	Leaf miners	nicotine
	Mealybugs	fatty acids
	Midges	deltamethrin, lambda-cyhalothrin + pirimicarb
	Scale insects	fatty acids
	Spider mites	fatty acids
	Weevils	alpha-cypermethrin, cypermethrin, deltamethrin, lambda-cyhalothrin, lambda-cyhalothrin + pirimicarb, zeta-cypermethrin
	Whiteflies	fatty acids
Plant growth regulation	Quality/yield control	chlormequat with di-1-p-menthene
Weeds	Broad-leaved weeds	bentazone, bentazone + MCPB, bentazone + pendimethalin, clomazone, cyanazine, cyanazine + pendimethalin, fomesafen + terbutryn (*spring sown*), glyphosate, MCPA + MCPB, MCPB, pendimethalin, terbuthylazine + terbutryn, trifluralin (*off-label*)
	Crops as weeds	cycloxydim, fluazifop-P-butyl, glyphosate, pendimethalin, quizalofop-P-ethyl, tepraloxydim
	Grass weeds	cyanazine, cyanazine + pendimethalin, cycloxydim, fluazifop-P-butyl, glyphosate, pendimethalin, propaquizafop, quizalofop-P-ethyl, tepraloxydim, terbuthylazine + terbutryn, tri-allate, trifluralin (*off-label*)
	Mayweeds	bentazone + pendimethalin
	Weeds, miscellaneous	glyphosate

Miscellaneous arable - Miscellaneous arable crops

Crop control	Pre-harvest desiccation	diquat (*off-label*)
Diseases	Botrytis	iprodione + thiophanate-methyl (*off-label*)
Pests	Slugs/snails	metaldehyde
	Whiteflies	alginate/polysaccaride
Weeds	Broad-leaved weeds	diquat, diquat (*between row treatment*), diquat + paraquat, glyphosate, paraquat, paraquat (*stale seedbed/inter-row*)
	Crops as weeds	diquat + paraquat, glyphosate, paraquat, paraquat (*stale seedbed/inter-row*)
	Grass weeds	diquat + paraquat, glyphosate, paraquat, paraquat (*stale seedbed/inter-row*)

Miscellaneous arable - Miscellaneous arable situations

Diseases	Soil-borne diseases	dazomet (*soil fumigation*)
Pests	Birds/mammals	aluminium ammonium sulphate (*seed treatment*), ziram
	Free-living nematodes	dazomet (*soil fumigation*)
	Slugs/snails	metaldehyde
	Soil pests	dazomet (*soil fumigation*)
Weeds	Broad-leaved weeds	amitrole, citronella oil, diquat + paraquat, glufosinate-ammonium, glufosinate-ammonium (*uncropped*), glyphosate, paraquat, paraquat (*minimum cultivation*), thifensulfuron-methyl (*in green cover*)
	Crops as weeds	amitrole (*barley stubble*), diquat + paraquat, fluazifop-P-butyl, glyphosate, paraquat, paraquat (*minimum cultivation*), tepraloxydim
	Grass weeds	amitrole, diquat + paraquat, fluazifop-P-butyl, glufosinate-ammonium, glufosinate-ammonium (*uncropped*), glyphosate, paraquat, paraquat (*minimum cultivation*), paraquat (*stubble treatment*), tepraloxydim
	Weeds, miscellaneous	2,4-D + dicamba + triclopyr, amitrole, cycloxydim, dazomet (*soil fumigation*), diquat + paraquat, fluazifop-P-butyl, glufosinate-ammonium, glyphosate, glyphosate (*wiper application*), metsulfuron-methyl, paraquat, tepraloxydim

Miscellaneous field vegetables - All vegetables

Diseases	Soil-borne diseases	dazomet (*soil fumigation*)
Pests	Aphids	natural plant extracts, nicotine, rotenone
	Capsid bugs	nicotine
	Free-living nematodes	dazomet (*soil fumigation*)
	Leafhoppers	natural plant extracts, nicotine
	Mites	natural plant extracts
	Sawflies	nicotine
	Soil pests	dazomet (*soil fumigation*)
	Thrips	natural plant extracts, nicotine
	Whiteflies	natural plant extracts
Weeds	Broad-leaved weeds	glufosinate-ammonium
	Grass weeds	glufosinate-ammonium

	Weeds, miscellaneous	ammonium sulphamate (*pre-planting*), dazomet (*soil fumigation*), glufosinate-ammonium

Oilseed crops - Linseed/flax

Crop control	Pre-harvest desiccation	diquat, glufosinate-ammonium, glyphosate
Diseases	Alternaria	iprodione (*seed treatment*)
	Botrytis	tebuconazole (*reduction*)
	Powdery mildew	tebuconazole
	Seed-borne diseases	prochloraz (*seed treatment*)
Pests	Beetles	zeta-cypermethrin
Weeds	Broad-leaved weeds	amidosulfuron, bentazone, bromoxynil, clopyralid, glyphosate, MCPA, metsulfuron-methyl, trifluralin, trifluralin (*off-label*)
	Crops as weeds	cycloxydim, fluazifop-P-butyl, glyphosate, quizalofop-P-ethyl, tepraloxydim
	Grass weeds	cycloxydim, fluazifop-P-butyl, glyphosate, propaquizafop, quizalofop-P-ethyl, tepraloxydim, trifluralin, trifluralin (*off-label*)
	Mayweeds	clopyralid, metsulfuron-methyl
	Weeds, miscellaneous	glyphosate

Oilseed crops - Miscellaneous oilseeds

Crop control	Pre-harvest desiccation	diquat (*off-label*), diquat (*off-label - for morphine production*), glyphosate (*off-label - for morphine production*)
Diseases	Alternaria	boscalid (*off-label - for morphine production*), iprodione (*off-label - for morphine production - seed treatment*)
	Botrytis	boscalid (*off-label - for morphine production*)
	Damping off	cymoxanil + fludioxonil + metalaxyl-M (*off-label - for morphine production - seed treatment*), thiram (*off-label - for morphine production - seed treatment*)
	Downy mildew	chlorothalonil + metalaxyl-M (*off-label - for morphine production*), cymoxanil + fludioxonil + metalaxyl-M (*off-label - for morphine production - seed treatment*), mancozeb (*off-label - for morphine production*)
	Powdery mildew	azoxystrobin (*off-label - for morphine production*)
	Pythium	cymoxanil + fludioxonil + metalaxyl-M (*off-label - for morphine production - seed treatment*), thiram (*off-label - for morphine production - seed treatment*)
	Sclerotinia stem rot	boscalid (*off-label - for morphine production*)
	White blister	propiconazole (*off-label*)
Pests	Aphids	deltamethrin (*off-label*), pirimicarb (*off-label*)
	Beetles	deltamethrin (*off-label*), lambda-cyhalothrin (*off-label - for morphine production*)
Weeds	Broad-leaved weeds	asulam (*off-label - for morphine production*), bentazone (*off-label*), chlorthal-dimethyl (*off-label - for morphine production*), clopyralid (*off-label*), clopyralid (*off-label - for morphine production*), ethofumesate (*off-label - for morphine production*), fluroxypyr (*off-label - for*

		morphine production), pendimethalin (*off-label*), trifluralin (*off-label*)
	Crops as weeds	fluroxypyr (*off-label - for morphine production*), propaquizafop (*off-label - for morphine production*)
	Grass weeds	asulam (*off-label - for morphine production*), ethofumesate (*off-label - for morphine production*), propaquizafop (*off-label - for morphine production*), propyzamide (*off-label*), trifluralin (*off-label*)
	Mayweeds	clopyralid (*off-label*), clopyralid (*off-label - for morphine production*), ethofumesate (*off-label - for morphine production*)
	Weeds, miscellaneous	diquat (*off-label - for morphine production*)

Oilseed crops - Oilseed rape

Crop control	Pre-harvest desiccation	diquat, glufosinate-ammonium, glyphosate
Diseases	Alternaria	difenoconazole, iprodione, iprodione (*seed treatment*), iprodione + thiophanate-methyl, metconazole, prochloraz, tebuconazole, vinclozolin
	Black scurf and stem canker	difenoconazole, iprodione + thiophanate-methyl, tebuconazole
	Botrytis	chlorothalonil, iprodione, iprodione + thiophanate-methyl, prochloraz, vinclozolin
	Canker	carbendazim + flusilazole (*reduction*), metconazole (*foliar disease - reduction*), metconazole (*reduction*), prochloraz, prochloraz + propiconazole, tebuconazole
	Damping off	thiram (*seed treatment*)
	Downy mildew	chlorothalonil, mancozeb
	Light leaf spot	carbendazim, carbendazim + flusilazole, cyproconazole, difenoconazole, flusilazole, iprodione + thiophanate-methyl, prochloraz, prochloraz + propiconazole, propiconazole (*reduction*), tebuconazole
	Phoma leaf spot	carbendazim + flusilazole, cyproconazole
	Powdery mildew	sulphur
	Ring spot	tebuconazole (*reduction*)
	Sclerotinia stem rot	boscalid, iprodione, iprodione + thiophanate-methyl, metconazole (*reduction*), prochloraz, prochloraz + tebuconazole, tebuconazole, vinclozolin
	White leaf spot	prochloraz
Pests	Aphids	deltamethrin, lambda-cyhalothrin, lambda-cyhalothrin + pirimicarb, nicotine (*off-label*), pirimicarb, tau-fluvalinate
	Beetles	alpha-cypermethrin, beta-cyfluthrin + imidacloprid (*seed treatment*), bifenthrin, cypermethrin, deltamethrin, lambda-cyhalothrin, lambda-cyhalothrin + pirimicarb, tau-fluvalinate, zeta-cypermethrin
	Birds/mammals	aluminium ammonium sulphate
	Midges	alpha-cypermethrin, cypermethrin, deltamethrin, lambda-cyhalothrin, lambda-cyhalothrin + pirimicarb, zeta-cypermethrin
	Slugs/snails	methiocarb, thiodicarb
	Weevils	alpha-cypermethrin, bifenthrin, cypermethrin, deltamethrin, lambda-cyhalothrin, lambda-cyhalothrin + pirimicarb, zeta-cypermethrin

Plant growth regulation	Growth control	tebuconazole
	Quality/yield control	sulphur
Weeds	Broad-leaved weeds	carbetamide, clomazone, clopyralid, clopyralid + picloram, cyanazine, glyphosate, metazachlor, metazachlor + quinmerac, napropamide, propachlor, propyzamide, pyridate (*off-label*), trifluralin
	Crops as weeds	carbetamide, cycloxydim, fluazifop-P-butyl, glyphosate, propyzamide, quizalofop-P-ethyl, tepraloxydim
	Grass weeds	carbetamide, cyanazine, cycloxydim, fluazifop-P-butyl, glyphosate, metazachlor, metazachlor + quinmerac, napropamide, propachlor, propaquizafop, propyzamide, quizalofop-P-ethyl, tepraloxydim, trifluralin
	Mayweeds	clopyralid, clopyralid + picloram
	Weeds, miscellaneous	glyphosate

Oilseed crops - Soya

Crop control	Pre-harvest desiccation	diquat (*off-label*)
Weeds	Broad-leaved weeds	fomesafen (*off-label*), trifluralin (*off-label*)
	Crops as weeds	fomesafen (*off-label*)
	Grass weeds	trifluralin (*off-label*)

Oilseed crops - Sunflowers

Crop control	Pre-harvest desiccation	diquat (*off-label*)
Pests	Slugs/snails	methiocarb
Weeds	Broad-leaved weeds	diquat (*off-label*), pendimethalin, trifluralin (*off-label*)
	Crops as weeds	pendimethalin
	Grass weeds	pendimethalin, trifluralin (*off-label*)

Root and tuber crops - Beet crops

Diseases	Black leg	hymexazol (*seed treatment*)
	Botrytis	iprodione (*off-label*)
	Canker	iprodione (*off-label*)
	Cercospora leaf spot	epoxiconazole + pyraclostrobin
	Damping off	cymoxanil + fludioxonil + metalaxyl-M (*off-label - seed treatment*)
	Powdery mildew	carbendazim + flusilazole, cyproconazole, epoxiconazole + pyraclostrobin, flusilazole, quinoxyfen, sulphur
	Ramularia leaf spots	cyproconazole, epoxiconazole + pyraclostrobin, propiconazole (*reduction*)
	Rhizoctonia	azoxystrobin (*off-label*), metalaxyl-M (*off-label*), propamocarb hydrochloride (*off-label*)

	Rust	carbendazim + flusilazole, cyproconazole, cyproconazole (*off-label*), epoxiconazole + pyraclostrobin, fenpropimorph (*off-label*), flusilazole, propiconazole
	Seed-borne diseases	thiram (*seed soak*)
Pests	Aphids	beta-cyfluthrin + clothianidin (*seed treatment*), carbosulfan, dimethoate, dimethoate (*excluding Myzus persicae*), imidacloprid (*seed treatment*), lambda-cyhalothrin + pirimicarb, nicotine, oxamyl, pirimicarb, pirimicarb (*off-label*), triazamate (*including Myzus persicae*)
	Beetles	benfuracarb, carbosulfan, chlorpyrifos, deltamethrin, imidacloprid (*seed treatment*), lambda-cyhalothrin, lambda-cyhalothrin + pirimicarb, oxamyl, tefluthrin (*seed treatment*)
	Birds/mammals	aluminium ammonium sulphate
	Caterpillars	cypermethrin, lambda-cyhalothrin (*off-label*), nicotine
	Cutworms	cypermethrin, lambda-cyhalothrin, lambda-cyhalothrin (*off-label*), lambda-cyhalothrin + pirimicarb, methiocarb (*reduction*), zeta-cypermethrin
	Flies	carbosulfan, dimethoate, imidacloprid (*seed treatment*), oxamyl
	Free-living nematodes	benfuracarb, carbosulfan, oxamyl
	Leaf miners	benfuracarb, dimethoate, lambda-cyhalothrin, lambda-cyhalothrin + pirimicarb, nicotine
	Leatherjackets	chlorpyrifos, methiocarb (*reduction*)
	Millipedes	benfuracarb, carbosulfan, imidacloprid (*seed treatment*), methiocarb (*reduction*), oxamyl, tefluthrin (*seed treatment*)
	Slugs/snails	methiocarb
	Springtails	benfuracarb, carbosulfan, imidacloprid (*seed treatment*), tefluthrin (*seed treatment*)
	Symphylids	benfuracarb, carbosulfan, imidacloprid (*seed treatment*), tefluthrin (*seed treatment*)
	Wireworms	carbosulfan
Plant growth regulation	Quality/yield control	sulphur
Weeds	Broad-leaved weeds	carbetamide, chloridazon, chloridazon + chlorpropham + metamitron, chloridazon + ethofumesate, chloridazon + metamitron, chloridazon + quinmerac, chlorpropham + metamitron, clopyralid, desmedipham + ethofumesate + phenmedipham, desmedipham + phenmedipham, diquat, ethofumesate, ethofumesate + lenacil + phenmedipham, ethofumesate + metamitron, ethofumesate + phenmedipham, glufosinate-ammonium, glyphosate, lenacil, lenacil + phenmedipham, metamitron, paraquat, phenmedipham, propyzamide, trifluralin, triflusulfuron-methyl, triflusulfuron-methyl (*off-label*)
	Crops as weeds	carbetamide, cycloxydim, fluazifop-P-butyl, glyphosate, glyphosate (*wiper application*), paraquat, propaquizafop (*off-label*), propyzamide, quizalofop-P-ethyl, sodium chloride (commodity substance), tepraloxydim
	Grass weeds	carbetamide, chloridazon, chloridazon + chlorpropham + metamitron, chloridazon + ethofumesate, chloridazon + metamitron, chloridazon + quinmerac, chlorpropham + metamitron, cycloxydim, desmedipham +

		ethofumesate + phenmedipham, ethofumesate, ethofumesate + metamitron, ethofumesate + phenmedipham, fluazifop-P-butyl, fluazifop-P-butyl (*off-label*), glufosinate-ammonium, glyphosate, lenacil, metamitron, paraquat, propaquizafop, propaquizafop (*off-label*), propyzamide, quizalofop-P-ethyl, tepraloxydim, tri-allate, trifluralin
	Mayweeds	chloridazon + quinmerac, clopyralid
	Polygonums	sodium chloride (commodity substance)
	Speedwells	chloridazon + quinmerac
	Weeds, miscellaneous	glufosinate-ammonium, glyphosate

Root and tuber crops - Carrots/parsnips/celeriac

Diseases	Alternaria	azoxystrobin, fenpropimorph (*off-label*), iprodione + thiophanate-methyl (*off-label*), tebuconazole
	Cavity spot	metalaxyl-M, metalaxyl-M (*off-label*)
	Crown rot	fenpropimorph (*off-label*), iprodione + thiophanate-methyl (*off-label*)
	Damping off	thiram (*seed treatment*)
	Powdery mildew	azoxystrobin, fenpropimorph (*off-label*), sulphur (*off-label*), tebuconazole
	Pythium	cymoxanil + fludioxonil + metalaxyl-M (*off-label - seed treatment*)
	Sclerotinia	azoxystrobin (*off-label*), tebuconazole
	Seed-borne diseases	thiram (*seed soak*)
	Septoria diseases	chlorothalonil (*off-label*)
Pests	Aphids	aldicarb, carbosulfan, deltamethrin (*off-label*), lambda-cyhalothrin + pirimicarb, nicotine, pirimicarb, pirimicarb (*off-label*)
	Birds/mammals	aluminium ammonium sulphate
	Caterpillars	deltamethrin (*off-label*), nicotine
	Cutworms	chlorpyrifos, cypermethrin (*off-label*), lambda-cyhalothrin, lambda-cyhalothrin (*off-label*), lambda-cyhalothrin + pirimicarb
	Flies	lambda-cyhalothrin (*off-label*), tefluthrin (*off-label - seed treatment*)
	Free-living nematodes	aldicarb, carbosulfan
	Leaf miners	nicotine
	Pests, miscellaneous	deltamethrin (*off-label*), dimethoate (*off-label*)
	Stem nematodes	oxamyl (*off-label*)
Plant growth regulation	Growth control	maleic hydrazide (*off-label*)
Weeds	Broad-leaved weeds	chlorpropham, chlorpropham + pentanochlor, clomazone, ioxynil (*off-label*), linuron, linuron (*off-label*), metoxuron, metribuzin (*off-label*), pendimethalin, pentanochlor, prometryn, trifluralin, trifluralin (*off-label*)
	Crops as weeds	cycloxydim, fluazifop-P-butyl, pendimethalin, tepraloxydim
	Grass weeds	chlorpropham, chlorpropham + pentanochlor, cycloxydim, fluazifop-P-butyl, fluazifop-P-butyl (*off-label*), linuron, metribuzin (*off-label*), pendimethalin,

		pentanochlor, prometryn, propaquizafop, tepraloxydim, trifluralin, trifluralin (*off-label*)
	Mayweeds	metoxuron
	Polygonums	chlorpropham

Root and tuber crops - Potatoes

Crop control	Pre-harvest desiccation	carfentrazone-ethyl, diquat, glufosinate-ammonium, sulphuric acid (commodity substance)
Diseases	Black dot	azoxystrobin (*off-label - for baby tuber production*)
	Black scurf and stem canker	flutolanil (*tuber treatment*), imazalil + pencycuron (*tuber treatment - reduction*), imazalil + thiabendazole (*tuber treatment - reduction*), pencycuron (*tuber treatment*)
	Blight	benalaxyl + mancozeb, Bordeaux mixture, chlorothalonil, chlorothalonil + mancozeb, chlorothalonil + propamocarb hydrochloride, copper oxychloride, cyazofamid, cymoxanil, cymoxanil + famoxadone, cymoxanil + mancozeb, dimethomorph + mancozeb, fenamidone + mancozeb, fenamidone + propamocarb hydrochloride, fluazinam, fluazinam + metalaxyl-M, mancozeb, mancozeb + metalaxyl-M, mancozeb + propamocarb hydrochloride, mancozeb + zoxamide
	Dry rot	imazalil, imazalil + thiabendazole (*tuber treatment - reduction*), thiabendazole (*off-label*), thiabendazole (*tuber treatment - post-harvest*)
	Gangrene	2-aminobutane, imazalil, imazalil + thiabendazole (*tuber treatment - reduction*), thiabendazole (*tuber treatment - post-harvest*)
	Rhizoctonia	flutolanil (*off-label - chitted seed treatment*), flutolanil (*tuber treatment*), imazalil + pencycuron (*tuber treatment*), iprodione (*seed treatment*), pencycuron (*tuber treatment*), tolclofos-methyl (*off-label - chitted seed treatment*), tolclofos-methyl (*tuber treatment*)
	Silver scurf	2-aminobutane, imazalil, imazalil + pencycuron (*tuber treatment - reduction*), imazalil + thiabendazole (*tuber treatment - reduction*), thiabendazole (*tuber treatment - post-harvest*)
	Skin spot	2-aminobutane, imazalil, imazalil + thiabendazole (*tuber treatment - reduction*), thiabendazole (*tuber treatment - post-harvest*)
Pests	Aphids	aldicarb, lambda-cyhalothrin, lambda-cyhalothrin + pirimicarb, nicotine, oxamyl, pirimicarb, pymetrozine
	Beetles	deltamethrin (*off-label*)
	Caterpillars	cypermethrin, nicotine
	Cutworms	chlorpyrifos, cypermethrin, lambda-cyhalothrin + pirimicarb, zeta-cypermethrin
	Cyst nematodes	1,3-dichloropropene, ethoprophos, fosthiazate, oxamyl
	Free-living nematodes	1,3-dichloropropene, aldicarb, oxamyl
	Leaf miners	nicotine
	Leatherjackets	methiocarb (*reduction*)
	Slugs/snails	methiocarb, thiodicarb
	Wireworms	ethoprophos, fosthiazate (*reduction*)
Plant growth regulation	Growth control	chlorpropham, chlorpropham (*fog*), chlorpropham (*thermal fog*), ethylene (commodity substance) (*in store*), maleic hydrazide

	Plant growth regulation, miscellaneous	maleic hydrazide
Weeds	Broad-leaved weeds	bentazone, clomazone, diquat, diquat + paraquat, flufenacet + metribuzin, glufosinate-ammonium, linuron, metribuzin, paraquat, pendimethalin, rimsulfuron
	Crops as weeds	cycloxydim, diquat + paraquat, metribuzin, paraquat, pendimethalin, rimsulfuron
	Grass weeds	cycloxydim, diquat + paraquat, flufenacet + metribuzin, glufosinate-ammonium, linuron, metribuzin, paraquat, pendimethalin, propaquizafop
	Weeds, miscellaneous	diquat, glufosinate-ammonium (*not for seed*)

Stem and bulb vegetables - Asparagus

Diseases	Phytophthora	metalaxyl-M (*off-label*)
	Rust	azoxystrobin
	Stemphylium	azoxystrobin, iprodione (*off-label*)
Pests	Aphids	nicotine
	Beetles	cypermethrin (*off-label*)
	Caterpillars	nicotine
	Leaf miners	nicotine
Weeds	Broad-leaved weeds	clopyralid (*off-label*), diuron (*off-label*), metamitron (*off-label*), metribuzin (*off-label*), simazine, simazine (*off-label*)
	Grass weeds	diuron (*off-label*), metamitron (*off-label*), metribuzin (*off-label*), simazine, simazine (*off-label*)
	Weeds, miscellaneous	diuron (*off-label*), glyphosate (*off-label*), metribuzin (*off-label*)

Stem and bulb vegetables - Celery/chicory

Diseases	Botrytis	azoxystrobin (*off-label*), iprodione (*off-label*)
	Damping off and foot rot	etridiazole (*seeds and seedlings*)
	Phytophthora	etridiazole (*transplants*), fosetyl-aluminium (*off-label - for forcing*), fosetyl-aluminium (*off-label - in forcing sheds*)
	Rhizoctonia	azoxystrobin (*off-label*)
	Sclerotinia	azoxystrobin (*off-label*), iprodione (*off-label*)
	Seed-borne diseases	thiram (*seed soak*)
	Septoria diseases	Bordeaux mixture, carbendazim, chlorothalonil (*qualified minor use*), copper ammonium carbonate, copper oxychloride
Pests	Aphids	deltamethrin (*off-label*), nicotine, pirimicarb, pirimicarb (*off-label*), pirimicarb (*off-label - for forcing*)
	Beetles	deltamethrin (*off-label*)
	Caterpillars	Bacillus thuringiensis (*off-label*), deltamethrin (*off-label*), lambda-cyhalothrin (*off-label*), nicotine
	Cutworms	lambda-cyhalothrin (*off-label - for forcing*)
	Flies	lambda-cyhalothrin (*off-label*)
	Leaf miners	cypermethrin (*off-label*), nicotine

	Leafhoppers	deltamethrin (*off-label*)
	Pests, miscellaneous	deltamethrin (*off-label*)
Weeds	Broad-leaved weeds	chlorpropham, chlorpropham + pentanochlor, linuron, pentanochlor, prometryn, propachlor (*off-label*), propachlor (*off-label - under crop covers*), propyzamide (*off-label*)
	Crops as weeds	propaquizafop (*off-label*)
	Grass weeds	chlorpropham, chlorpropham + pentanochlor, linuron, pentanochlor, prometryn, propachlor (*off-label*), propachlor (*off-label - under crop covers*), propaquizafop (*off-label*), propyzamide (*off-label*)
	Polygonums	chlorpropham

Stem and bulb vegetables - Globe artichokes/cardoons

Pests	Aphids	nicotine
	Caterpillars	nicotine
	Leaf miners	nicotine

Stem and bulb vegetables - Onions/leeks/garlic

Diseases	Bacterial blight	copper oxychloride (*off-label*)
	Botrytis	chlorothalonil, iprodione, propamocarb hydrochloride (*off-label*), thiabendazole + thiram (*off-label - seed treatment*)
	Collar rot	iprodione
	Damping off	thiram (*seed treatment*)
	Downy mildew	azoxystrobin, azoxystrobin (*off-label*), chlorothalonil + metalaxyl-M (*qualified minor use*), dimethomorph + mancozeb (*off-label*), fosetyl-aluminium (*off-label*), mancozeb (*off-label*), mancozeb + metalaxyl-M (*off-label*), metalaxyl-M (*off-label*), propamocarb hydrochloride (*off-label*)
	Fungus diseases	mancozeb + metalaxyl-M (*off-label*)
	Fusarium diseases	thiabendazole + thiram (*off-label - seed treatment*)
	Phytophthora	propamocarb hydrochloride
	Powdery mildew	azoxystrobin (*off-label*), tebuconazole (*off-label*)
	Purple blotch	azoxystrobin
	Pythium	propamocarb hydrochloride
	Rhynchosporium	propiconazole (*off-label*)
	Ring spot	azoxystrobin (*off-label*), prochloraz (*off-label*)
	Rust	azoxystrobin, azoxystrobin (*off-label*), cyproconazole, fenpropimorph, propiconazole (*off-label*), tebuconazole, tebuconazole (*off-label*)
	Storage rots	copper oxychloride (*off-label*)
	White rot	tebuconazole (*off-label*)
	White tip	chlorothalonil + metalaxyl-M (*qualified minor use*)
Pests	Aphids	deltamethrin (*off-label*), nicotine
	Caterpillars	Bacillus thuringiensis (*off-label*), deltamethrin (*off-label*), nicotine
	Cutworms	chlorpyrifos
	Flies	tefluthrin (*off-label - seed treatment*)

	Leaf miners	nicotine
	Pests, miscellaneous	deltamethrin (*off-label*), dimethoate (*off-label*), dimethoate (*off-label - seedlings*)
	Stem nematodes	aldicarb, oxamyl (*off-label*)
	Thrips	lambda-cyhalothrin (*off-label*)
Plant growth regulation	Growth control	maleic hydrazide
Weeds	Broad-leaved weeds	bentazone (*off-label*), chloridazon (*off-label*), chloridazon + propachlor (*off-label*), chlorpropham, chlorthal-dimethyl, chlorthal-dimethyl (*off-label*), chlorthal-dimethyl + propachlor, clopyralid, clopyralid (*off-label*), cyanazine, cyanazine (*fen soils only*), ethofumesate (*off-label*), fluroxypyr (*off-label*), glyphosate, ioxynil, ioxynil (*off-label*), linuron (*off-label*), metamitron (*off-label*), paraquat (*off-label*), pendimethalin, pendimethalin (*off-label*), prometryn, prometryn (*off-label*), propachlor, propachlor (*off-label*), pyridate, sodium monochloroacetate, trifluralin (*off-label*)
	Crops as weeds	cycloxydim, fluazifop-P-butyl, fluroxypyr (*off-label*), glyphosate, pendimethalin, propaquizafop (*off-label*), tepraloxydim
	Grass weeds	chloridazon (*off-label*), chloridazon + propachlor (*off-label*), chlorpropham, chlorthal-dimethyl + propachlor, cyanazine, cyanazine (*fen soils only*), cycloxydim, ethofumesate (*off-label*), fluazifop-P-butyl, fluazifop-P-butyl (*off-label*), glyphosate, paraquat (*off-label*), pendimethalin, prometryn, prometryn (*off-label*), propachlor, propachlor (*off-label*), propaquizafop, propaquizafop (*off-label*), tepraloxydim, tepraloxydim (*off-label*), trifluralin (*off-label*)
	Mayweeds	clopyralid, clopyralid (*off-label*)
	Weeds, miscellaneous	glyphosate

Flowers and ornamentals

Flowers - Bedding plants, general

Diseases	Alternaria	iprodione (*seed treatment*)
	Botrytis	carbendazim (*off-label*), thiram
	Petal blight	mancozeb
	Phytophthora	propamocarb hydrochloride
	Powdery mildew	bupirimate, carbendazim (*off-label*), copper ammonium carbonate, dinocap
	Pythium	propamocarb hydrochloride
	Rust	azoxystrobin (*off-label - in pots*), mancozeb, propiconazole (*off-label*), thiram
Pests	Aphids	imidacloprid, malathion, pirimicarb
	Flies	imidacloprid
	Weevils	imidacloprid
	Whiteflies	imidacloprid
Plant growth regulation	Flowering control	paclobutrazol

	Growth control	chlormequat, daminozide, paclobutrazol
Weeds	Broad-leaved weeds	chlorpropham, chlorpropham + pentanochlor, chlorthal-dimethyl, pentanochlor
	Grass weeds	chlorpropham, chlorpropham + pentanochlor, pentanochlor
	Polygonums	chlorpropham

Flowers - Bulbs/corms

Crop control	Pre-harvest desiccation	sulphuric acid (commodity substance)
Diseases	Botrytis	carbendazim (*off-label*), chlorothalonil (*off-label - for galanthamine production*), tebuconazole (*off-label - for galanthamine production*), thiram, vinclozolin (*off-label - for galanthamine production*)
	Crown rot	metalaxyl-M (*off-label*)
	Fire	mancozeb, thiram
	Fungus diseases	formaldehyde (commodity substance) (*dip*)
	Fusarium diseases	carbendazim (*off-label*), thiabendazole (*bulb dip or spray*)
	Ink disease	chlorothalonil (*qualified minor use*)
	Penicillium rot	carbendazim (*off-label*)
	Phytophthora	etridiazole, propamocarb hydrochloride
	Powdery mildew	bupirimate
	Pythium	propamocarb hydrochloride
	Sclerotinia	carbendazim (*off-label*)
	Stagonospora	carbendazim (*off-label*)
	White mould	chlorothalonil (*off-label*), mancozeb (*off-label - for galanthamine production*)
Pests	Aphids	malathion, nicotine
	Birds/mammals	aluminium ammonium sulphate (*bulb treatment*), aluminium ammonium sulphate (*corm treatment*)
	Capsid bugs	nicotine
	Flies	chlorpyrifos (*off-label*)
	Leaf miners	nicotine
	Leafhoppers	nicotine
	Sawflies	nicotine
	Stem nematodes	1,3-dichloropropene
	Thrips	nicotine
Plant growth regulation	Flowering control	paclobutrazol
	Growth control	2-chloroethylphosphonic acid, chlormequat, paclobutrazol
Weeds	Broad-leaved weeds	bentazone, chlorpropham, chlorpropham + pentanochlor, cyanazine, diquat, diquat + paraquat, diuron (*off-label - for galanthamine production*), paraquat, pendimethalin (*off-label - for galanthamine production*), pentanochlor
	Crops as weeds	cycloxydim, pendimethalin (*off-label - for galanthamine production*)

	Grass weeds	chlorpropham, chlorpropham + pentanochlor, cyanazine, cycloxydim, diquat + paraquat, paraquat, pentanochlor
	Speedwells	pendimethalin (*off-label - for galanthamine production*)

Flowers - Miscellaneous flowers

Diseases	Black spot	captan, kresoxim-methyl, mancozeb, myclobutanil
	Powdery mildew	bupirimate, dinocap, dodemorph, fenarimol, kresoxim-methyl, myclobutanil
	Rust	mancozeb, myclobutanil, penconazole
Pests	Aphids	imidacloprid, malathion, pirimicarb, rotenone
	Flies	imidacloprid
	Leaf miners	abamectin
	Sawflies	rotenone
	Spider mites	abamectin
	Thrips	abamectin
	Weevils	imidacloprid
	Whiteflies	imidacloprid
Plant growth regulation	Flowering control	paclobutrazol
	Growth control	paclobutrazol
Weeds	Broad-leaved weeds	chlorthal-dimethyl, dichlobenil, pentanochlor, propyzamide, simazine
	Grass weeds	dichlobenil, pentanochlor, propyzamide, simazine
	Weeds, miscellaneous	dichlobenil

Flowers - Pot plants

Diseases	Botrytis	carbendazim (*off-label*), iprodione
	Phytophthora	fosetyl-aluminium, propamocarb hydrochloride
	Powdery mildew	carbendazim (*off-label*)
	Pythium	propamocarb hydrochloride
	Root rot	fosetyl-aluminium
	Rust	mancozeb
Pests	Aphids	deltamethrin, imidacloprid, pirimicarb
	Caterpillars	deltamethrin
	Flies	imidacloprid
	Mealybugs	deltamethrin, petroleum oil
	Scale insects	deltamethrin, petroleum oil
	Spider mites	petroleum oil
	Weevils	imidacloprid
	Whiteflies	deltamethrin, imidacloprid
Plant growth regulation	Flowering control	2-chloroethylphosphonic acid, paclobutrazol
	Fruiting control	paclobutrazol
	Growth control	2-chloroethylphosphonic acid, chlormequat, chlormequat with choline chloride, daminozide, paclobutrazol

Flowers - Protected bulbs/corms

Pests	Aphids	malathion, pirimicarb

Flowers - Protected flowers

Diseases	Powdery mildew	imazalil
	Rust	azoxystrobin (*off-label*), propiconazole (*off-label*)
Pests	Aphids	malathion, nicotine, pirimicarb, Verticillium lecanii
	Leaf miners	abamectin
	Spider mites	abamectin, tebufenpyrad
	Thrips	abamectin, imidacloprid (*off-label*)
Plant growth regulation	Growth control	2-chloroethylphosphonic acid

Miscellaneous flowers and ornamentals - Miscellaneous uses

Pests	Birds/mammals	aluminium ammonium sulphate, bone oil
	Caterpillars	diflubenzuron
	Leaf miners	deltamethrin (*off-label*)
	Slugs/snails	metaldehyde, methiocarb (*outdoor only*)
	Thrips	deltamethrin (*off-label*)
	Whiteflies	alginate/polysaccaride
Plant growth regulation	Growth control	maleic hydrazide
Weeds	Broad-leaved weeds	amitrole, diquat + paraquat, glyphosate
	Crops as weeds	glyphosate
	Grass weeds	amitrole, diquat + paraquat, glyphosate
	Weeds, miscellaneous	amitrole, glyphosate, glyphosate (*wiper application*)

Miscellaneous flowers and ornamentals - Soils

Diseases	Soil-borne diseases	metam-sodium
Pests	Free-living nematodes	metam-sodium
	Leaf miners	deltamethrin (*off-label*)
	Soil pests	metam-sodium
	Thrips	deltamethrin (*off-label*)
Weeds	Weeds, miscellaneous	metam-sodium

Ornamentals - Nursery stock

Diseases	Alternaria	iprodione (*seed treatment*)
	Black root rot	carbendazim (*off-label*)
	Black spot	myclobutanil
	Botrytis	chlorothalonil, thiram (*except Hydrangea*)
	Canker	azaconazole + imazalil
	Damping off	copper ammonium carbonate, tolclofos-methyl

	Damping off and foot rot	etridiazole (*seeds and seedlings*)
	Foot rot	tolclofos-methyl
	Fungus diseases	dichlorophen, prochloraz
	Phytophthora	etridiazole, etridiazole (*rooted cuttings and transplants*), fosetyl-aluminium, propamocarb hydrochloride
	Powdery mildew	imazalil, myclobutanil
	Pythium	propamocarb hydrochloride
	Root rot	tolclofos-methyl
	Rust	myclobutanil, oxycarboxin
	Silver leaf	azaconazole + imazalil
Pests	Ants	pirimiphos-methyl
	Aphids	cypermethrin, deltamethrin, dimethoate, imidacloprid, natural plant extracts, nicotine, pirimicarb, pirimiphos-methyl, pymetrozine, rotenone, Verticillium lecanii
	Birds/mammals	ziram
	Capsid bugs	cypermethrin, deltamethrin, pirimiphos-methyl
	Caterpillars	Bacillus thuringiensis, cypermethrin, deltamethrin, diflubenzuron, nicotine, teflubenzuron
	Cutworms	cypermethrin
	Earwigs	pirimiphos-methyl
	Flies	imidacloprid
	Leaf miners	abamectin, cypermethrin (*off-label*), deltamethrin (*off-label*), dimethoate, nicotine, oxamyl (*off-label*), pirimiphos-methyl, thiacloprid (*off-label*)
	Leafhoppers	natural plant extracts, nicotine
	Mealybugs	deltamethrin, petroleum oil
	Mites	natural plant extracts
	Sawflies	pirimiphos-methyl
	Scale insects	deltamethrin, petroleum oil
	Slugs/snails	metaldehyde
	Spider mites	abamectin, bifenthrin, dimethoate, fenbutatin oxide, petroleum oil, pirimiphos-methyl
	Thrips	abamectin, cypermethrin, deltamethrin, deltamethrin (*off-label*), natural plant extracts, nicotine, pirimiphos-methyl, spinosad, thiacloprid (*off-label*)
	Weevils	chlorpyrifos, chlorpyrifos (*conifers*), fipronil, imidacloprid
	Whiteflies	buprofezin, cypermethrin, deltamethrin, imidacloprid, natural plant extracts, nicotine, pirimiphos-methyl, teflubenzuron, thiacloprid (*off-label*), Verticillium lecanii
Plant growth regulation	Growth control	2-(1-naphthyl)acetic acid, 4-indol-3-yl-butyric acid, 4-indol-3-yl-butyric acid + 2-(1-naphthyl)acetic acid with dichlorophen, chlormequat, copper hydroxide, daminozide, indol-3-ylacetic acid
Weeds	Broad-leaved weeds	clopyralid, diquat, diquat + paraquat, diuron, diuron (*not container grown*), diuron + paraquat, glufosinate-ammonium, isoxaben, isoxaben + trifluralin (*outdoor stock only*), napropamide, oxadiazon, paraquat (*off-label - inter-row directed treatment*), pentanochlor, propachlor, propyzamide, simazine, simazine (*for resale only*), trifluralin (*off-label*)
	Crops as weeds	cycloxydim

	Grass weeds	cycloxydim, diquat + paraquat, diuron, diuron + paraquat, fluazifop-P-butyl (*off-label*), glufosinate-ammonium, isoxaben + trifluralin (*outdoor stock only*), napropamide, oxadiazon, paraquat (*off-label - inter-row directed treatment*), pentanochlor, propachlor, propyzamide, simazine, simazine (*for resale only*), trifluralin (*off-label*)
	Mayweeds	clopyralid
	Mosses	dichlorophen
	Weeds, miscellaneous	ammonium sulphamate, ammonium sulphamate (*pre-planting*), glyphosate

Ornamentals - Trees and shrubs

Diseases	Fungus diseases	prochloraz
	Powdery mildew	penconazole
	Scab	penconazole
	Verticillium wilt	chloropicrin (*soil fumigation*)
Pests	Aphids	deltamethrin
	Capsid bugs	deltamethrin
	Caterpillars	Bacillus thuringiensis, deltamethrin, diflubenzuron
	Mealybugs	deltamethrin
	Scale insects	deltamethrin
	Thrips	deltamethrin
	Whiteflies	deltamethrin
Plant growth regulation	Growth control	maleic hydrazide
Weeds	Broad-leaved weeds	ammonium sulphamate, asulam, carbetamide + diflufenican + oxadiazon, chlorthal-dimethyl, dichlobenil, diuron, diuron + glyphosate, glufosinate-ammonium, isoxaben, isoxaben + trifluralin, metazachlor, paraquat, propyzamide, simazine
	Grass weeds	carbetamide + diflufenican + oxadiazon, cycloxydim (*off-label*), dichlobenil, diuron, diuron + glyphosate, glufosinate-ammonium, isoxaben + trifluralin, metazachlor, paraquat, propyzamide, simazine
	Weeds, miscellaneous	ammonium sulphamate, dichlobenil, diquat + paraquat, diuron, glyphosate, glyphosate (*wiper application*)
	Woody weeds/scrub	asulam, glyphosate

Ornamentals - Woody ornamentals

Diseases	Fungus diseases	prochloraz
Pests	Aphids	fatty acids
	Mealybugs	fatty acids
	Scale insects	fatty acids
	Spider mites	fatty acids
	Whiteflies	fatty acids
Weeds	Aquatic weeds	propyzamide
	Bindweeds	oxadiazon

	Broad-leaved weeds	dichlobenil, diuron, diuron + paraquat, glufosinate-ammonium, oxadiazon, paraquat, propyzamide, simazine
	Grass weeds	dichlobenil, diuron, diuron + paraquat, glufosinate-ammonium, oxadiazon, paraquat, propyzamide, simazine
	Weeds, miscellaneous	dichlobenil, glyphosate, propyzamide

Forestry

Forest nurseries, general - Forest nurseries

Pests	Aphids	pirimicarb
	Birds/mammals	aluminium ammonium sulphate
	Weevils	carbosulfan (*off-label*)
Weeds	Broad-leaved weeds	cyanazine (*off-label*), diquat + paraquat, napropamide, paraquat, paraquat (*stale seedbed*)
	Grass weeds	cyanazine (*off-label*), cycloxydim, diquat + paraquat, napropamide, paraquat (*stale seedbed*), propaquizafop

Forestry plantations, general - Forestry plantations

Crop control	Chemical stripping/ thinning	glyphosate
Pests	Beetles	chlorpyrifos
	Birds/mammals	aluminium ammonium sulphate, warfarin, ziram
	Caterpillars	diflubenzuron
	Weevils	carbosulfan, chlorpyrifos
Weeds	Aquatic weeds	glyphosate, propyzamide
	Broad-leaved weeds	2,4-D, 2,4-D + dicamba + triclopyr, ammonium sulphamate, atrazine, clopyralid (*off-label*), diquat + paraquat, glufosinate-ammonium, glyphosate, glyphosate (*wiper application*), isoxaben, metamitron (*off-label*), metazachlor, paraquat, propyzamide, triclopyr
	Crops as weeds	diquat + paraquat
	Grass weeds	ammonium sulphamate, atrazine, cycloxydim, diquat + paraquat, glufosinate-ammonium, glyphosate, metamitron (*off-label*), metazachlor, paraquat, propaquizafop, propyzamide
	Weeds, miscellaneous	ammonium sulphamate, glufosinate-ammonium, glyphosate, paraquat, propyzamide
	Woody weeds/scrub	2,4-D, 2,4-D + dicamba + triclopyr, ammonium sulphamate, asulam, asulam (*off-label*), dicamba, glyphosate, triclopyr

Miscellaneous forestry situations - Cut logs/timber

Pests	Beetles	chlorpyrifos
Weeds	Grass weeds	propaquizafop

Woodland on farms - Woodland

Diseases	Dutch elm disease	thiabendazole
Pests	Aphids	lambda-cyhalothrin
	Beetles	lambda-cyhalothrin
	Birds/mammals	aluminium phosphide
	Sawflies	lambda-cyhalothrin
Weeds	Broad-leaved weeds	2,4-D + dicamba + triclopyr, atrazine, cyanazine (*off-label*), lenacil (*off-label*), metazachlor, pendimethalin (*off-label*)
	Crops as weeds	fluazifop-P-butyl
	Grass weeds	atrazine, cyanazine (*off-label*), cycloxydim, fluazifop-P-butyl, glyphosate, lenacil (*off-label*), metazachlor, propaquizafop
	Weeds, miscellaneous	glyphosate
	Woody weeds/scrub	2,4-D + dicamba + triclopyr, glyphosate

Fruit and hops

Bush fruit - Bilberries/blueberries/cranberries

Diseases	Botrytis	pyrimethanil (*off-label*)
Pests	Caterpillars	Bacillus thuringiensis (*off-label*)
	Gall, rust and leaf & bud mites	tebufenpyrad (*off-label*)
	Scale insects	thiacloprid (*off-label*)
Weeds	Broad-leaved weeds	asulam (*off-label*), propachlor (*off-label*)
	Grass weeds	propachlor (*off-label*)
	Weeds, miscellaneous	glufosinate-ammonium, glyphosate (*off-label*)

Bush fruit - Currants

Diseases	Botrytis	carbendazim, chlorothalonil, chlorothalonil (*qualified minor use*), fenhexamid, fenhexamid + tolylfluanid, pyrimethanil (*off-label*), tolylfluanid
	Leaf spots	Bordeaux mixture, chlorothalonil, chlorothalonil (*qualified minor use*), copper ammonium carbonate, dodine, mancozeb
	Powdery mildew	bupirimate, carbendazim, chlorothalonil, chlorothalonil (*qualified minor use*), fenarimol, kresoxim-methyl, myclobutanil, penconazole
	Rust	copper oxychloride, thiram
Pests	Aphids	chlorpyrifos, pirimicarb
	Capsid bugs	chlorpyrifos, fenpropathrin
	Caterpillars	Bacillus thuringiensis (*off-label*), chlorpyrifos, diflubenzuron
	Gall, rust and leaf & bud mites	fenpropathrin, sulphur, tebufenpyrad (*off-label*)
	Midges	fenpropathrin
	Sawflies	fenpropathrin

	Scale insects	thiacloprid (*off-label*)
	Spider mites	chlorpyrifos, clofentezine (*off-label*), fenpropathrin
Weeds	Bindweeds	oxadiazon
	Broad-leaved weeds	asulam, asulam (*off-label*), chlorpropham, chlorthal-dimethyl, dichlobenil, diuron (*off-label*), glufosinate-ammonium, isoxaben, MCPB, napropamide, oxadiazon, paraquat, pendimethalin, propachlor (*off-label*), propyzamide
	Crops as weeds	fluazifop-P-butyl, pendimethalin
	Grass weeds	chlorpropham, dichlobenil, fluazifop-P-butyl, glufosinate-ammonium, napropamide, oxadiazon, paraquat, pendimethalin, propachlor (*off-label*), propyzamide
	Polygonums	chlorpropham
	Weeds, miscellaneous	dichlobenil, diuron (*off-label*), glufosinate-ammonium, glyphosate (*off-label*)

Bush fruit - Gooseberries

Diseases	Botrytis	carbendazim, chlorothalonil, chlorothalonil (*qualified minor use*), fenhexamid, fenhexamid + tolylfluanid, pyrimethanil (*off-label*), tolylfluanid
	Leaf spots	chlorothalonil, chlorothalonil (*qualified minor use*), mancozeb
	Powdery mildew	bupirimate, carbendazim, chlorothalonil, chlorothalonil (*qualified minor use*), fenarimol, myclobutanil, sulphur
Pests	Aphids	chlorpyrifos, pirimicarb
	Capsid bugs	chlorpyrifos
	Caterpillars	Bacillus thuringiensis (*off-label*), chlorpyrifos
	Gall, rust and leaf & bud mites	tebufenpyrad (*off-label*)
	Sawflies	nicotine, rotenone
	Scale insects	thiacloprid (*off-label*)
	Spider mites	chlorpyrifos
Weeds	Bindweeds	oxadiazon
	Broad-leaved weeds	asulam (*off-label*), chlorpropham, chlorthal-dimethyl, dichlobenil, isoxaben, MCPB, napropamide, oxadiazon, paraquat, pendimethalin, propachlor (*off-label*), propyzamide
	Crops as weeds	fluazifop-P-butyl, pendimethalin
	Grass weeds	chlorpropham, dichlobenil, fluazifop-P-butyl, napropamide, oxadiazon, paraquat, pendimethalin, propachlor (*off-label*), propyzamide
	Polygonums	chlorpropham
	Weeds, miscellaneous	dichlobenil

Bush fruit - Miscellaneous bush fruit

Pests	Aphids	nicotine
	Birds/mammals	aluminium ammonium sulphate
	Capsid bugs	nicotine
	Leafhoppers	nicotine
	Mealybugs	petroleum oil

	Sawflies	nicotine
	Scale insects	petroleum oil
	Spider mites	petroleum oil
Weeds	Broad-leaved weeds	diquat + paraquat, glufosinate-ammonium
	Crops as weeds	diquat + paraquat
	Grass weeds	diquat + paraquat, glufosinate-ammonium

Bush fruit - Protected bush fruit

Pests	Aphids	pymetrozine (*off-label*)

Cane fruit - Outdoor cane fruit

Crop control	Sucker/shoot control	sodium monochloroacetate (*off-label*)
Diseases	Botrytis	carbendazim, chlorothalonil, fenhexamid, fenhexamid + tolylfluanid, iprodione, thiram, tolylfluanid
	Cane blight	carbendazim (*off-label*)
	Cane spot	Bordeaux mixture, carbendazim, chlorothalonil, copper ammonium carbonate, copper oxychloride, thiram
	Phytophthora	metalaxyl-M (*off-label*)
	Powdery mildew	azoxystrobin (*off-label*), bupirimate (*outdoor only*), chlorothalonil, fenarimol, fenpropimorph (*off-label*)
	Purple blotch	copper oxychloride
	Spur blight	Bordeaux mixture, carbendazim, thiram
Pests	Aphids	1,3-dichloropropene, chlorpyrifos, pirimicarb
	Beetles	chlorpyrifos, deltamethrin, rotenone
	Birds/mammals	aluminium ammonium sulphate
	Capsid bugs	thiacloprid (*off-label*)
	Caterpillars	Bacillus thuringiensis
	Free-living nematodes	1,3-dichloropropene
	Mealybugs	petroleum oil
	Midges	chlorpyrifos
	Scale insects	petroleum oil
	Spider mites	bifenthrin (*off-label*), chlorpyrifos, clofentezine (*off-label*), petroleum oil, tebufenpyrad (*off-label*)
	Weevils	bifenthrin (*off-label*)
Plant growth regulation	Growth control	sodium monochloroacetate (*off-label*)
Weeds	Bindweeds	oxadiazon
	Broad-leaved weeds	asulam (*off-label*), chlorthal-dimethyl, dichlobenil, diquat + paraquat, glufosinate-ammonium, isoxaben, MCPB, napropamide, oxadiazon, paraquat, pendimethalin, propachlor (*off-label*), propyzamide, propyzamide (*England only*), trifluralin
	Crops as weeds	diquat + paraquat, fluazifop-P-butyl, pendimethalin
	Grass weeds	dichlobenil, diquat + paraquat, fluazifop-P-butyl, glufosinate-ammonium, napropamide, oxadiazon, paraquat, pendimethalin, propachlor (*off-label*), propyzamide, propyzamide (*England only*), trifluralin

	Weeds, miscellaneous	dichlobenil, glufosinate-ammonium

Cane fruit - Protected cane fruit

Diseases	Botrytis	pyrimethanil (*off-label*)
	Cane blight	carbendazim (*off-label*)
	Powdery mildew	myclobutanil (*off-label*)
	Root rot	fluazinam (*off-label*)
Pests	Aphids	pymetrozine (*off-label*)
	Spider mites	bifenthrin (*off-label*), clofentezine (*off-label*)
	Weevils	bifenthrin (*off-label*)

Hops, general - Hops

Crop control	Chemical stripping/ thinning	diquat, diquat + paraquat, paraquat, sodium monochloroacetate (*off-label*)
	Pre-harvest desiccation	pymetrozine (*off-label*)
Diseases	Downy mildew	Bordeaux mixture, chlorothalonil, copper oxychloride, fosetyl-aluminium, metalaxyl-M (*off-label*)
	Powdery mildew	bupirimate, fenpropimorph (*off-label*), myclobutanil (*off-label*), penconazole, sulphur
Pests	Aphids	1,3-dichloropropene, bifenthrin, cypermethrin, deltamethrin, imidacloprid, tebufenpyrad
	Free-living nematodes	1,3-dichloropropene
	Mealybugs	petroleum oil
	Scale insects	petroleum oil
	Spider mites	bifenthrin, petroleum oil, tebufenpyrad
Weeds	Bindweeds	oxadiazon
	Broad-leaved weeds	asulam, diquat + paraquat, isoxaben, oxadiazon, paraquat, pendimethalin
	Crops as weeds	diquat + paraquat, fluazifop-P-butyl, pendimethalin
	Grass weeds	diquat + paraquat, fluazifop-P-butyl, oxadiazon, paraquat, pendimethalin, propyzamide (*off-label*)

Miscellaneous fruit situations - Fruit crops, general

Pests	Aphids	fatty acids, natural plant extracts
	Leafhoppers	natural plant extracts
	Mealybugs	fatty acids, petroleum oil
	Mites	natural plant extracts
	Scale insects	fatty acids, petroleum oil
	Spider mites	fatty acids, petroleum oil
	Thrips	natural plant extracts
	Whiteflies	fatty acids, natural plant extracts
Plant growth regulation	Fruiting control	ethylene (commodity substance) (*in store*)
Weeds	Broad-leaved weeds	glufosinate-ammonium
	Grass weeds	glufosinate-ammonium

Miscellaneous fruit situations - Fruit nursery stock

Weeds	Broad-leaved weeds	metazachlor, trifluralin (*off-label*)
	Grass weeds	metazachlor, trifluralin (*off-label*)

Miscellaneous fruit situations - Orchards

Weeds	Broad-leaved weeds	paraquat
	Grass weeds	paraquat

Miscellaneous fruit situations - Protected miscellaneous fruit

Diseases	Botrytis	iprodione (*off-label*)
Pests	Aphids	natural plant extracts, nicotine
	Leafhoppers	natural plant extracts
	Mites	natural plant extracts
	Thrips	natural plant extracts
	Whiteflies	natural plant extracts

Miscellaneous nuts - Nuts

Diseases	Bacterial canker	copper oxychloride (*off-label*)
	Blight	copper oxychloride (*off-label*)
Pests	Aphids	lambda-cyhalothrin (*off-label*)
	Caterpillars	lambda-cyhalothrin (*off-label*)
Weeds	Broad-leaved weeds	glufosinate-ammonium
	Grass weeds	glufosinate-ammonium
	Weeds, miscellaneous	glufosinate-ammonium, glyphosate (*off-label*)

Other fruit - Rhubarb

Diseases	Downy mildew	mancozeb + metalaxyl-M (*off-label*)
Weeds	Broad-leaved weeds	dichlobenil (*off-label - established*), paraquat (*off-label*), propyzamide, propyzamide (*outdoor*), simazine, simazine (*off-label*)
	Grass weeds	dichlobenil (*off-label - established*), paraquat (*off-label*), propyzamide, propyzamide (*outdoor*), simazine, simazine (*off-label*)
	Weeds, miscellaneous	glyphosate (*off-label*)

Other fruit - Vines

Diseases	Botrytis	fenhexamid, iprodione (*off-label*), pyrimethanil (*off-label*)
	Bunch rots	iprodione (*off-label*)
	Downy mildew	copper oxychloride, fosetyl-aluminium (*off-label*), mancozeb + zoxamide, metalaxyl-M (*off-label*)
	Fungus diseases	mancozeb (*off-label*)
	Powdery mildew	dinocap (*off-label*), fenbuconazole (*off-label*), myclobutanil (*off-label*), sulphur

Pests	Aphids	nicotine
	Mealybugs	petroleum oil
	Scale insects	petroleum oil
	Spider mites	petroleum oil
Weeds	Bindweeds	oxadiazon
	Broad-leaved weeds	dichlobenil (*off-label*), glufosinate-ammonium, isoxaben, oxadiazon, paraquat
	Grass weeds	glufosinate-ammonium, oxadiazon, paraquat, propyzamide (*off-label*)
	Weeds, miscellaneous	glufosinate-ammonium, glyphosate (*off-label*)

Soft fruit - Miscellaneous soft fruit

Pests	Aphids	nicotine, rotenone
	Capsid bugs	nicotine
	Leafhoppers	nicotine
	Sawflies	nicotine

Soft fruit - Strawberries

Crop control	Sucker/shoot control	paraquat
Diseases	Black spot	azoxystrobin (*off-label*)
	Botrytis	boscalid + pyraclostrobin, captan, carbendazim, chlorothalonil, chlorothalonil (*off-label*), chlorothalonil (*outdoor crops only*), fenhexamid, fenhexamid + tolylfluanid, iprodione, mepanipyrim, pyrimethanil, thiram, tolyifluanid
	Crown rot	chloropicrin (*soil fumigation*), fosetyl-aluminium (*off-label*)
	Powdery mildew	bupirimate, dinocap, fenarimol, fenpropimorph (*off-label*), kresoxim-methyl, myclobutanil, sulphur
	Red core	chloropicrin (*soil fumigation*), fosetyl-aluminium, fosetyl-aluminium (*off-label*)
	Verticillium wilt	chloropicrin (*soil fumigation*)
Pests	Aphids	1,3-dichloropropene, chlorpyrifos, nicotine, pirimicarb, pymetrozine (*off-label*)
	Beetles	methiocarb
	Birds/mammals	aluminium ammonium sulphate
	Capsid bugs	thiacloprid (*off-label*)
	Caterpillars	Bacillus thuringiensis, chlorpyrifos
	Free-living nematodes	1,3-dichloropropene, chloropicrin (*soil fumigation*)
	Slugs/snails	methiocarb
	Spider mites	abamectin (*off-label*), bifenthrin, chlorpyrifos, clofentezine (*off-label*), fenbutatin oxide, tebufenpyrad
	Stem nematodes	1,3-dichloropropene
	Tarsonemid mites	abamectin (*off-label - in propagation*)
	Weevils	chlorpyrifos
Weeds	Broad-leaved weeds	asulam (*off-label*), chlorpropham, chlorthal-dimethyl, clopyralid, diquat + paraquat, ethofumesate (*off-label*), glufosinate-ammonium, isoxaben, napropamide,

		pendimethalin, phenmedipham, propachlor, propachlor (*off-label*), propyzamide, trifluralin
	Crops as weeds	cycloxydim, diquat + paraquat, ethofumesate (*off-label*), fluazifop-P-butyl, pendimethalin
	Grass weeds	chlorpropham, cycloxydim, diquat + paraquat, ethofumesate (*off-label*), fluazifop-P-butyl, glufosinate-ammonium, napropamide, pendimethalin, propachlor, propachlor (*off-label*), propyzamide, trifluralin
	Mayweeds	clopyralid
	Polygonums	chlorpropham
	Weeds, miscellaneous	glufosinate-ammonium

Tree fruit - Miscellaneous tree fruit

Diseases	Verticillium wilt	chloropicrin (*soil fumigation*)
Pests	Aphids	nicotine, rotenone
	Birds/mammals	aluminium ammonium sulphate, ziram
	Capsid bugs	nicotine
	Sawflies	nicotine
Weeds	Broad-leaved weeds	paraquat
	Grass weeds	paraquat
	Weeds, miscellaneous	glyphosate

Tree fruit - Pome fruit

Crop control	Sucker/shoot control	glyphosate
Diseases	Blossom wilt	vinclozolin
	Botrytis	iprodione (*off-label*)
	Botrytis fruit rot	thiram
	Canker	Bordeaux mixture, copper oxychloride
	Collar rot	copper oxychloride (*off-label*), fosetyl-aluminium
	Crown rot	fosetyl-aluminium
	Phytophthora	mancozeb + metalaxyl-M (*off-label - applied to orchard floor*)
	Powdery mildew	bupirimate, carbendazim, dinocap, fenarimol, fenbuconazole (*reduction*), kresoxim-methyl (*reduction*), myclobutanil, penconazole, sulphur
	Scab	Bordeaux mixture, captan, carbendazim, carbendazim (*off-label*), dithianon, dodine, fenarimol, fenbuconazole, kresoxim-methyl, mancozeb, myclobutanil, pyrimethanil, sulphur, thiram, tolylfluanid
	Storage rots	captan, carbendazim, carbendazim (*off-label*), metalaxyl-M (*off-label*), thiram
Pests	Aphids	chlorpyrifos, cypermethrin, deltamethrin, nicotine, pirimicarb, thiacloprid, triazamate
	Capsid bugs	chlorpyrifos, cypermethrin, deltamethrin, nicotine
	Caterpillars	Bacillus thuringiensis (*off-label*), chlorpyrifos, cypermethrin, deltamethrin, diflubenzuron, fenoxycarb, methoxyfenozide
	Gall, rust and leaf & bud mites	amitraz, diflubenzuron, tolylfluanid (*reduction*)

	Sawflies	chlorpyrifos, cypermethrin, deltamethrin, rotenone
	Spider mites	amitraz, bifenthrin, chlorpyrifos, clofentezine, fenpyroximate, tebufenpyrad, tolylfluanid (*reduction*)
	Suckers	amitraz, chlorpyrifos, cypermethrin, deltamethrin, diflubenzuron, lambda-cyhalothrin
	Weevils	chlorpyrifos
Plant growth regulation	Fruiting control	1-methylcyclopropene (*post-harvest use*), 2-chloroethylphosphonic acid (*off-label - for cider making*), paclobutrazol
	Growth control	gibberellins (*off-label*)
	Plant growth regulation, miscellaneous	paclobutrazol
	Quality/yield control	gibberellins
Weeds	Bindweeds	oxadiazon
	Broad-leaved weeds	2,4-D, 2,4-D + dichlorprop-P + MCPA + mecoprop-P, asulam, asulam (*off-label*), clopyralid (*off-label*), dicamba + MCPA + mecoprop-P, dichlobenil, diquat + paraquat, diuron, fluroxypyr (*off-label*), glufosinate-ammonium, isoxaben, napropamide, oxadiazon, pendimethalin, propyzamide
	Crops as weeds	diquat + paraquat, pendimethalin
	Grass weeds	amitrole, dichlobenil, diquat + paraquat, diuron, glufosinate-ammonium, napropamide, oxadiazon, pendimethalin, propyzamide
	Mayweeds	dicamba + MCPA + mecoprop-P
	Polygonums	dicamba + MCPA + mecoprop-P
	Weeds, miscellaneous	amitrole, amitrole (*off-label*), dichlobenil, glufosinate-ammonium, glyphosate

Tree fruit - Stone fruit

Crop control	Sucker/shoot control	glyphosate
Diseases	Bacterial canker	Bordeaux mixture, copper oxychloride
	Blossom wilt	fenbuconazole (*off-label*), myclobutanil (*off-label*)
	Botrytis	fenhexamid (*off-label*)
	Leaf curl	Bordeaux mixture, copper ammonium carbonate, copper oxychloride
	Rust	myclobutanil (*off-label*)
	Sclerotinia	fenbuconazole (*off-label*)
Pests	Aphids	chlorpyrifos, cypermethrin, deltamethrin, nicotine, pirimicarb, pirimicarb (*off-label*), tebufenpyrad (*off-label*), thiacloprid (*off-label*), thiacloprid (*off-label - under temporary covers*)
	Caterpillars	Bacillus thuringiensis (*off-label*), chlorpyrifos, cypermethrin, deltamethrin, diflubenzuron
	Gall, rust and leaf & bud mites	diflubenzuron
	Sawflies	deltamethrin
	Spider mites	chlorpyrifos, clofentezine, fenpyroximate (*off-label*)
Plant growth regulation	Growth control	paclobutrazol (*off-label*)

Weeds	Broad-leaved weeds	asulam, asulam (*off-label*), diquat + paraquat, glufosinate-ammonium, isoxaben, napropamide, pendimethalin, propyzamide
	Crops as weeds	diquat + paraquat, pendimethalin
	Grass weeds	diquat + paraquat, glufosinate-ammonium, napropamide, pendimethalin, propyzamide, propyzamide (*off-label*)
	Weeds, miscellaneous	amitrole (*off-label*), glufosinate-ammonium, glyphosate

Grain/crop store uses

Stored produce - Food/produce storage

Diseases	Alternaria	iprodione
	Aspergillus diseases	imazalil
	Botrytis	iprodione
	Cladosporium diseases	imazalil
	Penicillium rot	imazalil

Stored seed - Stored grain/rape/linseed

Pests	Birds/mammals	aluminium ammonium sulphate
	Food/grain storage pests	bifenthrin + malathion, chlorpyrifos-methyl, d-phenothrin + tetramethrin, pirimiphos-methyl
	Pests, miscellaneous	aluminium phosphide, magnesium phosphide

Grass

Grass seed - Grass seed crops

Diseases	Crown rust	propiconazole
	Damping off	thiram (*seed treatment*)
	Drechslera leaf spot	propiconazole
	Powdery mildew	propiconazole
	Rhynchosporium	propiconazole
Pests	Aphids	deltamethrin (*off-label*), dimethoate
Plant growth regulation	Growth control	trinexapac-ethyl
Weeds	Broad-leaved weeds	2,4-D, 2,4-D + mecoprop-P, bromoxynil + ethofumesate + ioxynil, clopyralid (*off-label*), dicamba + MCPA + mecoprop-P, ethofumesate, MCPA, mecoprop-P
	Grass weeds	bromoxynil + ethofumesate + ioxynil, ethofumesate
	Mayweeds	dicamba + MCPA + mecoprop-P
	Polygonums	dicamba + MCPA + mecoprop-P

Grassland - Leys

Diseases	Crown rust	propiconazole

	Drechslera leaf spot	propiconazole
	Powdery mildew	propiconazole
	Rhynchosporium	propiconazole
Pests	Flies	chlorpyrifos, cypermethrin
	Leatherjackets	chlorpyrifos, methiocarb (*seed admixture*)
	Slugs/snails	methiocarb (*seed admixture*)
Weeds	Broad-leaved weeds	2,4-D, 2,4-D + dicamba + triclopyr, 2,4-D + MCPA, 2,4-DB + linuron + MCPA, 2,4-DB + MCPA, amidosulfuron, bentazone + MCPA + MCPB, bromoxynil + ethofumesate + ioxynil, bromoxynil + ioxynil + mecoprop-P, clopyralid, clopyralid + fluroxypyr + triclopyr, clopyralid + triclopyr, dicamba, dicamba + MCPA + mecoprop-P, dicamba + mecoprop-P, ethofumesate, fluroxypyr, fluroxypyr + triclopyr, MCPA, MCPA + MCPB, MCPB, mecoprop-P
	Crops as weeds	fluroxypyr
	Grass weeds	bromoxynil + ethofumesate + ioxynil, ethofumesate
	Mayweeds	clopyralid, dicamba + MCPA + mecoprop-P, dicamba + mecoprop-P
	Polygonums	2,4-DB + MCPA, dicamba + MCPA + mecoprop-P, dicamba + mecoprop-P
	Weeds, miscellaneous	diquat + paraquat, glyphosate

Grassland - Permanent pasture

Pests	Aphids	pirimicarb
	Birds/mammals	aluminium ammonium sulphate, aluminium phosphide, strychnine hydrochloride (commodity substance) (*areas of restricted public access*)
	Flies	chlorpyrifos
	Leatherjackets	chlorpyrifos
Plant growth regulation	Quality/yield control	sulphur
Weeds	Aquatic weeds	glyphosate, MCPA, triclopyr
	Broad-leaved weeds	2,4-D, 2,4-D + dicamba + triclopyr, 2,4-D + MCPA, 2,4-D + mecoprop-P, amidosulfuron, asulam, citronella oil, clopyralid, clopyralid + 2,4-D + MCPA, clopyralid + fluroxypyr + triclopyr, clopyralid + triclopyr, dicamba (*wiper application*), dicamba + MCPA + mecoprop-P, dicamba + mecoprop-P, ethofumesate, fluroxypyr, fluroxypyr + triclopyr, glyphosate (*wiper application*), MCPA, MCPA + MCPB, MCPB, mecoprop-P, paraquat (*sward destruction/direct drilling*), thifensulfuron-methyl, triclopyr
	Crops as weeds	fluroxypyr
	Grass weeds	ethofumesate, glyphosate, paraquat (*sward destruction/ direct drilling*)
	Mayweeds	clopyralid, dicamba + MCPA + mecoprop-P, dicamba + mecoprop-P
	Polygonums	dicamba + MCPA + mecoprop-P, dicamba + mecoprop-P
	Weeds, miscellaneous	diquat + paraquat, glufosinate-ammonium, glyphosate

	Woody weeds/scrub	asulam, clopyralid + triclopyr, dicamba, glyphosate, triclopyr

Turf/amenity grass - Amenity grassland

Diseases	Brown patch	iprodione
	Dollar spot	iprodione
	Fusarium diseases	azoxystrobin, iprodione
	Melting out	iprodione
	Red thread	iprodione
	Snow mould	iprodione
Pests	Flies	chlorpyrifos
	Leatherjackets	chlorpyrifos
	Slugs/snails	metaldehyde
Plant growth regulation	Growth control	dicamba + maleic hydrazide + MCPA, maleic hydrazide, trinexapac-ethyl
Weeds	Broad-leaved weeds	2,4-D, 2,4-D + dicamba + triclopyr, 2,4-D + picloram, asulam, asulam (*not fine turf*), clopyralid + triclopyr, dicamba (*wiper application*), dicamba + maleic hydrazide + MCPA, dicamba + MCPA + mecoprop-P, MCPA
	Weeds, miscellaneous	glyphosate, glyphosate (*wiper application*)
	Woody weeds/scrub	2,4-D + picloram, clopyralid + triclopyr, dicamba

Turf/amenity grass - Managed amenity turf

Crop control	Miscellaneous non-selective situations	glufosinate-ammonium
Diseases	Anthracnose	carbendazim + iprodione, chlorothalonil
	Brown patch	iprodione
	Cladosporium diseases	carbendazim + iprodione
	Dollar spot	carbendazim, chlorothalonil, fenarimol, iprodione, pyraclostrobin (*reduction*), thiophanate-methyl
	Fungus diseases	dichlorophen
	Fusarium diseases	azoxystrobin, carbendazim, carbendazim + epoxiconazole, carbendazim + iprodione, chlorothalonil, fenarimol, iprodione, pyraclostrobin, thiophanate-methyl
	Melting out	iprodione
	Red thread	carbendazim + iprodione, chlorothalonil, dichlorophen, fenarimol, iprodione, pyraclostrobin, thiophanate-methyl
	Snow mould	iprodione
Pests	Birds/mammals	aluminium phosphide
	Earthworms	carbendazim, thiophanate-methyl
	Flies	chlorpyrifos
	Leatherjackets	chlorpyrifos
	Slugs/snails	metaldehyde
Plant growth regulation	Growth control	maleic hydrazide, trinexapac-ethyl
Weeds	Broad-leaved weeds	2,4-D, 2,4-D + dicamba, 2,4-D + dicamba + fluroxypyr, 2,4-D + mecoprop-P, chlorthal-dimethyl, clopyralid + diflufenican + MCPA, clopyralid + fluroxypyr + MCPA,

		dicamba + dichlorprop-P + ferrous sulphate + MCPA, dicamba + dichlorprop-P + MCPA, dicamba + MCPA + mecoprop-P, dichlorprop-P + ferrous sulphate + MCPA, dichlorprop-P + MCPA, fluroxypyr + mecoprop-P, MCPA, MCPA + mecoprop-P, mecoprop-P
	Crops as weeds	2,4-D + dicamba + fluroxypyr, dicamba + MCPA + mecoprop-P, dichlorprop-P + MCPA, mecoprop-P
	Grass weeds	tepraloxydim (*off-label*)
	Mayweeds	dicamba + MCPA + mecoprop-P
	Mosses	dicamba + dichlorprop-P + ferrous sulphate + MCPA, dichlorophen, dichlorophen + ferrous sulphate, dichlorprop-P + ferrous sulphate + MCPA, ferrous sulphate
	Polygonums	dicamba + MCPA + mecoprop-P
	Weeds, miscellaneous	diquat + paraquat, glyphosate (*pre-establishment*)

Non-crop pest control

Farm buildings/yards - Farm buildings

Diseases	Fungus diseases	formaldehyde (commodity substance)
Pests	Birds/mammals	alphachloralose (*indoor use*), aluminium ammonium sulphate, bone oil, brodifacoum, brodifacoum (*indoor use only*), bromadiolone, chlorophacinone, cholecalciferol, cholecalciferol + difenacoum, difenacoum, warfarin, zinc phosphide
	Flies	diflubenzuron, d-phenothrin + tetramethrin, pyrethrins, tetramethrin
	Food/grain storage pests	alpha-cypermethrin
	Pests, miscellaneous	alpha-cypermethrin
	Wasps	d-phenothrin + tetramethrin
Weeds	Weeds, miscellaneous	glyphosate

Farm buildings/yards - Farmyards

Pests	Birds/mammals	bromadiolone, difenacoum

Farmland pest control - Farmland situations

Pests	Birds/mammals	aluminium phosphide, bone oil
Weeds	Crops as weeds	dichlobenil (*blight prevention*)

Miscellaneous non-crop pest control - Manure/rubbish

Pests	Flies	diflubenzuron

Miscellaneous non-crop pest control - Miscellaneous pest control situations

Pests	Birds/mammals	carbon dioxide (commodity substance), paraffin oil (commodity substance) (*egg treatment*), strychnine hydrochloride (commodity substance) (*areas of restricted public access*)
	Wasps	resmethrin + tetramethrin

Protected salad and vegetable crops

Protected brassicas - Protected brassica vegetables

Diseases	Black rot	copper oxychloride (*off-label*)
	Damping off	fosetyl-aluminium (*off-label*)
	Downy mildew	fosetyl-aluminium (*off-label*), fosetyl-aluminium (*off-label - seedlings*)
	Spear rot	copper oxychloride (*off-label*)
Pests	Caterpillars	Bacillus thuringiensis (*off-label*)
	Pests, miscellaneous	dimethoate (*off-label*)

Protected brassicas - Protected salad brassicas

Diseases	Black rot	copper oxychloride (*off-label*)
	Damping off	fosetyl-aluminium (*off-label*)
	Downy mildew	fosetyl-aluminium (*off-label*), fosetyl-aluminium (*off-label - seedlings*)
	Spear rot	copper oxychloride (*off-label*)
Pests	Aphids	cypermethrin (*off-label*), deltamethrin (*off-label*), pymetrozine (*off-label - for baby leaf production*), thiacloprid (*off-label - for baby leaf production*)
	Beetles	deltamethrin (*off-label - baby leaf production*), deltamethrin (*off-label - for baby leaf production*)
	Caterpillars	Bacillus thuringiensis (*off-label*), cypermethrin (*off-label*), deltamethrin (*off-label*), deltamethrin (*off-label - baby leaf production*), deltamethrin (*off-label - for baby leaf production*)
	Pests, miscellaneous	deltamethrin (*off-label*), dimethoate (*off-label*)

Protected crops, general - All protected crops

Diseases	Soil-borne diseases	dazomet (*soil fumigation*)
Pests	Aphids	nicotine, rotenone
	Capsid bugs	nicotine
	Free-living nematodes	dazomet (*soil fumigation*)
	Leaf miners	nicotine
	Leafhoppers	nicotine
	Sawflies	nicotine
	Slugs/snails	metaldehyde
	Soil pests	dazomet (*soil fumigation*)

	Thrips	nicotine
	Whiteflies	alginate/polysaccaride
Weeds	Weeds, miscellaneous	dazomet (*soil fumigation*)

Protected crops, general - Glasshouses

Diseases	Fungus diseases	dichlorophen, formaldehyde (commodity substance) (*spray, dip or fumigant*)
	Phytophthora	fosetyl-aluminium
	Root rot	fosetyl-aluminium
	Soil-borne diseases	metam-sodium
Pests	Free-living nematodes	metam-sodium
	Soil pests	metam-sodium
Weeds	Mosses	dichlorophen
	Weeds, miscellaneous	metam-sodium

Protected crops, general - Protected vegetables, general

Diseases	White blister	propamocarb hydrochloride (*off-label*)
Pests	Aphids	natural plant extracts
	Leafhoppers	natural plant extracts
	Mites	natural plant extracts
	Thrips	natural plant extracts
	Whiteflies	natural plant extracts

Protected crops, general - Soils and compost

Diseases	Soil-borne diseases	formaldehyde (commodity substance) (*drench*), metam-sodium
Pests	Free-living nematodes	metam-sodium
	Pests, miscellaneous	dimethoate (*off-label - for watercress propagation*)
	Soil pests	metam-sodium
Weeds	Weeds, miscellaneous	metam-sodium

Protected fruiting vegetables - Protected aubergines

Pests	Aphids	pymetrozine (*off-label*)
	Leaf miners	deltamethrin (*off-label*), thiacloprid (*off-label*)
	Thrips	deltamethrin (*off-label*), thiacloprid (*off-label*)
	Whiteflies	thiacloprid (*off-label*)

Protected fruiting vegetables - Protected cucurbits

Diseases	Botrytis	chlorothalonil, iprodione (*off-label*)
	Damping off	propamocarb hydrochloride (*off-label*)
	Downy mildew	azoxystrobin (*off-label*), metalaxyl-M (*off-label*)
	Phytophthora	propamocarb hydrochloride

	Powdery mildew	azoxystrobin (*off-label*), bupirimate, chlorothalonil, fenarimol (*off-label*), sulphur (*off-label*)
	Pythium	propamocarb hydrochloride
	Root diseases	carbendazim (*off-label*)
	Root rot	propamocarb hydrochloride (*off-label*)
Pests	Ants	pirimiphos-methyl
	Aphids	deltamethrin, nicotine, pirimicarb (*off-label*), pirimiphos-methyl, pymetrozine, pymetrozine (*off-label*), Verticillium lecanii
	Capsid bugs	pirimiphos-methyl
	Caterpillars	Bacillus thuringiensis, deltamethrin
	Earwigs	pirimiphos-methyl
	Leaf miners	cypermethrin (*off-label*), deltamethrin (*off-label*), oxamyl (*off-label*), pirimiphos-methyl, thiacloprid (*off-label*)
	Leafhoppers	nicotine
	Mealybugs	deltamethrin, petroleum oil
	Sawflies	pirimiphos-methyl
	Scale insects	deltamethrin, petroleum oil
	Spider mites	abamectin, abamectin (*off-label*), clofentezine (*off-label*), fenbutatin oxide, fenbutatin oxide (*off-label*), petroleum oil, pirimiphos-methyl
	Thrips	abamectin, deltamethrin, deltamethrin (*off-label*), nicotine, pirimiphos-methyl, spinosad, thiacloprid (*off-label*)
	Whiteflies	buprofezin, cypermethrin, deltamethrin, nicotine, pirimiphos-methyl, thiacloprid (*off-label*), Verticillium lecanii
Weeds	Broad-leaved weeds	paraquat (*off-label - inter-row directed treatment*)
	Grass weeds	paraquat (*off-label - inter-row directed treatment*), propyzamide (*off-label*)

Protected fruiting vegetables - Protected tomatoes

Diseases	Blight	azoxystrobin (*off-label*), chlorothalonil, copper oxychloride, propamocarb hydrochloride
	Botrytis	azoxystrobin (*off-label*), chlorothalonil, iprodione, pyrimethanil (*off-label*), thiram
	Damping off	copper oxychloride, propamocarb hydrochloride (*off-label*)
	Damping off and foot rot	etridiazole (*seeds and seedlings*)
	Didymella stem rot	azoxystrobin (*off-label*), carbendazim (*off-label*)
	Foot rot	copper oxychloride
	Leaf mould	chlorothalonil, copper ammonium carbonate
	Phytophthora	copper oxychloride, etridiazole (*transplants*), propamocarb hydrochloride
	Powdery mildew	azoxystrobin (*off-label*), bupirimate (*off-label*), fenarimol (*off-label*), sulphur (*off-label*)
	Pythium	propamocarb hydrochloride
	Root diseases	etridiazole (*off-label*)
	Root rot	propamocarb hydrochloride (*off-label*)

Pests	Ants	pirimiphos-methyl
	Aphids	deltamethrin, fatty acids, nicotine, pirimicarb, pirimiphos-methyl, Verticillium lecanii
	Capsid bugs	pirimiphos-methyl
	Caterpillars	Bacillus thuringiensis, deltamethrin, deltamethrin (*off-label*), nicotine
	Earwigs	pirimiphos-methyl
	Leaf miners	abamectin, abamectin (*off-label*), deltamethrin (*off-label*), nicotine, oxamyl (*off-label*), pirimiphos-methyl, thiacloprid (*off-label*)
	Leafhoppers	nicotine
	Mealybugs	deltamethrin, fatty acids, petroleum oil
	Sawflies	pirimiphos-methyl
	Scale insects	deltamethrin, fatty acids, petroleum oil
	Spider mites	abamectin, fatty acids, fenbutatin oxide, petroleum oil, pirimiphos-methyl
	Thrips	abamectin, deltamethrin (*off-label*), nicotine, pirimiphos-methyl, thiacloprid (*off-label*)
	Whiteflies	buprofezin, deltamethrin, fatty acids, nicotine, pirimiphos-methyl, pymetrozine (*off-label*), spiromesifen, thiacloprid (*off-label*), Verticillium lecanii
Plant growth regulation	Fruiting control	(2-naphthyloxy)acetic acid, 2-chloroethylphosphonic acid
Weeds	Broad-leaved weeds	paraquat (*off-label - inter-row directed treatment*)
	Grass weeds	paraquat (*off-label - inter-row directed treatment*)

Protected herb crops - Protected herbs

Diseases	Botrytis	propamocarb hydrochloride (*off-label*)
	Downy mildew	fosetyl-aluminium (*off-label*), propamocarb hydrochloride (*off-label*)
	Fungus diseases	mancozeb + metalaxyl-M (*off-label*)
	Powdery mildew	sulphur (*off-label*)
	Rhizoctonia	azoxystrobin (*off-label*)
	Ring spot	prochloraz (*off-label*)
Pests	Aphids	nicotine, pymetrozine (*off-label*), thiacloprid (*off-label*)
	Beetles	deltamethrin, deltamethrin (*off-label*)
	Caterpillars	Bacillus thuringiensis (*off-label*)
	Leafhoppers	deltamethrin, deltamethrin (*off-label*), nicotine
	Thrips	nicotine
	Whiteflies	nicotine
Weeds	Broad-leaved weeds	prometryn, prometryn (*off-label*)
	Grass weeds	prometryn, prometryn (*off-label*), propyzamide (*off-label*)

Protected leafy vegetables - Mustard and cress

Diseases	Botrytis	propamocarb hydrochloride (*off-label*)
	Damping off and foot rot	etridiazole
	Downy mildew	propamocarb hydrochloride (*off-label*)

Pests	Aphids	nicotine
	Beetles	deltamethrin (*off-label*)
	Caterpillars	nicotine
	Leaf miners	nicotine
	Leafhoppers	deltamethrin (*off-label*)
Weeds	Broad-leaved weeds	propachlor (*off-label*)
	Grass weeds	propachlor (*off-label*)

Protected leafy vegetables - Protected leafy vegetables

Diseases	Big vein	carbendazim (*off-label*)
	Botrytis	iprodione, iprodione (*off-label*), propamocarb hydrochloride (*off-label*), pyrimethanil (*off-label*), thiram
	Downy mildew	fosetyl-aluminium, mancozeb, propamocarb hydrochloride (*off-label*), thiram
	Fungus diseases	mancozeb + metalaxyl-M (*off-label*)
	Rhizoctonia	azoxystrobin (*off-label*), tolclofos-methyl
	Ring spot	prochloraz (*off-label*)
Pests	Aphids	nicotine, pirimicarb, pirimicarb (*off-label*), pymetrozine (*off-label*), thiacloprid (*off-label*), Verticillium lecanii
	Beetles	deltamethrin (*off-label*)
	Caterpillars	Bacillus thuringiensis (*off-label*)
	Leaf miners	abamectin
	Leafhoppers	deltamethrin (*off-label*), nicotine
	Thrips	abamectin, nicotine, pirimicarb (*off-label*)
	Whiteflies	cypermethrin, nicotine, Verticillium lecanii
Weeds	Broad-leaved weeds	chlorpropham with cetrimide, paraquat (*off-label - inter-row directed treatment*)
	Grass weeds	chlorpropham with cetrimide, paraquat (*off-label - inter-row directed treatment*), propyzamide (*off-label*)
	Polygonums	chlorpropham with cetrimide

Protected leafy vegetables - Protected spinach

Diseases	Downy mildew	metalaxyl-M (*off-label*)
Pests	Aphids	cypermethrin (*off-label*), pirimicarb (*off-label*)
	Caterpillars	cypermethrin (*off-label*)
	Pests, miscellaneous	dimethoate (*off-label*)

Protected legumes - Protected peas and beans

Pests	Aphids	Verticillium lecanii
	Leaf miners	oxamyl (*off-label*)
	Whiteflies	Verticillium lecanii

Protected root and tuber vegetables - Protected carrots/ parsnips/celeriac

Pests	Pests, miscellaneous	dimethoate (*off-label*)

Protected root and tuber vegetables - Protected potatoes

Diseases	Blight	benalaxyl + mancozeb
Pests	Aphids	nicotine
	Leafhoppers	nicotine
	Thrips	nicotine
	Whiteflies	nicotine

Protected root and tuber vegetables - Protected root brassicas

Diseases	Botrytis	propamocarb hydrochloride (*off-label*)
	Downy mildew	propamocarb hydrochloride (*off-label*)
	Rhizoctonia	tolclofos-methyl (*off-label*)
	White blister	mancozeb + metalaxyl-M (*off-label*)
Pests	Aphids	pirimicarb (*off-label*)

Protected stem and bulb vegetables - Protected asparagus

Pests	Aphids	nicotine

Protected stem and bulb vegetables - Protected celery/chicory

Diseases	Botrytis	azoxystrobin (*off-label*), carbendazim (*off-label*), iprodione (*off-label*)
	Rhizoctonia	azoxystrobin (*off-label*), tolclofos-methyl (*off-label*)
	Sclerotinia	azoxystrobin (*off-label*)
Pests	Aphids	deltamethrin (*off-label*), nicotine
	Beetles	deltamethrin (*off-label*)
	Caterpillars	Bacillus thuringiensis (*off-label*), deltamethrin (*off-label*)
	Leafhoppers	deltamethrin (*off-label*)
	Pests, miscellaneous	deltamethrin (*off-label*)
	Whiteflies	cypermethrin
Weeds	Broad-leaved weeds	paraquat (*off-label - inter-row directed treatment*), prometryn
	Grass weeds	paraquat (*off-label - inter-row directed treatment*), prometryn, propyzamide (*off-label*)

Protected stem and bulb vegetables - Protected onions/leeks/garlic

Pests	Aphids	deltamethrin (*off-label*), nicotine
	Caterpillars	deltamethrin (*off-label*)
	Leafhoppers	nicotine
	Pests, miscellaneous	deltamethrin (*off-label*), dimethoate (*off-label - seedlings*)
	Stem nematodes	oxamyl (*off-label*)
	Thrips	nicotine
	Whiteflies	nicotine

Weeds	Broad-leaved weeds	paraquat (*off-label - inter-row directed treatment*)
	Grass weeds	paraquat (*off-label - inter-row directed treatment*)

Total vegetation control

Aquatic situations, general - Aquatic situations

Plant growth regulation	Growth control	maleic hydrazide
Weeds	Aquatic weeds	2,4-D, dichlobenil, glyphosate
	Broad-leaved weeds	2,4-D
	Grass weeds	glyphosate
	Weeds, miscellaneous	glyphosate, terbutryn

Non-crop areas, general - Farm buildings/yards

Plant growth regulation	Growth control	maleic hydrazide

Non-crop areas, general - Miscellaneous non-crop situations

Pests	Birds/mammals	warfarin
Plant growth regulation	Growth control	maleic hydrazide
Weeds	Broad-leaved weeds	amitrole, asulam, MCPA, triclopyr
	Grass weeds	amitrole
	Weeds, miscellaneous	amitrole, glyphosate
	Woody weeds/scrub	glyphosate, triclopyr

Non-crop areas, general - Non-crop farm areas

Weeds	Aquatic weeds	triclopyr
	Broad-leaved weeds	2,4-D + dicamba + triclopyr, amitrole, diquat + paraquat, diuron, diuron + glyphosate, glufosinate-ammonium, glyphosate, MCPA, paraquat, picloram, triclopyr
	Crops as weeds	diquat + paraquat
	Grass weeds	amitrole, diquat + paraquat, diuron, diuron + glyphosate, glufosinate-ammonium, glyphosate, paraquat
	Weeds, miscellaneous	amitrole, amitrole + 2,4-D + diuron, dichlobenil, diuron, diuron + paraquat, glufosinate-ammonium, glyphosate, sodium chlorate
	Woody weeds/scrub	asulam, dicamba, glyphosate, picloram, triclopyr, triclopyr (*directed treatment*)

Non-crop areas, general - Paths/roads etc

Diseases	Fungus diseases	dichlorophen
Pests	Slugs/snails	metaldehyde
Weeds	Broad-leaved weeds	2,4-D + dicamba + triclopyr, diquat + paraquat, diuron, glyphosate, MCPA

Grass weeds	diquat + paraquat, diuron, glyphosate
Mosses	dichlorophen
Weeds, miscellaneous	dichlobenil, dichlorophen, diquat + paraquat, diuron, diuron + paraquat, glyphosate, sodium chlorate
Woody weeds/scrub	2,4-D + dicamba + triclopyr, glyphosate

SECTION 2
PESTICIDE PROFILES

1 abamectin

A selective acaricide and insecticide for use in ornamentals

Products

Dynamec	Syngenta Bioline	18 g/l	EC	08701

Uses

- Leaf miner in ***miscellaneous flowers, ornamental specimens, protected cherry tomatoes*** *(off-label)*, ***protected flowers, protected lettuce, protected ornamentals, protected tomatoes***
- Red spider mites in ***protected peppers*** *(off-label)*, ***protected strawberries*** *(off-label)*
- Tarsonemid mites in ***strawberries*** *(off-label - in propagation)*
- Two-spotted spider mite in ***miscellaneous flowers, ornamental specimens, protected cucumbers, protected flowers, protected ornamentals, protected tomatoes***
- Western flower thrips in ***miscellaneous flowers, ornamental specimens, protected cucumbers, protected flowers, protected lettuce, protected ornamentals, protected tomatoes***

Specific Off-Label Approvals (SOLAs)

- ***protected cherry tomatoes, protected peppers, protected strawberries*** *(OLA 031128) Dec 2008* [1]
- ***strawberries*** *(in propagation) (OLA 040165) Dec 2008* [1]

Efficacy guidance

- Treat at first sign of infestation. Repeat sprays may be required
- For effective control total cover of all plant surfaces is essential, but avoid run-off
- Target pests quickly become immobilised but 3-5 d may be required for maximum mortality

Restrictions

- Number of treatments 6 on protected tomatoes and cucumbers (only 4 of which can be made when flowers or fruit present); 5 on protected peppers; 4 on protected lettuce; 3 on strawberries; not restricted on flowers but rotation with other products advised
- Maximum concentration must not exceed 50 ml per 100 l water
- Do not mix with wetters, stickers or other adjuvants
- Do not use on ferns (*Adiantum* spp) or Shasta daisies
- Consult manufacturer for list of plant varieties tested for safety
- There is insufficient evidence to support product compatibility with integrated and biological pest control programmes
- Unprotected persons must be kept out of treated areas until the spray has dried

Crop-specific information

- On tomato or cucumber crops that are in flower, or have started to set fruit, treat only between 1 Mar and 31 Oct. Seedling tomatoes or cucumbers that have not started to flower or set fruit may be treated at any time. Do not treat cherry tomatoes
- Apply to lettuce only between 1 Mar and 31 Oct
- Some spotting or staining may occur on carnation, kalanchoe and begonia foliage
- HI 14 d for protected lettuce; 3 d for other protected edible crops

Environmental safety

- High risk to bees. Do not apply to crops in flower or to those in which bees are actively foraging. Do not apply when flowering weeds are present
- Extremely dangerous to fish or other aquatic life. Do not contaminate surface waters or ditches with chemical or used container
- Where bumble bees are used in tomatoes as pollinators, keep them out for 24 h after treatment
- Permissible in organic systems

Hazard classification and safety precautions

Hazard H03, H04
Risk phrases R20, R21, R22a, R36
Operator protection A, C, D, H, K, M; U02a, U05a, U09a, U19a, U20a
Environmental protection E12a, E13a, E15a, E31b, E34
Storage and disposal E01, E04, E26, E30b
Medical advice M04

SEE SECTION 3 FOR PRODUCTS ALSO REGISTERED

2 aldicarb

A soil-applied, systemic carbamate insecticide and nematicide

Products

1	Me2 New Aldee	Me2	10% w/w	GR	10434
2	Temik 10G	Bayer CropScience	10% w/w	GR	10021

Uses

- Aphids in ***carrots, parsnips*** [1, 2]; ***early potatoes, maincrop potatoes*** [1]; ***potatoes*** [2]
- Free-living nematodes in ***carrots, parsnips*** [1, 2]; ***early potatoes, maincrop potatoes*** [1]; ***potatoes*** [2]
- Spraing vectors in ***early potatoes, maincrop potatoes*** [1]; ***potatoes*** [2]
- Stem and bulb nematodes in ***bulb onions*** [1, 2]
- Stem nematodes in ***bulb onions*** [1, 2]

Approval information

- In March 2003 the E U Council decided not to include aldicarb in Annex I under Directive 91/414. This meant that all approved products containing aldicarb have to be withdrawn. However for certain uses it was decided that this withdrawal can take place over a longer period than is usually provided, in order to allow time for the development of alternatives.
- In the UK, for uses on sugar beet, approvals for sale and supply were revoked in September 2003. Off-label approval for use on outdoor and protected sweetcorn seedlings and outdoor leeks was revoked on 18 September 2004.
- Products containing this active ingredient have been granted derogations for specified 'Essential Uses' for use until 31 December 2007. Sale and supply must cease by 30 June 2007 but growers have no guarantee that the products will continue to be available until then.
  For more information see 'The Review Programme' under 'Pesticide Legislation' in Section 5

Efficacy guidance

- Persistence and activity may be reduced in very wet soils or where pH exceeds 8.0. Do not use within 14 d of liming
- Use in potatoes reduces incidence of spraing disease

Restrictions

- Aldicarb is subject to the Poisons Rules 1982 and the Poisons Act 1972. See notes in Section 5
- This product contains an anticholinesterase carbamate compound. Do not use if under medical advice not to work with such compounds
- Maximum number of treatments 1 per crop
- Must be incorporated into soil by physical means. See label for details of application rates, timing, suitable applicators and techniques of incorporation
- No edible crops other than those listed (see label) should be planted into treated soil for at least 8 wk after application
- Keep unprotected persons out of treated glasshouses for at least 1 d

Crop-specific information

- Latest use: at planting, sowing, drilling or transplanting
- HI potatoes 8 wk; carrots, parsnips 12 wk
- Do not harvest spring sown bulb onions until mature bulb stage

Environmental safety

- Keep in original container, tightly closed, in a safe place, under lock and key
- Dangerous for the environment
- Toxic to aquatic organisms
- Dangerous to livestock. Keep all livestock out of treated areas/away from treated water for at least 13 wk. Bury or remove spillages
- Dangerous to game, wild birds and animals

Hazard classification and safety precautions

Hazard H01, H11

Risk phrases R21, R28, R51, R58

Operator protection A, B, H, K, M [1, 2]; C [1, 2] (or D+E); U02a, U04a, U05a, U07, U09a, U12, U13, U19a, U20a

Environmental protection E06b (13 wk); E10a, E13b, E32a, E32d, E33, E34, E36, E38

Storage and disposal E01, E04, E30b
Medical advice M02, M04

3 alginate/polysaccaride

A contact insecticide that works by physical action

Products

Agri-50E	Fargro	-	SC	-

Uses

- Whitefly in ***all edible crops, all non-edible crops, protected crops***

Approval information

- Product not controlled by Control of Pesticides Regulations/Plant Protection Products Regulations because it acts by physical means only

Efficacy guidance

- Product acts by blocking insect spiracles and inhibiting respiration by physical action
- For maximum effectiveness direct spray contact with target insects is essential. Ensure thorough and uniform spray coverage
- Best results obtained from treatments in early morning (outdoor crops only) or late afternoon when pests less active
- Frequency of treatment should be adjusted to maintain effective level of control
- Product most effective when used in water between pH 5.0 and 8.5. Water should be buffered if necessary, especially in areas of very hard water
- Product may be used in conjunction with biological control agents

Restrictions

- No restrictions on number of treatments
- Do use any tank mixture without first consulting supplier or distributor
- Do not mix with heavy metal products, sulphur, mineral oils or strongly anionic products
- Do not mix with anti-foaming agents

Crop-specific information

- Harvest interval 0 days for all edible crops
- Carry out plant safety check before large scale treatment of flowering ornamentals

Hazard classification and safety precautions

Operator protection U05a, U10, U19a, U20c
Storage and disposal E01, E04
Medical advice M04

4 alphachloralose

A narcotic glucofuranose rodenticide used to kill mice

Products

Alphamouse	Killgerm	100%	CB	H6692

Uses

- Mice in ***farm buildings*** *(indoor use)*

Efficacy guidance

- Place baits where mice are active, in runs or harbourages. Best results obtained where temperature is below 15°C
- Leave in position for 7-10 days only. If mouse activity remains continue treatment with a rodenticide containing a different active ingredient

Restrictions

- Alphachloralose is subject to the Poisons Rules 1982 and Poisons Act 1972. See notes in Section 5
- May be used only by persons instructed or trained in the use of alphachloralose
- Harmful by inhalation and if swallowed

- Do not use outdoors. Products must be used in situations where baits are placed within a building or other enclosed structure, and the target is living or feeding predominantly within that building or structure
- Wash out all mixing equipment thoroughly at the end of every operation
- Prevent access to baits by children, birds and non-target animals
- Do not prepare or lay baits where food, feed or water could become contaminated
- A suitable warning dye must always be used

Environmental safety

- Remove all remains of bait and bait containers after treatment and burn or bury. Do not dispose of in refuse sacks or on open rubbish tips
- Search for and burn or bury all rodent bodies. Do not place in refuse bins or on rubbish tips
- Take all precautions to prevent domestic animals or livestock having access to prepared bait

Hazard classification and safety precautions

Hazard H03
Risk phrases R20, R22a
Operator protection A, D, H; U05a, U13, U20b
Environmental protection E15a, E32d, E34
Storage and disposal E30b
Vertebrate/rodent control products V01b, V02, V03a, V04a
Medical advice M04

5 alpha-cypermethrin

A contact and ingested pyrethroid insecticide for use in arable crops

Products

1	Alphaguard 100 EC	Interfarm	100 g/l	EC	10772
2	Alphathrin	Nufarm UK	100 g/l	EC	11163
3	Antec Durakil 1.5 SC	Antec Biosentry	15 g/l	SC	H7560
4	Antec Durakil 6SC	Antec Biosentry	60 g/l	SC	H7559
5	Contest	BASF	15% w/w	WG	10216
6	Fernpath Dart	AgriGuard	100 g/l	EC	11970
7	I T Alpha-Cyper	I T Agro	100 g/l	EC	10483

Uses

- Brassica pod midge in ***winter oilseed rape*** [1, 2, 5-7]
- Cabbage seed weevil in ***spring oilseed rape***, ***winter oilseed rape*** [1, 2, 5-7]
- Cabbage stem flea beetle in ***winter oilseed rape*** [1, 2, 5-7]
- Cabbage white butterfly in ***salad brassicas*** *(off-label - for baby leaf production)* [5]
- Caterpillars in ***broccoli***, ***brussels sprouts***, ***cabbages***, ***calabrese***, ***cauliflowers***, ***kale*** [1, 2, 5-7]
- Cereal aphid in ***spring barley***, ***spring wheat***, ***winter barley***, ***winter wheat*** [1, 2, 5-7]
- Diamond-back moth in ***salad brassicas*** *(off-label - for baby leaf production)* [5]
- Flea beetle in ***broccoli***, ***brussels sprouts***, ***cabbages***, ***calabrese***, ***cauliflowers***, ***kale*** [1, 2, 5-7]
- Lesser mealworm in ***poultry houses*** [3, 4]
- Pea and bean weevil in ***broad beans***, ***combining peas***, ***spring field beans***, ***vining peas***, ***winter field beans*** [5-7]
- Pea aphid in ***combining peas***, ***vining peas*** [6, 7]; ***combining peas*** *(reduction)*, ***vining peas*** *(reduction)* [5]
- Pea moth in ***combining peas***, ***vining peas*** [5-7]
- Pollen beetle in ***spring oilseed rape***, ***winter oilseed rape*** [1, 2, 5-7]
- Poultry red mite in ***poultry houses*** [3, 4]
- Rape winter stem weevil in ***winter oilseed rape*** [1, 2, 5-7]
- Small white butterfly in ***salad brassicas*** *(off-label - for baby leaf production)* [5]
- Yellow cereal fly in ***winter barley***, ***winter wheat*** [5]

Specific Off-Label Approvals (SOLAs)

- ***salad brassicas*** *(for baby leaf production) (OLA 011221) Dec 2008* [5]

Efficacy guidance

- For cabbage stem flea beetle control spray oilseed rape when adult or larval damage first seen and about 1 mth later

- For flowering pests on oilseed rape apply at any time during flowering, on pollen beetle best results achieved at green to yellow bud stage (GS 3,3-3,7), on seed weevil between 20 pods set stage and 80% petal fall (GS 4,7-5,8)
- Spray cereals in autumn for control of cereal aphids, in spring/summer for grain aphids. (See label for details)
- For flea beetle, caterpillar and cabbage aphid control on brassicas apply when the pest or damage first seen or as a preventive spray. Repeat if necessary
- For pea and bean weevil control in peas and beans apply when pest attack first seen and repeat as necessary
- For lesser mealworm control in poultry houses apply a coarse, low-pressure spray as routine treatment after clean-out and before each new crop. Spray vertical surfaces and ensure an overlap onto ceilings. It is not necessary to treat the floor [3, 4]
- Use the higher recommended concentration where extreme residual action is required or surfaces are dirty or highly absorbent [3, 4]

Restrictions

- Maximum number of treatments 4 per crop on edible brassicas, 3 per crop on winter oilseed rape, peas, 2 on beans, spring oilseed rape (only 1 after yellow bud stage - GS 3,7)
- Maximum number of applications in animal husbandry use 2 per crop when used in premises that are occupied by poultry [3, 4]
- Apply up to 2 sprays on cereals in autumn and spring, 1 in summer between 1 Apr and 31 Aug. See label for details of rates
- Only 1 aphicide treatment may be applied in cereals between 1 Apr and 31 Aug in any one year and spray volume must not be reduced in this period
- Do not apply to a cereal crop if any product containing a pyrethroid or dimethoate has been applied after the start of ear emergence (GS 51)

Crop-specific information

- Latest use: before the end of flowering for oilseed rape; before 31 Mar in year of harvest for cereals (autumn and spring application); before early dough stage (GS 83) for cereals (summer application).
- HI vining peas 1 d; brassicas 7 d; combining peas, broad beans, field beans 11 d
- For summer cereal application do not spray within 6 m from edge of crop and do not reduce volume when used after 31 Mar

Environmental safety

- Dangerous for the environment [1, 2, 5-7]
- Very toxic to aquatic organisms [1, 2, 5-7]
- Dangerous to bees. Do not apply to crops in flower, or to those in which bees are actively foraging, except as directed on oilseed rape, wheat and barley. Do not apply when flowering weeds are present
- Extremely dangerous to fish or other aquatic life. Do not contaminate surface waters or ditches with chemical or used container [3, 4]
- Risk to non-target insects or other arthropods [6, 7]
- LERAP Category A (except [3, 4])
- Where possible spray oilseed rape crops in the late evening or early morning or in dull weather
- Do not spray within 6 m of the edge of a cereal crop after 31 Mar in yr of harvest
- Do not apply directly to poultry; collect eggs before application [3, 4]

Hazard classification and safety precautions

Hazard H03, H11 [1, 2, 5-7]; H08 [1, 2, 6, 7]

Risk phrases R21, R37, R38, R41, R67 [1, 2, 6, 7]; R22a [5-7]; R22b, R25, R66 [1, 2]; R50 [1, 2, 5, 7]; R51 [6]; R58 [1, 2, 5-7]

Operator protection A, H [1-7]; C [1, 2, 6, 7]; U02a [3, 4, 6, 7]; U04a, U11 [1, 2, 6, 7]; U05a, U10 [1, 2, 5-7]; U14 [6, 7]; U19a [1-4, 6, 7]; U20b [5-7]

Environmental protection E02a, E05a, E12e, E13a, E32a [3, 4]; E12e [6] (oilseed rape, wheat, barley); E12e [7]; E15a [1-7]; E16c, E16d [1, 2, 5-7]; E19b [1, 2]; E22c, E34 [6, 7]; E31b [5-7]; E32d, E38 [5]; E32e [1, 2, 7]

Consumer protection C09, C11 [3, 4]

Storage and disposal E01 [5, 6]; E04, E30a [1, 2, 5-7]; E26 [5-7]

Medical advice M03, M05b [1, 2, 6, 7]; M05a [5]

SEE SECTION 3 FOR PRODUCTS ALSO REGISTERED

6 aluminium ammonium sulphate

An inorganic bird and animal repellent

Products

1	Curb Crop Spray Powder	Sphere	88% w/w	WP	02480
2	Guardsman STP Seed Dressing Powder	Sphere	88% w/w	DS	03606
3	Liquid Curb Crop Spray	Sphere	83 g/l	SC	03164

Uses

- Animal repellent in ***agricultural premises, all top fruit, broad beans, bush fruit, cane fruit, carrots, flowerhead brassicas, forest nursery beds, forestry plantations, grain stores, grassland, leaf brassicas, peas, spring barley, spring field beans, spring oats, spring oilseed rape, spring wheat, strawberries, sugar beet, winter barley, winter field beans, winter oats, winter oilseed rape, winter wheat*** [1, 3]; ***amenity areas*** [1]
- Bird repellent in ***agricultural premises, all top fruit, broad beans, bush fruit, cane fruit, carrots, flowerhead brassicas, forest nursery beds, forestry plantations, grain stores, grassland, leaf brassicas, peas, spring barley, spring field beans, spring oats, spring oilseed rape, spring wheat, strawberries, sugar beet, winter barley, winter field beans, winter oats, winter oilseed rape, winter wheat*** [1, 3]; ***amenity areas*** [1]
- Birds in ***corms*** *(corm treatment)*, ***flower bulbs*** *(bulb treatment)*, ***seeds*** *(seed treatment)* [2]
- Damaging mammals in ***corms*** *(corm treatment)*, ***flower bulbs*** *(bulb treatment)*, ***seeds*** *(seed treatment)* [2]

Efficacy guidance

- Apply as overall spray to growing crops before damage starts or mix powder with seed depending on type of protection required
- Spray deposit protects growth present at spraying but gives little protection to new growth
- Product must be sprayed onto dry foliage to be effective and must dry completely before dew or frost forms. In winter this may require some wind

Restrictions

- Maximum number of treatments 1 per batch for corms, flower bulbs, seeds [2]

Crop-specific information

- Latest use: no restriction

Hazard classification and safety precautions

Operator protection U05a, U20a [1, 3]; U20b [2]
Environmental protection E15a, E32a [1-3]; E19b, E32d [1, 3]
Storage and disposal E01, E04, E26 [1, 3]; E30a [2]
Treated seed S02, S05 [2]
Medical advice M03 [1, 3]

7 aluminium phosphide

A phosphine generating compound used against vertebrates and grain store pests

Products

1	Degesch Fumigation Tablets	Rentokil	56% w/w	GE	09313
2	Luxan Talunex	Luxan	57% w/w	GE	06563
3	Phostoxin	Rentokil	56% w/w	GE	09315

Uses

- Insect pests in ***stored grain*** [1]
- Moles in ***farm woodland, grassland, lawns, managed amenity turf*** [2]; ***farmland*** [2, 3]
- Rabbits in ***farm woodland, grassland, lawns, managed amenity turf*** [2]; ***farmland*** [2, 3]
- Rats in ***farm woodland, grassland, lawns, managed amenity turf*** [2]; ***farmland*** [2, 3]

Efficacy guidance

- Product releases poisonous hydrogen phosphide gas in contact with moisture
- Place fumigation tablets in grain stores as directed [1]

- Place pellets in burrows or runs and seal hole by heeling in or covering with turf. Do not cover pellets with soil. Inspect daily and treat any new or re-opened holes [2, 3]
- Apply pellets by means of Luxan Topex Applicator [2]

Restrictions

- Aluminium phosphide is subject to the Poisons Rules 1982 and the Poisons Act 1972. See notes in Section 5
- Only to be used by operators instructed or trained in the use of aluminium phosphide and familiar with the precautionary measures to be taken. See label and HSE Guidance Notes for full precautions
- Only open container outdoors [2, 3] and for immediate use. Keep away from liquid or water as this causes immediate release of gas. Do not use in wet weather
- Do not use within 3 m of human or animal habitation. Before application ensure that no humans or domestic animals are in adjacent buildings or structures. Allow a minimum airing-off period of 4 h before re-admission

Environmental safety

- Product liberates very toxic, highly flammable gas
- Keep in original container, tightly closed, in a safe place, under lock and key
- Dangerous for the environment [2]
- Very toxic to aquatic organisms [2]
- Prevent access by livestock, pets and other non-target mammals and birds to buildings under fumigation and ventilation [1]
- Dangerous to fish or other aquatic life. Do not contaminate surface waters or ditches with chemical or used container [1, 3]
- Pellets must never be placed or allowed to remain on ground surface
- Do not use adjacent to watercourses
- Take particular care to avoid gassing non-target animals, especially those protected under the Wildlife and Countryside Act (e.g. badgers, polecat, reptiles, natterjack toads, most birds). Do not use in burrows where there is evidence of badger or fox activity, or when burrows might be occupied by birds
- Dust remaining after decomposition is harmless and of no environmental hazard
- Dispose of empty containers as directed on label

Hazard classification and safety precautions

Hazard H01, H07 [1-3]; H03 [3]; H11 [2]
Risk phrases R21, R26, R28 [1-3]; R36, R37, R50 [2]
Operator protection A, H [1-3]; D [1, 3]; G [2]; U01, U13, U19a, U20a [1-3]; U05a [2, 3]; U05b [1]; U07 [1, 3]; U14, U15 [2]; U18 [1, 2]
Environmental protection E02a [1] (4 h min); E02b [1]; E13b [1, 3]; E15a [2]; E32b, E34 [1-3]
Storage and disposal E01, E04, E30b [1-3]; E29a [1, 3]
Vertebrate/rodent control products V04a [2, 3]
Medical advice M04

8 amidosulfuron

A post-emergence sulfonylurea herbicide for cleavers and other broad-leaved weed control in cereals

Products

1	Eagle	Bayer CropScience	75% w/w	WG	07318
2	Landgold Amidosulfuron	Landgold	75% w/w	WG	09021
3	Pursuit	Bayer CropScience	75% w/w	WG	07333
4	Squire	Bayer CropScience	50% w/w	WG	08715

Uses

- Annual dicotyledons in ***durum wheat, spring barley, spring oats, spring rye, spring wheat, triticale, winter barley, winter oats, winter rye, winter wheat*** [1-3]; ***linseed*** [1, 3]
- Charlock in ***permanent pasture, rotational grassland*** [4]
- Cleavers in ***durum wheat, spring barley, spring oats, spring rye, spring wheat, triticale, winter barley, winter oats, winter rye, winter wheat*** [1-3]; ***linseed*** [1, 3]; ***permanent pasture, rotational grassland*** [4]
- Docks in ***permanent pasture, rotational grassland*** [4]

- Forget-me-not in ***permanent pasture, rotational grassland*** [4]
- Shepherd's purse in ***permanent pasture, rotational grassland*** [4]

Efficacy guidance

- For best results apply in spring (from 1 Feb) in warm weather when soil moist and weeds growing actively. When used in grassland following cutting or grazing, docks should be allowed to regrow before treatment
- Weed kill is slow, especially under cool, dry conditions. Weeds may sometimes only be stunted but will have little or no competitive effect on crop
- May be used on all soil types unless certain sequences are used on linseed. See label
- Spray is rainfast after 1 h
- Cleavers controlled from emergence to flower bud stage. If present at application charlock (up to flower bud), shepherds purse (up to flower bud) and field forget-me-not (up to 6 leaves) will also be controlled
- Amidosulfuron is a member of the ALS-inhibitor group of herbicides and products should be used in a planned Resistance Management strategy. See Section 5 for more information

Restrictions

- Maximum number of treatments 1 per crop, or 1 per yr when used on grass
- Use after 1 Feb and do not apply to rotational grass after 30 Jun, or to permanent grassland after 15 Oct
- Do not apply to crops undersown or due to be undersown with clover or lucerne
- Do not use on swards containing red clover but may be used on swards with white clover. Where clover has been newly established apply from 3-leaf stage of grass and 1-2 trifoliate leaves of clover [4]
- Do not spray crops under stress, suffering drought, waterlogged, grazed, lacking nutrients or if soil compacted
- Do not spray if frost expected
- Do not roll or harrow within 1 wk of spraying
- Certain mixtures or sequences with other sulfonylurea products are permitted under explicitly detailed conditions. See label for details. There are no recommendations for mixtures with metsulfuron-methyl products on linseed
- Certain mixtures with fungicides are expressly forbidden. See label for details

Crop-specific information

- Latest use: before first spikelets just visible (GS 51) for cereals; before flower buds visible for linseed; 15 Oct for grassland
- Broadcast crops should be sprayed post-emergence after plants have a well established root system
- Hay or silage from treated grass crops must not be cut for at least 21 d after treatment [4]

Following crops guidance

- If a treated crop fails cereals may be sown after 15 d and thorough cultivation
- After normal harvest of a treated crop only cereals, winter oilseed rape, mustard, turnips, winter field beans or vetches may be sown in the same year as treatment and these must be preceded by ploughing or thorough cultivation
- Only cereals may be sown within 12 mth of application to grassland [4]
- Cereals or potatoes must be sown as the following crop after use of permitted mixtures or sequences with other sulfonyl urea herbicides in cereals. Only cereals may be sown after the use of such sequences in linseed

Environmental safety

- Dangerous for the environment [4]
- Very toxic to aquatic organisms [4]
- Keep livestock out of treated areas for at least 7d following treatment and until poisonous weeds, such as ragwort, have died down and become unpalatable [4]
- Dangerous to fish or other aquatic life. Do not contaminate surface waters or ditches with chemical or used container [1-4]
- Take care to wash out sprayers thoroughly. See label for details
- Avoid drift onto neighbouring broad-leaved plants or onto surface waters or ditches

Hazard classification and safety precautions

Hazard H04, H11 [4]

Risk phrases R36, R50, R58 [4]

Operator protection C [4]; U05a [4]; U20a [1, 2]; U20b [3, 4]
Environmental protection E07a [4] (7 d); E13b, E31a [1-4]; E32d, E38 [4]
Storage and disposal E01, E04 [4]

9 amidosulfuron + iodosulfuron-methyl-sodium

A post-emergence sulfonylurea herbicide mixture for cereals

Products

Chekker	Bayer CropScience	12.5:1.25% w/w	WG	10955

Uses

- Annual dicotyledons in ***spring barley, spring rye, spring wheat, triticale, winter barley, winter rye, winter wheat***
- Chickweed in ***spring barley, spring rye, spring wheat, triticale, winter barley, winter rye, winter wheat***
- Cleavers in ***spring barley, spring rye, spring wheat, triticale, winter barley, winter rye, winter wheat***
- Mayweeds in ***spring barley, spring rye, spring wheat, triticale, winter barley, winter rye, winter wheat***
- Volunteer oilseed rape in ***spring barley, spring rye, spring wheat, triticale, winter barley, winter rye, winter wheat***

Efficacy guidance

- Best results obtained from treatment in warm weather when soil is moist and the weeds are growing actively
- Weeds must be present at application to be controlled
- Dry conditions resulting in moisture stress may reduce effectiveness
- Weed control is slow especially under cool dry conditions
- Occasionally weeds may only be stunted but they will normally have little or no competitive effect on the crop
- Amidosulfuron and iodosulfuron are members of the ALS-inhibitor group of herbicides and products should be used in a planned Resistance Management strategy. See Section 5 for more information

Restrictions

- Maximum number of treatments 1 per crop
- Must only be applied between 1 Feb in yr of harvest and specified latest time of application
- Do not apply to crops undersown or to be undersown with grass, clover or alfalfa
- Do not roll or harrow within 1 wk of spraying
- Do not spray crops under stress from any cause or if the soil is compacted
- Do not spray if frost expected
- Do not apply in mixture or in sequence with any other ALS inhibitor

Crop-specific information

- Treat drilled crops after the 2-leaf stage; treat broadcast crops after the plants have a well-established root system
- Latest use: before first spikelet of inflorescence just visible (GS 51)

Following crops guidance

- Cereals, winter oilseed rape and winter field beans may be sown in the same yr as treatment provided they are preceded by ploughing or thorough cultivation. Any crop may be sown in the spring of the yr following treatment
- A minimum of 3 mth must elapse between treatment and sowing winter oilseed rape

Environmental safety

- Dangerous for the environment
- Toxic to aquatic organisms
- LERAP Category B
- Take extreme care to avoid damage by drift onto broad-leaved plants outside the target area or onto ponds, waterways and ditches
- Observe carefully label instructions for sprayer cleaning

Hazard classification and safety precautions

Hazard H04, H11

Risk phrases R36, R51, R58

Operator protection A, C, H; U05a, U08, U11, U14, U15, U20b

Environmental protection E15a, E16a, E16b, E31a, E32d, E38

Storage and disposal E01, E04

10 2-aminobutane

A fumigant alkylamine fungicide permitted for use only on stored seed potatoes

Products

Certis 2-Aminobutane	Certis	720 g/l	VP	11182

Uses

- Gangrene in ***seed potatoes***
- Silver scurf in ***seed potatoes***
- Skin spot in ***seed potatoes***

Approval information

- Products containing this active ingredient have been granted derogations for specified 'Essential Uses' for use until 31 December 2007. Sale and supply must cease by 30 June 2007 but growers have no guarantee that the products will continue to be available until then.
  For more information see 'The Review Programme' under 'Pesticide Legislation' in Section 5

Efficacy guidance

- Product used for treatment of seed potato tubers by fumigation in appropriate premises (see label)

Restrictions

- Maximum number of treatments 1 per batch. Maximum quantity to be fumigated in a single stack must not exceed 2000 tonnes
- Do not treat immature tubers. Allow period of healing before treating damaged tubers
- Treatment must only be carried out by trained operators in suitable fumigation chambers under licence from the British Technology Group
- Fumigate within 21 d of lifting
- This product must not be used on any crops other than those listed, including any extrapolations that would normally be permissible under the Long Term Arrangements for Extension of Use (see Section 5)

Environmental safety

- Dangerous for the environment
- Very toxic to aquatic organisms
- Keep in original container, tightly closed, in a safe place, under lock and key
- Do not empty into drains
- Do not supply treated potatoes for consumption by humans or lactating dairy cows
- Use must be in accordance with approved Code of Practice for the Control of Substances Hazardous to Health: Fumigation Operations

Hazard classification and safety precautions

Hazard H03, H05, H07, H11

Risk phrases R20, R22a, R35, R50

Operator protection A, C; U02a, U04a, U05b, U10, U11, U13, U19a, U20b

Environmental protection E15a, E19b, E31a, E34

Storage and disposal E01, E04, E29b, E30b

Medical advice M04

11 amitraz

An amidine acaricide and insecticide for use in top fruit

Products

Mitac HF	Bayer CropScience	200 g/l	EC	07358

Uses

- Pear sucker in ***pears***
- Red spider mites in ***apples*, *pears***
- Rust mite in ***apples***

Approval information

- Use on apples is permitted until Aug 2005. Products containing this active ingredient have been granted derogations for specified 'Essential Uses' for use until 31 December 2007. Sale and supply must cease by 30 June 2007 but growers have no guarantee that the products will continue to be available until then.
  For more information see 'The Review Programme' under 'Pesticide Legislation' in Section 5

Efficacy guidance

- Best results achieved in dry conditions, do not spray if rain imminent

Restrictions

- Maximum total dose equivalent to two full dose treatments

Crop-specific information

- HI apples, pears 2 wk
- For red spider mites on apples and pears spray at 60-80% egg hatch and repeat 3 wk later
- For pear sucker control spray when significant numbers of nymphs have hatched but before there is significant contamination of the fruit with honeydew, normally Jun/Jul

Environmental safety

- Dangerous for the environment
- Very toxic to aquatic organisms
- Keep in original container, tightly closed, in a safe place, under lock and key

Hazard classification and safety precautions

Hazard H03, H08, H11
Risk phrases R22a, R22b, R38, R43, R50, R58
Operator protection A, C, H; U02a, U04a, U05a, U08, U13, U14, U19a, U20a
Environmental protection E15a, E31b, E32d, E34, E38
Consumer protection C02 (2 wk)
Storage and disposal E01, E04, E26, E30b
Medical advice M03, M05b

12 amitrole

A translocated, foliar-acting, non-selective triazole herbicide

Products

1	Aminotriazole Technical	Nufarm UK	92.5% w/w	TC	11137
2	Weedazol-TL	Nufarm UK	225 g/l	SL	11430

Uses

- Annual and perennial weeds in ***apricots*** *(off-label)*, ***cherries*** *(off-label)*, ***peaches*** *(off-label)*, ***plums*** *(off-label)*, ***quinces*** *(off-label)* [2]
- Annual weeds in ***fallows*, *field margins*, *headlands*, *stubbles*** [2]
- Barren brome in ***apple orchards*, *pear orchards*** [2]
- Couch in ***amenity areas*, *industrial sites*, *non-crop areas*** [1]; ***apple orchards*, *fallows*, *field margins*, *headlands*, *pear orchards*, *stubbles*** [2]
- Creeping bent in ***amenity areas*, *industrial sites*, *non-crop areas*** [1]
- Creeping thistle in ***amenity areas*, *industrial sites*, *non-crop areas*** [1]; ***fallows*, *stubbles*** [2]
- Docks in ***amenity areas*, *industrial sites*, *non-crop areas*** [1]; ***fallows*, *field margins*, *headlands*, *stubbles*** [2]
- General weed control in ***apple orchards*, *pear orchards*** [2]
- Perennial weeds in ***apple orchards*, *fallows*, *field margins*, *headlands*, *pear orchards*** [2]
- Total vegetation control in ***amenity areas*, *industrial sites*, *non-crop areas*** [1]
- Volunteer potatoes in ***stubbles*** *(barley stubble)* [2]

Specific Off-Label Approvals (SOLAs)

- ***apricots*, *cherries*, *peaches*, *plums*, *quinces*** *(OLA 031869) Dec 2008* [2]

SEE SECTION 3 FOR PRODUCTS ALSO REGISTERED

Approval information

- Amitrole included in Annex I under EC Directive 91/414

Efficacy guidance

- In non-crop land may be applied at any time from Apr to Oct. Best results achieved in spring or early summer when weeds growing actively. For coltsfoot, hogweed and horsetail summer and autumn applications are preferred
- Uptake is via foliage and heavy rain immediately after application will reduce efficacy. Amitrole is less affected by drought than some residual herbicides and remains effective for up to 2 mth

Restrictions

- Keep off suckers or foliage of desirable trees or shrubs
- Do not spray areas into which the roots of adjacent trees or shrubs extend
- Do not spray on sloping ground when rain imminent and run-off may occur
- Do mix product with acids

Environmental safety

- Harmful to aquatic organisms
- Keep livestock out of treated areas for at least two weeks following treatment and until poisonous weeds, such as ragwort, have died down and become unpalatable

Hazard classification and safety precautions

Hazard H03 [2]
Risk phrases R22a, R40, R48, R52, R58 [2]
Operator protection A [1, 2]; C [2]; U05a, U20b [2]; U08, U19a [1, 2]; U20a [1]
Environmental protection E07a, E15a [2]; E13c [1]; E31b [1, 2]
Storage and disposal E01, E04, E26 [2]; E30a [1, 2]

13 amitrole + 2,4-D + diuron

A total herbicide mixture of translocated and residual chemicals

Products

Trik	Nufarm UK	26.6:11.2:46.4% w/w	WP	11441

Uses

- Total vegetation control in ***land not intended to bear vegetation***

Approval information

- Amitrole and 2,4-D included in Annex I under EC Directive 91/414

Efficacy guidance

- Apply in spring or late summer/early autumn when weeds are growing actively and have sufficient leaf area to absorb chemical
- Apply maintenance treatment if necessary at lower rate when weeds 7-10 cm high
- Increase dose rate on areas of peat or high carbon content

Restrictions

- Maximum number of treatments 1 per yr for land not intended to bear vegetation
- Do not use on ground under which roots of valuable trees or shrubs are growing

Environmental safety

- Dangerous for the environment
- Very toxic to aquatic organisms
- Keep livestock out of treated areas for at least two weeks following treatment and until poisonous weeds, such as ragwort, have died down and become unpalatable

Hazard classification and safety precautions

Hazard H03, H11
Risk phrases R22a, R37, R48, R50, R58, R68
Operator protection A, C, D, H, M; U05a, U08, U14, U15, U19a, U20a
Environmental protection E07a, E15a, E32a, E32e
Storage and disposal E01, E04, E30a
Medical advice M03, M05a

14 ammonium sulphamate

A non-selective, inorganic, general purpose herbicide and tree-killer

Products

1	Amcide	B H & B	99.5% w/w	CR	04246
2	Root-Out	Dax	98.5% w/w	CR	03510

Uses
- Annual weeds in ***forest***, ***ornamental specimens*** *(pre-planting)*, ***vegetables*** *(pre-planting)* [1]; ***ornamental specimens***, ***trees and shrubs*** [2]
- Perennial dicotyledons in ***forest*** [1]; ***trees and shrubs*** [2]
- Perennial grasses in ***forest*** [1, 2]
- Perennial weeds in ***ornamental specimens*** [2]; ***ornamental specimens*** *(pre-planting)*, ***vegetables*** *(pre-planting)* [1]
- Rhododendrons in ***forest*** [1, 2]
- Woody weeds in ***forest*** [1, 2]

Efficacy guidance
- Apply as spray to low scrub and herbaceous weeds from Apr to Sep in dry weather when rain unlikely and cultivate after 3-8 wk
- Apply as crystals in frills or notches in trunks of standing trees at any time of year
- Apply as concentrated solution or crystals to stump surfaces within 48 h of cutting. Rhododendrons must be cut level with ground and sprayed to cover cut surface, bark and immediate root area
- Stainless steel or plastic sprayers are recommended. Solutions are corrosive to mild steel, galvanised iron, brass and copper

Restrictions
- Keep spray at least 30 cm from growing plants. Low doses may be used under mature trees with undamaged bark

Following crops guidance
- Allow 8-12 wk after treatment before replanting

Environmental safety
- Harmful to fish. Do not contaminate surface waters or ditches with chemical or used container

Hazard classification and safety precautions

Operator protection U09b, U20b [1]; U11, U14, U19a [1, 2]; U15, U20a [2]

Environmental protection E13c [1, 2]; E32a [1]

Storage and disposal E01, E30a

15 asulam

A translocated carbamate herbicide for control of docks and bracken

Products

1	Asulox	Bayer CropScience	400 g/l	SL	09969
2	Greencrop Frond	Greencrop	400 g/l	SL	11912
3	I T Asulam	I T Agro	400 g/l	SL	10186
4	Inter Asulam	I T Agro	400 g/l	SL	10929
5	Spitfire	AgriGuard	400 g/l	SL	12002

Uses
- Bracken in ***amenity vegetation*** [2-5]; ***forest***, ***permanent pasture***, ***rough grazing*** [1-5]; ***forest*** *(off-label)*, ***non-crop areas*** [1]
- Brome grasses in ***poppies*** *(off-label - for morphine production)* [1]
- Docks in ***amenity grass*** *(not fine turf)*, ***apple orchards***, ***blackcurrants***, ***cherries***, ***hops***, ***pear orchards***, ***permanent pasture***, ***plums*** [1-5]; ***amenity vegetation***, ***blackberries*** *(off-label)*, ***blueberries*** *(off-label)*, ***clover seed crops*** *(off-label)*, ***cranberries*** *(off-label)*, ***damsons*** *(off-label)*, ***gooseberries*** *(off-label)*, ***loganberries*** *(off-label)*, ***mint*** *(off-label)*, ***nectarines*** *(off-label)*, ***parsley*** *(off-label)*, ***poppies*** *(off-label - for morphine production)*, ***quinces*** *(off-label)*, ***raspberries*** *(off-label)*,

redcurrants *(off-label)*, ***road verges***, ***strawberries*** *(off-label)*, ***tarragon*** *(off-label)*, ***waste ground***, ***whitecurrants*** *(off-label)* [1]
- Meadow grasses in ***poppies*** *(off-label - for morphine production)* [1]

Specific Off-Label Approvals (SOLAs)
- ***blackberries***, ***blueberries***, ***cranberries***, ***damsons***, ***gooseberries***, ***loganberries***, ***nectarines***, ***quinces***, ***raspberries***, ***redcurrants***, ***whitecurrants*** *(OLA 001892) Dec 2008* [1]
- ***clover seed crops*** *(OLA 001894) Dec 2008* [1]
- ***forest*** *(OLA 001898) Dec 2008* [1]
- ***mint***, ***parsley***, ***tarragon*** *(OLA 001900) Dec 2008* [1]
- ***poppies*** *(for morphine production) (OLA 040471) Dec 2008* [1]
- ***strawberries*** *(OLA 001896) Dec 2008* [1]

Approval information
- May be applied through CDA equipment
- Approved for aerial application on bracken in agricultural grassland, amenity grassland, forestry and rough upland intended for grazing [1-3, 5]. See notes in Section 5
- Approved for use near surface waters. See notes in Section 5

Efficacy guidance
- Spray bracken when fronds fully expanded but not senescent, usually Jul-Aug; docks in full leaf before flower stem emergence
- Bracken fronds must not be damaged by stock, frost or cutting before treatment
- Uptake and reliability of bracken control may be improved by use of specified additives - see label. Additives not recommended on forestry land
- To allow adequate translocation do not cut or admit stock for 14 d after spraying bracken or 7 d after spraying docks. Preferably leave undisturbed until late autumn
- Complete bracken control rarely achieved by one treatment. Survivors should be sprayed when they recover to full green frond, which may be in the ensuing year but more likely in the second year following initial application

Restrictions
- Maximum number of treatments 1 per crop or 1 per yr
- Do not apply in drought or hot, dry conditions
- Do not use in pasture before mowing for hay

Crop-specific information
- Harvest interval: 10 wk for tarragon; 6 wk for mint, parsley
- Latest Use: 1st treatment between Mar and Apr; 2nd treatment between Sep and Nov for cane and soft fruit; Aug for amenity vegetation, forest, non-crop areas, grass; before 31 Mar in yr of harvest for strawberries; flower buds emerging for poppies
- In forestry areas some young trees may be checked if sprayed directly (see label)
- In fruit crops apply as a directed spray
- Do not treat blackcurrant cuttings, hop sets or weak hills
- Some grasses and herbs will be damaged by full dose. Most sensitive are cocksfoot, Yorkshire fog, timothy, bents, annual meadow-grass, daisies, docks, plantains, saxifrage
- Apply as spot treatment in parsley, mint and tarragon, not directly to crop

Following crops guidance
- Allow at least 6 wk between spraying and planting any crop

Environmental safety
- Dangerous for the environment
- Very toxic to aquatic organisms
- Keep livestock out of treated areas for at least two weeks following treatment and until poisonous weeds, such as ragwort, have died down and become unpalatable
- The use of asulam near surface waters has been considered by PSD. Whilst every care should be taken to avoid contamination, any that does occur during the normal course of spraying should offer no harm to operators, to users and consumers of the water, to domestic and farm animals and to wildlife. Before spraying such areas the appropriate regulatory authority should be notified

Hazard classification and safety precautions

Hazard H04, H11

Risk phrases R43 [1-4]; R50 [1-5]; R58 [2-5]

Operator protection A, C, D, H [1-5]; M [1] (for ULV application); M [2-5]; U05a, U20a [5]; U08, U20b [2-4]; U14 [1-4]; U19a, U20b [1] (ULV); U19a [2-5]; U20c [1]
Environmental protection E07a, E15a, E31a [1-5]; E32d, E38 [1]
Storage and disposal E01, E04 [5]; E26 [2-5]; E30a [1-5]

16 atrazine

A triazine herbicide with residual and foliar activity, with restricted permitted uses

Products

1	Aconite 50	DAPT	500 g/l	SC	10642
2	Alpha Atrazine 50 SC	Makhteshim	500 g/l	SC	04877
3	Atrazol	Sipcam	500 g/l	SC	07598
4	DG90	Sipcam	90% w/w	WG	11200
5	Gesaprim	Syngenta	500 g/l	SC	08411
6	Greencrop Amaize	Greencrop	500 g/l	SC	11973

Uses
- Annual dicotyledons in ***conifer plantations*** [3]; ***farm forestry*** [2, 6]; ***forest*** [2, 3, 6]; ***maize, sweetcorn*** [1-6]
- Annual grasses in ***conifer plantations*** [3]; ***farm forestry*** [2, 6]; ***forest*** [2, 3, 6]; ***maize, sweetcorn*** [1-3, 5, 6]
- Perennial grasses in ***farm forestry, forest*** [2, 6]

Approval information
- Atrazine was reviewed in 1992 and approvals for aerial use, and use on non-crop land (excluding home garden use) revoked. Restrictions were also placed on the number of applications that could be made to crops
- Use on maize is permitted until Sep 2005. Products containing this active ingredient have been granted derogations for specified 'Essential Uses' for use until 31 December 2007. Sale and supply must cease by 30 June 2007 but growers have no guarantee that the products will continue to be available until then.
  For more information see 'The Review Programme' under 'Pesticide Legislation' in Section 5

Efficacy guidance
- Root activity enhanced by rainfall soon after application and reduced on high organic soils. Foliar activity effective on weeds up to 3 cm high
- Resistant weed strains may develop with repeated use of atrazine or other triazines

Restrictions
- Maximum number of applications (including other atrazine/simazine products) 1 per crop (or lower doses to 3.0 l/ha total) for maize, sweetcorn; 1 per season for conifer plantations, raspberries, roses
- Maximum total dose equivalent to one full dose treatment
- Not recommended for use on soils with more than 10% organic matter
- Do not apply to raspberries in season of planting
- Do not use on Christmas trees

Crop-specific information
- Harvest interval: 7 mth for maize, sweetcorn
- Latest use: before cane emergence for raspberries
- May be used pre- or early post-weed emergence in maize and sweetcorn
- In conifers apply as overall spray in Feb-Apr. May be used in first spring after planting
- Apply to raspberries in spring before new cane emergence, not in season of planting
- Application rates vary with crop, soil type and weed problem. See label for details

Following crops guidance
- Latest use: 7 mth before a succeeding crop for maize, sweetcorn; Apr for conifer plantations; before cane emergence for raspberries; before weeds at 3 cm for roses

Environmental safety
- Dangerous for the environment [5]
- Very toxic to aquatic organisms. May cause long-term adverse effects in the aquatic environment
- Dangerous to fish or other aquatic life and aquatic higher plants. Do not contaminate surface waters or ditches with chemical or used container

SEE SECTION 3 FOR PRODUCTS ALSO REGISTERED

- LERAP Category B
- Use must be restricted to one product containing atrazine or simazine, and either to a single application at the maximum approved rate or (subject to any existing maximum permitted number of treatments) to several applications at lower doses up to the maximum approved rate for a single application
- On slopes heavy rainfall soon after application may cause surface run-off
- To reduce soil run-off, especially from forest plantations, users are advised to plant grass strips 6 m wide between treated areas and surface waters

Hazard classification and safety precautions

Hazard H03, H11 [2, 5, 6]; H04 [4]
Risk phrases R22a, R43 [2, 6]; R48, R50, R58 [2, 5, 6]
Operator protection A, C, D, H, M [1-6]; B [3-5]; U05a [2, 4]; U08, U19a [6]; U14 [2, 6]; U15, U20b [2]; U20c [1, 3-6]
Environmental protection E13b [1, 3, 4]; E14b [1]; E15a [2, 5, 6]; E16a, E16b [1-6]; E31a [3, 4]; E31b [1, 2, 6]; E31c, E32d, E38 [5]; E32e [2]
Storage and disposal E01, E04 [2, 4, 5]; E26 [1-4, 6]; E30a [1-6]
Medical advice M05a [6]

17 azaconazole

A conazole available only in mixture

18 azaconazole + imazalil

A fungicide mixture for use in horticulture

Products

Nectec Paste	Certis	1:2% w/w	PA	08510

Uses

- Canker in ***ornamental plant production***
- Silver leaf in ***ornamental plant production***

Approval information

- Imazalil included in Annex I under EC Directive 91/414
- Products containing azaconazole have been granted derogations for specified 'Essential Uses' for use until 31 December 2007. Sale and supply must cease by 30 June 2007 but growers have no guarantee that the products will continue to be available until then.
  For more information see 'The Review Programme' under 'Pesticide Legislation' in Section 5

Efficacy guidance

- Paint pruning cuts immediately. If necessary clean wounds and cut back any loose bark
- Ensure whole of cut area is fully covered beyond wound to surrounding healthy bark
- Cut back established cankers to sound healthy wood before treatment. Work paste into all crevices

Restrictions

- Maximum number of treatments 1 per wound per yr
- Treat only during dry weather and not in frosty conditions
- Use only during dormant periods
- Do not apply to grafting cuts

Environmental safety

- Harmful to fish or other aquatic life. Do not contaminate surface waters or ditches with chemical or used container

Hazard classification and safety precautions

Risk phrases R52, R58
Operator protection A; U05a
Environmental protection E13c, E32e
Storage and disposal E01, E04

19 azoxystrobin

A systemic translaminar and protectant strobilurin fungicide for cereals

Products

1	Amistar	Syngenta	250 g/l	SC	10443
2	Heritage	Scotts	50% w/w	WG	11383
3	Landgold Strobilurin 250	Landgold	250 g/l	SC	09595
4	Me2 Azoxystrobin	Me2	250 g/l	SC	09654
5	Standon Azoxystrobin	Standon	250 g/l	SC	09515

Uses

- Alternaria in ***carrots*** [1]
- Ascochyta in ***broad beans*** *(off-label)*, ***combining peas***, ***vining peas*** [1]
- Black dot in ***potatoes*** *(off-label - for baby tuber production)* [1]
- Black rot in ***broccoli*** *(off-label)*, ***brussels sprouts*** *(off-label)*, ***cabbages*** *(off-label)*, ***calabrese*** *(off-label)*, ***cauliflowers*** *(off-label)* [1]
- Black spot in ***protected strawberries*** *(off-label)*, ***strawberries*** *(off-label)* [1]
- Botrytis in ***broad beans*** *(off-label)*, ***celery*** *(off-label)*, ***dwarf beans*** *(off-label)*, ***navy beans*** *(off-label)*, ***protected celery*** *(off-label)*, ***runner beans*** *(off-label)* [1]
- Brown rust in ***spring barley***, ***spring wheat***, ***winter barley***, ***winter wheat*** [1, 3-5]; ***spring rye*** *(off-label)*, ***triticale***, ***winter rye*** [1]
- Crown rust in ***spring oats***, ***winter oats*** [1]
- Dark leaf spot in ***salad brassicas*** *(off-label - for baby leaf production)* [1]
- Didymella in ***inert substrate aubergines*** *(off-label)*, ***inert substrate tomatoes*** *(off-label)* [1]
- Didymella stem rot in ***aubergines*** *(off-label)*, ***protected tomatoes*** *(off-label)* [1]
- Downy mildew in ***bulb onions***, ***inert substrate courgettes*** *(off-label)*, ***inert substrate cucumbers*** *(off-label)*, ***inert substrate gherkins*** *(off-label)*, ***protected courgettes*** *(off-label)*, ***protected cucumbers*** *(off-label)*, ***protected gherkins*** *(off-label)*, ***salad brassicas*** *(off-label - for baby leaf production)*, ***salad onions*** *(off-label)* [1]
- Fusarium patch in ***amenity grass***, ***managed amenity turf*** [2]
- Glume blotch in ***spring wheat***, ***winter wheat*** [1, 3-5]
- Grey mould in ***aubergines*** *(off-label)*, ***inert substrate aubergines*** *(off-label)*, ***inert substrate tomatoes*** *(off-label)*, ***protected tomatoes*** *(off-label)* [1]
- Late blight in ***aubergines*** *(off-label)*, ***inert substrate aubergines*** *(off-label)*, ***inert substrate tomatoes*** *(off-label)*, ***protected tomatoes*** *(off-label)* [1]
- Late ear diseases in ***spring wheat***, ***winter wheat*** [1, 3-5]
- Net blotch in ***spring barley***, ***winter barley*** [1, 3-5]
- Powdery mildew in ***aubergines*** *(off-label)*, ***carrots***, ***chives*** *(off-label)*, ***herbs (see appendix 6)*** *(off-label)*, ***inert substrate aubergines*** *(off-label)*, ***inert substrate courgettes*** *(off-label)*, ***inert substrate cucumbers*** *(off-label)*, ***inert substrate gherkins*** *(off-label)*, ***inert substrate tomatoes*** *(off-label)*, ***parsley*** *(off-label)*, ***poppies*** *(off-label - for morphine production)*, ***protected courgettes*** *(off-label)*, ***protected cucumbers*** *(off-label)*, ***protected gherkins*** *(off-label)*, ***protected peppers*** *(off-label)*, ***protected tomatoes*** *(off-label)*, ***raspberries*** *(off-label)*, ***spring oats***, ***triticale***, ***winter oats***, ***winter rye*** [1]; ***spring barley***, ***winter barley*** [1, 3-5]; ***spring wheat***, ***winter wheat*** [5]
- Purple blotch in ***leeks*** [1]
- Rhizoctonia in ***celery*** *(off-label)*, ***protected celery*** *(off-label)*, ***protected chives*** *(off-label)*, ***protected herbs (see appendix 6)*** *(off-label)*, ***protected lettuce*** *(off-label)*, ***protected parsley*** *(off-label)*, ***swedes*** *(off-label)*, ***turnips*** *(off-label)* [1]
- Rhynchosporium in ***spring barley***, ***winter barley*** [1, 3-5]; ***spring rye*** *(off-label)*, ***triticale***, ***winter rye*** [1]
- Ring spot in ***chives*** *(off-label)*, ***herbs (see appendix 6)*** *(off-label)*, ***parsley*** *(off-label)* [1]
- Root malformation disorder in ***red beet*** *(off-label)* [1]
- Rust in ***asparagus***, ***chives*** *(off-label)*, ***herbs (see appendix 6)*** *(off-label)*, ***leeks***, ***parsley*** *(off-label)*, ***spring field beans***, ***winter field beans*** [1]
- Sclerotinia in ***celeriac*** *(off-label)*, ***celery*** *(off-label)*, ***lettuce*** *(off-label)*, ***protected celery*** *(off-label)* [1]
- Septoria leaf spot in ***spring wheat***, ***winter wheat*** [1, 3-5]
- Stemphylium in ***asparagus*** [1]

- White rust in ***chrysanthemums*** *(off-label - in pots)*, ***protected chrysanthemums*** *(off-label)* [1]
- Yellow rust in ***spring wheat***, ***winter wheat*** [1, 3-5]

Specific Off-Label Approvals (SOLAs)

- ***aubergines***, ***protected courgettes***, ***protected cucumbers***, ***protected gherkins***, ***protected tomatoes*** *(OLA 021533) Jul 2008* [1]
- ***broad beans***, ***dwarf beans***, ***navy beans***, ***runner beans*** *(OLA 032311) Jul 2008* [1]
- ***broccoli***, ***calabrese*** *(OLA 040985) Jul 2008* [1]
- ***brussels sprouts*** *(OLA 031995) Jul 2008* [1]
- ***cabbages***, ***cauliflowers*** *(OLA 040615) Jul 2008* [1]
- ***celeriac*** *(OLA 031862) Jul 2008* [1]
- ***celery***, ***protected celery*** *(OLA 001041) Jul 2008* [1]
- ***chives***, ***herbs (see appendix 6)***, ***parsley*** *(OLA 021293) Jul 2008* [1]
- ***chrysanthemums*** *(in pots) (OLA 011684) Jul 2008* [1]
- ***inert substrate aubergines***, ***inert substrate courgettes***, ***inert substrate cucumbers***, ***inert substrate gherkins***, ***inert substrate tomatoes*** *(OLA 011685) Jul 2008* [1]
- ***lettuce*** *(OLA 011465) Jul 2008* [1]
- ***poppies*** *(for morphine production) (OLA 031137) Jul 2008* [1]
- ***potatoes*** *(for baby tuber production) (OLA 030068) Dec 2008* [1]
- ***protected chives***, ***protected herbs (see appendix 6)***, ***protected lettuce***, ***protected parsley*** *(OLA 030659) Jul 2008* [1]
- ***protected chrysanthemums*** *(OLA 001536) May 2008* [1]
- ***protected chrysanthemums*** *(OLA 011684) Jul 2008* [1]
- ***protected peppers*** *(OLA 021295) Jul 2008* [1]
- ***protected strawberries***, ***strawberries*** *(OLA 021294) Jul 2008* [1]
- ***raspberries*** *(OLA 030365) Jul 2008* [1]
- ***red beet***, ***swedes***, ***turnips*** *(OLA 040614) Dec 2008* [1]
- ***salad brassicas*** *(for baby leaf production) (OLA 011465) Jul 2008* [1]
- ***salad onions*** *(OLA 021687) Jul 2008* [1]

Approval information

- Azoxystrobin included in Annex I under EC Directive 91/414

Efficacy guidance

- Best results obtained from use as a protectant or during early stages of disease establishment or when a predictive assessment indicates a risk of disease development
- Azoxystrobin inhibits fungal respiration and should always be used in mixture with fungicides with other modes of action
- In cereals control of established infections can be improved by appropriate tank mixtures or application as part of a programme
- Treatment can reduce the severity of take-all in cereals
- Treatment under poor growing conditions may give less reliable results
- In turf use product at full dose rate in a disease control programme, alternating with fungicides of different modes of action
- For good control of *Fusarium* patch in amenity turf and grass repeat treamtment at minimum intervals of 2 wk
- Azoxystrobin is a member of the QoI cross resistance group. Product should be used preventatively and not relied on for its curative potential
- Use product in cereals as part of an Integrated Crop Management strategy incorporating other methods of control, including where appropriate other fungicides with a different mode of action. Do not apply more than two foliar applications of QoI containing products to any cereal crop
- There is a significant risk of widespread resistance occurring in *Septoria tritici* populations in UK. Failure to follow resistance management action may result in reduced levels of disease control
- On cereal crops product must always be used in mixture with another product, recommended for control of the same target disease, that contains a fungicide from a different cross resistance group and is applied at a dose that will give robust control
- Strains of barley powdery mildew resistant to QoIs are common in the UK

Restrictions

- Maximum total dose ranges from 2-4 times the single full dose depending on crop and product. See labels for details

- On turf the maximum number of treatments is 4 per yr but they must not exceed one third of the total number of fungicide treatments applied
- Do not use where there is risk of spray drift onto neighbouring apple crops
- The same spray equipment should not be used to treat apples

Crop-specific information

- Latest use: grain watery ripe (GS 71) for cereals; before senescence for asparagus
- HI 3 d for inert substrate vegetables; 10 d for carrots, 14 d for onions, celery, lettuce, salad brassicas, vining peas; 21 d for leeks; 35 d for field beans, red beet, swedes, turnips; 36 d for combining peas
- Always use in mixture with another product from a different cross-resistance group on cereals
- All crops should be treated when not under stress. Check leaf wax on peas if necessary
- Consult processor before treating any crops for processing
- Treat asparagus after the harvest season. Where a new bed is established do not treat within 3 wk of transplanting out the crowns
- Do not apply to turf when ground is frozen or during drought

Environmental safety

- Dangerous for the environment [1, 3-5]
- Very toxic to aquatic organisms [1, 3-5]
- Extremely dangerous to fish or other aquatic life. Do not contaminate surface waters or ditches with chemical or used container [2]
- Avoid spray drift onto surrounding areas or crops, especially apples, plums or privet

Hazard classification and safety precautions

Hazard H11 [1, 3-5]
Risk phrases R50, R58 [1, 3-5]
Operator protection U05a [1, 3]; U09a, U19a, U20b [1, 3-5]; U20c [2]
Environmental protection E13a [2]; E15a [1, 3-5]; E31b [2-5]; E31c, E32d, E38 [1]
Storage and disposal E01, E04 [1, 3]; E24 [2]; E26, E30a [1-5]
Medical advice M05a [2]

20 azoxystrobin + chlorothalonil

A preventative and systemic fungicide mixture for cereals

Products

Amistar Opti	Syngenta	100:500 g/l	SC	11863

Uses

- Brown rust in ***spring barley***, ***winter barley***, ***winter wheat***
- Glume blotch in ***winter wheat***
- Net blotch in ***spring barley***, ***winter barley***
- Rhynchosporium in ***spring barley***, ***winter barley***
- Septoria leaf spot in ***winter wheat***
- Yellow rust in ***winter wheat***

Approval information

- Azoxystrobin included in Annex I under EC Directive 91/414

Efficacy guidance

- Best results obtained from applications made as a protectant treatment or in earliest stages of disease development. Further applications may be needed if disease attack is prolonged
- When used to control listed diseases treatment at first or second node stage can reduce the severity of take-all
- Control of *Septoria* and rust diseases may be improved by mixture with a triazole fungicide
- Azoxystrobin is a member of the QoI cross resistance group. Product should be used preventatively and not relied on for its curative potential
- Use product in cereals as part of an Integrated Crop Management strategy incorporating other methods of control, including where appropriate other fungicides with a different mode of action. Do not apply more than two foliar applications of QoI containing products to any cereal crop
- There is a significant risk of widespread resistance occurring in *Septoria tritici* populations in UK. Failure to follow resistance management action may result in reduced levels of disease control

SEE SECTION 3 FOR PRODUCTS ALSO REGISTERED

Restrictions

- Maximum number of treatments 2 per crop
- Maximum total dose on barley equivalent to one full dose treatment
- Do not use where there is risk of spray drift onto neighbouring apple crops
- The same spray equipment should not be used to treat apples

Crop-specific information

- Latest Use: before beginning of heading (GS 51) for barley; before caryopsis watery ripe (GS 71) for wheat

Environmental safety

- Dangerous for the environment
- Very toxic to aquatic organisms
- LERAP Category B

Hazard classification and safety precautions

Hazard H02, H11
Risk phrases R22a, R26, R37, R40, R41, R43, R50, R58
Operator protection A, C, H; U02a, U05a, U09a, U11, U14, U15, U19a, U20b
Environmental protection E15a, E16a, E31c, E32d, E34, E38
Storage and disposal E01, E04, E30a
Medical advice M04

21 azoxystrobin + cyproconazole

A contact and systemic broad spectrum fungicide mixture for cereals

Products

Priori Xtra	Syngenta	200:80 g/l	SC	11518

Uses

- Brown rust in ***spring barley, spring rye, spring wheat, winter barley, winter rye, winter wheat***
- Crown rust in ***spring oats, winter oats***
- Eyespot in ***spring barley*** *(reduction)*, ***spring wheat*** *(reduction)*, ***winter barley*** *(reduction)*, ***winter wheat*** *(reduction)*
- Net blotch in ***spring barley, winter barley***
- Powdery mildew in ***spring barley, spring oats, spring rye, spring wheat, winter barley, winter oats, winter rye, winter wheat***
- Rhynchosporium in ***spring barley*** *(moderate control)*, ***spring rye*** *(moderate control)*, ***winter barley*** *(moderate control)*, ***winter rye*** *(moderate control)*
- Septoria diseases in ***spring wheat, winter wheat***
- Yellow rust in ***spring wheat, winter wheat***

Approval information

- Azoxystrobin included in Annex I under EC Directive 91/414

Efficacy guidance

- Best results obtained from treatment during the early stages of disease development
- A second application may be needed if disease attack is prolonged
- Azoxystrobin is a member of the QoI cross resistance group. Product should be used preventatively and not relied on for its curative potential
- Use product as part of an Integrated Crop Management strategy incorporating other methods of control, including where appropriate other fungicides with a different mode of action. Do not apply more than two foliar applications of QoI containing products to any cereal crop
- There is a significant risk of widespread resistance occurring in *Septoria tritici* populations in UK. Failure to follow resistance management action may result in reduced levels of disease control
- Strains of wheat and barley powdery mildew resistant to QoIs are common in the UK. Control of wheat powdery mildew can only be relied upon from the triazole component
- On cereal crops product must always be used in mixture with another product, recommended for control of the same target disease, that contains a fungicide from a different cross resistance group and is applied at a dose that will give robust control

Restrictions

- Maximum number of treatments 2 per crop
- Do not use where there is risk of spray drift onto neighbouring apple crops
- The same spray equipment should not be used to treat apples

Crop-specific information

- Latest use: up to and including anthesis complete (GS 69) for rye and wheat; up to and including emergence of ear complete (GS 59) for barley and oats

Environmental safety

- Dangerous for the environment
- Very toxic to aquatic organisms

Hazard classification and safety precautions

Hazard H03, H11
Risk phrases R22a, R50, R58, R63
Operator protection A; U05a, U09a, U19a, U20b
Environmental protection E15a, E31c, E32d, E34, E38
Storage and disposal E01, E04, E30a
Medical advice M03

22 azoxystrobin + fenpropimorph

A protectant and eradicant fungicide mixture

Products

Amistar Pro	Syngenta	100:280 g/l	SE	10513

Uses

- Brown rust in ***spring barley, spring wheat, winter barley, winter wheat***
- Late ear diseases in ***spring wheat, winter wheat***
- Net blotch in ***spring barley, winter barley***
- Powdery mildew in ***spring barley, winter barley***
- Rhynchosporium in ***spring barley, winter barley***
- Septoria diseases in ***spring wheat, winter wheat***
- Yellow rust in ***spring wheat, winter wheat***

Approval information

- Azoxystrobin included in Annex I under EC Directive 91/414

Efficacy guidance

- Best results obtained from application before infection following a disease risk assessment, or when disease first seen in crop
- Results may be less reliable when used on crops under stress
- Treatments for protection against ear disease should be made at ear emergence
- Azoxystrobin is a member of the QoI cross resistance group. Product should be used preventatively and not relied on for its curative potential
- Use product as part of an Integrated Crop Management strategy incorporating other methods of control, including where appropriate other fungicides with a different mode of action. Do not apply more than two foliar applications of QoI containing products to any cereal crop
- There is a significant risk of widespread resistance occurring in *Septoria tritici* populations in UK. Failure to follow resistance management action may result in reduced levels of disease control
- Strains of barley powdery mildew resistant to QoIs are common in the UK
- On cereal crops product must always be used in mixture with another product, recommended for control of the same target disease, that contains a fungicide from a different cross resistance group and is applied at a dose that will give robust control

Restrictions

- Maximum number of treatments 2 per crop
- Do not use where there is risk of spray drift onto neighbouring apple crops
- The same spray equipment should not be used to treat apples

Crop-specific information

- Latest use: before early milk stage (GS 73)
- HI 5 wk

SEE SECTION 3 FOR PRODUCTS ALSO REGISTERED

Environmental safety

- Dangerous for the environment
- Very toxic to aquatic organisms

Hazard classification and safety precautions

Hazard H03, H11
Risk phrases R20, R38, R43, R50, R58, R63
Operator protection A, H; U02a, U05a, U09a, U14, U15, U19a, U20b
Environmental protection E15a, E31c, E32d, E38
Storage and disposal E01, E04, E26, E30a
Medical advice M03

23 Bacillus thuringiensis

A bacterial insecticide for control of caterpillars

Products

Dipel DF	Fargro	32000 IU/mg	WG	11184

Uses

- Caterpillars in ***amenity vegetation***, ***apples*** *(off-label)*, ***broccoli***, ***brussels sprouts***, ***cabbages***, ***calabrese*** *(off-label)*, ***cauliflowers***, ***celery*** *(off-label)*, ***cherries*** *(off-label)*, ***chinese cabbage*** *(off-label)*, ***chives*** *(off-label)*, ***collards*** *(off-label)*, ***herbs (see appendix 6)*** *(off-label)*, ***kale*** *(off-label)*, ***lettuce*** *(off-label)*, ***ornamental specimens***, ***parsley*** *(off-label)*, ***pears*** *(off-label)*, ***protected broccoli*** *(off-label)*, ***protected brussels sprouts*** *(off-label)*, ***protected cabbages*** *(off-label)*, ***protected calabrese*** *(off-label)*, ***protected cauliflowers*** *(off-label)*, ***protected celery*** *(off-label)*, ***protected chinese cabbage*** *(off-label)*, ***protected chives*** *(off-label)*, ***protected cucumbers***, ***protected herbs (see appendix 6)*** *(off-label)*, ***protected kale*** *(off-label)*, ***protected lettuce*** *(off-label)*, ***protected ornamentals***, ***protected parsley*** *(off-label)*, ***protected peppers***, ***protected tomatoes***, ***protected watercress*** *(off-label)*, ***raspberries***, ***strawberries***, ***watercress*** *(off-label)*
- Winter moth in ***bilberries*** *(off-label)*, ***blackcurrants*** *(off-label)*, ***blueberries*** *(off-label)*, ***cranberries*** *(off-label)*, ***gooseberries*** *(off-label)*, ***redcurrants*** *(off-label)*, ***vaccinium spp.*** *(off-label)*, ***whitecurrants*** *(off-label)*

Specific Off-Label Approvals (SOLAs)

- ***apples***, ***cherries***, ***chives***, ***herbs (see appendix 6)***, ***lettuce***, ***parsley***, ***pears***, ***protected chives***, ***protected herbs (see appendix 6)***, ***protected lettuce***, ***protected parsley*** *(OLA 022577) Dec 2008* [1]
- ***bilberries***, ***blackcurrants***, ***blueberries***, ***cranberries***, ***gooseberries***, ***redcurrants***, ***vaccinium spp.***, ***whitecurrants*** *(OLA 040358) Dec 2008* [1]
- ***calabrese***, ***celery***, ***chinese cabbage***, ***collards***, ***kale***, ***protected broccoli***, ***protected brussels sprouts***, ***protected cabbages***, ***protected calabrese***, ***protected cauliflowers***, ***protected celery***, ***protected chinese cabbage***, ***protected kale*** *(OLA 040739) Dec 2008* [1]
- ***protected watercress***, ***watercress*** *(OLA 031071) Dec 2008* [1]

Efficacy guidance

- Pest control achieved by ingestion by caterpillars of the treated plant vegetation. Caterpillars cease feeding and die in 2-3 d
- Apply as soon as larvae appear on crop and repeat every 7-10 d until the end of the hatching period
- Good coverage is essential, especially of undersides of leaves. Spray onto dry foliage and do not apply if rain expected within 6 h

Restrictions

- No restriction on number of treatments on edible crops
- Apply spray mixture as soon as possible after preparation

Crop-specific information

- HI zero

Environmental safety

- Store out of direct sunlight

Hazard classification and safety precautions

Operator protection A, C; U05a, U15, U20c

Environmental protection E15a, E32a
Storage and disposal E01, E04, E26, E30a

24 benalaxyl

A phenylamide (acylalanine) fungicide available only in mixtures

25 benalaxyl + mancozeb

A systemic and protectant fungicide mixture

Products

1	Galben M	Sipcam	8:65% w/w	WP	07220
2	Intro Plus	Interfarm	8:65% w/w	WP	11630
3	Tairel	Sipcam	8:65% w/w	WB	07767

Uses
- Blight in ***potatoes, protected potatoes***

Approval information
- Approved for aerial application on potatoes [1, 3]. See notes in Section 5

Efficacy guidance
- Apply to potatoes at blight warning prior to crop becoming infected and repeat at 10-21 d intervals depending on risk of infection
- Spray irrigated potatoes after irrigation and at 14 d intervals, crops in polythene tunnels at 10 d intervals
- When active potato growth ceases use a non-systemic fungicide to end of season, starting not more than 10 d after last application
- For reduction of downy mildew in oilseed rape apply at seedling to 7-leaf stage, before end Nov as soon as infection seen

Restrictions
- Maximum number of treatments 5 per crop (including other phenylamide-based fungicides) for potatoes
- Do not treat potatoes showing active blight infection

Crop-specific information
- HI 7 d

Environmental safety
- Dangerous for the environment
- Very toxic to aquatic organisms

Hazard classification and safety precautions
Hazard H04 [1-3]; H11 [2]
Risk phrases R36 [1, 3]; R37, R43, R50, R58 [2]
Operator protection A; U05a [1-3]; U08, U20b [1, 3]; U14, U15, U19a [2]
Environmental protection E13b [1, 3]; E15a, E32e [2]; E32a [1]
Consumer protection C02 [3] (7 d)
Storage and disposal E01, E04 [1-3]; E29a, E30a [1, 3]
Medical advice M04 [2]

26 benfuracarb

A soil-applied carbamate insecticide and nematicide for beet crops

Products

Oncol 10G	Nufarm UK	10% w/w	GR	08249

Uses
- Docking disorder vectors in ***fodder beet, mangels, sugar beet***
- Leaf miner in ***fodder beet, mangels, sugar beet***
- Millipedes in ***fodder beet, mangels, sugar beet***
- Pygmy beetle in ***fodder beet, mangels, sugar beet***

- Springtails in ***fodder beet, mangels, sugar beet***
- Symphylids in ***fodder beet, mangels, sugar beet***

Efficacy guidance

- Apply at sowing with suitable applicator so that the granules are mixed with the moving soil closing the seed furrow. See label for recommended applicators and calibration

Restrictions

- Maximum number of treatments 1 per crop

Crop-specific information

- Latest use: at planting

Environmental safety

- Dangerous for the environment
- Toxic to aquatic organisms
- Dangerous to game, wild birds and animals

Hazard classification and safety precautions

Hazard H03, H11
Risk phrases R20, R36, R38, R51, R58
Operator protection A, B, C, D, E, H, K, M; U02a, U05a, U13, U19a, U20a
Environmental protection E15a, E19b, E32a, E32d
Storage and disposal E01, E04, E30a
Medical advice M02, M05a

27 bentazone

A post-emergence contact diazinone herbicide

Products

1	Agriguard Bentazone 480	AgriGuard	480 g/l	SL	11569
2	Basagran SG	BASF	87% w/w	SG	08360
3	Landgold Bentazone SL	Landgold	480 g/l	SL	10841
4	Standon Bentazone S	Standon	87% w/w	SG	10124

Uses

- Annual dicotyledons in ***broad beans, linseed, navy beans, peas, potatoes, runner beans, spring field beans, winter field beans*** [1-4]; ***bulb onions*** *(off-label)*, ***evening primrose*** *(off-label)*, ***leeks*** *(off-label)*, ***salad onions*** *(off-label)* [2]; ***dwarf beans*** [1, 3]; ***french beans*** [2, 4]; ***narcissi*** [1, 2, 4]
- Common storksbill in ***leeks*** *(off-label)* [2]
- Fool's parsley in ***leeks*** *(off-label)* [2]

Specific Off-Label Approvals (SOLAs)

- ***bulb onions, salad onions*** *(OLA 971234) Dec 2008* [2]
- ***evening primrose*** *(OLA 971232) Dec 2008* [2]
- ***leeks*** *(OLA 021551) Dec 2008* [2]

Approval information

- Bentazone included in Annex I under EC Directive 91/414

Efficacy guidance

- Most effective control obtained when weeds are growing actively and less than 5 cm high or across. Good spray cover is essential
- Various recommendations are made for use in spray programmes with other herbicides or in tank mixes. See label for details
- The addition of specified ajuvant oils is recommended for use on some crops to improve fat hen control. Do not use under hot or humid conditions. See label for details
- Split dose application may be made in all recommended crops except peas and generally gives better weed control. See label for details

Restrictions

- Maximum number of treatments varies with crop and product - see labels
- Crops must be treated at correct stage of growth to avoid danger of scorch. See labels for details
- Not all varieties of recommended crops are fully tolerant. Use only on tolerant varieties named in labels. Do not use on forage or mange-tout varieties of peas

- Do not use on crops which have been affected by drought, waterlogging, frost or other stress conditions
- Do not apply insecticides within 7 d of treatment [1, 3]
- Leave 7 d after using a post-emergence grass herbicide or 14 d where treatment precedes the grass herbicide and carry out a leaf wax test where relevant [1, 3]
- Do not spray at temperatures above 21°C. Delay spraying until evening if necessary
- Do not apply if rain or frost expected, if unseasonably cold, if foliage wet or in drought
- A minimum of 6 h (preferably 12 h) free from rain is required after application
- May be used on selected varieties of maincrop and second early potatoes (see label for details), not on seed crops or first earlies

Crop-specific information

- Latest use: before shoots exceed 15 cm high for potatoes and spring field beans (or 6-7 leaf pairs), 4 leaf pairs (6 pairs or 15 cm high with split dose) for broad beans, before flower buds visible for French, navy, runner, soya and winter field beans and linseed, before flower buds can be found enclosed in terminal shoot for peas
- HI onions 3 wk
- Best results in narcissi obtained by using a suitable pre-emergence herbicide first
- Consult processor before using on crops for processing
- A satisfactory wax test must be carried out before use on peas
- Do not treat narcissi during flower bud initiation

Environmental safety

- Dangerous for the environment [1]
- Harmful to aquatic organisms

Hazard classification and safety precautions

Hazard H03 [1-4]; H11 [1]
Risk phrases R22a, R43 [1-4]; R36 [1]; R41 [2, 4]; R52, R58 [2-4]
Operator protection A [1-4]; C [1, 2, 4]; U05a, U08, U19a, U20b [1-4]; U11 [2, 4]; U14 [2, 3]
Environmental protection E15a [1-4]; E31b [1-3]; E32a [4]; E34 [1, 3]
Storage and disposal E01, E04, E30a [1-4]; E26 [1, 3, 4]; E29a [1]
Medical advice M03 [3]; M05a [2]

28 bentazone + MCPA + MCPB

A post-emergence herbicide for undersown spring cereals and grass

Products

Acumen	BASF	200:80:200 g/l	SL	00028

Uses

- Annual dicotyledons in ***newly sown grass***, ***undersown spring cereals***

Approval information

- Bentazone included in Annex I under EC Directive 91/414

Efficacy guidance

- Best results when weeds small and actively growing provided crop is at correct growth stage
- A minimum of 6 h free from rain is required after treatment

Restrictions

- Maximum number of treatments 1 per crop on undersown cereals; 1 per yr on grassland
- Do not treat red clover after 3-trifoliate leaf stage
- Do not use on crops suffering from herbicide damage or physical stress or in temperatures above 21°C
- Do not use on seed crops or on cereals undersown with lucerne or seed mixtures containing lucerne
- Do not roll or harrow for 7 d before or after spraying
- Clovers may be scorched and undersown crop checked but effects likely to be outgrown. Clover stand in grassland may be reduced
- Do not apply if frost expected, if crop wet or when temperatures at or above 21°C

SEE SECTION 3 FOR PRODUCTS ALSO REGISTERED

Crop-specific information

- Latest use: before 1st node detectable for undersown cereals (GS 31), 2 wk before grazing for grassland
- Apply to cereals from 2-fully expanded leaf stage but before first node detectable (GS 12-30) provided clover has reached 1-trifoliate leaf stage
- On first year grass leys use on seedling stage weeds before end of Sep
- Apply to newly sown leys after grass has reached 2-leaf stage provided clovers have at least 1 trifoliate leaf and red clover has not passed 3-trifoliate leaf stage

Environmental safety

- Dangerous for the environment
- Toxic to aquatic organisms
- Keep livestock out of treated areas for at least 2 wk and until foliage of poisonous weeds such as ragwort has died and become unpalatable

Hazard classification and safety precautions

Hazard H03, H11
Risk phrases R22a, R36, R38, R43, R51, R58
Operator protection A, C; U05a, U08, U14, U19a, U20b
Environmental protection E07a, E15a, E31c, E32d, E34, E38
Storage and disposal E01, E04, E29b, E30a
Medical advice M03, M05a

29 bentazone + MCPB

A post-emergence herbicide mixture for use in peas

Products

Pulsar	BASF	200:200 g/l	SL	04002

Uses

- Annual dicotyledons in ***peas***

Approval information

- Bentazone included in Annex I under EC Directive 91/414

Efficacy guidance

- To be used in tank mix with cyanazine
- Best results achieved when weeds small and actively growing provided crop at correct stage. Good spray cover is essential
- May be applied as single treatment or as split dose treatment applying first spray when susceptible weeds are not beyond the 2 leaf stage
- Do not apply if rain or frost expected or if foliage wet. A minimum of 24 h free from rain required after treatment
- Do not apply during drought or unseasonably cold weather

Restrictions

- Maximum number of treatments 1 per crop or 2 per crop with split dose
- Do not treat forage pea cultivars or mange-tout peas or other varieties specifically listed in the label
- Apply after a satisfactory wax test. Early drilled crops or crops affected by frost or abrasion may not have sufficiently waxy cuticle
- Allow 7 d after spraying before using a grass weed herbicide, or wait 14 d afterwards and test leaf wax before treating
- Do not use as tank mix with any other product than cyanazine, nor after use of TCA
- Do not apply insecticides within 7 d of treatment
- Do not apply to any crop that may have been subjected to stress conditions, where foliage damaged or under hot, sunny conditions when temperature exceeds 21°C
- Do not spray if foliage is wet or if frost or rain expected within 24 h
- Consult processors before use on crops for processing

Crop-specific information

- Latest use: before enclosed bud stage of crop (GS 10x)
- Apply only to listed pea cultivars (see label) from 3 fully expanded leaf (for full dose) or from 2 node stage (for split dose) to before flower buds can be found enclosed in terminal shoot (GS 102 or 103 to before GS 201)

Environmental safety

- Dangerous for the environment
- Toxic to aquatic organisms
- Keep livestock out of treated areas for 14 d and until foliage of any poisonous weeds such as ragwort has died and become unpalatable

Hazard classification and safety precautions

Hazard H04, H11
Risk phrases R36, R38, R43, R51, R58
Operator protection A, C; U05a, U08, U14, U19a, U20a
Environmental protection E07a, E15a, E31c, E32d, E34, E38
Storage and disposal E01, E04, E30a
Medical advice M05a

30 bentazone + pendimethalin

A contact and residual herbicide mixture for combining peas

Products

Impuls	BASF	480:400 g/l	KL	11780

Uses

- Annual dicotyledons in ***combining peas***
- Chickweed in ***combining peas***
- Cleavers in ***combining peas***
- Knotgrass in ***combining peas***
- Mayweeds in ***combining peas***

Approval information

- Bentazone and pendimethalin included in Annex I under EC Directive 91/414

Efficacy guidance

- Best results achieved by treating weeds when they are small and ensuring maximum spray coverage
- Moist soil required for maximum efficacy. Loose or cloddy seedbeds must be consolidated before spraying
- Residual control reduced on soils with more than 6% OM
- Minimum 6 hr free from rain required for optimum foliar activity. Do not spray when foliage wet

Restrictions

- Maximum total dose equivalent to one full dose treatment
- Severe crop damage may occur if mixed with any other product. Leave minimum 7 d before, or 14 d after, using a post-emergence grass herbicide. Do not apply insecticides within 7 d of use. Do not apply if TCA has been used
- Do not use on soils prone to waterlogging
- If day time temperatures likely to exceed 21°C spray in the evening
- Do not treat crops under stress from any cause
- Equipment must be washed out thoroughly after use. Traces of product may damage susceptible crops sprayed later

Crop-specific information

- Latest use: before 3rd node stage (GS 103)
- Crystal violet leaf wax test must be carried out before use. Transient leaf margin scorch may occur after treatment
- Crop damage may occur if heavy rain follows application on stony or gravelly soils

Following crops guidance

- Land must be ploughed to 150 mm before any succeeding crop except cereals
- In the event of crop failure spring barley may be sown after ploughing to 150 mm

SEE SECTION 3 FOR PRODUCTS ALSO REGISTERED

Environmental safety

- Dangerous for the environment
- Very toxic to aquatic organisms

Hazard classification and safety precautions

Hazard H03, H11
Risk phrases R22a, R43, R50, R58
Operator protection A; U05a, U08, U13, U19a, U20b
Environmental protection E15a, E31c, E32d, E34, E38
Storage and disposal E01, E04, E29b, E30a
Medical advice M03, M05a

31 beta-cyfluthrin

A non-systemic pyrethroid insecticide available only in mixtures

32 beta-cyfluthrin + clothianidin

An insecticidal seed treatment mixture for beet

Products

Poncho Beta	Bayer CropScience	53:400 g/l	EC	11999

Uses

- Beet virus yellows vectors in ***fodder beet*** *(seed treatment)*, ***sugar beet*** *(seed treatment)*

Approval information

- Beta-cyfluthrin included in Annex I under EC Directive 91/414

Efficacy guidance

- In addition to control of aphid virus vectors, product improves crop establishment by reducing damage caused by symphylids, springtails, millipedes, wireworms and leatherjackets
- Additional control measures should be taken where very high populations of soil pests are present
- Product reduces direct feeding damage by foliar pests such as pygmy mangold beetle, beet flea beetle, mangold fly
- Product is not active against nematodes
- Treatment does not alter physical characteristics of pelleted seed and no change to standard drill settings should be necessary

Restrictions

- Maximum number of treatments: 1 per seed batch
- Product must be co-applied with a colouring dye
- Product must only be applied as a coating to pelleted seed using special treatment machinery

Crop-specific information

- Latest use: pre-drilling

Environmental safety

- Extremely dangerous to fish or other aquatic life. Do not contaminate surface waters or ditches with chemical or used container

Hazard classification and safety precautions

Hazard H03
Risk phrases R22a
Operator protection A, H; U04a, U05a, U07, U13, U14, U20b, U24
Environmental protection E13a, E33, E34, E36
Storage and disposal E01, E04, E26, E30a
Treated seed S02, S03, S04a, S04b, S05, S06a, S06b, S07
Medical advice M03

33 beta-cyfluthrin + imidacloprid

An insecticide mixture for seed treatment of oilseed rape

Products

1 Chinook Blue	Bayer CropScience	100:100 g/l	LS	11262
2 Chinook Colourless	Bayer CropScience	100:100 g/l	LS	11206

Uses

- Cabbage stem flea beetle in ***spring oilseed rape*** *(seed treatment)* [1]; ***winter oilseed rape*** *(seed treatment)* [1, 2]
- Flea beetle in ***spring oilseed rape*** *(seed treatment)*, ***winter oilseed rape*** *(seed treatment)* [1]

Approval information

- Beta-cyfluthrin included in Annex I under EC Directive 91/414

Efficacy guidance

- Drill treated seed as soon as possible after treatment
- Product will reduce damage by early attacks of flea beetle and may also reduce subsequent larval damage. However a follow-up spray treatment may be required if pest activity is heavy and prolonged

Restrictions

- Maximum number of treatments 1 per batch of seed
- Do not use on seed with more than 9% moisture content, on sprouted seed or on cracked, split or otherwise damaged seed
- Product must be co-applied with a colouring agent [2]
- Must be applied using seed treatment machine recommended by manufacturer
- Allow treated seed to dry before packaging. Drying can be assisted by subsequent application of talc
- Product can cause a transient tingling or numbing sensation to exposed skin. Avoid skin contact with the product, treated seed or dust throughout all operations in the treatment plant and during drilling

Crop-specific information

- Latest use: before drilling

Environmental safety

- Dangerous for the environment
- Toxic to aquatic organisms
- Do not use treated seed as food or feed
- Dangerous to birds, game and other wildlife. Treated seed should not be left on the soil surface. Bury spillages
- Do not broadcast treated seed

Hazard classification and safety precautions

Hazard H03, H11
Risk phrases R22a, R51, R58
Operator protection A, H; U04a, U05a, U07, U20b
Environmental protection E13a, E32d, E33, E34, E36, E38
Storage and disposal E01, E04, E26, E30a
Treated seed S01, S02, S03, S04c, S05, S06a, S06b, S07, S08
Medical advice M03

34 bifenox

A diphenyl ether herbicide for cereals

Products

Fox	Makhteshim	480 g/l	SC	11981

Uses

- Cleavers in ***winter barley***, ***winter rye***, ***winter wheat***
- Field pansy in ***winter barley***, ***winter rye***, ***winter wheat***
- Forget-me-not in ***winter barley***, ***winter rye***, ***winter wheat***

- Poppies in ***winter barley, winter rye, winter wheat***
- Red dead-nettle in ***winter barley, winter rye, winter wheat***
- Speedwells in ***winter barley, winter rye, winter wheat***

Efficacy guidance

- Bifenox is absorbed by foliage and emerging roots of susceptible species
- Best results obtained when weeds are growing actively with adequate soil moisture

Restrictions

- Maximum number of treatments: one per crop
- Do not apply to crops suffering from stress from whatever cause
- Do not apply if the crop is wet or if rain or frost is expected
- Avoid drift onto broad-leaved plants outside the target area

Crop-specific information

- Latest use: before 2nd node detectable (GS 32) for all crops

Environmental safety

- Dangerous for the environment
- Very toxic to aquatic organisms

Hazard classification and safety precautions

Hazard H11
Risk phrases R50, R58
Operator protection A; U05a, U08, U20b
Environmental protection E15a, E19b, E32d, E34, E38
Storage and disposal E01, E04, E30a
Medical advice M03

35 bifenthrin

A contact and residual pyrethroid acaricide/insecticide for use in agricultural and horticultural crops

Products

1	Starion	Belchim	100 g/l	EC	11641
2	Talstar	Certis	100 g/l	EC	11325

Uses

- Aphids in ***broccoli, brussels sprouts, cabbages, calabrese, cauliflowers*** [1, 2]; ***winter barley, winter oats, winter wheat*** [1]
- Cabbage stem flea beetle in ***winter oilseed rape*** [1]
- Caterpillars in ***broccoli, brussels sprouts, cabbages, calabrese, cauliflowers*** [1, 2]
- Clay-coloured weevil in ***protected raspberries*** *(off-label)*, ***raspberries*** *(off-label)* [2]
- Damson-hop aphid in ***hops*** [2]
- Fruit tree red spider mite in ***apples, pears*** [2]
- Rape winter stem weevil in ***winter oilseed rape*** [1]
- Two-spotted spider mite in ***hops, ornamental specimens, protected raspberries*** *(off-label)*, ***raspberries*** *(off-label)*, ***strawberries*** [2]
- Virus vectors in ***winter barley, winter oats, winter wheat*** [1]
- Whitefly in ***broccoli, brussels sprouts, cabbages, calabrese, cauliflowers*** [1, 2]

Specific Off-Label Approvals (SOLAs)

- ***protected raspberries, raspberries*** *(OLA 030872) Jun 2006* [2]

Efficacy guidance

- Timing of application varies with crop and pest. See label for details
- Good spray cover of upper and lower plant surfaces essential to achieve effective pest control

Restrictions

- Maximum number of treatments 5 per yr for hops; 2 per yr for all other crops

Crop-specific information

- Latest use: before 31 Mar in yr of harvest for cereals; before 30 Nov in yr of sowing for winter oilseed rape
- HI 2 d for brassicas, raspberries; zero for all other crops

Environmental safety

- Dangerous for the environment

- Very toxic to aquatic organisms
- Extremely dangerous to bees. Do not apply to crops in flower or to those in which bees are actively foraging. Do not apply when flowering weeds are present
- Keep in original container, tightly closed, in a safe place, under lock and key
- Broadcast air assisted LERAP (18 m) and LERAP Category A

Hazard classification and safety precautions

Hazard H03, H08, H11
Risk phrases R20, R22a, R22b, R36, R37, R50, R58
Operator protection A, C, H; U02a, U04a, U05a, U08, U11, U20b [1, 2]; U19a [2]
Environmental protection E12c, E16c, E31b, E34, E38 [1, 2]; E15a, E16d [2]; E15b, E32d [1]; E17b [1, 2] (18 m)
Storage and disposal E01, E04, E30b [1, 2]; E26 [2]
Medical advice M03, M05b

36 bifenthrin + malathion

An insecticide mixture for pest control in stored grain and grain stores

Products

1	Prostore 157 UL	Nickerson	7.5:150 g/l	UL	12017
2	Prostore 420 EC	Nickerson	20:400 g/l	EC	12036

Uses

- Grain storage pests in ***grain stores*** [2]; ***stored grain*** [1, 2]

Efficacy guidance

- Best results in stored grain obtained from treatment at the start of storage after high-temperature drying. Cool warm grain as soon as possible to prevent moisture migration and ensure optimum protection from reinfestation
- Grain stores should be cleaned before treatment and treated 3-4 wk before filling. Treat all surfaces, especially any cracks and crevices [2]
- Controls a wide range of stored grain insect pests and cosmopolitan food mites
- Under normal storage conditions a single treatment will give 3-6 mth protection
- Knock-down is complete within 24 h at normal harvest grain teperatures
- Control may not be satisfactory if pest strains resistant to bifenthrin or malathion are present

Restrictions

- Maximum number of treatments 1 per store [2] or batch of grain
- Do not treat wheat (including durum), oats, rye or triticale intended for use as seed
- Must only be applied with appropriate equipment designed to treat grain
- Do not apply with hand-held equipment

Environmental safety

- Dangerous for the environment

Hazard classification and safety precautions

Hazard H03, H11
Risk phrases R22a, R38, R41, R43 [2]; R22b, R50, R58, R66, R67 [1, 2]
Operator protection A, C, H [1, 2]; D, J, M [2]; U05a, U23a [1, 2]; U14 [2]
Environmental protection E15a, E32a, E34, E38
Storage and disposal E01, E04
Medical advice M03, M05b

37 bitertanol

A conazole fungicide available only in mixtures

38 bitertanol + fuberidazole

A broad spectrum fungicide mixture for seed treatment in cereals

Products

Sibutol	Bayer CropScience	375:23 g/l	FS	11305

Uses

- Bunt in ***spring wheat*** *(seed treatment)*, ***winter wheat*** *(seed treatment)*
- Fusarium root rot in ***spring oats*** *(seed treatment)*, ***spring rye*** *(seed treatment)*, ***spring wheat*** *(seed treatment)*, ***triticale*** *(seed treatment)*, ***winter oats*** *(seed treatment)*, ***winter rye*** *(seed treatment)*, ***winter wheat*** *(seed treatment)*
- Loose smut in ***spring wheat*** *(seed treatment - reduction)*, ***winter wheat*** *(seed treatment - reduction)*
- Septoria seedling blight in ***spring wheat*** *(seed treatment - reduction)*, ***winter wheat*** *(seed treatment - reduction)*

Efficacy guidance

- Treated seed should preferably be drilled in the same season
- Control of loose smut may be inadequate for use on seed for multiplication

Restrictions

- Maximum number of treatments 1 per batch of seed
- Do not use on seed with more than 16% moisture content, or on sprouted, cracked or skinned seed
- Must be applied simultaneously with water in the ratio 1 part product to 2 parts water in a recommended seed treatment machine

Crop-specific information

- Before drilling

Environmental safety

- Dangerous for the environment
- Toxic to aquatic organisms
- Do not use treated seed as food or feed
- Treated seed harmful to game and wildlife

Hazard classification and safety precautions

Hazard H11
Risk phrases R51, R58
Operator protection A, H; U07, U20a
Environmental protection E03, E15a, E32d, E33, E34, E36, E38
Storage and disposal E26, E30a
Treated seed S01, S02, S03, S04a, S05, S06a, S07

39 bitertanol + fuberidazole + imidacloprid

A broad spectrum fungicide and insecticide seed treatment for winter wheat and winter oats

Products

Sibutol Secur	Bayer CropScience	140:8.6:87.5 g/l	LS	11308

Uses

- Bunt in ***winter wheat*** *(seed treatment)*
- Fusarium root rot in ***winter oats*** *(seed treatment)*, ***winter wheat*** *(seed treatment)*
- Loose smut in ***winter wheat*** *(seed treatment - reduction)*
- Seedling blight and foot rot in ***winter wheat*** *(seed treatment - reduction)*
- Virus vectors in ***winter oats*** *(seed treatment)*, ***winter wheat*** *(seed treatment)*
- Wireworm in ***winter oats*** *(seed treatment - reduction of damage)*, ***winter wheat*** *(seed treatment - reduction of damage)*

Efficacy guidance

- Best applied through recommended seed treatment machines
- Evenness of seed cover improved by simultaneous application of equal volumes of product and water or dilution of product with an equal volume of water

- Drill treated seed in the same season. Use minimum 125 kg treated seed per ha. Lower drilling rates and/or early drilling affect duration of BYDV protection needed and may require follow-up aphicide treatment
- In high risk areas where aphid activity is heavy and prolonged a follow-up aphicide treatment may be required
- Protection against foliar air-borne and splash-borne diseases later in the season will require appropriate fungicide follow-up sprays

Restrictions
- Maximum number of treatments 1 per batch of seed
- Do not use on seed with more than 16% moisture content, or on sprouted, cracked or skinned seed

Crop-specific information
- Latest use: before drilling
- Slightly delayed and reduced emergence may occur but this is normally outgrown
- Field emergence which is delayed for any reason may be accentuated by treatment

Environmental safety
- Dangerous for the environment
- Toxic to aquatic organisms
- Do not use treated seed as food or feed
- Dangerous to birds, game and other wildlife. Treated seed should not be left on the soil surface. Bury spillages
- Treated seed should not be broadcast, but drilled to a depth of 4 cm in a well prepared seedbed
- If seed is left on the soil surface the field should be harrowed and rolled to ensure good incorporation

Hazard classification and safety precautions

Hazard H11
Risk phrases R51, R58
Operator protection A, H; U04a, U05a, U13, U20b
Environmental protection E03, E15a, E32d, E34, E38
Storage and disposal E01, E04, E26, E30a
Treated seed S01, S02, S03, S04c, S05, S06a, S06b, S07
Medical advice M03

40 bone oil

A ready-to-use animal repellent

Products

Renardine 72-2	Roebuck Eyot	30% w/w	AL	06769

Uses
- Badgers in ***amenity areas***, ***farmland***
- Cats in ***agricultural premises***, ***amenity areas***, ***farmland***
- Dogs in ***agricultural premises***, ***amenity areas***, ***farmland***
- Foxes in ***amenity areas***, ***farmland***
- Moles in ***amenity areas***, ***farmland***
- Rabbits in ***amenity areas***, ***farmland***

Efficacy guidance
- Soak pieces of stick, rags or sand in product and distribute around area to be protected as directed
- Repeat treatment weekly or after heavy rainfall

Environmental safety
- Do not apply directly to animals or crops
- Used as directed product is purely deterrent and does not harm animals

41 Bordeaux mixture

A protectant copper sulphate/lime complex fungicide

Products

Wetcol 3	Ford Smith	30 g/l (copper)	SC	02360

Uses

- Bacterial canker in ***cherries***
- Blight in ***outdoor tomatoes***, ***potatoes***
- Cane spot in ***loganberries***, ***raspberries***
- Canker in ***apples***, ***pears***
- Celery leaf spot in ***celery***
- Currant leaf spot in ***blackcurrants***
- Downy mildew in ***hops***
- Leaf curl in ***apricots***, ***nectarines***, ***peaches***
- Scab in ***apples***, ***pears***
- Spur blight in ***raspberries***

Efficacy guidance

- Spray interval normally 7-14 d but varies with disease and crop. See label for details
- Commence spraying potatoes before crop meets in row or immediately first blight period occurs
- For canker control spray monthly from Aug to Oct
- For peach leaf curl control spray at leaf fall in autumn and again in Feb
- Spray when crop foliage dry. Do not spray if rain imminent

Restrictions

- Do not use on copper sensitive cultivars, including Doyenne du Comice pears

Crop-specific information

- HI 7 d

Environmental safety

- Harmful to fish or other aquatic life. Do not contaminate surface waters or ditches with chemical or used container
- Harmful to livestock. Keep all livestock out of treated areas for at least 3 wk

Hazard classification and safety precautions

Hazard H03, H04
Risk phrases R22a, R36, R37, R38
Operator protection U13, U15, U20a
Environmental protection E06c, E13c, E32a
Storage and disposal E30a

42 boscalid

A translocated and translaminar anilide fungicide for oilseed rape

Products

1	Filan	BASF	50% w/w	WG	11449
2	Me2 Succotash	Me2	50% w/w	WG	11815

Uses

- Alternaria in ***poppies*** *(off-label - for morphine production)* [1]
- Botrytis in ***poppies*** *(off-label - for morphine production)* [1]
- Sclerotinia stem rot in ***poppies*** *(off-label - for morphine production)* [1]; ***spring oilseed rape***, ***winter oilseed rape*** [1, 2]

Specific Off-Label Approvals (SOLAs)

- ***poppies*** *(for morphine production) (OLA 040163) Dec 2008* [1]

Efficacy guidance

- For best results apply in high disease risk situations at flowering stage of crop but before symptoms are visible
- Ensure adequate spray penetration and good coverage

Restrictions

- Maximum total dose equivalent to two full dose treatments
- Avoid spray drift onto neighbouring crops

Crop-specific information

- Latest use: up to and including full flower

Environmental safety

- Dangerous for the environment [1, 2]
- Toxic to aquatic organisms
- Product represents minimal hazard to bees when used as directed. However local bee-keepers should be notified if crops are to be sprayed when in flower

Hazard classification and safety precautions

Hazard H11
Risk phrases R51, R58
Operator protection U05a, U20c
Environmental protection E15a, E31c, E32d, E38
Storage and disposal E01, E04, E26, E30a

43 boscalid + pyraclostrobin

A protectant and systemic fungicide mixture

Products

Signum	BASF	26.7:6.7% w/w	WG	11450

Uses

- Dark leaf spot in ***cabbages, cauliflowers***
- Grey mould in ***strawberries***
- Ring spot in ***brussels sprouts, cabbages, cauliflowers***
- White blister in ***brussels sprouts, cabbages*** *(qualified minor use),* ***chinese cabbage*** *(off-label),* ***collards*** *(off-label),* ***kale*** *(off-label),* ***komatsuna*** *(off-label),* ***pak choi*** *(off-label),* ***salad brassicas*** *(off-label - baby leaf production)*

Specific Off-Label Approvals (SOLAs)

- ***chinese cabbage, collards, kale, komatsuna, pak choi*** *(OLA 031595) Nov 2005* [1]
- ***salad brassicas*** *(baby leaf production) (OLA 031595) Nov 2005* [1]

Efficacy guidance

- On brassicas apply as a protectant spray or at the first sign of disease and repeat at 3-4 wk intervals depending on disease pressure
- Ensure adequate spray penetration and coverage by increasing water volume in dense crops
- For best results on strawberries apply as a protectant spray at the white bud stage. Applications should be made in sequence with other products as part of a fungicide spray programme during flowering at 7-10 day intervals
- Pyraclostrobin is a member of the QoI cross-resistance group of fungicides and should be used in programmes with fungicides with a different mode of action

Restrictions

- Maximum total dose equivalent to two full dose treatments on all crops
- On brassicas use a maximum of three applications per yr
- On strawberries do not use consecutive treatments and use a maximum of one QoI containing product in three fungicide applications
- Consult processor before use on crops for processing

Crop-specific information

- HI 14 for brassica crops; 3 d for strawberries

Environmental safety

- Dangerous for the environment
- Very toxic to aquatic organisms
- LERAP Category B

Hazard classification and safety precautions

Hazard H03, H11

Risk phrases R22a, R50, R58
Operator protection A; U05a, U20c
Environmental protection E15a, E16a, E31c, E32d, E38
Storage and disposal E01, E04, E26, E30a
Medical advice M05a

44 brodifacoum

An anticoagulant coumarin rodenticide

Products

1	Klerat	Sorex	0.005% w/w	RB	H6701
2	Klerat Wax Blocks	Sorex	0.005% w/w	BB	H6703
3	Sorexa Checkatube	Sorex	0.2% w/w	XX	H6702
4	Talon Rat & Mouse Bait (Cut Wheat)	Killgerm	0.005% w/w	RB	H6709
5	Talon Rat & Mouse Bait (Whole Wheat)	Killgerm	0.005% w/w	RB	H6710

Uses

- Mice in ***agricultural premises*** [4, 5]; ***farm buildings*** [1, 2]; ***farm buildings*** *(indoor use only)* [3]
- Rats in ***agricultural premises*** [4, 5]; ***farm buildings*** [1, 2]

Efficacy guidance

- Effective against rodents resistant to other commonly used anticoagulants
- A single feed representing a fraction of the pest's normal daily food requirement can be lethal
- Ready-to-use in a baiting programme
- Cover baits by placing in bait boxes, drain pipes or under boards, or use bait tubes [3]
- Inspect bait sites frequently and top up as long as there is evidence of feeding

Restrictions

- For use only by professional pest contractors
- Do not use outdoors. Products must be used in situations where baits are placed within a building or other enclosed structure, and the target is living or feeding predominantly within that building or structure

Environmental safety

- Prevent access to baits by children, birds and other animals
- Do not prepare or lay baits where food, feed or water could become contaminated
- Remove all remains of bait or bait containers after use and burn or bury
- Search for and burn or bury all rodent bodies. Do not place in refuse bins or on rubbish tips
- Contains human taste deterrent denatonium benzoate [1, 2]

Hazard classification and safety precautions

Operator protection U13 [1-5]; U20a [4, 5]; U20b [1-3]
Environmental protection E32a [1, 2, 4, 5]
Storage and disposal E04 [1, 2, 4, 5]; E30b [1-5]
Vertebrate/rodent control products V01a [1, 2]; V01b, V02 [3-5]; V03a, V04a [1-3]; V03b, V04b [4, 5]
Medical advice M03 [4, 5]; M05b [1-3]

45 bromadiolone

An anti-coagulant coumarin-derivative rodenticide

Products

1	Endorats Premium Rat Killer	Irish Drugs	0.005% w/w	GB	H6725
2	Sakarat Bromabait	Killgerm	0.005% w/w	RB	H7902
3	Tomcat 2	Antec Biosentry	0.005% w/w	PT	H6736
4	Tomcat 2 Blox	Antec Biosentry	0.005% w/w	BB	H6731

Uses

- Mice in ***farm buildings*** [2-4]; ***farmyards*** [2]
- Rats in ***farm buildings*** [1-4]; ***farmyards*** [2]

Efficacy guidance

- Ready-to-use baits are formulated on a mould-resistant, whole-wheat base
- Use in baiting programme. Place baits in protected situations, sufficient for continuous feeding between treatments
- Chemical is effective against warfarin- and coumatetralyl-resistant rats and mice and does not induce bait shyness
- Use bait bags where loose baiting inconvenient (eg behind ricks, silage clamps etc)
- The resiatance status of the rodent population should be assessed when considering the choice of product to use

Restrictions

- For use only by professional operators

Environmental safety

- Access to baits by children, birds and animals, particularly cats, dogs, pigs and poultry, must be prevented
- Baits must not be placed where food, feed or water could become contaminated
- Remains of bait and bait containers must be removed after treatment and burned or buried
- Rodent bodies must be searched for and burned or buried. They must not be placed in refuse bins or on rubbish tips
- Take extreme care to prevent domestic animals having access to the bait

Hazard classification and safety precautions

Operator protection U13 [1-4]; U20b [2-4]
Environmental protection E15a [3, 4]; E32a [2-4]
Storage and disposal E26, E29a [2]; E30a [1-4]
Treated seed S06a [1]
Vertebrate/rodent control products V01a, V03a, V04a [1, 3, 4]; V01b, V03b, V04b [2]; V02 [1-4]
Medical advice M03 [2-4]

46 bromoxynil

A contact acting HBN herbicide

Products

1	Alpha Bromolin 225 EC	Makhteshim	225 g/l	LI	08255
2	Alpha Bromotril P	Makhteshim	240 g/l	LI	07099
3	Barclay Mutiny	Barclay	250 g/l	SC	11379
4	Bravado	DAPT	225 g/l	LI	10487
5	Emblem	Nufarm UK	20% w/w	WB	11559
6	Flagon 400 EC	Makhteshim	400 g/l	EC	08875
7	Greencrop Tassle	Greencrop	240 g/l	SC	09659

Uses

- Annual dicotyledons in ***forage maize*** [3, 5, 7]; ***linseed, spring barley, spring oats, spring wheat, winter barley, winter oats, winter wheat*** [1, 4, 6]; ***maize, sweetcorn*** *(off-label)* [2]

Specific Off-Label Approvals (SOLAs)

- ***sweetcorn*** *(OLA 972230) Dec 2008* [2]

Approval information

- Bromoxynil was reviewed in 1995 and approvals for home garden use, and most hand held applications revoked

Efficacy guidance

- Spray when main weed flush has germinated and the largest are at the 4 leaf stage
- Weed control can be enhanced by using a split treatment spraying each application when the weeds are seedling to 2 true leaves. Apply the second treatment before the crop canopy covers the ground [2, 3]

Restrictions

- Maximum number of treatments 1 per crop or yr or maximum total dose equivalent to one full dose treatment
- Do not apply with oils or other adjuvants
- Do not apply using hand-held equipment or at concentrations higher than those recommended

- Do not apply during frosty weather, drought, when soil is waterlogged, when rain expected within 4 h or to crops under any stress
- Take particular care to avoid drift onto neighbouring susceptible crops or open water surfaces

Crop-specific information

- Latest use: before 10 fully expanded leaf stage of crop for maize, sweetcorn; before crop 20 cm tall and before flower buds visible for linseed; before 2nd node detectable (GS 32) for cereals
- Apply to spring sown maize from 2 fully expanded leaves up to 9 fully expanded leaves [2, 3]
- Apply to spring sown linseed from 1 fully expanded leaf to 20 cm tall [1, 4]
- Foliar scorch, which rapidly disappears without affecting growth, will occur if treatment made in hot weather or during rapid growth

Environmental safety

- Dangerous for the environment
- Very toxic to aquatic organisms
- Keep livestock out of treated areas for at least 6 wk after treatment
- High risk to bees. Do not apply to crops in flower or to those in which bees are actively foraging
- Dangerous to fish or other aquatic life. Do not contaminate surface waters or ditches with chemical or used container [2, 4]
- LERAP Category B [5]

Hazard classification and safety precautions

Hazard H03 [1-7]; H04 [4]; H08 [1, 4, 6]; H11 [1, 3, 5-7]
Risk phrases R20 [1, 6]; R22a [1-7]; R22b [1, 4, 6]; R36 [1-4, 6, 7]; R38 [6]; R43 [4, 6]; R50 [1, 5, 6]; R51 [7]; R52 [3]; R58 [1, 3, 5-7]; R63 [1, 2, 5-7]
Operator protection A, C [1-7]; H [1, 4, 6]; U02a [1-4, 7]; U05a, U19a [1-7]; U08, U13, U20a [1-4, 6, 7]; U10 [6]; U11 [2, 6]; U14, U15 [1, 4-6]; U22a [5]; U23a [1-3, 5-7]
Environmental protection E06a [1, 4, 6] (6 wk); E12a, E16a, E32d, E38 [5]; E12f, E31b, E34 [1-4, 6, 7]; E13b [2, 4]; E15a [1, 3, 5-7]; E32e [1, 2, 6]
Consumer protection C02 [1, 4, 6] (6 wk)
Storage and disposal E01, E04, E30a [1-7]; E26 [1, 3, 4, 6, 7]
Medical advice M03 [2-4, 7]; M04 [1, 2, 6]; M05b [6]

47 bromoxynil + diflufenican + ioxynil

A selective contact and translocated herbicide for cereals

Products

Capture	Bayer CropScience	300:50:200 g/l	SC	09982

Uses

- Annual dicotyledons in ***spring barley, spring wheat, triticale, winter barley, winter rye, winter wheat***
- Chickweed in ***spring barley, spring wheat, triticale, winter barley, winter rye, winter wheat***
- Knotgrass in ***spring barley, spring wheat, triticale, winter barley, winter rye, winter wheat***
- Mayweeds in ***spring barley, spring wheat, triticale, winter barley, winter rye, winter wheat***
- Speedwells in ***spring barley, spring wheat, triticale, winter barley, winter rye, winter wheat***

Efficacy guidance

- Best results obtained on small weeds and when competition removed early
- Good spray coverage essential for good activity

Restrictions

- Maximum number of treatments 1 per crop
- Do not treat crops undersown or to be undersown
- Do not treat frosted crops or those that are under stress from any cause
- Do not apply by hand-held equipment or at concentrations higher than those recommended

Crop-specific information

- Latest use: before 2nd node detectable (GS 32)

Following crops guidance

- In the event of crop failure winter wheat may be drilled immediately after normal cultivations. Winter barley may be re-drilled after ploughing

- Land must be ploughed and an interval of 12 wk elapse after treatment before planting spring crops of wheat, barley, oilseed rape, peas, field beans, sugar beet, potatoes, carrots, edible brassicas or onions
- Successive treatments of any products containing diflufenican can lead to soil build-up and inversion ploughing to a depth of 15 cm must precede sowing any following non-cereal crop. Even where ploughing occurs some crops may be damaged

Environmental safety

- Dangerous for the environment
- Toxic to aquatic organisms
- Keep livestock out of treated areas for at least 2 wk after treatment
- Harmful to bees. Do not apply to crops in flower or to those in which bees are actively foraging
- LERAP Category B

Hazard classification and safety precautions

Hazard H03, H11
Risk phrases R22a, R51, R58, R63
Operator protection A, C, H; U05a, U08, U13, U19a, U20b
Environmental protection E06a (2 wk); E12f, E15a, E16a, E31b, E32d, E34, E38
Consumer protection C02 (2 wk)
Storage and disposal E01, E04, E26, E30a
Medical advice M03

48 bromoxynil + ethofumesate + ioxynil

A post-emergence herbicide for new grass leys

Products

Leyclene	Bayer CropScience	50:200:25 g/l	EC	07263

Uses

- Annual dicotyledons in ***grass seed crops***, ***seedling leys***
- Annual meadow grass in ***grass seed crops***, ***seedling leys***

Approval information

- Ethofumesate included in Annex I under EC Directive 91/414

Efficacy guidance

- Best results when weeds small and growing actively in a vigorous crop, soil moist and further rain within 10 d. Mid Oct to end Dec normally suitable
- Annual meadow grass controlled during early crop establishment. Spray weed grasses before fully tillered
- Ash or trash should be burned, buried or removed before spraying

Restrictions

- Maximum number of treatments 2 per yr
- Do not apply by hand-held equipment or at concentrations higher than those recommended
- Do not use on crops under stress, during periods of very dry weather, prolonged frost or waterlogging
- Do not use where clovers or other legumes are valued components of ley
- Do not roll for 7 d before or after spraying
- Do not spray in cold conditions or when heavy rain or frost imminent
- Do not use on soils with more than 10% organic matter
- Do not cut for 14 d after spraying or graze in Jan-Feb after spraying in Oct-Dec

Crop-specific information

- Latest use: 6 wk before cutting or grazing
- Apply to healthy ryegrasses or tall fescue after 2-3 leaf stage, to cocksfoot, timothy and meadow fescue at least 60 d after emergence and after 2-3 leaf stage

Following crops guidance

- Any crop may be sown 5 mth after application following mould board ploughing to at least 15 cm

Environmental safety

- Dangerous for the environment
- Toxic to aquatic organisms

SEE SECTION 3 FOR PRODUCTS ALSO REGISTERED

- Keep livestock out of treated areas for at least 6 wk after treatment
- Harmful to bees. Do not apply to crops in flower or to those in which bees are actively foraging

Hazard classification and safety precautions

Hazard H04, H08, H11

Risk phrases R20, R21, R22a, R38, R51, R58, R63

Operator protection A, C; U05a, U08, U13, U19a, U20b

Environmental protection E06a (6 wk); E12f, E15a, E31b, E32d, E34, E38

Consumer protection C02 (6 wk)

Storage and disposal E01, E04, E30a

Medical advice M03

49 bromoxynil + ioxynil

A contact acting post-emergence HBN herbicide for cereals

Products

1	Alpha Briotril 24/16	Makhteshim	240:160 g/l	EC	04876
2	Alpha Briotril Plus 19/19	Makhteshim	190:190 g/l	EC	04740
3	Biotite 380	DAPT	190:190 g/l	EC	10647
4	Mextrol Biox	Nufarm UK	200:200 g/l	EC	09470
5	Oxytril CM	Bayer CropScience	200:200 g/l	EC	10005

Uses

- Annual dicotyledons in ***spring barley, spring oats, spring wheat, winter barley, winter oats, winter wheat*** [1-5]; ***spring rye, undersown spring cereals, undersown winter cereals*** [4, 5]; ***triticale, winter rye*** [1, 2, 4, 5]; ***undersown barley, undersown oats, undersown rye, undersown triticale, undersown wheat*** [1, 2]

Efficacy guidance

- Best results achieved on young weeds growing actively in a highly competitive crop
- Do not apply during periods of drought or when rain imminent (some labels say 'if likely within 4 or 6 h')
- Recommended for tank mixture with hormone herbicides to extend weed spectrum. See labels for details

Restrictions

- Maximum number of treatments 1 per crop
- Do not spray crops stressed by drought, waterlogging or other factors
- Do not roll or harrow for several days before or after spraying. Number of days specified varies with product. See label for details
- Do not apply by hand-held equipment or at concentrations higher than those recommended

Crop-specific information

- Latest use: before 2nd node detectable stage (GS 31)
- HI (animal consumption) 6 wk
- Apply to winter or spring cereals from 1-2 fully expanded leaf stage, but before second node detectable (GS 32)
- Spray oats in spring when danger of frost past. Do not spray winter oats in autumn
- Apply to undersown cereals pre-sowing or pre-emergence of legume provided cover crop is at correct stage. Only spray trefoil pre-sowing
- On crops undersown with grasses alone or on direct re-seeds apply from 2-leaf stage of grass

Environmental safety

- Dangerous for the environment [1, 2, 4, 5]
- Very toxic to aquatic organisms
- Keep livestock out of treated areas for at least 6 wk after treatment
- Harmful to bees. Do not apply to crops in flower or to those in which bees are actively foraging. Do not apply when flowering weeds are present [1-5]
- LERAP Category B [1-3]

Hazard classification and safety precautions

Hazard H03 [1-5]; H04 [3]; H08 [1-3]; H11 [1, 2, 4, 5]

Risk phrases R20, R38 [1, 2]; R21 [1]; R22a, R22b [1-5]; R36 [1-3]; R43, R66, R67 [4, 5]; R50 [1, 4, 5]; R51 [2]; R58, R63 [1, 2, 4, 5]

Operator protection A, C; U02a, U14, U15, U20a [1-3]; U05a, U08, U13, U19a [1-5]; U10, U11 [1, 2]; U20b [4, 5]; U23a [1, 2, 4, 5]
Environmental protection E06a [1, 2] (6 wk); E07a, E32a, E32d, E38 [4, 5]; E07b [3] (6 wk); E12f, E34 [1-5]; E13b [3, 5]; E15a [1, 2, 4]; E16a, E31b [1-3]; E16b [3]; E32e [1, 2]
Consumer protection C02 [1-3] (6 wk); C02 [4, 5] (14 d)
Storage and disposal E01, E04, E30a [1-5]; E26 [1-3]
Medical advice M03, M05b [1-5]; M04 [1, 2]

50 bromoxynil + ioxynil + mecoprop-P

A post-emergence contact and translocated herbicide

Products

Swipe P	Nufarm UK	56:56:224 g/l	EW	11955

Uses

- Annual dicotyledons in ***durum wheat***, ***ryegrass***, ***spring barley***, ***spring oats***, ***spring wheat***, ***triticale***, ***winter barley***, ***winter oats***, ***winter wheat***

Approval information

- Mecoprop-P included in Annex I under EC Directive 91/414

Efficacy guidance

- Best results when weeds small and growing actively in a strongly competitive crop
- Do not spray in rain or when rain imminent. Control may be reduced by rain within 6 h
- Application to wet crops or weeds may reduce control. Rainfall in the 6 hour period following spraying may reduce weed control
- To achieve optimum control of large over-wintered weeds or in advanced crops increase water volume to aid spray penetration and cover
- Recommended for tank-mixing with approved MCPA-amine for hemp nettle control

Restrictions

- Maximum number of treatments 1 per crop. The total amount of mecoprop-P applied in a single yr must not exceed the maximum total dose approved for any single product for the crop/situation
- Do not treat durum wheat or winter oats in autumn
- Do not spray crops undersown with legumes or use on winter oats or durum wheat in autumn
- Winter and spring crops should not be sprayed in the spring until the risk of frost is over
- Do not spray crops under stress from frost, waterlogging, drought or other causes
- Do not roll within 5 d after spraying
- Do not mix with manganese sulphate
- Do not apply by hand-held equipment or at concentrations higher than those recommended

Crop-specific information

- Latest use:before first node detectable (GS31) for spring oats; before second node detectable (GS32) for durum wheat, spring barley, spring wheat, triticale, winter barley, winter oats, winter wheat
- HI (animal consumption) 6 wk
- Spray cereals from 3 leaves unfolded (GS 13) to before second node detectable (GS 32) for winter sown cereals, spring wheat and spring barley and before first node detectable (GS 31) for spring oats
- Apply to direct sown ryegrass or cereals undersown with ryegrass from 2-3 leaf stage of grass
- Yield of barley may be reduced if frost occurs within 3-4 wk of treatment of low vigour crops on light soils or subject to stress
- Some crop yellowing may follow treatment but yield not normally affected

Environmental safety

- Dangerous for the environment
- Toxic to aquatic organisms
- Keep livestock out of treated areas for at least two weeks following treatment and until poisonous weeds, such as ragwort, have died down and become unpalatable
- Harmful to bees. Do not apply to crops in flower or to those in which bees are actively foraging. Do not apply when flowering weeds are present
- LERAP Category B

- Do not harvest crops for animal consumption for at least 6 wk after last application
- Avoid drift onto neighbouring crops, especially beans, beet, brassicas (including oilseed rape), carrots, lettuce, legumes, tomatoes and other horticultural crops, which are all very susceptible

Hazard classification and safety precautions

Hazard H03, H11
Risk phrases R21, R22a, R51, R58, R63
Operator protection A, C, H, M; U05a, U08, U13, U14, U15, U19a, U20a, U23a
Environmental protection E07a, E12f, E15a, E16a, E31b, E31c, E32d, E34, E38
Consumer protection C02 (6 wk)
Storage and disposal E01, E04, E26, E30a
Medical advice M03

51 bromoxynil + prosulfuron

A contact and residual herbicide mixture for maize and sweetcorn

Products

Jester	Syngenta	60:3 % w/w	WG	08681

Uses

- Annual dicotyledons in ***maize***
- Black bindweed in ***maize***
- Chickweed in ***maize***
- Hemp-nettle in ***maize***
- Knotgrass in ***maize***
- Mayweeds in ***maize***

Approval information

- Prosulfuron included in Annex I under EC Directive 91/414

Efficacy guidance

- Prosulfuron is a member of the ALS-inhibitor group of herbicides and products should be used in a planned Resistance Management strategy. See Section 5 for more information

Restrictions

- Maximum total dose equivalent to one full dose treatment
- Do not treat crops grown for seed production
- Consult before use on crops intended for processing
- Do not use in frosty weather or on crops under stress
- Do not tank mix with organophosphate insecticides or apply in tank mix or sequence with any other sulfonylurea
- Do not apply by knapsack sprayer or in volumes less than those recommended
- Take care to wash out sprayers thoroughly. See label for details

Crop-specific information

- Latest use: before 5 crop leaves unfolded for maize
- Apply post-emergence up to when the crop has four unfolded leaves
- Product must be used with Agral
- Apply in cool conditions, or during the evening, to avoid scorch

Following crops guidance

- Winter or spring cereals, winter or spring beans, spring sown peas or oilseed rape may be sown following normal harvest of a treated crop. Mould-board ploughing to 20 cm is recommended in some cases

Environmental safety

- Dangerous for the environment
- Toxic to aquatic organisms
- LERAP Category B
- Take special care to avoid drift outside the target area

Hazard classification and safety precautions

Hazard H02, H11
Risk phrases R22a, R23, R36, R51, R58, R63
Operator protection A, C; U05a, U11, U15, U20c, U23a

Environmental protection E15a, E16a, E31c, E32d, E38
Storage and disposal E01, E04, E30a

52 bupirimate

A systemic pyrimidine fungicide active against powdery mildew

Products

Nimrod	Makhteshim	250 g/l	EC	10563

Uses

- Powdery mildew in ***apples***, ***begonias***, ***blackcurrants***, ***chrysanthemums***, ***courgettes*** *(outdoor only)*, ***gooseberries***, ***hops***, ***marrows*** *(outdoor only)*, ***pears***, ***protected cucumbers***, ***protected strawberries***, ***protected tomatoes*** *(off-label)*, ***pumpkins*** *(off-label)*, ***raspberries*** *(outdoor only)*, ***roses***, ***squashes*** *(off-label)*, ***strawberries***

Specific Off-Label Approvals (SOLAs)

- ***protected tomatoes*** *(OLA 010910) Dec 2008* [1]
- ***pumpkins***, ***squashes*** *(OLA 010909) Dec 2008* [1]

Efficacy guidance

- On apples during periods that favour disease development lower doses applied weekly give better results than higher rates fortnightly
- Not effective in protected crops against strains of mildew resistant to bupirimate

Restrictions

- Maximum number of treatments or maximum total dose depends on crop and variety (see label for details)

Crop-specific information

- HI depends on crop and variety (see label for details)
- Apply before or at first signs of disease and repeat at 7-14 d intervals. Timing and maximum dose vary with crop. See label for details
- With apples, hops and ornamentals cultivars may vary in sensitivity to spray. See label for details
- If necessary to spray cucurbits in winter or early spring spray a few plants 10-14 d before spraying whole crop to test for likelihood of leaf spotting problem
- On roses some leaf puckering may occur on young soft growth in early spring or under low light intensity. Avoid use of high rates or wetter on such growth
- Never spray flowering begonias (or buds showing colour) as this can scorch petals
- Do not mix with other chemicals for application to begonias, cucumbers or gerberas

Environmental safety

- Dangerous for the environment
- Toxic to aquatic organisms
- Product has negligible effect on *Phytoseiulus* and *Encarsia* and may be used in conjunction with biological control of red spider mite

Hazard classification and safety precautions

Hazard H03, H08, H11
Risk phrases R20, R22b, R38, R42, R51, R58
Operator protection A, C; U05a, U20b
Environmental protection E15a, E31c
Storage and disposal E01, E04, E30a
Medical advice M05b

53 buprofezin

A moulting inhibitor, thiadiazine insecticide for whitefly control

Products

Applaud	Certis	250 g/l	SC	11532

SEE SECTION 3 FOR PRODUCTS ALSO REGISTERED

Uses

- Glasshouse whitefly in ***aubergines, protected cucumbers, protected ornamentals, protected peppers, protected tomatoes***
- Tobacco whitefly in ***aubergines, protected cucumbers, protected ornamentals, protected peppers, protected tomatoes***

Efficacy guidance

- Product has contact, residual and some vapour activity
- Whitefly most susceptible at larval stages but residual effect can also kill nymphs emerging from treated eggs and application to pupae reduces emergence
- Adult whitefly not directly affected. Resistant strains of tobacco whitefly are known and where present control likely to be reduced or ineffective

Restrictions

- Maximum number of treatments 8 per crop for tomatoes and cucumbers; 4 per crop on protected ornamentals; 2 per crop for aubergines and peppers
- Do not apply more than 2 sprays within a 65 d period on tomatoes, or within a 45 d period on cucumbers
- Do not treat *Dieffenbachia* or *Closmoplictrum*
- Do not apply to crops under stress
- Do not leave spray liquid in sprayer for long periods
- Do not apply as fog or mist

Crop-specific information

- HI edible crops 3 d
- In IPM programme apply as single application and allow at least 60 d before re-applying
- In All Chemical programme apply twice at 7-14 d interval and allow at least 60 d before re-applying
- See label for list of ornamentals successfully treated but small scale test advised to check varietal tolerance. This is especially important if spraying flowering ornamentals with buds showing colour

Environmental safety

- Product may be used either in IPM programme in association with *Encarsia formosa* or in All Chemical programme

Hazard classification and safety precautions

Hazard H04
Risk phrases R36, R58
Operator protection U05a, U14, U15, U20c
Environmental protection E15a, E32a, E34
Storage and disposal E01, E04, E30a

54 captan

A protectant dicarboximide fungicide with horticultural uses

Products

1	Alpha Captan 80 WDG	Makhteshim	80% w/w	WG	07096
2	Alpha Captan 83 WP	Makhteshim	83% w/w	WP	04806
3	PP Captan 80-WG	Tomen	80% w/w	WG	11006

Uses

- Black spot in ***roses*** [1-3]
- Botrytis in ***strawberries*** [1, 2]
- Gloeosporium in ***apples*** [3]
- Gloeosporium rot in ***apples*** [1, 2]
- Scab in ***apples, pears*** [1-3]

Restrictions

- Maximum number of treatments 12 per yr on apples and pears as pre-harvest sprays
- Product must not be used as dip or drench on apples or pears [3]
- Do not use on apple cultivars Bramley, Monarch, Winston, King Edward, Spartan, Kidd's Orange or Red Delicious or on pear cultivar D'Anjou
- Do not mix with alkaline materials or oils
- Do not use on fruit for processing
- Powered visor respirator with hood and neck cape must be used when handling concentrate

Crop-specific information

- HI apples, pears 14 d; strawberries 7 d
- For control of scab apply at bud burst and repeat at 10-14 d intervals until danger of scab infection ceased
- For suppression of fruit storage rots apply from late Jul and repeat at 2-3 wk intervals
- For black spot control in roses apply after pruning with 3 further applications at 14 d intervals or spray when spots appear and repeat at 7-10 d intervals
- For grey mould in strawberries spray at first open flower and repeat every 7-10 d [1]
- Do not leave diluted material for more than 2 h. Agitate well before and during spraying

Environmental safety

- Dangerous for the environment
- Very toxic to aquatic organisms

Hazard classification and safety precautions

Hazard H02 [2]; H03 [1]; H04 [3]; H11 [1, 2]
Risk phrases R23, R37, R38 [2]; R36, R43 [1-3]; R40, R50, R58 [1, 2]; R41 [1]
Operator protection A [1-3]; C, K, M [3]; D, E, H [1, 3]; U05a [1-3]; U09a [1]; U11, U19a [1, 2]; U14, U15, U20b [2]; U20c [1, 3]
Environmental protection E13c, E31b, E34 [3]; E15a, E32a, E32d [1, 2]
Storage and disposal E01, E04, E30a

55 carbendazim

A systemic benzimidazole fungicide with curative and protectant activity

Products

1	AgriGuard Pro-Turf	AgriGuard	500 g/l	SC	11837
2	Ashlade Carbendazim Flowable	Nufarm UK	500 g/l	SC	06213
3	Delsene 50 Flo	Nufarm UK	500 g/l	SC	11452
4	Mascot Systemic	Rigby Taylor	500 g/l	SC	11335
5	Nuturf Carbendazim	Nufarm UK	500 g/l	SC	11469
6	Ringer	SumiAgro Amenity	500 g/l	SC	10692
7	Tripart Defensor FL	Tripart	500 g/l	SC	02752
8	Turfclear	Scotts	500 g/l	SC	07506

Uses

- American gooseberry mildew in ***blackcurrants***, ***gooseberries*** [7]
- Big vein in ***lettuce*** *(off-label)*, ***protected lettuce*** *(off-label)* [3]
- Black root rot in ***ornamental plant production*** *(off-label)*, ***protected ornamentals*** *(off-label)* [3]
- Botrytis in ***bedding plants*** *(off-label)*, ***bulbs/corms*** *(off-label)*, ***chrysanthemums*** *(off-label)*, ***pot plants*** *(off-label)*, ***protected celery*** *(off-label)* [3]; ***blackcurrants***, ***gooseberries***, ***outdoor tomatoes***, ***raspberries***, ***strawberries*** [7]
- Cane blight in ***blackberries*** *(off-label)*, ***loganberries*** *(off-label)*, ***protected blackberries*** *(off-label)*, ***protected loganberries*** *(off-label)*, ***protected raspberries*** *(off-label)*, ***protected rubus hybrids*** *(off-label)*, ***raspberries*** *(off-label)*, ***rubus hybrids*** *(off-label)* [3]
- Cane spot in ***raspberries*** [7]
- Celery leaf spot in ***celery*** [7]
- Chocolate spot in ***broad beans***, ***spring field beans***, ***winter field beans*** [7]
- Didymella stem rot in ***protected tomatoes*** *(off-label)* [3]
- Dollar spot in ***managed amenity turf*** [1, 2, 4-6, 8]
- Eye rot in ***apples*** [7]
- Eyespot in ***spring barley***, ***winter barley***, ***winter wheat*** [2, 3]
- Fusarium in ***bulbs/corms*** *(off-label)*, ***freesias*** *(off-label)* [3]
- Fusarium patch in ***managed amenity turf*** [1, 2, 4-6, 8]
- Gloeosporium in ***apples*** [7]
- Light leaf spot in ***spring oilseed rape***, ***winter oilseed rape*** [2, 3, 7]
- Penicillium rot in ***bulbs/corms*** *(off-label)* [3]
- Powdery mildew in ***apples*** [7]; ***bedding plants*** *(off-label)*, ***chrysanthemums*** *(off-label)*, ***pot plants*** *(off-label)* [3]
- Rhynchosporium in ***spring barley***, ***winter barley*** [2, 3, 7]

SEE SECTION 3 FOR PRODUCTS ALSO REGISTERED

- Root diseases in ***inert substrate cucumbers*** *(off-label)* [3]
- Scab in ***apples*** [7]; ***pears*** *(off-label)* [3]
- Sclerotinia in ***bulbs/corms*** *(off-label)* [3]
- Spur blight in ***raspberries*** [7]
- Stagonospora in ***bulbs/corms*** *(off-label)* [3]
- Storage rots in ***pears*** *(off-label)* [3]
- Trichoderma in ***mushroom compost*** *(off-label - spawn treatment)* [3]
- Wormcast formation in ***managed amenity turf*** [1, 4-6, 8]

Specific Off-Label Approvals (SOLAs)

- ***bedding plants, bulbs/corms, chrysanthemums, freesias, ornamental plant production, pot plants, protected ornamentals*** *(OLA 041004) Dec 2008* [3]
- ***blackberries, loganberries, protected blackberries, protected loganberries, protected raspberries, protected rubus hybrids, raspberries, rubus hybrids*** *(OLA 041009) Jul 2008* [3]
- ***inert substrate cucumbers*** *(OLA 041005) Dec 2008* [3]
- ***lettuce, protected lettuce*** *(OLA 041011) Dec 2008* [3]
- ***mushroom compost*** *(spawn treatment) (OLA 041007) Dec 2008* [3]
- ***pears*** *(OLA 041008) Dec 2008* [3]
- ***protected celery*** *(OLA 041006) Dec 2008* [3]
- ***protected tomatoes*** *(OLA 041010) Dec 2008* [3]

Approval information

- Following implementation of Directive 98/82/EC, approval for use of carbendazim on numerous crops was revoked in 1999. UK approvals for use of carbendazim on strawberries revoked in 2001 as a result of implementation of the MRL Directives

Efficacy guidance

- Products vary in the diseases listed as controlled for several crops. Labels must be consulted for full details and for rates and timings
- Mostly applied as spray or drench. Spray treatments normally applied at first sign of disease and repeated after 1 mth if required
- Apply as drench rather than spray where red spider mite predators are being used
- Do not apply during drought conditions. Rain or irrigation after treatment may improve control
- Leave 4 mth intervals where used only for worm cast control [1, 4-6, 8]
- To delay appearance of resistant strains alternate treatment with non-MBC fungicide. Eyespot in cereals and *Botrytis cinerea* in many crops are now widely resistant

Restrictions

- Maximum number of treatments (including applications of any product containing benomyl, carbendazim or thiophanate-methyl) varies with crop treated and product used - see labels for details
- Do not treat crops or turf suffering from drought or other physical or chemical stress
- Consult processors before using on crops for processing
- Not compatible with alkaline products such as lime sulphur

Crop-specific information

- Latest use varies with crop and product used. See labels for details
- HI 2 d for inert substrate cucumbers, protected tomatoes; 7-14 d (depends on dose) for apples, pears; 14 d for mushrooms, protected celery; 21 d for oilseed rape, dwarf beans, lettuce, 6 mth for cane fruit
- On turf apply as preventative treatment in spring or autumn during periods of high disease risk
- Apply as a drench to control soil-borne diseases in cucumbers and tomatoes and as a pre-planting dip treatment for bulbs

Environmental safety

- Dangerous for the environment [1, 7, 8]
- Very toxic to aquatic organisms [8]
- Harmful to aquatic organisms [7]
- Harmful to fish or other aquatic life. Do not contaminate surface waters or ditches with chemical or used container [4-6]
- After use dipping suspension must not be discharged directly into ditches or drains

Hazard classification and safety precautions

Hazard H03 [1-8]; H11 [1, 7, 8]

Risk phrases R22a [1, 7]; R50 [8]; R52 [7]; R58 [7, 8]; R68 [2-6, 8]

Operator protection A, C, H, M; U19a [8]; U20a [1]; U20b [2, 3, 5, 7]; U20c [4, 6, 8]
Environmental protection E13c [4-6]; E15a [1-3, 7, 8]; E31a [4]; E31b [1, 6-8]; E32a [2, 3, 5]; E32d [4, 8]; E32e, E34 [6]; E38 [8]
Storage and disposal E01, E04 [2, 3, 5, 7]; E24 [8]; E26 [1, 4, 6, 8]; E29a [1, 7]; E29b [6]; E30a [1-8]

56 carbendazim + epoxiconazole

A systemic fungicide mixture for managed amenity turf

Products

Capricorn	Bayer Environ.	125:125 g/l	SC	10035

Uses
- Fusarium patch in ***managed amenity turf***

Efficacy guidance
- Treat at the first sign of disease and repeat as necessary
- Apply after cutting and delay further mowing for at least 48 h to allow systemic distribution within the plant
- Only moderate control of Fusarium patch claimed

Restrictions
- Maximum number of treatments 2 per yr
- Do not apply during drought conditions or to frozen turf
- Use only on established turf

Environmental safety
- Harmful to fish or other aquatic life. Do not contaminate surface waters or ditches with chemical or used container
- LERAP Category B

Hazard classification and safety precautions
Hazard H04
Risk phrases R43
Operator protection A, C, H, M; U04a, U05a, U09b, U20c
Environmental protection E13c, E16a, E16b, E31b, E34
Storage and disposal E01, E04, E26, E30a

57 carbendazim + flusilazole

A broad-spectrum systemic and protectant fungicide for cereals

Products

1	Contrast	DuPont	125:250 g/l	SC	06150
2	Punch C	DuPont	125:250 g/l	SC	06801

Uses
- Brown rust in ***spring barley***, ***spring wheat***, ***winter barley***, ***winter wheat***
- Canker in ***spring oilseed rape*** *(reduction)*, ***winter oilseed rape*** *(reduction)*
- Eyespot in ***spring barley***, ***spring wheat***, ***winter barley***, ***winter wheat***
- Light leaf spot in ***spring oilseed rape***, ***winter oilseed rape***
- Phoma leaf spot in ***spring oilseed rape***, ***winter oilseed rape***
- Powdery mildew in ***spring barley***, ***spring wheat***, ***sugar beet***, ***winter barley***, ***winter wheat***
- Rust in ***sugar beet***
- Septoria in ***spring wheat***, ***winter wheat***
- Yellow rust in ***spring barley***, ***spring wheat***, ***winter barley***, ***winter wheat***

Efficacy guidance
- Apply at early stage of disease development or in routine preventive programme
- Most effective timing of treatment on cereals varies with disease. See label for details
- Higher rate active against both MBC-sensitive and MBC-resistant eyespot
- Rain occurring within 2-3 h of spraying may reduce effectiveness
- To prevent build-up of resistant strains of cereal mildew tank mix with approved morpholine fungicide

SEE SECTION 3 FOR PRODUCTS ALSO REGISTERED

Restrictions

- Maximum number of treatments 1 per crop on sugar beet; 2 per crop on cereals and oilseed rape
- Do not apply to crops under stress or during frosty weather

Crop-specific information

- Latest use on cereals varies with crop and dose used - see label for details; before first flower opened stage for oilseed rape
- HI 7 wk for sugar beet
- Treat oilseed rape in autumn when leaf lesions first appear and spring from the start of stem extension when disease appears

Environmental safety

- Dangerous for the environment
- Very toxic to aquatic organisms

Hazard classification and safety precautions

Hazard H02, H11
Risk phrases R36, R40, R46, R58, R60, R61 [1, 2]; R50 [1]; R51 [2]
Operator protection A, C [1, 2]; H, M [1]; U05a, U11, U19a, U20b
Environmental protection E15a, E31b, E32d, E34, E38
Storage and disposal E01, E04, E26, E30a
Medical advice M04

58 carbendazim + iprodione

A systemic and contact fungicide mixture

Products

Vitesse	Bayer Environ.	87.5:175 g/l	SC	10042

Uses

- Anthracnose in ***managed amenity turf***
- Fusarium patch in ***managed amenity turf***
- Pink patch in ***managed amenity turf***
- Red thread in ***managed amenity turf***
- Timothy leaf spot in ***managed amenity turf***

Approval information

- Iprodione included in Annex I under EC Directive 91/414

Efficacy guidance

- Maximum efficacy against Anthracnose in turf achieved by treatment at early disease development stage. Curative treatments for well-established Anthracnose are not recommended

Crop-specific information

- May be used on turf all year round, but is best suited for spring or late summer/early autumn application
- Where grass is being mown, apply after mowing. Delay further mowing for at least 48 h after treatment

Environmental safety

- Harmful to fish or other aquatic life. Do not contaminate surface waters or ditches with chemical or used container

Hazard classification and safety precautions

Operator protection A, C, H, M; U20c
Environmental protection E13c, E31b
Storage and disposal E26, E30a

59 carbetamide

A residual pre- and post-emergence carbamate herbicide for a range of field crops

Products

1	Carbetamex	Makhteshim	70% w/w	WP	11150
2	Crawler	Makhteshim	60% w/w	WG	11840

Uses

- Annual grasses in ***cabbage seed crops, collards, fodder rape seed crops, kale seed crops, lucerne, red clover, sainfoin, spring cabbage, sugar beet seed crops, swede seed crops, turnip seed crops, white clover, winter field beans, winter oilseed rape***
- Some annual dicotyledons in ***cabbage seed crops, collards, fodder rape seed crops, kale seed crops, lucerne, red clover, sainfoin, spring cabbage, sugar beet seed crops, swede seed crops, turnip seed crops, white clover, winter field beans, winter oilseed rape***
- Volunteer cereals in ***cabbage seed crops, collards, fodder rape seed crops, kale seed crops, lucerne, red clover, sainfoin, spring cabbage, sugar beet seed crops, swede seed crops, turnip seed crops, white clover, winter field beans, winter oilseed rape***

Efficacy guidance

- Best results pre- or early post-emergence of weeds under cool, moist conditions. Adequate soil moisture is essential
- Dicotyledons controlled include chickweed, cleavers and speedwell
- Weed growth stops rapidly after treatment but full effects may take 6-8 wk to develop
- Various tank mixes are recommended to broaden the weed spectrum. See label for details
- Always follow WRAG guidelines for preventing and managing herbicide resistant weeds. Section 5 for more information

Restrictions

- Maximum number of treatments 1 per crop for all crops
- Do not treat any crop on waterlogged soil
- Do not use on soils with more than 10% organic matter as residual activity is impaired
- Do not apply during prolonged periods of cold weather when weeds are dormant

Crop-specific information

- HI 6 wk for all crops
- Apply to brassicas from mid-Oct to end-Feb provided crop has at least 4 true leaves (spring cabbage, spring greens), 3-4 true leaves (seed crops, oilseed rape)
- Apply to established lucerne or sainfoin from Nov to end-Feb
- Apply to established red or white clover from Feb to mid-Mar

Following crops guidance

- Succeeding crops may be sown 2 wk after treatment for brassicas, field beans, 8 wk after treatment for peas, runner beans, 16 wk after treatment for cereals, maize
- Ploughing is not necessary before sowing subsequent crops

Environmental safety

- Do not graze crops for at least 6 wk after treatment

Hazard classification and safety precautions

Operator protection U19a [1]; U20c [1, 2]
Environmental protection E15a, E32a
Storage and disposal E30a

60 carbetamide + diflufenican + oxadiazon

A residual herbicide mixture for amenity vegetation

Products

Helmsman	Bayer Environ.	1:0.1:2% w/w	GR	09934

Uses

- Annual dicotyledons in ***amenity trees and shrubs, amenity vegetation***
- Annual meadow grass in ***amenity trees and shrubs, amenity vegetation***
- Perennial dicotyledons in ***amenity trees and shrubs, amenity vegetation***

Efficacy guidance

- Apply to soil surface in late winter or early spring using hand shaker or gravity fed applicator. Uniform application is essential for satisfactory weed control
- Chemical absorbed by emerging shoots of susceptible weeds as they grow through the treated soil
- Adequate moisture is essential. In dry conditions irrigate after application. Effectiveness may be reduced if drought conditions follow application

- Do not disturb treated soil by hoeing otherwise weed control may be impaired
- In some cases heavy rain shortly after application may splash granules onto the lower leaves of shrubs which may result in transient damage

Restrictions

- Maximum number of treatments 1 per yr
- Do not apply to plants grown in containers
- Do not apply to areas under-planted with bulbs or annual plants
- Not to be used on food crops

Environmental safety

- Harmful to aquatic organisms
- Do not apply on slopes where heavy rain soon after application could cause surface run-off

Hazard classification and safety precautions

Hazard H04
Risk phrases R36, R52, R58
Operator protection A; U05a, U19a, U20b
Environmental protection E15a, E32a, E32d
Consumer protection C01
Storage and disposal E01, E04, E30a

61 carbon dioxide (commodity substance)

A gas for the control of trapped rodents and other vertebrates

Products

carbon dioxide	various	99.9%	GA

Uses

- Birds in ***traps***
- Mice in ***traps***
- Rats in ***traps***

Approval information

- Approval for the use of carbon dioxide as a commodity substance was granted on 8 October 1993 by Ministers under regulation 5 of the Control of Pesticides Regulations 1986
- Only to be used where a licence has been issued in accordance with Section 16(1) of the Wildlife and Countryside Act 1981

Efficacy guidance

- Use to destroy trapped rodent pests
- Use to control birds covered by general licences issued by the Agriculture and Environment Departments under Section 16(1) of the Wildlife and Countryside Act (1981) for the control of opportunistic bird species, where birds have been trapped or stupefied with alphachloralose/seconal

Restrictions

- Operators must wear self-contained breathing apparatus when carbon dioxide levels are greater than 0.5% v/v
- Operators must be suitably trained and competent

Environmental safety

- Unprotected persons and non-target animals must be excluded from the treatment enclosures and surrounding areas unless the carbon dioxide levels are below 0.5% v/v

Hazard classification and safety precautions

Operator protection G

62 carbosulfan

A systemic carbamate insecticide for control of soil and stem pests

Products

1	Marshal Soil Insecticide suSCon CR granules	Fargro	10% w/w	GR	06978
2	Marshal Soil Insecticide suSCon CR Sachets	Fargro	10% w/w	WB	11754
3	Posse 10G	Belchim	10% w/w	GR	11640

Uses

- Aphids in ***broccoli, brussels sprouts, cabbages, calabrese, carrots, cauliflowers, collards, fodder beet, kale, mangels, parsnips, sugar beet*** [3]
- Cabbage root fly in ***broccoli, brussels sprouts, cabbages, calabrese, cauliflowers, collards, kale*** [3]
- Cabbage stem weevil in ***broccoli, brussels sprouts, cabbages, calabrese, cauliflowers, collards, kale*** [3]
- Flea beetle in ***broccoli, brussels sprouts, cabbages, calabrese, cauliflowers, collards, fodder beet, kale, mangels, sugar beet*** [3]
- Free-living nematodes in ***carrots, fodder beet, mangels, parsnips, sugar beet*** [3]
- Large pine weevil in ***forest*** [1, 2]; ***forest nurseries*** *(off-label)* [1]
- Mangold fly in ***fodder beet, mangels, sugar beet*** [3]
- Millipedes in ***fodder beet, mangels, sugar beet*** [3]
- Pygmy mangold beetle in ***fodder beet, mangels, sugar beet*** [3]
- Springtails in ***fodder beet, mangels, sugar beet*** [3]
- Symphylids in ***fodder beet, mangels, sugar beet*** [3]
- Wireworm in ***fodder beet, mangels, sugar beet*** [3]

Specific Off-Label Approvals (SOLAs)

- ***forest nurseries*** *(OLA 992383) Dec 2008* [1]

Efficacy guidance

- At recommended rates seed of drilled crops is not damaged by contact with product
- Products formulated as controlled release granules to give pine weevil control throughout the 2-year establishment phase [1, 2]

Restrictions

- This product contains an anticholinesterase carbamate compound. Do not use if under medical advice not to work with such compounds
- Maximum number of treatments 1 per crop or tree

Crop-specific information

- Latest use: before or at planting for forestry uses
- HI 100 d for beet crops, carrots, parsnips; 12 wk for broccoli, cabbages, calabrese, cauliflowers; 14 wk for Brussels sprouts; 23 wk for collards
- Apply to drilled crops with suitable granule applicator feeding directly into seed furrow or immediately behind drill coulter (behind seed drill boot for brassicas) or use bow-wave technique [3]
- See label for details of suitable applicators and settings. Correct calibration is essential [3]
- Where used in forestry, granules should be placed in the planting hole by hand or metered applicator before or after placing the tree. Sachets should be placed in the planting hole by hand before placement of the tree
- Safe to Douglas Fir and Sitka Spruce. Other conifers may vary in their sensitivity [1, 2]
- Forest trees may take 10-15 d to achieve full protection. An additional pre-planting insecticide dip or spray is advised, but post-planting sprays during the 2-yr establishment should not be necessary [1, 2]

Environmental safety

- Dangerous for the environment [2, 3]
- Very toxic to aquatic organisms
- Dangerous to game, wild birds and animals [1-3]
- Keep in original container, tightly closed, in a safe place, under lock and key

Hazard classification and safety precautions

Hazard H03 [2]; H04 [3]; H11 [2, 3]

Risk phrases R20, R51 [2]; R43, R50 [3]; R58 [2, 3]

Operator protection A [1, 2]; B, K, M [1, 3]; C, D, E [1]; C [3] (or D+E); H; U02a, U04a, U05a, U13, U20a [1-3]; U09a, U14, U19a [3]; U10 [1, 2]; U11 [1]; U12, U22b [2]

Environmental protection E10a [1-3]; E13b [1]; E15a, E32d, E38 [2, 3]; E32a, E34 [3]

Storage and disposal E01 [1-3]; E04 [1, 3]; E30a [1, 2]; E30b [3]

Medical advice M02 [1-3]; M03, M05a [3]

63 carboxin

A carboxamide fungicide available only in mixtures

64 carboxin + thiram

A fungicide seed dressing for cereals

Products

Anchor	Crompton	200:200 g/l	FS	08684

Uses

- Bunt in ***spring wheat*** *(seed treatment)*, ***winter wheat*** *(seed treatment)*
- Covered smut in ***spring barley*** *(seed treatment)*, ***spring oats*** *(seed treatment)*, ***winter barley*** *(seed treatment)*, ***winter oats*** *(seed treatment)*
- Fusarium foot rot and seedling blight in ***spring barley*** *(seed treatment)*, ***spring oats*** *(seed treatment)*, ***spring rye*** *(seed treatment)*, ***spring wheat*** *(seed treatment)*, ***triticale*** *(seed treatment)*, ***winter barley*** *(seed treatment)*, ***winter oats*** *(seed treatment)*, ***winter rye*** *(seed treatment)*, ***winter wheat*** *(seed treatment)*
- Leaf stripe in ***spring barley*** *(seed treatment)*, ***winter barley*** *(seed treatment - reduction)*
- Loose smut in ***spring barley*** *(seed treatment - reduction)*, ***spring oats*** *(seed treatment - reduction)*, ***winter barley*** *(seed treatment - reduction)*, ***winter oats*** *(seed treatment - reduction)*
- Net blotch in ***spring barley*** *(seed treatment)*
- Septoria seedling blight in ***spring wheat*** *(seed treatment)*, ***winter wheat*** *(seed treatment)*

Efficacy guidance

- Apply through suitable liquid flowable seed treating equipment of the batch treatment or continuous flow type where a secondary mixing auger is fitted
- Drill flow may be affected by treatment. Always re-calibrate seed drill before use

Restrictions

- Maximum number of treatments 1 per batch of seed
- Do not treat seed with moisture content above 16%
- Do not apply to cracked, split or sprouted seed
- Do not store treated seed from one season to the next

Crop-specific information

- Latest use: pre-drilling

Environmental safety

- Dangerous for the environment
- Very toxic to aquatic organisms
- Do not use treated seed as food or feed
- Treated seed harmful to game and wildlife

Hazard classification and safety precautions

Hazard H03, H11

Risk phrases R22a, R48, R50, R58

Operator protection A, D, H; U05a, U09a, U20c

Environmental protection E15a, E31a, E34, E38

Storage and disposal E01, E04, E26, E27, E30a

Treated seed S01, S02, S04b, S05, S06a, S07

Medical advice M03

65 carfentrazone-ethyl

A triazolinone contact herbicide for cereals and potato haulm destruction

Products

1	Aurora 50 WG	Belchim	50% w/w	WG	11613
2	Spotlight 24 EC	Belchim	240 g/l	EC	11617

Uses

- Cleavers in ***durum wheat***, ***spring barley***, ***spring oats***, ***spring wheat***, ***triticale***, ***winter barley***, ***winter oats***, ***winter wheat*** [1]
- Haulm destruction in ***seed potatoes***, ***ware potatoes*** [2]
- Ivy-leaved speedwell in ***durum wheat***, ***spring barley***, ***spring oats***, ***spring wheat***, ***triticale***, ***winter barley***, ***winter oats***, ***winter wheat*** [1]

Approval information

- Carfentrazone-ethyl included in Annex I under EC Directive 91/414

Efficacy guidance

- Best weed control results achieved from good spray cover applied to small actively growing weeds [1]
- Carfentrazone-ethyl acts by contact only; see label for optimum timing on specified weeds. Weeds emerging after application will not be controlled [1]
- For weed control use as two spray programme with one application in autumn and one in spring [1]
- Efficacy of haulm destruction will be reduced where flailed haulm covers the stems at application [2]
- For potato crops with very dense vigorous haulm or where regrowth occurs following a single application a second application may be necessary to achieve satisfactory desiccation. A minimum interval between applications of 7 d should be observed to achieve optimum performance [2]

Restrictions

- Maximum number of treatments 2 per crop for cereals (1 in Autumn and 1 in Spring) [1]; 2 per crop for ware potato haulm destruction [2]; 1 per crop for seed potato haulm destruction [2]
- Do not treat crops cereal under stress from drought, waterlogging, cold, pests, diseases, nutrient or lime deficiency or any factors reducing plant growth [1]
- Do not treat cereals undersown with clover or other legumes [1]
- Allow at least 2-3 wk between application and lifting potatoes to allow skins to set if potatoes are to be stored [2]
- Follow label instructions for sprayer cleaning

Crop-specific information

- Latest use: before 3rd node detectable (GS 33) on cereals [1]
- HI 7 d for potatoes [2]
- For weed control in cereals treat from 2 leaf stage [1]
- For potato haulm destruction ware crops should be treated at the onset of senescence; seed crops should be flailed when tubers have reached the desired size and then treated once the flailed haulm is clear of the stems that remain [2]

Following crops guidance

- No restrictions apply on the planting of succeeding crops 1 mth after application to potatoes for haulm destruction or 3 mth after application to cereals for weed control
- In the event of failure of a treated cereal crop, all cereals, ryegrass, maize, oilseed rape, peas, sunflowers, *Phacelia*, vetches, carrots or onions may be planted within 1 mth of treatment

Environmental safety

- Dangerous for the environment
- Very toxic to aquatic organisms
- Some non-target crops are sensitive. Avoid drift onto broad-leaved plants outside the treated area, or onto ponds waterways or ditches

Hazard classification and safety precautions

Hazard H03 [2]; H04 [1]; H11 [1, 2]

Risk phrases R22b [2]; R43 [1]; R50, R58 [1, 2]

Operator protection A, H [1]; U05a, U08, U13, U14 [1]
Environmental protection E15a, E31b, E32d, E34, E38
Storage and disposal E01, E04, E30a
Medical advice M05a [1]; M05b [2]

66 carfentrazone-ethyl + mecoprop-P

A foliar applied herbicide for cereals

Products

Platform S	Belchim	1.5:60% w/w	WG	11611

Uses

- Charlock in ***spring barley, spring oats, spring wheat, winter barley, winter oats, winter wheat***
- Chickweed in ***spring barley, spring oats, spring wheat, winter barley, winter oats, winter wheat***
- Cleavers in ***spring barley, spring oats, spring wheat, winter barley, winter oats, winter wheat***
- Red dead-nettle in ***spring barley, spring oats, spring wheat, winter barley, winter oats, winter wheat***
- Speedwells in ***spring barley, spring oats, spring wheat, winter barley, winter oats, winter wheat***

Approval information

- Carfentrazone-ethyl and mecoprop-P included in Annex I under EC Directive 91/414

Efficacy guidance

- Best results obtained when weeds have germinated and growing vigorously in warm moist conditions
- Treatment of large weeds and poor spray coverage may result in reduced weed control

Restrictions

- Maximum number of treatments 2 per crop. The total amount of mecoprop-P applied in a single yr must not exceed the maximum total dose approved for any single product for the crop/situation
- Do not treat crops suffering from stress from any cause
- Do not treat crops undersown or to be undersown

Crop-specific information

- Latest use: before 3rd node detectable (GS 33)
- Can be used on all varieties of wheat and barley in autumn or spring from the beginning of tillering
- Early sown crops may be prone to damage if treated after period of rapid growth in autumn

Following crops guidance

- In the event of crop failure, any cereal, maize, oilseed rape, peas, vetches or sunflowers may be sown 1 mth after a spring treatment. Any crop may be planted 3 mth after treatment

Environmental safety

- Dangerous for the environment
- Very toxic to aquatic organisms
- Keep livestock out of treated areas for at least two weeks following treatment and until poisonous weeds, such as ragwort, have died down and become unpalatable

Hazard classification and safety precautions

Hazard H03, H11
Risk phrases R22a, R41, R43, R50, R58
Operator protection A, C, H, M; U05a, U08, U11, U13, U14, U20b
Environmental protection E07a, E15a, E31b, E32d, E34, E38
Storage and disposal E01, E04, E30a
Medical advice M05a

67 carfentrazone-ethyl + metsulfuron-methyl

A foliar applied herbicide mixture for cereals

Products

Ally Express	DuPont	40:10% w/w	WG	08640

Uses

- Annual dicotyledons in ***durum wheat***, ***spring barley***, ***spring oats***, ***spring wheat***, ***triticale***, ***winter barley***, ***winter oats***, ***winter wheat***

Approval information

- Carfentrazone-ethyl and metsulfuron-methyl included in Annex I under EC Directive 91/414

Efficacy guidance

- Best results achieved from applications made in good growing conditions
- Good spray cover of weeds must be obtained
- Growth of weeds is inhibited within hours of treatment but the time taken for visible colour changes to appear will vary according to species and weather
- Product has short residual life in soil. Under normal moisture conditions susceptible weeds germinating soon after treatment will be controlled
- Metsulfuron-methyl is a member of the ALS-inhibitor group of herbicides and products should be used in a planned Resistance Management strategy. See Section 5 for more information

Restrictions

- Maximum number of treatments 1 per crop
- Product must only be used after 1 Feb
- Do not use on any crop suffering stress from drought, waterlogging, cold, pest or disease attack, or nutrient deficiency
- Do not use on crops undersown with grass or legumes or on any broad-leaved crop
- Do not apply within 7 d of rolling
- Do not apply to a crop already treated with any sulfonylurea product except those containing flupyrsulfuron alone or in mixture with carfentrazone-ethyl or thifensulfuron-methyl

Crop-specific information

- Latest use: before 3rd node detectable (GS 33)
- Can be used on all soil types
- Apply in the spring from the 3-leaf stage on all crops
- Slight necrotic spotting of crops can occur under certain crop and soil conditions. Recovery is rapid and there is no effect on grain yield or quality

Following crops guidance

- Only cereals, oilseed rape, field beans or grass may be sown as a following crop in the same calendar yr. Any crop may follow in the next spring
- In the event of failure of a treated crop, only winter wheat may be sown 1-3 mth later after ploughing and cultivating to at least 15 cm

Environmental safety

- Extremely dangerous to fish or other aquatic life. Do not contaminate surface waters or ditches with chemical or used container
- LERAP Category B
- Take extreme care to avoid drift onto broad-leaved plants outside the target area or onto ponds, waterways or ditches, or onto land intended for cropping
- Spraying equipment should not be drained or flushed onto land planted, or to be planted, with trees or crops other than cereals and should be thoroughly cleansed after use - see label for instructions

Hazard classification and safety precautions

Hazard H04, H11
Risk phrases R43, R50, R58
Operator protection A, H; U05a, U08, U14, U19a, U20b
Environmental protection E16a, E16b, E32a
Storage and disposal E01, E04, E30a

SEE SECTION 3 FOR PRODUCTS ALSO REGISTERED

68 chloridazon

A residual pyridazinone herbicide for beet crops

Products

1	Better DF	Sipcam	65% w/w	SG	06250
2	Better Flowable	Sipcam	430 g/l	SC	04924
3	Burex 430 SC	Interfarm	430 g/l	SC	09494
4	Pyramin DF	BASF	65% w/w	WG	03438
5	Sculptor	Sipcam	430 g/l	SC	08836
6	Takron	BASF	430 g/l	ZZ	11627
7	Tripart Gladiator 2	Tripart	430 g/l	SC	06618

Uses

- Annual dicotyledons in ***bulb onions*** *(off-label)* [4]; ***fodder beet***, ***mangels***, ***sugar beet*** [1-7]; ***leeks*** *(off-label)*, ***salad onions*** *(off-label)* [1, 4]; ***onions*** *(off-label)* [1]
- Annual meadow grass in ***bulb onions*** *(off-label)* [4]; ***fodder beet***, ***mangels***, ***sugar beet*** [1, 2, 4-7]; ***leeks*** *(off-label)*, ***salad onions*** *(off-label)* [1, 4]; ***onions*** *(off-label)* [1]

Specific Off-Label Approvals (SOLAs)

- ***bulb onions***, ***leeks***, ***salad onions*** *(OLA 970732) Dec 2008* [4]
- ***leeks***, ***onions***, ***salad onions*** *(OLA 012570) Dec 2008* [1]

Efficacy guidance

- Absorbed by roots of germinating weeds and best results achieved pre-emergence of weeds or crop when soil moist and adequate rain falls after application
- Application rate depends on soil type. See label for details

Restrictions

- Maximum number of treatments generally 1 per crop (pre-emergence) for fodder beet and mangels; 1 (pre-emergence) + 3 (post-emergence) per crop for sugar beet, but labels vary slightly. Check labels for maximum total dose for the crop to be treated
- Maximum total dose for onions and leeks equivalent to one full dose recommended for these crops [4]
- Do not use on Coarse Sands, Sands or Fine Sands or where organic matter exceeds 5%

Crop-specific information

- Latest use: pre-emergence for fodder beet and mangels; normally before leaves of crop meet in row for sugar beet, but labels vary slightly; up to and including second true leaf stage for onions and leeks [4]
- Where used pre-emergence spray as soon as possible after drilling in mid-Mar to mid-Apr on fine, firm, clod-free seedbed
- Where crop drilled after mid-Apr or soil dry apply pre-drilling and incorporate to 2.5 cm immediately afterwards
- Various tank mixes recommended on sugar beet for pre- and post-emergence use and as repeated low dose treatments. See label for details
- Crop vigour may be reduced by treatment of crops growing under unfavourable conditions including poor tilth, drilling at incorrect depth, soil capping, physical damage, pest or disease damage, excess seed dressing, trace-element deficiency or a sudden rise in temperature after a cold spell

Following crops guidance

- Winter cereals or any spring sown crop may follow a treated beet crop harvested at the normal time and after ploughing
- In the event of failure of a treated crop only a beet crop or maize may be drilled, after cultivation

Environmental safety

- Dangerous for the environment [3, 4, 6, 7]
- Very toxic to aquatic organisms [3, 4, 6]
- Harmful to fish or other aquatic life. Do not contaminate surface waters or ditches with chemical or used container [1, 2, 5]

Hazard classification and safety precautions

Hazard H03 [4]; H04 [3, 5-7]; H11 [3, 4, 6, 7]

Risk phrases R22a [4]; R36, R38 [5, 7]; R37 [7]; R43 [3, 6, 7]; R50, R58 [3, 4, 6]

Operator protection A [3, 5-7]; U05a [4-7]; U08 [1-7]; U14 [3, 5-7]; U15 [5, 7]; U19a [5-7]; U20a [2, 3]; U20b [1, 4-7]
Environmental protection E13c [1, 2, 5]; E15a [3, 4, 6, 7]; E31b [3, 5-7]; E32a [1, 2, 4]; E32d, E38 [4, 6]; E32e, E34 [3]
Storage and disposal E01, E04 [3-7]; E26 [3, 5, 7]; E30a [1-7]
Medical advice M05a [4, 6]

69 chloridazon + chlorpropham + metamitron

A residual herbicide mixture for beet crops

Products

Newtron	Nufarm UK	83:42:333 g/l	SE	11208

Uses
- Annual dicotyledons in ***fodder beet*, *mangels*, *sugar beet***
- Annual meadow grass in ***fodder beet*, *mangels*, *sugar beet***

Efficacy guidance
- Apply as part of a spray programme. For best results a full programme of pre- and post-emergence treatments is required
- Pre-emergence weed control will be reduced under very dry conditions but subsequent rain may activate the chemical
- Best post-emergence performance requires tank mxture with phenmedipham. See label for details
- Correct post-emergence timing important for good control. First treatment should be made when weeds at early cotyledon stage and subsequent applications made when new weed flushes appear

Restrictions
- Subject to the maximum total dose, maximum number of treatments 1 pre-emergence followed by 3 post-emergence, or 5 post-emergence treatments unless another product has been used pre-emergence, in which case 4 post-emergence sprays may be applied
- Must be tank mixed when used post-emergence. See label
- Do not roll or harrow treated soil. Spray after rolling

Crop-specific information
- Latest use: before crop leaves meet between rows

Following crops guidance
- Only sugar beet, fodder beet or mangels should be sown within 4 mth of last application
- After harvesting treated crops land should be mould-board ploughed to 15 cm and any spring crop may then be drilled

Environmental safety
- Dangerous for the environment
- Very toxic to aquatic organisms

Hazard classification and safety precautions
Hazard H03, H11
Risk phrases R22a, R36, R38, R43, R50, R58
Operator protection A, C; U05a, U08, U11, U14, U15, U19a, U20a
Environmental protection E15a, E31c, E32e, E34
Storage and disposal E01, E04, E30a

70 chloridazon + ethofumesate

A contact and residual herbicide for beet crops

Products

1	Gremlin	Sipcam	285:176 g/l	SC	09468
2	Magnum	BASF	275:170 g/l	SC	11727

Uses

- Annual dicotyledons in ***fodder beet*** [2]; ***sugar beet*** [1, 2]
- Annual meadow grass in ***fodder beet*** [2]; ***sugar beet*** [1, 2]

Approval information

- Ethofumesate included in Annex I under EC Directive 91/414

Efficacy guidance

- Best results achieved on a fine firm clod-free seedbed when soil moist and adequate rain falls after spraying. Efficacy and crop safety may be reduced if heavy rain falls just after incorporation
- Effectiveness may be reduced under conditions of low pH
- May be applied by conventional or repeat low dose method. See label for details

Restrictions

- Maximum number of treatments 1 pre-emergence for fodder beet, mangels; 1 pre-emergence plus 3 post-emergence for sugar beet [2]; 1 pre-emergence or 3 post-emergence on sugar beet [1]
- May be used on soil classes Loamy Sand - Silty Clay Loam [2]. Additional restrictions apply for some tank mixtures
- Do not treat beet post-emergence with recommended tank mixtures when temperature is, or is likely to be, above 21°C on day of spraying or under conditions of high light intensity
- If a mixture with phenmedipham is applied to a crop previously treated with a pre-emergence herbicide and the crop is suffering from stress from whatever cause, no further such post-emergence applications may be made [2]

Crop-specific information

- Latest use: pre-crop emergence for fodder beet, mangels; before crop leaves meet between rows for sugar beet
- Apply up to cotyledon stage of weeds
- Crop vigour may be reduced by treatment of crops growing under unfavourable conditions including poor tilth, drilling at incorrect depth, soil capping, physical damage, excess nitrogen, excess seed dressing, trace element deficiency or a sudden rise in temperature after a cold spell. Frost after pre-emergence treatment may check crop growth

Following crops guidance

- In the event of crop failure only sugar beet, fodder beet or mangels may be re-drilled
- Any crop may be sown 3 mth after spraying following ploughing to 15 cm

Environmental safety

- Dangerous for the environment [2]
- Very toxic to aquatic organisms [2]
- Harmful to fish or other aquatic life. Do not contaminate surface waters or ditches with chemical or used container [1]

Hazard classification and safety precautions

Hazard H03 [1, 2]; H11 [2]
Risk phrases R22a [1, 2]; R50, R58 [2]
Operator protection A, H [1, 2]; C [1]; U05a [2]; U08, U19a, U20a [1, 2]
Environmental protection E13c, E34 [1]; E15a, E32d, E38 [2]; E31b [1, 2]
Storage and disposal E01, E04, E30a
Medical advice M05a [2]

71 chloridazon + metamitron

A contact and residual herbicide mixture for sugar beet

Products

Volcan Combi	Sipcam	300:280 g/l	SC	10256

Uses

- Annual dicotyledons in ***sugar beet***
- Annual meadow grass in ***sugar beet***

Efficacy guidance

- Best results achieved from a sequential programmme of sprays
- Pre-emergence use improves efficacy of post-emergence programme. Best results obtained from application to a moist seed bed

FOR FULL CONDITIONS OF USE ALWAYS READ THE PRODUCT LABEL

- First post-emergence treatment should be made when weeds at early cotyledon stage and subsequent applications made when new weed flushes reach this stage
- Weeds surviving an earlier treatment should be treated again after 7-10 d even if no new weeds have appeared

Restrictions
- Maximum number of treaments 1 pre-emergence followed by 3 post-emergence
- Take advice if light soils have a high proportion of stones

Crop-specific information
- Latest use: when leaves of crop meet in rows
- Tolerance of crops growing under stress from any cause may be reduced
- Crops treated pre-emergence and subsequently subjected to frost may be checked and recovery may not be complete

Following crops guidance
- After the last application only sugar beet or mangels may be sown within 4 mth; cereals may be sown after 16 wk. Land should be mouldboard ploughed to 15 cm and thoroughly cultivated before any succeeding crop

Environmental safety
- Harmful to fish or other aquatic life. Do not contaminate surface waters or ditches with chemical or used container

Hazard classification and safety precautions

Hazard H04
Risk phrases R36, R38
Operator protection U08, U14, U15, U20a
Environmental protection E13c, E31b, E34
Storage and disposal E30a

72 chloridazon + propachlor

A residual pre-emergence herbicide for use in onions and leeks

Products

Ashlade CP	Nufarm UK	86:400 g/l	SC	06481

Uses
- Annual dicotyledons in ***bulb onions*** *(off-label)*, ***chives*** *(off-label)*, ***leeks*** *(off-label)*, ***salad onions*** *(off-label)*
- Annual grasses in ***bulb onions*** *(off-label)*, ***chives*** *(off-label)*, ***leeks*** *(off-label)*, ***salad onions*** *(off-label)*

Specific Off-Label Approvals (SOLAs)
- ***bulb onions, leeks*** *(OLA 981362) Dec 2008* [1]
- ***chives, salad onions*** *(OLA 951233) Dec 2008* [1]

Efficacy guidance
- Best results achieved from application to firm, moist, weed-free seedbed when adequate rain falls afterwards

Restrictions
- Maximum number of treatments 1 per crop
- Do not use on soils with more than 10% organic matter
- Ensure crops are drilled to 20 mm depth

Crop-specific information
- Latest use: pre-emergence of crop; before 2 true leaf stage for onions and leeks
- HI 12 wk for chives, salad onions
- Apply pre-emergence of sown crops, preferably soon after drilling, before weeds emerge. Loose or fluffy seedbeds must be consolidated before application
- Apply to transplanted crops when soil has settled after planting
- Crops stressed by nutrient deficiency, pests or diseases, poor growing conditions or pesticide damage may be checked by treatment, especially on sandy or gravelly soils

Following crops guidance

- In the event of crop failure only onions, leeks or maize should be planted
- Any crop can follow a treated onion or leek crop harvested normally as long as the ground is cultivated thoroughly before drilling

Environmental safety

- Harmful to fish or other aquatic life. Do not contaminate surface waters or ditches with chemical or used container

Hazard classification and safety precautions

Hazard H03
Risk phrases R22a, R36, R43
Operator protection A, C; U02a, U04a, U05a, U08, U11, U13, U14, U19a, U20a
Environmental protection E15a, E31b
Storage and disposal E01, E04, E30a
Medical advice M03

73 chloridazon + quinmerac

A herbicide mixture for use in beet crops

Products

Fiesta T	BASF	360:60 g/l	SC	11734

Uses

- Annual dicotyledons in ***fodder beet, mangels, sugar beet***
- Annual meadow grass in ***fodder beet, mangels, sugar beet***
- Chickweed in ***fodder beet, mangels, sugar beet***
- Cleavers in ***fodder beet, mangels, sugar beet***
- Mayweeds in ***fodder beet, mangels, sugar beet***
- Poppies in ***fodder beet, mangels, sugar beet***
- Speedwells in ***fodder beet, mangels, sugar beet***

Efficacy guidance

- Best results obtained when adequate soil moisture is present at application and afterwards to form an active herbicidal layer in the soil
- A programme of pre-emergence treatment followed by post-emergence application(s) in mixture with a contact herbicide optimises weed control and is essential for some difficult weed species such as cleavers
- Treatment pre-emergence only may not provide sufficient residual activity to give season-long weed control
- Effectiveness may be reduced under conditions of low pH
- Always follow WRAG guidelines for preventing and managing herbicide resistant weeds. Section 5 for more information

Restrictions

- Maximum total dose on sugar beet equivalent to one treatment pre-emergence plus two treatments post-emergence, all at maximum individual dose; maximum total dose on fodder beet and mangels equivalent to one full dose pre-emergence treatment
- Heavy rain shortly after treatment may check crop growth particularly when it leaves water standing in surface depressions
- Treatment of stressed crops or those growing in unfavourable conditions may depress crop vigour and possibly reduce stand
- Do not use on Sands or soils of high organic matter content
- Where rates of nitrogen higher than those generally recommended are considered necessary, apply at least 3 wk before drilling

Crop-specific information

- Latest use: before plants meet between the rows for sugar beet; pre-emergence for fodder beet, mangels

Following crops guidance

- Winter cereals or any spring sown crop may follow a treated beet crop harvested at the normal time and after ploughing
- In the event of failure of a treated crop only a beet crop may be drilled, after cultivation

Environmental safety

- Dangerous for the environment
- Toxic to aquatic organisms
- To reduce movement to groundwater do not apply to dry soil or when heavy rain is forecast

Hazard classification and safety precautions

Hazard H04, H11
Risk phrases R43, R51, R58
Operator protection A; U05a, U08, U14, U19a, U20b
Environmental protection E15a, E31c, E32d, E38
Storage and disposal E01, E04, E29b, E30a
Medical advice M05a

74 chlormequat

A plant-growth regulator for reducing stem growth and lodging

Products

1	Adjust	Mandops	620 g/l	SL	05589
2	Agriguard Chlormequat 720	AgriGuard	720 g/l	SL	09919
3	Barleyquat B	Mandops	620 g/l	SL	07051
4	BASF 3C Chlormequat 720	BASF	720 g/l	SL	06514
5	Bettaquat B	Mandops	620 g/l	SL	07050
6	Fargro Chlormequat	Fargro	460 g/l	SL	02600
7	Greencrop Carna	Greencrop	600 g/l	SL	09403
8	Hive	Nufarm UK	730 g/l	SL	11392
9	K2	Mandops	620 g/l	SL	10370
10	Mandops Chlormequat 700	Mandops	700 g/l	SL	06002
11	Manipulator	Mandops	620 g/l	SL	05871
12	Mirquat	Nufarm UK	730 g/l	SL	10604
13	Sigma PCT	Nufarm UK	460 g/l	SL	11209
14	Stabilan 700	Nufarm UK	700 g/l	SL	11393

Uses

- Increasing yield in ***spring barley*** [1, 3, 9-11]; ***winter barley*** [1, 3, 4, 7, 9-11]
- Lodging control in ***spring barley*** [1, 3, 9, 11]; ***spring oats, winter oats*** [2, 4, 7, 8, 10, 12, 14]; ***spring rye, triticale*** [4]; ***spring wheat, winter wheat*** [1, 2, 4, 5, 7-14]; ***winter barley*** [1-4, 8, 9, 11-14]; ***winter rye*** [2, 4]
- Stem shortening in ***bedding plants, camellias, hibiscus trionum, lilies*** [6, 8, 12, 14]; ***geraniums, poinsettias*** [4, 6, 8, 12, 14]

Approval information

- Approved for aerial application on wheat and oats [4, 7, 10, 12, 13]; on winter barley [06154, 7]; on triticale, rye [4]. See notes in Section 5

Efficacy guidance

- Most effective results on cereals normally achieved from application from Apr onwards, on wheat and rye from leaf sheath erect to first node detectable (GS 30-31), on oats at second node detectable (GS 32), on winter barley from mid-tillering to leaf sheath erect (GS 25-30). However, recommendations vary with product. See label for details
- Results on barley can be variable
- In tank mixes with other pesticides optimum timing for herbicide action may differ from that for growth reduction. See label for details of tank mix recommendations
- Addition of approved non-ionic wetter recommended for products approved on oats

Restrictions

- Maximum number of treatments 1 per crop for cereals (2 per crop for split applications on winter wheat or winter barley); varies with species for ornamentals [4]
- Maximum total dose equivalent to one full dose treatment [1, 5, 9]

- Do not use on very late sown spring wheat or oats or on crops under stress
- Do not use straw from treated cereals as horticultural growth medium or mulch
- Mixtures with liquid nitrogen fertilizers may cause scorch and are specifically excluded on some labels
- At least 6 h, preferably 24 h, required before rain for maximum effectiveness. Do not apply to wet crops
- Check labels for tank mixtures known to be incompatible

Crop-specific information

- Latest use varies with product. See label for details
- May be used on cereals undersown with grass or clovers
- Ornamentals to be treated must be well established and growing vigorously. Do not treat in strong sunlight or when temperatures are likely to fall below 10°C
- Temporary yellow spotting may occur on poinsettias. It can be minimised by use of a non-ionic wetting agent - see label

Environmental safety

- Dangerous for the environment [2, 14]
- Harmful to aquatic organisms [6, 8, 14]
- Wash equipment thoroughly with water and wetting agent immediately after use and spray out. Spray out again before storing or using for another product. Traces can cause harm to susceptible crops sprayed later

Hazard classification and safety precautions

Hazard H03 [1-14]; H11 [2, 14]

Risk phrases R21 [2]; R22a [1-14]; R52, R58 [6, 8, 14]

Operator protection A; U05a [1-11, 13, 14]; U05b [12]; U08, U19a [1-14]; U13 [6, 10]; U20a [10]; U20b [1-7, 9, 11-14]

Environmental protection E08 [10]; E15a, E34 [1-14]; E31a [10, 14]; E31b [1-3, 5-9, 11-13]; E31c [4]

Consumer protection C01 [10]

Storage and disposal E01, E04, E30a [1-14]; E26 [2, 6, 7, 10, 14]

Medical advice M03 [1-7, 9-14]; M05a [4, 6, 8, 14]

75 chlormequat + 2-chloroethylphosphonic acid

A plant growth regulator for use in cereals

Products

1	Greencrop Tycoon	Greencrop	305:155 g/l	SL	09571
2	Strate	Bayer CropScience	360:180 g/l	SL	10020
3	Sypex	BASF	305:155 g/l	SL	04650
4	Upgrade	Bayer CropScience	360:180 g/l	SL	10029

Uses

- Lodging control in ***spring barley***, ***winter barley***, ***winter wheat***

Efficacy guidance

- Best results obtained when crops growing vigorously
- Recommended dose varies with growth stage. See labels for details and recommendations for use of sequential treatments

Restrictions

- 2-chloroethylphosphonic acid is an anticholinesterase organophosphorus compound. Do not use if under medical advice not to work with such compounds
- Maximum number of treatments 1 per crop; maximum total dose equivalent to one full dose treatment
- Product must always be used with specified non-ionic wetter - see labels
- Do not use on any crop in sequence with any other product containing 2-chloroethylphosphonic acid
- Do not spray when crop wet or rain imminent
- Do not spray during cold weather or periods of night frost, when soil is very dry, when crop diseased or suffering pest damage, nutrient deficiency or herbicide stress
- If used on seed crops grown for certification inform seed merchant beforehand

- Do not use on wheat variety Moulin or on any winter varieties sown in spring [1, 3]
- Do not use on spring barley Triumph [1]
- Do not treat barley on soils with more than 10% organic matter [1, 3]
- Do not use straw from treated cereals as a horticultural growth medium or as a mulch [1, 3]
- Do not use in programme with any other product containing 2-chloroethylphosphonic acid [1]
- Only crops growing under conditions of high fertility should be treated

Crop-specific information

- Latest use: before flag leaf ligule/collar just visible (GS 39) or 1st spikelet visible (GS 51) for wheat or barley at top dose; or before flag leaf sheath opening (GS 47) for winter wheat at reduced dose
- Apply before lodging has started

Environmental safety

- Dangerous for the environment [2]
- Harmful to aquatic organisms [3]
- Harmful to fish or other aquatic life. Do not contaminate surface waters or ditches with chemical or used container [2, 4]

Hazard classification and safety precautions

Hazard H03 [1-4]; H11 [2]
Risk phrases R22a, R37 [1-4]; R41 [1, 2, 4]; R52 [3]
Operator protection A [1-4]; C [2, 4]; U05a, U08, U19a, U20b [1-4]; U11 [1, 2, 4]; U13 [2, 4]
Environmental protection E13c [2, 4]; E15a [1, 3]; E31b, E34 [1-4]
Storage and disposal E01, E04, E30a [1-4]; E26 [1, 2, 4]
Medical advice M01, M03 [1-4]; M05a [3]

76 chlormequat + 2-chloroethylphosphonic acid + mepiquat chloride

A plant growth regulator for reducing lodging in cereals

Products

Cyclade	BASF	230:155:75 g/l	SL	08958

Uses

- Lodging control in ***spring barley***, ***winter barley***, ***winter wheat***

Efficacy guidance

- Best results achieved in a vigorous, actively growing crop with adequate fertility and moisture
- Optimum timing on all crops is from second node detectable stage (GS 32)
- Recommended for use as part of an intensive growing system which includes provision for optimum fertilizer treatment and disease control

Restrictions

- 2-chloroethylphosphonic acid is an anticholinesterase organophosphorus compound. Do not use if under medical advice not to work with such compounds
- Maximum number of treatments 1 per crop
- Maximum total dose depends on spraying regime adopted. See label
- Must be used with a non-ionic wetting agent
- Do not apply to stressed crops or those on soils of low fertility unless receiving adequate dressings of fertilizer
- Do not apply in temperatures above 21°C or if crop is wet or if rain expected
- Do not treat variety Moulin nor any winter varieties sown in spring
- Do not use in a programme with any other product containing 2-chloroethylphosphonic acid
- Do not apply to barley on soils with more than 10% organic matter (winter wheat may be treated)
- Notify seed merchant in advance if use on a seed crop is proposed

Crop-specific information

- Latest use: before first spikelet of ear visible (GS 51) using reduced dose on winter barley; before flag leaf sheath opening (GS 47) using reduced dose on winter wheat; before flag leaf just visible on spring barley
- May be applied to crops undersown with grasses or clovers
- Treatment may cause some delay in ear emergence

Environmental safety

- Harmful to aquatic organisms
- Do not use straw from treated crops as a horticultural growth medium

Hazard classification and safety precautions

Hazard H03
Risk phrases R22a, R37, R52
Operator protection A; U05a, U08, U19a, U20b
Environmental protection E15a, E31c, E34
Storage and disposal E01, E04, E30a
Medical advice M01, M03, M05a

77 chlormequat + mepiquat chloride

A plant growth regulator for reducing lodging in cereals

Products

Stronghold	BASF	345:115 g/l	SL	09134

Uses

- Lodging control in ***winter wheat***

Efficacy guidance

- Optimum timing is when leaf sheaths erect (GS 30)
- Benefit will vary according to crop and stage of growth at application

Restrictions

- Maximum total dose equivalent to one full dose treatment
- Do not apply to stressed crops or those on soils of low fertility unless receiving adequate dressings of fertilizer
- Do not treat crops where significant foot diseases, especially Take-all, are expected
- Do not treat crops on soils of low fertility
- Do not apply in temperatures above 21°C or if crop is wet or if rain expected
- Do not treat any winter varieties sown in spring
- Notify seed merchant in advance if use on a seed crop is proposed

Crop-specific information

- Latest use: before 3rd node detectable (GS 33)
- Apply during good growing conditions at the correct timings - see label
- May be applied to crops undersown with grasses or clovers
- Treatment may cause some delay in ear emergence
- Mixtures with liquid fertilizers may cause scorching in some circumstances

Environmental safety

- Do not use straw from treated crops as a horticultural growth medium or mulch

Hazard classification and safety precautions

Hazard H03
Risk phrases R22a
Operator protection A; U05a, U08, U19a, U20b
Environmental protection E15a, E31c, E34
Storage and disposal E01, E04, E30a
Medical advice M03, M05a

78 chlormequat with choline chloride

A plant growth regulator for use in cereals and certain ornamentals

Products

1	Barclay Take 5	Barclay	645 g/l	SL	11368
2	Cropsafe 5C Chlormequat	Certis	645 g/l	SL	11179
3	New 5C Cycocel	BASF	645 g/l	SL	01482
4	New 5C Quintacel	Nufarm UK	645 g/l	SL	12074

Uses

- Increasing yield in ***winter barley*** [3, 4]
- Lodging control in ***spring oats, spring wheat, triticale, winter oats, winter rye, winter wheat*** [1, 3, 4]; ***spring rye*** [3, 4]; ***winter barley*** [1]
- Stem shortening in ***geraniums, poinsettias*** [2-4]

Approval information

- Approved for aerial application on wheat, oats, rye, triticale, winter barley [1, 3, 4]. See notes in Section 5

Efficacy guidance

- Influence on growth varies with crop and growth stage. Risk of lodging reduced by application at early stem extension. Root development and yield can be improved by earlier treatment
- Most effective results normally achieved from spring application. On winter barley an autumn treatment may also be useful. Timing of spray is critical and recommendations vary with product. See label for details
- Often used in tank-mixes with pesticides. Recommendations for mixtures and sequential treatments vary with product. See label for details

Restrictions

- Maximum number of treatments 1 per crop for spring wheat, oats, rye, triticale; 1 or 2 per crop for winter wheat and winter barley (depending on dose); 1-3 per yr for ornamentals. See label for details
- Do not spray very late sown spring crops, crops on soils of low fertility, crops under stress from any cause or if frost expected
- Do not use on spring barley
- Add authorised non-ionic wetter when spraying oats
- At least 6 h required before rain for maximum effectiveness. Do not apply to wet crops
- Do not use straw from treated cereals as horticultural growth medium or mulch

Crop-specific information

- Latest use: before 3rd node detectable (GS 33) for oats, winter wheat, autumn drilled spring wheat; before 2nd node detectable (GS 32) for rye; before 1st node detectable (GS 31) for spring wheat, winter barley, triticale; after potting for Poinsettias
- Mixtures with liquid nitrogen fertilizers may cause scorch
- May be used on cereals undersown with grass or clovers

Environmental safety

- Wash equipment thoroughly with water and wetting agent immediately after use and spray out. Traces can cause harm to susceptible crops sprayed later

Hazard classification and safety precautions

Hazard H03
Risk phrases R22a
Operator protection A; U05a, U08, U19a, U20b
Environmental protection E15a, E34 [1-4]; E31b [1, 2]; E31c [2-4]
Storage and disposal E01, E04, E30a [1-4]; E26 [1]; E29b [2]
Medical advice M03, M05a

79 chlormequat with choline chloride + imazaquin

A plant growth regulator mixture for winter wheat

Products

1	Meteor	BASF	368:28:0.8 g/l	SL	10403
2	Standon Imazaquin 5C	Standon	368:0.8 g/l	SL	08813

Uses

- Increasing yield in ***winter wheat***
- Lodging control in ***winter wheat***

Efficacy guidance

- Apply to crops during good growing conditions or to those at risk from lodging
- On soils of low fertility, best results obtained where adequate nitrogen fertilizer used

SEE SECTION 3 FOR PRODUCTS ALSO REGISTERED

Restrictions

- Maximum number of applications 1 per crop (2 per crop at split dose)
- Do not treat durum wheat
- Do not apply to undersown crops
- Do not use straw from treated cereals as horticultural growth medium or mulch
- Do not apply when crop wet or rain imminent

Crop-specific information

- Latest use: before second node detectable (GS 31)
- Apply as single dose from leaf sheath lengthening up to and including 1st node detectable or as split dose, the first from tillers formed to leaf sheath lengthening, the second from leaf sheath erect up to and including 1st node detectable

Hazard classification and safety precautions

Hazard H03 [1, 2]; H11 [1]
Risk phrases R22a, R36 [1, 2]; R51, R58 [1]
Operator protection A, C; U05a, U08, U13, U19a, U20b
Environmental protection E15a, E31b, E34
Storage and disposal E01, E04, E26, E30a
Medical advice M03

80 chlormequat with di-1-p-menthene

A plant-growth regulator for reducing stem growth and lodging

Products

Podquat	Mandops	470 g/l	SL	03003

Uses

- Increasing yield in ***broad beans***, ***combining peas***, ***spring field beans***, ***winter field beans***

Restrictions

- Do not apply to plants covered by frost
- Do not use in tank mixture with herbicides on pea crops
- Do not use in tank mixture with liquid nitorgen fertilizer

Crop-specific information

- On beans either apply in autumn as soon as possible after 3-leaf stage (GS 1,3) but before growth ceases followed by spring treatment in mid-Mar to early Apr, or use a single spring spray during stem elongation. See label for details
- On peas treat when standing crop reaches 7-10 cm above ground level
- May be used at temperatures down to 1°C provided spray dries on leaves before rain, frost or snow occurs

Hazard classification and safety precautions

Hazard H03
Risk phrases R22a
Operator protection A; U05a, U08, U19a, U20a
Environmental protection E15a, E31a, E34
Storage and disposal E01, E04, E30a
Medical advice M03

81 2-chloroethylphosphonic acid

A plant growth regulator for cereals and various horticultural crops

See also *chlormequat + 2-chloroethylphosphonic acid*
chlormequat + 2-chloroethylphosphonic acid + mepiquat chloride

Products

1	Agriguard Cerusite	AgriGuard	480 g/l	SL	11494
2	Cerone	Bayer CropScience	480 g/l	SL	09985
3	Ethrel C	Certis	480 g/l	SL	11387
4	Pan Ethephon	Pan Agriculture	480 g/l	SL	11020

Uses

- Basal bud stimulation in ***protected roses*** [3]
- Fruit ripening in ***apples*** *(off-label - for cider making)*, ***outdoor tomatoes***, ***protected tomatoes*** [3]
- Increasing branching in ***geraniums*** [3]
- Inducing flowering in ***bromeliads*** [3]
- Lodging control in ***spring barley*** [1, 2]; ***triticale***, ***winter barley***, ***winter rye***, ***winter wheat*** [1, 2, 4]
- Stem shortening in ***narcissi*** [3]

Specific Off-Label Approvals (SOLAs)

- ***apples*** *(for cider making) (OLA 030875) Dec 2008* [3]

Approval information

- Approved for aerial application on winter barley [1, 2]. See notes in Section 5

Efficacy guidance

- Best results achieved on crops growing vigorously under conditions of high fertility
- Optimum timing varies between crops and products. See labels for details
- Do not spray crops when wet or if rain imminent
- Best results on horticultural crops when temperature does not fall below 10°C [3]
- Use on tomatoes 17 d before planned pulling date [3]
- Apply as drench to daffodils when stems average 15 cm [3]
- Apply to glasshouse roses when new growth started after pruning [3]

Restrictions

- 2-chloroethylphosphonic acid is an anticholinesterase organophosphorus compound. Do not use if under medical advice not to work with such compounds
- Maximum number of treatments 1 per crop or yr
- Do not spray crops suffering from stress caused by any factor, during cold weather or period of night frost nor when soil very dry
- Do not apply to cereals within 10 d of herbicide or liquid fertilizer application
- Do not spray wheat or triticale where the leaf sheaths have split and the ear is visible

Crop-specific information

- Latest use: before 1st spikelet visible (GS 51) for spring barley, winter barley, winter rye; before flag leaf sheath opening (GS 47) for triticale, winter wheat
- HI cider apples, tomatoes 5 d

Environmental safety

- Dangerous for the environment [1, 4]
- Harmful to aquatic organisms
- Avoid accidental deposits on painted objects such as cars, trucks, aircraft

Hazard classification and safety precautions

Hazard H04 [1-4]; H11 [1, 4]
Risk phrases R36 [1]; R37, R41, R52, R58 [2-4]; R38 [1-4]
Operator protection A, C; U02a, U09a, U20a [3]; U05a, U13 [1-4]; U08, U20b [1, 2, 4]; U11 [2, 3]
Environmental protection E13c, E32d, E38 [2, 3]; E15a [1, 4]; E31b [1-4]; E34 [1, 3]
Storage and disposal E01, E04, E30a
Medical advice M01 [2-4]

82 2-chloroethylphosphonic acid + mepiquat chloride

A plant growth regulator for reducing lodging in cereals

Products

1	Barclay Banshee	Barclay	155:305 g/l	SL	11343
2	Guilder	Nufarm UK	155:305 g/l	SL	11894
3	Me2 Terpitz	Me2	155:305 g/l	SL	09634
4	Standon Mepiquat Plus	Standon	155:305 g/l	SL	09373
5	Terpal	BASF	155:305 g/l	SL	02103

Uses

- Increasing yield in ***winter barley*** *(low lodging situations)* [1-3, 5]
- Lodging control in ***spring barley*** [1-3, 5]; ***triticale***, ***winter barley***, ***winter rye***, ***winter wheat*** [1-5]

SEE SECTION 3 FOR PRODUCTS ALSO REGISTERED

Efficacy guidance

- Best results achieved on crops growing vigorously under conditions of high fertility
- Recommended dose and timing vary with crop, cultivar, growing conditions, previous treatment and desired degree of lodging control. See label for details
- May be applied to crops undersown with grass or clovers
- Do not apply to crops if wet or rain expected as efficacy will be impaired

Restrictions

- 2-chloroethylphosphonic acid is an anticholinesterase organophosphorus compound. Do not use if under medical advice not to work with such compounds
- Maximum number of treatments 2 per crop
- Add an authorised non-ionic wetter to spray solution. See label for recommended product and rate
- Do not treat crops damaged by herbicides or stressed by drought, waterlogging etc
- Do not treat crops on soils of low fertility unless adequately fertilized
- Do not use in a programme with any other product containing 2-chloroethylphosphonic acid
- Do not apply to winter cultivars sown in spring or treat winter barley, triticale or winter rye on soils with more than 10% organic matter (winter wheat may be treated)
- Do not apply at temperatures above 21°C

Crop-specific information

- Latest use: before ear visible (GS 49) for winter barley, spring barley, winter wheat and triticale; flag leaf just visible (GS 37) for winter rye
- Late tillering may be increased with crops subject to moisture stress and may reduce quality of malting barley

Environmental safety

- Harmful to aquatic organisms [1]
- Do not use straw from treated cereals as a mulch or growing medium

Hazard classification and safety precautions

Hazard H03 [2, 5]; H04 [1, 3, 4]
Risk phrases R22a [5]; R36, R37 [1-5]; R52 [1]; R58 [1, 5]
Operator protection A, C [2, 5]; U05a [3, 4]; U11 [2-4]; U20b [1-5]
Environmental protection E15a [1-5]; E31b [1, 3, 4]; E31c [2, 5]
Storage and disposal E01, E04 [3-5]; E26 [1, 3, 4]; E29b [2, 5]; E30a [1-5]
Medical advice M01 [1-5]; M05a [5]

83 chlorophacinone

An anticoagulant rodenticide

Products

1	Drat Rat Bait	B H & B	0.005% w/w	RB	H6743
2	Endorats	Irish Drugs	0.005% w/w	RB	H6744
3	Karate Ready to Use Rat Bait	JohnsonDiversey	0.006% w/w	RB	H6745
4	Karate Ready to Use Rodenticide Sachets	JohnsonDiversey	0.006%w/w	RB	H6746

Uses

- Rats in ***farm buildings*** [1-4]
- Voles in ***farm buildings*** [1]

Efficacy guidance

- Chemical formulated with oil, thus improving weather resistance of bait
- Use in baiting programme
- Bait stations should be sited where rats active, by rat holes, along runs or in harbourages. Place bait in suitable containers
- Replenish baits every few days and remove unused bait when take ceases or after 7-10 d

Restrictions

- For use only by professional operators
- Resistance status of target population should be taken into account when considering choice of rodenticide
- Bait stations may be sited conveniently but bait should be inaccessible to non-target animals and protected from prevailing weather

Environmental safety

- Prevent access to baits by children, domestic animals and birds; see label for other precautions required
- Harmful to game, wild birds and animals
- Store unused sachets in a safe place. Do not store half-used sachets [4]

Hazard classification and safety precautions

Operator protection U13 [1-4]; U20a [2]; U20b [1, 3]; U20c [4]
Environmental protection E31a [1]; E32a [2-4]
Storage and disposal E30a [1-3]
Vertebrate/rodent control products V01a, V03a, V04a [1, 2]; V01b, V03b, V04b [3, 4]; V02 [1-4]
Medical advice M03 [3, 4]

84 chloropicrin

A highly toxic horticultural soil fumigant

Products

1	Chloropicrin Fumigant	Dewco-Lloyd	99.5% w/w	LI	04216
2	K & S Chlorofume	K & S Fumigation	99.3% w/w	VP	08722

Uses

- Crown rot in ***strawberries*** *(soil fumigation)*
- Nematodes in ***strawberries*** *(soil fumigation)*
- Red core in ***strawberries*** *(soil fumigation)*
- Replant disease in ***all top fruit*** *(soil fumigation)*, ***hardy ornamentals*** *(soil fumigation)*
- Verticillium wilt in ***strawberries*** *(soil fumigation)*

Efficacy guidance

- Treat pre-planting
- Apply with specialised injection equipment
- For treating small areas or re-planting a single tree, a hand-operated injector may be used. Mark the area to be treated and inject to 22 cm at intervals of 22 cm
- Polythene sheeting (150 gauge) should be progressively laid over soil as treatment proceeds. The margin of the sheeting around the treated area must be embedded or covered with treated soil. Remove progressively after at least 4 d provided good air movement conditions prevail. Aerate soil for 15 d before planting

Restrictions

- Chloropicrin is subject to the Poisons Rules 1982 and the Poisons Act 1972. See notes in Section 5
- Before use, consult the code of practice for the fumigation of soil with chloropicrin. 2 fumigators must be present
- Remove contaminated gloves, boots or other clothing immediately and ventilate them in the open air until all odour is eliminated

Crop-specific information

- Latest use: at least 20 d before planting

Following crops guidance

- Carry out a cress test before replanting treated soil

Environmental safety

- Avoid treatment or vapour release when persistent still air conditions prevail
- Dangerous to game, wild birds and animals [1]
- Dangerous to bees [1]
- Dangerous to fish. Do not contaminate surface waters or ditches with chemical or used container [1]
- Dangerous to livestock. Keep all livestock out of treated areas until advised otherwise by fumigator

Hazard classification and safety precautions

Hazard H01, H04 [1, 2]; H05 [2]
Risk phrases R26, R27, R28, R37 [1, 2]; R34 [2]; R36, R38 [1]
Operator protection A, G, H, K, M; U04a, U05a, U10, U19a, U20a [1, 2]; U13, U14, U15 [1]

Environmental protection E02a [1, 2] (until advised); E06b [1] (until advised); E06b, E33, E36 [2]; E10a, E12d, E13b, E32a [1]; E15a, E34 [1, 2]
Storage and disposal E01 [1]; E04, E30b [1, 2]
Medical advice M04 [1]

85 chlorothalonil

A protectant chlorophenyl fungicide for use in many crops and turf

See also azoxystrobin + chlorothalonil

Products

1	Agriguard Chlorothalonil	AgriGuard	500 g/l	SC	09390
2	Bombardier FL	Unicrop	500 g/l	SC	07910
3	Bravo 500	Syngenta	500 g/l	SC	10518
4	Clortosip 500	Sipcam	500 g/l	SC	09320
5	Daconil Turf	Scotts	500 g/l	SC	09265
6	Flute	Sipcam	720 g/l	SC	08953
7	Fusonil Turf	Rigby Taylor	500 g/l	SC	09695
8	Greencrop Orchid	Greencrop	500 g/l	SC	09566
9	Inzacur	Nufarm UK	500 g/l	SC	12014
10	Joules	Nufarm UK	500 g/l	SC	11047
11	Marnoch Chlorothalonil	Marnoch	500 g/l	SC	09763
12	Repulse	Certis	500 g/l	SC	11328
13	Rover	Sipcam	500 g/l	SC	09848
14	Scorpio 500 SC	DAPT	500 g/l	SC	10927
15	Sipcam Echo 75	Sipcam	75% w/w	WG	08302

Uses

- Anthracnose in ***managed amenity turf*** [5, 7, 9]
- Ascochyta in ***combining peas*** [1-4, 6, 8, 11-13, 15]
- Blight in ***potatoes*** [1-4, 6, 8, 10-15]; ***protected tomatoes*** [2-4, 12, 13]
- Botrytis in ***blackberries***, ***raspberries*** [1-4, 11, 13, 15]; ***blackcurrants***, ***strawberries*** *(outdoor crops only)* [4, 11, 13]; ***broccoli***, ***brussels sprouts***, ***cabbages***, ***cauliflowers*** [1-4, 6, 11-13, 15]; ***calabrese*** [6, 15]; ***combining peas*** [1, 4, 11-13, 15]; ***daffodils*** *(off-label - for galanthamine production)* [3]; ***gooseberries***, ***redcurrants*** [11]; ***gooseberries*** *(qualified minor use)*, ***redcurrants*** *(qualified minor use)* [4, 13]; ***onions*** [1, 2, 4, 11-13]; ***protected cucumbers***, ***protected ornamentals*** [1-4, 12, 13, 15]; ***protected tomatoes*** [1-3, 12, 15]; ***spring oilseed rape***, ***winter oilseed rape*** [1, 3, 11, 13, 15]; ***strawberries*** [3, 12, 15]; ***strawberries*** *(off-label)* [3, 4]
- Cane spot in ***blackberries***, ***raspberries*** [3, 12]
- Celery leaf spot in ***celery*** *(qualified minor use)* [1-4, 12, 13, 15]
- Chocolate spot in ***spring field beans***, ***winter field beans*** [1-4, 6, 8, 10, 11, 13-15]
- Currant leaf spot in ***blackcurrants*** [2-4, 6, 8, 11-13, 15]; ***gooseberries***, ***redcurrants*** [11]; ***gooseberries*** *(qualified minor use)*, ***redcurrants*** *(qualified minor use)* [2-4, 12, 13, 15]
- Dark leaf spot in ***broccoli***, ***brussels sprouts***, ***cabbages***, ***calabrese***, ***cauliflowers*** [8, 15]
- Dollar spot in ***managed amenity turf*** [5, 7, 9]
- Downy mildew in ***brassica seed beds*** [3]; ***broccoli***, ***brussels sprouts***, ***cabbages***, ***cauliflowers*** [1-4, 6, 11-13, 15]; ***calabrese*** [6, 15]; ***edible brassicas*** [12]; ***hops*** [1-4, 11-13, 15]; ***spring oilseed rape***, ***winter oilseed rape*** [1, 3, 4, 11, 13, 15]
- Fusarium patch in ***managed amenity turf*** [5, 7, 9]
- Glume blotch in ***spring wheat*** [4, 6, 8, 10, 13-15]; ***winter wheat*** [1-4, 6, 8, 10, 11, 13-15]
- Ink disease in ***irises*** *(qualified minor use)* [2-4, 13, 15]
- Late ear diseases in ***winter wheat*** [1, 11]
- Leaf mould in ***protected tomatoes*** [1, 2, 4, 12, 13, 15]
- Leaf rot in ***onions*** [1-4, 11-13, 15]
- Mycosphaerella in ***combining peas*** [1-4, 10-15]
- Neck rot in ***onions*** [1-4, 11-13, 15]
- Powdery mildew in ***blackberries*** [3]; ***blackcurrants*** [4, 11, 13]; ***gooseberries***, ***redcurrants*** [11]; ***gooseberries*** *(qualified minor use)*, ***redcurrants*** *(qualified minor use)* [4, 13]; ***protected cucumbers*** [15]; ***raspberries*** [3, 15]
- Red thread in ***managed amenity turf*** [5, 7, 9]
- Rhynchosporium in ***spring barley***, ***winter barley*** [3, 10, 14]

- Ring spot in ***broccoli*, *brussels sprouts*, *cabbages*, *cauliflowers*** [3, 6, 12, 15]; ***calabrese*** [6, 15]; ***edible brassicas*** [12]
- Septoria leaf spot in ***celeriac*** *(off-label)* [3]; ***celery*** *(qualified minor use)* [11]; ***spring wheat*** [4, 6, 8, 10, 13-15]; ***winter wheat*** [1-4, 6, 8, 10, 11, 13-15]
- White mould in ***daffodils*** *(off-label)* [3]

Specific Off-Label Approvals (SOLAs)

- ***celeriac*** *(OLA 032083) Dec 2008* [3]
- ***daffodils*** *(for galanthamine production) (OLA 041518) Dec 2008* [3]
- ***daffodils*** *(OLA 041518) Dec 2008* [3]
- ***strawberries*** *(OLA 011635) Dec 2008* [3]
- ***strawberries*** *(OLA 021253) Dec 2008* [4]
- ***strawberries*** *(OLA 992452) Dec 2008* [3]

Approval information

- Approval for aerial spraying on potatoes [1-4, 6, 8, 12, 15]. See notes in Section 5
- Following implementation of Directive 98/82/EC, approval for use of chlorothalonil on numerous crops was revoked in 1999, although uses on blackcurrants, redcurrants and gooseberries were subsequently reinstated for some products. Uses on loganberries revoked in 2001 following implementation of MRL Directives

Efficacy guidance

- For some crops products differ in diseases listed as controlled. See label for details and for application rates, timing and number of sprays
- Apply as protective spray or as soon as disease appears and repeat as directed
- Activity against Septoria may be reduced where serious mildew or rust present. In such conditions mix with suitable mildew or rust fungicide
- May be used for preventive and curative treatment of turf [5, 7]

Restrictions

- Maximum number of treatments 2 per crop for brassicas; 3 per crop for celery, peas, cereals; 2 per spawning for mushrooms
- Maximum total dose for other crops varies with crop and product - see labels for details
- Operators must use a vehicle fitted with a cab and a forced air filtration unit with a pesticide filter complying with HSE Guidance Note PM74 or to an equivalent or higher standard when making broadcast or air-assisted applications
- Must only be applied by a pedestrian controlled sprayer or vehicle mounted/drawn equipment [5, 7]
- Do not mow or water turf for 24 h after treatment. Do not add surfactant or mix with liquid fertilizer [5, 7]

Crop-specific information

- Latest time of application to cereals varies with product. Consult label
- Latest use for winter oilseed rape before flowering; end Aug in yr of harvest for blackcurrants, gooseberries, redcurrants, blueberries, loganberries, raspberries
- HI 8 wk for field beans; 6 wk for combining peas; 28 d or before 31 Aug for post-harvest treatment for blackcurrants, gooseberries, redcurrants; 14 d for onions, strawberries; 10 d for hops; 7-14 d for broccoli, Brussels sprouts, cabbages, cauliflowers, celery, onions, potatoes; 3 d for hops, cane fruit; 12-48 h for protected cucumbers, protected tomatoes; 24 h for mushrooms
- For Botrytis control in strawberries important to start spraying early in flowering period and repeat at least 3 times at 10 d intervals
- On strawberries some scorching of calyx may occur with protected crops

Environmental safety

- Dangerous for the environment [1-3, 5, 7-9, 11, 12]
- Very toxic to aquatic organisms [3, 5, 7-9, 11, 12]
- Dangerous to fish or other aquatic life. Do not contaminate surface waters or ditches with chemical or used container [4, 6, 10, 13-15]
- Do not spray from the air within 250 m horizontal distance of surface waters or ditches
- LERAP Category B
- Broadcast air-assisted LERAP (18 m) [1-4, 7-9, 11, 13, 15]

Hazard classification and safety precautions

Hazard H03 [1-3, 5, 7-9, 12]; H04 [4, 6, 10, 11, 13-15]; H11 [1-3, 5, 7-9, 11, 12]

Risk phrases R20 [3, 5, 7, 12]; R36 [3-5, 7-10, 13, 14]; R37 [1, 2, 4, 5, 8, 11-13, 15]; R38 [1-4, 7, 8, 10-14]; R40, R50, R58 [3, 5, 7-9, 11, 12]; R41 [1, 2, 4, 6, 8, 10, 12-15]; R43 [3-8, 11-13]
Operator protection A, C, H [1-15]; D [15]; J [1-4, 6, 8, 11, 13]; M [1-9, 11-13, 15]; U02a [3, 4, 6-8, 10, 11, 13-15]; U05a, U09a, U19a, U20a [1-15]; U13, U15 [4, 6, 8-10, 13-15]; U14 [4, 6, 8-10, 12-15]
Environmental protection E13b [4, 6, 10, 13-15]; E15a [1-3, 5, 7-9, 11, 12]; E16a [1-15]; E16b [1-8, 10-15]; E17b [1-4, 7-9, 11] (18 m); E17b [12]; E17b [13, 15] (18 m); E18 [1-4, 6-8, 11, 13, 15]; E31a [5, 9, 12]; E31b [1, 2, 4, 6, 8, 10, 13-15]; E31c [3, 7, 11]; E32d, E38 [3, 5, 7, 12]; E34 [1, 2, 5, 9, 10, 14]
Consumer protection C02 [10, 14] (field beans 14 d; potatoes 7 d)
Storage and disposal E01, E04, E30a [1-15]; E26 [1-3, 5-11, 14]
Medical advice M05a [8]

86 chlorothalonil + cyproconazole

A systemic protectant and curative fungicide mixture for cereals

Products

Alto Elite	Syngenta	375:40 g/l	SC	08467

Uses

- Brown rust in ***spring barley***, ***winter barley***, ***winter wheat***
- Chocolate spot in ***spring field beans***, ***winter field beans***
- Eyespot in ***winter barley***, ***winter wheat***
- Grey mould in ***combining peas***
- Net blotch in ***spring barley***, ***winter barley***
- Powdery mildew in ***spring barley***, ***winter barley***, ***winter wheat***
- Rhynchosporium in ***spring barley***, ***winter barley***
- Rust in ***spring field beans***, ***winter field beans***
- Septoria diseases in ***winter wheat***
- Yellow rust in ***spring barley***, ***winter barley***, ***winter wheat***

Efficacy guidance

- Apply at first signs of infection or as soon as disease becomes active
- A repeat application may be made if re-infection occurs
- For established mildew tank-mix with an approved mildewicide
- When applied prior to third node detectable (GS 33) a useful reduction of eyespot will be obtained

Restrictions

- Maximum total dose equivalent to 2 full dose treatments
- Do not apply at concentrations higher than recommended

Crop-specific information

- Latest use: completion of ear emergence in wheat and barley
- HI peas, field beans 6 wk
- If applied to winter wheat in spring at GS 30-33 straw shortening may occur but yield is not reduced

Environmental safety

- Dangerous for the environment
- Very toxic to aquatic organisms
- LERAP Category B

Hazard classification and safety precautions

Hazard H03, H11
Risk phrases R20, R37, R40, R41, R43, R50, R58
Operator protection A, C, H, M; U05a, U11, U19a, U20a
Environmental protection E15a, E16a, E16b, E31c, E32d, E38
Storage and disposal E01, E04, E30a
Medical advice M04

87 chlorothalonil + flutriafol

A systemic eradicant and protectant fungicide for winter wheat

Products

1	Halo	Headland	375:47 g/l	SC	11546
2	Impact Excel	Headland	300:47 g/l	SC	11547

Uses
- Brown rust in ***winter wheat***
- Late ear diseases in ***winter wheat***
- Powdery mildew in ***winter wheat***
- Septoria in ***winter wheat***
- Yellow rust in ***winter wheat***

Efficacy guidance
- Generally disease control and yield benefit will be optimised when application is made at an early stage of disease development
- Apply as soon as disease is seen establishing in the crop

Restrictions
- Maximum number of treatments 2 per crop (including other products containing flutriafol)

Crop-specific information
- Latest use: before early milk stage (GS 73)
- On certain cultivars with erect leaves high transpiration can result in flag leaf tip scorch. This may be increased by treatment but does not affect yield

Environmental safety
- Dangerous for the environment
- Very toxic to aquatic organisms
- LERAP Category B

Hazard classification and safety precautions

Hazard H02, H11 [2]; H04 [1]
Risk phrases R23, R40, R50, R58 [2]; R38 [1]; R41, R43 [1, 2]
Operator protection A, C, H; U02a, U05a, U09a, U19a, U20b [1, 2]; U11 [2]
Environmental protection E13b [1]; E15a, E38 [2]; E16a, E31b [1, 2]
Storage and disposal E01, E04, E30a

88 chlorothalonil + mancozeb

A multi-site protectant fungicide mixture

Products

1	Adagio	Interfarm	201:274 g/l	SC	10796
2	Guru	Interfarm	286: 194 g/l	SC	10801

Uses
- Blight in ***potatoes*** [1]
- Septoria diseases in ***winter wheat*** [2]

Efficacy guidance
- Start spray treatments on potatoes immediately after a blight warning or just before the haulm meets in the row [1]
- It is essential to start the blight spray programme before the disease appears in the crop [1]
- Repeat treatments for blight at 7, 10 or 14 d intervals depending on blight risk and continue until haulm is to be destroyed [1]
- Best results on winter wheat achieved from protective applications to the flag leaf. If disease already present on lower leaves treat as soon as flag leaf is just visible (GS 37) [2]
- Activity on Septoria may be reduced in presence of severe mildew infection [2]

Restrictions
- Maximum number of treatments 5 per crop for potatoes [1]
- Maximum total dose on winter wheat equivalent to 1.5 full dose treatments [2]

- Do not treat crops under stress from frost, drought, waterlogging, trace element deficiency or pest attack [2]
- Broadcast air assisted applications must only be made by equipment fitted with a cab with a forced air filtration unit plus a pesticide filter complying with HSE Guidance Note PM 74 or an equivalent or higher standard

Crop-specific information
- Before grain watery ripe (GS 71) for winter wheat
- HI potatoes 7 d
- Irrigated potato crops should be sprayed immediately after irrigation [1]

Environmental safety
- Dangerous for the environment
- Very toxic to aquatic organisms
- LERAP Category B

Hazard classification and safety precautions

Hazard H04, H11
Risk phrases R37, R43, R50, R58 [1, 2]; R40 [2]
Operator protection A, C, H, M; U02a, U05a, U08, U10, U11, U13, U14, U19a [1, 2]; U20a [1]; U20b [2]
Environmental protection E15a [1]; E16a, E16b, E31b, E32d, E38 [1, 2]
Storage and disposal E01, E04, E26, E30a
Medical advice M04 [1]

89 chlorothalonil + metalaxyl-M

A systemic and protectant fungicide for various crops

Products

Folio Gold	Syngenta	500:37.5 g/l	SC	10704

Uses
- Alternaria in ***brussels sprouts*** *(moderate control)*, ***calabrese*** *(moderate control)*, ***cauliflowers*** *(moderate control)*
- Downy mildew in ***broad beans***, ***brussels sprouts***, ***bulb onions*** *(qualified minor use)*, ***calabrese***, ***cauliflowers***, ***poppies*** *(off-label - for morphine production)*, ***shallots*** *(qualified minor use)*, ***spring field beans***, ***winter field beans***
- Ring spot in ***brussels sprouts*** *(reduction)*, ***calabrese*** *(reduction)*, ***cauliflowers*** *(reduction)*
- White blister in ***brussels sprouts***, ***calabrese***, ***cauliflowers***
- White tip in ***leeks*** *(qualified minor use)*

Specific Off-Label Approvals (SOLAs)
- ***poppies*** *(for morphine production) (OLA 040469) Dec 2008* [1]

Approval information
- Metalaxyl-M included in Annex I under EC Directive 91/414

Efficacy guidance
- Apply at first signs of disease or when weather conditions favourable for disease pressure
- Best results obtained when used in a full and well-timed programme. Repeat treatment at 14-21 d intervals if necessary
- Treatment of established disease will be less effective
- Evidence of effectiveness in bulb onions, shallots and leeks is limited

Restrictions
- Maximum total dose equivalent to 2 full doses on broad beans, field beans, cauliflowers, calabrese; 3 full doses on Brussels sprouts, leeks, onions and shallots

Crop-specific information
- HI 14 d for all crops

Environmental safety
- Dangerous for the environment
- Very toxic to aquatic organisms
- LERAP Category B

Hazard classification and safety precautions

Hazard H03, H11
Risk phrases R20, R36, R37, R38, R40, R43, R50, R58
Operator protection A, C, H; U05a, U09a, U15, U20a
Environmental protection E15a, E16a, E31c, E32d, E38
Consumer protection C02 (14 d)
Storage and disposal E01, E04, E30a

90 chlorothalonil + propamocarb hydrochloride

A contact and systemic fungicide mixture for blight control in potatoes

Products

1	Merlin	Bayer CropScience	375:375 g/l	SC	07943
2	Pan Magician	Pan Agriculture	375:375 g/l	SC	11992

Uses

- Blight in ***potatoes***

Efficacy guidance

- Commence treatment early in the season as soon as there is risk of infection
- In the absence of a blight warning treatment should start just before potatoes meet along the row
- Use only as a protectant. Stop use when blight readily visible (1% leaf area destroyed)
- Repeat sprays at 10-14 d intervals depending on blight infection risk. See label for details
- Complete blight spray programme after end Aug up to haulm destruction with protectant fungicides preferably fentin based

Restrictions

- Maximum number of treatments 5 per crop
- Apply to dry foliage. Do not apply if rainfall or irrigation imminent

Crop-specific information

- HI 7 d

Environmental safety

- Dangerous for the environment
- Very toxic to aquatic organisms
- LERAP Category B

Hazard classification and safety precautions

Hazard H03, H11
Risk phrases R20, R40, R41, R43, R50, R58
Operator protection A, C, H; U05a, U08, U11, U14, U19a, U20a
Environmental protection E15a, E16a, E16b, E31b, E32d, E34, E38
Storage and disposal E01, E04, E26, E30a

91 chlorothalonil + tetraconazole

A systemic protectant and curative fungicide mixture for cereals

Products

Voodoo	Sipcam	250:62.5 g/l	SE	09414

Uses

- Brown rust in ***spring barley***, ***spring barley*** *(autumn sown)*, ***spring wheat***, ***spring wheat*** *(autumn sown)*, ***winter barley***, ***winter wheat***
- Crown rust in ***spring oats***, ***winter oats***
- Net blotch in ***spring barley***, ***spring barley*** *(autumn sown)*, ***winter barley***
- Powdery mildew in ***spring barley***, ***spring barley*** *(autumn sown)*, ***spring wheat***, ***spring wheat*** *(autumn sown)*, ***winter barley***, ***winter wheat***
- Rhynchosporium in ***spring barley***, ***spring barley*** *(autumn sown)*, ***winter barley***
- Septoria diseases in ***spring wheat***, ***spring wheat*** *(autumn sown)*, ***winter wheat***

SEE SECTION 3 FOR PRODUCTS ALSO REGISTERED

- Sooty moulds in ***spring wheat, winter wheat***
- Yellow rust in ***spring barley, spring barley*** *(autumn sown)*, ***spring wheat, spring wheat*** *(autumn sown)*, ***winter barley, winter wheat***

Efficacy guidance

- For best results apply before onset of disease attack, or at an early stage of disease development. Further treatments may be necessary if disease pressure remains high
- May be used in a programme in combination with a number of other fungicides to improve overall control and reduce potential for development of resistance

Restrictions

- Maximum total dose equivalent to three full dose treatments on wheat; two full dose treatments on barley

Crop-specific information

- Latest use: before early flowering (GS 63) in wheat; before end of ear emergence (GS 59) in barley

Environmental safety

- Harmful to fish or other aquatic life. Do not contaminate surface waters or ditches with chemical or used container
- LERAP Category B

Hazard classification and safety precautions

Hazard H04
Risk phrases R36, R38
Operator protection A, C; U05a, U19a, U20a
Environmental protection E13c, E16a, E16b, E31b, E34
Storage and disposal E01, E04, E30a
Medical advice M03

92 chlorotoluron

A contact and residual urea herbicide for cereals

Products

1	Alpha Chlortoluron 500	Makhteshim	500 g/l	SC	04848
2	Atol	Nufarm UK	700 g/l	SC	11138
3	Copal 500	DAPT	500 g/l	SC	10633
4	Headland Tolerate	Headland	500 g/l	SC	10774
5	Lentipur CL 500	Nufarm UK	500 g/l	SC	08743
6	Tolugan 700	Makhteshim	700 g/l	SC	08064
7	Tolurex 90 WDG	Makhteshim	90% w/w	WG	11403
8	Tripart Culmus	Tripart	500 g/l	SC	06619

Uses

- Annual dicotyledons in ***durum wheat*** [5, 7]; ***triticale*** [1-7]; ***winter barley, winter wheat*** [1-8]
- Annual grasses in ***durum wheat*** [5, 7]; ***triticale*** [1-7]; ***winter barley, winter wheat*** [1-8]
- Blackgrass in ***durum wheat*** [5, 7]; ***triticale*** [1-7]; ***winter barley, winter wheat*** [1-8]
- Rough meadow grass in ***durum wheat*** [5, 7]; ***triticale*** [1-7]; ***winter barley, winter wheat*** [1-8]
- Wild oats in ***durum wheat*** [5, 7]; ***triticale*** [1-7]; ***winter barley, winter wheat*** [1-8]

Approval information

- May be applied through CDA equipment. See labels for details [2, 3, 5]
- Approved for aerial application on winter wheat, winter barley, durum wheat, triticale [5]. See notes in Section 5

Efficacy guidance

- Best results achieved by application soon after drilling. Application in autumn controls most weeds germinating in early spring
- For wild oat control apply within 1 wk of drilling, not after 2-leaf stage. Blackgrass and meadow grasses controlled to 5 leaf, ryegrasses to 3 leaf stage
- Any trash or burnt straw should be buried and dispersed during seedbed preparation
- Control may be reduced if prolonged dry conditions follow application
- Harrowing after treatment may reduce weed control
- Always follow WRAG guidelines for preventing and managing herbicide resistant weeds. Section 5 for more information

Restrictions
- Maximum number of treatments 1 per crop
- Use only on listed crop varieties. See label. Ensure seed well covered at drilling
- Apply only as pre-emergence spray in durum wheat, pre- or post-emergence in wheat, barley or triticale
- Do not apply pre-emergence to crops sown after 30 Nov
- Do not apply to crops severely checked by waterlogging, pests, frost or other factors
- Do not use on undersown crops or those due to be undersown
- Do not apply post-emergence in mixture with liquid fertilizers
- Do not roll for 7 d before or after application to an emerged crop
- Do not use on soils with more than 10% organic matter
- Crops on stony or gravelly soils may be damaged, especially after heavy rain

Crop-specific information
- Latest use: pre-emergence for durum wheat. Post emergence timings on other crops vary - see labels

Environmental safety
- Dangerous for the environment
- Very toxic to aquatic organisms. May cause long-term adverse effects in the aquatic environment

Hazard classification and safety precautions

Hazard H03 [1-5, 8]; H04 [1, 3]; H11 [1, 2, 4-8]
Risk phrases R22a, R36, R38 [1, 3, 4, 8]; R50, R58 [1, 2, 4-8]; R63, R68 [2, 5]
Operator protection A, C [1-7]; U05a [1-4, 6, 7]; U08, U19a [1-8]; U10, U11 [4]; U20a [2, 6]; U20b [1, 3-5, 7, 8]
Environmental protection E13c [3]; E15a [1, 2, 4-8]; E31b [1-8]; E32d [2, 4, 5]; E32e [1, 6, 7]; E34 [1-4, 6, 7]
Storage and disposal E01, E04 [1-7]; E26 [4, 8]; E30a [1-8]
Medical advice M03 [1-4, 6, 7]

93 chlorotoluron + isoproturon

A urea herbicide mixture for cereals

Products

Tolugan Extra	Makhteshim	300:300 g/l	SC	09393

Uses
- Annual dicotyledons in ***winter barley, winter wheat***
- Annual grasses in ***winter barley, winter wheat***
- Blackgrass in ***winter barley, winter wheat***
- Rough meadow grass in ***winter barley, winter wheat***
- Wild oats in ***winter barley, winter wheat***

Approval information
- Isoproturon included in Annex I under EC Directive 91/414

Efficacy guidance
- Best results achieved by application soon after drilling. Application in autumn controls most weeds germinating in early spring
- For wild oat control apply within 1 wk of drilling, not after 2-leaf stage. Blackgrass and meadow grasses controlled to 5 leaf, ryegrasses to 3 leaf stage
- Any trash or burnt straw should be buried and dispersed during seedbed preparation
- Control may be reduced if prolonged dry conditions follow application
- Harrowing after treatment may reduce weed control
- Where strains of herbicide-resistant blackgrass occur control may be unsatisfactory
- Always follow WRAG guidelines for preventing and managing herbicide resistant weeds. Section 5 for more information

Restrictions
- Maximum number of treatments 1 per crop
- Use only on listed crop varieties. See label. Ensure seed well covered at drilling
- Apply only as post-emergence treatment

- Do not apply to crops severely checked by waterlogging, pests, frost or other factors
- Do not use on undersown crops or those due to be undersown
- Do not apply post-emergence in mixture with liquid fertilizers
- Do not roll for 7 d before or after application to an emerged crop
- Do not use on soils with more than 10% organic matter
- Crops on stony or gravelly soils may be damaged, especially after heavy rain

Crop-specific information

- Latest use: before end of tillering (GS 29)

Environmental safety

- Dangerous for the environment
- Very toxic to aquatic organisms
- Do not apply to dry, cracked or waterlogged soils where heavy rain may lead to contamination of drains by isoproturon

Hazard classification and safety precautions

Hazard H03, H11
Risk phrases R40, R43, R50, R58
Operator protection A, C, H; U05a, U08, U19a, U20b
Environmental protection E15a, E31c, E32e, E34
Storage and disposal E01, E04, E26, E30a

94 chlorpropham

A residual carbamate herbicide and potato sprout suppressant

See also chloridazon + chlorpropham + metamitron

Products

1	BL 500	Whyte Agrochemicals	500 g/l	HN	00279
2	Comrade	United Phosphorus	400 g/l	EC	10181
3	Luxan Gro-Stop Fog	Luxan	300 g/l	HN	09388
4	Luxan Gro-Stop HN	Luxan	300 g/l	HN	07689
5	MSS CIPC 5 G	Whyte Agrochemicals	5% w/w	GR	01402
6	MSS CIPC 50 M	Whyte Agrochemicals	500 g/l	EC	11165
7	MSS Sprout Nip	Whyte Agrochem.	100% w/w	HN	11804
8	Warefog 25	Whyte Agrochemicals	600 g/l	HN	06776

Uses

- Annual dicotyledons in ***acroclinium, african marigolds, blackcurrants, calendula, carrots, celery, china asters, chrysanthemums, coreopsis, flower bulbs, french marigolds, gooseberries, lettuce, lucerne, onions, parsley, straw flower, strawberries, sweet sultan*** [2]
- Annual grasses in ***acroclinium, african marigolds, blackcurrants, calendula, carrots, celery, china asters, chrysanthemums, coreopsis, flower bulbs, french marigolds, gooseberries, lettuce, lucerne, onions, parsley, straw flower, strawberries, sweet sultan*** [2]
- Chickweed in ***acroclinium, african marigolds, blackcurrants, calendula, carrots, celery, china asters, chrysanthemums, coreopsis, french marigolds, gooseberries, lettuce, lucerne, parsley, straw flower, strawberries, sweet sultan*** [2]
- Polygonums in ***acroclinium, african marigolds, blackcurrants, calendula, carrots, celery, china asters, chrysanthemums, coreopsis, french marigolds, gooseberries, lettuce, lucerne, parsley, straw flower, strawberries, sweet sultan*** [2]
- Sprout suppression in ***potatoes*** [5, 6]; ***potatoes*** *(thermal fog)* [7]; ***ware potatoes*** *(fog)* [8]; ***ware potatoes*** *(thermal fog)* [1, 3, 4]

Approval information

- Some products are formulated for application by thermal fogging. See labels for details [1, 3, 4, 7, 8]

Efficacy guidance

- Apply weed control sprays to freshly cultivated soil. Adequate rainfall must occur after spraying. Activity is greater in cold, wet than warm, dry conditions
- For sprout suppression apply with suitable fogging or rotary atomiser equipment or sprinkle granules over dry tubers before sprouting commences. Repeat applications may be needed. See labels for details

- Best results on potatoes obtained in purpose-built box stores with suitable forced draft ventilation. Potatoes in bulk stores should not be stacked more than 3 m high
- It is important to treat potatoes before the eyes open to obtain best results
- Effectiveness of fogging reduced in non-dedicated stores without proper insulation and temperature controls. Best results obtained at 5-10°C and 75-80% humidity

Restrictions

- Maximum number of treatments 1 per batch for sprout suppresion of potatoes; 2 per yr for flower bulbs; 1 per yr for grass seed crops [2]; 1 per crop for other named crops
- Maximum total dose equivalent to the above multiples of full dose treatments
- Not to be used on grass seed crops if grass to be grazed or cut for fodder before 31 May following treatment [2]
- Excess rainfall after application may result in crop damage
- Do not use on Sands, Very Light soils or soils low in organic matter
- Poor conditions at drilling or planting, soil compaction, surface capping, waterlogging or attack by pests may result in crop damage
- On crops under glass high temperatures and poor ventilation may cause crop damage
- Only clean, mature, disease-free potatoes should be treated for sprout suppression. Use of chlorpropham can inhibit tuber wound healing and the severity of skin spot infection in store may be increased if damaged tubers are treated
- Do not fog potatoes with a high level of skin spot
- Do not use on potatoes for seed. Do not handle, dry or store seed potatoes or any other seed or bulbs in boxes or buildings in which potatoes are being or have been treated
- Do not remove potatoes for sale or processing for at least 21 d (4 wk [4]) after application
- A minimum interval of 45 d must elapse between applications

Crop-specific information

- Latest use: 3 d before drilling carrots; 7d before planting out celery; 2 wk before planting or at least 10 d after planting for chrysanthemums [2]; 21 d before removal for sale or processing for potatoes [1, 5, 6, 8]; before tulip leaves unfurl or flower bulbs 5 cm high
- Cure potatoes according to label instructions before treatment and allow 3 wk between completion of loading into store and first treatment
- Apply weed control sprays to seeded crops pre-emergence of crop or weeds, to onions as soon as first crop seedlings visible, to planted crops a few days before planting, to bulbs immediately after planting, to fruit crops in late autumn-early winter. See label for further details

Environmental safety

- Dangerous for the environment
- Toxic to aquatic organisms
- Keep unprotected persons out of treated stores for at least 24 h after application

Hazard classification and safety precautions

Hazard H02, H07 [1, 6, 7]; H03 [2-4, 8]; H08 [2]; H11 [1, 3-8]

Risk phrases R20, R21, R22a [2]; R23, R24, R39 [1, 6, 7]; R25 [1]; R36 [1, 6-8]; R38 [2, 8]; R40, R42, R43 [3, 4]; R48 [8]; R51 [1, 3, 4, 6, 8]; R52 [5]; R58 [1, 3-6, 8]

Operator protection A [1-4, 6, 7]; C [1, 2, 6, 7]; D, E, H, M [1, 3, 4, 6-8]; J [1, 6-8]; U02a, U04a, U20a [3, 4]; U05a [1-8]; U08 [1-4, 6-8]; U09a [2]; U14, U15 [3, 4, 8]; U19a [1, 3, 4, 6-8]; U20b [1, 2, 5-7]; U20c [8]

Environmental protection E02a [1, 6] (24 h); E02a [7]; E02a [8] (12 h); E15a, E34 [1-8]; E31a [1, 6-8]; E31b [2]; E32a [3-5]; E32d [3, 4]; E32e [1, 2, 5-8]

Consumer protection C12 [3, 4]

Storage and disposal E01, E04 [1-8]; E26 [1-4, 6, 7]; E28 [1, 2, 6, 7]; E29a [3, 4]; E29b [2]; E30a [2, 5, 8]; E30b [1, 3, 4, 6, 7]

Medical advice M03 [2]; M04 [1, 3, 4, 6, 7]

95 chlorpropham with cetrimide

A soil-acting herbicide for lettuce under cold glass

Products

Croptex Pewter	Certis	80:80 g/l	SC	11181

SEE SECTION 3 FOR PRODUCTS ALSO REGISTERED

Uses

- Annual dicotyledons in ***protected lettuce***
- Annual grasses in ***protected lettuce***
- Chickweed in ***protected lettuce***
- Polygonums in ***protected lettuce***

Efficacy guidance

- Adequate irrigation must be applied before or after treatment
- Best results achieved on firm soil of fine tilth, free from clods and weeds
- Control of susceptible weeds is achieved at, or shortly after, germination and residual control lasts for 6-8 wk depending on weather conditions

Restrictions

- Maximum number of treatments 1 per crop
- Apply within 24 h of drilling
- Do not apply to crop foliage or use where seed has germinated
- Excess irrigation may cause temporary check to crop under certain circumstances
- Do not apply where tomatoes, brassicas or other sensitive crops are growing in the same house

Crop-specific information

- Latest use: before crop emergence or pre-planting
- Apply to drilled lettuce under cold glass within 24 h post-drilling, to transplanted crops pre-planting and treat up to 7 d later

Following crops guidance

- In the event of crop failure only lettuce should be grown within 2 mth
- After cutting a normally harvested crop the treated area should be cultivated to a minimum depth of 15 cm before sowing any succeeding crop

Environmental safety

- Dangerous for the environment
- Toxic to aquatic organisms
- Highly flammable

Hazard classification and safety precautions

Hazard H03, H07, H11
Risk phrases R21, R22a, R36, R37, R38, R51, R58
Operator protection A, C; U05a, U08, U14, U15, U19a, U20a
Environmental protection E15a, E32d, E34
Storage and disposal E01, E04
Medical advice M03

96 chlorpropham + fenuron

A residual herbicide for runner beans and spinach

Products

Croptex Chrome	Certis	80:15 g/l	EC	11180

Uses

- Annual dicotyledons in ***leaf spinach***, ***runner beans*** *(off-label)*
- Annual grasses in ***leaf spinach***
- Chickweed in ***leaf spinach***

Specific Off-Label Approvals (SOLAs)

- ***runner beans*** *(OLA 030871) Dec 2008* [1]

Approval information

- Products containing fenuron have been granted derogations for specified 'Essential Uses' for use until 31 December 2007. Sale and supply must cease by 30 June 2007 but growers have no guarantee that the products will continue to be available until then.
  For more information see 'The Review Programme' under 'Pesticide Legislation' in Section 5

Efficacy guidance

- Apply to soil freshly cultivated and free of established weeds. Adequate rainfall must occur after spraying. Activity is greater in cold, wet than warm, dry conditions

Restrictions
- Maximum number of treatments 1 per crop
- Do not use on Sands, Very Light soils or soils low in organic matter
- Poor conditions at drilling or planting, soil compaction, surface capping, waterlogging or attack by pests may result in crop damage
- This product must not be used on any crops other than those listed, including any extrapolations that would normally be permissible under the Long Term Arrangements for Extension of Use (see Section 5)

Crop-specific information
- Latest use: pre crop emergence

Following crops guidance
- In the event of crop failure only recommended crops should be replanted in treated soil
- Plough or cultivate to 15 cm after harvest to dissipate any residues

Environmental safety
- Dangerous for the environment
- Very toxic to aquatic organisms

Hazard classification and safety precautions

Hazard H03, H08, H11
Risk phrases R21, R22a, R36, R37, R38, R50, R58
Operator protection A, C; U05a, U08, U11, U13, U14, U19a, U20b
Environmental protection E15a, E19b, E31a, E32e, E34
Storage and disposal E01, E04, E30a
Medical advice M03, M05a

97 chlorpropham + metamitron

A contact and residual herbicide mixture for beet crops

Products

Golmet	Nufarm UK	24.5:280 g/l	SC	10684

Uses
- Annual dicotyledons in ***fodder beet, mangels, sugar beet***
- Annual meadow grass in ***fodder beet, mangels, sugar beet***

Efficacy guidance
- May be used pre-emergence alone or post-emergence in tank mixture or with an authorised adjuvant oil
- Use as part of a spray programme. For optimum efficacy a full programme of pre- and post-emergence sprays is recommended
- Ideally one pre-emergence application should be followed by a repeat overall low dose programme on mineral soils. On organic soils a programme of up to five post-emergence sprays is likely to be most effective
- Product combines residual and contact activity. Adequate soil moisture and a fine firm seedbed are essential for optimum residual activity

Restrictions
- Maximum number of treatments one pre-emergence plus three post-emergence or, where no pre-emergence treatment has been applied, up to five post-emergence treatments with an adjuvant oil

Crop-specific information
- Latest use: before crop foliage meets across the rows

Following crops guidance
- Only sugar beet, fodder beet or mangels should be sown as following crops within 4 mth of treatment. After normal harvest of a treated crop land should be mouldboard ploughed to 15 cm after which any spring crops may be sown or planted

Environmental safety
- Dangerous for the environment
- Very toxic to aquatic organisms

SEE SECTION 3 FOR PRODUCTS ALSO REGISTERED

Hazard classification and safety precautions

Hazard H03, H11

Risk phrases R22a, R36, R38, R50, R58

Operator protection A, C; U05a, U08, U11, U14, U15, U19a, U20a

Environmental protection E15a, E31c, E34

Storage and disposal E01, E04, E26, E30a

98 chlorpropham + pentanochlor

A contact and residual herbicide for horticultural crops

Products

Atlas Brown	Nufarm UK	150:300 g/l	EC	07703

Uses

- Annual dicotyledons in ***carrots, celeriac, celery, chrysanthemums, narcissi, parsley, parsnips, tulips***
- Annual meadow grass in ***carrots, celeriac, celery, chrysanthemums, narcissi, parsley, parsnips, tulips***

Approval information

- Products containing pentanochlor have been granted derogations for specified 'Essential Uses' for use until 31 December 2007. Sale and supply must cease by 30 June 2007 but growers have no guarantee that the products will continue to be available until then.
  For more information see 'The Review Programme' under 'Pesticide Legislation' in Section 5

Efficacy guidance

- Apply as pre- or post-weed emergence spray
- Best results by application to weeds up to 2-leaf stage on fine, firm, moist seedbed
- Greatest contact action achieved under warm, moist conditions, the short residual action greatest in earlier part of year

Restrictions

- Maximum number of treatments 2 per crop
- This product must not be used on any crops other than those listed, including any extrapolations that would normally be permissible under the Long Term Arrangements for Extension of Use (see Section 5), except herbs and ornamentals

Crop-specific information

- Latest use: pre-emergence for narcissi, tulips
- HI 28 d for carrots, parsnips
- Apply to carrots and related crops pre- or post-emergence after fully expanded cotyledon stage, to narcissi and tulips at any time before emergence
- Apply to chrysanthemums either pre-planting or after planting as carefully directed spray, avoiding foliage

Following crops guidance

- Any crop may be sown or planted after 4 wk following ploughing and cultivation

Environmental safety

- Dangerous for the environment
- Toxic to aquatic organisms

Hazard classification and safety precautions

Hazard H03, H11

Risk phrases R21, R22a, R22b, R36, R48, R51, R58

Operator protection C; U05a, U09a, U14, U15, U19a, U20a

Environmental protection E15a, E31b, E32e, E38

Consumer protection C02 (carrots, parsnips 28 d)

Storage and disposal E01, E04, E30a

Medical advice M04, M05b

99 chlorpyrifos

A contact and ingested organophosphorus insecticide and acaricide

Products

	Product	Company	Conc.	Form.	Reg. No.
1	Alpha Chlorpyrifos 48 EC	Makhteshim	480 g/l	EC	04821
2	Ballad	Headland	480 g/l	EC	11659
3	Chevron 48	DAPT	480 g/l	EC	10645
4	Crossfire 480	Bayer Environ.	480 g/l	EC	09929
5	Cyren	Cheminova	480 g/l	EC	08358
6	Cyren	Headland	480 g/l	EC	11028
7	Dursban WG	Dow	75% w/w	WG	09153
8	Equity	Dow	480 g/l	EC	11520
9	Greencrop Pontoon	Greencrop	480 g/l	EC	09667
10	Lorsban T	Rigby Taylor	480 g/l	EC	07813
11	Lorsban WG	Landseer	75% w/w	WG	11962
12	Maraud	Scotts	480 g/l	EC	09274
13	Pyrinex 48EC	Makhteshim	480 g/l	EC	08644
14	Spannit	SumiAgro Amenity	480 g/l	EC	08744
15	Spannit Granules	SumiAgro	6% w/w	GR	10935
16	Standon Chlorpyrifos 48	Standon	480 g/l	EC	08286
17	suSCon Green Soil Insecticide	Fargro	10% w/w	CG	06312

Uses

- Ambrosia beetle in ***cut logs*** [1, 3, 5-7, 9, 10, 13]
- Aphids in ***apples, gooseberries, pears, plums, raspberries, strawberries*** [1, 3, 5, 6, 8, 9, 11, 13, 14, 16]; ***blackcurrants*** [1, 11, 13]; ***broccoli, cabbages, calabrese, cauliflowers, chinese cabbage*** [1, 3, 5-9, 13, 16]; ***brussels sprouts*** [1, 13, 16]; ***currants*** [3, 5, 6, 8, 9, 14, 16]; ***edible brassicas*** [14]; ***redcurrants, whitecurrants*** [9, 11]; ***spring barley, winter barley*** [1, 3, 7, 8, 13, 14]; ***spring oats, spring wheat, winter oats, winter wheat*** [1, 3, 5-9, 13, 14, 16]
- Apple blossom weevil in ***apples*** [1, 3, 5, 6, 8, 9, 11, 13, 14, 16]
- Apple sucker in ***apples*** [16]
- Bean seed fly in ***green beans*** *(off-label - harvested as a dry pulse)* [7]
- Black pine beetle in ***forestry transplants*** [9, 10]
- Cabbage root fly in ***broccoli, cabbages, cauliflowers*** [1, 3, 5-9, 13, 15, 16]; ***brussels sprouts*** [1, 7, 8, 13, 16]; ***calabrese, chinese cabbage*** [1, 3, 5-9, 13, 16]; ***collards*** *(off-label)*, ***kale*** *(off-label)* [1, 7]; ***edible brassicas*** [14]; ***kohlrabi*** *(off-label)*, ***mooli*** *(off-label)*, ***radishes*** *(off-label)* [7]
- Capsids in ***apples, gooseberries, pears*** [1, 3, 5, 6, 8, 9, 11, 13, 14, 16]; ***blackcurrants*** [1, 11, 13]; ***currants*** [3, 5, 6, 8, 9, 14, 16]; ***redcurrants, whitecurrants*** [9, 11]
- Caterpillars in ***apples*** [1, 3, 5, 6, 8, 9, 13, 14]; ***blackcurrants*** [1, 11, 13]; ***broccoli, cabbages, calabrese, cauliflowers, chinese cabbage*** [1, 3, 5-9, 13, 16]; ***brussels sprouts*** [1, 13, 16]; ***currants*** [3, 5, 6, 8, 9, 14, 16]; ***edible brassicas*** [14]; ***gooseberries*** [1, 3, 5, 6, 8, 9, 11, 13, 14, 16]; ***pears*** [1, 3, 5, 6, 8, 9, 11, 13, 14]; ***redcurrants, whitecurrants*** [9, 11]
- Codling moth in ***apples, pears*** [1, 3, 5, 6, 8, 9, 11, 13, 14, 16]
- Cutworms in ***broccoli, cabbages, calabrese, cauliflowers, chinese cabbage*** [1, 3, 5-9, 13, 16]; ***brussels sprouts*** [1, 13, 16]; ***carrots, potatoes*** [1, 5-9, 13, 14, 16]; ***edible brassicas*** [14]; ***onions*** [1, 5-9, 13, 16]
- Damson-hop aphid in ***plums*** [1, 9, 11, 13, 16]
- Elm bark beetle in ***cut logs*** [3, 5, 6, 10]
- Frit fly in ***amenity grass*** [2, 4-6, 12, 16]; ***amenity turf*** [9, 16]; ***golf courses*** [5, 6, 14]; ***grassland, maize, spring oats, spring wheat, winter oats, winter wheat*** [1, 3, 5-9, 13, 14, 16]; ***managed amenity turf*** [1-6, 10, 12, 13]; ***rotational grassland*** [1, 13]; ***spring barley*** [8, 14]; ***winter barley*** [8]
- Larch shoot beetle in ***cut logs*** [1, 3, 5-10, 13]
- Leatherjackets in ***amenity grass*** [2, 4-6, 12, 16]; ***amenity turf*** [9, 16]; ***broccoli, cabbages, calabrese, cauliflowers, chinese cabbage*** [3, 5-8, 16]; ***brussels sprouts*** [16]; ***edible brassicas*** [14]; ***golf courses*** [5, 6, 14]; ***grassland, spring oats, spring wheat, sugar beet, winter oats, winter wheat*** [1, 3, 5-9, 13, 14, 16]; ***managed amenity turf*** [1-6, 10, 12, 13]; ***rotational grassland*** [1, 13]; ***spring barley*** [1, 3, 7, 8, 13]; ***winter barley*** [1, 3, 7, 8, 13, 14]
- Mealy aphid in ***plums*** [1, 9, 11, 13]
- Narcissus fly in ***flower bulbs*** *(off-label)* [1]
- Pear sucker in ***pears*** [16]

- Pine shoot beetle in ***cut logs*** [1, 3, 5-10, 13]
- Pine weevil in ***forestry transplants*** [9, 10]
- Pygmy mangold beetle in ***sugar beet*** [3, 5-8, 14]
- Raspberry beetle in ***raspberries*** [1, 3, 5, 6, 8, 9, 11, 13, 14, 16]
- Raspberry cane midge in ***raspberries*** [1, 3, 5, 6, 8, 9, 11, 13, 14, 16]
- Red spider mites in ***apples, gooseberries, pears, plums, raspberries, strawberries*** [1, 3, 5, 6, 8, 9, 11, 13, 14, 16]; ***blackcurrants*** [1, 11, 13]; ***currants*** [3, 5, 6, 8, 9, 14, 16]; ***redcurrants, whitecurrants*** [9, 11]
- Sawflies in ***apples*** [1, 3, 5, 6, 8, 9, 11, 13, 14, 16]; ***pears*** [8, 14]
- Strawberry blossom weevil in ***strawberries*** [8, 11]
- Suckers in ***apples, pears*** [1, 3, 5, 6, 8, 9, 11, 13, 14]
- Summer-fruit tortrix moth in ***apples*** [16]
- Thrips in ***spring oats, spring wheat, winter oats, winter wheat*** [3, 5, 6, 14]; ***winter barley*** [14]
- Tortrix moths in ***apples, pears, plums, strawberries*** [1, 3, 5, 6, 8, 9, 11, 13, 14, 16]
- Vine weevil in ***conifers*** [5-10]; ***container-grown ornamentals, ornamental plant production*** [17]; ***ornamental plant production*** *(conifers)* [1, 13]; ***strawberries*** [1, 3, 5, 6, 8, 9, 11, 13, 16]
- Wheat bulb fly in ***spring barley, winter barley*** [7, 8]; ***spring oats, spring wheat, winter oats, winter wheat*** [1, 3, 5-9, 13, 14, 16]
- Wheat-blossom midge in ***spring oats, spring wheat, winter oats, winter wheat*** [1, 3, 5-9, 13, 14, 16]
- Whitefly in ***broccoli, cabbages, calabrese, cauliflowers, chinese cabbage*** [3, 5-8]
- Winter moth in ***apples, pears, plums*** [1, 3, 5, 6, 8, 9, 11, 13, 14, 16]
- Woolly aphid in ***apples*** [1, 3, 5, 6, 8, 9, 11, 13, 14, 16]; ***pears*** [3, 5, 6, 8, 14]
- Yellow cereal fly in ***spring wheat, winter wheat*** [7, 8]

Specific Off-Label Approvals (SOLAs)

- ***collards, kale*** *(OLA 021688) Dec 2008* [1]
- ***collards, kale*** *(OLA 031390) Dec 2008* [7]
- ***flower bulbs*** *(OLA 021458) Dec 2008* [1]
- ***green beans*** *(harvested as a dry pulse) (OLA 031391) Dec 2008* [7]
- ***kohlrabi, mooli, radishes*** *(OLA 031391) Dec 2008* [7]

Approval information

- Following implementation of Directive 98/82/EC, approval for use of chlorpyrifos on numerous crops was revoked in 1999

Efficacy guidance

- Brassicas raised in plant-raising beds may require retreatment at transplanting
- Activity may be reduced when soil temperature below 5°C or on organic soils
- In dry conditions the effect of granules applied as a surface band may be reduced [15]
- Where pear suckers resistant to chlorpyrifos occur control is unlikely to be satisfactory
- Efficacy against frit fly and leatherjackets in grass may be reduced if applied during periods of frost

Restrictions

- Maximum number of treatments and timing vary with crop, product and pest. See label for details
- On carrots the maximum total dose applied per crop must not exceed the equivalent of 3 (on mineral soils) or 4 (on organic soils) full dose applications
- Do not treat potatoes under severe drought stress. The variety Desiree is particularly susceptible
- Do not apply to sugar beet under stress or within 4 d of applying a herbicide
- Do not use product in growing media used for aquatic or semi-aquatic plants, and not for propagation of any edible crop
- Do not graze lactating cows on treated pasture for 14 d after treatment

Crop-specific information

- Latest use 4d after transplanting edible brassica crops; end Jul for sugar beet; varies for other crops. See individual labels
- HI range from 7 d to 6 wk depending on crop. See labels
- For vine weevil control in hardy ornamental nursery stock incorporate in growing medium when plants first potted from rooted cutting stage. Treat the fresh growing medium when plants are potted on into larger containers [17]
- For turf pests apply from Nov where high larval populations detected or damage seen [2, 4-6, 10, 12]

- On lettuce apply only to strong well developed plants when damage first seen, or on professional advice
- In apples use pre-blossom up to pink/white bud and post-blossom after petal fall
- Test tolerance of glasshouse ornamentals before widescale use for propagating unrooted cuttings, or potting unusual plants and new species, and when using any media with a high content of non-peat ingredients

Environmental safety

- Dangerous for the environment
- Very toxic to aquatic organisms
- Toxic to aquatic organisms [15]
- Keep livestock out of treated areas for at least 14 d after treatment
- High risk to bees. Do not apply to crops in flower or to those in which bees are actively foraging
- Broadcast air-assisted LERAP (18 m) and LERAP Category A (all except [17])

Hazard classification and safety precautions

Hazard H03 [1-16]; H04 [3, 14]; H08 [1, 3, 4, 8, 9, 12-14, 16]; H11 [1, 2, 4-13, 15, 16]
Risk phrases R20 [1, 2, 5, 6, 13]; R21 [4, 9, 14-16]; R22a [1-16]; R22b [1-6, 8, 9, 12, 13, 16]; R36 [1, 13, 14]; R37, R67 [4, 8-10, 12, 16]; R38 [1-6, 8-10, 12-14, 16]; R40 [1, 13]; R41, R43 [1, 3, 13]; R50 [1, 2, 4-13, 16]; R51 [15]; R58 [1, 2, 4-13, 15, 16]
Operator protection A [1-17]; C [1-3, 5, 6, 13, 14]; H [1-3, 5, 6, 9-11, 13, 14, 16]; P [8]; U02a [1, 3, 13-15, 17]; U05a [1-6, 8-10, 12-14, 16]; U08, U19a [1-14, 16]; U14 [1, 2, 5, 6, 13]; U15 [1, 13]; U20a [14, 15]; U20b [1-13, 16, 17]
Environmental protection E06a [1-3, 5-8, 10-13] (14 d); E12a [2, 4-6, 11]; E12d [1, 3, 7-10, 12-14, 16]; E13a [1, 3, 13, 17]; E13b [14]; E15a [2, 4-12, 15, 16]; E16c, E16d, E34 [1-14, 16]; E17b [1]; E17b [3-7] (18 m); E17b [8]; E17b [9] (18 m); E17b [10]; E17b [11, 12] (18 m); E17b [13]; E17b [16] (18 m); E31b [1-3, 5-14, 16]; E31c [4]; E32a [14, 15, 17]; E32d, E38 [4, 7, 8, 10-12, 15]; E32e [2, 5, 6]
Consumer protection C01 [17]; C02 [14]; C02 [15] (6 wk)
Storage and disposal E01 [1-10, 12-17]; E04 [1-10, 12-14, 16]; E26 [1, 3, 9, 16, 17]; E30a [1-17]
Medical advice M01 [1-7, 9-17]; M03 [1-14, 16]; M05b [1-6, 8-10, 12, 13, 16]

100 chlorpyrifos-methyl

An organophosphorus insecticide and acaricide for grain store use

Products

1	Greencrop Storeclean 225	Greencrop	225 g/l	EC	11098
2	Reldan 22	Dow	225 g/l	EC	08191

Uses

- Grain storage pests in ***stored grain***
- Pre-harvest hygiene in ***grain stores***

Approval information

- Approvals for chlorpyrifos-methyl allowed to continue following the UK Review programme of cholinesterase compounds

Efficacy guidance

- Apply to grain after drying to moisture content below 14%, cooling and cleaning
- Insecticide may become depleted at grain surface if grain is being cooled by continuous extraction of air from the base leading to reduced control of grain store pests especially mites
- Resistance to organophosphorus compounds sometimes occurs in insect and mite pests of stored products

Restrictions

- This product contains an anticholinesterase organophosphorus compound. Do not use if under medical advice not to work with such compounds
- Maximum number of treatments 1 per batch or 1 per store, prior to storage
- Only treat grain in good condition
- Do not treat grain intended for sowing
- Minimum of 90 d must elapse between treatment of grain and removal from store for consumption or processing

SEE SECTION 3 FOR PRODUCTS ALSO REGISTERED

Crop-specific information

- May be applied pre-harvest to surfaces of empty store and grain handling machinery and as admixture with grain
- May be used in wheat, barley, oats, rye or triticale stores

Environmental safety

- Dangerous for the environment
- Very toxic to aquatic organisms

Hazard classification and safety precautions

Hazard H03, H11 [1]; H04 [2]
Risk phrases R22b, R41 [1, 2]; R50, R58 [1]
Operator protection A, C, H, M; U05a, U08, U19a, U20b
Environmental protection E13a [2]; E15a [1]; E31b, E34 [1, 2]
Storage and disposal E01, E04, E26, E30a
Medical advice M01, M05b [1, 2]; M03 [1]

101 chlorthal-dimethyl

A residual benzoic acid herbicide for use in horticulture

Products

Dacthal W-75	Certis	75% w/w	WP	11323

Uses

- Annual dicotyledons in ***bedding plants***, ***blackcurrants***, ***broccoli***, ***brussels sprouts***, ***bulb onions***, ***cabbages***, ***calabrese***, ***cauliflowers***, ***chives*** *(off-label)*, ***fodder rape***, ***gooseberries***, ***herbs (see appendix 6)*** *(off-label)*, ***kale***, ***leeks***, ***parsley*** *(off-label)*, ***raspberries***, ***roses***, ***runner beans***, ***sage***, ***shrubs***, ***strawberries***, ***swedes***, ***trees***, ***turnips***
- Black nightshade in ***dwarf beans*** *(off-label)*, ***navy beans*** *(off-label)*, ***poppies*** *(off-label - for morphine production)*
- Fat hen in ***poppies*** *(off-label - for morphine production)*
- General weed control in ***courgettes*** *(off-label)*, ***marrows*** *(off-label)*
- Slender speedwell in ***managed amenity turf***
- Volunteer potatoes in ***dwarf beans*** *(off-label)*, ***navy beans*** *(off-label)*

Specific Off-Label Approvals (SOLAs)

- ***chives***, ***herbs (see appendix 6)***, ***parsley*** *(OLA 030873) Dec 2008* [1]
- ***courgettes***, ***marrows*** *(OLA 041561) Dec 2008* [1]
- ***dwarf beans***, ***navy beans*** *(OLA 030874) Dec 2008* [1]
- ***poppies*** *(for morphine production) (OLA 032407) Dec 2008* [1]

Efficacy guidance

- Best results on fine firm weed-free soil when adequate rain or irrigation follows
- Apply after drilling or planting prior to weed emergence. Rates and timing vary with crop and soil type. See label for details
- For control of slender speedwell in turf apply when weeds growing actively. Do not mow for at least 3 d after treatment

Restrictions

- Maximum number of treatments 1 per crop or 1 per season for all crops
- Do not apply mixture with propachlor to newly planted strawberries after rolling or application of other herbicides
- Do not use on strawberries between flowering and harvest
- Many types of ornamental have been treated successfully. See label for details. For species of unknown susceptibility treat a small number of plants first
- Do not use on turf where bent grasses form a major constituent of sward
- Do not plant lettuce within 6 mth of application, seeded turf within 2 mth, other crops within 3 mth. In the event of crop failure deep plough before re-drilling or planting
- Do not use on dwarf French beans
- Do not use on organic soils

Crop-specific information

- Latest use: before crop emergence on runner beans, sage, other herbs and poppies; end of flowering for strawberries. Some restrictions apply to other crops. See label for details
- Recommended alone on roses, runner beans, various ornamentals, strawberries and turf, in tank-mix with propachlor on brassicas, onions, leeks, sage, ornamentals, established strawberries and newly planted soft fruit

Following crops guidance

- The interval between trreatment and planting a following crop should be 6 mth for lettuce, 2 mth for seeded turf, 3 mth for other crops
- In the event of failure of a treated crop the land must be deep mould-board ploughed before re-drilling or re-planting

Hazard classification and safety precautions

Hazard H03
Risk phrases R40, R58
Operator protection U19a, U20a
Environmental protection E15a, E32a
Storage and disposal E01, E04, E30a
Medical advice M04

102 chlorthal-dimethyl + propachlor

A residual herbicide mixture for a range of vegetable crops

Products

Decimate	Certis	225:216 g/l	SC	11008

Uses

- Annual dicotyledons in ***broccoli, brussels sprouts, bulb onions, cabbages, calabrese, cauliflowers, fodder rape, kale, leeks, mustard, salad onions, swedes, turnips***
- Annual meadow grass in ***broccoli, brussels sprouts, bulb onions, cabbages, calabrese, cauliflowers, fodder rape, kale, leeks, mustard, salad onions, swedes, turnips***

Efficacy guidance

- Product may be applied pre- or post-emergence of the crop but pre-emergence of the weeds. Emerged weeds are not controlled
- Soils should be fine, firm and free of weed growth
- For best results apply to moist soil and before rain. In dry weather irrigate lightly after application and before seedling weeds appear
- Weed control may be reduced under very wet, dry or windy conditions
- Do not use on soils with more than 10% organic matter as weed control will be impaired

Restrictions

- Maximum number of treatments 1 per crop for all crops

Crop-specific information

- Use on brassica crops pre-crop emergence may result in a check to growth but this is usually outgrown within 8 wk without effect on yield or quality
- Post-emergence treatment on brassica crops should not be made until they have 3-4 true leaves. Post-emergence treatment on onions should be made post-crook stage
- Young transplanted plants, particularly cabbages and cauliflowers, raised under glass should be hardened off before treatment
- Care should be taken with modular transplants to ensure that that roots are not exposed to spray

Following crops guidance

- The interval between treatment and planting succeeding crops should be 2 mth for seeded turf, 6 mth for lettuce, 3 mth for all other crops
- In the event of failure of a treated crop the land must be deep mouldboard ploughed before redrilling or replanting

Environmental safety

- Dangerous for the environment
- Toxic to aquatic organisms

Hazard classification and safety precautions

Hazard H03, H11
Risk phrases R36, R40, R43, R51, R58
Operator protection A, C; U02a, U04a, U05a, U09b, U11, U14, U15, U19a, U20a
Environmental protection E15a, E31a
Storage and disposal E01, E04, E30a
Medical advice M04

103 cholecalciferol

A hypercalcaemic rodenticide

Products

Sorex Fatal	Sorex	0.1% w/w	RB	H7087

Uses

- Rats in ***farm buildings/yards***

Efficacy guidance

- Carry out pre-baiting with blank pellets at each proposed baiting point in runs, burrows and where droppings seen, for at least 3 d
- Inspect baiting points daily and replenish for up to 5 d
- Remove all remaining bait and containers after 5 d to prevent bait shyness
- If rat activity continues switch to a rodenticide with a different active ingredient

Environmental safety

- Prevent access to bait by children, birds and non-target animals particularly dogs, cats, pigs and poultry

Hazard classification and safety precautions

Operator protection A; U13, U20b
Environmental protection E32a
Storage and disposal E30a
Vertebrate/rodent control products V01a, V02, V03b, V04a
Medical advice M03

104 cholecalciferol + difenacoum

A mixture of rodenticides with different modes of action

Products

Sorexa CD Mouse Bait	Sorex	0.075:0.0025% w/w	RB	H6751

Uses

- Mice in ***farm buildings/yards***

Efficacy guidance

- Lay baits in mouse runs in many locations throughout infested area
- It is important to lay many small baits as mice are sporadic feeders
- Cover baits to protect from moisture in outdoor situations
- Inspect baits frequently and replace or top up with fresh material as necessary

Environmental safety

- Protect baits from access by children, birds and non-target animals, particularly dogs, cats, pigs and poultry
- Product contains Bitrex to help prevent accidental human consumption

Hazard classification and safety precautions

Operator protection A; U13, U20b
Environmental protection E32a
Storage and disposal E30a
Vertebrate/rodent control products V01b, V02, V03b, V04b
Medical advice M03

105 choline chloride

A plant growth regulator available only in mixtures

See also *chlormequat with choline chloride + imazaquin*

106 citronella oil

A natural plant extract herbicide

Products

Barrier H	Barrier	22.9% w/w	EW	10136

Uses
- Ragwort in ***grassland, land temporarily removed from production***

Efficacy guidance
- Best results obtained from spot treatment of ragwort in the rosette stage, during dry still conditions
- Aerial growth of ragwort is rapidly destroyed. Longer term control depends on overall management strategy
- Check for regrowth after 28 d and re-apply as necessary

Crop-specific information
- Contact with grasses will result in transient scorch which is outgrown in good growing conditions

Environmental safety
- Apply away from bees
- Harmful to fish or other aquatic life. Do not contaminate surface waters or ditches with chemical or used container
- Keep livestock out of treated areas for at least 2 wk and until foliage of any poisonous weeds such as ragwort has died and become unpalatable

Hazard classification and safety precautions

Hazard H04
Risk phrases R36, R38
Operator protection A, C; U05a, U20c
Environmental protection E07b (2 wk); E12g, E13c, E32a
Storage and disposal E01, E04, E26, E30a

107 clodinafop-propargyl

A contact acting herbicide for annual grass weed control in wheat, triticale and rye

Products

1	Greencrop Boulevard	Greencrop	240 g/l	EC	09960
2	Landgold Clodinafop	Landgold	240 g/l	EC	10172
3	Marathon	Me2	240 g/l	EC	09959
4	Standon Clodinafop 240	Standon	240 g/l	EC	10174
5	Topik	Syngenta	240 g/l	EC	08461

Uses
- Blackgrass in ***durum wheat, spring rye, spring wheat, triticale, winter rye, winter wheat***
- Rough meadow grass in ***durum wheat, spring rye, spring wheat, triticale, winter rye, winter wheat***
- Wild oats in ***durum wheat, spring rye, spring wheat, triticale, winter rye, winter wheat***

Efficacy guidance
- Spray when majority of weeds have germinated but before competition reduces yield
- Products contain a herbicide safener (cloquintocet-mexyl) that improves crop tolerance to clodinafop-propargyl
- Optimum control achieved when all grass weeds emerged. Wait for delayed germination on dry or cloddy seedbed
- A mineral oil additive is recommended to give more consistent control of very high blackgrass populations or for late season treatments. See label for details

- Weed control not affected by soil type, organic matter or straw residues
- Control may be reduced if rain falls within 1 h of treatment
- Always follow WRAG guidelines for preventing and managing herbicide resistant weeds. Section 5 for more information

Restrictions
- Maximum total dose equivalent to one full dose treatment
- Do not use on barley or oats
- Do not treat crops under stress or suffering from waterlogging, pest attack, disease or frost
- Do not treat crops undersown with grass mixtures
- Do not mix with products containing MCPA, mecoprop-P, 2,4-D or 2,4-DB
- MCPA, mecoprop, 2,4-D or 2,4-DB should not be applied within 21 d before, or 7 d after, treatment

Crop-specific information
- Latest use: before second node detectable stage (GS 32) for durum wheat, triticale, rye; before flag leaf sheath extending (GS 41) for wheat
- Spray in autumn, winter or spring from 1 true leaf stage (GS 11) to before second node detectable (GS 32) on durum, rye, triticale; before flag leaf sheath extends (GS 41) on wheat

Following crops guidance
- Only a broad leaved crop may be sown after failure of a treated crop. After normal harvest any broad leaved crop or wheat, durum wheat, rye, triticale or barley should be sown

Environmental safety
- Dangerous for the environment
- Very toxic to aquatic organisms

Hazard classification and safety precautions

Hazard H03 [2-5]; H04 [1]; H11 [1-5]
Risk phrases R36, R38, R43, R50, R58
Operator protection A, C, H, K; U02a, U05a, U09a [1-5]; U19a, U20b [1]; U20a [2-5]
Environmental protection E15a [1-5]; E31c, E32d, E38 [4, 5]; E32a [1-3]
Storage and disposal E01, E04, E30a [1-5]; E26 [1-4]

108 clodinafop-propargyl + trifluralin

A contact and residual herbicide for winter wheat

Products

Hawk	Syngenta	12:383 g/l	EC	08417

Uses
- Annual dicotyledons in ***winter wheat***
- Blackgrass in ***winter wheat***
- Rough meadow grass in ***winter wheat***
- Wild oats in ***winter wheat***

Efficacy guidance
- Product contains a herbicide safener (cloquintocet-mexyl) that improves crop tolerance to clodinafop-propargyl
- Optimum weed control obtained when most grass weeds emerged and broad leaved weeds at susceptible stage before they compete with the crop. Pre-emergence control of annual meadow-grass and broad leaved weeds may not be satisfactory on medium and heavy soils
- Delay treatment if dry or cloddy seedbeds favour late weed germination
- Optimum weed control obtained in crops growing in a fine, firm seedbed cleared of trash and straw
- Wild oats and other weeds germinating from beneath treated zone not controlled
- Speed of action depends on temperature and growing conditions and may appear slow in dry or cold weather
- Rain within 1 h of application may reduce grass weed control
- Always follow WRAG guidelines for preventing and managing herbicide resistant weeds. Section 5 for more information

Restrictions
- Maximum number of treatments 1 per crop
- Do not use on barley or oats
- Do not treat crops under stress or suffering from waterlogging, pest attack, disease or frost
- Do not treat crops undersown with grass mixtures
- Do not mix with products containing MCPA, mecoprop-P, 2,4-D or 2,4-DB
- MCPA, mecoprop, 2,4-D or 2,4-DB should not be applied within 21 d before, or 7 d after, treatment
- Do not use on Sands, stony or gravelly soils or soils with over 10% organic matter
- Product must be applied with specified adjuvant - see label

Crop-specific information
- Latest use: before ear at 1 cm stage (GS 30)
- Spray in autumn, winter or spring from first to third leaf unfolded stage (GS 11-13) but before ear at 1 cm stage (GS 30)

Following crops guidance
- After normal harvest any broad leaved crop (except sugar beet), barley, wheat, durum wheat, rye or triticale may be sown. Before drilling or planting subsequent crops soil must be mouldboard ploughed to 15 cm
- See label for details of crops that may be safely drilled or planted within 5 mth of treatment. If a treated crop should fail after 5 mth but before normal harvest, sow only a broad-leaved crop (not sugar beet)
- 12 mth must elapse after treatment before sugar beet is drilled

Environmental safety
- Dangerous for the environment
- Very toxic to aquatic organisms
- Crops must not be treated if there is risk of run-off (eg from frozen foliage)

Hazard classification and safety precautions

Hazard H04, H11
Risk phrases R36, R43, R50, R58
Operator protection A, C, H, K; U02a, U05a, U09a, U20a
Environmental protection E15a, E31c, E32d, E38
Storage and disposal E01, E04, E30a

109 clofentezine

A selective ovicidal tetrazine acaricide for use in top fruit

Products

Apollo 50 SC	Makhteshim	500 g/l	SC	10590

Uses
- Red spider mites in ***apples***, ***cherries***, ***pears***, ***plums***
- Spider mites in ***blackcurrants*** *(off-label)*, ***protected blackberries*** *(off-label)*, ***protected peppers*** *(off-label)*, ***protected raspberries*** *(off-label)*, ***protected strawberries*** *(off-label)*, ***raspberries*** *(off-label)*, ***strawberries*** *(off-label)*

Specific Off-Label Approvals (SOLAs)
- ***blackcurrants***, ***raspberries***, ***strawberries*** *(OLA 010912) Dec 2008* [1]
- ***blackcurrants***, ***raspberries***, ***strawberries*** *(OLA 012269) Dec 2008* [1]
- ***protected blackberries***, ***protected raspberries*** *(OLA 010911) Dec 2008* [1]
- ***protected blackberries***, ***protected raspberries*** *(OLA 012268) Dec 2008* [1]
- ***protected peppers*** *(OLA 011431) Jul 2007* [1]
- ***protected strawberries*** *(OLA 012271) Dec 2008* [1]

Efficacy guidance
- Acts on eggs and early motile stages of mites. For effective control total cover of plants is essential, particular care being needed to cover undersides of leaves

Restrictions
- Maximum number of treatments 1 per yr for top fruit; 2 per crop on protected peppers, 3 per yr for soft and cane fruit

SEE SECTION 3 FOR PRODUCTS ALSO REGISTERED

Crop-specific information

- HI protected cane fruit 7 d; blackcurrants, raspberries, strawberries 14 d; apples, pears 28 d, plums, cherries 8 wk
- For red spider mite control spray apples and pears between bud burst and pink bud, plums and cherries between white bud and first flower. Rust mite is also suppressed
- On established infestations apply in conjunction with an adult acaricide

Environmental safety

- Harmful to aquatic organisms
- Product safe on predatory mites, bees and other predatory insects

Hazard classification and safety precautions

Risk phrases R52
Operator protection U05a, U08, U20b
Environmental protection E15a, E32a
Storage and disposal E01, E04, E26, E30a

110 clomazone

An isoxazolidinone residual herbicide for oilseed rape and field beans

Products

Centium 360 CS	Belchim	360 g/l	CS	11607

Uses

- Chickweed in ***carrots, combining peas, potatoes, spring field beans, spring oilseed rape, vining peas, winter field beans, winter oilseed rape***
- Cleavers in ***carrots, combining peas, potatoes, spring field beans, spring oilseed rape, vining peas, winter field beans, winter oilseed rape***
- Fool's parsley in ***carrots, combining peas, potatoes, spring field beans, spring oilseed rape, vining peas***
- Red dead-nettle in ***carrots, combining peas, potatoes, spring field beans, spring oilseed rape, vining peas, winter field beans, winter oilseed rape***
- Shepherd's purse in ***carrots, combining peas, potatoes, spring field beans, spring oilseed rape, vining peas, winter field beans, winter oilseed rape***

Efficacy guidance

- Best results obtained from application as soon as possible after sowing crop and before emergence of crop or weeds
- Uptake is via roots and shoots. Seedbeds should be firm, level and free from clods. Loose puffy seedbeds should be consolidated before spraying
- Efficacy is reduced on organic soils and on dry cloddy seedbeds
- Clomazone acts by inhibiting synthesis of chlorophyll pigments. Susceptible weeds emerge but are chlorotic and die shortly afterwards
- Season-long control of weeds may not be achieved

Restrictions

- Maximum number of treatments one per crop
- Crops must be covered by a minimum of 20 mm settled soil. Do not apply to broadcast crops. Direct-drilled crops should be harrowed across the slits to cover seed before spraying
- Do not treat two consecutive crops of carrots in one calendar yr
- Do not use on Sands or Very Light soils

Crop-specific information

- Latest use: pre-emergence of crop
- Crop plants emerged at time of treatment may be severely damaged
- Some transient crop bleaching may occur under certain climatic conditions and can be severe where heavy rain follows application. This is normally rapidly outgrown and has no effect on final crop yield. Overlapping spray swaths may cause severe damage to field beans

Following crops guidance

- Following normal harvest of a spring or autumn treated crop, cereals, oilseed rape, field beans, combining peas, potatoes, maize, turnips, linseed or sugar beet may be sown

- In the event of failure of an autumn treated crop, winter cereals or winter beans may be sown in the autumn if 6 wk have elapsed since treatment. In the spring following crop failure combining peas, field beans or potatoes may be sown if 6 wk have elapsed since treatment, and spring cereals, maize, turnips, onions, carrots or linseed may be sown if 7 mth have elapsed since treatment
- In the event of a failure of a spring treated crop a wide range of crops may be sown provided intervals of 6-9 wk have elapsed since treatment. See label for details
- Prior to resowing any listed replacement crop the soil should be ploughed and cultivated to 15 cm

Environmental safety

- Take extreme care to avoid drift outside the target area, or on to ponds, waterways or ditches as considerable damage may occur

Hazard classification and safety precautions

Hazard H04
Risk phrases R43
Operator protection A, H; U05a, U14, U20b
Environmental protection E15a, E31b, E32d, E34
Storage and disposal E01, E04, E30a

111 clopyralid

A foliar translocated picolinic herbicide for a wide range of crops

Products

1	Dow Shield	Dow	200 g/l	SL	10988
2	Fernpath Torate	AgriGuard	200 g/l	SL	11033
3	Greencrop Champion	Greencrop	200 g/l	SL	11755
4	Loncid	I T Agro	200 g/l	SL	10832
5	Pirlid	AgriGuard	200 g/l	SL	11946

Uses

- Annual dicotyledons in ***broccoli, brussels sprouts, cabbages, calabrese, fodder beet, fodder rape, kale, linseed, maize, mangels, onions, ornamental plant production, permanent pasture, red beet, spring oilseed rape, strawberries, sugar beet, swedes, sweetcorn, turnips, winter barley, winter oats, winter oilseed rape*** [1-5]; ***cauliflowers, conifers and broadleaved trees*** *(off-label)* [1]; ***rotational grassland*** [3]; ***spring barley, spring oats, spring wheat, winter wheat*** [1, 3, 4]
- Corn marigold in ***broccoli, brussels sprouts, cabbages, calabrese, cauliflowers, fodder beet, fodder rape, kale, linseed, maize, mangels, onions, ornamental plant production, permanent pasture, red beet, spring oilseed rape, strawberries, sugar beet, swedes, sweetcorn, turnips, winter barley, winter oats, winter oilseed rape*** [1-5]; ***rotational grassland*** [3]; ***spring barley, spring oats, spring wheat, winter wheat*** [1, 3, 4]
- Creeping thistle in ***apples*** *(off-label)*, ***pears*** *(off-label)* [1]; ***broccoli, brussels sprouts, cabbages, calabrese, cauliflowers, fodder beet, fodder rape, kale, linseed, maize, mangels, onions, ornamental plant production, permanent pasture, red beet, spring oilseed rape, strawberries, sugar beet, swedes, sweetcorn, turnips, winter barley, winter oats, winter oilseed rape*** [1-5]; ***rotational grassland*** [3]; ***spring barley, spring oats, spring wheat, winter wheat*** [1, 3, 4]
- Groundsel in ***chives*** *(off-label)*, ***herbs (see appendix 6)*** *(off-label)*, ***leaf spinach*** *(off-label)*, ***parsley*** *(off-label)*, ***spinach beet*** *(off-label)* [1]
- Mayweeds in ***broccoli, brussels sprouts, cabbages, calabrese, cauliflowers, fodder beet, fodder rape, kale, linseed, maize, mangels, onions, ornamental plant production, permanent pasture, red beet, spring oilseed rape, strawberries, sugar beet, swedes, sweetcorn, turnips, winter barley, winter oats, winter oilseed rape*** [1-5]; ***chives*** *(off-label)*, ***herbs (see appendix 6)*** *(off-label)*, ***honesty*** *(off-label)*, ***leaf spinach*** *(off-label)*, ***parsley*** *(off-label)*, ***poppies*** *(off-label - for morphine production)*, ***spinach beet*** *(off-label)* [1]; ***rotational grassland*** [3]; ***spring barley, spring oats, spring wheat, winter wheat*** [1, 3, 4]
- Thistles in ***asparagus*** *(off-label)*, ***grass seed crops*** *(off-label)*, ***honesty*** *(off-label)*, ***poppies*** *(off-label - for morphine production)* [1]

Specific Off-Label Approvals (SOLAs)

- ***apples, pears*** *(OLA 002082) Dec 2008* [1]

- ***asparagus*** *(OLA 000520) Dec 2008* [1]
- ***chives, herbs (see appendix 6), parsley*** *(OLA 010377) Dec 2008* [1]
- ***conifers and broadleaved trees*** *(OLA 920757) Dec 2008* [1]
- ***grass seed crops*** *(OLA 920662) Dec 2008* [1]
- ***honesty*** *(OLA 930480) Dec 2008* [1]
- ***leaf spinach, spinach beet*** *(OLA 010094) Dec 2008* [1]
- ***poppies*** *(for morphine production) (OLA 040161) Dec 2008* [1]

Efficacy guidance

- Best results achieved by application to young actively growing weed seedlings. Treat creeping thistle at rosette stage and repeat 3-4 wk later as directed
- High activity on weeds of Compositae family. For most crops recommended for use in tank mixes. See label for details

Restrictions

- Maximum total dose varies between the equivalent of one and two full dose treatments, depending on the crop treated. See labels for details
- Do not apply to cereals later than the second node detectable stage (GS 32)
- Do not apply when crop damp or when rain expected within 6 h
- Do not use straw from treated cereals in compost or any other form for glasshouse crops. Straw may be used for strawing down strawberries
- Straw from treated grass seed crops or linseed should be baled and carted away. If incorporated do not plant winter beans in same year
- Do not use on onions at temperatures above 20°C or when under stress
- Do not treat maiden strawberries or runner beds or apply to early leaf growth during blossom period or within 4 wk of picking. Aug or early Sep sprays may reduce yield

Crop-specific information

- Latest use: before 3rd node detectable (GS 33) for cereals; before flower buds visible from above for oilseed rape, linseed and honesty; post final harvest of spears and pre-emergence of fern for asparagus
- HI grassland 7 d; apples, pears, strawberries 4 wk; maize, sweetcorn, onions, Brussels sprouts, broccoli, cabbage, cauliflowers, calabrese, kale, fodder rape, oilseed rape, swedes, turnips, sugar beet, red beet, fodder beet, mangels, sage, honesty 6 wk
- Timing of application varies with weed problem, crop and other ingredients of tank mixes. See labels for details
- Apply as directed spray in woody ornamentals, avoiding leaves, buds and green stems. Do not apply in root zone of families Compositae or Leguminosae

Following crops guidance

- Do not plant susceptible autumn-sown crops in same year as treatment. Do not apply later than Jul where susceptible crops are to be planted in spring. See label for details

Environmental safety

- Dangerous for the environment
- Harmful to aquatic organism
- Wash spray equipment thoroughly with water and detergent immediately after use. Traces of product can damage susceptible plants sprayed later
- Keep livestock out of treated areas for at least 7 d and until foliage of any poisonous weeds such as ragwort has died and become unpalatable

Hazard classification and safety precautions

Hazard H04 [4]; H11 [2, 5]
Risk phrases R41 [4]; R52, R58 [1, 4]
Operator protection A, C; U08, U19a [1-5]; U11 [4]; U20a [3]; U20b [1, 2, 4, 5]
Environmental protection E07a [1-5] (7 d); E15a, E31b [1-5]; E32d [1]
Storage and disposal E01 [1, 4]; E26, E30a [1-5]

112 clopyralid + 2,4-D + MCPA

A translocated herbicide mixture for grassland

Products

Lonpar	Dow	35:150:175 g/l	SL	08686

Uses

- Creeping thistle in ***established grassland***

Approval information

- 2,4-D included in Annex I under EC Directive 91/414

Efficacy guidance

- Treatment must be made when weeds and grass are actively growing
- Important to ensure sufficient leaf area for uptake especially on established thistles with extensive root system. Treat at rosette stage
- On large well established thistles and where there is a large soil seed reservoir further treatment in the following year may be needed
- To allow maximum translocation do not cut grass for 28 d after treatment

Restrictions

- Maximum number of treatments 1 or 2 per yr depending on dose used - see label
- Do not spray in drought, very hot or very cold weather or if rain expected within 6 h
- Do not treat grass less than 1 yr old or sports or amenity turf
- Product kills or severely checks clover and should not be used where clover is an important constituent of the sward
- Do not drill grass, grass mixtures, kale, swedes or turnips into the sward within 6 wk of spraying

Crop-specific information

- Very occasionally some yellowing of sward may occur after treatment which is quickly outgrown

Following crops guidance

- Susceptible crops (see label) must not be planted in the same calendar yr as treatment. Ensure that remains of treated crop have completely decayed before planting a subsequent susceptible crop. Where a susceptible crop is to be planted in the spring, product must not be sprayed later than end Jul in previous yr

Environmental safety

- Harmful to fish or other aquatic life. Do not contaminate surface waters or ditches with chemical or used container
- Keep livestock out of treated areas for at least 1 wk and until foliage of any poisonous weeds, such as ragwort, has died and become unpalatable
- Wash spray equipment thoroughly with water and detergent immediately after use. Traces of product can damage susceptible plants sprayed later

Hazard classification and safety precautions

Hazard H03, H04
Risk phrases R22a, R41
Operator protection A, C, H, M; U05a, U15
Environmental protection E07b (1 wk); E13c, E34
Storage and disposal E01, E04, E26

113 clopyralid + diflufenican + MCPA

A selective herbicide for use in established turf

Products

Spearhead	Bayer Environ.	20:15:300 g/l	SL	09941

Uses

- Annual dicotyledons in ***managed amenity turf***
- Perennial dicotyledons in ***managed amenity turf***

Efficacy guidance

- Best results achieved by application when grass and weeds are actively growing
- Treatment during early part of the season is recommended, but not during drought

Restrictions

- Maximum number of treatments 1 per yr
- Only use when sward is satisfactorily established and regular mowing has begun
- Turf sown in spring or early summer may be ready for treatment after 2 mth. Later sown turf should not be sprayed until growth is resumed in the following spring

- Avoid mowing within 3-4 d before or after treatment
- Do not use cuttings from treated area as a mulch for any crop

Environmental safety

- Extremely dangerous to fish or other aquatic life. Do not contaminate surface waters or ditches with chemical or used container
- Keep livestock out of treated areas/away from treated water for at least 1 wk and until foliage of any poisonous weeds, such as ragwort, has died and become unpalatable
- LERAP Category B
- Avoid drift. Small amounts of spray can cause serious injury to herbaceous plants, vegetables, fruit and glasshouse crops
- Keep livestock out of treated areas

Hazard classification and safety precautions

Hazard H03, H04
Risk phrases R20, R21, R22a, R36, R38
Operator protection A, C; U02a, U05a, U08, U19a, U20b
Environmental protection E07b, E13a, E16a, E16b, E31b, E34
Storage and disposal E01, E04, E26, E30a
Medical advice M03

114 clopyralid + fluroxypyr + MCPA

A translocated herbicide mixture for use in sports and amenity turf

Products

Greenor	Rigby Taylor	20:40:200 g/l	ME	10909

Uses

- Annual dicotyledons in ***managed amenity turf***

Approval information

- Fluroxypyr included in Annex I under EC Directive 91/414

Efficacy guidance

- Best results achieved when weeds actively growing and turf grass competitive
- Treatment should normally be between Apr-Sep when the soil is moist
- Do not apply during drought unless irrigation is applied
- Allow 3 d before or after mowing established turf to ensure sufficient weed leaf surface present to allow uptake and movement

Restrictions

- Maximum number of treatments 2 per yr
- Do not treat grass under stress from frost, drought, waterlogging, trace element deficiency, disease or pest attack
- Do not treat if night temperatures are low, when frost is imminent or during prolonged cold weather

Crop-specific information

- Treat young turf only in spring when at least 2 mth have elapsed since sowing
- Allow 5 d after mowing young turf before treatment
- Product selective on a number of turf grass species (see label) but consultation or testing recommended before treatment of any cultivar

Environmental safety

- Dangerous for the environment
- Very toxic to aquatic organisms
- Wash spray equipment thoroughly with water and detergent immediately after use. Traces of product can damage susceptible plants sprayed later

Hazard classification and safety precautions

Hazard H04, H11
Risk phrases R36, R43, R50, R58
Operator protection A, C; U05a, U08, U14, U19a, U20b
Environmental protection E15a, E31b, E32d, E38
Storage and disposal E01, E04, E26, E30a

115 clopyralid + fluroxypyr + triclopyr

A foliar acting herbicide mixture for grassland

Products

Pastor	Dow	50:75:100 g/l	EC	11168

Uses

- Annual dicotyledons in ***newly sown grass***
- Docks in ***established grassland, newly sown grass***
- Stinging nettle in ***established grassland***
- Thistles in ***established grassland***

Approval information

- Fluroxypyr included in Annex I under EC Directive 91/414

Efficacy guidance

- Apply in spring or autumn depending on weeds present or as a split treatment in spring followed by autumn
- Treatment must be made when weeds and grass are actively growing
- It is important to ensure sufficient leaf area is present for uptake especially on established docks and thistles
- On large well established docks and where there is a large soil seed reservoir further treatment in the following year may be needed
- To allow maximum translocation in the weeds do not cut grass for 4 wk after treatment

Restrictions

- Maximum number of treatments 1 per yr at full dose or 2 per yr at half dose
- Do not spray in drought, very hot or very cold weather
- Do not treat sports or amenity turf
- Product kills or severely checks clover and should not be used where clover is an important constituent of the sward
- Do not roll or harrow 10 d before or 7 d after treatment
- Do not allow spray or drift to reach other crops, amenity plantings, gardens, ponds, lakes or water courses

Crop-specific information

- HI 7 d
- Application during active growth ensures minimal check to grass. Newly sown grass may be treated from the third leaf visible stage
- Product may be used in established grassland which is under non-rotational setaside arrangements
- Very occasionally some yellowing of the sward may occur after treatment but is quickly outgrown

Following crops guidance

- Residues in incompletely decayed plant tissue may affect succeeding susceptible crops such as peas, beans and other legumes, carrots and related crops, potatoes, tomatoes, lettuce
- Do not plant susceptible autumn-sown crops in the same yr as treatment with product. Spring sown crops may follow if treatment was before end Jul in the previous yr

Environmental safety

- Dangerous for the environment
- Toxic to aquatic organisms
- Keep livestock out of treated areas for at least 7 d following treatment and until foliage of poisonous weeds such as ragwort has died and become unpalatable
- Wash spray equipment thoroughly with water and detergent immediately after use. Traces of product can damage susceptible plants sprayed later

Hazard classification and safety precautions

Hazard H03, H08, H11
Risk phrases R22b, R37, R38, R41, R43, R51, R58, R67
Operator protection A, C; U02a, U05a, U08, U11, U14, U19a, U20b
Environmental protection E07c (7 d); E15a, E31b, E32d, E34, E38
Consumer protection C01

SEE SECTION 3 FOR PRODUCTS ALSO REGISTERED

Storage and disposal E01, E04, E26, E30a
Medical advice M05b

116 clopyralid + picloram

A post-emergcne herbicide mixture for oilseed rape

Products

Galera	Dow	267:67 g/l	SL	11961

Uses
- Cleavers in ***winter oilseed rape***
- Mayweeds in ***winter oilseed rape***

Efficacy guidance
- Best results obtained from treatment when weeds are small and actively growing
- Cleavers that germinate after treatment will not be controlled

Restrictions
- Maximum total dose equivalent to one full dose treatment
- Do not treat crops under stress from cold, drought, pest damage, nutrient deficiency or any other cause
- Do not roll or harrow for 7 d before or after spraying
- Do not apply through CDA applicators
- Do not use any treated plant material for composting or mulching
- Do not use manure from animals fed on treated crops for composting
- Chop and incorporate all treated plant remains in early autumn, or as soon as possible after harvest, to release any residues into the soil. Ensure that all treated plant remains have completely decayed before planting susceptible crops

Following crops guidance
- Any crop may be sown in the calenday yr following treatment
- Ploughing or thorough cultivation should be carried out before planting leguminous crops
- Do not attempt to plant peas, beans, other legumes, carrots, other umbelliferous crops, potatoes, lettuce, other Compositae, or any glasshouse or protected crops if treated crop remains have not fully decayed by the time of planting
- In the event of failure of an autumn treated crop only oilseed rape, wheat, barley, oats, maize or ryegrass may be sown in the spring and only after ploughing or thorough cultivation

Environmental safety
- Take extreme care to avoid drift onto crops and non-target plants outside the target area

Hazard classification and safety precautions

Risk phrases R58
Operator protection A, C; U05a
Storage and disposal E01, E04, E30a

117 clopyralid + triclopyr

A perennial and woody weed herbicide for use in grassland

Products

1	Blaster	Headland Amenity	60:240 g/l	EC	10571
2	Grazon 90	Dow	60:240 g/l	EC	05456
3	Thistlex	Dow	200:200 g/l	EC	11533

Uses
- Brambles in ***amenity grass*** [1]; ***established grassland*** [2]
- Broom in ***amenity grass*** [1]; ***established grassland*** [2]
- Creeping thistle in ***permanent pasture, rotational grassland*** [3]
- Docks in ***amenity grass*** [1]; ***established grassland*** [2]
- Gorse in ***amenity grass*** [1]; ***established grassland*** [2]
- Perennial dicotyledons in ***amenity grass*** [1]; ***established grassland*** [2]
- Stinging nettle in ***amenity grass*** [1]; ***established grassland*** [2]
- Thistles in ***amenity grass*** [1]; ***established grassland*** [2]

Efficacy guidance

- Must be applied to actively growing weeds
- Correct timing crucial for good control. Spray stinging nettle before flowering, docks in rosette stage in spring, creeping thistle before flower stems 15-20 cm high, brambles, broom and gorse in Jun-Aug
- Allow 2-3 wk regrowth after grazing or mowing before spraying perennial weeds
- Where there is a large reservoir of weed seed in the soil further treatment in the following yr may be needed

Restrictions

- Maximum number of treatments 1 per yr
- Only use on permanent pasture or rotational grassland established for at least 1 yr
- Do not apply where clover is an important constituent of sward
- Do not roll or harrow within 10 d before or 7 d after spraying
- Do not cut grass for 21 d before or 28 d after spraying
- Do not use any treated plant material for composting or mulching, and do not use manure for composting from animals fed on treated crops
- Do not apply by hand-held rotary atomiser equipment
- Do not allow drift onto other crops, amenity plantings or gardens, ponds, lakes or water courses. All conifers, especially pine and larch, are very sensitive

Crop-specific information

- Latest use: 7 d before grazing or cutting
- Some transient yellowing of treated swards may occur but is quickly outgrown

Following crops guidance

- Residues in plant tissues which have not completely decayed may affect succeeding susceptible crops such as peas, beans, other legumes, carrots, parsnips, potatoes, tomatoes, lettuce, glasshouse and protected crops
- Do not plant susceptible autumn-sown crops (eg winter beans) in same year as treatment
- Do not direct drill kale, swedes, turnips, grass or grass mixtures within 6 wk of spraying
- Do not spray after end Jul where susceptible crops to be planted next spring

Environmental safety

- Dangerous for the environment
- Very toxic to aquatic organisms
- Keep livestock out of treated areas for at least 7 d after spraying and until foliage of any poisonous weeds such as ragwort or buttercup has died down and become unpalatable

Hazard classification and safety precautions

Hazard H03 [1, 2]; H04 [3]; H11 [1-3]
Risk phrases R22a, R38, R43, R50 [1, 2]; R41, R58 [1-3]; R52 [3]
Operator protection A, C [1-3]; H, M [1, 2]; U02a, U05a, U11, U20b [1-3]; U08, U14, U19a, U23b [1, 2]; U15, U23a [3]
Environmental protection E07a [1-3] (7 d); E15a, E31b, E32d [1-3]; E23, E34, E38 [1, 2]
Consumer protection C01 [1, 2]
Storage and disposal E01, E04, E26, E30a
Medical advice M03 [3]

118 copper ammonium carbonate

A protectant copper fungicide

Products

Croptex Fungex	Certis	8% w/w (copper)	SL	11049

Uses

- Blight in ***outdoor tomatoes***
- Cane spot in ***loganberries***, ***raspberries***
- Celery leaf spot in ***celery***
- Currant leaf spot in ***blackcurrants***
- Damping off in ***seedlings of ornamentals***
- Leaf curl in ***peaches***

SEE SECTION 3 FOR PRODUCTS ALSO REGISTERED

- Leaf mould in ***protected tomatoes***
- Powdery mildew in ***chrysanthemums***, ***cucumbers***

Efficacy guidance

- Apply spray to both sides of foliage
- With protected crops keep foliage dry before and after spraying
- Ventilate glasshouse immediately after spraying

Restrictions

- Maximum number of treatments 5 per crop for celery; 3 per yr for blackcurrants and loganberries; 2 per yr for peaches and raspberries; 1 per tray when used as a drench for ornamental seedlings
- Do not spray plants which are dry at the roots

Environmental safety

- Harmful to fish or other aquatic life. Do not contaminate surface waters or ditches with chemical or used container
- Keep all livestock out of treated areas for at least 3 wk. Bury or remove spillages

Hazard classification and safety precautions

Hazard H03
Risk phrases R22a, R41
Operator protection U05a, U11, U15, U20c
Environmental protection E06c (3 wk); E13c, E31a, E32d, E34
Storage and disposal E01, E04, E30a
Medical advice M03

119 copper hydroxide

An inorganic root pruning and root development agent

Products

Spin Out	Fargro	71 g/l	LI	07610

Uses

- Root control in ***container-grown stock***

Efficacy guidance

- Product should be painted on inner surfaces of containers prior to planting
- Used containers should be free from any loose soil or dirt
- Apply using conventional painting techniques such as brush, sponge, pad or airless paint sprayer
- Thorough coverage with a single coating to a minimum thickness of 0.075 mm is sufficient for root control

Restrictions

- Maximum number of treatments 1 per container
- Tested on a wide range of ornamental species (see label). Efficacy and tolerance should be tested on a small scale for species not listed
- Performance characteristics may vary under some conditions such as when substrates with a pH of less than 5.0 are used
- Product must not be diluted
- Knapsack sprayer or similar equipment is not suitable

Crop-specific information

- Latest use: before planting
- Can be used in production of container grown forestry seedlings, hardy ornamental and herbaceous plant species
- Use at any stage of plant development from seedlings to large container grown trees

Environmental safety

- Harmful to fish or other aquatic life. Do not contaminate surface waters or ditches with chemical or used container

Hazard classification and safety precautions

Operator protection A, D; U11, U12, U20b
Environmental protection E13c, E32a
Storage and disposal E26, E30a

120 copper oxychloride

A protectant copper fungicide and bactericide

Products

1	Cuprokylt	Unicrop	50% w/w (copper)	WP	00604
2	Cuprokylt FL	Unicrop	270 g/l (copper)	SC	08299
3	Headland Copper	Headland	256 g/l (copper)	SC	07799

Uses

- Bacterial canker in ***cherries*** [1, 3]; ***cob nuts*** *(off-label)*, ***hazel nuts*** *(off-label)*, ***walnuts*** *(off-label)* [2, 3]; ***plums*** [1, 2]
- Bacterial rot in ***bulb onions*** *(off-label)*, ***garlic*** *(off-label)*, ***leeks*** *(off-label)*, ***salad onions*** *(off-label)*, ***shallots*** *(off-label)* [1, 3]
- Black rot in ***broccoli*** *(off-label)*, ***brussels sprouts*** *(off-label)*, ***cabbages*** *(off-label)*, ***calabrese*** *(off-label)*, ***cauliflowers*** *(off-label)*, ***chinese cabbage*** *(off-label)*, ***collards*** *(off-label)*, ***kale*** *(off-label)* [1, 3]; ***protected brassica seedlings*** *(off-label)* [2]; ***protected broccoli*** *(off-label)*, ***protected brussels sprouts*** *(off-label)*, ***protected cabbages*** *(off-label)*, ***protected calabrese*** *(off-label)*, ***protected cauliflowers*** *(off-label)*, ***protected chinese cabbage*** *(off-label)*, ***protected collards*** *(off-label)*, ***protected kale*** *(off-label)* [3]
- Blight in ***cob nuts*** *(off-label)*, ***hazel nuts*** *(off-label)*, ***walnuts*** *(off-label)* [2, 3]; ***outdoor tomatoes***, ***potatoes*** [1-3]; ***protected tomatoes*** [1, 2]
- Buck-eye rot in ***outdoor tomatoes***, ***protected tomatoes*** [1-3]
- Cane spot in ***loganberries***, ***raspberries*** [1, 2]
- Canker in ***apples***, ***pears*** [1-3]
- Celery leaf spot in ***celery*** [1-3]
- Collar rot in ***apples*** *(off-label)* [2]
- Damping off in ***outdoor tomatoes***, ***protected tomatoes*** [1, 2]
- Downy mildew in ***grapevines*** [3]; ***hops*** [1-3]
- Foot rot in ***outdoor tomatoes***, ***protected tomatoes*** [1, 2]
- Leaf curl in ***peaches*** [1, 2]
- Pseudomonas storage rots in ***bulb onions*** *(off-label)*, ***garlic*** *(off-label)*, ***leeks*** *(off-label)*, ***salad onions*** *(off-label)*, ***shallots*** *(off-label)* [1, 3]
- Purple blotch in ***blackberries*** [3]
- Pythium in ***watercress*** *(off-label)* [3]; ***watercress*** *(off-label - during propagation)* [2]
- Rhizoctonia in ***watercress*** *(off-label)* [3]; ***watercress*** *(off-label - during propagation)* [2]
- Rust in ***blackcurrants*** [1-3]
- Spear rot in ***broccoli*** *(off-label)*, ***brussels sprouts*** *(off-label)*, ***cabbages*** *(off-label)*, ***calabrese*** *(off-label)*, ***cauliflowers*** *(off-label)*, ***chinese cabbage*** *(off-label)*, ***collards*** *(off-label)*, ***kale*** *(off-label)* [1, 3]; ***protected brassica seedlings*** *(off-label)* [2]; ***protected broccoli*** *(off-label)*, ***protected brussels sprouts*** *(off-label)*, ***protected cabbages*** *(off-label)*, ***protected calabrese*** *(off-label)*, ***protected cauliflowers*** *(off-label)*, ***protected chinese cabbage*** *(off-label)*, ***protected collards*** *(off-label)*, ***protected kale*** *(off-label)* [3]

Specific Off-Label Approvals (SOLAs)

- ***apples*** *(OLA 982336) Dec 2008* [2]
- ***broccoli***, ***brussels sprouts***, ***cabbages***, ***calabrese***, ***cauliflowers***, ***chinese cabbage***, ***collards***, ***kale*** *(OLA 010115) Dec 2008* [1]
- ***broccoli***, ***brussels sprouts***, ***bulb onions***, ***cabbages***, ***calabrese***, ***cauliflowers***, ***chinese cabbage***, ***cob nuts***, ***collards***, ***garlic***, ***hazel nuts***, ***kale***, ***leeks***, ***protected broccoli***, ***protected brussels sprouts***, ***protected cabbages***, ***protected calabrese***, ***protected cauliflowers***, ***protected chinese cabbage***, ***protected collards***, ***protected kale***, ***salad onions***, ***shallots***, ***walnuts***, ***watercress*** *(OLA 020415) Dec 2008* [3]
- ***bulb onions***, ***garlic***, ***leeks***, ***salad onions***, ***shallots*** *(OLA 991127) Dec 2008* [1]
- ***cob nuts***, ***hazel nuts***, ***walnuts*** *(OLA 990385) Dec 2008* [2]
- ***protected brassica seedlings*** *(OLA 010117) Dec 2008* [2]
- ***watercress*** *(during propagation) (OLA 001538) Dec 2008* [2]

Approval information

- Approved for aerial application on potatoes [1]. See notes in Section 5

Efficacy guidance

- Spray crops at high volume when foliage dry but avoid run off. Do not spray if rain expected soon

- Spray interval commonly 10-14 d but varies with crop, see label for details
- If buck-eye rot occurs, spray soil surface and lower parts of tomato plants to protect unaffected fruit [1, 2]
- A follow-up spray in the following spring should be made to top fruit severely infected with bacterial canker

Restrictions

- Maximum number of treatments 3 per crop for apples, blackberries, grapevines, pears; 2 per crop for watercress. Not specified for other crops
- Some peach cultivars are sensitive to copper. Treat non-sensitive varieties only [1, 2]

Crop-specific information

- Latest use: before bud burst for apples and pears; before planting out protected brassica seedlings
- HI calabrese 3 d; hops 7 d; onions, garlic, leeks, shallots 14 d; nuts 3 mth
- Slight damage may occur to leaves of cherries and plums [1, 2]

Environmental safety

- Dangerous for the environment
- Very toxic to aquatic organisms
- Keep all livestock out of treated areas for at least 3 wk

Hazard classification and safety precautions

Hazard H03, H11 [2]
Risk phrases R22a, R50, R58 [2]
Operator protection A [3]; U20a [1, 3]; U20c [2]
Environmental protection E06a [1-3] (3 wk); E13c [1, 3]; E15a, E31c, E32d [2]; E31a [1]; E31b, E34 [3]
Storage and disposal E26 [2]; E30a [1-3]

121 copper sulphate

See Bordeaux Mixture

122 cyanazine

A contact and residual triazine herbicide

Products

1	Fortrol	Makhteshim	500 g/l	SC	11174
2	I T Cyanazine	I T Agro	500 g/l	SC	10179

Uses

- Annual dicotyledons in ***broad beans, peas, salad onions*** [2]; ***broad beans*** *(Scotland only)*, ***broccoli*** *(off-label)*, ***cabbages*** *(off-label)*, ***calabrese*** *(off-label)*, ***cauliflowers*** *(off-label)*, ***collards*** *(off-label)*, ***combining peas, farm forestry*** *(off-label)*, ***forest nurseries*** *(off-label)*, ***kale*** *(off-label)*, ***salad onions*** *(fen soils only)*, ***vining peas*** [1]; ***bulb onions, narcissi, winter oilseed rape*** [1, 2]
- Annual grasses in ***broad beans, peas, salad onions*** [2]; ***broad beans*** *(Scotland only)*, ***combining peas, farm forestry*** *(off-label)*, ***forest nurseries*** *(off-label)*, ***salad onions*** *(fen soils only)*, ***vining peas*** [1]; ***bulb onions, narcissi, winter oilseed rape*** [1, 2]

Specific Off-Label Approvals (SOLAs)

- ***broccoli, cabbages, calabrese, cauliflowers, collards, kale*** *(OLA 031074) Dec 2007* [1]
- ***farm forestry*** *(OLA 031454) Dec 2007* [1]
- ***forest nurseries*** *(OLA 031455) Dec 2007* [1]

Approval information

- Products containing this active ingredient have been granted derogations for specified 'Essential Uses' for use until 31 December 2007. Sale and supply must cease by 30 June 2007 but growers have no guarantee that the products will continue to be available until then.
  For more information see 'The Review Programme' under 'Pesticide Legislation' in Section 5

Efficacy guidance

- Weeds controlled before emergence or at young seedling stage. Soil should be moist and some rain should follow application

- Best results achieved when applied during mild, bright weather. Avoid applications in dull, cold or wet conditions
- Numerous tank mixtures recommended to broaden weed spectrum. See label for details of tank mix partners and timings
- Some strains of blackgrass have developed resistance to many blackgrass herbicides which may lead to poor control

Restrictions

- Maximum number of treatments 1 per crop for edible crops
- Maximum total dose equivalent to one full dose treatment on onions
- Variety restrictions apply in peas. See label for details of tolerant varieties
- Do not treat crops suffering from physical damage or stress from pest attack
- Treatment directly after conditions that reduce wax formation may lead to damage
- Do not use as pre-emergence treatment on soils with more than 10% organic matter
- Do not use on Sands, Very Light or stony soils
- These products must not be used on any crops other than those listed, including any extrapolations that would normally be permissible under the Long Term Arrangements for Extension of Use (see Section 5)

Crop-specific information

- Latest use: pre-emergence for broad beans; before 31 Jan for winter oilseed rape; pre-emergence (high dose) or before flower buds visible for peas; before flower buds visible for field beans
- HI 8 wk for onions, salad onions; 11 wk for brassica crops
- In early drillings of peas in late Feb or Mar application may be delayed until immediately prior to emergence to increase effective duration of control
- Peas should be drilled to give 20-25 mm of settled soil above the seed. On light soils crops may be damaged if heavy rain falls soon after spraying
- Apply to oilseed rape after 1 Nov from 5-leaf stage when winter hardened. Do not apply after 31 Jan
- Apply post-emergence to onions after 2 true leaf stage only on fen soils with more than 10% organic matter

Environmental safety

- Dangerous for the environment
- Very toxic to aquatic organisms

Hazard classification and safety precautions

Hazard H03, H11
Risk phrases R22a, R50, R58
Operator protection A, C; U05a, U08, U19a, U20b
Environmental protection E15a, E19b, E31c, E32d, E34
Storage and disposal E01, E04, E26, E30a
Medical advice M03, M05a

123 cyanazine + pendimethalin

A contact and residual herbicide mixture

Products

Bullet	Makhteshim	150:264 g/l	SC	11204

Uses

- Annual dicotyledons in ***combining peas, spring field beans***
- Annual meadow grass in ***combining peas, spring field beans***
- Rough meadow grass in ***combining peas, spring field beans***

Approval information

- Pendimethalin included in Annex I under EC Directive 91/414
- Products containing cyanazine have been granted derogations for specified 'Essential Uses' for use until 31 December 2007. Sale and supply must cease by 30 June 2007 but growers have no guarantee that the products will continue to be available until then.
  For more information see 'The Review Programme' under 'Pesticide Legislation' in Section 5

Efficacy guidance

- Best results achieved in crops growing on firm, fine seedbeds, where light rain falls soon after treatment. Do not disturb soil after application. Weed control may be reduced if prolonged dry conditions follow application
- Weeds present before crop emergence should be controlled with a contact herbicide as a tank-mix or in sequence
- Weed control may be reduced on soils with a high Kd factor, where OM exceeds 6% or ash content is high. Surface organic crop residues should be dispersed
- Weeds germinating more than 2 mth after spraying may not be controlled

Restrictions

- Maximum number of treatments 1 per crop
- Do not use on very light, very stony or gravelly soils or those with more than 10% organic matter
- Do not apply where crop root growth is likely to be restricted
- Seed should be covered by a minimum of 25 mm of settled soil after seedbed consolidation
- Do not apply later than when the growing point of the crop is within 13 mm of the soil surface
- Do not treat pea variety Vedette

Crop-specific information

- Latest use: pre-emergence of crop

Following crops guidance

- After a dry season land must be ploughed to 150 mm before drilling ryegrass
- Before drilling a winter crop plough or cultivate to 150 mm
- In the event of crop failure land must be ploughed or cultivated to 150 mm and then an interval of 8 wk must elapse before drilling any crop

Environmental safety

- Dangerous for the environment
- Very toxic to aquatic organisms

Hazard classification and safety precautions

Hazard H03, H11
Risk phrases R22a, R50, R58
Operator protection A; U05a, U08, U13, U19a, U20b
Environmental protection E15a, E19b, E31b, E32d, E34
Storage and disposal E01, E04, E26, E30a
Medical advice M03

124 cyazofamid

A cyanoimidazolesulfonamide protectant fungicide for potatoes

Products

Ranman Twinpack	Belchim	400 g/l	KL	11851

Uses

- Blight in ***potatoes***

Approval information

- Cyazofamid included in Annex I under EC Directive 91/414

Efficacy guidance

- Apply as a protectant treatment before blight enters the crop and repeat every 7-10 d depending on severity of disease pressure
- Commence spray programme immediately the risk of blight in the locality occurs, usually when the crop meets along the rows
- Product must always be used with organosilicone adjuvant provided in the twin pack
- To minimise the chance of development of resistance no more than three applications should be made consecutively (out of a permissible total of six) in the blight control programme. For more information on Resistance Management see Section 5

Restrictions

- Maximum number of treatments 6 per crop (no more than three of which should be consecutive)
- Mixed product must not be allowed to stand overnight
- Consult processor before using on crops intended for processing

Crop-specific information

- HI 7 d

Environmental safety

- Dangerous for the environment
- Very toxic to aquatic organisms

Hazard classification and safety precautions

Hazard H03, H11
Risk phrases R20, R36, R48 (adjuvant); R41, R50, R58
Operator protection A, C; U02a, U05a, U11, U15, U20a; U19a (adjuvant)
Environmental protection E15a, E31c, E32d, E34, E38
Storage and disposal E01, E04, E30a
Medical advice M03

125 cycloxydim

A translocated post-emergence oxime herbicide for grass weed control

Products

1	Greencrop Pomeroy	Greencrop	200 g/l	EC	11769
2	Greencrop Valentia	Greencrop	200 g/l	EC	10197
3	Landgold Cycloxydim	Landgold	200 g/l	EC	06269
4	Laser	BASF	200 g/l	EC	05251
5	Marnoch Clodim	Me2	200 g/l	EC	11460
6	Standon Cycloxydim	Standon	200 g/l	EC	08830

Uses

- Annual grasses in ***amenity vegetation*** *(off-label)* [4]; ***brussels sprouts, cabbages, cauliflowers, fodder beet, mangels, spring field beans, sugar beet, swedes, winter field beans, winter oilseed rape*** [1-6]; ***bulb onions, carrots, dwarf beans, farm forestry, forest, leeks, linseed, parsnips, salad onions, spring oilseed rape, strawberries*** [1, 2, 4-6]; ***calabrese, peas*** [2]; ***combining peas, vining peas*** [3-5]; ***early potatoes*** [1, 3, 4]; ***flower bulbs*** [2, 4-6]; ***forest nurseries, ornamental plant production*** [1]; ***maincrop potatoes*** [2-5]
- Annual meadow grass in ***combining peas, maincrop potatoes, vining peas*** [1, 6]
- Black bent in ***brussels sprouts, cabbages, cauliflowers, fodder beet, maincrop potatoes, mangels, spring field beans, sugar beet, swedes, winter field beans, winter oilseed rape*** [1-6]; ***bulb onions, carrots, dwarf beans, flower bulbs, leeks, linseed, parsnips, salad onions, spring oilseed rape, strawberries*** [1, 2, 4-6]; ***calabrese, peas*** [2]; ***combining peas, vining peas*** [1, 3-6]; ***early potatoes*** [1, 3, 4]; ***ornamental plant production*** [1]
- Blackgrass in ***brussels sprouts, cabbages, cauliflowers, fodder beet, maincrop potatoes, mangels, spring field beans, sugar beet, swedes, winter field beans, winter oilseed rape*** [1-6]; ***bulb onions, carrots, dwarf beans, flower bulbs, leeks, linseed, parsnips, salad onions, spring oilseed rape, strawberries*** [1, 2, 4-6]; ***calabrese, peas*** [2]; ***combining peas, vining peas*** [1, 3-6]; ***early potatoes*** [1, 3, 4]; ***ornamental plant production*** [1]
- Couch in ***brussels sprouts, cabbages, cauliflowers, fodder beet, maincrop potatoes, mangels, spring field beans, sugar beet, swedes, winter field beans, winter oilseed rape*** [1-6]; ***bulb onions, carrots, dwarf beans, flower bulbs, leeks, linseed, parsnips, salad onions, spring oilseed rape, strawberries*** [1, 2, 4-6]; ***calabrese, peas*** [2]; ***combining peas, vining peas*** [1, 3-6]; ***early potatoes*** [1, 3, 4]; ***ornamental plant production*** [1]
- Creeping bent in ***brussels sprouts, cabbages, cauliflowers, fodder beet, maincrop potatoes, mangels, spring field beans, sugar beet, swedes, winter field beans, winter oilseed rape*** [1-6]; ***bulb onions, carrots, dwarf beans, flower bulbs, leeks, linseed, parsnips, salad onions, spring oilseed rape, strawberries*** [1, 2, 4-6]; ***calabrese, peas*** [2]; ***combining peas, vining peas*** [1, 3-6]; ***early potatoes*** [1, 3, 4]; ***ornamental plant production*** [1]
- Green cover in ***land temporarily removed from production*** [1, 2, 4, 5]
- Onion couch in ***brussels sprouts, bulb onions, cabbages, carrots, cauliflowers, dwarf beans, flower bulbs, fodder beet, leeks, linseed, maincrop potatoes, mangels, parsnips, salad onions, spring field beans, spring oilseed rape, strawberries, sugar beet, swedes, winter field beans, winter oilseed rape*** [1, 2, 4-6]; ***calabrese, peas*** [2]; ***combining peas, vining peas*** [1, 4-6]; ***early potatoes*** [1, 4]; ***ornamental plant production*** [1]

- Perennial grasses in ***amenity vegetation*** *(off-label)* [4]; ***farm forestry, forest*** [1, 2, 4-6]; ***forest nurseries*** [1]
- Volunteer cereals in ***brussels sprouts, cabbages, cauliflowers, fodder beet, maincrop potatoes, mangels, spring field beans, sugar beet, swedes, winter field beans, winter oilseed rape*** [1-6]; ***bulb onions, carrots, dwarf beans, flower bulbs, leeks, linseed, parsnips, salad onions, spring oilseed rape, strawberries*** [1, 2, 4-6]; ***calabrese, peas*** [2]; ***combining peas, vining peas*** [1, 3-6]; ***early potatoes*** [1, 3, 4]; ***ornamental plant production*** [1]
- Wild oats in ***brussels sprouts, cabbages, cauliflowers, fodder beet, maincrop potatoes, mangels, spring field beans, sugar beet, swedes, winter field beans, winter oilseed rape*** [1-6]; ***bulb onions, carrots, dwarf beans, flower bulbs, leeks, linseed, parsnips, salad onions, spring oilseed rape, strawberries*** [1, 2, 4-6]; ***calabrese, peas*** [2]; ***combining peas, vining peas*** [1, 3-6]; ***early potatoes*** [1, 3, 4]; ***ornamental plant production*** [1]

Specific Off-Label Approvals (SOLAs)

- ***amenity vegetation*** *(OLA 962585) Dec 2008* [4]

Efficacy guidance

- Best results achieved when weeds small and have not begun to compete with crop. Effectiveness reduced by drought, cool conditions or stress. Weeds emerging after application are not controlled
- Foliage death usually complete after 3-4 wk but longer under cool conditions, especially late treatments to winter oilseed rape
- Perennial grasses should have sufficient foliage to absorb spray and should not be cultivated for at least 14 d after treatment
- On established couch pre-planting cultivation recommended to fragment rhizomes and encourage uniform emergence
- Split applications to volunteer wheat and barley at GS 12-14 will often give adequate control in winter oilseed rape. See label for details
- Apply to dry foliage when rain not expected for at least 2 h

Restrictions

- Maximum number of treatments 1 per crop for spring oilseed rape, early potatoes; 2 per yr on amenity vegetation; 2 per crop (the second at reduced dose) for other crops. See label for details
- Must be used with Actipron (see label)
- Do not apply to crops damaged or stressed by adverse weather, pest or disease attack or other pesticide treatment
- Prevent drift onto other crops, especially cereals and grass

Crop-specific information

- HI cabbage, cauliflower, calabrese, salad onions 4 wk; peas, dwarf beans 5 wk; bulb onions, carrots, parsnips, strawberries 6 wk; sugar and fodder beet, leeks, mangels, potatoes, field beans, swedes, Brussels sprouts, winter field beans 8 wk; oilseed rape, soya beans, linseed 12 wk
- Recommended time of application varies with crop. See label for details
- On peas a crystal violet wax test should be done if leaf wax likely to have been affected by weather conditions or other chemical treatment. The wax test is essential if other products are to be sprayed before or after treatment
- May be used on ornamental bulbs when crop 5-10 cm tall. Product has been used on tulips, narcissi, hyacinths and irises but some subjects may be more sensitive and growers advised to check tolerance on small number of plants before treating the rest of the crop
- May be applied to land temporarily removed from production where the green cover is made up predominantly of tolerant crops listed on label. Use on industrial crops of linseed and oilseed rape on land temporarily removed from production also permitted.

Following crops guidance

- Guideline intervals for sowing succeeding crops after failed treated crop: field beans, peas, sugar beet, rape, kale, swedes, radish, white clover, lucerne 1 wk; dwarf French beans 4 wk; wheat, barley, maize 8 wk
- Oats should not be sown after failure of a treated crop

Environmental safety

- Dangerous for the environment
- Toxic to aquatic organisms

Hazard classification and safety precautions

Hazard H03, H11 [1-4, 6]; H04 [5]

Risk phrases R22b, R51, R58 [1-4, 6]; R36 [1-3, 5, 6]; R38 [1-6]

Operator protection A, C; U05a, U08, U20b [1-6]; U19a [1, 2]
Environmental protection E13c [5]; E15a [1-4, 6]; E31b [1-3, 6]; E31c [4, 5]; E32d, E38 [4]
Storage and disposal E01, E04, E30a [1-6]; E26 [1-3, 6]
Medical advice M05b [1-4, 6]

126 cymoxanil

A urea fungicide for potatoes

Products

1	Option	DuPont	60% w/w	WG	11834
2	Sipcam C 50	Sipcam	50% w/w	WP	10610

Uses
- Blight in ***potatoes***

Efficacy guidance
- Product to be used in mixture with recommended partner containing mancozeb or fluazinam in order to combine systemic and protective activity
- Commence spray programme before infection appears as soon as weather conditions favourable for disease development occur. At latest the first treatment should be made as the foliage meets along the rows
- Repeat treatments at 7-10 day intervals according to disease incidence and weather conditions

Restrictions
- Maximum number of treatments 4 per crop

Crop-specific information
- HI 14 d [1], 7 d [2]

Environmental safety
- Dangerous for the environment
- Very toxic to aquatic organisms

Hazard classification and safety precautions

Hazard H03 [1, 2]; H11 [1]
Risk phrases R22a [1, 2]; R50, R58 [1]
Operator protection A; U05a, U08, U20b [1, 2]; U19a [1]
Environmental protection E13c [2]; E15a, E32d, E34, E38 [1]; E32a [1, 2]
Storage and disposal E01, E04, E30a
Medical advice M03

127 cymoxanil + famoxadone

A preventative and curative fungicide mixture for potatoes

Products

Tanos	DuPont	25:25% w/w	WG	10677

Uses
- Blight in ***potatoes***

Approval information
- Famoxadone included in Annex I under EC Directive 91/414

Efficacy guidance
- Commence spray programme before infection appears as soon as weather conditions favourable for disease development occur. At latest the first treatment should be made as the foliage meets along the rows
- Repeat treatments at 7-10 day intervals according to disease incidence and weather conditions
- Reduce spray interval if conditions are conducive to the spread of blight
- Spray as soon as possible after irrigation
- Increase water volume in dense crops

- Product combines active substances with different modes of action and is effective against strains of potato blight that are insensitive to phenylamide fungicides
- To minimise the likelihood of development of resistance to strobilurin fungicides these products should be used in a planned Resistance Management strategy. See Section 5 for more information

Restrictions

- Maximum number of treatments 12 per crop
- Consult processors before using on crops for processing

Crop-specific information

- HI 14 d for potatoes

Environmental safety

- Dangerous for the environment
- Very toxic to aquatic organisms
- LERAP Category B

Hazard classification and safety precautions

Hazard H03, H11
Risk phrases R22a, R50, R58
Operator protection A, D; U05a, U20b
Environmental protection E15a, E16a, E16b, E31b, E32d, E34, E38
Storage and disposal E01, E04, E30a
Medical advice M03

128 cymoxanil + fludioxonil + metalaxyl-M

A fungicide seed dressing for peas

Products

Wakil XL	Syngenta	10:5:17.5% w/w	WS	10562

Uses

- Ascochyta in ***combining peas*** *(seed treatment)*, ***vining peas*** *(seed treatment)*
- Damping off in ***combining peas*** *(seed treatment)*, ***poppies*** *(off-label - for morphine production - seed treatment)*, ***red beet*** *(off-label - seed treatment)*, ***vining peas*** *(seed treatment)*
- Downy mildew in ***broad beans*** *(off-label - seed treatment)*, ***combining peas*** *(seed treatment)*, ***poppies*** *(off-label - for morphine production - seed treatment)*, ***spring field beans*** *(off-label - seed treatment)*, ***vining peas*** *(seed treatment)*, ***winter field beans*** *(off-label - seed treatment)*
- Pythium in ***carrots*** *(off-label - seed treatment)*, ***parsnips*** *(off-label - seed treatment)*, ***poppies*** *(off-label - for morphine production - seed treatment)*

Specific Off-Label Approvals (SOLAs)

- ***broad beans***, ***spring field beans***, ***winter field beans*** *(seed treatment) (OLA 023932) Dec 2008* [1]
- ***carrots***, ***parsnips*** *(seed treatment) (OLA 021191) Dec 2008* [1]
- ***poppies*** *(for morphine production - seed treatment) (OLA 040683) Dec 2008* [1]
- ***red beet*** *(seed treatment) (OLA 032313) Dec 2008* [1]

Approval information

- Metalaxyl-M included in Annex I under EC Directive 91/414

Efficacy guidance

- Apply through continuous flow seed treaters which should be calibrated before use

Restrictions

- Max no of treatments 1 per seed batch
- Ensure moisture content of treated seed satisfactory and store in a dry place
- Check calibration of seed drill with treated seed before drilling and sow as soon as possible after treatment
- Consult before using on crops for processing

Crop-specific information

- Latest use: pre-drilling

Environmental safety

- Harmful to aquatic organisms
- Do not use treated seed as food or feed

Hazard classification and safety precautions

Risk phrases R52, R58
Operator protection A, D, H; U02a, U05a, U20b
Environmental protection E03, E15a, E32a, E32d
Storage and disposal E01, E04, E30a
Treated seed S02, S04b, S05, S07

129 cymoxanil + mancozeb

A protectant and systemic fungicide for potato blight control

Products

1	Agriguard Cymoxanil Plus	AgriGuard	4.5:68% w/w	WP	10893
2	Curzate M68	DuPont	4.5:68% w/w	WP	08072
3	Globe	Sipcam	6:70 % w/w	WP	10339
4	Me2 Cymoxeb	Me2	4.5:68% w/w	WP	09486
5	Rhythm	Interfarm	4.5:68% w/w	WP	11669
6	Solace	Nufarm UK	4.5:68% w/w	WP	11936

Uses

- Blight in ***potatoes***

Efficacy guidance

- Apply immediately after blight warning or as soon as local conditions dictate and repeat at 7-14 d intervals until haulm dies down or is burnt off
- Spray interval should not be more than 7-10 d in irrigated crops (see product label). Apply treatment after irrigation
- To minimise the likelihood of development of resistance these products should be used in a planned Resistance Management strategy. See Section 5 for more information

Restrictions

- Maximum number of treatments 6 per crop
- Do not allow packs to become wet during storage

Crop-specific information

- HI 7 d [3]
- Destroy and remove any haulm that remains after harvest of early varieties to reduce blight pressure on neighbouring maincrop potatoes

Environmental safety

- Dangerous for the environment
- Very toxic to aquatic organisms
- Keep product away from fire or sparks

Hazard classification and safety precautions

Hazard H04 [1-6]; H11 [1, 2, 4-6]
Risk phrases R36, R37, R42 [1, 2, 4-6]; R38 [1, 4, 6]; R43 [1-6]; R50, R58 [2, 5]
Operator protection A, C [1-6]; H [3]; U02a [3]; U05a, U08 [1-6]; U11 [2, 5]; U14 [5]; U19a [1, 2, 4-6]; U20a [1, 4, 6]; U20b [2, 3, 5]
Environmental protection E13b [3]; E15a, E32a [1-6]; E32d, E38 [2, 5]
Storage and disposal E01, E04, E30a

130 cypermethrin

A contact and stomach acting pyrethroid insecticide

Products

1	Cyperguard 100 EC	Nufarm UK	100 g/l	EC	12039
2	Cyperkill 5	Mitchell Cotts	50 g/l	EC	00625
3	I T Cyper	I T Agro	100 g/l	EC	10557
4	Jundi 100 EC	DAPT	100 g/l	EC	10848

Products – continued

5 Permasect C	Nufarm UK	100 g/l	EC	11121
6 Toppel 10	United Phosphorus	100 g/l	EC	08772

Uses

- Aphids in ***apples, vining peas*** [1-6]; ***broccoli, brussels sprouts, cabbages, calabrese, cauliflowers, durum wheat, hops, lettuce, pears, triticale, winter oats, winter rye*** [1, 3, 4, 6]; ***cherries, ornamental plant production, plums*** [3, 4, 6]; ***chinese cabbage*** *(off-label)*, ***leaf spinach*** *(off-label)*, ***protected chinese cabbage*** *(off-label)*, ***protected spinach*** *(off-label)* [6]; ***lettuce*** *(outdoor crops)* [2]; ***ornamental specimens*** [1]; ***winter barley, winter wheat*** [1-4, 6]
- Asparagus beetle in ***asparagus*** *(off-label)* [6]
- Barley yellow dwarf vectors in ***triticale, winter oats, winter rye*** [5]
- Barley yellow dwarf virus vectors in ***durum wheat, triticale, winter oats, winter rye*** [1, 3, 4, 6]; ***spring barley*** *(autumn sown)*, ***spring wheat*** *(autumn sown)* [5]; ***winter barley, winter wheat*** [1-6]
- Bladder pod midge in ***spring oilseed rape, winter oilseed rape*** [1, 3, 4, 6]
- Cabbage stem flea beetle in ***winter oilseed rape*** [1-6]
- Capsids in ***apples*** [1, 3, 4, 6]; ***ornamental plant production*** [3, 4, 6]; ***ornamental specimens*** [1]; ***pears*** [1-6]
- Caterpillars in ***apples*** [1-4, 6]; ***broccoli, brussels sprouts, cabbages, calabrese, cauliflowers*** [1-6]; ***cherries, lettuce, pears, plums*** [1, 3, 4, 6]; ***chinese cabbage*** *(off-label)*, ***leaf spinach*** *(off-label)*, ***protected chinese cabbage*** *(off-label)*, ***protected spinach*** *(off-label)* [6]; ***fodder beet, kale, mangels, potatoes, red beet, sugar beet*** [2, 5]; ***lettuce*** *(outdoor crops)* [2]; ***ornamental plant production*** [3, 4, 6]; ***ornamental specimens*** [1]
- Codling moth in ***apples*** [1-6]
- Cutworms in ***carrots*** *(off-label)*, ***parsnips*** *(off-label)* [6]; ***fodder beet, mangels, red beet*** [2, 5]; ***lettuce*** [1, 3, 4, 6]; ***lettuce*** *(outdoor crops)* [2]; ***ornamental plant production*** [3, 4, 6]; ***ornamental specimens*** [1]; ***potatoes, sugar beet*** [1-6]
- Frit fly in ***grass re-seeds*** [1-6]
- Leaf miner in ***chicory*** *(off-label)*, ***endives*** *(off-label)*, ***lettuce*** *(off-label)*, ***ornamental specimens*** *(off-label)*, ***protected courgettes*** *(off-label)*, ***protected cucumbers*** *(off-label)*, ***protected gherkins*** *(off-label)* [2]
- Pea and bean weevil in ***spring field beans, vining peas, winter field beans*** [1-6]
- Pea moth in ***vining peas*** [1-6]
- Pod midge in ***spring oilseed rape, winter oilseed rape*** [2, 5]
- Pollen beetle in ***spring oilseed rape, winter oilseed rape*** [1-6]
- Rape winter stem weevil in ***winter oilseed rape*** [1, 3, 4, 6]
- Sawflies in ***apples*** [1, 3, 4, 6]
- Seed weevil in ***spring oilseed rape, winter oilseed rape*** [1-6]
- Suckers in ***apples*** [5]; ***pears*** [1, 3, 4, 6]
- Thrips in ***ornamental plant production*** [3, 4, 6]; ***ornamental specimens*** [1]
- Tortrix moths in ***apples*** [1-6]; ***pears*** [2, 5]
- Whitefly in ***broccoli, brussels sprouts, cabbages, calabrese, cauliflowers, protected cucumbers, protected lettuce, protected ornamentals*** [1, 3, 4, 6]; ***ornamental plant production*** [3, 4, 6]; ***ornamental specimens, protected celery*** [1]
- Winter moth in ***apples, pears*** [5]; ***cherries, plums*** [1]
- Yellow cereal fly in ***durum wheat*** [1, 3, 4, 6]; ***spring barley, spring wheat*** [2]; ***spring barley*** *(autumn sown)*, ***spring wheat*** *(autumn sown)* [5]; ***triticale, winter oats, winter rye*** [1, 3-6]; ***winter barley, winter wheat*** [1-6]

Specific Off-Label Approvals (SOLAs)

- ***asparagus*** *(OLA 032223) Dec 2008* [6]
- ***carrots, parsnips*** *(OLA 982184) Dec 2008* [6]
- ***chicory, endives, lettuce, ornamental specimens, protected courgettes, protected cucumbers, protected gherkins*** *(OLA 992260) Dec 2008* [2]
- ***chinese cabbage, leaf spinach, protected chinese cabbage, protected spinach*** *(OLA 983133) Dec 2008* [6]

Approval information

- Following implementation of Directive 98/82/EC, approval for use of cypermethrin on numerous crops was revoked in 1999

Efficacy guidance

- Products combine rapid action, good persistence, and high activity on Lepidoptera.
- As effect is mainly via contact good coverage is essential for effective action. Spray volume should be increased on dense crops
- A repeat spray after 10-14 d is needed for some pests of outdoor crops, several sprays at shorter intervals for whitefly and other glasshouse pests
- Rates and timing of sprays vary with crop and pest. See label for details
- Add a non-ionic wetter to improve results on leaf brassicas. In Brussels sprouts use of a drop-leg sprayer may be beneficial
- Where aphids in hops, pear suckers or glasshouse whitefly resistant to cypermethrin occur control is unlikely to be satisfactory

Restrictions

- Maximum number of treatments varies with crop and product. See label or approval notice for details

Crop-specific information

- Latest use: normally before end Mar in yr of harvest, but varies with product and crop. See labels for details
- HI apples, pears 14 d; vining peas 7 d; chicory, lettuce, courgettes, cucumbers, gherkins, endives, asparagus 1 d; other crops 0 d
- Test spray sample of new or unusual ornamentals before committing whole batches

Environmental safety

- Dangerous for the environment
- Very toxic to aquatic organisms
- Flammable
- LERAP Category A
- Broadcast air-assisted LERAP (18 m) [4, 5]
- Do not spray cereals after 31 Mar within 6 m of the edge of the growing crop

Hazard classification and safety precautions

Hazard H03, H08 [1-6]; H04 [2, 4]; H11 [1, 3, 5, 6]

Risk phrases R20, R41, R51 [3]; R21 [1, 6]; R22a [1-3, 5, 6]; R22b [1-6]; R36 [2, 4]; R37 [2, 3, 5]; R38 [1, 2, 4, 6]; R43 [2]; R50, R66, R67 [1, 5, 6]; R58 [1, 3, 5, 6]

Operator protection A, C; U02a, U04a [2, 5]; U05a, U19a, U20b [1-6]; U08 [3]; U10 [1, 2, 4-6]; U11 [2, 3, 5]

Environmental protection E13a [2, 4]; E15a [1, 3, 5, 6]; E16c, E16d, E31b [1-6]; E17b [4, 5] (18 m); E19b [1, 6]; E32d [3, 5]; E34 [2, 5]

Storage and disposal E01, E04, E26, E30a

Medical advice M03 [1, 2, 6]; M04 [5]; M05b [1, 3-6]

131 cyproconazole

A contact and systemic conazole fungicide for cereals and other field crops

See also *azoxystrobin + cyproconazole*
chlorothalonil + cyproconazole

Products

1	Cabaret	Bayer CropScience	240 g/l	EC	11915
2	Caddy 240 EC	Bayer CropScience	240 g/l	EC	11256

Uses

- Brown rust in ***red beet*** *(off-label)* [2]; ***spring barley***, ***spring rye***, ***winter barley***, ***winter rye***, ***winter wheat*** [1, 2]
- Chocolate spot in ***spring field beans*** *(with chlorothalonil)*, ***winter field beans*** *(with chlorothalonil)* [1, 2]
- Crown rust in ***spring oats***, ***winter oats*** [1, 2]
- Eyespot in ***winter barley***, ***winter wheat*** [1, 2]
- Late ear diseases in ***winter wheat*** [1, 2]
- Light leaf spot in ***winter oilseed rape*** [1, 2]
- Net blotch in ***spring barley***, ***winter barley*** [1, 2]

SEE SECTION 3 FOR PRODUCTS ALSO REGISTERED

- Phoma leaf spot in ***winter oilseed rape*** [1, 2]
- Powdery mildew in ***spring barley***, ***spring oats***, ***spring rye***, ***sugar beet***, ***winter barley***, ***winter oats***, ***winter rye***, ***winter wheat*** [1, 2]
- Ramularia leaf spots in ***sugar beet*** [1, 2]
- Rhynchosporium in ***spring barley***, ***winter barley*** [1, 2]
- Rust in ***leeks***, ***spring field beans***, ***sugar beet***, ***winter field beans*** [1, 2]; ***red beet*** *(off-label)* [1]
- Septoria diseases in ***winter wheat*** [1, 2]
- Yellow rust in ***spring barley***, ***winter barley***, ***winter wheat*** [1, 2]

Specific Off-Label Approvals (SOLAs)

- ***red beet*** *(OLA 031883) Dec 2008* [2]
- ***red beet*** *(OLA 041024) Dec 2008* [1]

Efficacy guidance

- Apply at start of disease development or as preventive treatment and repeat as necessary
- Most effective time of treatment varies with disease and use of tank mixes may be desirable. See label for details
- Product alone gives useful reduction of cereal eyespot. Where high infections probable a tank mix with prochloraz recommended
- On oilseed rape a two spray autumn/spring programme recommended for high risk situations and on susceptible varieties

Restrictions

- Maximum total dose equivalent to 1-3 full dose treatments depending on crop treated. See labels for details

Crop-specific information

- HI 6 wk for field beans; 14 d for sugar beet, red beet, leeks
- Latest use: completion of flowering for rye, wheat; emergence of ear complete (GS 59) for barley, oats; before lowest pods more than 2 cm long (GS 5,1) for winter oilseed rape
- Application to winter wheat in spring between the start of stem elongation and the third node detectable stage (GS 30-33) may cause straw shortening, but does not cause loss of yield

Environmental safety

- Dangerous for the environment
- Toxic to aquatic organisms

Hazard classification and safety precautions

Hazard H03, H11
Risk phrases R22b, R36, R38, R51, R58, R63
Operator protection A, C, H; U05a, U08, U20b
Environmental protection E15a, E31b, E32d, E38
Storage and disposal E01, E04, E26, E30a
Medical advice M05b

132 cyproconazole + cyprodinil

A broad spectrum fungicide mixture for cereals

Products

Radius	Syngenta	5.33:40 % w/w	WG	09387

Uses

- Brown rust in ***winter wheat***
- Eyespot in ***spring barley***, ***winter barley***, ***winter wheat***
- Net blotch in ***spring barley***, ***winter barley***
- Powdery mildew in ***spring barley***, ***winter barley***, ***winter wheat***
- Rhynchosporium in ***spring barley***, ***winter barley***
- Septoria diseases in ***winter wheat***
- Yellow rust in ***winter wheat***

Efficacy guidance

- Best results obtained from treatment at early stages of development. Later treatments may be required if disease pressure remains high
- Eyespot is fully controlled by treatment in spring during stem extension. Control may be reduced when very dry conditions follow application

Restrictions

- Maximum total dose equivalent to two full dose treatments

Crop-specific information

- Latest use: before first spikelet of inflorescence visible (GS 51) for barley; before caryopsis watery ripe (GS 71) for winter wheat

Environmental safety

- Dangerous for the environment
- Very toxic to aquatic organisms
- LERAP Category B

Hazard classification and safety precautions

Hazard H03, H11

Risk phrases R38, R43, R50, R58, R63

Operator protection A, H; U05a, U20a

Environmental protection E15a, E16a, E16b, E32d, E38

Storage and disposal E01, E04, E30a

133 cyproconazole + propiconazole

A broad-spectrum mixture of conazole fungicides for wheat and barley

Products

1	Alto Xtra	Syngenta	160:250 g/l	EC	11839
2	Menara	Syngenta	160:250 g/l	EC	09321

Uses

- Brown rust in ***spring barley***, ***winter barley***, ***winter wheat*** [1, 2]
- Glume blotch in ***winter wheat*** [1]
- Late ear diseases in ***winter wheat*** [1, 2]
- Net blotch in ***spring barley***, ***winter barley*** [1, 2]
- Powdery mildew in ***spring barley***, ***winter barley*** [1, 2]; ***winter wheat*** [1]
- Rhynchosporium in ***spring barley***, ***winter barley*** [1, 2]
- Septoria diseases in ***winter wheat*** [2]
- Septoria leaf spot in ***winter wheat*** [1]
- Yellow rust in ***spring barley***, ***winter barley*** [1]; ***winter wheat*** [1, 2]

Approval information

- Propiconazole included in Annex I under EC Directive 91/414

Efficacy guidance

- Treatment should be made at first sign of disease
- Useful reduction of eyespot when applied in spring during stem extension (GS 30-32)

Restrictions

- Maximum total dose equivalent to one full dose treatment

Crop-specific information

- Latest use: before beginning of anthesis (GS 60) for barley; before grain watery ripe (GS 71) for wheat
- Application to winter wheat in spring may cause straw shortening but does not cause loss of yield

Environmental safety

- Dangerous for the environment
- Very toxic to aquatic organisms

Hazard classification and safety precautions

Hazard H03, H11

Risk phrases R36, R38, R50, R58, R63, R67

Operator protection A, C, H; U05a, U08, U20b

SEE SECTION 3 FOR PRODUCTS ALSO REGISTERED

Environmental protection E15a, E31c, E32d, E38
Storage and disposal E01, E04, E26, E30a

134 cyproconazole + trifloxystrobin

A conazole and strobilurin fungicide mixture for wheat and barley

Products

Sphere	Bayer CropScience	80:187.5 g/l	EC	11429

Uses

- Brown rust in ***spring barley**, **winter barley**, **winter wheat***
- Net blotch in ***spring barley**, **winter barley***
- Powdery mildew in ***spring barley**, **winter barley**, **winter wheat***
- Rhynchosporium in ***spring barley**, **winter barley***
- Septoria diseases in ***winter wheat***
- Yellow rust in ***spring barley**, **winter barley**, **winter wheat***

Approval information

- Trifloxystrobin included in Annex I under EC Directive 91/414

Efficacy guidance

- Best results obtained from treatment at early stages of disease development. Further treatment may be needed if disease attack is prolonged
- Trifloxystrobin is a member of the QoI cross resistance group. Product should be used preventatively and not relied on for its curative potential
- Use product as part of an Integrated Crop Management strategy incorporating other methods of control, including where appropriate other fungicides with a different mode of action. Do not apply more than two foliar applications of QoI containing products to any cereal crop
- There is a significant risk of widespread resistance occurring in *Septoria tritici* populations in UK. Failure to follow resistance management action may result in reduced levels of disease control
- Strains of barley powdery mildew resistant to QoIs are common in the UK

Restrictions

- Maximum total dose per crop equivalent to two full dose treatments

Crop-specific information

- HI 35 d

Environmental safety

- Dangerous for the environment
- Very toxic to aquatic organisms

Hazard classification and safety precautions

Hazard H03, H11
Risk phrases R36, R50, R58, R63
Operator protection A, C, H; U05a, U09a, U19a, U20b
Environmental protection E15a, E31b, E32d, E38
Storage and disposal E01, E04, E30a

135 cyprodinil

An anilinopyrimidine systemic broad spectrum fungicide for cereals

See also cyproconazole + cyprodinil

Products

Unix	Syngenta	75% w/w	WG	11512

Uses

- Eyespot in ***winter barley**, **winter wheat***
- Net blotch in ***spring barley**, **winter barley***
- Powdery mildew in ***spring barley**, **spring wheat**, **winter barley**, **winter wheat***
- Rhynchosporium in ***spring barley**, **winter barley***

Efficacy guidance
- Best results obtained from treatment at early stages of disease development
- For best control of eyespot spray before or during the period of stem extension in spring. Control may be reduced if very dry conditions follow treatment

Restrictions
- Maximum number of treatments equal to one and two thirds full dose on wheat and twice full dose on barley

Crop-specific information
- Latest use: up to and including first awns visible (GS 49) for spring barley, winter barley; up to and including grain watery ripe (GS 71) for spring wheat, winter wheat

Environmental safety
- Dangerous for the environment
- Very toxic to aquatic organisms
- LERAP Category B

Hazard classification and safety precautions

Hazard H11
Risk phrases R50, R58
Operator protection A, H; U05a, U20a
Environmental protection E15a, E16a, E16b, E32d, E38
Storage and disposal E01, E04, E30a

136 cyprodinil + picoxystrobin

A broad spectrum fungicide mixture for cereals

Products

Acanto Prima	Syngenta	30:8 % w/w	WG	11864

Uses
- Brown rust in ***spring barley, spring wheat, winter barley, winter wheat***
- Eyespot in ***spring barley, spring wheat*** *(moderate control)*, ***winter barley, winter wheat*** *(moderate control)*
- Glume blotch in ***spring wheat, winter wheat***
- Net blotch in ***spring barley, winter barley***
- Powdery mildew in ***spring barley, winter barley***
- Rhynchosporium in ***spring barley, winter barley***
- Septoria leaf spot in ***spring wheat, winter wheat***
- Yellow rust in ***spring wheat, winter wheat***

Efficacy guidance
- Best results obtained from use as a protectant treatment or in the earliest stages of disease development. Further applications may be needed if disease attack is prolonged
- Picoxystrobin is a member of the QoI cross resistance group. Product should be used preventatively and not relied on for its curative potential
- Use product as part of an Integrated Crop Management strategy incorporating other methods of control, including where appropriate other fungicides with a different mode of action. Do not apply more than two foliar applications of QoI containing products to any cereal crop
- There is a significant risk of widespread resistance occurring in *Septoria tritici* populations in UK. Failure to follow resistance management action may result in reduced levels of disease control
- In wheat product must always be used in mixture with another product, recommended for control of the same target disease, that contains a fungicide from a different cross resistance group and is applied at a dose that will give robust control
- Strains of barley powdery mildew resistant to QoIs are common in the UK

Restrictions
- Maximum number of treatments 2 per crop

Crop-specific information
- Latest use: before early milk stage (GS 73) wheat; before ear just visible (GS 51) for barley

SEE SECTION 3 FOR PRODUCTS ALSO REGISTERED

Environmental safety

- Dangerous for the environment
- Very toxic to aquatic organisms

Hazard classification and safety precautions

Hazard H11
Risk phrases R50, R58
Operator protection A, C, H; U05a, U09a, U11, U14, U15, U20a
Environmental protection E15a, E31c, E32d, E34, E38
Storage and disposal E01, E04, E30a

137 2,4-D

A translocated phenoxy herbicide for cereals, grass and amenity use

See also *amitrole + 2,4-D + diuron*
clopyralid + 2,4-D + MCPA

Products

1	Agricorn D II	FCC	490 g/l	SL	09415
2	Depitox	Nufarm UK	490 g/l	SL	11149
3	Dicotox Extra	Bayer Environ.	400 g/l	EC	09930
4	Dioweed 50	United Phosphorus	500 g/l	SL	08050
5	Dormone	Bayer Environ.	465 g/l	SL	09932
6	Headland Staff	Headland	470 g/l	SL	07189
7	Herboxone	Headland	500 g/l	SL	10032
8	HY-D Super	Agrichem	500 g/l	SL	11618
9	Syford	Vitax	500 g/l	SL	02062

Uses

- Annual dicotyledons in ***amenity grass, apple orchards, pear orchards*** [1, 2]; ***amenity turf*** [3, 5, 9]; ***conifer plantations, forest*** [3]; ***established grassland*** [1, 2, 4, 6-9]; ***grass seed crops*** [9]; ***managed amenity turf*** [1, 2, 4, 5]; ***rotational grassland*** [4]; ***spring barley, spring wheat, winter barley, winter rye, winter wheat*** [1, 2, 4, 6-8]; ***spring rye*** [1, 2, 4]; ***winter oats*** [1, 2, 6-8]
- Aquatic weeds in ***aquatic situations*** [2]; ***water or waterside areas*** [5]
- Heather in ***conifer plantations, forest*** [3]
- Perennial dicotyledons in ***amenity grass, apple orchards, pear orchards*** [1, 2]; ***amenity turf*** [3, 5, 9]; ***conifer plantations, forest*** [3]; ***established grassland*** [1, 2, 4, 6-9]; ***grass seed crops*** [9]; ***managed amenity turf*** [1, 2, 4, 5]; ***rotational grassland*** [4]; ***spring barley, spring wheat, winter barley, winter rye, winter wheat*** [1, 2, 4, 6-8]; ***spring rye*** [1, 2, 4]; ***water or waterside areas*** [5]; ***winter oats*** [1, 2, 6-8]
- Willows in ***conifer plantations, forest*** [3]
- Woody weeds in ***conifer plantations, forest*** [3]

Approval information

- 2,4-D included in Annex I under EC Directive 91/414
- May be applied through CDA equipment (see label for details) [9]. See notes in Section 5 on ULV application
- Approved for aquatic weed control [2, 5]. See notes in Section 5 on use of herbicides in or near water

Efficacy guidance

- Best results achieved by spraying weeds in seedling to young plant stage when growing actively in a strongly competing crop
- Most effective stage for spraying perennials varies with species. See label for details
- Spray aquatic weeds when in active growth between May and Sep

Restrictions

- Maximum number of treatments normally 1 per crop and in forestry, 2 per yr in grassland and 2 or 3 per yr in amenity turf. Check individual labels
- Do not use on newly sown leys containing clover
- Do not spray grass seed crops after ear emergence
- Do not spray within 6 mth of laying turf or sowing fine grass
- Do not dump surplus herbicide in water or ditch bottoms
- Do not plant conifers until at least 1 mth after treatment

- Do not spray crops stressed by cold weather or drought or if frost expected
- Do not roll or harrow within 7 d before or after spraying
- Do not spray if rain falling or imminent
- Do not mow amenity grass 2 d before or 1 d after spraying. Do not mow grassland or graze for at least 10 d after spraying

Crop-specific information

- Latest use: before 1st node detectable in cereals; end Aug for conifer plantations; before established grassland 25 cm high
- Spray winter cereals in spring when leaf-sheath erect but before first node detectable (GS 31), spring cereals from 5-leaf stage to before first node detectable (GS 15-31)
- Cereals undersown with grass and/or clover, but not with lucerne, may be treated
- Selective treatment of resistant conifers can be made in Aug when growth ceased and plants hardened off, spray must be directed if applied earlier. See label for details

Following crops guidance

- Do not use shortly before or after sowing any crop
- Do not direct drill brassicas or grass/clover mixtures within 3 wk of application

Environmental safety

- Dangerous for the environment [3-9]
- Toxic (very toxic [3]) to aquatic organisms [4-9]
- 2,4-D is active at low concentrations. Take extreme care to avoid drift onto neighbouring crops, especially beet crops, brassicas, most market garden crops including lettuce and tomatoes under glass, pears and vines
- May be used to control aquatic weeds in presence of fish if used in strict accordance with directions for waterweed control and precautions needed for aquatic use [2, 5]
- Keep livestock out of treated areas for at least 2 wk following treatment and until poisonous weeds, such as ragwort, have died down and become unpalatable
- Dangerous to aquatic higher plants. Do not contaminate surface waters or ditches with chemical or used container [4, 6, 7]
- Water containing the herbicide must not be used for irrigation purposes within 3 wk of treatment or until the concentration in water is below 0.05 ppm

Hazard classification and safety precautions

Hazard H03 [1-9]; H11 [3-9]

Risk phrases R20 [8]; R21 [3, 5, 8]; R22a, R58 [1-9]; R22b, R36, R50, R66, R67 [3]; R38 [1-3]; R41 [1, 2, 4-9]; R43 [3-9]; R51 [4-9]

Operator protection A, C, H, M [1-9]; D [4]; U05a, U08 [1-9]; U11 [1, 2, 4-9]; U14 [3-9]; U15 [4, 6-8]; U20a [1, 4]; U20b [2, 3, 5-9]

Environmental protection E07a, E34 [1-9]; E13b [1]; E13c [5, 8]; E14b [4, 6, 7]; E15a [2-5, 7, 9]; E21 [5] (3 wk); E31a [1, 6-8]; E31b [2, 4, 9]; E31c [3]; E32d [3, 5, 9]; E38 [3, 5-9]

Storage and disposal E01, E04, E30a [1-9]; E26 [2-4, 6-8]

Medical advice M03 [1-9]; M05a [4, 6-8]; M05b [3]

138 2,4-D + dicamba

A translocated herbicide for use on turf

Products

New Estermone	Vitax	200:35 g/l	EC	06336

Uses

- Annual dicotyledons in ***managed amenity turf***
- Perennial dicotyledons in ***managed amenity turf***

Approval information

- 2,4-D included in Annex I under EC Directive 91/414

Efficacy guidance

- Best results achieved by application when weeds growing actively in spring or early summer (later with irrigation and feeding)
- More resistant weeds may need repeat treatment after 3 wk
- Do not use during drought conditions or mow for 3 d before or after treatment

SEE SECTION 3 FOR PRODUCTS ALSO REGISTERED

Restrictions

- Maximum number of treatments 3 per yr
- Do not treat newly sown or turfed areas
- Avoid spray drift onto cultivated crops or ornamentals
- Do not re-seed for 6 wk after application

Environmental safety

- Dangerous for the environment
- Toxic to aquatic organisms
- Keep livestock out of treated areas for at least two weeks following treatment and until poisonous weeds, such as ragwort, have died down and become unpalatable

Hazard classification and safety precautions

Hazard H04, H11
Risk phrases R51, R58
Operator protection A, C, H, M; U08, U11, U20b
Environmental protection E07a, E15a, E31a, E32d, E38
Storage and disposal E30a

139 2,4-D + dicamba + fluroxypyr

A translocated and contact herbicide mixture for amenity turf

Products

Holster	SumiAgro Amenity	285:52.2:105 g/l	SL	10593

Uses

- Annual dicotyledons in ***managed amenity turf***
- Buttercups in ***managed amenity turf***
- Chickweed in ***managed amenity turf***
- Clover in ***managed amenity turf***
- Dandelions in ***managed amenity turf***

Approval information

- 2,4-D and fluroxypyr included in Annex I under EC Directive 91/414

Efficacy guidance

- Apply when weeds actively growing (normally between Apr and Sep) and when soil is moist
- Best results obtained from treatment in spring or early summer before weeds begin to flower
- Do not apply if turf is wet or if rainfall expected within 4 h of treatment. Both circumstances will reduce weed control

Restrictions

- Maximum number of treatments 2 per yr
- Do not mow for 3 d before or after treatment
- Avoid overlapping or overdosing, especially on newly sown turf
- Do not spray in drought conditions or if turf under stress from frost, waterlogging, trace element deficiency, pest or disease attack

Crop-specific information

- Latest Use: normally Sep for managed amenity turf
- New turf may be treated in spring provided at least 2 mth have elapsed since sowing

Environmental safety

- Dangerous for the environment
- Toxic to aquatic organisms
- Keep livestock out of treated areas for at least two weeks following treatment and until poisonous weeds, such as ragwort, have died down and become unpalatable

Hazard classification and safety precautions

Hazard H03, H11
Risk phrases R22a, R22b, R36, R38, R51, R58
Operator protection A, C, H, M; U05a, U08, U19a, U20b
Environmental protection E07a, E15a, E31b, E32d, E38
Storage and disposal E01, E04, E26, E30a
Medical advice M03

140 2,4-D + dicamba + triclopyr

A translocated herbicide for perennial and woody weed control

Products

1 Broadsword	United Phosphorus	200:85:65 g/l	EC	09140
2 Greengard	SumiAgro Amenity	200:85:65 g/l	EC	11715
3 Nu-Shot	Nufarm UK	200:85:65 g/l	EC	11148

Uses

- Annual dicotyledons in ***amenity grass*** [2, 3]; ***land not intended to bear vegetation*** [1]; ***permanent pasture, rotational grassland*** [1, 3]
- Brambles in ***farm forestry, forest, natural surfaces not intended to bear vegetation*** [1-3]
- Docks in ***amenity grass*** [2, 3]; ***land not intended to bear vegetation*** [1]; ***permanent pasture, rotational grassland*** [1, 3]
- Gorse in ***farm forestry, forest, natural surfaces not intended to bear vegetation*** [1-3]
- Green cover in ***land temporarily removed from production*** [1]
- Japanese knotweed in ***farm forestry, forest, natural surfaces not intended to bear vegetation*** [1-3]
- Perennial dicotyledons in ***amenity grass*** [2, 3]; ***farm forestry, forest, natural surfaces not intended to bear vegetation*** [1-3]; ***land not intended to bear vegetation*** [1]; ***permanent pasture, rotational grassland*** [1, 3]
- Plantains in ***amenity grass*** [2, 3]; ***land not intended to bear vegetation*** [1]; ***permanent pasture, rotational grassland*** [1, 3]
- Rhododendrons in ***farm forestry, forest, natural surfaces not intended to bear vegetation*** [1-3]
- Stinging nettle in ***amenity grass*** [2, 3]; ***permanent pasture, rotational grassland*** [1, 3]
- Thistles in ***amenity grass*** [2, 3]; ***land not intended to bear vegetation*** [1]; ***permanent pasture, rotational grassland*** [1, 3]
- Woody weeds in ***farm forestry, forest, natural surfaces not intended to bear vegetation*** [1-3]

Approval information

- 2,4-D included in Annex I under EC Directive 91/414

Efficacy guidance

- Apply as foliar spray to herbaceous or woody weeds. Timing and growth stage for best results vary with species. See label for details
- Dilute with water for stump treatment and apply after felling up to the start of regrowth. Treat any regrowth with a spray to the growing foliage
- May be applied at 1/3 dilution in weed wipers or 1/8 dilution with ropewick applicators

Restrictions

- Maximum total dose per yr equivalent to two full dose treatments
- Do not use on pasture established less than 1 yr or on grass grown for seed
- Where clover a valued constituent of sward only use as a spot treatment
- Do not graze for 7 d or mow for 14 d after treatment
- Do not direct drill grass, clover or brassicas for at least 6 wk after grassland treatment
- Do not plant trees for 1-3 mth after spraying depending on dose applied. See label
- Avoid spray drift into greenhouses or onto crops or ornamentals. Vapour drift may occur in hot conditions

Crop-specific information

- Latest Use: before weed flower buds open
- Sprays may be applied in pines, spruce and fir providing drift is avoided. Optimum time is mid-autumn when tree growth ceased but weeds not yet senescent

Environmental safety

- Dangerous for the environment
- Toxic (very toxic [1]) to aquatic organisms [2, 3]
- Docks and other weeds may become increasingly palatable after treatment and may be preferentially grazed

SEE SECTION 3 FOR PRODUCTS ALSO REGISTERED

- Keep livestock out of treated areas for at least two weeks following treatment and until poisonous weeds, such as ragwort, have died down and become unpalatable
- Take extreme care to avoid drift onto neighbouring crops, especially beet crops, brassicas, most market garden crops including lettuce and tomatoes under glass, pears and vines

Hazard classification and safety precautions

Hazard H03, H08, H11

Risk phrases R22a, R22b, R36, R58 [1-3]; R37 [1, 3]; R38 [1, 2]; R43 [3]; R50, R67 [1]; R51 [2, 3]

Operator protection A, C, H, M [1-3]; D [1, 3]; U02a, U04a, U13, U20a [1, 2]; U05a, U08, U11, U19a [1-3]; U14, U20b [3]

Environmental protection E07a, E31b, E34 [1-3]; E15a [2, 3]; E15b [1]; E32d [1, 3]; E38 [3]

Consumer protection C01 [1, 2]

Storage and disposal E01, E04, E26, E30a

Medical advice M03, M05b

141 2,4-D + dichlorprop-P + MCPA + mecoprop-P

A translocated herbicide for use in apple and pear orchards

Products

UPL Camppex	United Phosphorus	34.5:66.6:52.8:82.1 g/l	SL	11661

Uses

- Annual dicotyledons in ***apple orchards***, ***pear orchards***
- Chickweed in ***apple orchards***, ***pear orchards***
- Cleavers in ***apple orchards***, ***pear orchards***
- Perennial dicotyledons in ***apple orchards***, ***pear orchards***

Approval information

- 2,4-D and mecoprop-P included in Annex I under EC Directive 91/414

Efficacy guidance

- Use on emerged weeds in established apple and pear orchards (from 1 yr after planting) as directed application
- Spray when weeds in active growth and at growth stage recommended on label
- Effectiveness may be reduced by rain within 12 h

Restrictions

- Maximum number of treatments 2 per yr
- Applications must be made around, and not directly to, trees. Do not allow drift onto trees
- Do not spray during blossom period. Do not spray to run-off
- Do not roll, harrow or cut grass crops on orchard floor within at least 3 d before or after spraying

Crop-specific information

- Latest use: before blossom period

Environmental safety

- Harmful to aquatic organisms
- Keep livestock out of treated areas for at least two weeks following treatment and until poisonous weeds, such as ragwort, have died down and become unpalatable
- Take extreme care to avoid drift onto neighbouring sensitive crops

Hazard classification and safety precautions

Hazard H03

Risk phrases R20, R21, R22a, R43, R52, R58

Operator protection A, C, H, M; U05a, U08, U20b

Environmental protection E07a, E13c, E31b, E34

Storage and disposal E01, E04, E26, E30a

Medical advice M03, M05a

142 2,4-D + MCPA

A translocated herbicide mixture for cereals and grass

Products

Headland Polo	Headland	360:315 g/l	SL	10283

Uses

- Annual dicotyledons in ***established grassland, rotational grassland, spring barley, spring wheat, winter barley, winter oats, winter wheat***
- Perennial dicotyledons in ***established grassland, rotational grassland, spring barley, spring wheat, winter barley, winter oats, winter wheat***

Approval information

- 2,4-D included in Annex I under EC Directive 91/414

Efficacy guidance

- Best results achieved by spraying weeds in seedling to young plant stage when growing actively in a strongly competing crop
- Most effective stage for spraying perennials varies with species. See label for details

Restrictions

- Maximum number of treatments 1 per crop in cereals and established grassland; 2 per yr in rotational grassland
- Do not spray if rain falling or imminent
- Do not cut grass or graze for at least 10 d after spraying
- Do not use on newly sown leys containing clover or other legumes
- Do not spray crops stressed by cold weather or drought or if frost expected
- Do not use shortly before or after sowing any crop
- Do not roll or harrow within 7 d before or after spraying

Crop-specific information

- Spray winter cereals in spring when leaf-sheath erect but before first node detectable (GS 31), spring cereals from 5-leaf stage to before first node detectable (GS 15-31)
- Latest use: before first node detectable (GS 31) in cereals

Environmental safety

- Dangerous for the environment
- Toxic to aquatic organisms
- Keep livestock out of treated areas for at least two weeks following treatment and until poisonous weeds, such as ragwort, have died down and become unpalatable
- 2,4-D and MCPA are active at low concentrations. Take extreme care to avoid drift onto neighbouring crops, especially beet crops, brassicas, most market garden crops including lettuce and tomatoes under glass, pears and vines

Hazard classification and safety precautions

Hazard H03, H11
Risk phrases R20, R21, R22a, R41, R43, R51, R58
Operator protection A, C, H, M; U05a, U08, U11, U14, U15, U20b
Environmental protection E07a, E15a, E31a, E34
Storage and disposal E01, E04, E26, E30a
Medical advice M03, M05a

143 2,4-D + mecoprop-P

A translocated herbicide for use in amenity turf

Products

1 Supertox 30	Bayer Environ.	93.5:95 g/l	SL	09946
2 Sydex	Vitax	125:125 g/l	SL	06412

Uses

- Annual dicotyledons in ***established grassland, grass seed crops*** [2]; ***managed amenity turf*** [1, 2]
- Perennial dicotyledons in ***established grassland, grass seed crops*** [2]; ***managed amenity turf*** [1, 2]

Approval information

- 2,4-D and mecoprop-P included in Annex I under EC Directive 91/414

Efficacy guidance

- May be applied from Apr to Sep, best results in May-Jun when weeds in active growth. Less susceptible weeds may require second treatment not less than 6 wk later
- For best results apply fertilizer 1-2 wk before treatment
- Do not spray during drought conditions or when rain imminent as efficacy will be impaired
- Do not close mow infrequently mown turf for 3-4 d before or after treatment. On fine turf do not mow for 24 h before or after treatment

Restrictions

- Maximum number of treatments varies with product. Consult label. The total amount of mecoprop-P applied in a single yr must not exceed the maximum total dose approved for any single product for the crop/situation
- Do not use first 4 mowings for mulching unless composted for at least 6 mth (do not use first 2 mowings for composting on some labels)
- Grass cuttings should not be used for mulching but may be composted and used 6 mth later
- Do not apply during frosty weather
- Do not roll or harrow within 7 d before or after treatment

Crop-specific information

- Newly laid turf or newly sown grass should not be treated for at least 6 mth

Environmental safety

- Dangerous for the environment [2]
- Harmful to aquatic organisms
- Keep livestock out of treated areas for at least two weeks following treatment and until poisonous weeds, such as ragwort, have died down and become unpalatable
- Take extreme care to avoid drift onto neighbouring crops, especially beet crops, brassicas, most market garden crops including lettuce and tomatoes under glass, pears and vines

Hazard classification and safety precautions

Hazard H03 [1, 2]; H04 [1]; H11 [2]
Risk phrases R22a [2]; R36, R41 [1]; R43, R52, R58 [1, 2]
Operator protection A, C, H, M; U05a, U08, U11, U14, U20b [1, 2]; U15 [1]
Environmental protection E07a, E15a, E32d, E34 [1, 2]; E23, E31c [1]; E31b, E38 [2]
Storage and disposal E01, E04, E30a [1, 2]; E26 [1]
Medical advice M03 [2]

144 2,4-D + picloram

A persistent translocated herbicide for non-crop land

Products

1	Atladox HI	Nomix Enviro	240:65 g/l	SL	05559
2	Tordon 101	Dow	240:65 g/l	SL	05816

Uses

- Annual dicotyledons in ***amenity grass, non-crop grass, road verges***
- Brambles in ***amenity grass, non-crop grass, road verges***
- Creeping thistle in ***amenity grass, non-crop grass, road verges***
- Docks in ***amenity grass, non-crop grass, road verges***
- Japanese knotweed in ***amenity grass, non-crop grass, road verges***
- Perennial dicotyledons in ***amenity grass, non-crop grass, road verges***
- Ragwort in ***amenity grass, non-crop grass, road verges***
- Scrub clearance in ***amenity grass, non-crop grass, road verges***
- Woody weeds in ***amenity grass, non-crop grass, road verges***

Approval information

- 2,4-D included in Annex I under EC Directive 91/414

Efficacy guidance

- Apply as overall foliar spray during period of active growth when foliage well developed

Restrictions

- Maximum number of treatments 1 per yr
- Do not apply around desirable trees or shrubs where roots may absorb chemical
- Prevent leaching into areas where desirable plants are present
- Avoid drift of spray onto desirable plants
- Do not use cuttings from treated grass for mulching or composting

Environmental safety

- Harmful to aquatic organisms
- Keep livestock out of treated areas for at least 2 wk following treatment and until poisonous weeds, such as ragwort, have died down and become unpalatable

Hazard classification and safety precautions

Hazard H04
Risk phrases R37, R43, R52, R58
Operator protection A, C; U05a, U08, U14, U15, U20b
Environmental protection E07a [2]; E13c [1]; E15a, E31a, E32d [1, 2]
Storage and disposal E01, E04, E30a

145 daminozide

A hydrazide plant growth regulator for use in certain ornamentals

Products

1	B-Nine	Certis	85% w/w	SP	11471
2	Dazide	Fine	85% w/w	SP	02691

Uses

- Internode reduction in ***azaleas***, ***bedding plants***, ***chrysanthemums***, ***hydrangeas*** [1, 2]; ***ornamental specimens***, ***poinsettias*** [1]; ***pot plants*** [2]

Approval information

- Sales of daminozide for use on food crops were halted worldwide by the manufacturer in Oct 1989. Sales for use on flower crops were not affected

Efficacy guidance

- Best results obtained by application in late afternoon when glasshouse has cooled down
- Spray when foliage dry
- A reduced rate tank mix with chlormequat + choline chloride is recommended for use on poinsettias. See label for details [1]

Restrictions

- Maximum number of treatments 2 per crop
- Apply only to turgid, well watered plants. Do not water for 24 h after spraying
- Do not use on chrysanthemum Fandango
- Do not mix with other spray chemicals except as recommended above

Crop-specific information

- Latest use: end Sep for Poinsettias

Environmental safety

- Harmful to fish or other aquatic life. Do not contaminate surface waters or ditches with chemical or used container [2]

Hazard classification and safety precautions

Hazard H04
Risk phrases R36 [2]; R41 [1]
Operator protection A, C [1, 2]; H [2]; U05a, U19a [1, 2]; U09a, U20a [2]; U11 [1]
Environmental protection E13c, E31a [2]; E15a [1]
Storage and disposal E01, E04, E30a

146 dazomet

A methyl isothiocyanate releasing soil fumigant

Products

Basamid	Certis	97% w/w	GR	11324

Uses

- Nematodes in ***field crops*** *(soil fumigation)*, ***protected crops*** *(soil fumigation)*, ***vegetables*** *(soil fumigation)*
- Soil insects in ***field crops*** *(soil fumigation)*, ***protected crops*** *(soil fumigation)*, ***vegetables*** *(soil fumigation)*
- Soil-borne diseases in ***field crops*** *(soil fumigation)*, ***protected crops*** *(soil fumigation)*, ***vegetables*** *(soil fumigation)*
- Weed seeds in ***field crops*** *(soil fumigation)*, ***protected crops*** *(soil fumigation)*, ***vegetables*** *(soil fumigation)*

Efficacy guidance

- Dazomet acts by releasing methyl isothiocyanate in contact with moist soil
- Soil sterilization is carried out after harvesting one crop and before planting the next
- The soil must be of fine tilth, free of clods and evenly moist to the depth of sterilization
- Soil moisture must not be less than 50% of water-holding capacity or oversaturated. If too dry, water at least 7-14 d before treatment
- In order to obtain short treatment times it is recommended to treat soils when soil temperature is above 7°C. Treatment should be used outdoors before winter rains make soil too wet to cultivate - usually early Nov
- For club root control treat only in summer when soil temperature above 10°C
- Where onion white rot a problem unlikely to give effective control where inoculum level high or crop under stress
- Apply granules with suitable applicators, mix into soil immediately to desired depth and seal surface with polythene sheeting, by flooding or heavy rolling. See label for suitable application and incorporation machinery
- With 'planting through' technique polythene seal is left in place to form mulch into which new crop can be planted

Restrictions

- Maximum number of treatments 1 per crop or batch of soil
- Do not treat ground where water table may rise into treated layer

Crop-specific information

- Latest use: pre-planting

Following crops guidance

- With 'planting through' technique no gas release cultivations are made and safety test with cress is particularly important. Conduct cress test on soil samples from centre as well as edges of bed. Observe minimum of 30 d from application to cress test in soils at 10°C or above, at least 50 d in soils below 10°C. See label for details
- In all other situations 14-28 d after treatment cultivate lightly to allow gas to disperse and conduct cress test after a further 14-28 d (timing depends on soil type and temperature). Do not treat structures containing live plants or any ground within 1 m of live plants

Environmental safety

- Dangerous for the environment
- Very toxic to aquatic organisms

Hazard classification and safety precautions

Hazard H03, H11
Risk phrases R22a, R50, R58
Operator protection A, M; U04a, U05a, U09a, U19a, U20a
Environmental protection E15a, E32a, E32d, E34, E38
Storage and disposal E01, E04, E29a, E30a
Medical advice M03, M05a

147 2,4-DB

A translocated phenoxy herbicide for use in lucerne

Products

DB Straight	United Phosphorus	300 g/l	SL	07523

Uses

- Annual dicotyledons in ***lucerne***, ***undersown spring cereals*** *(undersown with lucerne)*, ***undersown winter cereals*** *(undersown with lucerne)*
- Thistles in ***lucerne***, ***undersown spring cereals*** *(undersown with lucerne)*, ***undersown winter cereals*** *(undersown with lucerne)*

Approval information

- 2,4-DB included in Annex I under EC Directive 91/414

Efficacy guidance

- Best results achieved on young seedling weeds under good growing conditions. Treatment less effective in cold weather and dry soil conditions
- Rain within 12 h may reduce effectiveness

Restrictions

- Do not allow spray drift onto neighbouring crops

Crop-specific information

- Latest use: before first node detectable (GS 31) for undersown cereals; fourth trifoliate leaf for lucerne
- In direct sown lucerne spray when seedlings have reached first trifoliate leaf stage. Optimum time 3-4 trifoliate leaves
- Do not treat any lucerne after fourth trifoliate leaf
- In spring barley and spring oats undersown with lucerne spray from when cereal has 1 leaf unfolded and lucerne has first trifoliate leaf
- In spring wheat undersown with lucerne spray from when cereal has 3 leaves unfolded and lucerne has first trifoliate leaf

Environmental safety

- Dangerous for the environment
- Toxic to aquatic organisms
- Keep livestock out of treated areas for at least two weeks following treatment and until poisonous weeds, such as ragwort, have died down and become unpalatable

Hazard classification and safety precautions

Hazard H03, H11
Risk phrases R22a, R38, R41, R51, R58
Operator protection U05a, U08, U20a
Environmental protection E07a, E15a, E19b, E31b, E34
Storage and disposal E26, E30a
Medical advice M03, M05a

148 2,4-DB + linuron + MCPA

A translocated herbicide for undersown cereals and grass

Products

Alistell	United Phosphorus	220:30:30 g/l	EC	11053

Uses

- Annual dicotyledons in ***seedling grassland***, ***undersown spring cereals*** *(undersown with clover)*, ***undersown winter cereals*** *(undersown with clover)*

Approval information

- 2,4-DB and linuron included in Annex I under EC Directive 91/414

Efficacy guidance

- Best results achieved on young seedling weeds growing actively in warm, moist weather
- May be applied at any time of year provided crop at correct stage and weather suitable
- Avoid spraying if rain falling or imminent

Restrictions

- Maximum number of treatments 1 per crop or yr
- Spray winter cereals when fully tillered but before first node detectable (GS 29-30)
- Spray spring wheat from 5-fully expanded leaf stage (GS 15), barley and oats from 2-fully expanded leaves (GS 12)
- Do not spray cereals undersown with lucerne, peas or beans
- Apply to clovers after 1-trifoliate leaf, to grasses after 2-fully expanded leaf stage
- Do not spray in conditions of drought, waterlogging or extremes of temperature
- In frosty weather clover leaf scorch may occur but damage normally outgrown
- Do not use on sand or soils with more than 10% organic matter
- Do not roll or harrow within 7 d before or after spraying
- Avoid drift of spray or vapour onto susceptible crops

Crop-specific information

- Latest use: before first node detectable (GS 31) for undersown cereals; 2 wk before grazing for grassland

Environmental safety

- Dangerous for the environment
- Toxic to aquatic organisms
- Keep livestock out of treated areas for at least two weeks following treatment and until poisonous weeds, such as ragwort, have died down and become unpalatable
- LERAP Category B
- Do not apply by hand-held sprayers

Hazard classification and safety precautions

Hazard H03, H11
Risk phrases R22a, R22b, R40, R41, R51, R58, R66
Operator protection A, C, H; U05a, U08, U11, U13, U19a, U20b
Environmental protection E07a, E15a, E16a, E31b, E32d, E34, E38
Storage and disposal E01, E04, E30a
Medical advice M03, M05a

149 2,4-DB + MCPA

A translocated herbicide for cereals, clovers and leys

Products

1	Agrichem DB Plus	Agrichem	243:40 g/l	SL	00044
2	Headland Cedar	Headland	240:40 g/l	SL	10489
3	Redlegor	United Phosphorus	244:44 g/l	SL	07519

Uses

- Annual dicotyledons in ***all cereals, rotational grassland*** [1]; ***grass re-seeds, spring barley, spring oats, undersown barley*** *(red or white clover)*, ***undersown oats*** *(red or white clover)*, ***undersown wheat*** *(red or white clover)*, ***winter barley, winter oats*** [2]; ***seedling leys*** [3]; ***spring wheat, winter wheat*** [2, 3]; ***undersown spring cereals, undersown winter cereals*** [1, 3]
- Perennial dicotyledons in ***all cereals, rotational grassland*** [1]; ***grass re-seeds, spring barley, spring oats, undersown barley*** *(red or white clover)*, ***undersown oats*** *(red or white clover)*, ***undersown wheat*** *(red or white clover)*, ***winter barley, winter oats*** [2]; ***seedling leys*** [3]; ***spring wheat, winter wheat*** [2, 3]; ***undersown spring cereals, undersown winter cereals*** [1, 3]
- Polygonums in ***seedling leys, spring wheat, undersown spring cereals, undersown winter cereals, winter wheat*** [3]

Approval information

- 2,4-DB included in Annex I under EC Directive 91/414

Efficacy guidance

- Best results achieved on young seedling weeds under good growing conditions
- Spray thistles and other perennials when 10-20 cm high provided clover at correct stage
- Effectiveness may be reduced by rain within 12 h, by very cold conditions or drought

Restrictions

- Maximum number of treatments 1 per crop
- Do not spray established clover crops or lucerne

- Do not roll or harrow within 7 d before or after spraying
- Do not spray immediately before or after sowing any crop
- Avoid drift onto neighbouring sensitive crops

Crop-specific information

- Latest use: before first node detectable (GS 31) for cereals; 4th trifoliate leaf stage of clover for grass re-seeds
- Apply in spring to winter cereals from leaf sheath erect stage, to spring barley or oats from 2-leaf stage (GS 12), to spring wheat from 5-leaf stage (GS 15)
- Spray clovers as soon as possible after first trifoliate leaf, grasses after 2-3 leaf stage
- Red clover may suffer temporary distortion after treatment

Environmental safety

- Harmful to aquatic organisms
- Keep livestock out of treated areas for at least two weeks following treatment and until poisonous weeds, such as ragwort, have died down and become unpalatable

Hazard classification and safety precautions

Hazard H03
Risk phrases R20, R21 [1]; R22a, R41, R52 [1-3]; R38 [2, 3]; R58 [1, 3]
Operator protection A, C [1, 2]; U05a, U08, U11, U20b [1-3]; U14 [3]; U15 [2, 3]
Environmental protection E07a, E13c [1-3]; E19b, E31c [1]; E31a, E38 [2]; E31b [3]; E34 [1, 2]
Storage and disposal E01, E30a [1, 2]; E04 [1]; E26 [1-3]
Medical advice M03 [1-3]; M05a [1, 3]

150 deltamethrin

A pyrethroid insecticide with contact and residual activity

Products

1	Bandu	Cheminova	25 g/l	EC	10629
2	Bandu	Headland	25 g/l	EC	10994
3	Decis	Bayer CropScience	25 g/l	EC	07172
4	Decis Protech	Bayer CropScience	15 g/l	EW	11502
5	Landgold Deltaland	Landgold	25 g/l	EC	09906
6	Pearl Micro	Bayer CropScience	6.25% w/w	GR	08620
7	Thripstick	Aquaspersions	0.125 g/l	AL	02134

Uses

- American serpentine leaf miner in ***aubergines*** *(off-label)*, ***non-edible ornamentals*** *(off-label)*, ***rooting beds*** *(off-label)* [1, 3]; ***non-edible glasshouse crops*** *(off-label)*, ***protected aubergines*** *(off-label)* [4]; ***protected courgettes*** *(off-label)*, ***protected cucumbers*** *(off-label)*, ***protected gherkins*** *(off-label)*, ***protected peppers*** *(off-label)*, ***protected tomatoes*** *(off-label)* [1, 3, 4]
- Aphids in ***amenity vegetation*** [1-4]; ***apples***, ***spring barley***, ***spring oats***, ***spring wheat***, ***winter barley***, ***winter oats***, ***winter wheat*** [1-6]; ***bulb onions*** *(off-label)*, ***carrots*** *(off-label)*, ***celery*** *(off-label)*, ***garlic*** *(off-label)*, ***leeks*** *(off-label)*, ***parsnips*** *(off-label)*, ***peas***, ***protected celery*** *(off-label)*, ***protected chinese cabbage*** *(off-label)*, ***protected salad onions*** *(off-label)*, ***salad onions*** *(off-label)* [6]; ***chinese cabbage*** *(off-label)*, ***lettuce*** *(off-label)* [3, 4, 6]; ***evening primrose*** *(off-label)* [4]; ***grass seed crops*** *(off-label)* [1, 3, 4, 6]; ***nursery stock*** [3, 6]; ***ornamental plant production***, ***protected cucumbers***, ***protected peppers***, ***protected tomatoes*** [1-4, 6]; ***protected ornamentals*** [1, 4]; ***protected pot plants*** [2-4, 6]; ***radicchio*** *(off-label)* [1, 3, 6]
- Apple sawfly in ***apples*** [6]
- Apple sucker in ***apples*** [1-6]
- Barley yellow dwarf vectors in ***winter barley***, ***winter wheat*** [1-4]
- Barley yellow dwarf virus vectors in ***spring barley***, ***spring oats***, ***spring wheat***, ***winter oats*** [6]; ***winter barley***, ***winter wheat*** [5, 6]
- Beet virus yellows vectors in ***winter oilseed rape*** [1-6]
- Brassica pod midge in ***mustard***, ***spring oilseed rape***, ***winter oilseed rape*** [6]
- Cabbage seed weevil in ***mustard***, ***spring oilseed rape***, ***winter oilseed rape*** [1-6]
- Cabbage stem flea beetle in ***winter oilseed rape*** [1-6]
- Cabbage stem weevil in ***mustard***, ***spring oilseed rape***, ***winter oilseed rape*** [1-5]
- Capsids in ***amenity vegetation*** [1-4]; ***apples*** [1-6]; ***nursery stock*** [3, 6]; ***ornamental plant production*** [1-4, 6]

- Caterpillars in ***amenity vegetation*** [1-4]; ***apples, broccoli, brussels sprouts, cabbages, cauliflowers, kale, plums, swedes, turnips*** [1-6]; ***bulb onions*** *(off-label)*, ***carrots*** *(off-label)*, ***celery*** *(off-label)*, ***garlic*** *(off-label)*, ***leeks*** *(off-label)*, ***parsnips*** *(off-label)*, ***protected celery*** *(off-label)*, ***protected chinese cabbage*** *(off-label)*, ***protected salad onions*** *(off-label)*, ***salad onions*** *(off-label)* [6]; ***chinese cabbage*** *(off-label)*, ***lettuce*** *(off-label)*, ***protected salad brassicas*** *(off-label - for baby leaf production)* [3, 4, 6]; ***marjoram*** *(off-label)*, ***sorrel*** *(off-label)* [4]; ***nursery stock*** [3, 6]; ***ornamental plant production, protected cucumbers, protected peppers, protected tomatoes*** [1-4, 6]; ***protected ornamentals*** [1, 4]; ***protected pot plants*** [2-4, 6]; ***protected salad brassicas*** *(off-label - baby leaf production)*, ***salad brassicas*** *(off-label - baby leaf production)* [2]; ***radicchio*** *(off-label)* [1, 3, 4, 6]; ***salad brassicas*** *(off-label)* [3]; ***salad brassicas*** *(off-label - for baby leaf production)* [4, 6]; ***tomato houses*** *(off-label)* [7]
- Codling moth in ***apples*** [1-6]
- Colorado beetle in ***potatoes*** *(off-label)* [3]
- Cutworms in ***lettuce*** [1-4, 6]
- Damson-hop aphid in ***hops, plums*** [1-6]
- Flea beetle in ***broccoli, brussels sprouts, cabbages, cauliflowers, kale, swedes, turnips*** [3, 6]; ***cress*** *(off-label)*, ***protected herbs (see appendix 6)***, ***salad brassicas*** *(off-label)* [3]; ***evening primrose*** *(off-label)*, ***sorrel*** *(off-label)* [1, 3, 4, 6]; ***marjoram*** *(off-label)*, ***protected lamb's lettuce*** *(off-label)*, ***protected scarole*** *(off-label)*, ***protected watercress*** *(off-label)*, ***radicchio*** *(off-label)* [4]; ***protected chives*** *(off-label)*, ***protected parsley*** *(off-label)* [2, 4]; ***protected herbs (see appendix 6)*** *(off-label)*, ***protected lettuce*** *(off-label)* [2, 4, 6]; ***protected radicchio*** *(off-label)*, ***salad brassicas*** *(off-label - for baby leaf production)* [4, 6]; ***protected salad brassicas*** *(off-label - baby leaf production)*, ***salad brassicas*** *(off-label - baby leaf production)* [2]; ***protected salad brassicas*** *(off-label - for baby leaf production)* [3, 4, 6]; ***sugar beet*** [1-6]
- Fruit tree tortrix moth in ***plums*** [1-5]
- Insect control in ***celery*** *(off-label)*, ***protected celery*** *(off-label)*, ***protected chinese cabbage*** *(off-label)*, ***protected salad onions*** *(off-label)*, ***salad onions*** *(off-label)* [2]
- Insect pests in ***bulb onions*** *(off-label)*, ***carrots*** *(off-label)*, ***garlic*** *(off-label)*, ***leeks*** *(off-label)* [3, 4]; ***celery*** *(off-label)*, ***protected celery*** *(off-label)*, ***protected spring onions*** *(off-label)*, ***salad onions*** *(off-label)* [1, 3, 4]; ***chinese cabbage*** *(off-label)* [3]; ***protected chinese cabbage*** *(off-label)* [1, 4]
- Leaf miner in ***tomato houses*** *(off-label)* [7]
- Leafhoppers in ***cress*** *(off-label)*, ***protected herbs (see appendix 6)*** [3]; ***marjoram*** *(off-label)* [1, 3, 4, 6]; ***protected chives*** *(off-label)*, ***protected parsley*** *(off-label)* [2, 4]; ***protected herbs (see appendix 6)*** *(off-label)*, ***protected lettuce*** *(off-label)* [2, 4, 6]; ***protected lamb's lettuce*** *(off-label)*, ***protected scarole*** *(off-label)*, ***protected watercress*** *(off-label)*, ***radicchio*** *(off-label)*, ***sorrel*** *(off-label)* [4]; ***protected radicchio*** *(off-label)* [4, 6]
- Mealybugs in ***amenity vegetation*** [1-4]; ***nursery stock*** [3, 6]; ***ornamental plant production, protected cucumbers, protected peppers, protected tomatoes*** [1-4, 6]; ***protected ornamentals*** [1, 4]; ***protected pot plants*** [2-4, 6]
- Pea and bean weevil in ***broad beans, peas, spring field beans, winter field beans*** [1-6]
- Pea midge in ***peas*** [3, 4, 6]
- Pea moth in ***peas*** [1-6]
- Pear sucker in ***pears*** [1-6]
- Phorid flies in ***mushrooms*** *(off-label)* [2-4]
- Plum fruit moth in ***plums*** [6]
- Plum sawfly in ***plums*** [6]
- Pollen beetle in ***evening primrose*** *(off-label)* [1, 3, 4, 6]; ***mustard, spring oilseed rape, winter oilseed rape*** [1-6]
- Rape winter stem weevil in ***winter oilseed rape*** [6]
- Raspberry beetle in ***raspberries*** [1-6]
- Sawflies in ***apples, plums*** [1-5]
- Scale insects in ***amenity vegetation*** [1-4]; ***nursery stock*** [3, 6]; ***ornamental plant production, protected cucumbers, protected peppers, protected tomatoes*** [1-4, 6]; ***protected ornamentals*** [1, 4]; ***protected pot plants*** [2-4, 6]
- Sciarid flies in ***mushrooms*** *(off-label)* [2-4]
- Thrips in ***amenity vegetation*** [1-4]; ***nursery stock*** [3, 6]; ***ornamental plant production*** [1-4, 6]; ***protected cucumbers*** [7]
- Tortrix moths in ***apples*** [1-6]

- Western flower thrips in ***aubergines*** *(off-label)*, ***non-edible ornamentals*** *(off-label)*, ***rooting beds*** *(off-label)* [1, 3]; ***non-edible glasshouse crops*** *(off-label)*, ***protected aubergines*** *(off-label)* [4]; ***protected courgettes*** *(off-label)*, ***protected cucumbers*** *(off-label)*, ***protected gherkins*** *(off-label)*, ***protected peppers*** *(off-label)*, ***protected tomatoes*** *(off-label)* [1, 3, 4]
- Whitefly in ***amenity vegetation*** [1-4]; ***nursery stock*** [3, 6]; ***ornamental plant production***, ***protected cucumbers***, ***protected peppers***, ***protected tomatoes*** [1-4, 6]; ***protected ornamentals*** [1, 4]; ***protected pot plants*** [2-4, 6]
- Yellow cereal fly in ***spring barley***, ***spring oats***, ***spring wheat***, ***winter barley***, ***winter oats***, ***winter wheat*** [6]

Specific Off-Label Approvals (SOLAs)

- ***aubergines***, ***non-edible ornamentals***, ***protected courgettes***, ***protected cucumbers***, ***protected gherkins***, ***protected peppers***, ***protected tomatoes***, ***rooting beds*** *(OLA 001839) Dec 2008* [3]
- ***aubergines***, ***celery***, ***evening primrose***, ***grass seed crops***, ***marjoram***, ***non-edible ornamentals***, ***protected celery***, ***protected chinese cabbage***, ***protected courgettes***, ***protected cucumbers***, ***protected gherkins***, ***protected peppers***, ***protected spring onions***, ***protected tomatoes***, ***radicchio***, ***rooting beds***, ***salad onions***, ***sorrel*** *(OLA 011755) Dec 2008* [1]
- ***bulb onions***, ***garlic*** *(OLA 001847) Dec 2008* [3]
- ***bulb onions***, ***carrots***, ***celery***, ***chinese cabbage***, ***evening primrose***, ***garlic***, ***grass seed crops***, ***leeks***, ***lettuce***, ***marjoram***, ***parsnips***, ***radicchio***, ***salad onions***, ***sorrel*** *(OLA 011795) Dec 2008* [6]
- ***bulb onions***, ***carrots***, ***celery***, ***chinese cabbage***, ***evening primrose***, ***garlic***, ***grass seed crops***, ***leeks***, ***lettuce***, ***marjoram***, ***protected celery***, ***protected chinese cabbage***, ***protected chives***, ***protected herbs (see appendix 6)***, ***protected lamb's lettuce***, ***protected lettuce***, ***protected parsley***, ***protected radicchio***, ***protected scarole***, ***protected spring onions***, ***protected watercress***, ***radicchio***, ***salad onions***, ***sorrel*** *(OLA 031140) Dec 2008* [4]
- ***carrots*** *(OLA 001845) Dec 2008* [3]
- ***celery***, ***chinese cabbage***, ***protected celery***, ***protected spring onions***, ***salad onions*** *(OLA 001849) Dec 2008* [3]
- ***celery***, ***protected celery***, ***protected chinese cabbage***, ***protected salad onions***, ***salad onions*** *(OLA 041187) Dec 2008* [2]
- ***chinese cabbage***, ***lettuce*** *(OLA 001843) Dec 2008* [3]
- ***cress*** *(OLA 022908) Dec 2008* [3]
- ***evening primrose***, ***grass seed crops***, ***marjoram***, ***radicchio***, ***sorrel*** *(OLA 001841) Dec 2008* [3]
- ***leeks*** *(OLA 001851) Dec 2008* [3]
- ***mushrooms*** *(OLA 030039) Dec 2008* [3]
- ***mushrooms*** *(OLA 030040) Dec 2008* [3]
- ***mushrooms*** *(OLA 031138) Dec 2008* [4]
- ***mushrooms*** *(OLA 041189) Dec 2008* [2]
- ***non-edible glasshouse crops***, ***protected aubergines***, ***protected courgettes***, ***protected cucumbers***, ***protected gherkins***, ***protected peppers***, ***protected tomatoes*** *(OLA 031139) Dec 2008* [4]
- ***potatoes*** *(OLA 001855) Dec 2008* [3]
- ***protected celery***, ***protected chinese cabbage***, ***protected salad onions*** *(OLA 011796) Dec 2008* [6]
- ***protected chives***, ***protected herbs (see appendix 6)***, ***protected lettuce***, ***protected parsley*** *(OLA 041188) Dec 2008* [2]
- ***protected herbs (see appendix 6)***, ***protected lettuce***, ***protected radicchio*** *(OLA 040740) Dec 2008* [6]
- ***protected salad brassicas***, ***salad brassicas*** *(baby leaf production) (OLA 041190) Dec 2008* [2]
- ***protected salad brassicas*** *(for baby leaf production) (OLA 001853) Dec 2008* [3]
- ***protected salad brassicas*** *(for baby leaf production) (OLA 011796) Dec 2008* [6]
- ***protected salad brassicas***, ***salad brassicas*** *(for baby leaf production) (OLA 031140) Dec 2008* [4]
- ***salad brassicas*** *(for baby leaf production) (OLA 011795) Dec 2008* [6]
- ***salad brassicas*** *(OLA 001853) Dec 2008* [3]
- ***tomato houses*** *(OLA 920180) Dec 2008* [7]

Approval information

- Deltamethrin included in Annex I under EC Directive 91/414

Efficacy guidance

- A contact and stomach poison with 3-4 wk persistence, particularly effective on caterpillars and sucking insects
- Normally applied at first signs of damage with follow-up treatments where necessary at 10-14 d intervals. Rates, timing and recommended combinations with other pesticides vary with crop and pest. See label for details
- Spray is rainfast within 1 h [3]
- May be applied in frosty weather provided foliage not covered in ice
- Temperatures above 35°C may reduce effectiveness or persistence
- Spray Thripstick without dilution onto polythene covering on glasshouse floor. Spray before planting cucumbers where thrips are a known problem or when pest problem seen. Repeat every 8 wk to ensure complete control [7]

Restrictions

- Maximum number of treatments varies with crop and pest, 4 per crop for wheat and barley, only 1 application between 1 Apr and 31 Aug. See label or off-label approval notice for other crops
- Do not apply more than 1 aphicide treatment to cereals in summer
- Do not spray crops suffering from drought or other physical stress
- Consult processer before treating crops for processing
- Do not apply to a cereal crop if any product containing a pyrethroid insecticide or dimethoate has been applied to that crop after the start of ear emergence (GS 51)
- Do not spray cereals after 31 Mar in the year of harvest within 6 m of the outside edge of the crop
- Reduced volume spraying must not be used on cereals after 31 Mar in yr of harvest

Crop-specific information

- Latest use: early dough (GS 83) for barley, oats, wheat; before flowering for mustard, oilseed rape, evening primrose; before 31 Mar for grass seed crops, marjoram, radicchio, sorrel
- HI mushrooms 2 d; leeks, protected spring onions 3 d; chinese cabbage, most herbs, protected celery, protected lettuce, protected salad brassicas, radicchio, sorrel 7 d; marjoram, radicchio, sorrel 14 d; carrots, parsnips 21 d

Environmental safety

- Dangerous for the environment
- Very toxic to aquatic organisms
- High risk to bees. Do not apply to crops in flower or to those in which bees are actively foraging. Do not apply when flowering weeds are present
- Do not apply in tank-mixture with a triazole-containing fungicide when bees are likely to be actively foraging in the crop
- High risk to non-target insects or other arthropods. Do not spray within 6 m of the field boundary
- Extremely dangerous to fish or other aquatic life. Do not contaminate surface waters or ditches with chemical or used container
- LERAP Category A (except [7])
- Broadcast air-assisted LERAP (18 m) [3-6]

Hazard classification and safety precautions

Hazard H02 [6]; H03, H08 [1-3, 5]; H11 [1-6]

Risk phrases R20 [1-3]; R21 [5]; R22a, R22b, R41 [1-3, 5]; R25, R36 [6]; R38, R51 [1-3, 5, 6]; R50 [4]; R58 [1-6]

Operator protection A, C, H; U02a, U05b, U06, U09a [7]; U04a, U08 [1-3, 5, 6]; U05a, U19a [1-3, 5-7]; U10 [1, 2]; U11, U14, U15 [4]; U20b [1-7]

Environmental protection E12b [3, 4, 6] (see label for guidance on cereals, oilseed rape, peas, beans); E12e [1, 2] (cereals, oilseed rape, peas, beans); E12e [5] (oilseed rape, cereals, peas, beans); E13a [6, 7]; E15a [1-5]; E16c, E16d [1-6]; E17b [3-6] (18 m); E22a [4]; E31a [7]; E31b [5]; E31c [1-4]; E32a [6]; E32d [3, 4, 6]; E34 [1-3, 5, 6]; E38 [1-4, 6]

Storage and disposal E01, E04, E30a [1-7]; E26 [1-3, 5-7]; E27 [7]; E29b [6]

Medical advice M03 [1-3, 5, 6]; M04 [6]; M05b [1-3, 5]

151 desmedipham

A contact carbamate herbicide available only in mixtures

152 desmedipham + ethofumesate + phenmedipham

A selective contact and residual herbicide for beet

Products

1	Betanal Expert	Bayer CropScience	25:151:75 g/l	EC	10592
2	Landgold Deputy	Landgold	25:151:75 g/l	EC	11975

Uses

- Annual dicotyledons in ***sugar beet***
- Annual meadow grass in ***sugar beet***

Approval information

- Ethofumesate included in Annex I under EC Directive 91/414

Efficacy guidance

- Product recommended for low-volume overall application in a planned spray programme
- Product acts mainly by contact action. A full programme also gives some residual control but this may be reduced on soils with more than 5% organic matter
- Best results achieved from treatments applied at fully expanded cotyledon stage of largest weeds present. Occasional larger weeds will usually be controlled from a full programme of sprays
- Where a pre-emergence band spray has been applied treatment must be timed according to size of the untreated weeds between the rows
- Susceptible weeds may not all be killed by the first spray. Repeat applications as each flush of weeds reaches cotyledon size normally necessary for season long control
- Sequential treatments should be applied when the previous one is still showing an effect on the weeds
- Various mixtures with other beet herbicides are recommended. See label for details

Restrictions

- Maximum total dose equivalent to three full dose treatments
- Do not spray crops stressed by nutrient deficiency, wind damage, pest or disease attack, or previous herbicide treatments. Stressed crops treated under conditions of high light intensity may be checked and not recover fully
- If temperature likely to exceed 21°C spray after 5 pm
- Crystallisation may occur if spray volume exceeds that recommended or spray mixture not used within 2 h, especially if the water temperature is below 5°C
- Before use, wash out sprayer to remove all traces of previous products, especially hormone and sulfonyl urea weedkillers

Crop-specific information

- Latest use: before crop meets between rows
- Apply first treatment when majority of crop plants have reached the fully expanded cotyledon stage
- Frost within 7 d of treatment may cause check from which the crop may not recover

Following crops guidance

- Beet crops may be sown at any time after treatment. Any other crop may be sown 3 mth after treatment following mouldboard ploughing to 15 cm minimum

Environmental safety

- Dangerous for the environment
- Toxic to aquatic organisms

Hazard classification and safety precautions

Hazard H11
Risk phrases R51, R58
Operator protection A, H; U08, U20a
Environmental protection E15a, E32d, E38 [1]; E31b [1, 2]
Storage and disposal E26, E30a [1, 2]; E29a [2]
Medical advice M03

153 desmedipham + phenmedipham

A mixture of contact herbicides for use in sugar beet

Products

Betanal Carrera	Bayer CropScience	40:125 g/l	SE	11029

Uses

- Annual dicotyledons in ***sugar beet***

Efficacy guidance

- Product is recommended for low volume overall spraying in a planned programme involving pre and/or post-emergence treatments at doses recommended for low volume programmes
- Best results obtained from treatment when earliest germinating weeds have reached cotyledon stage
- Further treatments must be applied as each flush of weeds reaches cotyledon stage but allowing a minimum of 7 d between each spray
- Where a pre-emergence band spray has been applied, the first treatment should be timed according to the size of the weeds in the untreated area between the rows
- Product is absorbed by leaves of emerged weeds which are killed by scorching action in 2-10 d. Apply overall as a fine spray to optimise weed cover and spray retention
- Various tank mixtures and sequences recommended to widen weed spectrum and add residual activity - see label for details

Restrictions

- Maximum total dose equivalent to three full dose treatments
- Do not spray crops stressed by nutrient deficiency, frost, wind damage, pest or disease attack, or previous herbicide treatments. Stressed crops may be checked and not recover fully
- If temperature likely to exceed 21°C spray after 5 pm
- Before use, wash out sprayer to remove all traces of previous products, especially hormone and sulfonyl urea weedkillers
- Crystallisation may occur if spray volume exceeds that recommended or spray mixture not used within 2 h, especially if the water temperature is below 5°C
- Product may cause non-reinforced PVC pipes and hoses to soften and swell. Wherever possible, use reinforced PVC or synthetic rubber hoses

Crop-specific information

- Latest use: before crop leaves meet between rows
- Product safe to use on all soil types

Environmental safety

- Dangerous for the environment
- Toxic to aquatic organisms
- Risk to certain non-target insects or other arthropods. Avoid spraying within 6 m of field boundary

Hazard classification and safety precautions

Hazard H04, H11
Risk phrases R43, R51, R58
Operator protection A, H; U05a, U08, U19a, U20b
Environmental protection E15a, E22b, E31b, E32d, E38
Storage and disposal E01, E04, E26, E30a

154 dicamba

A translocated benzoic herbicide for control of bracken and perennial weeds

See also *2,4-D + dicamba*
2,4-D + dicamba + fluroxypyr
2,4-D + dicamba + triclopyr

Products

I T Dicamba	I T Agro	480 g/l	SL	10976

Uses
- Bracken in ***amenity grass*, *forest*, *grassland*, *land not intended to bear vegetation*, *permanent pasture***
- Docks in ***amenity grass*** *(wiper application)*, ***permanent pasture*** *(wiper application)*, ***rotational grassland***

Efficacy guidance
- For weed control in established leys and pasture apply when weeds actively growing before flowering shoots appear. Large well-established docks may require a second treatment in the following yr
- For dock control in grass using a ropewick applicator treat vigorously growing weeds before flower stems senesce
- Bracken is controlled by application to the ground litter from Mar to early May before any new frond growth appears. Subsequent rainfall is necessary to wash the treatment down to the rhizomes. Any bracken with rhizomes in the area to which the treatment is carried are controlled or severely repressed
- Bracken control may be reduced where rhizomes are broken up by cultivation or ploughing

Restrictions
- Maximum total dose equivalent to one full dose per yr
- Avoid drift onto susceptible crops. Tomatoes may be affected by vapour drift at a considerable distance
- Do not roll crops within 1 wk of spraying. Do not treat crops where clover forms an important part of the sward
- Do not apply in the root zone of new trees or mature trees standing on site and do not allow any spray to contact tree foliage

Crop-specific information
- Trees may be treated in a band between the rows before planting or in the yr of planting. Delay treatment until the yr after planting on trees planted on ploughed lines

Environmental safety
- Keep livestock out of treated areas for at least 2-3 wk and until foliage of any poisonous weeds such as ragwort and buttercups has died and become unpalatable

Hazard classification and safety precautions

Hazard H04

Risk phrases R36, R52

Operator protection A, C; U05a, U08, U11, U19a, U20b

Environmental protection E07a, E15a, E31b

Storage and disposal E01, E04, E26, E30a

155 dicamba + dichlorprop-P + ferrous sulphate + MCPA

A herbicide/fertilizer combination for moss and weed control in turf

Products

Renovator 2	Scotts	0.03:0.22:16.3: 0.22% w/w	GR	11411

Uses
- Annual dicotyledons in ***amenity turf*, *golf courses***
- Moss in ***amenity turf*, *golf courses***
- Perennial dicotyledons in ***amenity turf*, *golf courses***

Efficacy guidance
- Apply from mid-Apr to mid-Aug when weeds are growing
- For best control of moss scarify vigorously after 2 wk to remove dead moss
- Where regrowth of moss or weeds occurs a repeat treatment may be made after 6 wk
- Avoid treatment of wet grass or during drought. If no rain falls within 48 h water in thoroughly

Restrictions
- Do not treat new turf until established for 6-9 mth
- The first 4 mowings after treatment should not be used to mulch cultivated plants unless composted at least 6 mth

SEE SECTION 3 FOR PRODUCTS ALSO REGISTERED

- Avoid walking on treated areas until it has rained or they have been watered
- Do not re-seed or turf within 8 wk of last treatment
- Do not cut grass for at least 3 d before and at least 4 d after treatment
- Do not apply during freezing conditions or when rain imminent

Crop-specific information

- Apply with a suitable calibrated fertilizer distributor

Hazard classification and safety precautions

Operator protection U09a, U20a
Environmental protection E15a, E32a
Storage and disposal E30a

156 dicamba + dichlorprop-P + MCPA

A translocated herbicide mixture for turf

Products

1	Cleanrun 3	Scotts	0.03:0.22:0.22% w/w	MG	11410
2	Intrepid 2	Scotts	20.8:167:167 g/l	SL	11594

Uses

- Annual dicotyledons in ***managed amenity turf***
- Perennial dicotyledons in ***managed amenity turf***

Efficacy guidance

- Apply as directed on label between Apr and Sep when weeds growing actively
- If no rain falls within 48 h of treatment irrigate thoroughly [1]
- Do not mow for at least 3 d before and 3-4 d after treatment so that there is sufficient leaf growth for spray uptake and sufficient time for translocation. See label for details
- If re-growth occurs or new weeds germinate re-treatment recommended

Restrictions

- Do not use during drought unless irrigation is carried out before and after treatment
- Do not apply during freezing conditions or when heavy rain is imminent
- Do not treat new turf until established for about 6 mth after seeding or turfing
- The first 4 mowings after treatment should not be used to mulch cultivated plants unless composted for at least 6 mth
- Avoid walking where possible on treated areas until after rain or irrigation [1]
- Do not re-seed turf within 8 wk of last treatment

Crop-specific information

- Latest use: end Sep for managed amenity turf

Environmental safety

- Keep livestock out of treated areas for at least 2 weeks and until foliage of any poisonous weeds such as ragwort has died and become unpalatable

Hazard classification and safety precautions

Hazard H03 [2]
Risk phrases R22a, R38, R41, R43 [2]
Operator protection A, C [2]; U05a, U08, U11, U19a, U20a [2]; U09a, U20b [1]
Environmental protection E15a [1, 2]; E31b, E34 [2]; E32a [1]
Storage and disposal E01, E04, E26 [2]; E30a [1, 2]
Medical advice M03, M05a [2]

157 dicamba + maleic hydrazide + MCPA

A herbicide/plant growth regulator mixture for amenity grass

Products

Mazide Selective	Vitax	6:200:75 g/l	SL	05753

Uses

- Annual dicotyledons in ***amenity grass***, ***road verges***

- Growth retardation in ***amenity grass, road verges***
- Perennial dicotyledons in ***amenity grass, road verges***

Approval information

- Maleic hydrazide included in Annex I under EC Directive 91/414

Efficacy guidance

- Best results achieved by application in Apr-May when grass and weeds growing actively but before weeds have started to flower
- May be used either as one annual spray or as spring spray repeated after 8-10 wk

Restrictions

- Do not use on fine turf

Environmental safety

- Keep livestock out of treated areas for at least 2 wk following treatment until foliage of any poisonous weeds such as ragwort has died and become unpalatable

Hazard classification and safety precautions

Hazard H04
Risk phrases R36
Operator protection U08, U11, U20b
Environmental protection E07a, E15a, E31a
Storage and disposal E26, E30a

158 dicamba + MCPA + mecoprop-P

A translocated herbicide for cereals, grassland, amenity grass and orchards

Products

1	Banlene Super	Bayer CropScience	18:252:42 g/l	SL	10053
2	Field Marshal	United Phosphorus	18:360:80	SL	08956
3	Greencrop Triathlon	Greencrop	25:200:200 g/l	SL	10956
4	Headland Relay P	Headland	25:200:200 g/l	SL	08580
5	Headland Relay Turf	Headland Amenity	25:200:200 g/l	SL	08935
6	Headland Transfer	Headland	18:315:50 g/l	SL	11010
7	Headland Trinity	Headland	18:315:50 g/l	SL	10842
8	Hycamba Plus	Agrichem	20.4:125.9:240.2 g/l	SL	10180
9	Hyprone-P	Agrichem	16:101:92 g/l	SL	09125
10	Hysward-P	Agrichem	16:101:92 g/l	SL	09052
11	Mircam Plus	Nufarm UK	19.5:245:43.3 g/l	SL	11525
12	Nocweed	Bayer Environ.	18:252:42 g/l	SL	10609
13	Outrun	SumiAgro Amenity	20.4:125.9:240.2 g/l	SL	10581
14	Pasturol Plus	FCC	25:200:200 g/l	SL	10278
15	Pierce	Nufarm UK	31:256:238 g/l	SL	11924
16	T2 Green	Nufarm UK	31: 256: 237 g/l	SL	11925
17	Tribute	Nomix Enviro	18:252:42 g/l	SL	06921
18	Tribute Plus	Nomix Enviro	18:252:42 g/l	SL	09493
19	Trireme	Nufarm UK	19.5:245:43.3 g/l	SL	11524
20	Tritox	Scotts	15:178:54 g/l	SL	07764
21	UPL Grassland Herbicide	United Phosphorus	25:200:200	SL	08934

Uses

- Annual dicotyledons in ***amenity grass*** [3, 7, 8, 10, 13, 14, 19]; ***amenity turf*** [5, 11, 17, 18, 20]; ***apple orchards*** [1, 7, 11, 12]; ***established grassland*** [1, 10, 12]; ***grass seed crops*** [1, 6, 7, 9, 11, 12, 15]; ***grassland*** [11]; ***managed amenity turf*** [3, 5, 7, 8, 10, 12-14, 16-20]; ***newly sown grass*** [9]; ***pear orchards, spring rye, winter rye*** [1, 7, 12]; ***permanent pasture*** [1, 3, 4, 6, 7, 12, 14, 15, 21]; ***rotational grassland*** [1-4, 6, 7, 10-12, 14, 15, 21]; ***spring barley, spring oats, spring wheat, winter barley, winter oats, winter wheat*** [1, 2, 6, 7, 9, 11, 12]; ***undersown barley, undersown oats, undersown wheat*** [9, 12]; ***undersown rye*** [12]; ***undersown spring cereals, undersown winter cereals*** [1, 7]; ***undersown spring cereals*** *(grass only)*, ***undersown winter cereals*** *(grass only)* [2, 6]
- Buttercups in ***managed amenity turf*** [12]
- Chickweed in ***amenity turf, apple orchards, grassland*** [11]; ***established grassland*** [1, 12]; ***grass seed crops*** [1, 6, 7, 9, 11, 12, 15]; ***newly sown grass*** [9]; ***permanent pasture*** [1, 6, 7, 12, 15, 21]; ***rotational grassland*** [1, 2, 6, 7, 11, 12, 15, 21]; ***spring barley, spring oats, spring wheat, winter barley, winter oats, winter wheat*** [1, 2, 6, 7, 9, 11, 12]; ***spring rye, winter rye***

[1, 7, 12]; ***undersown barley, undersown oats, undersown wheat*** [9, 12]; ***undersown rye*** [12]; ***undersown spring cereals, undersown winter cereals*** [1, 7]; ***undersown spring cereals*** *(grass only)*, ***undersown winter cereals*** *(grass only)* [2, 6]
- Cleavers in ***amenity turf, apple orchards, grassland*** [11]; ***grass seed crops*** [6, 9, 11, 15]; ***newly sown grass*** [9]; ***permanent pasture*** [6, 15, 21]; ***rotational grassland*** [11, 15, 21]; ***spring barley, spring oats, spring wheat, winter barley, winter oats, winter wheat*** [1, 2, 6, 7, 9, 11, 12]; ***spring rye, winter rye*** [1, 7, 12]; ***undersown barley, undersown oats, undersown wheat*** [9, 12]; ***undersown rye*** [12]; ***undersown spring cereals, undersown winter cereals*** [1, 7]
- Clover in ***managed amenity turf*** [12]
- Daisies in ***managed amenity turf*** [12]
- Docks in ***amenity grass, managed amenity turf*** [10]; ***apple orchards, pear orchards*** [1, 7, 12]; ***established grassland*** [1, 10, 12]; ***grass seed crops*** [1, 7, 9, 12, 15]; ***permanent pasture*** [1, 7, 12, 15, 21]; ***rotational grassland*** [1, 7, 10, 12, 15, 21]
- Mayweeds in ***amenity turf, apple orchards, grassland*** [11]; ***established grassland*** [1, 12]; ***grass seed crops*** [1, 6, 7, 9, 11, 12, 15]; ***newly sown grass*** [9]; ***permanent pasture*** [1, 6, 7, 12, 15, 21]; ***rotational grassland*** [1, 2, 6, 7, 11, 12, 15, 21]; ***spring barley, spring oats, spring wheat, winter barley, winter oats, winter wheat*** [1, 2, 6, 7, 9, 11, 12]; ***spring rye, winter rye*** [1, 7, 12]; ***undersown barley, undersown oats, undersown wheat*** [9, 12]; ***undersown rye*** [12]; ***undersown spring cereals, undersown winter cereals*** [1, 7]; ***undersown spring cereals*** *(grass only)*, ***undersown winter cereals*** *(grass only)* [2, 6]
- Perennial dicotyledons in ***amenity grass*** [3, 7, 8, 10, 13, 14, 19]; ***amenity turf*** [5, 11, 17, 18, 20]; ***apple orchards*** [1, 7, 11, 12]; ***established grassland*** [1, 10, 12]; ***grass seed crops*** [1, 6, 7, 9, 11, 12, 15]; ***grassland*** [11]; ***managed amenity turf*** [3, 5, 7, 8, 10, 13, 14, 16-20]; ***newly sown grass*** [9]; ***pear orchards, spring rye, winter rye*** [1, 7, 12]; ***permanent pasture*** [1, 3, 4, 6, 7, 12, 14, 15, 21]; ***rotational grassland*** [1-4, 6, 7, 10-12, 14, 15, 21]; ***spring barley, spring oats, spring wheat, winter barley, winter oats, winter wheat*** [1, 2, 6, 7, 9, 11, 12]; ***undersown barley, undersown oats, undersown wheat*** [9, 12]; ***undersown rye*** [12]; ***undersown spring cereals, undersown winter cereals*** [1, 7]; ***undersown spring cereals*** *(grass only)*, ***undersown winter cereals*** *(grass only)* [2, 6]
- Plantains in ***managed amenity turf*** [12]
- Polygonums in ***amenity turf, apple orchards, grassland*** [11]; ***grass seed crops*** [6, 9, 11, 15]; ***newly sown grass*** [9]; ***permanent pasture*** [6, 15, 21]; ***rotational grassland*** [1, 2, 6, 7, 11, 12, 15, 21]; ***spring barley, spring oats, spring wheat, winter barley, winter oats, winter wheat*** [1, 2, 6, 7, 9, 11, 12]; ***spring rye, winter rye*** [1, 7, 12]; ***undersown barley, undersown oats, undersown wheat*** [9, 12]; ***undersown rye*** [12]; ***undersown spring cereals, undersown winter cereals*** [1, 7]; ***undersown spring cereals*** *(grass only)*, ***undersown winter cereals*** *(grass only)* [2, 6]
- Yarrow in ***managed amenity turf*** [12]

Approval information
- Mecoprop-P included in Annex I under EC Directive 91/414

Efficacy guidance
- Treatment should be made when weeds growing actively. Weeds hardened by winter weather may be less susceptible
- For best results apply in fine warm weather, preferably when soil is moist. Do not spray if rain expected within 6 h or in drought
- Application of fertilizer 1-2 wk before spraying aids weed control in turf
- Where a second treatment later in the season is needed in amenity situations and on grass allow 4-6 wk between applications to permit sufficient foliage regrowth for uptake

Restrictions
- Maximum number of treatments (including other mecoprop-P products) or maximum total dose varies with crop and product. See label for details. The total amount of mecoprop-P applied in a single yr must not exceed the maximum total dose approved for any single product for the crop/situation
- Do not apply to cereals after the first node is detectable (GS 31), or to grass under stress from drought or cold weather
- Do not spray cereals undersown with clovers or legumes, to be undersown with grass or legumes or grassland where clovers or other legumes are important
- Do not spray leys established less than 18 mth or orchards established less than 3 yr

- Do not roll or harrow within 7 d before or after treatment, or graze for at least 7 d afterwards (longer if poisonous weeds present)
- Do not use on turf or grass in year of establishment. Allow 6-8 wk after treatment before seeding bare patches
- The first mowings after use should not be used for mulching unless composted for 6 mth
- Turf should not be mown for 24 h before or after treatment (3-4 d for closely mown turf)
- Avoid drift onto all broad-leaved plants outside the target area

Crop-specific information

- Latest use: before first node detectable (GS 31) for cereals; 5-6 wk before head emergence for grass seed crops; mid-Oct for established grass
- HI 7-14 d before cutting or grazing for leys, permanent pasture
- Apply to winter cereals from the leaf sheath erect stage (GS 30), and to spring cereals from the 5 expanded leaf stage (GS 15)
- Spray grass seed crops 4-6 wk before flower heads begin to emerge (timothy 6 wk)
- Turf containing bulbs may be treated once the foliage has died down completely [4, 10, 18, 20]

Environmental safety

- Harmful to aquatic organisms
- Keep livestock out of treated areas for at least 2 wk following treatment and until poisonous weeds, such as ragwort, have died down and become unpalatable

Hazard classification and safety precautions

Hazard H03 [1-8, 11-19, 21]; H04 [9, 10, 20]

Risk phrases R20 [4, 6-8, 13, 14, 21]; R21 [1-7, 11, 12, 14, 17-19, 21]; R22a [1-8, 11-16, 19, 21]; R36 [2-4, 9-11, 14, 17-20]; R38 [6, 7, 11, 13, 19, 21]; R41 [1, 4-8, 12-16, 19, 21]; R43 [21]; R52 [4-7, 14, 17, 18, 20, 21]; R58 [6, 7, 20, 21]

Operator protection A, C [1-21]; H, M [2-11, 13-21]; U05a [1-21]; U08 [1-11, 13, 14, 17-21]; U09a [12]; U11 [1, 4, 6-16, 19-21]; U15 [4, 14]; U19a [3, 8-10, 13, 20]; U20b [1-14, 17-21]

Environmental protection E07a [1-4, 6-21]; E13c [1-10, 12-16, 19, 21]; E15a [11, 15-20]; E23, E32d [17, 18]; E31a [6, 7, 20]; E31b [1-5, 11, 12, 14-19, 21]; E31c [8-10, 13]; E34 [2-7, 11, 12, 14-21]; E38 [4, 14, 17, 18]

Storage and disposal E01, E04, E30a [1-21]; E26 [1-19, 21]

Medical advice M03 [2-7, 11, 14-16, 19, 21]; M05a [7-10, 13, 21]

159 dicamba + mecoprop-P

A translocated post-emergence herbicide for cereals and grassland

Products

	Product	Company	Content	Form	Reg. no.
1	Camber	Headland	42:319 g/l	SL	09901
2	Di-Farmon R	Headland	42:319 g/l	SL	08472
3	Dockmaster	Nufarm UK	18.7:150 g/l	SL	11651
4	Foundation	Syngenta	84:600 g/l	SL	08475
5	Foundation	Headland	84:600 g/l	SL	11708
6	Headland Saxon	Headland	84:600 g/l	SL	11947
7	High Load Mircam	Nufarm UK	80:600 g/l	SL	11930
8	Hyban-P	Agrichem	18.7:150 g/l	SL	09129
9	Hygrass-P	Agrichem	18.7:150 g/l	SL	09130
10	Mircam	Nufarm UK	18.7:150 g/l	SL	11707
11	Prompt	Headland	84:600 g/l	SL	11948

Uses

- Annual dicotyledons in ***established grassland*** [1, 2, 4-7, 11]; ***permanent pasture*** [9, 10]; ***rotational grassland*** [7, 9, 10]; ***spring barley***, ***spring oats***, ***spring wheat***, ***winter barley***, ***winter oats***, ***winter wheat*** [1-8, 10, 11]; ***spring rye***, ***triticale***, ***undersown rye***, ***winter rye*** [8]; ***undersown barley***, ***undersown oats***, ***undersown wheat*** [3, 8]
- Buttercups in ***permanent pasture*** [3, 8]; ***rotational grassland*** [8]
- Chickweed in ***established grassland*** [1, 2]; ***permanent pasture***, ***undersown barley***, ***undersown oats***, ***undersown wheat*** [3, 8]; ***rotational grassland*** [7, 8]; ***spring barley***, ***spring oats***, ***spring wheat***, ***winter barley***, ***winter oats***, ***winter wheat*** [1-8, 10, 11]; ***spring rye***, ***triticale***, ***undersown rye***, ***winter rye*** [8]

SEE SECTION 3 FOR PRODUCTS ALSO REGISTERED

- Cleavers in ***established grassland*** [1, 2]; ***rotational grassland*** [7]; ***spring barley, spring oats, spring wheat, winter barley, winter oats, winter wheat*** [1-8, 10, 11]; ***spring rye, triticale, undersown rye, winter rye*** [8]; ***undersown barley, undersown oats, undersown wheat*** [3, 8]
- Creeping thistle in ***permanent pasture*** [3, 8]; ***rotational grassland*** [8]
- Docks in ***permanent pasture*** [3, 8-10]; ***rotational grassland*** [8-10]
- Mayweeds in ***established grassland*** [1, 2]; ***rotational grassland*** [7]; ***spring barley, spring oats, spring wheat, winter barley, winter oats, winter wheat*** [1-8, 10, 11]; ***spring rye, triticale, undersown rye, winter rye*** [8]; ***undersown barley, undersown oats, undersown wheat*** [3, 8]
- Perennial dicotyledons in ***established grassland*** [4-7, 11]; ***permanent pasture, rotational grassland*** [9, 10]; ***spring barley, spring oats, spring wheat, winter barley, winter oats, winter wheat*** [3, 8, 10]; ***spring rye, triticale, undersown rye, winter rye*** [8]; ***undersown barley, undersown oats, undersown wheat*** [3, 8]
- Polygonums in ***established grassland*** [1, 2]; ***rotational grassland*** [7]; ***spring barley, spring oats, spring wheat, winter barley, winter oats, winter wheat*** [1-8, 10, 11]; ***spring rye, triticale, undersown rye, winter rye*** [8]; ***undersown barley, undersown oats, undersown wheat*** [3, 8]
- Scentless mayweed in ***permanent pasture*** [3, 8]; ***rotational grassland*** [8]
- Stinging nettle in ***permanent pasture, rotational grassland*** [8]
- Thistles in ***permanent pasture, rotational grassland*** [9, 10]

Approval information

- Mecoprop-P included in Annex I under EC Directive 91/414

Efficacy guidance

- Best results by application in warm, moist weather when weeds are actively growing

Restrictions

- Maximum number of treatments 1 per crop for cereals and 1 or 2 per yr on grass depending on label. The total amount of mecoprop-P applied in a single yr must not exceed the maximum total dose approved for any single product for the crop/situation
- Do not spray in cold or frosty conditions
- Do not spray if rain expected within 6 h
- Do not treat undersown grass until tillering begins
- Do not spray cereals undersown with clover or legume mixtures
- Do not roll, harrow within 7 d before or after spraying
- Do not treat crops suffering from stress from any cause
- Use product immediately following dilution; do not allow diluted product to stand before use [6]
- Avoid treatment when drift may damage neighbouring susceptible crops

Crop-specific information

- Latest use: before 1st node detectable for cereals; 7 d before cutting or 14 d before grazing grass
- Apply to winter sown crops from 5 expanded leaf stage (GS 15)
- Apply to spring sown cereals from 5 expanded leaf stage but before first node is detectable (GS 15-31)
- Treat grassland just before perennial weeds flower
- Transient crop prostration may occur after spraying but recovery is rapid

Environmental safety

- Dangerous for the environment [1, 2, 4-6, 11]
- Toxic to aquatic organisms
- Keep livestock out of treated areas for at least 2 wk and until foliage of poisonous weeds such as ragwort has died and become unpalatable

Hazard classification and safety precautions

Hazard H03 [1-7, 10, 11]; H04 [8, 9]; H11 [1, 2, 4-6, 11]

Risk phrases R21 [3, 7]; R22a [1-7, 10, 11]; R36 [3, 10]; R38 [1-6, 10, 11]; R41 [1, 2, 4-9, 11]; R51 [1, 2, 4-6, 11]; R52 [10]; R58 [1, 2, 4-6, 10, 11]

Operator protection A, C [1-11]; H [1-3, 5, 6, 8-11]; M [1-3, 5, 6, 9-11]; U05a [1-11]; U08 [1-8, 10, 11]; U09a [9]; U11 [1-9, 11]; U19a [8-10]; U20a [2]; U20b [1, 3-11]

Environmental protection E07a [1-9, 11]; E13c, E31c [8, 9]; E15a, E31b [1-7, 10, 11]; E34 [1-11]; E38 [1, 2, 4, 5]

Storage and disposal E01, E04, E26, E30a

Medical advice M03 [1-11]; M05a [8-10]

160 dichlobenil

A residual benzonitrile herbicide for woody crops and non-crop uses

Products

1	Casoron G	Nomix Enviro	6.75% w/w	GR	09023
2	Casoron G	Scotts	6.75% w/w	GR	10709
3	Casoron G	Certis	6.75% w/w	GR	11632
4	Casoron G4	Scotts	4% w/w	GR	10708
5	Embargo G	SumiAgro Amenity	6.75% w/w	GR	10367
6	Luxan Dichlobenil Granules	Luxan	6.75% w/w	GR	09250
7	Midstream GSR	Scotts	20% w/w	GR	11674

Uses

- Annual dicotyledons in ***grapevines*** *(off-label)*, ***rhubarb*** *(off-label - established)* [3]
- Annual grasses in ***rhubarb*** *(off-label - established)* [3]
- Annual weeds in ***apple orchards***, ***blackberries***, ***blackcurrants***, ***gooseberries***, ***loganberries***, ***pear orchards***, ***raspberries***, ***redcurrants*** [3, 5, 6]; ***established woody ornamentals***, ***roses*** [1, 2, 4-6]; ***ornamental trees*** [4, 5]
- Aquatic weeds in ***enclosed waters***, ***land immediately adjacent to aquatic areas***, ***open waters*** [1-3, 6, 7]
- Perennial dicotyledons in ***apple orchards***, ***blackberries***, ***blackcurrants***, ***gooseberries***, ***loganberries***, ***pear orchards***, ***raspberries***, ***redcurrants*** [3, 5, 6]; ***established woody ornamentals***, ***roses*** [1, 2, 4-6]; ***grapevines*** *(off-label)*, ***rhubarb*** *(off-label - established)* [3]; ***ornamental trees*** [4, 5]
- Perennial grasses in ***apple orchards***, ***blackberries***, ***blackcurrants***, ***gooseberries***, ***loganberries***, ***pear orchards***, ***raspberries***, ***redcurrants*** [3, 5, 6]; ***established woody ornamentals***, ***roses*** [1, 2, 4-6]; ***ornamental trees*** [4, 5]; ***rhubarb*** *(off-label - established)* [3]
- Total vegetation control in ***hard surfaces***, ***natural surfaces not intended to bear vegetation***, ***permeable surfaces overlying soil*** [1-3, 5, 6]; ***non-crop areas*** [1, 2]
- Volunteer potatoes in ***potato dumps*** *(blight prevention)* [3, 5, 6]

Specific Off-Label Approvals (SOLAs)

- ***grapevines*** *(OLA 041405) Dec 2008* [3]
- ***rhubarb*** *(established) (OLA 041404) Dec 2008* [3]

Approval information

- Approved for aquatic weed control [1-3, 6, 7]. See notes in Section 5 on use of herbicides in or near water

Efficacy guidance

- Best results achieved by application in winter to moist soil during cool weather, particularly if rain follows soon after. Do not disturb treated soil by hoeing
- In aquatic situations apply as soon as active growth commences to reduce the likelihood of subsequent water deoxygenation caused by die-back of heavy weed stands. Later partial treatment for rooted species may be effective [1, 3, 6, 7]
- Where mulching is practised, apply beforehand
- Lower rates control annuals, higher rates perennials, see label for details
- Residual activity lasts 3-6 mth with selective, up to 12 mth with non-selective rates
- Do not use on fen peat or moss soils as efficacy will be severely impaired
- For control of emergent, floating and submerged aquatics apply to water surface in early spring. Intended for use in still or sluggish flowing water [1, 3, 6, 7]
- Control in water is effective where flow does not exceed 90 m/hr (2.5 cm/second). Treatment of faster flowing water will only be effective if the flow can be checked for a minimum of 7 d after application

Restrictions

- Maximum number of treatments 1 per yr
- Maximum total dose equivalent to one full dose treatment
- Do not treat crops established for less than 2 yr. See label for lists of resistant and sensitive species. Do not treat stone fruit or Norway spruce (Christmas trees)
- Do not apply within 300 m of glasshouses or hops or near areas underplanted with bulbs, annuals or herbaceous stock
- Do not apply to frozen, snow-covered or waterlogged ground or when crop foliage wet

- Do not apply to sites less than 18 mth before replanting or sowing
- Store well away from corms, bulbs, tubers and seed

Crop-specific information

- Latest use: before end of spring for aquatic situations; 4 wk before budding for grapevines; before onset of spring growth for potato dumps and all other crops. See labels for precise details
- HI 200 d for grapevines
- Apply to crops in dormant period. See label for details of timing and rates
- Apply evenly by hand applicator or suitable mechanical spreader
- Apply to potato dump sites before emergence of potatoes for blight prevention

Environmental safety

- Dangerous for the environment
- Harmful to aquatic organisms
- If fish are known to be present in a water body intended for treatment for aquatic weed control, partial treatment should be adopted. See label for details [7]
- Do not dump surplus herbicide in water or ditch bottoms [1-3, 6, 7]
- Use not permitted on reservoirs that form part of a water supply system
- Do not use treated water for irrigation purposes within 2 wk of treatment or until concentration of dichlobenil falls below 0.3 ppm [7]
- Use as an aquatic herbicide subject to requirements set out in *Guidelines for the use of herbicides in or near watercourses or lakes* available from Defra

Hazard classification and safety precautions

Hazard H03, H11 [5]; H04 [6]
Risk phrases R21, R51 [5]; R43 [6]; R52 [1, 3, 4, 7]; R58 [1, 3-5, 7]
Operator protection A, H [5, 6]; U05a [5, 6]; U20b [2]; U20c [1, 3-7]
Environmental protection E13c, E19a [1-3, 6, 7]; E15a [4, 5]; E21 [2, 7]; E32a [1-7]; E34 [6]; E38 [1, 3, 4]
Storage and disposal E01, E04 [5, 6]; E25 [2]; E29a [2, 4, 7]; E30a [1-7]
Medical advice M03 [6]; M05b [5]

161 dichlorophen

A chlorophenol moss-killer, fungicide, bactericide and algicide

Products

1	50/50 Liquid Mosskiller	Vitax	360 g/l	SL	07191
2	Enforcer	Scotts	360 g/l	SL	09288
3	Fungo	Dax	170 g/l	SL	H4768
4	Mossicide	SumiAgro Amenity	360 g/l	SL	09606
5	Nomix Mosskiller	Nomix Enviro	360 g/l	SL	12080
6	Panacide M	Coalite	360 g/l	SL	05611
7	Panacide TS	Coalite	480 g/l	SL	05612
8	Super Mosstox	Bayer Environ.	360 g/l	SL	09942

Uses

- Algae in ***hard surfaces, paths and drives*** [3]
- Fungus diseases in ***glasshouses, ornamental specimens*** [2]; ***hard surfaces*** [1, 2, 5]; ***managed amenity turf*** [1]
- Moss in ***glasshouses, ornamental specimens*** [2]; ***hard surfaces*** [1-5, 8]; ***managed amenity turf*** [1, 2, 4-8]; ***paths and drives*** [3, 8]
- Red thread in ***managed amenity turf*** [8]

Approval information

- Certain products containing this active ingredient have been granted derogations for specified 'Essential Uses' for use until 31 December 2007. Sale and supply must cease by 30 June 2007 but growers have no guarantee that the products will continue to be available until then.
  For more information see 'The Review Programme' under 'Pesticide Legislation' in Section 5

Efficacy guidance

- For moss control in turf spray at any time when moss is growing but best results obtained from treatment in spring or early summer. Supplement spraying with other measures to improve fertility and drainage
- Rake out dead moss 2-3 wk after treatment

- Treat hard surfaces when rain not expected and brush away dead material when treatment fully dried
- Efficacy will be impaired if used during freezing, frosty or drought conditions

Restrictions

- To be used only by professional operators
- Keep unprotected persons out of treated areas for 48 h or until surfaces are dry

Environmental safety

- Dangerous for the environment
- Very toxic to aquatic organisms
- Prevent any surface run-off from entering storm drains

Hazard classification and safety precautions

Hazard H03 [1]; H04 [3, 6-8]; H05 [2, 4, 5, 8]; H11 [1, 2, 4, 5, 8]
Risk phrases R22a, R58 [1, 2, 4, 5, 8]; R34, R51 [2]; R35 [4, 5, 8]; R36, R38 [2, 3, 5-8]; R41 [1, 3]; R50 [1, 4, 5, 8]
Operator protection A, C, H [1-3, 5-8]; M [1, 2, 5-8]; U02a [1, 6, 7]; U02b [1, 2, 4, 5]; U04a [1, 2, 4-7]; U05a [1-8]; U08, U15 [1, 5]; U09a [2-4, 6-8]; U10 [2, 4, 8]; U11 [1, 2, 4]; U14 [1, 2, 5]; U19a [2]; U20a [4, 6, 7]; U20b [1-3, 5]; U20c [8]
Environmental protection E02a [1, 2, 4, 5] (48 h); E02a [6, 7]; E13b [3, 6, 7]; E15a [1-3, 5, 8]; E20 [1-6]; E31a [1, 2, 4-8]; E32d [4, 5]; E32e [2]; E34 [5]; E38 [4]
Consumer protection C09 [1-7]; C12 [2, 5]
Storage and disposal E01, E04, E30a [1-8]; E26 [4, 8]
Medical advice M03 [5-7]; M04 [2, 4, 8]; M05b [3]

162 dichlorophen + ferrous sulphate

A mosskiller/fertilizer mixture for use on turf

Products

SHL Lawn Sand Plus	Sinclair	0.3:5.4% w/w	SA	04439

Uses

- Moss in ***amenity turf***

Efficacy guidance

- Apply to established turf from late spring to early autumn when soil moist but not when grass wet or damp with dew
- Heavy infestations may need a repeat treatment

Restrictions

- Maximum number of treatments 2 per yr
- Do not treat newly sown grass or freshly laid turf for the first year
- Do not apply during drought or freezing conditions or when rain imminent
- Avoid walking on treated areas until it has rained or turf has been watered
- Do not mow for 3-4 d before or after application

Environmental safety

- Harmful to fish or other aquatic life. Do not contaminate surface waters or ditches with chemical or used container

Hazard classification and safety precautions

Operator protection U20c
Environmental protection E13d, E32a
Storage and disposal E01, E30a

163 1,3-dichloropropene

A halogenated hydrocarbon soil nematicide

Products

Telone II	Dow	94% w/w	VP	05749

Uses

- Free-living nematodes in ***potatoes***

SEE SECTION 3 FOR PRODUCTS ALSO REGISTERED

- Nematodes in ***hops, raspberries, strawberries***
- Potato cyst nematode in ***potatoes***
- Stem and bulb nematodes in ***narcissi***
- Stem nematodes in ***strawberries***
- Virus vectors in ***hops, raspberries, strawberries***

Efficacy guidance

- Before treatment soil should be in friable seed bed condition above 5°C, with adequate moisture and all crop remains decomposed. Tilling to 30 cm improves results
- Apply with sub-surface A-blade injector combined with soil-sealing roller or, in hops, with a hollow tined injector. See label for details of suitable machinery
- Leave soil undisturbed for 21 d after treatment. Wet or cold soils need longer exposure
- For control of potato-cyst nematode contact firm's representative, check nematode level and use in integrated control programme
- For use in hops apply in May or Jun following the yr of grubbing and leave as long as possible before replanting
- Do not use water to clean out apparatus nor use aluminium or magnesium alloy containers which may corrode

Restrictions

- Maximum number of treatments 2 per yr for hops, 1 per yr for other crops
- Do not drill or plant until odour of fumigant is eliminated
- Do not use fertilizer containing ammonium salts after treatment. Use only nitrate nitrogen fertilizers until crop well established and soil temperature above 18°C
- Do not use on extremely heavy clays or soils with many large stones

Crop-specific information

- Latest use: 6 wk before planting for raspberries, strawberries; pre-planting of crop for hops, narcissi, potatoes

Following crops guidance

- Allow at least 6 wk after treatment before planting raspberries or strawberries

Environmental safety

- Dangerous for the environment
- Toxic to aquatic organisms

Hazard classification and safety precautions

Hazard H02, H08, H11
Risk phrases R20, R21, R22b, R24, R25, R37, R38, R43, R51, R58
Operator protection A, C, H, M; U02a, U04a, U05a, U09a, U14, U15, U19a, U20a
Environmental protection E15a, E32d, E33, E34, E38
Storage and disposal E01, E04, E24, E28, E30b
Medical advice M04, M05b

164 dichlorprop-P

A translocated phenoxy herbicide for use in cereals

See also 2,4-D + dichlorprop-P + MCPA + mecoprop-P
dicamba + dichlorprop-P + ferrous sulphate + MCPA
dicamba + dichlorprop-P + MCPA

Products

Headland Link	Headland	500 g/l	SL	11091

Uses

- Annual dicotyledons in ***spring barley, spring oats, spring wheat, winter barley, winter oats, winter wheat***
- Charlock in ***spring barley, spring oats, spring wheat, winter barley, winter oats, winter wheat***
- Chickweed in ***spring barley, spring oats, spring wheat, winter barley, winter oats, winter wheat***
- Fat hen in ***spring barley, spring oats, spring wheat, winter barley, winter oats, winter wheat***

- Perennial dicotyledons in ***spring barley, spring oats, spring wheat, winter barley, winter oats, winter wheat***
- Redshank in ***spring barley, spring oats, spring wheat, winter barley, winter oats, winter wheat***

Efficacy guidance

- Best results achieved by application to young seedling weeds in good growing conditions in a strongly competing crop
- Where cleavers are to be controlled increase water volume if crop or weed is dense
- Control of redshank is best achieved in warm temperatures even if this means allowing the weed to grow beyond the seedling stage

Restrictions

- Maximum number of treatments 1 per crop
- Do not apply during cold weather, drought, rain or when rain is expected
- Do not roll or harrow within 7 d before or after spraying
- Do not apply immediately before sowing any crop

Crop-specific information

- Latest use: before second node detectable (GS 32)
- Spray winter cereals in the spring from the leaf sheath erect stage; spray spring cereals from the one leaf stage

Following crops guidance

- Keep livestock out of treated areas until foliage of any poisonous weeds such as ragwort has died and become unpalatable

Environmental safety

- Harmful to fish or other aquatic life. Do not contaminate surface waters or ditches with chemical or used container
- Keep livestock out of treated areas for at least 2 wk and until foliage of any poisonous weeds, such as ragwort, has died and become unpalatable
- Do not spray in windy conditions when drift may cause damage to neighbouring sensitive crops

Hazard classification and safety precautions

Hazard H03, H04
Risk phrases R22a, R41, R43
Operator protection A, C, H; U05a, U08, U20b
Environmental protection E07b, E13c, E31a, E34
Storage and disposal E01, E04, E26, E30a
Medical advice M03

165 dichlorprop-P + ferrous sulphate + MCPA

A herbicide/fertilizer combination for moss and weed control in turf

Products

1	SHL Granular Feed, Weed & Mosskiller	Sinclair	0.2:10.9:0.3% w/w	GR	10972
2	SHL Turf Feed, Weed & Mosskiller	Sinclair	0.2:10.9:0.3% w/w	DP	10973

Uses

- Annual dicotyledons in ***managed amenity turf***
- Buttercups in ***managed amenity turf***
- Moss in ***managed amenity turf***
- Perennial dicotyledons in ***managed amenity turf***

Efficacy guidance

- Apply between Mar and Sep when grass in active growth and soil moist
- A repeat treatment may be needed after 4-6 wk to control perennial weeds or if moss regrows
- Water in if rainfall does not occur within 48 h [2]

Restrictions

- Do not treat newly sown grass for 6 mth after establishment [1]
- Do not apply during drought or freezing conditions or when rain imminent
- Avoid walking on treated areas until it has rained or turf has been watered

SEE SECTION 3 FOR PRODUCTS ALSO REGISTERED

- Do not mow for 3-4 d before or after application
- Do not use first 4 mowings after treatment for mulching. Mowings should be composted for 6 mth before use
- Do not treat areas of fine turf such as golf or bowling greens [1]
- Avoid contact with tarmac surfaces as staining may occur

Environmental safety

- Avoid drift onto nearby plants and borders [1]

Hazard classification and safety precautions

Operator protection U20c
Environmental protection E32a
Storage and disposal E01, E30a

166 dichlorprop-P + MCPA

A translocated herbicide for use in cereals and turf

Products

1	SHL Granular Feed and Weed	Sinclair	0.2:0.3% w/w	GR	10970
2	SHL Turf Feed and Weed	Sinclair	0.2:0.31% w/w	DP	10963

Uses

- Annual dicotyledons in ***managed amenity turf*** [1, 2]
- Buttercups in ***managed amenity turf*** [1, 2]
- Clovers in ***managed amenity turf*** [2]
- Daisies in ***managed amenity turf*** [2]
- Dandelions in ***managed amenity turf*** [2]
- Perennial dicotyledons in ***managed amenity turf*** [1, 2]

Efficacy guidance

- Best results achieved by application to young seedling weeds in active growth between Mar and Sep
- Water in if rainfall does not occur within 48 h [2]

Restrictions

- Do not use on newly sown turf for 6 mth after establishment
- Do not cut or mow for 2-3 d before and after treatment
- Do not spray during cold weather, if rain or frost expected, if crop wet or in drought
- The first four mowings should not be used for mulching and should be composted for 6 mth before use
- Avoid close mowing for 3-4 d before or after treatment
- Do not treat areas of fine turf such as golf or bowling greens [1]

Environmental safety

- Avoid drift onto nearby plants and borders [2]
- Avoid contact with tarmac surfaces as staining may occur

Hazard classification and safety precautions

Operator protection U20c
Environmental protection E32a
Storage and disposal E01, E30a

167 dichlorprop-P + MCPA + mecoprop-P

A translocated herbicide mixture for winter and spring cereals

Products

Hymec Triple	Agrichem	310:160:130 g/l	SL	09949

Uses

- Annual dicotyledons in ***durum wheat***, ***spring barley***, ***spring oats***, ***spring wheat***, ***winter barley***, ***winter oats***, ***winter wheat***

- Chickweed in ***durum wheat, spring barley, spring oats, spring wheat, winter barley, winter oats, winter wheat***
- Cleavers in ***durum wheat, spring barley, spring oats, spring wheat, winter barley, winter oats, winter wheat***
- Field pansy in ***durum wheat, spring barley, spring oats, spring wheat, winter barley, winter oats, winter wheat***
- Mayweeds in ***durum wheat, spring barley, spring oats, spring wheat, winter barley, winter oats, winter wheat***
- Poppies in ***durum wheat, spring barley, spring oats, spring wheat, winter barley, winter oats, winter wheat***

Efficacy guidance

- Best results obtained if application is made while majority of weeds are at seedling stage but not if temperatures are too low
- Optimum results achieved by spraying when temperature is above 10°C. If temperatures are lower delay spraying until growth becomes more active

Restrictions

- Maximum number of treatments 1 per crop
- Do not spray in windy conditions where spray drift may cause damage to neighbouring crops, especially sugar beet, oilseed rape, peas, turnips and most horticultural crops including lettuce and tomatoes under glass

Crop-specific information

- Latest use: before second node detectable (GS 32) for all crops

Environmental safety

- Dangerous for the environment
- Very toxic to aquatic organisms

Hazard classification and safety precautions

Hazard H03, H11
Risk phrases R20, R21, R22a, R41, R50, R58
Operator protection A, C, H, M; U05a, U08, U11, U20b
Environmental protection E13c, E31c, E32d, E34
Storage and disposal E01, E04, E26, E30a
Medical advice M03, M05a

168 diclofop-methyl

A translocated phenoxypropionic herbicide available only in mixtures

169 diclofop-methyl + fenoxaprop-P-ethyl

A foliar acting herbicide mixture for grass control in wheat and barley

Products

1	Corniche	Bayer CropScience	250:20 g/l	EW	08947
2	Tigress Ultra	Bayer CropScience	250:20 g/l	EW	08946

Uses

- Blackgrass in ***spring barley, winter barley, winter wheat***
- Ryegrass in ***spring barley, winter barley, winter wheat***
- Wild oats in ***spring barley, winter barley, winter wheat***

Efficacy guidance

- Apply from emergence of crop up to before second node detectable (GS 32)
- Wild oats and blackgrass controlled from 2 fully expanded leaves to end of tillering but before 1st node detectable
- Ryegrass from seed controlled from 2 fully expanded leaves up to 3 tillers
- Spray is rainfast from 1 h after spraying. Do not apply to wet or icy foliage
- Any conditions resulting in moisture stress may reduce effectiveness especially with later applications to spring barley

SEE SECTION 3 FOR PRODUCTS ALSO REGISTERED

- Performance not affected by soils with high OM content or high Kd factor
- Always follow WRAG guidelines for preventing and managing herbicide resistant weeds. Section 5 for more information

Restrictions

- Maximum number of treatments 1 per crop
- Do not apply to undersown crops or those to be undersown
- Broadcast crops should be sprayed post-emergence after root system well established
- Do not roll or harrow within 1 wk
- Do not spray crops under stress from drought, waterlogging etc
- Do not spray immediately before or after a sudden drop in temperature, a period of warm days/ cold nights or when extremely low temperatures forecast
- Interval of 7 d must elapse before or after application of any other product
- Must not be mixed with weedkillers containing hormones or bifenox

Crop-specific information

- Latest use: before 2nd node detectable (GS 32)
- Can be sprayed in frosty weather provided crop hardened off
- Applications may be followed by transient leaf discoloration

Environmental safety

- Dangerous for the environment
- Very toxic to aquatic organisms
- Keep livestock out of treated areas for at least 7 d

Hazard classification and safety precautions

Hazard H11
Risk phrases R50, R58
Operator protection A, C, H; U05a, U20b
Environmental protection E06a (7 d); E13b, E31b, E32d, E38
Consumer protection C02 (6 wk)
Storage and disposal E01, E04, E26, E30a

170 difenacoum

An anticoagulant coumarin rodenticide

See also cholecalciferol + difenacoum

Products

1	Neosorexa	Sorex	0.005% w/w	AB	H6773
2	Neosorexa Ratpacks	Sorex	0.005% w/w	AB	H6782
3	Sakarat D (Whole Wheat)	Killgerm	0.005% w/w	RB	H7109
4	Sakarat D Wax Bait	Killgerm	0.005% w/w	RB	H7489
5	Sorexa Gel	Sorex	0.005% w/w	RB	H6790

Uses

- Mice in ***farm buildings***, ***farmyards*** [1-5]
- Rats in ***farm buildings***, ***farmyards*** [1-3]

Efficacy guidance

- Difenacoum is a chronic poison and rodents need to feed several times before accumulating a lethal dose. Effective against rodents resistant to other commonly used anticoagulants
- Ready-to-use in baiting programme for indoor and outdoor use
- Lay small baits about 1 m apart throughout infested areas for mice, larger baits for rats near holes and along runs; place 2-4 blocks 5-10 m apart depending on severity of infestation
- Cover baits by placing in bait boxes, drain pipes or under boards
- Inspect bait sites frequently and top up as long as there is evidence of feeding
- Product formulated for application through a skeleton or caulking gun [4]

Restrictions

- Only for use by farmers, horticulturists and other professional users
- When working in rodent infested areas wear synthetic rubber/PVC gloves to protect against rodent-borne diseases

Environmental safety

- Cover bait to prevent access by children, animals or birds

Hazard classification and safety precautions

Operator protection A [5]; U13 [1-5]; U20a [2]; U20b [1, 3-5]
Environmental protection E32a [2-5]
Storage and disposal E30a
Vertebrate/rodent control products V01a, V03a, V04a [1, 2, 5]; V01b, V03b, V04b [4]; V02 [1, 2, 4, 5]
Medical advice M03 [3, 4]

171 difenoconazole

A diphenyl-ether triazole protectant and curative fungicide

Products

Plover	Syngenta	250 g/l	EC	11763

Uses

- Alternaria in ***broccoli, brussels sprouts, cabbages, calabrese, cauliflowers, spring oilseed rape, winter oilseed rape***
- Brown rust in ***winter wheat***
- Light leaf spot in ***spring oilseed rape, winter oilseed rape***
- Ring spot in ***broccoli, brussels sprouts, cabbages, calabrese, cauliflowers***
- Septoria in ***winter wheat***
- Stem canker in ***spring oilseed rape, winter oilseed rape***
- Yellow rust in ***winter wheat***

Efficacy guidance

- For most effective control of Septoria, apply as part of a programme of sprays which includes a suitable flag leaf treatment
- Adequate control of yellow rust may require an earlier appropriate treatment
- Improved control of established infections on oilseed rape achieved by mixture with carbendazim. See label
- In cabbages a 3-spray programme should be used starting at the first sign of disease and repeated at 14-21 d intervals
- Product is fully rainfast 2 h after application

Restrictions

- Maximum number of treatments 3 per crop for cabbages; 2 per crop for oilseed rape; 1 per crop for wheat
- Apply to wheat any time from ear fully emerged stage but before early milk-ripe stage (GS 59-73)

Crop-specific information

- Latest use: before grain early milk-ripe stage (GS 73) for cereals; end of flowering for oilseed rape
- HI brassicas 21 d
- Recommended for use on all varieties of winter and spring sown oilseed rape and cabbage
- Treat oilseed rape in autumn from 4 expanded true leaf stage (GS 1,4). A repeat spray may be made in spring at the beginning of stem extension (GS 2,0) if visible symptoms develop

Environmental safety

- Dangerous for the environment
- Very toxic to aquatic organisms

Hazard classification and safety precautions

Hazard H04, H11
Risk phrases R38, R41, R43, R50, R58
Operator protection A, C, H; U05a, U09a, U11, U20b
Environmental protection E15a, E31a, E32d, E38
Storage and disposal E01, E04, E30a

172 diflubenzuron

A selective, persistent, contact and stomach acting insecticide

Products

1	Dimilin Flo	Crompton	480 g/l	SC	08769
2	Dimilin Flo	Certis	480 g/l	SC	11056

Uses

- Browntail moth in ***amenity trees and shrubs, hedges, nursery stock, ornamental specimens*** [2]
- Bud moth in ***apples, pears*** [2]
- Carnation tortrix moth in ***amenity trees and shrubs, hedges, nursery stock, ornamental specimens*** [2]
- Caterpillars in ***broccoli, brussels sprouts, cabbages, calabrese, cauliflowers*** [2]
- Clouded drab moth in ***apples, pears*** [2]
- Codling moth in ***apples, pears*** [2]
- Fruit tree tortrix moth in ***apples, pears*** [2]
- Houseflies in ***livestock houses, manure heaps, refuse tips*** [2]
- Lackey moth in ***amenity trees and shrubs, hedges, nursery stock, ornamental specimens*** [2]
- Oak leaf roller moth in ***forest*** [2]
- Pear sucker in ***pears*** [2]
- Pine beauty moth in ***forest*** [2]
- Pine looper in ***forest*** [2]
- Plum fruit moth in ***plums*** [2]
- Rust mite in ***apples, pears, plums*** [2]
- Sciarid flies in ***mushrooms*** [1]
- Small ermine moth in ***amenity trees and shrubs, hedges, nursery stock, ornamental specimens*** [2]
- Tortrix moths in ***plums*** [2]
- Winter moth in ***amenity trees and shrubs, apples, blackcurrants, forest, hedges, nursery stock, ornamental specimens, pears, plums*** [2]

Approval information

- Approved for aerial application in forestry when average wind velocity does not exceed 18 knots and gusts do not exceed 20 knots [2]. See notes in Section 5

Efficacy guidance

- Most active on young caterpillars and most effective control achieved by spraying as eggs start to hatch
- Dose and timing of spray treatments vary with pest and crop. See label for details
- Addition of wetter recommended for use on brassicas and for pear sucker control in pears

Restrictions

- Maximum number of treatments 3 per yr for apples, pears; 2 per yr for plums, blackcurrants; 2 per crop for brassicas; 1 per spawning for mushrooms [1]
- Before treating ornamentals check varietal tolerance on a small sample
- Do not use as a compost drench or incorporated treatment on ornamental crops
- Do not spray protected plants in flower or with flower buds showing colour
- For use only on the food crops specified on the label
- Do not apply directly to livestock/poultry

Crop-specific information

- HI apples, pears, plums, blackcurrants, brassicas 14 d
- Apply as casing mixing treatment (use immediately) on mushrooms or as post-casing drench [1]

Environmental safety

- Dangerous for the environment
- Very toxic to aquatic organisms
- LERAP Category B
- Broadcast air-assisted LERAP (20 m in orchards, 10 m in blackcurrants forestry or ornamentals)

Hazard classification and safety precautions

Hazard H11

Risk phrases R50

Operator protection U20c
Environmental protection E05a, E16a, E16b, E18, E32e [2]; E15a, E32a, E38 [1, 2]; E17b [2] (20 m when used in orchards; 10 m when used in blackcurrants, forestry or ornamentals)
Storage and disposal E26 [2]; E30a [1, 2]

173 diflufenican

A shoot absorbed pyridinecarboxamide herbicide available only in mixtures

See also *bromoxynil + diflufenican + ioxynil*
carbetamide + diflufenican + oxadiazon
clopyralid + diflufenican + MCPA

174 diflufenican + flufenacet

A contact and residual herbicide mixture for cereals

Products

Liberator	Bayer CropScience	100:400 g/l	SC	12032

Uses
- Annual meadow grass in ***winter barley***, ***winter wheat***
- Blackgrass in ***winter barley***, ***winter wheat***
- Chickweed in ***winter barley***, ***winter wheat***
- Cleavers in ***winter barley***, ***winter wheat***
- Field pansy in ***winter barley***, ***winter wheat***
- Mayweeds in ***winter barley***, ***winter wheat***
- Speedwells in ***winter barley***, ***winter wheat***

Efficacy guidance
- Best results obtained when there is moist soil at and after application and rain falls within 7 d
- Residual control may be reduced under prolonged dry conditions
- Activity may be slow under cool conditions and final level of weed control may take some time to appear
- Good weed control depends on burying any trash or straw before or during seedbed preparation
- Established perennial grasses and broad-leaved weeds will not be controlled
- Do not use as a stand-alone treatment for blackgrass control. Always follow WRAG guidelines for preventing and managing herbicide resistant weeds. Section 5 for more information

Restrictions
- Maximum number of treatments 1 per crop
- Do not treat undersown cereals or those to be undersown
- Do not use on waterlogged soils or soils prone to waterlogging
- Do not use on Sands or Very Light soils or on soils containing more than 10% organic matter
- Do not treat broadcast crops and treat shallow-drilled crops post-emergence only
- Do not incorporate into the soil or disturb the soil after application by rolling or harrowing
- Avoid treating crops under stress from whatever cause and avoid treating during periods of prolonged or severe frosts

Crop-specific information
- Latest use: before 31 Dec in yr of sowing and before 3rd tiller stage (GS 23) for wheat or 4th tiller stage (GS 24) for barley
- For pre-emergence treatments the seed should be covered with a minimum of 32 mm settled soil

Following crops guidance
- In the event of crop failure wheat, barley or potatoes may be sown provided the soil is ploughed to 15 cm, and and minimum of 12 weeks elapse between treatment and sowing spring wheat or spring barley

- After normal harvest wheat, barley or potatoes my be sown without special cultivations. Soil must be ploughed or cultivated to 15 cm before sowing oilseed rape, field beans, peas, sugar beet, carrots, onions or edible brassicae
- Successive treatments of any products containing diflufenican can lead to soil build-up and inversion ploughing must precede sowing any following non-cereal crop. Even where ploughing occurs some crops may be damaged

Environmental safety

- Dangerous for the environment
- Very toxic to aquatic organisms
- Risk to non-target insects or other arthropods. Avoid spraying within 6 m of the field boundary
- LERAP Category B

Hazard classification and safety precautions

Hazard H03, H11
Risk phrases R22a, R43, R48, R50, R58
Operator protection A, H; U05a, U14
Environmental protection E16a, E22c, E31b, E32d, E34, E38
Storage and disposal E01, E04, E30a
Medical advice M03, M05a

175 diflufenican + flurtamone

A contact and residual herbicide mixture for cereals

Products

1	Bacara	Bayer CropScience	100:250 g/l	SC	10744
2	Graduate	Bayer CropScience	80:400 g/l	SC	10776

Uses

- Annual dicotyledons in ***winter barley***, ***winter wheat*** [1, 2]
- Annual meadow grass in ***winter barley***, ***winter wheat*** [1, 2]
- Blackgrass in ***winter barley***, ***winter wheat*** [1]
- Loose silky bent in ***winter barley***, ***winter wheat*** [1]
- Volunteer oilseed rape in ***winter barley***, ***winter wheat*** [1, 2]

Efficacy guidance

- Apply pre-emergence or from when crop has first leaf unfolded before susceptible weeds pass recommended size
- Best results obtained on firm, fine seedbeds with adequate soil moisture present at and after application. Increase water volume where crop or weed foliage is dense
- Good weed control requires ash, trash and burnt straw to be buried during seed bed preparation
- Loose fluffy seedbeds should be rolled before application and the final seed bed should be fine and firm without large clods
- Speed of control depends on weather conditions and activity can be slow under cool conditions
- Always follow WRAG guidelines for preventing and managing herbicide resistant weeds. Section 5 for more information

Restrictions

- Maximum number of treatments 1 per crop
- Crops should be drilled to a normal depth of 25 mm and the seed well covered. Do not treat broadcast crops as uncovered seed may be damaged
- Do not treat spring sown cereals, durum wheat, oats, undersown cereals or those to be undersown
- Do not treat frosted crops or when frost is imminent. Severe frost after application, or any other stress, may lead to transient discoloration or scorch
- Do not use on Sands or Very Light soils or those that are very stony or gravelly
- Do not use on waterlogged soils, or on crops subject to temporary waterlogging by heavy rainfall, as there is risk of persistent crop damage which may result in yield loss
- Do not use on soils with more than 10% organic matter
- Do not harrow at any time after application and do not roll autumn treated crops until spring

Crop-specific information

- Latest use: before 2nd node detectable (GS32)
- Take particular care to match spray swaths otherwise crop discoloration and biomass reduction may occur which may lead to yield reduction

Following crops guidance

- In the event of crop failure winter wheat may be redrilled immediately after normal cutlivation, and winter barley may be sown after ploughing. Fields must be ploughed to a depth of 15 cm and 20 wk must elapse before sowing spring crops of wheat, barley, oilseed rape, peas, field beans or potatoes
- After normal harvest autumn cereals can be drilled after ploughing. Thorough mixing of the soil must take place before drilling field beans, leaf brassicae or winter oilseed rape. For sugar beet seed crops and winter onions complete inversion of the furrow slice is essential
- Do not broadcast or direct drill oilseed rape or other brassica crops as a following crop on treated land. See label for detailed advice on preparing land for subsequent autumn cropping in the normal rotation
- Successive treatments of any products containing diflufenican can lead to soil build-up and inversion ploughing must precede sowing any following non-cereal crop. Even where ploughing occurs some crops may be damaged

Environmental safety

- Dangerous for the environment
- Very toxic to aquatic organisms
- LERAP Category B

Hazard classification and safety precautions

Hazard H11
Risk phrases R50, R58
Operator protection A, H; U20c
Environmental protection E13a, E16a, E16b, E31b, E32d, E38
Storage and disposal E26, E30a

176 diflufenican + flurtamone + isoproturon

A contact and residual herbicide for winter cereals

Products

Ingot	Bayer CropScience	27:67:400 g/l	SC	09997

Uses

- Annual dicotyledons in ***winter barley***, ***winter wheat***
- Annual meadow grass in ***winter barley***, ***winter wheat***
- Blackgrass in ***winter barley***, ***winter wheat***
- Loose silky bent in ***winter barley***, ***winter wheat***
- Volunteer oilseed rape in ***winter barley***, ***winter wheat***

Approval information

- Isoproturon included in Annex I under EC Directive 91/414

Efficacy guidance

- Apply from when crop has first leaf unfolded before susceptible weeds pass recommended size
- Best results obtained on firm, fine seedbeds with adequate soil moisture. Good spray coverage of soil and weeds is essential
- Speed of control depends on weather conditions and activity will be reduced during prolonged dry periods especially in spring. Weed control, especially of grasses, may also be reduced in wet seasons or where heavy rain falls shortly after treatment
- Always follow WRAG guidelines for preventing and managing herbicide resistant weeds. Section 5 for more information

Restrictions

- Maximum number of treatments 1 per crop. Maximum total dose of isoproturon 2.5 kg a.i./ha per crop
- Crops should be drilled to a normal depth of 25 mm and the seed well covered. Do not treat broadcast crops

- Do not apply pre-emergence to wheat or barley
- Do not use on crops undersown or to be undersown
- Do not treat frosted crops or when frost is imminent. Severe frost after application may cause transient discoloration or scorch
- Do not use on Sands or Very Light soils or those that are very stony or gravelly as there is risk of crop damage
- Do not treat on soils with more than 10% organic matter
- Do not use on waterlogged soils, or on crops subject to temporary waterlogging by heavy rainfall, as there is risk of persistent crop damage which may result in yield loss
- Do not roll autumn treated crops until the spring and do not harrow at any time after application
- Take particular care to match spray swaths otherwise crop discoloration and biomass reduction may occur which may lead to yield reduction

Crop-specific information

- Latest use: before 2nd node detectable (GS32)

Following crops guidance

- In the event of crop failure winter wheat may be redrilled immediately after normal cutlivation, and winter barley may be sown after ploughing. Fields must be ploughed to a depth of 15 cm and 20 wk must elapse before sowing spring crops of wheat, barley, oilseed rape, peas, field beans or potatoes
- After normal harvest autumn cereals can be drilled after ploughing. Thorough mixing of the soil must take place before drilling field beans, leaf brassicae or winter oilseed rape. For sugar beet seed crops and winter onions complete inversion of the furrow slice is essential
- Do not broadcast or direct drill oilseed rape or other brassica crops as a following crop on treated land. See label for detailed advice on preparing land for subsequent autumn cropping in the normal rotation
- Successive treatments of any products containing diflufenican can lead to soil build-up and inversion ploughing must precede sowing any following non-cereal crop. Even where ploughing occurs some crops may be damaged

Environmental safety

- Dangerous for the environment
- Very toxic to aquatic organisms
- LERAP Category B
- Do not apply to dry, cracked or waterlogged soils where heavy rain may lead to contamination of drains by isoproturon

Hazard classification and safety precautions

Hazard H03, H11
Risk phrases R40, R50, R58
Operator protection A, C, H; U20c
Environmental protection E13a, E16a, E16b, E31b, E32d, E38
Storage and disposal E26, E30a

177 diflufenican + isoproturon

A contact and residual herbicide for use in winter cereals

Products

1	Fernpath Ipex	AgriGuard	50:500 g/l	SC	11391
2	Javelin	Bayer CropScience	62.5:500 g/l	SC	10440
3	Javelin Gold	Bayer CropScience	20:500 g/l	SC	09999
4	Me2 Sylvester	Me2	50:500 g/l	SC	12021
5	Panther	Bayer CropScience	50:500 g/l	SC	10745
6	Standon Diflufenican 625	Standon	62.5:500 g/l	SC	10753
7	Standon Diflufenican-IPU	Standon	50:500 g/l	SC	11109

Uses

- Annual dicotyledons in ***triticale, winter barley, winter rye, winter wheat***
- Annual grasses in ***triticale, winter barley, winter rye, winter wheat***
- Blackgrass in ***triticale, winter barley, winter rye, winter wheat***
- Wild oats in ***triticale, winter barley, winter rye, winter wheat***

Approval information

- Isoproturon included in Annex I under EC Directive 91/414

Efficacy guidance

- May be applied in autumn or spring (but only post-emergence on wheat or barley). Best control normally achieved by early post-emergence treatment
- Best results by application to fine, firm seedbed moist at or after application
- Weeds controlled from before emergence to 6 true leaf stage
- Apply to moist, but not waterlogged, soils
- Any trash or ash should be buried during seedbed preparation
- Always follow WRAG guidelines for preventing and managing herbicide resistant weeds. Section 5 for more information

Restrictions

- Maximum number of treatments 1 per crop. Maximum total dose of isoproturon 2.5 kg a.i./ha per yr
- On triticale and winter rye only treat named varieties and apply as pre-emergence spray
- Do not use on other cereals, broadcast or undersown crops or crops to be undersown
- Do not use on Sands or Very Light soils, or those that are very stony or gravelly or on soils with more than 10% organic matter
- Do not spray when heavy rain is forecast or on crops suffering from stress, frost, deficiency, pest or disease attack
- Do not harrow after application nor roll autumn-treated crops until spring

Crop-specific information

- Latest use: before 2nd node detectable (GS 32) (low dose) or before end Feb (high dose) for wheat and barley; pre-emergence of crop for triticale and rye
- Spray winter wheat and barley post-emergence before end Feb
- Drill crop to normal depth (25 mm) and ensure seed well covered

Following crops guidance

- In the event of crop failure winter wheat may be redrilled immediately after normal cutlivation, and winter barley may be sown after ploughing. Fields must be ploughed to a depth of 15 cm and 20 wk must elapse before sowing spring crops of wheat, barley, oilseed rape, peas, field beans, sugar beet, potatoes, carrots, edible brassicas or onions
- After normal harvest autumn cereals can be drilled after ploughing. Thorough mixing of the soil must take place before drilling field beans, leaf brassicae or winter oilseed rape. For sugar beet seed crops and winter onions complete inversion of the furrow slice is essential
- Do not broadcast or direct drill oilseed rape or other brassica crops as a following crop on treated land. See label for detailed advice on preparing land for subsequent autumn cropping in the normal rotation
- Successive treatments of any products containing diflufenican can lead to soil build-up and inversion ploughing must precede sowing any following non-cereal crop. Even where ploughing occurs some crops may be damaged

Environmental safety

- Dangerous for the environment
- Very toxic to aquatic organisms
- Do not apply to dry, cracked or waterlogged soils where heavy rain may lead to contamination of drains by isoproturon
- LERAP Category B

Hazard classification and safety precautions

Hazard H03, H11
Risk phrases R40, R50, R58
Operator protection A, C, H [1, 2, 4-7]; U20a [1]; U20b [3]; U20c [2, 4-7]
Environmental protection E13a, E32d, E38 [2, 3, 5]; E15a [1, 4, 6, 7]; E16a, E16b, E31b [1-7]
Storage and disposal E26 [1-3, 6]; E30a [1-7]

SEE SECTION 3 FOR PRODUCTS ALSO REGISTERED

178 diflufenican + trifluralin

A contact and residual herbicide for use in winter cereals

Products

Ardent	Bayer CropScience	40:400 g/l	SC	09968

Uses

- Annual dicotyledons in ***triticale***, ***winter barley***, ***winter rye***, ***winter wheat***
- Annual meadow grass in ***winter barley***, ***winter wheat***

Efficacy guidance

- Weeds controlled pre-emergence up to 4 leaves (2 leaves for annual meadow grass)
- Good weed control depends on trash being buried before or during seedbed preparation
- Spring germinating weeds are not controlled
- Best results achieved on firm, fine moist seedbeds. Seed should be well covered
- Rolling after autumn treatment will reduce weed control. Roll in spring if necessary
- Speed of control depends on temperature and growing conditions; activity can be slow under cool conditions
- Always follow WRAG guidelines for preventing and managing herbicide resistant weeds. Section 5 for more information

Restrictions

- Maximum number of treatments 1 per crop
- Ensure that crop is evenly drilled and seed well covered. Do not treat broadcast crops or those that are frosted or stressed
- Do not treat undersown crops or those to be undersown
- Do not harrow after application
- Do not use on oats or any spring sown cereals
- Do not use on Sands or Very Light soils, or those that are very stony or gravelly
- Do not use on soils with more than 10% organic matter; this may include newly ploughed grassland for a time

Crop-specific information

- Latest use: before 2nd tiller stage (GS 22), or end Nov, whichever is sooner
- Apply pre- or early post-emergence of crop up to maximum recommended weed size
- Crops occasionally show transient leaf discoloration after treatment. Symptoms are quickly outgrown and yield not affected

Following crops guidance

- In the event of crop failure treated soil cannot be redrilled with wheat or barley until the following autumn; sugar beet should not be drilled in the spring following autumn treatment; 5 mth must elapse between treatment and sowing spring oilseed rape, carrots or edible brassicas
- After normal harvest of a treated crop soils must be ploughed to 15 cm before drilling field beans, leaf brassicas, winter oilseed rape, autumn sown cereals or winter onions
- Successive treatments of any products containing diflufenican can lead to soil build-up and inversion ploughing must precede sowing any following non-cereal crop. Even where ploughing occurs some crops may be damaged

Environmental safety

- Dangerous for the environment
- Very toxic to aquatic organisms
- LERAP Category B

Hazard classification and safety precautions

Hazard H04, H11
Risk phrases R36, R43, R50, R66, R67
Operator protection A, C; U05a, U08, U13, U14, U20b
Environmental protection E13a, E16a, E31b, E32d, E38
Storage and disposal E01, E04, E26, E30a

179 dimethoate

A contact and systemic organophosphorus insecticide and acaricide

Products

1	BASF Dimethoate 40	BASF	400 g/l	EC	00199
2	Danadim	Headland	400 g/l	EC	11550
3	Rogor L40	Interfarm	400 g/l	EC	07611

Uses

- Aphids in ***broccoli***, ***brussels sprouts***, ***calabrese***, ***cauliflowers***, ***ornamental specimens*** [2]; ***fodder beet***, ***mangels***, ***red beet*** [3]; ***fodder beet*** *(excluding Myzus persicae)*, ***lettuce***, ***mangels*** *(excluding Myzus persicae)*, ***red beet*** *(excluding Myzus persicae)* [1, 2]; ***grass seed crops***, ***ornamental plant production*** [1]; ***mangel seed crops***, ***spring rye***, ***spring wheat***, ***sugar beet seed crops*** *(excluding Myzus persicae)* [1, 3]; ***sugar beet*** *(excluding Myzus persicae)*, ***triticale***, ***winter rye***, ***winter wheat*** [1-3]
- Cabbage aphid in ***broccoli***, ***brussels sprouts***, ***calabrese***, ***cauliflowers*** [1]
- Insect pests in ***bulb onions*** *(off-label)*, ***bulb onions*** *(off-label - seedlings)*, ***garlic*** *(off-label)*, ***garlic*** *(off-label - seedlings)*, ***leeks*** *(off-label)*, ***leeks*** *(off-label - seedlings)*, ***protected garlic*** *(off-label - seedlings)*, ***protected leeks*** *(off-label - seedlings)*, ***protected onions*** *(off-label - seedlings)*, ***protected roscoff cauliflowers*** *(off-label)*, ***protected salad onions*** *(off-label - seedlings)*, ***protected savoy cabbage*** *(off-label)*, ***protected shallots*** *(off-label - seedlings)*, ***roscoff cauliflowers*** *(off-label)*, ***salad onions*** *(off-label - seedlings)*, ***savoy cabbage*** *(off-label)*, ***shallots*** *(off-label)*, ***shallots*** *(off-label - seedlings)* [1]; ***calabrese*** *(off-label)*, ***celeriac*** *(off-label)*, ***chinese cabbage*** *(off-label)*, ***collards*** *(off-label)*, ***kale*** *(off-label)*, ***kohlrabi*** *(off-label)*, ***protected broccoli*** *(off-label)*, ***protected brussels sprouts*** *(off-label)*, ***protected cabbages*** *(off-label)*, ***protected calabrese*** *(off-label)*, ***protected cauliflowers*** *(off-label)*, ***protected celeriac*** *(off-label)*, ***protected chinese cabbage*** *(off-label)*, ***protected collards*** *(off-label)*, ***protected kale*** *(off-label)*, ***protected kohlrabi*** *(off-label)*, ***protected spinach*** *(off-label)* [1, 2]; ***compost*** *(off-label - for watercress propagation)* [2]
- Leaf miner in ***fodder beet***, ***mangels***, ***red beet***, ***sugar beet*** [1-3]; ***mangel seed crops***, ***sugar beet seed crops*** [3]; ***ornamental plant production*** [1]; ***ornamental specimens*** [2]
- Mangold fly in ***fodder beet***, ***mangels***, ***red beet***, ***sugar beet*** [2]
- Red spider mites in ***ornamental plant production*** [1]
- Wheat bulb fly in ***spring rye***, ***triticale***, ***winter rye*** [3]; ***spring wheat*** [1, 3]; ***winter wheat*** [1-3]

Specific Off-Label Approvals (SOLAs)

- ***bulb onions***, ***garlic***, ***leeks***, ***shallots*** *(OLA 001511) May 2005* [1]
- ***bulb onions***, ***garlic***, ***leeks***, ***protected garlic***, ***protected leeks***, ***protected onions***, ***protected salad onions***, ***protected shallots***, ***salad onions***, ***shallots*** *(seedlings) (OLA 001511) May 2005* [1]
- ***calabrese***, ***celeriac***, ***chinese cabbage***, ***collards***, ***kale***, ***kohlrabi***, ***protected broccoli***, ***protected brussels sprouts***, ***protected cabbages***, ***protected calabrese***, ***protected cauliflowers***, ***protected celeriac***, ***protected chinese cabbage***, ***protected collards***, ***protected kale***, ***protected kohlrabi***, ***protected spinach*** *(OLA 041505) Dec 2008* [2]
- ***calabrese***, ***celeriac***, ***chinese cabbage***, ***collards***, ***kale***, ***kohlrabi***, ***protected broccoli***, ***protected brussels sprouts***, ***protected cabbages***, ***protected calabrese***, ***protected cauliflowers***, ***protected celeriac***, ***protected chinese cabbage***, ***protected collards***, ***protected kale***, ***protected kohlrabi***, ***protected roscoff cauliflowers***, ***protected savoy cabbage***, ***protected spinach***, ***roscoff cauliflowers***, ***savoy cabbage*** *(OLA 940389) Dec 2008* [1]
- ***compost*** *(for watercress propagation) (OLA 041502) Dec 2008* [2]

Approval information

- Approved for aerial application on cereals, sugar beet [1, 3]. See notes in Section 5

Efficacy guidance

- Chemical has quick knock-down effect and systemic activity lasts for up to 14 d
- With some crops, products differ in range of pests listed as controlled. Uses section above provides summary. See labels for details
- For most pests apply when pest first seen and repeat 2-3 wk later or as necessary. Timing and number of sprays varies with crop and pest. See labels for details
- Best results achieved when crop growing vigorously. Systemic activity reduced when crops suffering from drought or other stress

- In hot weather apply in early morning or late evening
- Where aphids or spider mites resistant to organophosphorus compounds occur control is unlikely to be satisfactory and repeat treatments may result in lower levels of control

Restrictions

- Maximum number of treatments 6 per crop for Brussels sprouts, broccoli, cauliflower, calabrese, lettuce; 4 per crop for grass seed crops; 1 per crop for beet crops (see also below), cereals, watercress
- Contains an anticholinesterase organophosphorus compound. Do not use if under medical advice not to work with such compounds
- On beet crops only one treatment per crop may be made for control of leaf miners (max 84 g a.i./ha) and black bean aphid (max 400 g a.i./ha)
- In beet crops resistant strains of peach-potato aphid (*Myzus persicae*) are common and dimethoate products must not be used to control this pest
- Test for varietal susceptibility on all unusual plants or new cultivars
- Do not tank mix with alkaline materials. See label for recommended tank-mixes
- Consult processor before spraying crops grown for processing
- Must not be applied to cereals if any product containing a pyrethroid insecticide or dimethoate has been sprayed after the start of ear emergence (GS 51)

Crop-specific information

- Latest use: before 30 Jun in yr of harvest for beet crops; flowering just complete stage (GS 30) for ground applications to cereals; before 31 Mar in yr of harvest for aerial application to cereals; before 6 true leaves for shallots; before transfer to cropping beds for watercress; before 7 leaf stage for brassicas and protected vegetables
- HI brassicas 21 d; grass seed crops, lettuce, cereals 14 d; shallots 7 d

Environmental safety

- Dangerous for the environment
- Very toxic to aquatic organisms
- Harmful to game, wild birds and animals
- Harmful to livestock. Keep all livestock out of treated areas for at least 7 d
- Likely to cause adverse effects on beneficial arthropods
- High risk to bees. Do not apply to crops in flower or to those in which bees are actively foraging. Do not apply when flowering weeds are present
- Surface residues may also cause bee mortality following spraying
- High risk to non-target insects or other arthropods
- LERAP Category A
- Keep in original container, tightly closed, in a safe place, under lock and key
- Do not treat cereals after 1 Apr within 6 m of edge of crop

Hazard classification and safety precautions

Hazard H03, H08 [1-3]; H11 [3]
Risk phrases R20, R22a, R43 [1-3]; R21 [1, 3]; R48 [1]; R50, R58 [3]; R52 [2]
Operator protection A, H, M [1-3]; C, J [1, 3]; U02a, U04a, U19a [1-3]; U05a, U10, U13, U20a [1, 3]; U08, U15, U20b [2]
Environmental protection E06c [1-3] (7 d); E10b, E16c, E16d, E34 [1-3]; E12a, E13b, E22a [1, 2]; E12d, E15a, E17a, E31b [3]; E17a [2] (18 m); E18 [1, 3]; E31c [1]; E32e [2]
Storage and disposal E01, E04 [1-3]; E26 [1]; E29b [1, 3]; E30a [3]; E30b [1, 2]
Treated seed S04b [3]
Medical advice M01 [1-3]; M03 [2, 3]; M05a [1]

180 dimethomorph

A cinnamic acid fungicide with translaminar activity available only in mixtures

181 dimethomorph + mancozeb

A systemic and protectant fungicide for potato blight control

Products

Invader	BASF	7.5:66.7% w/w	WG	11978

Uses
- Blight in ***potatoes***
- Downy mildew in ***bulb onions*** *(off-label)*, ***garlic*** *(off-label)*, ***shallots*** *(off-label)*

Specific Off-Label Approvals (SOLAs)
- ***bulb onions, garlic, shallots*** *(OLA 011446) Dec 2008* [1]

Efficacy guidance
- Commence treatment as soon as there is a risk of blight infection
- In the absence of a warning treatment should start before the crop meets along the rows
- Repeat treatments every 7-14 d depending in the degree of infection risk
- Irrigated crops should be regarded as at high risk and treated every 7 d
- For best results good spray coverage of the foliage is essential
- To minimise the likelihood of development of resistance these products should be used in a planned Resistance Management strategy. See Section 5 for more information

Restrictions
- Maximum total dose equivalent to eight full dose treatments

Crop-specific information
- HI 7 d for all crops

Environmental safety
- Dangerous for the environment
- Very toxic to aquatic organisms
- LERAP Category B

Hazard classification and safety precautions

Hazard H04, H11
Risk phrases R36, R37, R38, R43, R50, R58
Operator protection A; U02a, U04a, U05a, U08, U13, U19a, U20b
Environmental protection E16a, E16b, E32d, E38
Storage and disposal E01, E04, E26, E30a
Medical advice M05a

182 dimoxystrobin

A protectant strobilurin fungicide for cereals available only in mixtures

183 dimoxystrobin + epoxiconazole

A protectant and curative fungicide mixture for cereals

Products

Swing Gold	BASF	133:50 g/l	SC	11658

Uses
- Brown rust in ***winter wheat***
- Fusarium ear blight in ***winter wheat***
- Septoria in ***winter wheat***
- Tan spot in ***winter wheat***

Efficacy guidance
- For best results apply from the start of ear emergence
- Dimoxystrobin is a member of the QoI cross resistance group. Product should be used preventatively and not relied on for its curative potential
- Use product as part of an Integrated Crop Management strategy incorporating other methods of control, including where appropriate other fungicides with a different mode of action. Do not apply more than two foliar applications of QoI containing products to any cereal crop
- There is a significant risk of widespread resistance occurring in *Septoria tritici* populations in UK. Failure to follow resistance management action may result in reduced levels of disease control

Restrictions
- Maximum number of treatments 1 per crop
- Do not apply before the start of ear emergence

Crop-specific information

- Latest use: up to and including flowering (GS 69)

Environmental safety

- Dangerous for the environment
- Very toxic to aquatic organisms
- LERAP Category B
- Avoid drift onto neighbouring crops

Hazard classification and safety precautions

Hazard H03, H11
Risk phrases R20, R22a, R40, R50, R58
Operator protection A; U05a, U20b
Environmental protection E15a, E16a, E31c, E32d, E34, E38
Storage and disposal E01, E04, E26, E30a
Medical advice M05a

184 dinocap

A protectant dinitrophenyl fungicide for powdery mildew control

Products

Karathane Liquid	Landseer	350 g/l	EC	09262

Uses

- Powdery mildew in ***apples***, ***chrysanthemums***, ***grapevines*** *(off-label)*, ***roses***, ***strawberries***

Specific Off-Label Approvals (SOLAs)

- ***grapevines*** *(OLA 021410) Dec 2008* [1]

Efficacy guidance

- Spray at 7-14 d intervals to maintain protective film
- Product must be applied to roses before disease becomes established
- Regular use on apples suppresses red spider mites and rust mites

Restrictions

- Maximum total dose equivalent to 10 full dose treatments on apples; 8 full dose treatments on grapevines
- Maximum number of treatments on strawberries 5 per yr
- When applied to apples during blossom period may cause spotting but no adverse effect on pollination or fruit set. Do not apply to Golden Delicious during blossom period
- Do not apply when temperature is above 24°C. Do not apply with white oils
- Certain chrysanthemum cultivars may be susceptible
- May cause petal spotting on white roses

Crop-specific information

- HI grapevines 21 d, apples 14 d, strawberries 7 d

Environmental safety

- Dangerous for the environment
- Very toxic to aquatic organisms
- LERAP Category B
- Broadcast air-assisted LERAP (18 m)

Hazard classification and safety precautions

Hazard H02, H08, H11
Risk phrases R20, R22b, R36, R38, R43, R48, R50, R58, R61, R67
Operator protection A, C, H, J, M; U04a, U05a, U10, U11, U19a, U20a
Environmental protection E15a, E16a, E31a, E32d, E34, E38; E17b (18 m)
Storage and disposal E01, E04, E26, E30a
Medical advice M04

185 diquat

A non-residual bipyridyl contact herbicide and crop desiccant

Products

1	Agriguard Diquat	AgriGuard	200 g/l	SL	10836
2	Greencrop Boomerang	Greencrop	200 g/l	SL	10691
3	Landgold Diquat	Landgold	200 g/l	SL	10649
4	Reglone	Syngenta	200 g/l	SL	10534
5	Standon Diquat SL	Standon	200 g/l	SL	10732
6	Waterloo	Interfarm	200 g/l	SL	11196

Uses

- Annual dicotyledons in ***flower bulbs***, ***sugar beet*** [1, 2, 6]; ***mustard*** *(off-label)*, ***sage*** *(off-label)*, ***sunflowers*** *(off-label)* [4]; ***ornamental plant production***, ***row crops*** [1]; ***potatoes*** [1, 2, 5]; ***row crops*** *(between row treatment)* [2, 4-6]
- Chemical stripping in ***hops*** [1, 2, 4, 6]
- General weed control in ***poppies*** *(off-label - for morphine production)*, ***potatoes*** [4]
- General weed growth in ***laid barley*** *(stockfeed only)*, ***laid oats*** *(stockfeed only)* [1-3, 5, 6]
- Haulm destruction in ***potatoes*** [3-5]
- Pre-harvest desiccation in ***borage*** *(off-label)*, ***echium plantaginium*** *(off-label)*, ***honesty*** *(off-label)*, ***laid barley*** *(stockfeed only)*, ***laid oats*** *(stockfeed only)*, ***mustard*** *(off-label)*, ***navy beans*** *(off-label)*, ***peas***, ***poppies*** *(off-label - for morphine production)*, ***sage*** *(off-label)*, ***soya beans*** *(off-label)*, ***sunflowers*** *(off-label)* [4]; ***clover seed crops*** [1, 2, 4, 6]; ***combining peas*** [1-3, 5, 6]; ***linseed***, ***spring field beans*** *(stock or pigeon feed only)*, ***spring oilseed rape***, ***winter field beans*** *(stock or pigeon feed only)*, ***winter oilseed rape*** [1-6]; ***potatoes*** [1, 2, 6]

Specific Off-Label Approvals (SOLAs)

- ***borage*** *(OLA 011654) Dec 2008* [4]
- ***echium plantaginium*** *(OLA 011650) Dec 2008* [4]
- ***honesty*** *(OLA 011655) Dec 2008* [4]
- ***mustard***, ***sage***, ***sunflowers*** *(OLA 040118) Dec 2008* [4]
- ***navy beans*** *(OLA 022039) Dec 2008* [4]
- ***poppies*** *(for morphine production) (OLA 032408) Dec 2008* [4]
- ***soya beans*** *(OLA 011653) Dec 2008* [4]

Approval information

- Diquat included in Annex I under EC Directive 91/414
- All approvals for aquatic weed control expired in 2004

Efficacy guidance

- Acts rapidly on green parts of plants and rainfast in 15 min
- Best results for potato desiccation achieved by spraying in bright light and low humidity conditions
- For weed control in row crops apply as overall spray before crop emergence or before transplanting or as an inter-row treatment using a spray guard
- Treatments for weed control normally require co-application with paraquat. Use of authorised non-ionic wetter for improved weed control essential for most uses except potato desiccation, treatment of hops and aquatic weed control (except *Lemna*)

Restrictions

- Maximum number of treatments 3 per crop on hops; 1 per crop in most other situations - see labels for details
- Do not apply potato haulm destruction treatment when soil dry. Tubers may be damaged if spray applied during or shortly after dry periods. See label for details of maximum allowable soil-moisture deficit and varietal drought resistance scores
- Do not add wetters to desiccant sprays for potatoes or for water weed control except for *Lemna* control.
- Consult processor before adding Agral or other wetter on peas. Staining may be increased
- Do not use treated straw or haulm as animal feed or bedding within 4 d of spraying
- Do not use on barley, oats and field beans intended for human consumption [2-5]

Crop-specific information

- Latest use varies with crop and product to ensure best results - see labels
- HI zero when used as a crop desiccant but see label for advisory intervals

- For pre-harvest desiccation, apply to potatoes when tubers the desired size and to other crops when mature or approaching maturity. See label for details of timing and of period to be left before harvesting potatoes and for timing of pea and bean desiccation sprays
- Spray linseed when seed matured evenly over whole field; direct combining can normally begin 10-20 d after spraying
- Apply as hop stripping treatment when shoots have reached top wire
- Apply to floating and submerged aquatic weeds in still or slow moving water [1]
- Treated laid barley and oats may be used only for stock feed; treated peas must be harvested dry; treated field beans may be used for pigeon and animal feed only
- If potato tubers are to be stored leave 14 d after treatment before lifting
- Use on thin or stressed crops of oilseed rape may increase seed losses and decrease oil yield to an unacceptable level
- Treatment of hops under stress or when leaves wet with a heavy dew may cause damage

Environmental safety

- Dangerous for the environment
- Very toxic to aquatic organisms
- Keep all livestock out of treated areas and away from treated water for at least 24 h
- Do not feed treated straw or haulm to livestock within 4 days of spraying
- Do not use on crops if the straw is to be used as animal feed/bedding
- Do not dump surplus herbicide in water or ditch bottoms

Hazard classification and safety precautions

Hazard H02, H11
Risk phrases R22a, R37, R50, R58 [1-6]; R36, R38 [1-3, 5]; R48 [2-6]
Operator protection A, C; U02a, U04a, U05a, U08, U19a, U20b
Environmental protection E06c [1, 4, 6] (24 h); E08 [4, 6] (4 d); E09 [1] (4 d); E15a, E31a, E34 [1-6]; E19a [1]; E32d, E38 [4, 6]
Storage and disposal E01, E04, E30a [1-6]; E26 [1-3, 5]
Medical advice M03 [1]; M04 [2-6]

186 diquat + paraquat

A non-selective non-residual bipyridyl contact herbicide

Products

1	ASAP	Me2	80:120 g/l	SC	11485
2	Fernpath Pronto	AgriGuard	80:120 g/l	SL	11988
3	PDQ	Syngenta	80:120 g/l	SL	10532
4	Speedway 2	Scotts	2.5:2.5% w/w	WG	10790

Uses

- Annual and perennial weeds in ***amenity vegetation, hard surfaces, natural surfaces not intended to bear vegetation, permeable surfaces overlying soil*** [4]
- Annual dicotyledons in ***all edible crops, all non-edible crops, forest nurseries, natural surfaces not intended to bear vegetation, ornamental plant production*** [1, 2]; ***apple orchards, bush fruit, cane fruit, cherries, cultivated land/soil, damsons, flower bulbs, non-crop areas, pear orchards, plums, row crops, strawberries, stubbles*** [3]; ***forest, hops, potatoes*** [1-3]
- Annual grasses in ***all edible crops, all non-edible crops, forest nurseries, natural surfaces not intended to bear vegetation, ornamental plant production*** [1, 2]; ***apple orchards, bush fruit, cane fruit, cherries, cultivated land/soil, damsons, flower bulbs, non-crop areas, pear orchards, plums, row crops, strawberries, stubbles*** [3]; ***forest, hops, potatoes*** [1-3]
- Chemical stripping in ***hops*** [3]
- Grass weeds in ***field margins*** [3]
- Green cover in ***land temporarily removed from production*** [1-3]
- Perennial non-rhizomatous grasses in ***apple orchards, bush fruit, cane fruit, cherries, cultivated land/soil, damsons, flower bulbs, forest, hops, non-crop areas, pear orchards, plums, potatoes, row crops, strawberries, stubbles*** [3]

- Sward destruction in ***grassland*** [3, 4]; ***managed amenity turf*** [4]; ***permanent pasture, rotational grassland*** [1, 2]
- Volunteer cereals in ***apple orchards, bush fruit, cane fruit, cherries, cultivated land/soil, damsons, forest, hops, non-crop areas, pear orchards, plums, potatoes, row crops, strawberries, stubbles*** [3]

Approval information

- Diquat and paraquat included in Annex I under EC Directive 91/414

Efficacy guidance

- Rapid kill obtained under bright conditions but most effective results on difficult weeds obtained from slower action in winter
- Apply to young emerged weeds less than 15 cm high, annual grasses must have at least 2 leaves when sprayed [3]
- Addition of approved non-ionic wetter recommended for control of certain species and with low dose rates. See label for details [3]
- Interval between spraying and cultivation varies. See label for details [3]
- For chemical stripping apply in Jul or after hops have reached top wire. Do not use on hops under drought conditions [3]
- Chemical rapidly inactivated in moist soil, activity reduced in dirty or muddy water
- Spray is rainfast in 10 min

Restrictions

- Paraquat is subject to the Poisons Rules 1982 and the Poisons Act 1972. See notes in Section 5
- Product for use only by local authorities and professional spray operators [4]
- Diquat may cause an allergic reaction
- Maximum number of treatments 2 per yr for grassland destruction
- In non-crop areas do not use around green bark [4]
- Do not put in a food or drinks container

Crop-specific information

- Latest use varies according to situation - see label
- Apply up to just before sown crops emerge or just before planting [3]
- In potatoes spray earlies up to 10% emergence, maincrop to 40% emergence, provided plants are less than 15 cm high. Do not use post-emergence on potatoes from diseased or small tubers or under very hot, dry conditions [3]
- Use guarded no-drift sprayers to kill inter-row weeds and strawberry runners [3]
- Apply to fruit crops as a directed spray, preferably in dormant season [3]
- If spraying bulbs at end of season ensure all crop foliage is detached from bulbs. Do not use on very sandy soils [3]

Following crops guidance

- On sandy or immature peat soils and on forest nursery seedbeds allow 3 d between spraying and planting [3]
- Where trash or dying weeds are left on surface allow at least 3 d before planting

Environmental safety

- Dangerous for the environment
- Very toxic to aquatic organisms
- Paraquat can be harmful to hares; spray stubbles early in the day
- Keep all livestock out of treated areas for at least 24 h
- Keep in original container, tightly closed, in a safe place, under lock and key

Hazard classification and safety precautions

Hazard H02 [1-3]; H03 [4]; H11 [1-4]

Risk phrases R21, R37, R38, R41, R50 [1-3]; R22a, R48, R58 [1-4]; R51 [4]

Operator protection A, C [1-4]; H, M [1-3]; U02a, U04a, U20a [2]; U04b, U04c, U19a, U20b [1, 3, 4]; U05a, U09a [1-4]; U14 [4]

Environmental protection E02a, E32e [4]; E06c [1-4] (24 h); E11 [1-3]; E15a, E34 [1-4]; E31a [2, 4]; E31c, E32d [1, 3]; E38 [1, 3, 4]

Storage and disposal E01, E04, E30b [1-4]; E26 [2, 4]

Medical advice M03 [2]; M04 [1, 3, 4]; M05c [1-3]

187 dithianon

A protectant and eradicant dicarbonitrile fungicide for scab control

Products

Dithianon Flowable	BASF	750 g/l	SC	10219

Uses

- Scab in ***apples, pears***

Efficacy guidance

- Apply at bud-burst and repeat every 10-14 d until danger of scab infection ceases
- Application at high rate within 48 h of a Mills period prevents new infection
- Spray programme also reduces summer infection with apple canker

Restrictions

- Maximum number of treatments 8 per crop
- Do not use on Golden Delicious apples after green cluster
- Do not mix with lime sulphur or highly alkaline products

Crop-specific information

- HI 4 wk

Environmental safety

- Dangerous for the environment
- Very toxic to aquatic organisms

Hazard classification and safety precautions

Hazard H03, H11
Risk phrases R20, R22a, R50, R58
Operator protection A, H; U05a, U08, U19a, U20a
Environmental protection E15a, E31b, E34, E38
Storage and disposal E01, E04, E26, E30a
Medical advice M03, M05a

188 diuron

A residual urea herbicide for non-crop areas and woody crops

See also *amitrole + 2,4-D + diuron*

Products

1	Diurex 50SC	Makhteshim	500 g/l	SC	11472
2	Diuron 80 WP	Bayer Environ.	80% w/w	WP	09931
3	Freeway	Bayer Environ.	500 g/l	SC	09933
4	Freeway	Bayer Environ.	500 g/l	SC	11129
5	Karmex	Headland	80% w/w	WP	09475
6	MSS Diuron 500 FL	Nufarm UK	500 g/l	SC	11493
7	MSS Diuron 80 WP	Nufarm UK	80% w/w	WP	11503
8	Nomix Diuron 80	Nomix Enviro	80% w/w	WP	12079
9	Nomix Diuron Flowable	Nomix Enviro	500 g/l	SC	12078
10	Unicrop Flowable Diuron	Unicrop	500 g/l	SC	02270

Uses

- Annual and perennial weeds in ***asparagus*** *(off-label)*, ***blackcurrants*** *(off-label)* [1]
- Annual dicotyledons in ***amenity trees and shrubs*** [1, 3, 4, 8]; ***amenity vegetation*** [6, 7, 9]; ***apple orchards, pear orchards*** [1, 3, 4, 10]; ***asparagus*** *(off-label)* [10]; ***established woody ornamentals*** [2]; ***hard surfaces*** [1, 6, 9]; ***land not intended to bear vegetation*** [2-9]; ***natural surfaces not intended to bear vegetation*** [1]; ***permeable surfaces overlying soil*** [1, 6, 7, 9]; ***trees and shrubs*** [3, 4]; ***woody nursery stock*** [3]; ***woody nursery stock*** *(not container grown)* [4]
- Annual grasses in ***amenity trees and shrubs*** [1, 3, 4]; ***apple orchards, pear orchards*** [1, 3, 4, 10]; ***asparagus*** *(off-label)* [10]; ***established woody ornamentals*** [2]; ***hard surfaces, natural surfaces not intended to bear vegetation, permeable surfaces overlying soil*** [1]; ***land not intended to bear vegetation*** [2-4]; ***trees and shrubs, woody nursery stock*** [3, 4]
- Annual meadow grass in ***amenity trees and shrubs, land not intended to bear vegetation*** [8]

- Black nightshade in ***daffodils*** *(off-label - for galanthamine production)* [10]
- Perennial weeds in ***amenity vegetation***, ***land not intended to bear vegetation***, ***permeable surfaces overlying soil*** [6, 7, 9]; ***hard surfaces*** [6, 9]
- Stinging nettle in ***daffodils*** *(off-label - for galanthamine production)* [10]
- Willowherb in ***blackcurrants*** *(off-label)* [1, 10]; ***daffodils*** *(off-label - for galanthamine production)* [10]

Specific Off-Label Approvals (SOLAs)

- ***asparagus***, ***blackcurrants*** *(OLA 032172) Dec 2008* [1]
- ***asparagus*** *(OLA 982352) Dec 2008* [10]
- ***blackcurrants*** *(OLA 951318) Dec 2008* [10]
- ***daffodils*** *(for galanthamine production) (OLA 041521) Dec 2008* [10]

Efficacy guidance

- Best results when applied to moist soil and rain falls soon afterwards
- Length of residual activity may be reduced on heavy or highly organic soils or those with ash substrates
- Selective rates must be applied to weed-free soil and activity persists for 2-3 mth
- Application for total vegetation control may be at any time of year, best results obtained in late winter to early spring. See label for list of resistant species
- Application to frozen ground not recommended

Restrictions

- Maximum number of treatments 1 per yr on land not intended for cropping (high rate), 2 per yr around trees and shrubs (low rate); one per year for apples, pears; 2 per yr (low rate) for blackcurrants [10]
- Do not treat trees and shrubs less than 5 cm tall or established less than 12 mth or in areas of container grown trees and shrubs. See label for list of sensitive ornamental species that should not be treated
- Do not use around trees and shrubs on Sands or Very Light soils or those that are gravelly or where less than 1% organic matter
- Do not use on lawns, grass tennis courts or similar areas of turf
- Do not apply non-selective rates on or near desirable plants where chemical may be washed into contact with roots
- Application for amenity use should only take place between the beginning of Feb and end of Apr (May in Scotland)

Crop-specific information

- Latest use: 30 April (31 May in Scotland) on non-porous surfaces; at any time on porous surfaces for land not intended to bear vegetation; before bud-burst for blackcurrants [10]; pre-emergence for asparagus
- Around trees and shrubs spray in late winter or early spring
- Treat hard surfaces between Feb and end May
- Apply to weed-free soil in apple and pear orchards established for at least 1 yr during Feb-Mar

Following crops guidance

- Succeeding crops should not be planted for at least 12 mth after treatment. Do not plant susceptible crops such as vegetables for at least 2 yr

Environmental safety

- Dangerous for the environment
- Very toxic to aquatic organisms
- Do not apply over drains or in drainage channels or gullies
- Avoid run-off when using on paved and similar surfaces

Hazard classification and safety precautions

Hazard H03 [1, 2, 4, 6-10]; H04 [2, 3, 5]; H11 [1, 2, 4-10]

Risk phrases R22a [1]; R36, R38 [2, 3, 5]; R37, R51 [5]; R40 [1, 2, 4, 8, 10]; R48, R50 [1, 2, 4, 6-10]; R58 [1, 2, 4-10]; R68 [6, 7, 9]

Operator protection A [1, 3-6, 8, 9]; C [5, 8]; H, M [1, 3, 4, 6, 9]; N [1, 6, 9]; U05a [1-10]; U08 [2-10]; U14, U15 [1]; U19a [2-5, 7, 10]; U20a [1, 6-9]; U20b [2-5, 10]

Environmental protection E13c [3]; E15a [1, 2, 4-10]; E31b [1, 3, 6, 9]; E31c [4, 10]; E32a [2, 5, 7]; E32d [2, 4, 6-10]; E32e [1]; E34 [2, 4, 6, 8-10]; E38 [2, 4-10]

Storage and disposal E01, E04, E30a [1-10]; E26 [4, 10]

Medical advice M03 [2, 4, 10]; M04 [1]; M05a [6-9]

SEE SECTION 3 FOR PRODUCTS ALSO REGISTERED

189 diuron + glyphosate

A non-selective residual herbicide mixture for non-crop and amenity use

Products

1	Touché	Nomix Enviro	217.6:109 g/l	RH	07913
2	Xanadu	Bayer Environ.	125:100 g/l	EW	09943

Uses

- Annual dicotyledons in ***amenity trees and shrubs, land not intended to bear vegetation***
- Annual grasses in ***amenity trees and shrubs, land not intended to bear vegetation***
- Perennial dicotyledons in ***amenity trees and shrubs, land not intended to bear vegetation***
- Perennial grasses in ***amenity trees and shrubs, land not intended to bear vegetation***

Approval information

- Glyphosate included in Annex I under EC Directive 91/414
- Approved for use through ULV applicators [1]

Efficacy guidance

- Best results achieved from treatment when weeds are green and actively growing. Symptoms may be slow to appear in poor growing conditions
- Perennial grasses are susceptible when tillering and making new rhizome growth, normally when plants have 4-5 new leaves
- Perennial dicotyledons most susceptible if treated at or near flowering but will be severely checked if treated at other times when growing actively
- Most species of germinating weeds are controlled. See label for more resistant species
- Weed control may be reduced on heavy or highly organic soils and when weeds are suffering stress in any situation
- At least 6 and preferably 24 h rain-free must follow spraying. Rain soon after treatment may reduce initial weed control
- Apply with manufacturer's specialist equipment [1]

Restrictions

- Maximum number of treatments 1 per yr
- Ornamental trees and shrubs should be established for at least 12 mth and be at least 50 mm tall
- Among ornamental trees and shrubs take care to avoid foliage or the stems of young plants
- Do not allow spray to contact desired plants or crops
- Do not use on tree and shrub nurseries or before planting
- Do not treat ornamental trees and shrubs on Sands or Very Light soils or those that are gravelly or with less than 1% organic matter [1]
- Do not decant, connect directly to Nomix applicator [1]

Crop-specific information

- Treat hard surfaces between Feb and end May

Environmental safety

- Dangerous for the environment
- Toxic to aquatic organisms
- Not to be used on food crops
- Do not apply over drains or in drainage channels, gullies or similar structures

Hazard classification and safety precautions

Hazard H03, H11
Risk phrases R40, R48, R51, R58
Operator protection A [2]; H, M; U02a [1]; U09a, U19a, U20b [1, 2]
Environmental protection E15a, E32a, E32d, E38 [1, 2]; E34 [2]
Consumer protection C01 [2]
Storage and disposal E01, E30a [1, 2]; E04 [2]; E26 [1]
Medical advice M03 [2]

190 diuron + paraquat

A total herbicide with contact and residual activity

Products

Dexuron	Nomix Enviro	300:100 g/l	SC	07169

Uses

- Annual dicotyledons in ***established woody ornamentals, woody nursery stock***
- Annual grasses in ***established woody ornamentals, woody nursery stock***
- Perennial dicotyledons in ***established woody ornamentals, woody nursery stock***
- Perennial grasses in ***established woody ornamentals, woody nursery stock***
- Total vegetation control in ***non-crop areas, paths***

Approval information

- Paraquat included in Annex I under EC Directive 91/414

Efficacy guidance

- Apply to emerged weeds at any time of year. Best results achieved by application in spring or early summer
- Effectiveness not reduced by rain soon after treatment

Restrictions

- Paraquat is subject to the Poisons Rules 1982 and the Poisons Act 1972. See notes in Section 5
- Maximum number of treatments 1 per yr
- Avoid contact of spray with green bark, buds or foliage of desirable trees or shrubs

Crop-specific information

- In nurseries use low dose rate as inter-row spray, not more than once per year

Environmental safety

- Dangerous for the environment
- Toxic to aquatic organisms
- Keep in original container, tightly closed, in a safe place, under lock and key
- Harmful to fish or other aquatic life. Do not contaminate surface waters or ditches with chemical or used container
- Keep livestock out of treated areas for at least 2 wk and until foliage of any poisonous weeds such as ragwort has died and become unpalatable

Hazard classification and safety precautions

Hazard H02, H11
Risk phrases R21, R25, R36, R38, R51, R68
Operator protection A, C; U02a, U04a, U05a, U09a, U11, U19a, U20b
Environmental protection E06c (24 h); E15a, E32a, E32d, E34, E38
Storage and disposal E01, E04, E30b
Medical advice M04

191 dodemorph

A systemic morpholine fungicide for powdery mildew control

Products

F238	BASF	385 g/l	EC	00206

Uses

- Powdery mildew in ***roses***

Efficacy guidance

- Spray roses every 10-14 d during mildew period or every 7 d and at increased dose if cleaning up established infection or if disease pressure high
- Add Citowett when treating rose varieties which are difficult to wet
- Product has negligible effect on *Phytoseiulus* spp being used to control red spider mites

Restrictions

- Do not use on seedling roses

SEE SECTION 3 FOR PRODUCTS ALSO REGISTERED

- Do not apply to roses under hot, sunny conditions, particularly under glass, but spray early in the morning or during the evening. Increase the humidity some hours before spraying
- Check tolerance of new varieties before treating rest of crop

Environmental safety

- Dangerous for the environment
- Toxic to aquatic organisms

Hazard classification and safety precautions

Hazard H04, H08, H11
Risk phrases R22a, R22b, R38, R41, R51, R58, R63
Operator protection A, C; U04a, U05a, U09a, U11, U14, U15, U20b
Environmental protection E15a, E31c, E32d, E38
Storage and disposal E01, E04, E26, E30a
Medical advice M05b

192 dodine

A protectant and eradicant guanidine fungicide

Products

1	Greencrop Budburst	Greencrop	450 g/l	SC	11042
2	Radspor FL	Truchem	450 g/l	SC	01685

Uses

- Currant leaf spot in ***blackcurrants***
- Scab in ***apples, pears***

Efficacy guidance

- Apply protective spray on apples and pears at bud-burst and at 10-14 d intervals until late Jun to early Jul
- Apply post-infection spray within 36 h of rain responsible for initiating infection. Where scab already present spray prevents production of spores
- On blackcurrants commence spraying at early grape stage and repeat at 2-3 wk intervals, and at least once after picking

Restrictions

- Do not apply in very cold weather (under 5°C) or under slow drying conditions to pears or dessert apples during bloom or immediately after petal fall
- Do not mix with lime sulphur or tetradifon
- Consult processors before use on crops grown for processing

Crop-specific information

- Latest use: early Jul for culinary apples; pre-blossom for dessert apples and pears

Environmental safety

- Dangerous for the environment
- Very toxic to aquatic organisms

Hazard classification and safety precautions

Hazard H03, H11
Risk phrases R21 [1]; R22a, R36, R38, R43, R50, R58 [1, 2]; R37, R41 [2]
Operator protection A, C; U05a, U08, U19a, U20b [1, 2]; U11 [2]
Environmental protection E15a, E31b, E34 [1, 2]; E38 [2]
Storage and disposal E01, E04, E30a [1, 2]; E26 [1]
Medical advice M03 [2]; M04 [1]

193 epoxiconazole

A systemic, protectant and curative triazole fungicide for use in cereals

See also *carbendazim + epoxiconazole*
dimoxystrobin + epoxiconazole

Products

1	Agriguard Epoxiconazole	AgriGuard	125 g/l	SC	09407
2	Epic	BASF	125 g/l	SC	12136
3	Landgold Epoxiconazole	Landgold	125 g/l	SC	09821
4	Me2 Puddy	Me2	125 g/l	SC	12164
5	Opus	BASF	125 g/l	SC	12057
6	Standon Epoxiconazole	Standon	125 g/l	SC	09517

Uses

- Brown rust in ***spring barley***, ***winter barley***, ***winter wheat*** [1-6]; ***spring rye***, ***spring wheat***, ***triticale***, ***winter rye*** [2-5]
- Eyespot in ***winter barley*** *(reduction)*, ***winter wheat*** *(reduction)* [1-6]
- Fusarium ear blight in ***spring wheat***, ***winter wheat*** [2-5]; ***winter wheat*** *(reduction)* [1, 6]
- Net blotch in ***spring barley***, ***winter barley*** [1-6]
- Powdery mildew in ***spring barley***, ***winter barley***, ***winter wheat*** [1-6]; ***spring oats***, ***spring rye***, ***spring wheat***, ***triticale***, ***winter oats***, ***winter rye*** [2-5]
- Rhynchosporium in ***spring barley***, ***winter barley*** [1-6]; ***spring rye***, ***winter rye*** [2-5]
- Septoria in ***spring wheat***, ***triticale*** [2-5]; ***winter wheat*** [1-6]
- Sooty moulds in ***spring wheat*** *(reduction)* [2-5]; ***winter wheat*** *(reduction)* [1-6]
- Yellow rust in ***spring barley***, ***winter barley***, ***winter wheat*** [1-6]; ***spring rye***, ***spring wheat***, ***triticale***, ***winter rye*** [2-5]

Efficacy guidance

- Apply at the start of foliar disease attack
- Optimum effect against eyespot achieved by spraying between leaf-sheath erect and second node detectable stages (GS 30-32)
- Best control of ear diseases of wheat obtained by treatment during ear emergence
- For Septoria spray after third node detectable stage (GS 33) when weather favouring disease development has occurred
- Mildew control improved by use of tank mixtures. See label for details

Restrictions

- Maximum total dose equivalent to two full dose treatments
- Product may cause damage to broad-leaved plant species
- Avoid spray drift onto neighbouring crops

Crop-specific information

- Latest use: up to and including flowering just complete (GS 69) in wheat, rye, triticale; up to and including emergence of ear just complete (GS 59) in barley, oats

Environmental safety

- Dangerous for the environment [1-6]
- Very toxic to aquatic organisms
- LERAP Category B

Hazard classification and safety precautions

Hazard H03 [2-6]; H11 [1-6]
Risk phrases R38, R40, R50, R62, R63 [2-6]; R51 [1]; R58 [1-6]
Operator protection A; U05a [1-6]; U20a [1]; U20b [2-6]
Environmental protection E15a, E16a [1-6]; E16b [1, 6]; E31b [1, 3, 6]; E31c, E32d, E38 [2, 4, 5]
Storage and disposal E01, E04, E26, E30a [1-6]; E29b [3]
Medical advice M05a [2-5]; M05b [6]

194 epoxiconazole + fenpropimorph

A systemic, protectant and curative fungicide mixture for cereals

Products

1	Eclipse	BASF	84:250 g/l	SE	11731
2	Greencrop Galore	Greencrop	84:250 g/l	SE	09561
3	Opus Team	BASF	84:250 g/l	SE	11759
4	Standon Epoxifen	Standon	84:250 g/l	SE	08972

SEE SECTION 3 FOR PRODUCTS ALSO REGISTERED

Uses

- Brown rust in ***spring barley*, *winter barley*, *winter wheat*** [1-4]; ***spring rye*, *spring wheat*, *triticale*, *winter rye*** [1, 3]
- Eyespot in ***winter barley*** *(reduction)*, ***winter wheat*** *(reduction)* [1-4]
- Fusarium ear blight in ***spring wheat*, *winter wheat*** [1, 3]; ***winter wheat*** *(reduction)* [2, 4]
- Net blotch in ***spring barley*, *winter barley*** [1-4]
- Powdery mildew in ***spring barley*, *winter barley*, *winter wheat*** [1-4]; ***spring oats*, *spring rye*, *spring wheat*, *triticale*, *winter oats*, *winter rye*** [1, 3]
- Rhynchosporium in ***spring barley*, *winter barley*** [1-4]; ***spring rye*, *winter rye*** [1, 3]
- Septoria in ***spring wheat*, *triticale*** [1, 3]; ***winter wheat*** [1-4]
- Sooty moulds in ***spring wheat*** *(reduction)* [1, 3]; ***winter wheat*** *(reduction)* [1-3]
- Yellow rust in ***spring barley*, *winter barley*, *winter wheat*** [1-4]; ***spring rye*, *spring wheat*, *triticale*, *winter rye*** [1, 3]

Efficacy guidance

- Apply at the start of foliar disease attack
- Optimum effect against eyespot achieved by spraying between leaf-sheath erect and second node detectable stages (GS 30-32)
- Best control of ear diseases obtained by treatment during ear emergence
- For Septoria spray after third node detectable stage (GS 33) when weather favouring disease development has occurred

Restrictions

- Maximum total dose equivalent to two full dose treatments
- Product may cause damage to broad-leaved plant species

Crop-specific information

- Latest use: up to and including flowering just complete (GS 69) in wheat, rye, triticale; up to and including emergence of ear just complete (GS 59) in barley, oats

Environmental safety

- Dangerous for the environment
- Toxic to aquatic organisms
- LERAP Category B
- Avoid spray drift onto neighbouring crops

Hazard classification and safety precautions

Hazard H03, H11
Risk phrases R20, R43 [2]; R38 [4]; R40, R62, R63 [1, 3, 4]; R51, R58 [1-4]
Operator protection A [1-4]; H [1, 3]; U05a, U20b [1-4]; U14 [2]; U19a [2, 4]
Environmental protection E15a, E16a [1-4]; E16b [2, 4]; E31a [2]; E31b [4]; E31c, E32d, E38 [1, 3]
Storage and disposal E01, E04, E26, E30a
Medical advice M05a [1, 3]; M05b [4]

195 epoxiconazole + fenpropimorph + kresoxim-methyl

A protectant, systemic and curative fungicide mixture for cereals

Products

1	Asana	BASF	125:150:125 g/l	SE	11934
2	BAS 493F	BASF	125:150:125 g/l	SE	11748
3	Mantra	BASF	125:150:125 g/l	SE	11728
4	Mastiff	BASF	125:150:125 g/l	SE	11747
5	Standon Kresoxim Super	Standon	125:150:125 g/l	SE	09794

Uses

- Brown rust in ***spring barley*, *winter barley*, *winter wheat*** [1-5]; ***spring rye*, *spring wheat*, *triticale*, *winter rye*** [1-4]
- Eyespot in ***spring rye*** *(reduction)*, ***triticale*** *(reduction)*, ***winter oats*** *(reduction)*, ***winter rye*** *(reduction)* [1-4]; ***winter barley*** *(reduction)*, ***winter wheat*** *(reduction)* [1-5]
- Fusarium ear blight in ***spring wheat*** *(reduction)* [1-4]; ***winter wheat*** *(reduction)* [1-5]
- Net blotch in ***spring barley*, *winter barley*** [1-5]
- Powdery mildew in ***spring barley*, *winter barley*** [1-5]; ***spring oats*, *spring rye*, *triticale*, *winter oats*, *winter rye*** [1-4]

- Rhynchosporium in ***spring barley*, *winter barley*** [1-5]; ***spring rye*, *winter rye*** [1-4]
- Septoria diseases in ***spring wheat*, *triticale*** [1-4]; ***winter wheat*** [1-5]
- Sooty moulds in ***winter wheat*** *(reduction)* [1-5]
- Yellow rust in ***spring barley*, *winter barley*, *winter wheat*** [1-5]; ***spring rye*, *spring wheat*, *triticale*, *winter rye*** [1-4]

Approval information

- Kresoxim-methyl included in Annex I under EC Directive 91/414

Efficacy guidance

- For best results spray at the start of foliar disease attack and repeat if infection conditions persist
- Optimum effect against eyespot obtained by treatment between leaf sheaths erect and first node detectable stages (GS 30-32)
- For protection against ear diseases apply during ear emergence
- Kresoxim-methyl is a member of the QoI cross resistance group. Product should be used preventatively and not relied on for its curative potential
- Use product as part of an Integrated Crop Management strategy incorporating other methods of control, including where appropriate other fungicides with a different mode of action. Do not apply more than two foliar applications of QoI containing products to any cereal crop
- There is a significant risk of widespread resistance occurring in *Septoria tritici* populations in UK. Failure to follow resistance management action may result in reduced levels of disease control
- Strains of barley powdery mildew resistant to QoIs are common in the UK

Restrictions

- Maximum total dose equivalent to two full dose treatments

Crop-specific information

- Latest use: mid flowering (GS 65) for wheat, rye, triticale; completion of ear emergence (GS 59) for barley, oats

Environmental safety

- Dangerous for the environment
- Very toxic to aquatic organisms
- LERAP Category B
- Avoid spray drift onto neighbouring crops. Product may damage broad-leaved species

Hazard classification and safety precautions

Hazard H03, H11
Risk phrases R40, R50, R58, R62, R63 [1-5]; R43 [5]
Operator protection A [1-5]; H [1-4]; U05a, U14, U20b
Environmental protection E15a, E16a [1-5]; E16b, E31b [5]; E31c, E32d, E38 [1-4]
Storage and disposal E01, E04, E26, E30a [1-5]; E29b [5]
Medical advice M05a [1-4]; M05b [5]

196 epoxiconazole + fenpropimorph + pyraclostrobin

A protectant and systemic fungicide mixture for cereals

Products

Diamant	BASF	42:214:114 g/l	SE	11557

Uses

- Brown rust in ***spring barley*, *spring wheat*, *winter barley*, *winter wheat***
- Crown rust in ***spring oats*, *winter oats***
- Fusarium ear blight in ***spring wheat*** *(reduction)*, ***winter wheat*** *(reduction)*
- Net blotch in ***spring barley*, *winter barley***
- Powdery mildew in ***spring barley*, *spring oats*, *winter barley*, *winter oats***
- Rhynchosporium in ***spring barley*, *winter barley***
- Septoria diseases in ***spring wheat*, *winter wheat***
- Yellow rust in ***spring barley*, *spring wheat*, *winter barley*, *winter wheat***

Efficacy guidance

- For best results spray at the start of foliar disease attack and repeat if infection conditions persist
- For protection against ear diseases apply during ear emergence

- Pyraclostrobin is a member of the QoI cross resistance group. Product should be used preventatively and not relied on for its curative potential
- Use product as part of an Integrated Crop Management strategy incorporating other methods of control, including where appropriate other fungicides with a different mode of action. Do not apply more than two foliar applications of QoI containing products to any cereal crop
- There is a significant risk of widespread resistance occurring in *Septoria tritici* populations in UK. Failure to follow resistance management action may result in reduced levels of disease control
- Strains of barley powdery mildew resistant to QoIs are common in the UK

Restrictions

- Maximum number of treatments 2 per crop

Crop-specific information

- Latest use: up to and including ear emergence just complete (GS 59) for barley, oats; up to and including flowering just complete (GS 69) for wheat

Environmental safety

- Dangerous for the environment
- Very toxic to aquatic organisms
- LERAP Category B
- Avoid spray drift onto neighbouring crops. Product may damage broad-leaved species

Hazard classification and safety precautions

Hazard H03, H11
Risk phrases R20, R22a, R38, R40, R50, R58, R63
Operator protection A; U05a, U14, U20b
Environmental protection E13b, E16a, E16b, E31c, E32d, E34, E38
Storage and disposal E01, E04, E26, E30a
Medical advice M03, M05a

197 epoxiconazole + kresoxim-methyl

A protectant, systemic and curative fungicide mixture for cereals

Products

1	Landmark	BASF	125:125 g/l	SC	11730
2	Me2 KME	Me2	125:125 g/l	SC	09594
3	Standon Kresoxim-Epoxiconazole	Standon	125:125 g/l	SC	09281

Uses

- Brown rust in ***spring barley***, ***winter barley***, ***winter wheat*** [1-3]; ***spring rye***, ***spring wheat***, ***triticale***, ***winter rye*** [1]
- Eyespot in ***spring rye*** *(reduction)*, ***triticale*** *(reduction)*, ***winter oats*** *(reduction)*, ***winter rye*** *(reduction)* [1]; ***winter barley*** *(reduction)*, ***winter wheat*** *(reduction)* [1-3]
- Fusarium ear blight in ***spring wheat*** *(reduction)* [1]; ***winter wheat*** *(reduction)* [1-3]
- Net blotch in ***spring barley***, ***winter barley*** [1-3]
- Powdery mildew in ***spring barley***, ***winter barley*** [1-3]; ***spring oats***, ***spring rye***, ***triticale***, ***winter oats***, ***winter rye*** [1]
- Rhynchosporium in ***spring barley***, ***winter barley*** [1-3]; ***spring rye***, ***winter rye*** [1]
- Septoria diseases in ***spring wheat***, ***triticale*** [1]; ***winter wheat*** [1-3]
- Sooty moulds in ***winter wheat*** *(reduction)* [1-3]
- Yellow rust in ***spring barley***, ***winter barley***, ***winter wheat*** [1-3]; ***spring rye***, ***spring wheat***, ***triticale***, ***winter rye*** [1]

Approval information

- Kresoxim-methyl included in Annex I under EC Directive 91/414

Efficacy guidance

- For best results spray at the start of foliar disease attack and repeat if infection conditions persist
- Optimum effect against eyespot obtained by treatment between leaf sheaths erect and first node detectable stages (GS 30-32)
- For protection against ear diseases apply during ear emergence
- Kresoxim-methyl is a member of the QoI cross resistance group. Product should be used preventatively and not relied on for its curative potential

- Use product as part of an Integrated Crop Management strategy incorporating other methods of control, including where appropriate other fungicides with a different mode of action. Do not apply more than two foliar applications of QoI containing products to any cereal crop
- There is a significant risk of widespread resistance occurring in *Septoria tritici* populations in UK. Failure to follow resistance management action may result in reduced levels of disease control
- Strains of barley powdery mildew resistant to QoIs are common in the UK

Restrictions

- Maximum total dose equivalent to two full dose treatments

Crop-specific information

- Latest use: mid flowering (GS 65) for wheat, rye, triticale; completion of ear emergence (GS 59) for barley, oats

Environmental safety

- Dangerous for the environment
- Very toxic to aquatic organisms
- LERAP Category B
- Avoid spray drift onto neighbouring crops. Product may damage broad-leaved species

Hazard classification and safety precautions

Hazard H03, H11
Risk phrases R40, R50, R58, R62, R63 [1-3]; R43 [2, 3]
Operator protection A; U05a, U14, U20b
Environmental protection E15a, E16a [1-3]; E16b, E31b [2, 3]; E31c, E32d, E38 [1]
Storage and disposal E01, E04, E26, E30a
Medical advice M05a [1]; M05b [2, 3]

198 epoxiconazole + kresoxim-methyl + pyraclostrobin

A protectant, systemic and curative fungicide mixture for cereals

Products

1	Covershield	BASF	50:67:133 g/l	SE	10900
2	Opponent	BASF	50:67:133 g/l	SE	10877

Uses

- Brown rust in ***spring barley, spring wheat, winter barley, winter wheat***
- Crown rust in ***spring oats, winter oats***
- Net blotch in ***spring barley, winter barley***
- Rhynchosporium in ***spring barley, winter barley***
- Septoria diseases in ***spring wheat, winter wheat***
- Yellow rust in ***spring barley, spring wheat, winter barley, winter wheat***

Approval information

- Kresoxim-methyl included in Annex I under EC Directive 91/414

Efficacy guidance

- For best results apply at the start of foliar disease attack
- Best results on Septoria leaf blotch when treated in the latent phase
- Good reduction of Fusarium ear blight can be obtained from treatment during flowering
- Yield response may be obtained in the absence of visual disease symptoms
- Kresoxim-methyl and pyraclostrobin are members of the QoI cross resistance group. Product should be used preventatively and not relied on for its curative potential
- Use product as part of an Integrated Crop Management strategy incorporating other methods of control, including where appropriate other fungicides with a different mode of action. Do not apply more than two foliar applications of QoI containing products to any cereal crop
- There is a significant risk of widespread resistance occurring in *Septoria tritici* populations in UK. Failure to follow resistance management action may result in reduced levels of disease control

Restrictions

- Maximum total dose equivalent to two full dose treatments

Crop-specific information

- Latest use: before grain watery ripe (GS 71) for wheat; up to and including emergence of ear just complete (GS 59) for barley and oats

SEE SECTION 3 FOR PRODUCTS ALSO REGISTERED

Environmental safety

- Dangerous for the environment
- Very toxic to aquatic organisms
- LERAP Category B

Hazard classification and safety precautions

Hazard H03, H11
Risk phrases R20, R22a, R40, R50, R58
Operator protection A; U05a, U14, U20b
Environmental protection E13b, E16a, E16b, E31c, E32d, E34, E38
Storage and disposal E01, E04, E26, E30a
Medical advice M05a

199 epoxiconazole + pyraclostrobin

A protectant, systemic and curative fungicide mixture for cereals

Products

1	Euro	Me2	50:133 g/l	SE	11511
2	Ibex	BASF	50:133 g/l	SE	10901
3	Opera	BASF	50:133 g/l	SE	10876

Uses

- Brown rust in ***spring barley***, ***spring wheat***, ***winter barley***, ***winter wheat*** [1-3]
- Cercospora leaf spot in ***sugar beet*** [2, 3]
- Crown rust in ***spring oats***, ***winter oats*** [1-3]
- Net blotch in ***spring barley***, ***winter barley*** [1-3]
- Powdery mildew in ***sugar beet*** [2, 3]
- Ramularia leaf spots in ***sugar beet*** [2, 3]
- Rhynchosporium in ***spring barley***, ***winter barley*** [1-3]
- Rust in ***sugar beet*** [2, 3]
- Septoria diseases in ***spring wheat***, ***winter wheat*** [1-3]
- Yellow rust in ***spring barley***, ***spring wheat***, ***winter barley***, ***winter wheat*** [1-3]

Efficacy guidance

- For best results apply at the start of foliar disease attack
- Good reduction of Fusarium ear blight can be obtained from treatment during flowering
- Yield response may be obtained in the absence of visual disease symptoms
- Pyraclostrobin is a member of the QoI cross resistance group. Product should be used preventatively and not relied on for its curative potential
- Use product as part of an Integrated Crop Management strategy incorporating other methods of control, including where appropriate other fungicides with a different mode of action. Do not apply more than two foliar applications of QoI containing products to any cereal crop
- There is a significant risk of widespread resistance occurring in *Septoria tritici* populations in UK. Failure to follow resistance management action may result in reduced levels of disease control

Restrictions

- Maximum total dose equivalent to two full dose treatments

Crop-specific information

- Latest use: before grain watery ripe (GS 71) for wheat; up to and including emergence of ear just complete (GS 59) for barley and oats
- HI 6 wk for sugar beet

Environmental safety

- Dangerous for the environment
- Very toxic to aquatic organisms
- LERAP Category B

Hazard classification and safety precautions

Hazard H03, H11
Risk phrases R20, R22a, R38, R40, R50, R58
Operator protection A; U05a, U14, U20b
Environmental protection E15a, E16a, E16b, E31c, E32d, E34, E38

Storage and disposal E01, E04, E26, E30a
Medical advice M03, M05a

200 esfenvalerate

A contact and ingested pyrethroid insecticide

Products

Sumi-Alpha	BASF	25 g/l	EC	10401

Uses
- Aphids in ***winter barley***, ***winter wheat***

Approval information
- Esfenvalerate included in Annex I under EC Directive 91/414

Efficacy guidance
- Crops at high risk (e.g. after grass or in areas with history of BYDV) should be treated when aphids first seen or by mid-Oct. Otherwise treat in late Oct-early Nov
- High risk crops will need a second treatment
- Product also recommended between onset of flowering and milky ripe stages (GS 61-73) for control of summer cereal aphids

Restrictions
- Maximum number of treatments 3 per crop of which 2 may be in autumn
- Do not use if another pyrethroid or dimethoate has been applied to crop after start of ear emergence (GS 51)

Crop-specific information
- Latest use: 31 Mar in yr of harvest (winter use)
- HI 20 d (summer use)

Environmental safety
- Dangerous for the environment
- Very toxic to aquatic organisms
- Extremely dangerous to bees. Do not apply to crops in flower or to those in which bees are actively foraging. Do not apply when flowering weeds are present
- High risk to non-target insects or other arthropods. Do not spray within 6 m of the field boundary
- LERAP Category A
- Store product in dark away from direct sunlight

Hazard classification and safety precautions
Hazard H03, H11
Risk phrases R20, R22a, R38, R41, R43, R50, R58
Operator protection A, C, H; U04a, U05a, U08, U11, U14, U19a, U20b
Environmental protection E12c, E15a, E16c, E16d, E22a, E31c, E34, E38
Storage and disposal E01, E04, E29b, E30a
Medical advice M03

201 ethofumesate

A benzofuran herbicide for grass weed control in various crops

See also bromoxynil + ethofumesate + ioxynil
chloridazon + ethofumesate
desmedipham + ethofumesate + phenmedipham

Products

1	Agriguard Ethofumesate 200	AgriGuard	200 g/l	EC	10698
2	Agriguard Ethofumesate Flo	AgriGuard	500 g/l	SC	09478
3	Ethosat 500	Makhteshim	500 g/l	SC	11501
4	Kubist Flo	Nufarm UK	500 g/l	SC	11167
5	Nortron Flo	Bayer CropScience	500 g/l	SC	08154

Uses

- Annual dicotyledons in ***bulb onions*** *(off-label)* [3]; ***fodder beet, mangels, red beet, sugar beet*** [1-5]; ***garlic*** *(off-label)*, ***horseradish*** *(off-label)*, ***strawberries*** *(off-label)* [3, 5]; ***onions*** *(off-label)* [5]
- Annual grasses in ***garlic*** *(off-label)*, ***horseradish*** *(off-label)*, ***onions*** *(off-label)*, ***strawberries*** *(off-label)* [5]; ***grass seed crops, permanent pasture, rotational grassland*** [1]
- Annual meadow grass in ***bulb onions*** *(off-label)*, ***horseradish*** *(off-label)*, ***strawberries*** *(off-label)* [3]; ***fodder beet, mangels, red beet, sugar beet*** [1-5]; ***garlic*** *(off-label)* [3, 4]; ***onions*** *(off-label)* [4]
- Blackgrass in ***bulb onions*** *(off-label)*, ***garlic*** *(off-label)*, ***horseradish*** *(off-label)*, ***strawberries*** *(off-label)* [3]; ***fodder beet, mangels, red beet, sugar beet*** [1-5]; ***grass seed crops, permanent pasture, rotational grassland*** [1]
- Chickweed in ***garlic*** *(off-label)*, ***onions*** *(off-label)* [4]; ***grass seed crops, permanent pasture, rotational grassland*** [1]
- Cleavers in ***grass seed crops, permanent pasture, rotational grassland*** [1]
- Clover in ***strawberries*** *(off-label)* [4]
- Fat hen in ***poppies*** *(off-label - for morphine production)* [5]
- General weed control in ***horseradish*** *(off-label)* [4]
- Grass weeds in ***poppies*** *(off-label - for morphine production)* [5]; ***strawberries*** *(off-label)* [4]
- Mayweeds in ***poppies*** *(off-label - for morphine production)* [5]

Specific Off-Label Approvals (SOLAs)

- ***bulb onions, garlic, horseradish, strawberries*** *(OLA 040153) Dec 2008* [3]
- ***garlic, horseradish, onions, strawberries*** *(OLA 001916) Dec 2008* [5]
- ***garlic, horseradish, onions, strawberries*** *(OLA 032082) Dec 2008* [4]
- ***poppies*** *(for morphine production) (OLA 040470) Dec 2008* [5]

Approval information

- Ethofumesate included in Annex I under EC Directive 91/414

Efficacy guidance

- Most products may be applied pre- or post-emergence of crop or weeds but some restricted to pre-emergence or post-emergence use only. Check label
- Volunteer cereals not well controlled pre-emergence, weed grasses should be sprayed before fully tillered
- Grass crops may be sprayed during rain or when wet. Not recommended in very dry conditions or prolonged frost

Restrictions

- Maximum number of treatments 1 pre- plus 1 or 2 post-emergence per crop or yr for beet crops and grass seed crops; 1 per crop for horseradish, strawberries; 2 per yr for amenity turf and established grassland
- Do not use on Sands or Heavy soils, Very Light soils containing a high percentage of stones, or soils with more than 5-10% organic matter (percentage varies according to label)
- Do not use on swards reseeded without ploughing
- Clovers will be killed or severely checked
- Do not graze or cut grass for 14 d after, or roll less than 7 d before or after spraying

Crop-specific information

- Latest use: before crops meet across rows for beet crops and mangels; 14 d before cutting or grazing for grass leys; pre-emergence for horseradish
- HI strawberries 11 wk; garlic, onions 3 mth [5]
- Apply in beet crops in tank mixes with other pre- or post-emergence herbicides. Recommendations vary for different mixtures. See label for details
- Safe timing on beet crops varies with other ingredient of tank mix. See label for details
- In grass crops apply to moist soil as soon as possible after sowing or post-emergence when crop in active growth, normally mid-Oct to mid-Dec. See label for details
- May be used in Italian, hybrid and perennial ryegrass, timothy, cocksfoot, meadow fescue and tall fescue. Apply pre-emergence to autumn-sown leys, post-emergence after 2-3 leaf stage. See label for details

Following crops guidance

- Any crop may be sown 3 mth after application of mixtures in beet crops following ploughing, 5 mth after application in grass crops

Environmental safety

- Dangerous for the environment
- Toxic to aquatic organisms
- Do not empty into drains

Hazard classification and safety precautions

Hazard H03, H08 [1]; H11 [1-5]
Risk phrases R20, R21, R22b, R36, R38 [1]; R51, R58 [3-5]
Operator protection A, C, H [1]; U05a [1]; U08 [1-3]; U09a [4, 5]; U19a, U20a [1-5]
Environmental protection E15a, E31b [1-4]; E19b [3]; E32d [3-5]; E34 [1, 3]; E38 [4, 5]
Storage and disposal E01, E04 [1]; E26, E30a [1-5]
Medical advice M03 [1]

202 ethofumesate + lenacil + phenmedipham

A contact and residual herbicide mixture for beet crops

Products

Agricola Lens Plus	Interfarm	65:50:71.3 g/l	EC	11587

Uses

- Annual dicotyledons in ***fodder beet***, ***mangels***, ***sugar beet***

Approval information

- Ethofumesate included in Annex I under EC Directive 91/414

Efficacy guidance

- Best results obtained from a programme of 'low dose' applications commencing when earliest germinating weeds have reached cotyledon stage
- Further treatments must be applied as each flush of weeds reaches cotyledon stage and before the largest weeds grow beyond the first true leaf stage, but allowing a minimum of 5 d between each spray
- Some species may not be completely killed until two or more applications have been made
- Control of some weeds can be improved by recommended tank mixtures. See label

Restrictions

- Maximum number of treatments 3 per crop for beet crops
- Do not spray when the temperature is above 21°C. In hot weather spray in the evening when temperatures begin to fall
- Do not treat crops under stress from whatever cause or under conditions of high light intensity
- Do not treat where rolling or harrowing has been carried out within 7 d before application

Crop-specific information

- Latest use: before crop leaves meet between rows for beet crops

Following crops guidance

- Any crop may follow normal harvest of a treated crop provided 4 mth have elapsed since treatment and the land has been ploughed to 15 cm and cultivated
- In the event of failure of a treated crop the land should be ploughed to 15 cm and then only sugar beet, fodder beet or mangels may be drilled until 4 mth have elapsed since treatment

Environmental safety

- Dangerous for the environment
- Toxic to aquatic organisms

Hazard classification and safety precautions

Hazard H03, H11
Risk phrases R21, R22a, R22b, R36, R37, R40, R41, R43, R51, R58
Operator protection A, C; U02a, U05a, U08, U11, U14, U19a, U20b
Environmental protection E15a, E31b, E34
Storage and disposal E01, E04, E30a
Medical advice M04, M05b

203 ethofumesate + metamitron

A contact and residual herbicide mixture for beet crops

Products

1	Galahad	Bayer CropScience	100:400 g/l	SC	10727
2	Torero	Makhteshim	150:350 g/l	SC	11158

Uses

- Annual dicotyledons in ***fodder beet***, ***sugar beet*** [1, 2]; ***mangels*** [1]
- Annual meadow grass in ***fodder beet***, ***sugar beet*** [1, 2]; ***mangels*** [1]

Approval information

- Ethofumesate included in Annex I under EC Directive 91/414

Efficacy guidance

- Best results obtained from a series of treatments applied as an overall fine spray commencing when earliest germinating weeds are no larger than fully expanded cotyledon and the majority of the crop at fully expanded cotyledon
- Apply subsequent sprays as each new flush of weeds reaches early cotyledon and continue until weed emergence ceases (maximum 3 sprays)
- Product may be used on all soil types but residual activity may be reduced on those with more than 5% organic matter

Restrictions

- Maximum number of treatments three per crop [2]
- Maximum total dose equivalent to three full dose treatments [1]

Crop-specific information

- Latest use: before crop leaves meet between rows
- Crop tolerance may be reduced by stress caused by growing conditions, effects of pests, disease or other pesticides, nutrient deficiency etc

Following crops guidance

- Beet crops may be sown at any time after treatment. Any other crop may be sown after mouldboard ploughing to 15 cm and a minimum interval of 3 mth after treatment

Environmental safety

- Dangerous for the environment
- Very toxic to aquatic organisms
- Do not empty into drains

Hazard classification and safety precautions

Hazard H03 [1]; H11 [1, 2]
Risk phrases R22a, R43, R50 [1]; R51 [2]; R58 [1, 2]
Operator protection A, H [1]; U05a, U08, U19a [1, 2]; U14, U15, U20a [2]; U20b [1]
Environmental protection E13c, E38 [1]; E15a, E19b [2]; E31b, E32d, E34 [1, 2]
Storage and disposal E01, E04, E26, E30a
Medical advice M03 [1]

204 ethofumesate + phenmedipham

A contact and residual herbicide for use in beet crops

Products

1	Powertwin	Makhteshim	200:200 g/l	SC	11187
2	Thunder	Makhteshim	200:200 g/l	SC	11186
3	Twin	Makhteshim	94:97 g/l	EC	11172

Uses

- Annual dicotyledons in ***fodder beet***, ***mangels***, ***sugar beet***
- Annual meadow grass in ***fodder beet***, ***mangels***, ***sugar beet***
- Blackgrass in ***fodder beet***, ***mangels***, ***sugar beet***

Approval information

- Ethofumesate included in Annex I under EC Directive 91/414

Efficacy guidance

- Best results achieved by repeat applications to cotyledon stage weeds. Larger susceptible weeds not killed by first treatment usually checked and controlled by second application
- Apply on all soil types at 7-10 d intervals
- On soils with more than 5-10% organic matter residual activity may be reduced

Restrictions

- Maximum number of treatments normally 3 per crop or maximum total dose equivalent to three full dose treatments - see labels for details
- Do not spray wet foliage or if rain imminent
- Spray must be applied low volume. See label for details
- Spray in evening if daytime temperatures above 21°C expected
- Avoid or delay treatment if frost expected within 7 d
- Avoid or delay treating crops under stress from wind damage, manganese or lime deficiency, pest or disease attack etc

Crop-specific information

- Latest use: before crop foliage meets in the rows
- Check from which recovery may not be complete may occur if treatment made during conditions of sharp diurnal temperature fluctuation

Following crops guidance

- Beet crops may be sown at any time after treatment. Any other crop may be sown after mouldboard ploughing to 15 cm and a minimum interval of 3 mth after treatment

Environmental safety

- Dangerous for the environment
- Toxic to aquatic organisms
- Do not empty into drains
- Extra care necessary to avoid drift because product is recommended for use as a fine spray

Hazard classification and safety precautions

Hazard H03 [3]; H04 [1, 2]; H11 [1-3]
Risk phrases R37, R40 [3]; R43, R51, R58 [1-3]
Operator protection A [1-3]; H [1, 2]; U05a, U08, U19a, U20a [1-3]; U14 [1, 2]
Environmental protection E15a, E19b, E31b, E32d, E34
Storage and disposal E01, E04, E30a [1-3]; E26 [1]
Medical advice M05a

205 ethoprophos

An organophosphorus nematicide and insecticide

Products

Mocap 10G	Bayer CropScience	10% w/w	GR	10003

Uses

- Potato cyst nematode in ***potatoes***
- Wireworm in ***potatoes***

Efficacy guidance

- Broadcast shortly before or during final soil preparation with suitable fertilizer spreader and incorporate immediately to 10-15 cm. Deeper incorporation may reduce efficacy. See label for details
- Treatment can be applied on all soil types. Control of pests reduced on organic soils
- Effectiveness dependent on soil moisture. Drought after application may reduce control
- Where pre-planting nematode populations are high, loss of crop vigour and subsequent loss of yield may occur in spite of treatment

Restrictions

- This product contains an anticholinesterase organophosphorous compound. Do not use if under medical advice not to work with such compounds
- Maximum number of treatments 1 per crop
- Do not harvest crops for human or animal consumption for at least 8 wk after application

SEE SECTION 3 FOR PRODUCTS ALSO REGISTERED

- Vehicles with a closed cab must be used when applying and incorporating
- Do not apply by hand or hand held equipment

Crop-specific information

- Latest use: pre-planting of crop
- Field scale application must only be through positive displacement type specialist granule applicators or dual purpose fertilizer applicators with microgranule setting

Environmental safety

- Dangerous for the environment
- Toxic to aquatic organisms
- Dangerous to livestock. Keep all livestock out of treated areas/away from treated water for at least 13 wk. Bury or remove spillages
- Dangerous to game, wild birds and animals
- Product supplied in returnable/refillable containers. Follow instructions on label

Hazard classification and safety precautions

Hazard H02, H11
Risk phrases R22a, R23, R43, R51, R58
Operator protection A, D, H; U02a, U04a, U05a, U07, U08, U13, U14, U19a, U20a
Environmental protection E06b (13 wk); E10a, E13b, E32d, E33, E34, E36, E38
Consumer protection C02 (8 wk)
Storage and disposal E01, E04, E30a
Medical advice M01, M03, M04

206 ethylene (commodity substance)

A gas used for fruit ripening and potato storage

Products

ethylene	various	-	GA

Uses

- Fruit ripening in ***fruit crops*** *(in store)*
- Sprout suppression in ***potatoes*** *(in store)*

Approval information

- Ethylene is a supported substance in the fourth stage of the EC Review Programme. If ethylene is included in Annex I under Directive 91/414, any commodity chemical approvals will no longer apply

Efficacy guidance

- For use in stored fruit or potatoes after harvest

Restrictions

- Handling and release of ethylene must only be undertaken by operators suitably trained and competent to carry out the work
- Operators must vacate treated areas immediately after ethylene introduction
- Unprotected persons must be excluded from the treated areas until atmospheres have been thoroughly ventilated for 15 minutes minimum before re-entry
- Ambient atmospheric ethylene concentration must not exceed 1000 ppm. Suitable self-contained breathing apparatus must be worn in atmospheres containing ethylene in excess of 1000 ppm
- A minimum 3 d post treatment period is required before removal of treated crop from storage
- Ethylene treatment must only be undertaken in fully enclosed storage areas that are air tight with appropriate air circulation and venting facilities

207 etridiazole

A protective thiadiazole fungicide for soil or compost incorporation

Products

1	Standon Etridiazole 35	Standon	35% w/w	WP	08778
2	Terrazole 35 WP	Scotts	35% w/w	WP	10468

Uses

- Damping off and foot rot in ***cabbages***, ***cauliflowers*** *(seeds and seedlings)*, ***celery*** *(seeds and seedlings)*, ***cucumbers*** *(seeds and seedlings)*, ***mustard and cress***, ***ornamental plant production*** *(seeds and seedlings)*, ***outdoor tomatoes*** *(seeds and seedlings)*, ***protected tomatoes*** *(seeds and seedlings)* [1, 2]
- Phytophthora in ***watercress*** *(off-label)* [1]
- Phytophthora root rot in ***cabbages***, ***cauliflowers*** *(transplants)*, ***celery*** *(transplants)*, ***cucumbers*** *(transplants)*, ***hardy ornamental nursery stock***, ***ornamental plant production*** *(rooted cuttings and transplants)*, ***outdoor tomatoes*** *(transplants)*, ***protected tomatoes*** *(transplants)*, ***tulips*** [1, 2]
- Phytophthora wilt in ***cabbages***, ***cauliflowers*** *(transplants)*, ***celery*** *(transplants)*, ***cucumbers*** *(transplants)*, ***hardy ornamental nursery stock***, ***ornamental plant production*** *(rooted cuttings and transplants)*, ***outdoor tomatoes*** *(transplants)*, ***protected tomatoes*** *(transplants)*, ***tulips*** [1, 2]
- Pythium in ***watercress*** *(off-label)* [1]
- Root diseases in ***inert substrate tomatoes*** *(off-label)* [1]

Specific Off-Label Approvals (SOLAs)

- ***inert substrate tomatoes***, ***watercress*** *(OLA 020409) Dec 2008* [1]

Efficacy guidance

- Best results obtained when used as prophylactic treatment
- Compost/soil incorporation more effective than when applied as drench
- After drenching wash spray residue from crop foliage
- When treating compost for blocking reduce dose by 50% to allow for compaction
- Uniform distribution and thorough incorporation essential for best results. Dilute powder with dry sand to facilitate handling
- Treat compost as soon as possible before use
- Do not apply to wet soil

Restrictions

- Maximum number of treatments 1 per crop or yr for compost incorporation. No restrictions on drenches
- Do not use on *Escallonia, Pyracantha, Gloxinia* spp., pansies or lettuces
- Do not drench seedlings until well established
- When using compost containing more than 20% inert material or a high proportion of young peat, check crop safety before widespread use
- Test on small numbers of plants in advance when treating subjects of unknown susceptibility or using compost with more than 20% inert material

Crop-specific information

- Latest use: normally before sowing or transplanting. 24 h after seeding watercress
- HI tomatoes, cucumbers, mustard and cress 3 d; inert substrate tomatoes 24 h
- Germination of lettuce in previously treated soil may be impaired

Environmental safety

- Dangerous for the environment
- Very toxic to aquatic organisms

Hazard classification and safety precautions

Hazard H03, H11
Risk phrases R22a, R43 [2]; R36, R40, R50, R58 [1, 2]; R38 [1]
Operator protection A, C; U02a, U04a, U05a, U08, U20b [1, 2]; U11, U14, U15, U19a [2]
Environmental protection E15a, E32a
Storage and disposal E01, E04, E30a
Medical advice M04 [2]

208 famoxadone

A strobilurin fungicide available only in mixtures

See also *cymoxanil + famoxadone*

209 famoxadone + flusilazole

A contact, preventative and curative fungicide mixture for cereals

Products

1	Charisma	DuPont	100:106.7 g/l	EC	10415
2	Medley	DuPont	100:106.7 g/l	EC	10933

Uses

- Brown rust in ***spring barley***, ***winter barley***, ***winter wheat***
- Net blotch in ***spring barley***, ***winter barley***
- Rhynchosporium in ***spring barley***, ***winter barley***
- Septoria diseases in ***winter wheat***
- Yellow rust in ***spring barley***, ***winter barley***, ***winter wheat***

Approval information

- Famoxadone included in Annex I under EC Directive 91/414

Efficacy guidance

- Best results obtained from application at early stage of disease development before infection spreads to new growth
- Famoxadone is a member of the QoI cross resistance group. Product should be used preventatively and not relied on for its curative potential
- Use product as part of an Integrated Crop Management strategy incorporating other methods of control, including where appropriate other fungicides with a different mode of action. Do not apply more than two foliar applications of QoI containing products to any cereal crop
- There is a significant risk of widespread resistance occurring in *Septoria tritici* populations in UK. Failure to follow resistance management action may result in reduced levels of disease control

Restrictions

- Maximum total dose equivalent to two full dose treatments in wheat and barley
- Do not apply to crops under stress
- Do not apply during frosty weather

Crop-specific information

- Latest use: before flowering (GS 60) for winter wheat; before quarter ear emergence (GS 53) for barley

Environmental safety

- Dangerous for the environment
- Very toxic to aquatic organisms
- LERAP Category B

Hazard classification and safety precautions

Hazard H02, H11
Risk phrases R36, R40, R50, R58, R61
Operator protection A, C; U05a, U19a, U20b
Environmental protection E15a, E16a, E16b, E31b, E32d, E38
Storage and disposal E01, E04, E26, E30a
Medical advice M04

210 fatty acids

A soap concentrate insecticide and acaricide

Products

Savona	Koppert	49% w/w	SL	06057

Uses

- Aphids in ***broad beans***, ***brussels sprouts***, ***cabbages***, ***cucumbers***, ***fruit trees***, ***lettuce***, ***outdoor tomatoes***, ***peas***, ***peppers***, ***protected tomatoes***, ***pumpkins***, ***runner beans***, ***woody ornamentals***
- Mealybugs in ***broad beans***, ***brussels sprouts***, ***cabbages***, ***cucumbers***, ***fruit trees***, ***lettuce***, ***outdoor tomatoes***, ***peas***, ***peppers***, ***protected tomatoes***, ***pumpkins***, ***runner beans***, ***woody ornamentals***

- Scale insects in ***broad beans, brussels sprouts, cabbages, cucumbers, fruit trees, lettuce, outdoor tomatoes, peas, peppers, protected tomatoes, pumpkins, runner beans, woody ornamentals***
- Spider mites in ***broad beans, brussels sprouts, cabbages, cucumbers, fruit trees, lettuce, outdoor tomatoes, peas, peppers, protected tomatoes, pumpkins, runner beans, woody ornamentals***
- Whitefly in ***broad beans, brussels sprouts, cabbages, cucumbers, fruit trees, lettuce, outdoor tomatoes, peas, peppers, protected tomatoes, pumpkins, runner beans, woody ornamentals***

Efficacy guidance

- Use only soft or rain water for diluting spray
- Pests must be sprayed directly to achieve any control. Spray all plant parts thoroughly to run off
- For glasshouse use apply when insects first seen and repeat as necessary. For scale insects apply several applications at weekly intervals after egg hatch
- To control whitefly spray when required and use biological control after 12 h

Restrictions

- Do not use on new transplants, newly rooted cuttings or plants under stress
- Do not use on specified susceptible shrubs. See label for details

Crop-specific information

- HI zero

Environmental safety

- Harmful to fish or other aquatic life. Do not contaminate surface waters or ditches with chemical or used container

Hazard classification and safety precautions

Operator protection U20c
Environmental protection E13c, E31a
Storage and disposal E26, E30a

211 fenamidone

A strobilurin fungicide available only in mixtures

212 fenamidone + mancozeb

A fungicide mixture for potatoes

Products

Sonata	Bayer CropScience	10:50% w/w	WG	11570

Uses

- Blight in ***potatoes***

Approval information

- Fenamidone included in Annex I under EC Directive 91/414

Efficacy guidance

- Apply as soon as there is risk of blight infection or immediately after a blight warning
- Fenamidone is a member of the QoI cross resistance group. To minimise the likelihood of development of resistance these products should be used in a planned Resistance Management strategy. In addition before application consult and adhere to the latest FRAG-UK resistance guidance on application of QoI fungicides to potatoes. See Section 5 for more information

Restrictions

- Maximum number of treatments (including any other QoI containing fungicide) 6 per yr but no more than three should be applied consecutively. Use in alternation with fungicides from a different cross-resistance group
- Observe a 7 d interval between treatments
- Use only as a protective treatment. Do not use when blight has become readily visible (1% leaf area destroyed)

Crop-specific information

- HI 7 d for potatoes

Environmental safety

- Dangerous for the environment
- Very toxic to aquatic organisms
- Risk to certain non-target insects or other arthropods. See directions for use
- LERAP Category B

Hazard classification and safety precautions

Hazard H04, H11
Risk phrases R36, R37, R50, R58
Operator protection D, E, H; U05a, U08, U11, U19a, U20a
Environmental protection E13b, E16a, E22b, E31b, E32d, E34, E38
Storage and disposal E01, E04, E26, E30a

213 fenamidone + propamocarb hydrochloride

A systemic fungicide mixture for potatoes

Products

Consento	Bayer CropScience	75:375 g/l	SC	11889

Uses

- Blight in ***potatoes***

Approval information

- Fenamidone included in Annex I under EC Directive 91/414

Efficacy guidance

- Apply as soon as there is risk of blight infection or immediately after a blight warning
- Fenamidone is a member of the QoI cross resistance group. To minimise the likelihood of development of resistance these products should be used in a planned Resistance Management strategy. In addition before application consult and adhere to the latest FRAG-UK resistance guidance on application of QoI fungicides to potatoes. See Section 5 for more information

Restrictions

- Maximum number of treatments (including any other QoI containing fungicide) 6 per yr but no more than three should be applied consecutively. Use in alternation with fungicides from a different cross-resistance group
- Observe a 7 d interval between treatments
- Use only as a protective treatment. Do not use when blight has become readily visible (1% leaf area destroyed)

Crop-specific information

- HI 7 d for potatoes

Environmental safety

- Dangerous for the environment
- Very toxic to aquatic organisms. May cause long-term adverse effects in the aquatic environment
- LERAP Category B

Hazard classification and safety precautions

Hazard H04, H11
Risk phrases R36, R50, R58
Operator protection A, C; U05a, U08, U11, U20a
Environmental protection E13b, E16a, E31b, E32d, E38
Storage and disposal E01, E04, E30a

214 fenarimol

A systemic curative and protective pyrimidine fungicide

Products

1	Rimidin	Rigby Taylor	120 g/l	SC	05907
2	Rubigan	Gowan	120 g/l	SC	11069

Uses

- Dollar spot in ***managed amenity turf*** [1]
- Fusarium patch in ***managed amenity turf*** [1]
- Powdery mildew in ***apples***, ***blackcurrants***, ***gooseberries***, ***marrows*** *(off-label)*, ***protected courgettes*** *(off-label)*, ***protected gherkins*** *(off-label)*, ***protected peppers*** *(off-label)*, ***protected tomatoes*** *(off-label)*, ***pumpkins*** *(off-label)*, ***raspberries***, ***roses***, ***squashes*** *(off-label)*, ***strawberries*** [2]
- Red thread in ***managed amenity turf*** [1]
- Scab in ***apples*** [2]

Specific Off-Label Approvals (SOLAs)

- ***marrows***, ***pumpkins***, ***squashes*** *(OLA 040673) Dec 2008* [2]
- ***protected courgettes***, ***protected gherkins*** *(OLA 040672) Dec 2008* [2]
- ***protected peppers***, ***protected tomatoes*** *(OLA 040674) Dec 2008* [2]

Efficacy guidance

- For turf disease control spray as preventive treatment and repeat as necessary. See label for details

Restrictions

- Maximum number of treatments 15 per yr on apples, 4 per yr on managed amenity turf, 3 per yr on raspberries and protected vegetables
- Do not mow within 24 h after treatment

Crop-specific information

- HI 14 d for apples, blackcurrants, gooseberries, raspberries, strawberries; 7 d for marrows, protected courgettes, protected gherkins, pumpkins, squashes; 2 d for protected peppers, protected tomatoes,

Environmental safety

- Harmful to aquatic organisms

Hazard classification and safety precautions

Hazard H03
Risk phrases R52, R62 [1, 2]; R58, R61 [2]; R63, R64 [1]
Operator protection A, C; U08, U19a [1, 2]; U20c [1]
Environmental protection E15a, E31b
Consumer protection C02 [2] (14 d)
Storage and disposal E01, E30a
Medical advice M05b [2]

215 fenbuconazole

A systemic protectant and curative triazole fungicide for top fruit and grapevines

Products

Indar 5EW	Landseer	50 g/l	EW	09518

Uses

- Blossom wilt in ***cherries*** *(off-label)*, ***mirabelles*** *(off-label)*, ***plums*** *(off-label)*
- Brown rot in ***cherries*** *(off-label)*, ***mirabelles*** *(off-label)*, ***plums*** *(off-label)*
- Powdery mildew in ***apples*** *(reduction)*, ***grapevines*** *(off-label)*, ***pears*** *(reduction)*
- Scab in ***apples***, ***pears***

Specific Off-Label Approvals (SOLAs)

- ***cherries***, ***mirabelles***, ***plums*** *(OLA 031372) Dec 2008* [1]
- ***grapevines*** *(OLA 032081) Dec 2008* [1]

Efficacy guidance

- Most effective when used as part of a routine preventative programme from bud burst to onset of petal fall
- After petal fall, tank mix with other protectant fungicides to enhance scab control
- See label for recommended spray intervals. In periods of rapid growth or high disease pressure, a 7 d interval should be used

Restrictions

- Maximum total dose on top fruit equivalent to ten full doses per yr on apples, pears; six full doses on grapevines; five full doses on plums; four full doses on cherries, mirabelles
- Consult processors before using on pears for processing
- Do not harvest for human or animal consumption for at least 4 wk after last application

Crop-specific information

- HI 28 d for apples, pears; 21 d for grapevines; 3 d for cherries, mirabelles, plums
- Safe to use on all main commercial varieties of apples and pears in UK

Environmental safety

- Dangerous for the environment
- Toxic to aquatic organisms

Hazard classification and safety precautions

Hazard H04, H11
Risk phrases R36, R41, R51, R58
Operator protection A, C; U05a, U08, U11, U20a
Environmental protection E15a, E32a, E32d, E38
Consumer protection C02 (4 wk)
Storage and disposal E01, E04, E26

216 fenbutatin oxide

A selective contact and ingested organotin acaricide

Products

Torq	Fargro	50% w/w	WP	08370

Uses

- Spider mites in ***protected peppers*** *(off-label)*
- Two-spotted spider mite in ***protected cucumbers***, ***protected ornamentals***, ***protected tomatoes***, ***tunnel grown strawberries***

Specific Off-Label Approvals (SOLAs)

- ***protected peppers*** *(OLA 022124) Dec 2008* [1]

Efficacy guidance

- Active on larvae and adult mites. Spray may take 7-10 d to effect complete kill but mites cease feeding and crop damage stops almost immediately
- Apply as soon as mites first appear and repeat as necessary
- On tunnel-grown strawberries apply when mites first appear, usually before flowering starts, and repeat 10-14 d later. A post-harvest spray is also recommended

Restrictions

- Maximum number of treatments 2 per crop pre-harvest + 1 post harvest for tunnel grown strawberries; 2 per crop for protected peppers
- Allow at least 10 d between spray applications and do not apply within 10 d of a previous spray
- Do not add wetters or mix with anything other than water
- Do not use white petroleum oil within 28 d of treatment, or any other pesticide within 7 d
- Do not apply to crops which are under stress for any reason
- On subjects of unknown susceptibility test treat on a small number of plants in advance

Crop-specific information

- HI glasshouse cucumbers, tomatoes, peppers 3 d; tunnel-grown strawberries 7 d

Environmental safety

- Dangerous for the environment
- Very toxic to aquatic organisms
- Suitable for use in IPM programmes with *Encarsia* or *Phytoseiulus* being used for biological control

Hazard classification and safety precautions

Hazard H01, H11
Risk phrases R26, R36, R38, R50, R58
Operator protection A, C, D, H, M; U05a, U08, U10, U19a, U20a
Environmental protection E15a, E32a, E32d, E34, E38

Storage and disposal E01, E04, E30a
Medical advice M04

217 fenhexamid

A protectant fungicide for soft fruit

Products

Teldor	Bayer CropScience	50% w/w	WG	11229

Uses

- Botrytis in ***blackberries***, ***blackcurrants***, ***cherries*** *(off-label)*, ***gooseberries***, ***grapevines***, ***loganberries***, ***plums*** *(off-label)*, ***raspberries***, ***redcurrants***, ***rubus hybrids***, ***strawberries***, ***whitecurrants***

Specific Off-Label Approvals (SOLAs)

- ***cherries***, ***plums*** *(OLA 031866) May 2011* [1]

Approval information

- Fenhexamid included in Annex I under EC Directive 91/414

Efficacy guidance

- Use as part of a programme of sprays throughout the flowering period to achieve effective control of Botrytis
- To minimise possibility of development of resistance, no more than two sprays of the product may be applied consecutively. Other fungicides from a different chemical group should then be used for at least two consecutive sprays. If only two applications are made on grapevines, only one may include fenhexamid
- Complete spray cover of all flowers and fruitlets throughout the blossom period is essential for successful control of Botrytis
- Spray programmes should normally start at the start of flowering

Restrictions

- Maximum number of treatments 2 per yr on grapevines; 4 per yr on other listed crops but no more than 2 sprays may be applied consecutively

Crop-specific information

- HI 1 d for strawberries, raspberries, loganberries, blackberries, Rubus hybrids; 3 d for cherries; 7 d for blackcurrants, redcurrants, whitecurrants, gooseberries; 21d for outdoor grapes

Environmental safety

- Dangerous for the environment

Hazard classification and safety precautions

Hazard H11
Risk phrases R58
Operator protection U08, U19a, U20b
Environmental protection E13c, E32a, E32d, E38
Storage and disposal E26, E30a

218 fenhexamid + tolylfluanid

A protectant fungicide mixture for soft fruit

Products

Talat	Certis	16.7:33.3% w/w	WG	11311

Uses

- Grey mould in ***blackberries***, ***blackcurrants***, ***gooseberries***, ***loganberries***, ***raspberries***, ***redcurrants***, ***rubus hybrids***, ***strawberries***, ***whitecurrants***

Approval information

- Fenhexamid included in Annex I under EC Directive 91/414

Efficacy guidance

- Effective control of grey mould in soft fruit requires a programme of treatment throughout the blossom period

- Complete even spray coverage of all flowers and developing fruitlets is essential for good results
- To minimise the likelihood of development of resistance use in a planned Resistance Management strategy. See Section 1 for more details

Restrictions
- Maximum total dose equivalent to four full dose treatments for strawberries, raspberries, loganberries, blackberries and other Rubus hybrids, and to two full dose treatments for currants and gooseberries
- Do not apply more than two treatments consecutively (including other products containing fenhexamid or tolylfluanid)
- Do not treat crops growing under glass or polythene
- Consult processor before using on crops for processing

Crop-specific information
- HI 14 days for blackberries, loganberries, raspberries, Rubus hybrids, strawberries; 21 days for currants, gooseberries

Environmental safety
- Dangerous for the environment
- Very toxic to aquatic organisms
- Risk to certain non-target insects or other arthropods. See directions for use
- LERAP Category B
- Broadcast air-assisted LERAP (7.5 m)

Hazard classification and safety precautions

Hazard H04, H11
Risk phrases R43, R50, R58
Operator protection A, H, M; U02a, U05a, U10, U14, U20b
Environmental protection E13b, E16a, E16b, E32a, E32d, E38; E17b (7.5 m)
Storage and disposal E01, E04, E26, E30a

219 fenoxaprop-P-ethyl

A phenoxypropionic acid herbicide for use in wheat

See also diclofop-methyl + fenoxaprop-P-ethyl

Products

1	Cheetah Super	Bayer CropScience	55 g/l	EW	08723
2	Triumph	Bayer CropScience	120 g/l	EC	10902

Uses
- Blackgrass in ***spring wheat***, ***winter wheat***
- Canary grass in ***spring wheat***, ***winter wheat***
- Rough meadow grass in ***spring wheat***, ***winter wheat***
- Wild oats in ***spring wheat***, ***winter wheat***

Efficacy guidance
- Treat weeds from 2 fully expanded leaves up to flag leaf ligule just visible; for awned canary-grass and rough meadow-grass from 2 leaves to the end of tillering
- A second application may be made in spring where susceptible weeds emerge after an autumn application
- Spray is rainfast 1 h after application
- Dry conditions resulting in moisture stress may reduce effectiveness
- Always follow WRAG guidelines for preventing and managing herbicide resistant weeds. Section 5 for more information

Restrictions
- Maximum total dose equivalent to one or two full dose treatments depending on product used
- Do not apply to barley, durum wheat, undersown crops or crops to be undersown
- Do not roll or harrow within 1 wk of spraying
- Do not spray crops under stress, suffering from drought, waterlogging or nutrient deficiency or those grazed or if soil compacted

- Avoid spraying immediately before or after a sudden drop in temperature or a period of warm days/cold nights
- Do not mix with hormone weedkillers

Crop-specific information

- Latest use: before flag leaf sheath extending (GS 41)
- Treat from crop emergence to flag leaf fully emerged (GS 41).
- Product may be sprayed in frosty weather provided crop hardened off but do not spray wet foliage or leaves covered with ice
- Broadcast crops should be sprayed post-emergence after plants have developed well-established root system

Environmental safety

- Dangerous for the environment [1, 2]
- Very toxic (toxic [1]) to aquatic organisms

Hazard classification and safety precautions

Hazard H04 [2]; H11 [1, 2]
Risk phrases R36, R38, R50 [2]; R51 [1]; R58 [1, 2]
Operator protection A, C, H; U05a, U20b [1, 2]; U08 [2]; U09a [1]
Environmental protection E13c, E31b, E32d, E38
Storage and disposal E01, E04, E30a

220 fenoxycarb

An insect specific growth regulator for top fruit

Products

Insegar WG	Syngenta	25% w/w	WG	09789

Uses

- Codling moth in ***apples***
- Summer-fruit tortrix moth in ***apples***, ***pears***

Efficacy guidance

- Best results from application at 5th instar stage before pupation. Product prevents transformation from larva to pupa
- Correct timing best identified from pest warnings
- Because of mode of action rapid knock-down of pest is not achieved and larvae continue to feed for a period after treatment
- Adequate water volume necessary to ensure complete coverage of leaves

Restrictions

- Maximum number of treatments 2 per crop
- Consult processors before use

Crop-specific information

- HI 42 d for apples, pears
- Use on all varieties of apples and pears

Environmental safety

- Dangerous for the environment
- Toxic to aquatic organisms
- High risk to bees. Do not apply to crops in flower or to those in which bees are actively foraging. Do not apply when flowering weeds are present
- Risk to certain non-target insects or other arthropods. See directions for use
- Broadcast air-assisted LERAP (8 m)
- Apply to minimise off-target drift to reduce effects on non-target organisms. Some margin of safety to beneficial arthropods is indicated.

Hazard classification and safety precautions

Hazard H11
Risk phrases R51, R58
Operator protection A, C, H; U05a, U20a, U23a
Environmental protection E12a, E15a, E22b, E32a, E32d, E38; E17b (8 m)
Storage and disposal E01, E04, E30a

SEE SECTION 3 FOR PRODUCTS ALSO REGISTERED

221 fenpropathrin

A contact and ingested pyrethroid acaricide and insecticide

Products

Meothrin	BASF	100 g/l	EC	10400

Uses

- Blackcurrant gall mite in ***blackcurrants***
- Blackcurrant leaf midge in ***blackcurrants***
- Capsids in ***blackcurrants***
- Currant sawfly in ***blackcurrants***
- Two-spotted spider mite in ***blackcurrants***

Approval information

- Products containing this active ingredient have been granted derogations for specified 'Essential Uses' for use until 31 December 2007. Sale and supply must cease by 30 June 2007 but growers have no guarantee that the products will continue to be available until then.
  For more information see 'The Review Programme' under 'Pesticide Legislation' in Section 5

Efficacy guidance

- Acts on motile stages of mites and gives rapid kill

Restrictions

- Maximum number of treatments 3 per yr for blackcurrants
- This product must not be used on any crops other than those listed, including any extrapolations that would normally be permissible under the Long Term Arrangements for Extension of Use (see Section 5)

Crop-specific information

- Latest use: 14 d after end of flowering for blackcurrants

Environmental safety

- Dangerous for the environment
- Very toxic to aquatic organisms
- High risk to bees. Do not apply to crops in flower or to those in which bees are actively foraging. Do not apply when flowering weeds are present
- Do not operate air-assisted sprayers within 18 m of surface water or ditches. Direct spray away from water
- LERAP Category A
- Broadcast air-assisted LERAP (18 m)

Hazard classification and safety precautions

Hazard H02, H04, H11
Risk phrases R25, R36, R38, R50, R58
Operator protection A, C, H, J; U02c, U04a, U05a, U08, U19a, U20b
Environmental protection E12a, E15a, E16c, E16d, E31c, E32e, E34; E17b (18 m)
Storage and disposal E01, E04, E29b, E30a
Medical advice M04, M05b

222 fenpropidin

A systemic, curative and protective piperidine (morpholine) fungicide

Products

Tern	Syngenta	750 g/l	EC	08660

Uses

- Brown rust in ***spring barley***, ***spring wheat***, ***winter barley***, ***winter wheat***
- Glume blotch in ***spring wheat***, ***winter wheat***
- Leaf blotch in ***spring barley***, ***winter barley***
- Powdery mildew in ***spring barley***, ***spring wheat***, ***winter barley***, ***winter wheat***
- Septoria leaf spot in ***spring wheat***, ***winter wheat***
- Yellow rust in ***spring barley***, ***spring wheat***, ***winter barley***, ***winter wheat***

Efficacy guidance

- Best results obtained when applied at early stage of disease development. See label for details of recommended timing alone and in mixtures
- Disease control enhanced by vapour-phase activity. Control can persist for 4-6 wk
- Alternate with triazole fungicides to discourage build-up of resistance

Restrictions

- Maximum number of treatments 3 per crop (up to 2 in yr of harvest) for winter crops; 2 per crop for spring crops

Crop-specific information

- Latest use: up to and including ear emergence complete (GS 59).
- HI 5 wk

Environmental safety

- Dangerous for the environment
- Very toxic to aquatic organisms

Hazard classification and safety precautions

Hazard H03, H11
Risk phrases R22a, R37, R50, R58
Operator protection A, C, H; U02a, U04a, U05a, U10, U20a
Environmental protection E15a, E32d, E34, E38
Storage and disposal E01, E04, E30a
Medical advice M03

223 fenpropimorph

A contact and systemic morpholine fungicide

See also *azoxystrobin + fenpropimorph*
epoxiconazole + fenpropimorph
epoxiconazole + fenpropimorph + kresoxim-methyl
epoxiconazole + fenpropimorph + pyraclostrobin

Products

1	Corbel	BASF	750 g/l	EC	00578
2	Landgold Fenpropimorph 750	Landgold	750 g/l	EC	10472
3	Marnoch Phorm	Me2	750 g/l	EC	11087
4	Standon Fenpropimorph 750	Standon	750 g/l	EC	08965

Uses

- Alternaria in ***carrots*** *(off-label)*, ***horseradish*** *(off-label)*, ***parsley root*** *(off-label)*, ***parsnips*** *(off-label)*, ***salsify*** *(off-label)* [1]
- Brown rust in ***spring barley***, ***spring wheat***, ***triticale***, ***winter barley***, ***winter wheat*** [1-4]
- Crown rot in ***carrots*** *(off-label)*, ***horseradish*** *(off-label)*, ***parsley root*** *(off-label)*, ***parsnips*** *(off-label)*, ***salsify*** *(off-label)* [1]
- Powdery mildew in ***carrots*** *(off-label)*, ***hops*** *(off-label)*, ***horseradish*** *(off-label)*, ***parsley root*** *(off-label)*, ***parsnips*** *(off-label)*, ***raspberries*** *(off-label)*, ***salsify*** *(off-label)*, ***strawberries*** *(off-label)* [1]; ***spring barley***, ***spring oats***, ***spring rye***, ***spring wheat***, ***winter barley***, ***winter oats***, ***winter rye***, ***winter wheat*** [1-4]
- Rhynchosporium in ***spring barley***, ***winter barley*** [1-4]
- Rust in ***leeks*** [3]; ***red beet*** *(off-label)*, ***sugar beet seed crops*** *(off-label)* [1]; ***spring field beans***, ***winter field beans*** [2-4]
- Yellow rust in ***spring barley***, ***spring wheat***, ***triticale***, ***winter barley***, ***winter wheat*** [1-4]

Specific Off-Label Approvals (SOLAs)

- ***carrots***, ***horseradish***, ***parsley root***, ***parsnips***, ***salsify*** *(OLA 023753) Dec 2008* [1]
- ***carrots***, ***horseradish***, ***parsley root***, ***parsnips***, ***salsify*** *(OLA 962483) Dec 2008* [1]
- ***hops*** *(OLA 010595) Dec 2008* [1]
- ***hops*** *(OLA 023759) Dec 2008* [1]
- ***raspberries***, ***strawberries*** *(OLA 040804) Dec 2008* [1]
- ***red beet*** *(OLA 023751) Dec 2008* [1]
- ***red beet*** *(OLA 941246) Dec 2008* [1]

- ***sugar beet seed crops*** *(OLA 023757) Dec 2008* [1]
- ***sugar beet seed crops*** *(OLA 961807) Dec 2008* [1]

Efficacy guidance

- On cereals spray at start of disease attack. See labels for recommended tank mixes. Follow-up treatments may be needed if disease pressure remains high
- On field beans a second application may be needed after 2-3 wk
- Product rainfast after 2 h

Restrictions

- Maximum number of treatments 2 per crop for spring cereals, field beans, red beet, sugar beet seed crops; 6 per crop for leeks, hops; 3 per crop for all other crops
- Consult processors before using on crops for processing
- An interval of at least 10 d on field beans and 14 d on leeks must elapse between applications

Crop-specific information

- Latest use: before 2 true leaves for sugar beet seed crops
- HI cereals, field beans 5 wk; carrots, horseradish, parsley root, parsnips, salsify 4 wk; leeks, red beet 3 wk; raspberries, strawberries 2 wk
- Scorch may occur if applied during frosty weather or in high temperatures

Environmental safety

- Dangerous for the environment
- Very toxic to aquatic organisms

Hazard classification and safety precautions

Hazard H03 [1, 2]; H04 [3, 4]; H11 [1, 2, 4]
Risk phrases R36 [2, 3]; R38 [1-4]; R50 [1]; R51 [2, 4]; R58 [1, 2, 4]; R63 [1, 2]
Operator protection A [1-4]; C [2-4]; U05a, U08, U14, U15, U19a [1-4]; U20a [3]; U20b [1, 2, 4]
Environmental protection E13b [3]; E15a [1, 2, 4]; E31b [3, 4]; E31c [1, 2]; E32d, E38 [1]
Storage and disposal E01, E04, E30a [1-4]; E26 [2-4]; E29b [1]
Medical advice M05a [1, 2, 4]

224 fenpropimorph + flusilazole

A broad-spectrum eradicant and protectant fungicide mixture for cereals

Products

1	Colstar	DuPont	375:160 g/l	EC	06783
2	Pluton	DuPont	375:160 g/l	EC	10957

Uses

- Brown rust in ***spring barley***, ***winter barley***, ***winter wheat***
- Net blotch in ***spring barley***, ***winter barley***
- Powdery mildew in ***spring barley***, ***winter barley***, ***winter wheat***
- Rhynchosporium in ***spring barley***, ***winter barley***
- Septoria diseases in ***winter wheat***
- Yellow rust in ***spring barley***, ***winter barley***, ***winter wheat***

Efficacy guidance

- Disease control is more effective if treatment made at an early stage of disease development
- Treat winter cereals in spring or early summer before diseases spread to new growth
- Spring barley should be treated when diseases are first evident
- Treatment may be repeated after 3-4 wk if necessary

Restrictions

- Maximum number of treatments (including other products containing flusilazole) 3 per crop (winter wheat), 2 per crop (winter or spring barley)
- Do not apply to crops under stress
- Do not apply during frosty weather

Crop-specific information

- Latest use: before beginning of anthesis (GS 60) for winter wheat; up to and including completion of ear emergence (GS 59) for barley

Environmental safety

- Dangerous for the environment
- Toxic to aquatic organisms

Hazard classification and safety precautions

Hazard H02, H11
Risk phrases R36, R40, R51, R58, R61
Operator protection A, C; U05a, U11, U19a, U20b
Environmental protection E15a, E31b, E32d, E38
Storage and disposal E01, E04, E26, E30a
Medical advice M04

225 fenpropimorph + kresoxim-methyl

A protectant and systemic fungicide mixture for cereals

Products

1	Ensign	BASF	300:150 g/l	SE	11729
2	Greencrop Monsoon	Greencrop	300:150 g/l	SE	09573
3	Standon Kresoxim FM	Standon	300:150 g/l	SE	08922

Uses

- Powdery mildew in ***spring barley***, ***winter barley*** [1-3]; ***spring oats***, ***spring rye***, ***triticale***, ***winter oats***, ***winter rye*** [1]
- Rhynchosporium in ***spring barley***, ***winter barley*** [1-3]; ***spring rye***, ***triticale***, ***winter rye*** [1]
- Septoria in ***spring wheat*** *(reduction)*, ***triticale*** *(reduction)* [1]; ***winter wheat*** *(reduction)* [1-3]

Approval information

- Kresoxim-methyl included in Annex I under EC Directive 91/414

Efficacy guidance

- For best results spray at the start of foliar disease attack and repeat if infection conditions persist
- Kresoxim-methyl is a member of the QoI cross resistance group. Product should be used preventatively and not relied on for its curative potential
- Use product as part of an Integrated Crop Management strategy incorporating other methods of control, including where appropriate other fungicides with a different mode of action. Do not apply more than two foliar applications of QoI containing products to any cereal crop
- There is a significant risk of widespread resistance occurring in *Septoria tritici* populations in UK. Strains of barley powdery mildew resistant to QoIs are common in UK. Failure to follow resistance management action may result in reduced levels of disease control
- Strains of barley powdery mildew resistant to QoIs are common in the UK

Restrictions

- Maximum total dose equivalent to two full dose treatments for all crops

Crop-specific information

- Latest use: completion of ear emergence (GS 59) for barley and oats; completion of flowering (GS 69) for wheat, rye and triticale

Environmental safety

- Dangerous for the environment
- Very toxic to aquatic organisms

Hazard classification and safety precautions

Hazard H03, H11
Risk phrases R40, R50, R58 [1-3]; R43 [2, 3]; R63 [1, 2]
Operator protection A [1-3]; H [1]; U05a, U14, U20b [1-3]; U19a [2]
Environmental protection E15a [1-3]; E31b [2, 3]; E31c, E32d, E38 [1]
Storage and disposal E01, E04, E26, E30a
Medical advice M05a [1, 3]

SEE SECTION 3 FOR PRODUCTS ALSO REGISTERED

226 fenpropimorph + pyraclostrobin

A protectant and curative fungicide mixture for cereals

Products

Jenton	BASF	375:100 g/l	EC	11898

Uses

- Brown rust in ***spring barley***, ***winter barley***
- Net blotch in ***spring barley***, ***winter barley***
- Powdery mildew in ***spring barley***, ***winter barley***
- Rhynchosporium in ***spring barley***, ***winter barley***
- Yellow rust in ***spring barley***, ***winter barley***

Efficacy guidance

- Best results obtained from treatment at the start of foliar disease attack
- Yield response may be obtained in the absence of visual disease
- Pyraclostrobin is a member of the QoI cross resistance group. Product should be used preventatively and not relied on for its curative potential
- Use product as part of an Integrated Crop Management strategy incorporating other methods of control, including where appropriate other fungicides with a different mode of action. Do not apply more than two foliar applications of QoI containing products to any cereal crop
- There is a significant risk of widespread resistance occurring in *Septoria tritici* populations in UK. Strains of barley powdery mildew resistant to QoIs are common in UK. Failure to follow resistance management action may result in reduced levels of disease control

Restrictions

- Maximum number of treatments 2 per crop

Crop-specific information

- Latest use: up to and including emergence of ear just complete (GS 59) for barley

Environmental safety

- Dangerous for the environment
- Very toxic to aquatic organisms
- LERAP Category B

Hazard classification and safety precautions

Hazard H03, H11
Risk phrases R20, R22a, R36, R38, R50, R58, R63
Operator protection A, C; U05a, U14, U20b
Environmental protection E15a, E16a, E31c, E32d, E34, E38
Storage and disposal E01, E04, E26, E30a
Medical advice M03, M05a

227 fenpropimorph + quinoxyfen

A systemic fungicide mixture for cereals

Products

Orka	Dow	250:66.7 g/l	EW	08879

Uses

- Powdery mildew in ***durum wheat***, ***spring barley***, ***spring oats***, ***spring rye***, ***spring wheat***, ***triticale***, ***winter barley***, ***winter oats***, ***winter rye***, ***winter wheat***

Approval information

- Quinoxyfen included in Annex I under Directive 91/414

Efficacy guidance

- For best results treat at early stage of disease development before infection spreads to new crop growth. Further treatment may be necessary if disease pressure remains high
- For control of established infections and broad spectrum disease control use in tank mixtures. See label
- Product rainfast after 1 h
- Systemic activity may be reduced in severe drought

Restrictions

- Maximum total dose 3.0 l product per ha
- Apply only in the spring from mid-tillering stage (GS 25)

Crop-specific information

- Latest use: when first awns visible (GS 49)
- Crop scorch may occur when treatment made in high temperatures

Environmental safety

- Dangerous for the environment
- Very toxic to aquatic organisms
- LERAP Category B

Hazard classification and safety precautions

Hazard H03, H11
Risk phrases R43, R50, R58, R61
Operator protection A, H; U05a, U14
Environmental protection E15a, E16a, E32d, E34, E38
Storage and disposal E01, E04

228 fenpyroximate

A mitochondrial electron transport inhibitor (METI) acaricide for apples

Products

Sequel	Certis	51.3 g/l	SC	11408

Uses

- Fruit tree red spider mite in ***apples***, ***plums*** *(off-label)*

Specific Off-Label Approvals (SOLAs)

- ***plums*** *(OLA 030869) Jan 2006* [1]

Efficacy guidance

- Kills motile stages of fruit tree red spider mite. Best results achieved if applied in warm weather
- Total spray cover of trees essential. Use higher volumes for large trees
- Apply when majority of winter eggs have hatched

Restrictions

- Maximum number of treatments 1 per yr or total dose equivalent to one full dose treatment
- Other mitochondrial electron transport inhibitor (METI) acaricides should not be applied to the same crop in the same calendar yr either separately or in mixture
- Do not apply when apple crops or pollinator are in flower
- Consult processor before use on crops for processing

Crop-specific information

- HI 2 wk

Environmental safety

- Dangerous for the environment
- Very toxic to aquatic organisms
- Risk to non-target insects or other arthropods
- Broadcast air-assisted LERAP (38 m)

Hazard classification and safety precautions

Hazard H03, H11
Risk phrases R20, R36, R41, R43, R50
Operator protection A, C, H; U05a, U08, U14, U15, U20b
Environmental protection E15a, E22c, E31c, E32e; E17b (38 m)
Storage and disposal E01, E04, E26, E30a

229 fenuron

A urea herbicide available only in mixtures

See also *chlorpropham + fenuron*

SEE SECTION 3 FOR PRODUCTS ALSO REGISTERED

230 ferrous sulphate

A herbicide/fertilizer combination for moss control in turf

See also dicamba + dichlorprop-P + ferrous sulphate + MCPA
dichlorophen + ferrous sulphate

Products

1	Elliott's Lawn Sand	Elliott	9.3% w/w	SA	04860
2	Elliott's Mosskiller	Elliott	24.6% w/w	GR	04909
3	Greenmaster Autumn	Scotts	18.2% w/w	GR	11185
4	Greenmaster Mosskiller	Scotts	24.1% w/w	GR	11576
5	SHL Lawn Sand	Sinclair	5.4% w/w	SA	05254
6	Taylors Lawn Sand	Rigby Taylor	4.1% w/w	SA	04451
7	Vitax Microgran 2	Vitax	10.9% w/w	MG	04541
8	Vitax Turf Tonic	Vitax	15% w/w	SA	04354

Uses
- Moss in ***managed amenity turf***

Efficacy guidance
- For best results apply when light showers expected, mow 3 d before treatment and do not mow for 3-4 d afterwards
- Water after 2 d if no rain
- Rake out dead moss thoroughly 7-14 d after treatment. See label
- Fertilizer component of autumn treatment encourages root growth, that of mosskiller formulations promotes tillering

Restrictions
- Maximum number of treatments - see labels
- Do not apply during drought or when heavy rain expected
- Do not apply in frosty weather or when the ground is frozen
- Do not walk on treated areas until well watered

Crop-specific information
- If spilt on paving, concrete, clothes etc brush off immediately to avoid discolouration
- Observe label restrictions for interval before cutting after treatment

Environmental safety
- Harmful to fish or other aquatic life. Do not contaminate surface waters or ditches with chemical or used container [3, 4]

Hazard classification and safety precautions

Operator protection U11 [4]; U20a [1, 2]; U20b [7, 8]; U20c [3-6]
Environmental protection E13c [3, 4]; E15a [1, 2, 6-8]; E32a [1-8]
Storage and disposal E01 [5, 8]; E30a [1-8]

231 fipronil

A phenylpyrazole insecticide for horticulture

Products

Vi-Nil	Certis	0.1% w/w	FG	11077

Uses
- Vine weevil in ***container-grown ornamentals***

Efficacy guidance
- Product may be used at any stage of plant propagation to potting on
- Treated compost will give control of vine weevil larvae into second yr after treatment
- Incorporate into fresh compost each time the subject is re-potted
- Control not guaranteed if untreated liners or plugs are potted into treated compost

Restrictions
- Maximum number of treatments 1 per batch of compost
- Check tolerance on sample plants before large scale treatment
- Do not use in compost for aquatic plants or marginals

Crop-specific information
- Latest use: before planting in treated compost
- Product may be used in all compost types but dose should be reduced in proportion if more than 20% inert material included

Environmental safety
- Harmful to fish or other aquatic life. Do not contaminate surface waters or ditches with chemical or used container

Hazard classification and safety precautions

Operator protection A, D; U13, U20b
Environmental protection E13c, E32a
Consumer protection C01
Storage and disposal E01, E04

232 florasulam

A triazolopyrimidine herbicide for cereals

Products

Boxer	Dow	50 g/l	SC	09819

Uses
- Annual dicotyledons in ***spring barley, spring oats, spring wheat, winter barley, winter oats, winter wheat***
- Charlock in ***spring barley, spring oats, spring wheat, winter barley, winter oats, winter wheat***
- Chickweed in ***spring barley, spring oats, spring wheat, winter barley, winter oats, winter wheat***
- Cleavers in ***spring barley, spring oats, spring wheat, winter barley, winter oats, winter wheat***
- Volunteer oilseed rape in ***spring barley, spring oats, spring wheat, winter barley, winter oats, winter wheat***

Approval information
- Florasulam included in Annex I under EC Directive 91/414

Efficacy guidance
- Best results obtained from treatment of small actively growing weeds in good conditions
- Apply in autumn or spring once crop has 3 leaves
- Product is mainly absorbed by leaves of weeds and is effective on all soil types
- Florasulam is a member of the ALS-inhibitor group of herbicides

Restrictions
- Maximum total dose on any crop equivalent to one full dose treatment
- Do not roll or harrow within 7 d before or after application
- Do not spray when crops under stress from cold, drought, pest damage, nutrient deficiency or any other cause
- Do not spray in tank mixture with a product containing any other ALS-inhibitor (eg sulfonylurea herbicides) except those containing metsulfuron-methyl, thifensulfuron-methyl or tribenuron-methyl. Product may be sprayed in sequence with any of the above ALS-inhibitors or flupyrsulfuron-methyl

Crop-specific information
- Latest use: before flag leaf sheath extending stage (GS 41) for all crops

Following crops guidance
- Only cereals, oilseed rape, field beans, grass or vegetable brassicas as transplants may be sown as a following crop in the same calendar yr as treatment. Oilseed rape may show some temporary reduction of vigour after a dry summer, but yields are not affected
- In addition to the above, linseed, peas, sugar beet, potatoes, maize, clover (for use in grass/clover mixtures) or carrots may be sown in the calendar yr following treatment
- In the event of failure of a treated crop in spring only spring wheat, spring barley, spring oats, maize or ryegrass may be sown

SEE SECTION 3 FOR PRODUCTS ALSO REGISTERED

Environmental safety

- Dangerous for the environment
- Very toxic to aquatic organisms
- See label for detailed instructions on tank cleaning

Hazard classification and safety precautions

Hazard H11
Risk phrases R50, R58
Operator protection U05a
Environmental protection E15a, E31b, E32d, E34, E38
Storage and disposal E01, E04, E26, E30a

233 florasulam + fluroxypyr

A post-emergence herbicide mixture for cereals

Products

1	GF 184	Dow	2.5:100 g/l	SE	10878
2	Hiker	Dow	1.0:100 g/l	SE	11451
3	Starane Gold	Dow	1.0:100 g/l	SE	10879
4	Starane XL	Dow	2.5:100 g/l	SE	10921

Uses

- Annual dicotyledons in ***spring barley, spring oats, spring wheat, winter barley, winter oats, winter wheat*** [1, 4]
- Chickweed in ***spring barley, spring oats, spring wheat, winter barley, winter oats, winter wheat*** [1-4]
- Cleavers in ***spring barley, spring oats, spring wheat, winter barley, winter oats, winter wheat*** [1-4]
- Mayweeds in ***spring barley, spring oats, spring wheat, winter barley, winter oats, winter wheat*** [1, 4]

Approval information

- Florasulam and fluroxypyr included in Annex I under EC Directive 91/414

Efficacy guidance

- Best results obtained when weeds are small and growing actively
- Products are mainly absorbed through weed foliage. Cleavers emerging after application will not be controlled
- Florasulam is a member of the ALS-inhibitor group of herbicides

Restrictions

- Maximum total dose equivalent to one full dose treatment
- Product must not be applied before 1 Mar in yr of harvest
- Do not roll or harrow 7 d before or after application
- Do not spray when crops are under stress from cold, drought, pest damage or nutrient deficiency
- Do not apply through CDA applicators
- These products may be applied in sequence with products containing only one of flupyrsulfuron-methyl, metsulfuron-methyl, thifensulfuron-methyl or tribenuron-methyl
- Apart from the above, these products must not be applied in tank mix or sequence with any other ALS-inhibitor (eg sulfonylurea herbicides)
- Only one other ALS-inhibitor product may be applied to the same crop

Crop-specific information

- Latest use: before flag leaf sheath extended for barley and wheat; before second node detectable for oats

Following crops guidance

- Cereals, oilseed rape, field beans or grass may follow treated crops in the same yr. Oilseed rape may suffer temporary vigour reduction after a dry summer
- In addition to the above, linseed, peas, sugar beet, potatoes, maize or clover may be sown in the calendar yr following treatment
- In the event of failure of a treated crop in the spring, only spring cereals, maize or ryegrass may be planted

Environmental safety

- Dangerous for the environment
- Toxic to aquatic organisms
- Take extreme care to avoid drift onto non-target crops or plants

Hazard classification and safety precautions

Hazard H04, H11
Risk phrases R36, R38, R51, R58, R67
Operator protection A, C; U05a, U11, U19a
Environmental protection E15a, E31c, E32d, E34, E38
Storage and disposal E01, E04, E30a

234 fluazifop-P-butyl

A phenoxypropionic acid grass herbicide for broadleaved crops

Products

Fusilade Max	Syngenta	125 g/l	EC	11519

Uses

- Annual grasses in ***blackcurrants***, ***broad beans*** *(off-label)*, ***carrots***, ***chinese cabbage*** *(off-label)*, ***collards*** *(off-label)*, ***combining peas***, ***farm forestry***, ***field margins***, ***flax***, ***flax for industrial use***, ***fodder beet***, ***garlic*** *(off-label)*, ***gooseberries***, ***haricot beans*** *(off-label)*, ***hops***, ***kale*** *(stockfeed only)*, ***linseed***, ***linseed for industrial use***, ***navy beans*** *(off-label)*, ***onions***, ***ornamental plant production*** *(off-label)*, ***parsnips*** *(off-label)*, ***protected ornamentals*** *(off-label)*, ***raspberries***, ***red beet*** *(off-label)*, ***spring field beans***, ***spring oilseed rape***, ***spring oilseed rape for industrial use***, ***strawberries***, ***sugar beet***, ***swedes*** *(stockfeed only)*, ***turnips*** *(stockfeed only)*, ***vining peas***, ***winter field beans***, ***winter oilseed rape***, ***winter oilseed rape for industrial use***
- Annual meadow grass in ***lucerne*** *(off-label)*
- Barren brome in ***field margins***
- Blackgrass in ***hops***, ***lucerne*** *(off-label)*, ***spring field beans***, ***winter field beans***
- Green cover in ***land temporarily removed from production***
- Perennial grasses in ***blackcurrants***, ***broad beans*** *(off-label)*, ***carrots***, ***chinese cabbage*** *(off-label)*, ***collards*** *(off-label)*, ***combining peas***, ***farm forestry***, ***flax***, ***flax for industrial use***, ***fodder beet***, ***garlic*** *(off-label)*, ***gooseberries***, ***haricot beans*** *(off-label)*, ***hops***, ***kale*** *(stockfeed only)*, ***linseed***, ***linseed for industrial use***, ***navy beans*** *(off-label)*, ***onions***, ***ornamental plant production*** *(off-label)*, ***parsnips*** *(off-label)*, ***protected ornamentals*** *(off-label)*, ***raspberries***, ***red beet*** *(off-label)*, ***spring field beans***, ***spring oilseed rape***, ***spring oilseed rape for industrial use***, ***strawberries***, ***sugar beet***, ***swedes*** *(stockfeed only)*, ***turnips*** *(stockfeed only)*, ***vining peas***, ***winter field beans***, ***winter oilseed rape***, ***winter oilseed rape for industrial use***
- Ryegrass in ***lucerne*** *(off-label)*
- Volunteer cereals in ***blackcurrants***, ***carrots***, ***combining peas***, ***farm forestry***, ***field margins***, ***flax***, ***flax for industrial use***, ***fodder beet***, ***gooseberries***, ***hops***, ***kale*** *(stockfeed only)*, ***linseed***, ***linseed for industrial use***, ***lucerne*** *(off-label)*, ***onions***, ***raspberries***, ***spring field beans***, ***spring oilseed rape***, ***spring oilseed rape for industrial use***, ***strawberries***, ***sugar beet***, ***swedes*** *(stockfeed only)*, ***turnips*** *(stockfeed only)*, ***vining peas***, ***winter field beans***, ***winter oilseed rape***, ***winter oilseed rape for industrial use***
- Wild oats in ***blackcurrants***, ***carrots***, ***combining peas***, ***farm forestry***, ***field margins***, ***flax***, ***flax for industrial use***, ***fodder beet***, ***gooseberries***, ***hops***, ***kale*** *(stockfeed only)*, ***linseed***, ***linseed for industrial use***, ***lucerne*** *(off-label)*, ***onions***, ***raspberries***, ***spring field beans***, ***spring oilseed rape***, ***spring oilseed rape for industrial use***, ***strawberries***, ***sugar beet***, ***swedes*** *(stockfeed only)*, ***turnips*** *(stockfeed only)*, ***vining peas***, ***winter field beans***, ***winter oilseed rape***, ***winter oilseed rape for industrial use***

Specific Off-Label Approvals (SOLAs)

- ***broad beans***, ***chinese cabbage***, ***collards***, ***garlic***, ***haricot beans***, ***navy beans***, ***ornamental plant production***, ***parsnips***, ***protected ornamentals***, ***red beet*** *(OLA 032138) Dec 2008* [1]
- ***lucerne*** *(OLA 041321) Dec 2008* [1]

Efficacy guidance

- Best results achieved by application when weed growth active under warm conditions with adequate soil moisture. Agral or other specified adjuvant must always be added to spray. See label

- Spray weeds from 2-expanded leaf stage to fully tillered, couch from 4 leaves when majority of shoots have emerged, with a second application if necessary
- Control may be reduced under dry conditions. Do not cultivate for 2 wk after spraying couch
- Annual meadow grass is not controlled
- May also be used to remove grass cover crops

Restrictions

- Maximum number of treatments normally 2 per crop for winter oilseed rape (including crops for industrial use); 2 per yr for farm forestry; 1 per crop or yr for spring oilseed rape (including crops for industrial use) and other crops
- Do not sow cereals for at least 8 wk after application of high rate or 2 wk after low rate
- Do not apply through CDA sprayer, with hand-held equipment or from air
- Avoid treatment before spring growth has hardened or when buds opening
- Do not treat bush and cane fruit or hops between flowering and harvest
- Oilseed rape, linseed and flax for industrial use must not be harvested for human or animal consumption nor grazed
- Do not use for forestry establishment on land not previously under arable cultivation or improved grassland
- Treated vegetation in field margins, land temporarily removed from production etc, must not be grazed or harvested for human or animal consumption and unprotected persons must be kept out of treated areas for at least 24 h

Crop-specific information

- Latest use: before 5 leaf stage for spring oilseed rape; before flowering for blackcurrants, gooseberries, hops, raspberries, strawberries; before flower buds visible for field beans, peas, linseed, flax, oilseed rape; 2 wk before sowing cereals or grass for field margins, land temporarily removed from production, oilseed rape for industrial use, grass crops for farm forestry
- HI beet crops, brassicas, carrots, parsnips, 8 wk; garlic, onions 4 wk
- Apply to sugar and fodder beet from 1-true leaf to 50% ground cover
- Apply to winter oilseed rape from 1-true leaf to established plant stage
- Apply to spring oilseed rape from 1-true leaf but before 5-true leaves
- Apply in fruit crops after harvest. See label for timing details on other crops
- Before using on onions or peas use crystal violet test to check that leaf wax is sufficient

Environmental safety

- Dangerous for the environment
- Very toxic to aquatic organisms

Hazard classification and safety precautions

Hazard H03, H11
Risk phrases R38, R50, R58, R61, R63
Operator protection A, C, H, M; U05a, U08, U20b
Environmental protection E15a, E31c, E32d, E38
Storage and disposal E01, E04, E30a

235 fluazinam

A pyridinamine fungicide for use in potatoes

Products

1	Greencrop Solanum	Greencrop	500 g/l	SC	11052
2	Landgold Fluazinam	Landgold	500 g/l	SC	08060
3	Shirlan	Syngenta	500 g/l	SC	10573

Uses

- Blight in ***potatoes*** [1-3]
- Root rot in ***protected blackberries*** *(off-label)*, ***protected raspberries*** *(off-label)*, ***protected rubus hybrids*** *(off-label)* [3]

Specific Off-Label Approvals (SOLAs)

- ***protected blackberries***, ***protected raspberries***, ***protected rubus hybrids*** *(OLA 032168) Dec 2008* [3]

Efficacy guidance

- Commence treatment at the first blight risk warning (before blight enters the crop). Products are rainfast within 1 h
- In the absence of a warning, treatment should start before foliage of adjacent plants meets in the rows
- Spray at 7-14 d intervals (5-10 d intervals [3]) depending on severity of risk (see label)
- Ensure complete coverage of the foliage and stems, increasing volume as haulm growth progresses, in dense crops and if blight risk increases

Restrictions

- Maximum number of treatments 10 per crop on potatoes; 2 per yr on protected cane fruit
- Do not use with hand-held sprayers

Crop-specific information

- HI 7 d (0 d [3])
- Latest Use: before end Mar in yr of harvest for protected blackberries, protected raspberries, protected Rubus hybrids

Environmental safety

- Dangerous for the environment
- Very toxic to aquatic organisms
- LERAP Category B

Hazard classification and safety precautions

Hazard H04, H11
Risk phrases R36, R43, R50, R58
Operator protection A, C, H; U02a, U04a, U05a, U08 [1-3]; U14, U15, U20a [2, 3]; U20b [1]
Environmental protection E15a, E16a [1-3]; E16b [2, 3]; E31b [1, 2]; E31c, E32d, E34, E38 [3]
Storage and disposal E01, E04, E30a [1-3]; E26 [1, 2]
Medical advice M03 [1-3]; M05a [3]

236 fluazinam + metalaxyl-M

A mixture of contact and systemic fungicides for potatoes

Products

Epok	Belchim	400:200 g/l	EC	11997

Uses

- Blight in ***potatoes***

Efficacy guidance

- Commence treatment at the first blight risk warning (before blight enters the crop). Products are rainfast within 1 h
- In the absence of a warning, treatment should start before foliage of adjacent plants meets in the rows
- Spray at 7-14 d intervals depending on severity of risk (see label)
- Ensure complete coverage of the foliage and stems, increasing volume as haulm growth progresses, in dense crops and if blight risk increases
- Metalaxyl-M works best on young actively growing foliage and efficacy declines with the onset of senescence. Therefore it is recommended that the product is only used for the first part of the blight control program, which should be completed with a reliable protectant fungicide

Restrictions

- Maximum number of treatments 5 per crop

Crop-specific information

- HI 7d for potatoes

Environmental safety

- Dangerous for the environment
- Very toxic to aquatic organisms
- LERAP Category B

Hazard classification and safety precautions

Hazard H03, H11

Risk phrases R20, R36, R38, R40, R43, R50, R58
Operator protection A, C, H; U05a, U07, U11, U14, U15, U19a, U20a
Environmental protection E15a, E16a, E16b, E31b, E32d, E34, E38
Storage and disposal E01, E04, E30a
Medical advice M05a

237 fludioxonil

A cyanopyrrole fungicide seed treatment for wheat and barley

See also cymoxanil + fludioxonil + metalaxyl-M

Products

Beret Gold	Syngenta	25 g/l	FS	11635

Uses
- Bunt in ***winter wheat*** *(seed treatment)*
- Covered smut in ***spring barley*** *(seed treatment)*, ***winter barley*** *(seed treatment)*
- Fusarium foot rot and seedling blight in ***spring barley*** *(seed treatment)*, ***spring oats*** *(seed treatment)*, ***spring wheat*** *(seed treatment)*, ***winter barley*** *(seed treatment)*, ***winter oats*** *(seed treatment)*, ***winter wheat*** *(seed treatment)*
- Leaf stripe in ***spring barley*** *(seed treatment - reduction)*, ***winter barley*** *(seed treatment - reduction)*
- Pyrenophora leaf spot in ***spring oats*** *(seed treatment)*, ***winter oats*** *(seed treatment)*
- Snow mould in ***spring barley*** *(seed treatment)*, ***winter barley*** *(seed treatment)*, ***winter wheat*** *(seed treatment)*

Efficacy guidance
- Apply direct to seed using conventional seed treatment equipment. Continuous flow treaters should be calibrated using product before use
- Effective against benzimidazole-resistant strains of *Fusarium nivale*

Restrictions
- Maximum number of treatments 1 per seed batch
- Do not apply to cracked, split or sprouted seed
- Sow treated seed within 6 mth

Crop-specific information
- Latest use: before drilling
- Product may reduce flow rate of seed through drill. Recalibrate with treated seed before drilling

Environmental safety
- Harmful to aquatic organisms
- Do not use treated seed as food or feed
- Treated seed harmful to game and wildlife

Hazard classification and safety precautions
Risk phrases R52, R58
Operator protection A, H; U05a, U20b
Environmental protection E03, E15a, E32a
Storage and disposal E01, E04, E30a
Treated seed S02, S05, S07

238 flufenacet

A broad spectrum oxyacetamide herbicide available only in mixtures

See also diflufenican + flufenacet

239 flufenacet + metribuzin

A herbicide mixture for potatoes

Products

Artist	Bayer CropScience	24:17.5% w/w	WP	11239

Uses
- Annual dicotyledons in ***early potatoes, maincrop potatoes***
- Annual meadow grass in ***early potatoes, maincrop potatoes***

Efficacy guidance
- Product acts through root uptake and needs sufficient soil moisture at and shortly after application
- Effectiveness is reduced under dry soil conditions
- Residual activity is reduced on mineral soils with a high organic matter content and on peaty or organic soils
- Ensure application is made evenly to both sides of potato ridges
- Perennial weeds are not controlled

Restrictions
- Maximum total dose equivalent to one full dose treatment
- Potatoes must be sprayed before emergence of crop and weeds
- See label for list of tolerant varieties. Do not treat Maris Piper grown on Sands or Very Light soils
- Do not use on Sands
- On stony or gravelly soils there is risk of crop damage especially if heavy rain falls soon after application

Crop-specific information
- Latest use: before potato crop emergence
- Consult processor before use on crops for processing

Following crops guidance
- Before drilling or planting any succeeding crop soil must be mouldboard ploughed to at least 15 cm as soon as possible after lifting and no later than end Dec
- In W Cornwall on soils with more than 5% organic matter treated early potatoes may be followed by summer planted brassica crops 14 wk after treatment and after mouldboard ploughing. Elsewhere cereals or winter beans may be grown in the same year if at least 16 wk have elapsed since treatment
- In the yr following treatment any crop may be grown except lettuce or radish, or vegetable brassica crops on silt soils in Lincs

Environmental safety
- Dangerous for the environment
- Very toxic to aquatic organisms
- Take care to avoid spray drift onto neighbouring crops, especially lettuce or brassicas

Hazard classification and safety precautions

Hazard H03, H11
Risk phrases R22a, R43, R48, R50, R58
Operator protection A, D, H; U05a, U08, U13, U14, U19a, U20b
Environmental protection E15a, E32a, E32d, E38
Storage and disposal E01, E04, E30a
Medical advice M03

240 flufenacet + pendimethalin

A broad spectrum residual and contact herbicide mixture for winter cereals

Products

1	Crystal	BASF	60:300 g/l	EC	10657
2	Ice	BASF	60:300 g/l	EC	10681
3	Standon FFA60 Plus	Standon	60:300 g/l	EC	11649
4	Trooper	BASF	60:300 g/l	EC	10682

SEE SECTION 3 FOR PRODUCTS ALSO REGISTERED

Uses

- Annual dicotyledons in ***winter barley, winter wheat***
- Annual grasses in ***winter barley, winter wheat***
- Annual meadow grass in ***winter barley, winter wheat***
- Blackgrass in ***winter barley, winter wheat***
- Chickweed in ***winter barley, winter wheat***
- Corn marigold in ***winter barley, winter wheat***
- Speedwells in ***winter barley, winter wheat***

Approval information

- Pendimethalin included in Annex I under EC Directive 91/414

Efficacy guidance

- Best results achieved when applied from pre-emergence of weeds to 2-leaf stage but post emergence treatment is not recommended on clay soils
- Product requires some soil moisture to be activated ideally from rain within 7 d of application. Prolonged dry conditions may reduce residual control
- Product is slow acting and final level of weed control may take some time to appear
- For effective weed control seed bed preparations should ensure even incorporation of any trash, straw and ash to 15 cm
- Efficacy may be reduced on soils with more than 6% organic matter
- Always follow WRAG guidelines for preventing and managing herbicide resistant weeds. Section 5 for more information

Restrictions

- Maximum total dose equivalent to one full dose treatment
- For pre-emergence treatments seed should be covered with min 32 mm settled soil. Shallow drilled crops should be treated post-emergence only
- Do not treat undersown crops
- Avoid spraying during periods of prolonged or severe frosts
- Do not use on stony or gravelly soils or those with more than 10% organic matter
- Pre-emergence treatment may only be used on crops drilled before 30 Nov. All crops must be treated before 31 Dec in yr of planting.
- Concentrated or diluted product may stain clothing or skin

Crop-specific information

- Latest use: before third tiller stage (GS 23) and before 31 Dec in yr of planting
- Very wet weather before and after treatment may result in loss of crop vigour and reduced yield, particularly where soils become waterlogged

Following crops guidance

- Any crop may follow a failed or normally harvested treated crop provided ploughing to at least 15 cm is carried out beforehand

Environmental safety

- Dangerous for the environment
- Very toxic to aquatic organisms
- Risk to certain non-target insects or other arthropods
- LERAP Category B
- Some products supplied in small volume returnable packs. Follow instructions for use

Hazard classification and safety precautions

Hazard H03, H11
Risk phrases R22a, R22b, R38, R40, R50, R58
Operator protection A, C, H; U02a, U05a, U20c
Environmental protection E15a, E16a, E16b, E31a, E34 [1-4]; E22b, E32d, E38 [1, 2, 4]; E22c [3]
Storage and disposal E01, E04, E30a
Medical advice M03, M05b

241 flupyrsulfuron-methyl

A sulfonylurea herbicide for winter wheat available only in mixtures

242 flupyrsulfuron-methyl + thifensulfuron-methyl

A sulfonylurea herbicide mixture for winter wheat

Products

Lexus Millenium	DuPont	10:40% w/w	WG	09206

Uses

- Annual dicotyledons in ***winter wheat***
- Blackgrass in ***winter wheat***

Approval information

- Flupyrsulfuron-methyl and thifensulfuron-methyl included in Annex I under EC Directive 91/414

Efficacy guidance

- Best results obtained when applied to small actively growing weeds
- Good spray cover of weeds must be obtained
- Increased degradation of active ingredient in high soil temperatures reduces residual activity
- Product has moderate residual life in soil. Under normal moisture conditions susceptible weeds germinating soon after treatment will be controlled
- Product may be used on all soil types but residual activity and weed control is reduced on highly alkaline soils
- Blackgrass should be treated from 1 leaf up to mid-tillering. Strains of blackgrass resistant to other herbicides may not be controlled
- Flupyrsulfuron-methyl and thifensulfuron-methyl are members of the ALS-inhibitor group of herbicides and products should be used in a planned Resistance Management strategy. See Section 5 for more information

Restrictions

- Maximum number of treatments 1 per crop
- Do not treat winter wheat at or after 1st node detectable stage (GS 31)
- Do not use on wheat undersown with grasses or legumes, or any other broad-leaved crop
- Do not apply within 7 d of rolling
- Do not treat any crop suffering from drought, waterlogging, pest or disease attack, nutrient deficiency, or any other stress factors
- Do not apply to a crop already treated with any other product containing thifensulfuron-methyl. Only specified products containing metsulfuron-methyl or tribenuron-methyl may follow treatment. See label

Crop-specific information

- Latest use: before 1st node detectable (GS 31)
- Slight chlorosis, speckling and stunting may occur in certain conditions. Recovery is rapid and yield not affected

Following crops guidance

- Only cereals, oilseed rape, field beans or grass may be sown in the yr of harvest of a treated crop. Any crop may be sown the following spring
- In the event of crop failure only winter wheat may be sown before normal harvest date. Land should be ploughed and cultivated to 15 cm minimum before resowing

Environmental safety

- Dangerous for the environment
- Very toxic to aquatic organisms
- LERAP Category B

Hazard classification and safety precautions

Hazard H04, H11
Risk phrases R43, R50, R58
Operator protection A, H; U05a, U08, U20b
Environmental protection E15a, E16a, E16b, E32a, E32d, E38
Storage and disposal E01, E04, E26, E30a

243 fluquinconazole

A protectant, eradicant and systematic triazole fungicide for winter wheat

Products

1	Flamenco	BASF	100 g/l	SC	11699
2	Galmano	Bayer CropScience	167 g/l	FS	11650
3	Jockey F	BASF	167 g/l	FS	11690
4	Sahara	Bayer CropScience	100 g/l	SC	11905

Uses

- Brown rust in ***winter wheat*** [1, 4]
- Bunt in ***winter wheat*** *(seed treatment)* [2, 3]
- Covered smut in ***winter barley*** *(seed treatment)* [3]
- Loose smut in ***winter barley*** *(seed treatment)* [3]
- Powdery mildew in ***winter wheat*** [1, 4]
- Septoria diseases in ***winter wheat*** [1, 4]
- Septoria leaf spot in ***winter wheat*** *(seed treatment)* [2, 3]
- Take-all in ***winter barley*** *(seed treatment - reduction)* [3]; ***winter wheat*** *(seed treatment - reduction)* [2, 3]
- Yellow rust in ***winter barley*** *(seed treatment)* [3]; ***winter wheat*** [1, 4]; ***winter wheat*** *(seed treatment)* [2, 3]

Efficacy guidance

- Best results obtained from spray treatments when disease first becomes active in crop but before infection spreads to younger leaves [1, 4]
- Adequate disease protection throughout the season will usually require a programme of at least two fungicide treatments [1, 4]
- May also be applied as a protectant at end of ear emergence (but no later) if crop still disease free [1, 4]
- With seed treatments ensure good even coverage of seed to obtain reliable disease control [2, 3]
- Seed treatment provides early control of *Septoria* leaf spot and yellow rust but may require foliar treatment for later infection [2, 3]

Restrictions

- Maximum total spray dose equivalent to two full doses [1, 4]
- Maximum number of seed treatments 1 per batch [2, 3]
- Do not apply spray treatments after the start of anthesis (GS 59) [1, 4]
- Do not mix with any other products [2, 3]
- Do not treat cracked, split or sprouted seed [2, 3]
- Do not use treated seed as food or feed [2, 3]

Crop-specific information

- Latest use: before beginning of anthesis (GS 59) for spray treatment; before drilling for seed treatment
- Ensure good spray coverage and increase volume in dense crops [1, 4]
- Delayed emergence may result when treated seed is drilled into heavy or poorly drained soils which then become wet or waterlogged [2, 3]

Environmental safety

- Dangerous for the environment
- Toxic to aquatic organisms
- Special PPE requirements and precautions apply where product supplied in returnable packs. Check label

Hazard classification and safety precautions

Hazard H02, H11

Risk phrases R22a, R48, R51, R58 [1-4]; R36, R43 [1, 4]

Operator protection A, H [1-4]; C [1, 4]; D [2, 3]; M [1]; U05a [1-4]; U07 [2]; U07 [3] (1000 l containers); U11, U14, U15 [1, 4]; U20b [2, 3]

Environmental protection E13b [2, 4]; E15a [1, 3]; E31c [1, 4]; E32a [3] (25-200 l containers); E32d, E34, E38 [1-4]; E33 [2]; E33 [3] (1000 l containers); E36 [2, 3]

Storage and disposal E01, E04 [1-4]; E26 [1, 2, 4]; E29b, E30a [2, 3]
Treated seed S01, S02, S03, S04b, S05, S06a, S07, S08 [2, 3]
Medical advice M03 [1-4]; M04 [1, 4]

244 fluquinconazole + prochloraz

A broad-spectrum triazole mixture for use as a spray and seed treatment in winter wheat

Products

1	Foil	BASF	54:174 g/l	SE	11700
2	Galmano Plus	Bayer CropScience	167:31 g/l	FS	11645
3	Jockey	BASF	167:31 g/l	FS	11689

Uses
- Brown rust in ***winter barley***, ***winter wheat*** [1]
- Bunt in ***winter wheat*** *(seed treatment)* [2, 3]
- Covered smut in ***winter barley*** *(seed treatment)* [3]
- Fusarium foot rot and seedling blight in ***winter barley*** *(seed treatment)* [3]
- Fusarium root rot in ***winter wheat*** *(seed treatment)* [2, 3]
- Loose smut in ***winter barley*** *(seed treatment)* [3]
- Powdery mildew in ***winter barley***, ***winter wheat*** *(moderate control)* [1]
- Rhynchosporium in ***winter barley*** [1]
- Septoria leaf spot in ***winter wheat*** [1]; ***winter wheat*** *(seed treatment)* [2, 3]
- Take-all in ***winter barley*** *(seed treatment - reduction)* [3]; ***winter wheat*** *(seed treatment - reduction)* [2, 3]
- Yellow rust in ***winter barley*** *(seed treatment)* [3]; ***winter wheat*** [1]; ***winter wheat*** *(seed treatment)* [2, 3]

Efficacy guidance
- Best results obtained from spay treatment when disease first becomes active in crop but before infection spreads to younger leaves [1]
- When treating seed ensure good even coverage of seed to obtain reliable disease control [2, 3]
- Seed treatment provides early control of *Septoria* leaf spot and yellow rust but may require foliar treatment for later infection [2, 3]

Restrictions
- Maximum number of treatments 1 per seed batch (seed dressing)
- Maximum total dose equivalent to two full doses (spray)
- Do not treat cracked, split or sprouted seed [2, 3]
- Do not use treated seed as food or feed [2, 3]

Crop-specific information
- Latest use: before 1st awns visible (GS 47) for winter barley; before beginning of anthesis (GS 59) for winter wheat; before drilling (seed treatment)
- Ensure good spray coverage and increase volume in dense crops [1]
- Delayed emergence may result when treated seed is drilled into heavy or poorly drained soils which then become wet or waterlogged [2, 3]

Environmental safety
- Dangerous for the environment
- Toxic to aquatic organisms
- Product supplied in returnable packs for which special PPE requirements and precautions apply. Check label

Hazard classification and safety precautions

Hazard H02 [2, 3]; H03, H04 [1]; H11 [1-3]
Risk phrases R22a, R48, R51, R58 [1-3]; R36 [1, 2]; R43, R66 [1]
Operator protection A, H [1-3]; C [1]; D [2, 3]; U05a [1-3]; U07 [2]; U07 [3] (1000 l containers only); U20b [2, 3]
Environmental protection E13b, E33, E36 [2]; E15a [1, 3]; E31c [1]; E32a [3] (25-200 l containers only); E32d, E34, E38 [1-3]; E33, E36 [3] (1000 l containers only)
Storage and disposal E01, E04, E30a [1-3]; E26 [1, 2]; E29b [2, 3]
Treated seed S01, S02, S03, S04b, S05, S06a, S07, S08 [2, 3]
Medical advice M03 [1, 2]; M04 [3]; M05b [1]

SEE SECTION 3 FOR PRODUCTS ALSO REGISTERED

245 fluroxypyr

A post-emergence aryloxyalkanoic acid herbicide

See also *2,4-D + dicamba + fluroxypyr*
clopyralid + fluroxypyr + MCPA
clopyralid + fluroxypyr + triclopyr
florasulam + fluroxypyr

Products

1	Agriguard Fluroxypyr	AgriGuard	200 g/l	EC	09298
2	Barclay Hurler	Barclay	200 g/l	EC	11356
3	Fernpath Hatchet	AgriGuard	200 g/l	EC	11510
4	Greencrop Reaper	Greencrop	200 g/l	EC	09359
5	Greencrop Reaper 2	Greencrop	200 g/l	EC	10060
6	Landgold Fluroxypyr	Landgold	200 g/l	EC	10490
7	Standon Fluroxypyr	Standon	200 g/l	EC	08923
8	Standon Fluroxypyr 200	Standon	200 g/l	EC	11534
9	Starane 2	Dow	200 g/l	EC	05496
10	Starane 2	Dow	200 g/l	EC	12018
11	Tomahawk	Makhteshim	200 g/l	EC	09249

Uses

- Annual dicotyledons in ***bulb onions*** *(off-label)*, ***leeks*** *(off-label)*, ***maize*** [9, 10]; ***durum wheat, spring barley, spring oats, spring rye, spring wheat, triticale, winter barley, winter oats, winter rye, winter wheat*** [1-11]; ***established grassland*** [1-5, 7-11]; ***forage maize*** [2, 4-8, 11]; ***newly sown grass*** [8]; ***seedling leys*** [1-5, 7, 9-11]
- Black bindweed in ***durum wheat, spring barley, spring oats, spring rye, spring wheat, triticale, winter barley, winter oats, winter rye, winter wheat*** [1-11]; ***established grassland*** [1-5, 7-11]; ***forage maize*** [2, 4-8, 11]; ***maize*** [9, 10]; ***newly sown grass*** [8]; ***seedling leys*** [1-5, 7, 9-11]
- Chickweed in ***durum wheat, spring barley, spring oats, spring rye, spring wheat, triticale, winter barley, winter oats, winter rye, winter wheat*** [1-11]; ***established grassland*** [1-5, 7-11]; ***forage maize*** [2, 4-8, 11]; ***maize*** [9, 10]; ***newly sown grass*** [8]; ***seedling leys*** [1-5, 7, 9-11]
- Cleavers in ***apple orchards*** *(off-label)*, ***maize, pear orchards*** *(off-label)*, ***poppies*** *(off-label - for morphine production)* [9, 10]; ***durum wheat, spring barley, spring oats, spring rye, spring wheat, triticale, winter barley, winter oats, winter rye, winter wheat*** [1-11]; ***established grassland*** [1-5, 7-11]; ***forage maize*** [2, 4-8, 11]; ***newly sown grass*** [8]; ***seedling leys*** [1-5, 7, 9-11]
- Docks in ***apple orchards*** *(off-label)*, ***maize, pear orchards*** *(off-label)* [9, 10]; ***durum wheat, spring barley, spring oats, spring rye, spring wheat, triticale, winter barley, winter oats, winter rye, winter wheat*** [1-11]; ***established grassland*** [1-5, 7-11]; ***forage maize*** [2, 4-8, 11]; ***newly sown grass*** [8]; ***seedling leys*** [1-5, 7, 9-11]
- Forget-me-not in ***durum wheat, spring barley, spring oats, spring rye, spring wheat, triticale, winter barley, winter oats, winter rye, winter wheat*** [1-11]; ***established grassland*** [1-5, 7-11]; ***forage maize*** [2, 4-8, 11]; ***maize*** [9, 10]; ***newly sown grass*** [8]; ***seedling leys*** [1-5, 7, 9-11]
- Hemp-nettle in ***durum wheat, spring barley, spring oats, spring rye, spring wheat, triticale, winter barley, winter oats, winter rye, winter wheat*** [1-11]; ***established grassland*** [1-5, 7-11]; ***forage maize*** [2, 4-8, 11]; ***maize*** [9, 10]; ***newly sown grass*** [8]; ***seedling leys*** [1-5, 7, 9-11]
- Poppies in ***poppies*** *(off-label - for morphine production)* [9, 10]
- Stinging nettle in ***apple orchards*** *(off-label)*, ***pear orchards*** *(off-label)* [9, 10]
- Volunteer potatoes in ***bulb onions*** *(off-label)*, ***leeks*** *(off-label)*, ***maize, poppies*** *(off-label - for morphine production)* [9, 10]; ***durum wheat, spring barley, spring oats, spring rye, spring wheat, triticale, winter barley, winter oats, winter rye, winter wheat*** [1-11]; ***established grassland, seedling leys*** [1-5, 7, 9-11]; ***forage maize*** [2, 4-8, 11]

Specific Off-Label Approvals (SOLAs)

- ***apple orchards, pear orchards*** *(OLA 002710) Dec 2005* [9, 10]
- ***bulb onions*** *(OLA 012137) Dec 2005* [9, 10]
- ***leeks*** *(OLA 023656) Dec 2005* [9, 10]
- ***poppies*** *(for morphine production) (OLA 032409) Dec 2005* [9, 10]

Approval information

- Fluroxypyr included in Annex I under EC Directive 91/414

Efficacy guidance

- Best results achieved under good growing conditions in a strongly competing crop
- A number of tank mixtures with HBN and other herbicides are recommended for use in autumn and spring to extend range of species controlled. See label for details
- Spray is rainfast in 1 h

Restrictions

- Maximum number of treatments 1 per crop or yr or maximum total dose equivalent to one full dose treatment
- Do not use on crops undersown with clovers or other legumes
- Do not treat crops suffering stress caused by any factor
- Do not roll or harrow for 7 d before or after treatment
- Do not spray if frost imminent

Crop-specific information

- Latest use: before flag leaf sheath opening (GS 47) for winter wheat and barley; before flag leaf sheath extending (GS 41) for spring wheat and barley; before second node detectable (GS 32) for oats, rye, triticale and durum wheat; before 7 leaves unfolded and before buttress roots appear for maize; before flower buds exposed for poppies
- HI apples, pears 4 wk; onions 11 wk
- Apply to new leys from 3 expanded leaf stage
- Timing varies in tank mixtures. See label for details
- Crops undersown with grass may be sprayed provided grasses are tillering

Following crops guidance

- Keep livestock out of treated areas for at least 3 d and until foliage of poisonous weeds such as ragwort has died and become unpalatable

Environmental safety

- Dangerous for the environment
- Toxic to aquatic organisms
- Keep livestock out of treated areas for at least 3 d following treatment and until poisonous weeds, such as ragwort, have died down and become unpalatable
- Wash spray equipment thoroughly with water and detergent immediately after use. Traces of product can damage susceptible plants sprayed later

Hazard classification and safety precautions

Hazard H03, H08, H11
Risk phrases R22b [1-11]; R36, R38 [1, 3]; R37 [1, 3, 4, 6-11]; R51, R58 [2, 4-11]; R67 [4-11]
Operator protection U08, U19a, U20b
Environmental protection E07a [1-11] (3 d); E15a, E31b [1-11]; E32d, E38 [9-11]; E34 [1, 3, 11]
Storage and disposal E26 [1-8, 11]; E30a [1-11]
Medical advice M05b

246 fluroxypyr + mecoprop-P

A post-emergence herbicide for broadleaved weeds in amenity turf

Products

Bastion T	Rigby Taylor	72:300 g/l	ME	06011

Uses

- Annual dicotyledons in ***amenity turf, managed amenity turf***
- Perennial dicotyledons in ***amenity turf, managed amenity turf***
- Slender speedwell in ***amenity turf, managed amenity turf***

Approval information

- Fluroxypyr included in Annex I under EC Directive 91/414

Efficacy guidance

- Best results achieved when soil moist and weeds in active growth, normally Apr-Sep

SEE SECTION 3 FOR PRODUCTS ALSO REGISTERED

Restrictions

- Maximum number of treatments 2 per yr. The total amount of mecoprop-P applied in a single yr must not exceed the maximum total dose approved for any single product for the crop/situation
- Do not treat turf under stress from any cause, if night temperatures are low, if ground frost imminent or during prolonged cold weather
- Do not apply in drought period unless irrigation applied
- Do not spray if turf wet

Crop-specific information

- Avoid mowing for 3 d before or after spraying (5 d before for young turf)
- Young turf must only be treated in spring provided that at least 2 mth have elapsed between sowing and application

Environmental safety

- Dangerous for the environment
- Toxic to aquatic organisms
- Wash spray equipment thoroughly with water and detergent immediately after use. Traces of product can damage susceptible plants sprayed later
- Avoid drift onto all broad-leaved plants outside the target area

Hazard classification and safety precautions

Hazard H03, H11
Risk phrases R21, R22a, R22b, R41, R51, R58
Operator protection A, C; U05a, U08, U11, U19a, U20b
Environmental protection E15a, E31b, E32d, E34, E38
Storage and disposal E01, E04, E26, E30a
Medical advice M03, M05a

247 fluroxypyr + triclopyr

A foliar acting herbicide for docks in grassland

Products

Doxstar	Dow	100:100 g/l	EC	11063

Uses

- Chickweed in ***newly sown grass***
- Docks in ***established grassland***, ***newly sown grass***

Approval information

- Fluroxypyr included in Annex I under EC Directive 91/414

Efficacy guidance

- Seedling docks in new leys are controlled up to 50 mm diameter. In established grass apply in spring or autumn or, at lower dose, in spring and autumn on docks up to 200 mm. A second application in the subsequent yr may be needed
- Allow 2-3 wk after cutting or grazing to allow sufficient regrowth of docks to occur before spraying
- Control may be reduced if rain falls within 2 h of application
- To allow maximum translocation to the roots of docks do not cut grass for 28 d after spraying

Restrictions

- Maximum number of treatments 2 per yr
- Do not roll or harrow for 10 d before or 7 d after spraying
- Do not spray in drought, very hot or very cold weather

Crop-specific information

- Latest use: 7 d before grazing or harvest of grass
- Grass less than one yr old may be treated from the third leaf visible stage
- Clover will be killed or severely checked by treatment

Following crops guidance

- Do not sow kale, turnips, swedes or grass mixtures containing clover by direct drilling or minimum cultivation techniques within 6 wk of application

Environmental safety

- Dangerous for the environment

- Toxic to aquatic organisms
- Keep livestock out of treated areas for at least 7 d following treatment and until poisonous weeds, such as ragwort, have died down and become unpalatable
- Do not allow drift to come into contact with crops, amenity plantings, gardens, ponds, lakes or watercourses
- Wash spray equipment thoroughly with water and detergent immediately after use. Traces of product can damage susceptible plants sprayed later

Hazard classification and safety precautions

Hazard H03, H08, H11
Risk phrases R22a, R22b, R37, R38, R43, R51, R58, R67
Operator protection A, C; U02a, U05a, U08, U14, U19a, U20b
Environmental protection E07a (7 d); E15a, E31b, E32d, E34, E38
Consumer protection C01
Storage and disposal E01, E04, E26, E30a
Medical advice M05b

248 flurtamone

A carotenoid synthesis inhibitor available only in mixtures

See also *diflufenican + flurtamone*
diflufenican + flurtamone + isoproturon

249 flusilazole

A systemic, protective and curative conazole fungicide for cereals and oilseed rape

See also *carbendazim + flusilazole*
famoxadone + flusilazole
fenpropimorph + flusilazole

Products

1	Capitan 40	DuPont	400 g/l	EC	10914
2	Genie 25	DuPont	250 g/l	EW	10285
3	Lyric	DuPont	250 g/l	EW	08252
4	Sanction 25	DuPont	250 g/l	EW	10284

Uses

- Brown rust in ***spring barley, winter barley, winter wheat*** [1-4]
- Eyespot in ***spring barley, winter barley, winter wheat*** [1-4]
- Light leaf spot in ***spring oilseed rape, winter oilseed rape*** [1-4]
- Net blotch in ***spring barley, winter barley*** [1-4]
- Powdery mildew in ***spring barley, winter barley, winter wheat*** [1-4]; ***sugar beet*** [2-4]
- Rhynchosporium in ***spring barley, winter barley*** [1-4]
- Rust in ***sugar beet*** [2-4]
- Septoria in ***spring barley, winter barley, winter wheat*** [1-4]
- Yellow rust in ***spring barley, winter barley, winter wheat*** [1-4]

Efficacy guidance

- On cereals use as a routine preventative spray or when disease first develops
- Best control of eyespot achieved by spraying between leaf-sheath erect and second node detectable stages (GS 30-32)
- Product active against both MBC-sensitive and MBC-resistant strains of eyespot
- Rain occurring within 2 h after application may reduce effectiveness
- See label for recommended tank-mixes to give broader spectrum control

Restrictions

- Maximum number of treatments (including other products containing flusilazole) on cereals and oilseed rape depends on dose and timing - see labels for details. High rate must not be used more than once in any crop
- Maximum total dose on sugar beet equivalent to one full dose
- Do not apply to crops under stress or during frosty weather

Crop-specific information

- Latest use: high dose before 3rd node detectable (GS 33) plus reduced dose before early milk stage (GS 72) on winter wheat and barley; before first flowers open (GS 4,0) on oilseed rape
- HI 7 wk for sugar beet
- Treat oilseed rape in autumn and/or spring at stem extension stage
- On sugar beet apply at an early stage of disease development, usually in early Aug

Environmental safety

- Dangerous for the environment
- Toxic to aquatic organisms

Hazard classification and safety precautions

Hazard H02, H11
Risk phrases R22a, R40, R51, R58, R61 [1-4]; R38 [2-4]
Operator protection A, C; U05a, U11, U19a [1-4]; U20a [2-4]; U20b [1]
Environmental protection E13c [3]; E15a [1, 2, 4]; E31b, E32d, E34, E38 [1-4]
Storage and disposal E01, E04, E30a [1-4]; E26 [2-4]
Medical advice M04

250 flutolanil

An oxathiin fungicide for treatment of potato seed tubers

Products

Rhino	Certis	460 g/l	FS	11802

Uses

- Black scurf in ***potatoes*** *(tuber treatment)*
- Rhizoctonia in ***potatoes*** *(off-label - chitted seed treatment)*
- Stem canker in ***potatoes*** *(tuber treatment)*

Specific Off-Label Approvals (SOLAs)

- ***potatoes*** *(chitted seed treatment) (OLA 041135) Dec 2008* [1]

Efficacy guidance

- Apply to clean tubers before chitting, prior to planting, or at planting
- Apply through canopied, hydraulic or spinning disc equipment (with or without electrostatics) mounted on a rolling conveyor or table
- May be diluted with water up to 2.0 l per tonne to improve tuber coverage. Disease in areas not covered by spray will not be controlled

Restrictions

- Maximum number of treatments 1 per batch of seed tubers

Crop-specific information

- Latest use: at planting
- HI 12 wk for potatoes
- Seed tubers should be of good quality and free from bacterial rots, physical damage or virus infection, and should not be sprouted to such an extent that mechanical damage to the shoots will occur during treatment or planting

Environmental safety

- Harmful to aquatic organisms

Hazard classification and safety precautions

Hazard H04
Risk phrases R43, R52
Operator protection A, H; U04a, U05a, U19a, U20a
Environmental protection E15a, E31a, E32d, E34
Storage and disposal E01, E04, E26, E30a
Treated seed S01, S03, S04a, S05
Medical advice M03

251 flutriafol

A broad-spectrum conazole fungicide for cereals

Products

1 Consul	Headland	125 g/l	SC	11523
2 Pointer	Headland	125 g/l	SC	11522

Uses

- Brown rust in ***spring barley***, ***winter barley***, ***winter wheat***
- Powdery mildew in ***spring barley***, ***winter barley***, ***winter wheat***
- Rhynchosporium in ***spring barley***, ***winter barley***
- Septoria leaf spot in ***winter wheat***
- Yellow rust in ***spring barley***, ***winter barley***, ***winter wheat***

Efficacy guidance

- Best results obtained from treatment in early stages of disease development. See label for detailed guidance on spray timing for specific diseases and the need for repeat treamtments
- Tank mix options available on the label to broaden activity spectrum
- Good spray coverage essential for optimum performance

Restrictions

- Maximum number of treatments 2 per crop (including other products containing flutriafol)

Crop-specific information

- Latest use: before early grain milky ripe stage (GS 73)
- Flag leaf tip scorch on wheat caused by stress may be increased by fungicide treatment

Environmental safety

- Harmful to aquatic organisms

Hazard classification and safety precautions

Hazard H03
Risk phrases R43 [1]; R48, R52 [1, 2]
Operator protection A, H; U05a, U09a, U19a, U20b
Environmental protection E13c, E31c, E38
Storage and disposal E01, E04, E30a

252 fomesafen

A contact and residual diphenyl ether herbicide for use in leguminous crops

Products

Flex	Syngenta	250 g/l	SL	08885

Uses

- Annual dicotyledons in ***dwarf beans***
- Black nightshade in ***soya beans*** *(off-label)*
- Volunteer oilseed rape in ***dwarf beans***, ***soya beans*** *(off-label)*

Specific Off-Label Approvals (SOLAs)

- ***soya beans*** *(OLA 023930) Nov 2006* [1]

Approval information

- Products containing this active ingredient have been granted derogations for specified 'Essential Uses' for use until 31 December 2007. Sale and supply must cease by 30 June 2007 but growers have no guarantee that the products will continue to be available until then.
  For more information see 'The Review Programme' under 'Pesticide Legislation' in Section 5

Efficacy guidance

- Fomesafen combines contact activity and residual soil uptake
- Best results obtained when seedling weeds are actively growing in warm moist conditions with adequate soil moisture
- Weeds should be treated at seedling stage. Larger weeds, and those hardened off by adverse conditions, may be less well controlled
- Residual activity gives control of later germinating weed seedlings

- Product may be used on all soil types but residual control of weeds germinating after treatment may be reduced on soils with high organic matter
- Weed spectrum may be widened by use in a programme with other herbicides. See label for details

Restrictions

- Maximum total dose equivalent to one full dose treatment
- Apply to healthy crops only. Crops growing under stress from drought, waterlogging etc should not be treated
- This product must not be used on any crops other than those listed, including any extrapolations that would normally be permissible under the Long Term Arrangements for Extension of Use (see Section 5)

Crop-specific information

- HI green beans 5 wk; soya beans 8 wk
- Some crop damage, such as leaf crinkling, may occur. Effect is transient and should not affect yield

Environmental safety

- Dangerous for the environment
- Very toxic to aquatic organisms

Hazard classification and safety precautions

Hazard H04, H11
Risk phrases R36, R43, R50, R58
Operator protection A, C, H; U04a, U05a, U11, U14, U19a, U20b
Environmental protection E15a, E31c, E32d, E38
Storage and disposal E01, E04, E30a
Medical advice M03

253 fomesafen + terbutryn

A residual herbicide for use in leguminous crops

Products

Reflex T	Syngenta	80:400 g/l	SC	08884

Uses

- Annual dicotyledons in ***broad beans*** *(spring sown)*, ***combining peas*** *(spring sown)*, ***spring field beans***, ***vining peas*** *(spring sown)*

Approval information

- Products containing fomesafen and terbutryn have been granted derogations for specified 'Essential Uses' for use until 31 December 2007. Sale and supply must cease by 30 June 2007 but growers have no guarantee that the products will continue to be available until then. For more information see 'The Review Programme' under 'Pesticide Legislation' in Section 5

Efficacy guidance

- Best results achieved when applied to moist, fine, firm tilth
- Rain soon after application is essential for optimum weed control
- Results may be unsatisfactory in dry, or excessively wet, soil conditions
- Weed control may be reduced on soils with more than 10% organic matter

Restrictions

- Maximum number of treatments 1 per crop
- Use only once every 5 yr
- All varieties of spring sown peas may be treated but forage varieties may be damaged from which recovery may not be complete. All varieties of spring sown field and broad beans may be treated
- Do not use on Sands
- This product must not be used on any crops other than those listed, including any extrapolations that would normally be permissible under the Long Term Arrangements for Extension of Use (see Section 5)

Crop-specific information

- Latest use: pre-emergence of crop

- Apply after planting, but before the crop emerges
- Emerged crop leaves may be severely scorched or killed

Following crops guidance

- Plough or cultivate to at least 150 mm before drilling or planting another crop. Only cereals should be planted in the calendar year of use with a minimum interval of 4 mth after treatment

Environmental safety

- Dangerous for the environment
- Very toxic to aquatic organisms

Hazard classification and safety precautions

Hazard H03, H11
Risk phrases R22a, R36, R43, R50, R58
Operator protection A, C, H; U04a, U05a, U08, U19a, U20b
Environmental protection E15a, E31c, E32d, E38
Storage and disposal E01, E04

254 formaldehyde (commodity substance)

An agricultural/horticultural and animal husbandry fungicide

Products

1	formaldehyde	various	38-40%	SL
2	paraformaldehyde	various	-	ZZ

Uses

- Fungus diseases in ***flower bulbs*** *(dip)*, ***glasshouses*** *(spray, dip or fumigant)*, ***mushroom houses*** *(spray or fumigant)* [1]; ***livestock houses*** [2]
- Soil-borne diseases in ***soil and compost*** *(drench)* [1]

Approval information

- Approval for the use of formaldehyde as a commodity substance was granted on 1 March 1991 by Ministers under regulation 5 of the Control of Pesticides Regulations 1986

Efficacy guidance

- Use as a dip to sterilize flower bulbs [1]
- Use as drench to sterilize soil and compost, indoors and outdoors [1]
- Use as spray or fumigant in mushroom houses [1]
- Use as spray, dip or fumigant for glasshouse hygiene [1]
- Use as fumigant to sterilize animal houses [2]

Restrictions

- Formaldehyde is subject to the Poisons Rules 1982 and the Poisons Act 1972. See notes in Section 5
- Operators must observe Occupational Exposure Standard as set out in HSE Guidance Note EH40/90 and ACOP 30 *Control of Substances Hazardous to Health in Fumigation*
- Operators must be supplied with a Section 6 (HSW) Safety Data Sheet before commencing work
- Observe maximum permitted concentrations. See PR 12, 1990, pp 9-10

Hazard classification and safety precautions

Operator protection A, D, P [1]

255 fosetyl-aluminium

A systemic phosphonic acid fungicide for various horticultural crops

Products

1	Aliette 80 WG	Certis	80% w/w	WG	11213
2	I T Fosetyl-AL	I T Agro	80% w/w	WP	11717
3	Standon Fosetyl-AL 80 WG	Standon	80% w/w	WG	10667

Uses

- Collar rot in ***apples*** [1-3]
- Crown rot in ***apples*** [1-3]; ***protected strawberries*** *(off-label)*, ***strawberries*** *(off-label)* [1]

SEE SECTION 3 FOR PRODUCTS ALSO REGISTERED

- Damping off in **broccoli** *(off-label)*, **brussels sprouts** *(off-label)*, **cabbages** *(off-label)*, **calabrese** *(off-label)*, **cauliflowers** *(off-label)*, **chinese cabbage** *(off-label)*, **collards** *(off-label)*, **kale** *(off-label)*, **protected broccoli** *(off-label)*, **protected brussels sprouts** *(off-label)*, **protected cabbages** *(off-label)*, **protected calabrese** *(off-label)*, **protected cauliflowers** *(off-label)*, **protected chinese cabbage** *(off-label)*, **protected collards** *(off-label)*, **protected kale** *(off-label)* [1]
- Downy mildew in **broad beans**, **hops**, **protected lettuce** [1-3]; **broccoli** *(off-label)*, **brussels sprouts** *(off-label)*, **cabbages** *(off-label)*, **calabrese** *(off-label)*, **cauliflowers** *(off-label)*, **chinese cabbage** *(off-label)*, **collards** *(off-label)*, **combining peas** *(off-label - seed treatment)*, **grapevines** *(off-label)*, **kale** *(off-label)*, **protected brassica seedlings** *(off-label)*, **protected broccoli** *(off-label)*, **protected brussels sprouts** *(off-label)*, **protected cabbages** *(off-label)*, **protected calabrese** *(off-label)*, **protected cauliflowers** *(off-label)*, **protected chinese cabbage** *(off-label)*, **protected collards** *(off-label)*, **protected kale** *(off-label)*, **salad onions** *(off-label)*, **vining peas** *(off-label - seed treatment)* [1]; **broccoli** *(off-label - seedlings)*, **brussels sprouts** *(off-label - seedlings)*, **cabbages** *(off-label - seedlings)*, **calabrese** *(off-label - seedlings)*, **cauliflowers** *(off-label - seedlings)*, **chinese cabbage** *(off-label - seedlings)*, **chives** *(off-label)*, **herbs (see appendix 6)** *(off-label)*, **kale** *(off-label - seedlings)*, **leaf spinach** *(off-label)*, **protected chives** *(off-label)*, **protected herbs (see appendix 6)** *(off-label)*, **spinach beet** *(off-label)*, **watercress** *(off-label - during propagation)* [1, 3]; **combining peas** *(off-label)*, **lettuce** *(off-label)*, **parsley** *(off-label)*, **protected broccoli** *(off-label - seedlings)*, **protected brussels sprouts** *(off-label - seedlings)*, **protected cabbages** *(off-label - seedlings)*, **protected calabrese** *(off-label - seedlings)*, **protected cauliflowers** *(off-label - seedlings)*, **protected chinese cabbage** *(off-label - seedlings)*, **protected kale** *(off-label - seedlings)*, **protected parsley** *(off-label)*, **vining peas** *(off-label)* [3]
- Phytophthora in **chicory** *(off-label - for forcing)* [1]; **chicory** *(off-label - in forcing sheds)* [3]; **watercress** *(off-label - during propagation)* [1, 3]
- Phytophthora root rot in **capillary benches** [1]; **protected pot plants** [1, 3]
- Phytophthora stem rot in **capillary benches** [1]; **protected pot plants** [1, 3]
- Phytophthora wilt in **hardy ornamental nursery stock** [1-3]
- Pythium in **watercress** *(off-label - during propagation)* [1, 3]
- Red core in **protected strawberries** *(off-label)*, **strawberries** *(off-label)* [1]; **strawberries** [1-3]
- Root rot in **capillary benches** [1]; **protected pot plants** [1, 3]

Specific Off-Label Approvals (SOLAs)

- ***broccoli, brussels sprouts, cabbages, calabrese, cauliflowers, chinese cabbage, collards, kale, protected broccoli, protected brussels sprouts, protected cabbages, protected calabrese, protected cauliflowers, protected chinese cabbage, protected collards, protected kale*** *(OLA 040149) Jul 2008* [1]
- ***broccoli, brussels sprouts, cabbages, calabrese, cauliflowers, chinese cabbage, kale, protected broccoli, protected brussels sprouts, protected cabbages, protected calabrese, protected cauliflowers, protected chinese cabbage, protected kale*** *(seedlings) (OLA 030366) Dec 2008* [3]
- ***broccoli, brussels sprouts, cabbages, calabrese, cauliflowers, chinese cabbage, kale*** *(seedlings) (OLA 030863) Dec 2008* [1]
- ***chicory*** *(for forcing) (OLA 030866) Dec 2008* [1]
- ***chicory*** *(in forcing sheds) (OLA 030366) Dec 2008* [3]
- ***chives, herbs (see appendix 6), leaf spinach, lettuce, parsley, protected chives, protected herbs (see appendix 6), protected parsley, spinach beet*** *(OLA 030366) Dec 2008* [3]
- ***chives, herbs (see appendix 6), protected chives, protected herbs (see appendix 6)*** *(OLA 030868) Dec 2008* [1]
- ***combining peas, vining peas*** *(OLA 030367) Dec 2008* [3]
- ***combining peas, vining peas*** *(seed treatment) (OLA 030861) Dec 2008* [1]
- ***grapevines*** *(OLA 030862) Dec 2008* [1]
- ***leaf spinach, spinach beet*** *(OLA 030864) Dec 2008* [1]
- ***protected brassica seedlings*** *(OLA 030863) Dec 2008* [1]
- ***protected strawberries, strawberries*** *(OLA 030579) Dec 2008* [1]
- ***salad onions*** *(OLA 023926) Dec 2007* [1]
- ***watercress*** *(during propagation) (OLA 030366) Dec 2008* [3]
- ***watercress*** *(during propagation) (OLA 030867) Dec 2008* [1]

Restrictions

- Maximum number of treatments 1 per seed batch for peas; 1 per batch of compost for lettuce; 1 per yr for strawberries (root dip or foliar spray); 1 per yr for apples (bark paste), 2 per yr for apples

(foliar spray); 2 per crop for broad beans, peas; 2 per yr for hops (basal spray); 6 per yr for hops (foliar spray); 1 per crop for brassicas and salad brassicas; 1 per crop during propagation, 2 per crop after planting out for lettuce; 7 per yr for grapevines
- Check tolerance of ornamental species before large-scale treatment

Crop-specific information
- Latest use: pre-planting for dipping autumn planted strawberry runners; up to 31 Dec for spraying autumn planted strawberry runners; pre-sowing for protected lettuce, peas; before transplanting or pre-emergence for brassicas; 24 h after seeding watercress
- HI apples 5 mth (bark paste) or 4 wk (spray); grapevines 35 d; chicory 21 d; broad beans 17 d; hops, strawberries, herbs and salad crops 14 d; leaf spinach, spinach beet 7 d
- Spray young orchards for crown rot protection after blossom when first leaves fully open and repeat after 4-6 wk. Apply as paste to bark of apples to control collar rot
- Apply to broad beans when infection appears (usually at flowering) and 14 d later. Consult before treating crops to be processed
- Use on strawberries only between harvest and 31 Dec
- Spray autumn-planted strawberry runners 2-3 wk after planting or use dip treatment at planting. Spray established crops in late summer/early autumn after picking and repeat annually
- Apply to hops as early season basal spray or as foliar spray every 10-14 d from when training is completed
- Use by wet incorporation in blocking compost for protected lettuce only from Sep to Apr and follow all directions carefully to avoid severe crop injury. Crop maturity may be delayed by a few days
- Apply as drench to rooted cuttings of hardy nursery stock after first potting and repeat mthly. Up to 6 applications may be needed
- See label for details of application to capillary benches and trays of young plants.

Environmental safety
- Harmful to aquatic organisms

Hazard classification and safety precautions

Risk phrases R52 [2, 3]
Operator protection A, C, H; U20c
Environmental protection E15a, E32a
Storage and disposal E30a

256 fosthiazate

An organophosphorus contact nematicide for potatoes

Products

Nemathorin 10G	Syngenta	10% w/w	FG	11003

Uses
- Potato cyst nematode in ***potatoes***
- Wireworm in ***potatoes*** *(reduction)*

Approval information
- Fosthiazate included in Annex I under EC Directive 91/414

Efficacy guidance
- Apply and incorporate granules in one operation to a uniform depth of 10-15 cm. Deeper incorporation will reduce control
- Application best achieved using equipment such as Horstine Farmery Microband Applicator, Matco or Stocks Micrometer applicators together with a rear mounted powered rotary cultivator
- Granules must not become wet or damp before use

Restrictions
- Contains an anticholinesterase organophosphorus compound. Do not use if under medical advice not to work with such compounds
- Maximum number of treatments 1 per crop
- Product must only be applied using tractor-mounted/drawn direct placement machinery. Do not use air assisted broadcast machinery
- Do not allow granules to stand overnight in the application hopper

- Do not apply more than once every four years on the same area of land
- Consult before using on crops intended for processing

Crop-specific information

- Latest use: at planting
- HI 17 wk for potatoes

Environmental safety

- Dangerous for the environment
- Toxic to aquatic organisms
- Dangerous to game, wild birds and animals
- Dangerous to livestock. Keep all livestock out of treated areas for at least 13 wk
- Risk of explosion if dust in the air
- Incorporation to 10-15 cm and ridging up of treated soil must be carried out immediately after application. Powered rotary cultivators are preferred implements for incorporation but discs, power, spring tine or Dutch harrows may be used provided two passes are made at right angles
- Failure completely to bury granules immediately after application is hazardous to wildlife

Hazard classification and safety precautions

Hazard H03, H11
Risk phrases R22a, R43, R51, R58
Operator protection A, E, G, H, K, M; U02a, U04a, U05a, U07, U09a, U13, U14, U20a
Environmental protection E06b (13 wk); E10a, E15a, E32a, E32d, E33, E34, E38
Consumer protection C02 (17 wk)
Storage and disposal E01, E04, E26, E30a
Medical advice M01, M03, M05a

257 fuberidazole

A benzimidazole (MBC) fungicide available only in mixtures

See also *bitertanol + fuberidazole*
bitertanol + fuberidazole + imidacloprid

258 fuberidazole + imidacloprid + triadimenol

A broad spectrum systemic fungicide and insecticide seed treatment for winter cereals

Products

Baytan Secur	Bayer CropScience	15:117:125 g/l	FS	11253

Uses

- Blue mould in ***winter wheat*** *(seed treatment)*
- Brown foot rot in ***winter barley*** *(seed treatment)*
- Brown rust in ***winter barley*** *(seed treatment)*, ***winter wheat*** *(seed treatment)*
- Bunt in ***winter wheat*** *(seed treatment)*
- Covered smut in ***winter barley*** *(seed treatment)*
- Fusarium foot rot and seedling blight in ***winter barley*** *(seed treatment - reduction)*, ***winter oats*** *(seed treatment - reduction)*, ***winter wheat*** *(seed treatment - reduction)*
- Lodging control in ***winter wheat*** *(seed treatment - reduction)*
- Loose smut in ***winter barley*** *(seed treatment)*, ***winter oats*** *(seed treatment)*, ***winter wheat*** *(seed treatment)*
- Net blotch in ***winter barley*** *(seed treatment - seed-borne only)*
- Powdery mildew in ***winter barley*** *(seed treatment)*, ***winter oats*** *(seed treatment)*, ***winter wheat*** *(seed treatment)*
- Pyrenophora leaf spot in ***winter oats*** *(seed treatment)*
- Septoria leaf spot in ***winter wheat*** *(seed treatment)*
- Septoria seedling blight in ***winter wheat*** *(seed treatment - reduction)*
- Snow rot in ***winter barley*** *(seed treatment - reduction)*
- Virus vectors in ***winter barley*** *(seed treatment)*, ***winter oats*** *(seed treatment)*, ***winter wheat*** *(seed treatment)*

- Wireworm in ***winter barley*** *(reduction of damage)*, ***winter oats*** *(reduction of damage)*, ***winter wheat*** *(reduction of damage)*
- Yellow rust in ***winter barley*** *(seed treatment)*, ***winter wheat*** *(seed treatment)*

Efficacy guidance

- Apply through recommended seed treatment machinery
- Evenness of seed cover improved by simultaneous application of equal volumes of product and water
- Calibrate drill for treated seed and drill at 4 cm into firm, well prepared seedbed
- Use minimum 125 kg treated seed per ha and increase seed rate as drilling season progresses. Lower drilling rates and/or early drilling affect duration of BYDV protection needed and may require follow-up aphicide treatment
- When aphid activity unusually late, or is heavy and prolonged in areas of high risk, and mild weather predominates, follow-up treatment may be required
- In addition to seed-borne diseases early attacks of various foliar, air-borne diseases are controlled or suppressed. See label for details

Restrictions

- Maximum number of treatments 1 per batch of seed
- Do not use on naked oats
- Do not drill treated winter wheat seed after end of Nov
- Do not handle seed unnecessarily
- Do not use treated seed as food or feed
- Treated seed should not be broadcast, but drilled to a depth of 4 cm in a well prepared seedbed
- Germination tests should be done on all batches of seed to be treated to ensure seed viability and suitability for treatment
- Do not use on seed with moisture content above 16%, on sprouted, cracked or skinned seed or on seed already treated with another seed treatment

Crop-specific information

- Latest use: before drilling
- Store treated seed in cool, dry, well-ventilated store and drill as soon as possible, preferably in season of purchase
- Treatment may accentuate effects of adverse seedbed conditions on crop emergence

Environmental safety

- Dangerous to fish or other aquatic life. Do not contaminate surface waters or ditches with chemical or used container
- Dangerous to birds, game and other wildlife. Treated seed should not be left on the soil surface. Bury spillages
- Seed should be drilled to a depth of 4 cm into a well-prepared seed bed
- If seed is present on the soil surface, or if spills have occurred, the field should be harrowed and rolled if conditions are appropriate to ensure good incorporation

Hazard classification and safety precautions

Risk phrases R58
Operator protection A, H; U04a, U05a, U07, U13, U14, U20b
Environmental protection E03, E13b, E32d, E36, E38
Storage and disposal E01, E04, E26, E30a
Treated seed S01, S02, S03, S04c, S05, S06a, S07, S08

259 fuberidazole + triadimenol

A broad spectrum systemic fungicide seed treatment for cereals

Products

Baytan Flowable	Makhteshim	22.5:187.5 g/l	FS	11714

Uses

- Blue mould in ***spring wheat*** *(seed treatment)*, ***triticale*** *(seed treatment)*, ***winter wheat*** *(seed treatment)*
- Brown foot rot in ***spring barley*** *(seed treatment)*, ***spring oats*** *(seed treatment)*, ***spring rye*** *(seed treatment)*, ***spring wheat*** *(seed treatment)*, ***triticale*** *(seed treatment)*, ***winter barley*** *(seed*

treatment), ***winter oats*** *(seed treatment)*, ***winter rye*** *(seed treatment)*, ***winter wheat*** *(seed treatment)*

- Brown rust in ***spring barley*** *(seed treatment)*, ***spring rye*** *(seed treatment)*, ***spring wheat*** *(seed treatment)*, ***winter barley*** *(seed treatment)*, ***winter rye*** *(seed treatment)*, ***winter wheat*** *(seed treatment)*
- Bunt in ***spring wheat*** *(seed treatment)*, ***winter wheat*** *(seed treatment)*
- Covered smut in ***spring barley*** *(seed treatment)*, ***winter barley*** *(seed treatment)*
- Crown rust in ***spring oats*** *(seed treatment)*, ***winter oats*** *(seed treatment)*
- Leaf stripe in ***spring barley*** *(seed treatment)*
- Loose smut in ***spring barley*** *(seed treatment)*, ***spring oats*** *(seed treatment)*, ***spring wheat*** *(seed treatment)*, ***winter barley*** *(seed treatment)*, ***winter oats*** *(seed treatment)*, ***winter wheat*** *(seed treatment)*
- Mildew in ***spring barley*** *(seed treatment)*, ***spring wheat*** *(seed treatment)*, ***winter barley*** *(seed treatment)*, ***winter wheat*** *(seed treatment)*
- Pyrenophora leaf spot in ***spring oats*** *(seed treatment)*, ***winter oats*** *(seed treatment)*
- Septoria in ***spring wheat*** *(seed treatment)*, ***winter wheat*** *(seed treatment)*
- Yellow rust in ***spring barley*** *(seed treatment)*, ***spring wheat*** *(seed treatment)*, ***winter barley*** *(seed treatment)*, ***winter wheat*** *(seed treatment)*

Efficacy guidance

- Apply through recommended seed treatment machinery
- Calibrate drill for treated seed and drill at 2.5-4 cm into firm, well prepared seedbed
- In addition to seed-borne diseases early attacks of various foliar, air-borne diseases are controlled or suppressed. See label for details

Restrictions

- Maximum number of treatments 1 per batch of seed
- Do not use on naked oats
- Do not drill treated winter wheat or rye seed after end of Nov. Seed rate should be increased as drilling season progresses
- Germination tests should be done on all batches of seed to be treated to ensure seed viability and suitability for treatment
- Do not use on seed with moisture content above 16%, on sprouted, cracked or skinned seed or on seed already treated with another seed treatment
- Do not use treated seed as food or feed

Crop-specific information

- Latest use: before drilling
- Treated spring wheat may be drilled in autumn up to end of Nov or from Feb onwards
- Store treated seed in cool, dry, well-ventilated store and drill as soon as possible, preferably in season of purchase
- Treatment may accentuate effects of adverse seedbed conditions on crop emergence

Environmental safety

- Dangerous for the environment
- Harmful to aquatic organisms

Hazard classification and safety precautions

Hazard H11
Risk phrases R52, R58
Operator protection A; U20b
Environmental protection E03, E15a, E32a, E34
Storage and disposal E26, E30a
Treated seed S01, S02, S03, S04a, S05, S06a, S07

260 gibberellins

A plant growth regulator for use in top fruit

Products

Novagib	Fine	10 g/l	SL	08954

Uses
- Growth regulation in ***pears*** *(off-label)*
- Reducing fruit russeting in ***apples***

Specific Off-Label Approvals (SOLAs)
- ***pears*** *(OLA 012756) Dec 2008* [1]

Efficacy guidance
- Apply to apples at completion of petal fall and repeat 3 or 4 times at 7-10 d intervals. Number of sprays and spray interval depend on dose (see labels)

Restrictions
- Maximum total dose on apples equivalent to four full dose treatments
- Good results achieved on Cox's Orange Pippin, Discovery, Golden Delicious and Karmijn apples. For other cultivars test on a small number of trees
- Return bloom may be reduced in yr following treatment
- Consult before treating apple crops grown for processing

Crop-specific information
- HI zero for apples

Hazard classification and safety precautions

Operator protection U20b
Environmental protection E15a, E31b
Storage and disposal E01, E24, E26, E29a, E30a

261 glufosinate-ammonium

A non-selective, non-residual phosphinic acid contact herbicide

Products

1	Challenge	Bayer CropScience	150 g/l	SL	07306
2	Challenge 60	Fargro	60 g/l	SL	08236
3	Finale	Bayer Environ.	120 g/l	SL	10092
4	Harvest	Bayer CropScience	150 g/l	SL	07321
5	Kaspar	Certis	150 g/l	SL	11214

Uses
- Annual and perennial weeds in ***apple orchards, bilberries, blueberries, cane fruit, cherries, cranberries, cultivated land/soil, currants, damsons, fallows, forest, grapevines, non-crop farm areas, pear orchards, plums, potatoes*** *(not for seed)*, ***strawberries, stubbles, sugar beet, tree nuts, vegetables*** [1, 4]
- Annual dicotyledons in ***apple orchards, cherries, currants, damsons, fallows, forest, grapevines, headlands*** *(uncropped)*, ***non-crop farm areas, pear orchards, plums, strawberries, sugar beet, tree nuts, vegetables, ware potatoes*** [5]; ***bush fruit, fruit trees, ornamental trees, shrubs*** [2]; ***cane fruit, cultivated land/soil*** [2, 5]; ***nursery stock, woody ornamentals*** [3]
- Annual grasses in ***apple orchards, cherries, currants, damsons, fallows, forest, grapevines, headlands*** *(uncropped)*, ***non-crop farm areas, pear orchards, plums, strawberries, sugar beet, tree nuts, vegetables, ware potatoes*** [5]; ***bush fruit, fruit trees, ornamental trees, shrubs*** [2]; ***cane fruit, cultivated land/soil*** [2, 5]; ***nursery stock, woody ornamentals*** [3]
- Green cover in ***land temporarily removed from production*** [1, 4, 5]
- Harvest management/desiccation in ***combining peas, ware potatoes*** [5]; ***combining peas*** *(not for seed)*, ***potatoes*** [1, 4]; ***linseed, spring field beans, spring oilseed rape, winter field beans, winter oilseed rape*** [1, 4, 5]
- Line marking preparation in ***managed amenity turf*** [3]
- Perennial dicotyledons in ***apple orchards, cherries, currants, damsons, fallows, forest, grapevines, headlands*** *(uncropped)*, ***land temporarily removed from production, non-crop farm areas, pear orchards, plums, strawberries, tree nuts*** [5]; ***bush fruit, cultivated land/soil, fruit trees, ornamental trees, shrubs*** [2]; ***cane fruit*** [2, 5]; ***nursery stock, woody ornamentals*** [3]
- Perennial grasses in ***apple orchards, cherries, currants, damsons, fallows, forest, grapevines, headlands*** *(uncropped)*, ***land temporarily removed from production, non-crop farm areas,***

pear orchards, plums, strawberries, tree nuts [5]; ***bush fruit, cultivated land/soil, fruit trees, ornamental trees, shrubs*** [2]; ***cane fruit*** [2, 5]; ***nursery stock, woody ornamentals*** [3]
- Sward destruction in ***grassland*** [1, 4, 5]

Efficacy guidance
- Activity quickest under warm, moist conditions. Light rainfall 3-4 h after application will not affect activity. Do not spray wet foliage or if rain likely within 6 h
- For weed control uses treat when weeds growing actively. Deep rooted weeds may require second treatment [2]
- On uncropped headlands apply in May/Jun to prevent weeds invading field
- Glufosinate kills all green tissue but does not harm mature bark

Restrictions
- Maximum number of treatments 4 per crop for potatoes (including 2 desiccant uses); 2 per crop (including 1 desiccant use) for oilseed rape, dried peas, field beans, linseed, wheat and barley; 3 (2 [2]) per yr for fruit and forestry; 2 per yr for strawberries, non-crop land and land temporarily removed from production; 1 per crop for sugar beet, vegetables and other crops; 1 per yr for grassland destruction, on cultivated land prior to planting edible crops
- Pre-harvest desiccation sprays should not be used on seed crops of wheat, barley, peas or potatoes but may be used on seed crops of oilseed rape, field beans and linseed [1, 4, 5]
- Do not desiccate potatoes in exceptionally wet weather or in saturated soil. See label for details
- Do not spray potatoes after emergence if grown from small or diseased seed or under very dry conditions
- Application for weed control and line-marking uses in sports turf must be between 1 Mar and 30 Sep [3]
- Do not allow spray to contact dormant or green buds, suckers, damaged or green bark and foliage of wanted plants [2]
- Do not use straw from treated crops as animal feed or bedding [1, 4]
- Do not spray hedge bottoms [1, 4]

Crop-specific information
- Latest use: 30 Sep for use on non-crop land, top fruit, soft fruit, cane fruit, bush fruit, forestry; pre-drilling, pre-planting or pre-emergence in sugar beet, vegetables and other crops; before winter dormancy for grassland destruction.
- HI potatoes, oilseed rape, combining peas, field beans, linseed 7 d; wheat, barley 14 d
- Ploughing or other cultivations can follow 4 h after spraying
- Crops can normally be sown/planted immediately after spraying or sprayed post-drilling. On sand, very light or immature peat soils allow at least 3 d before sowing/planting or expected emergence
- For weed control in potatoes apply pre-emergence or up to 10% emergence on earlies and seed crops, up to 40% on maincrop, on plants up to 15 cm high
- In sugar beet and vegetables apply just before crop emergence, using stale seedbed technique
- In top and soft fruit, grapevines and forestry apply between 1 Mar and 30 Sep as directed sprays
- For grass destruction apply before winter dormancy occurs. Heavily grazed fields should show active regrowth. Plough from the day after spraying
- Apply pre-harvest desiccation treatments 10-21 d before harvest (14-21 d for oilseed rape). See label for timing details on individual crops
- For potato haulm desiccation apply to listed varieties (not seed crops) at onset of senescence, 14-21 d before harvest

Environmental safety
- Harmful to fish or other aquatic life. Do not contaminate surface waters or ditches with chemical or used container
- Keep livestock out of treated areas until foliage of any poisonous weeds such as ragwort has died and become unpalatable
- Treated pea haulm may be fed to livestock from 7 d after spraying, treated grain from 14 d [1, 4]
- Product supplied in small volume returnable container - see label for filling and mixing instructions [4]

Hazard classification and safety precautions

Hazard H03 [1, 4, 5]; H04 [1, 3]

Risk phrases R21, R22a [1, 4, 5]; R36 [1, 3]; R41 [4, 5]

Operator protection A [1, 2, 4, 5]; B [3]; C, H, M; U02a [2]; U04a, U13, U14, U15, U19a [1, 4, 5]; U05a [1, 3-5]; U08 [1-5]; U11 [4, 5]; U20a [1, 2, 4, 5]; U20b [3]

Environmental protection E07a, E31b [3-5]; E07b [1, 2]; E09 [1]; E13c [1-5]; E22b [5]; E22c [4]; E31a [2]; E32d [4, 5]; E34 [1, 4, 5]
Storage and disposal E01, E04 [1, 3-5]; E26 [3]; E30a [1-5]
Medical advice M03 [1, 4, 5]

262 glyphosate

A translocated non-residual phosphonic acid herbicide

Products

1	Amega Pro TMF	Nufarm UK	480 g/l	SL	09630
2	Asteroid	Headland	360 g/l	SL	11118
3	Azural	Cardel	360 g/l	SL	11668
4	Barclay Barbarian	Barclay	360 g/l	SL	11362
5	Barclay Gallup 360	Barclay	360 g/l	SL	11369
6	Barclay Gallup Amenity	Barclay	360 g/l	SL	06753
7	Barclay Gallup Biograde 360	Barclay	360 g/l	SL	11370
8	Barclay Gallup Biograde Amenity	Barclay	360 g/l	SL	11363
9	Barclay Gallup Hi-Aktiv	Barclay	490 g/l	SL	11371
10	Buggy SG	Sipcam	36% w/w	WG	08573
11	CDA Vanquish Biactive	Bayer Environ.	120 g/l	RH	10950
12	Clinic	Nufarm UK	360 g/l	SL	09378
13	Dow Agrosciences Glyphosate 360	Dow	360 g/l	SL	11552
14	Envision	Headland	450 g/l	SL	10569
15	Flame	Dow	360 g/l	SL	11583
16	Glyfos	Cheminova	360 g/l	SL	07109
17	Glyfos	Headland	360 g/l	SL	10995
18	Glyfos Gold	Nomix Enviro	360 g/l	SL	10570
19	Glyfos ProActive	Nomix Enviro	360 g/l	SL	07800
20	Glyphogan	Makhteshim	360 g/l	SL	05784
21	Glyphosate 360	SumiAgro Amenity	360 g/l	SL	11060
22	Glyphosate 360	Monsanto	360 g/l	SL	11726
23	Greenaway Gly-490	Greenaway	490 g/l	SL	11064
24	Greencrop Gypsy	Greencrop	360 g/l	SL	09432
25	Habitat	SumiAgro Amenity	360 g/l	SL	10498
26	Hilite	Nomix Enviro	144 g/l	RH	06261
27	I T Glyphosate	I T Agro	360 g/l	SL	07212
28	Kernel	Cheminova	480 g/l	SL	10616
29	Kernel	Headland	480 g/l	SL	10993
30	KN 540	Monsanto	540 g/l	SL	12009
31	Landgold Glyphosate 360	Landgold	360 g/l	SL	05929
32	Manifest	Headland	360 g/l	SL	11041
33	Nufosate	Nufarm UK	360 g/l	SL	10096
34	Reliance	Cardel	360 g/l	SL	11667
35	Romany	Greencrop	360 g/l	SL	11000
36	Roundup Amenity	Monsanto	360 g/l	SL	10319
37	Roundup Biactive	Monsanto	360 g/l	SL	10320
38	Roundup Gold	Monsanto	450 g/l	SL	10975
39	Roundup Max	Monsanto	68% w/w	SG	10231
40	Roundup Pro Biactive	Monsanto	360 g/l	SL	10330
41	Roundup Pro-Green	Monsanto	450 g/l	SL	11907
42	Roundup ProVide	Monsanto	68% w/w	SL	11721
43	Roundup Rail	Monsanto	360 g/l	SL	11874
44	Roundup Ultra ST	Monsanto	450 g/l	SL	10199
45	Samurai	Monsanto	360 g/l	SL	10334
46	Scorpion	Cardel	360 g/l	SL	11656
47	Spasor	Bayer Environ.	360 g/l	SL	09945
48	Spasor Biactive	Bayer Environ.	360 g/l	SL	09940
49	Stacato	Sipcam	360 g/l	SL	05892
50	Sting ECO	Monsanto	120 g/l	SL	10337
51	Stirrup	Nomix Enviro	144 g/l	RH	06132
52	Tangent	Headland Amenity	450 g/l	SL	11872
53	Touchdown	Syngenta	330 g/l	SL	10538
54	Touchdown Quattro	Syngenta	360 g/l	SL	10608
55	Trustee Elite	Barclay	450 g/l	SL	11556
56	Tumbleweed Pro-Active	Scotts	450 g/l	SL	11963
57	Typhoon 360	Makhteshim	360 g/l	SL	11175

Uses

- Annual and perennial weeds in ***all top fruit, hedges, managed amenity turf*** *(pre-establishment)*, ***ornamental plant production*** [26, 51]; ***amenity areas*** [6-9, 23, 40, 48]; ***amenity areas*** *(wiper application)*, ***broad-leaved trees, conifers, farm buildings/yards, land clearance*** [40]; ***amenity grass, amenity grass*** *(wiper application)*, ***amenity trees and shrubs*** *(wiper application)*, ***woody ornamentals*** [40, 47]; ***amenity trees and shrubs*** [26, 40, 47, 51]; ***amenity vegetation*** [11, 26, 40, 51]; ***apple orchards, cherries, damsons, pear orchards, plums*** [2-5, 7, 9, 10, 12-17, 20, 22-24, 28-30, 32, 34, 35, 37, 38, 44-46, 55, 57]; ***asparagus*** *(off-label)*, ***blackcurrants*** *(off-label)*, ***blueberries*** *(off-label)*, ***chestnuts*** *(off-label)*, ***cob nuts*** *(off-label)*, ***grapevines*** *(off-label)*, ***hazel nuts*** *(off-label)*, ***rhubarb*** *(off-label)*, ***walnuts*** *(off-label)* [4]; ***combining peas, durum wheat, spring barley, spring field beans, spring oilseed rape, spring wheat, winter barley, winter field beans, winter oilseed rape, winter wheat*** [2, 4, 54]; ***cultivated land/soil*** [1-3, 12, 13, 15, 22, 24, 35, 45]; ***farm forestry*** [42]; ***fencelines, walls*** [40, 48]; ***field crops*** *(wiper application)* [3, 12, 13, 15, 22, 24, 35, 45]; ***forest*** [4-10, 12-14, 16-21, 23-26, 28, 29, 32, 33, 35, 36, 40-42, 51, 52, 55-57]; ***grassland*** [1, 3, 10, 12, 13, 15, 16, 20, 22, 28, 34, 35, 37, 38, 45, 46]; ***hard surfaces*** [18, 23, 40-43, 48, 52, 54, 56]; ***industrial sites*** [6-9, 18, 19, 21, 25, 33, 36, 40]; ***land immediately adjacent to aquatic areas*** [23, 41, 42, 56]; ***land not intended to bear vegetation*** [3-11, 13-17, 19, 21-23, 25, 26, 28, 29, 32-37, 39, 40, 44-46, 51, 53, 55, 57]; ***land prior to cultivation*** [52]; ***land temporarily removed from production*** [2, 10, 30, 37, 38, 54]; ***linseed*** [2, 54]; ***mustard*** [54]; ***natural surfaces not intended to bear vegetation, permeable surfaces overlying soil*** [18, 23, 41, 42, 52, 54, 56]; ***non-crop areas*** [38, 40, 48]; ***non-crop farm areas*** [3-5, 12, 13, 22, 24, 35, 45, 49]; ***paths and drives*** [18, 25, 33, 36, 40, 48]; ***railway tracks*** [43]; ***road verges*** [18, 19, 25, 33, 36, 40, 48]; ***spring oats, winter oats*** [2, 4]; ***stubbles*** [1-4, 10, 12, 13, 15, 16, 20, 22-24, 27, 28, 30, 31, 34, 35, 37-39, 45, 46, 49, 53, 54]
- Annual dicotyledons in ***all edible crops, all non-edible crops, hard surfaces, land not intended to bear vegetation, permeable surfaces overlying soil*** [50]; ***combining peas, durum wheat, leeks, linseed, mustard, onions, spring field beans, spring oilseed rape, sugar beet, swedes, turnips, vining peas, winter field beans, winter oilseed rape*** [39]; ***cultivated land/soil*** [4, 5, 7, 9, 23]; ***spring barley, spring oats, spring wheat, winter barley, winter oats, winter wheat*** [14, 17, 29, 32, 39, 55, 57]; ***stubbles*** [5, 7, 9, 44, 50]
- Annual grasses in ***all edible crops, all non-edible crops, hard surfaces, land not intended to bear vegetation, permeable surfaces overlying soil*** [50]; ***combining peas, durum wheat, leeks, linseed, mustard, onions, spring barley, spring field beans, spring oats, spring oilseed rape, spring wheat, sugar beet, swedes, turnips, vining peas, winter barley, winter field beans, winter oats, winter oilseed rape, winter wheat*** [39]; ***cultivated land/soil*** [4, 5, 7, 9, 14, 17, 23, 29, 32, 44, 55, 57]; ***stubbles*** [5, 7, 9, 44, 50]
- Annual weeds in ***durum wheat, leeks, linseed, mustard, onions, spring barley, spring field beans, spring oats, spring oilseed rape, spring wheat, sugar beet, swedes, turnips, winter barley, winter field beans, winter oats, winter oilseed rape, winter wheat*** [30, 37, 38]; ***grassland*** [30, 34, 37, 38, 46, 49]; ***peas*** [37, 38]; ***vining peas*** [30]
- Aquatic weeds in ***open waters*** [23, 41, 42, 56]
- Black bent in ***combining peas, spring field beans, spring oilseed rape, stubbles, winter field beans, winter oilseed rape*** [4, 5, 7, 9, 14, 17, 23, 29, 32, 44, 55, 57]; ***durum wheat, spring barley, spring oats, spring wheat, winter barley, winter oats, winter wheat*** [4, 5, 7, 9, 23, 44]; ***linseed*** [7, 9, 14, 17, 29, 32, 44, 55, 57]
- Bolters in ***sugar beet*** *(wiper application)* [3-5, 7, 9, 12, 13, 15, 22-24, 30, 34, 35, 37, 38, 44-46]
- Bracken in ***amenity trees and shrubs, fencelines, land clearance, non-crop areas, paths and drives*** [40]; ***farm forestry*** [42]; ***forest*** [4-10, 12-14, 16-21, 23-25, 28, 29, 32, 33, 35, 36, 40-42, 52, 55-57]; ***grassland*** [53]; ***tolerant conifers*** [10, 16, 19, 28, 40]
- Brambles in ***tolerant conifers*** [10, 16, 19, 28, 40]
- Chemical thinning in ***forest*** [6-9, 12, 14, 17, 18, 20, 21, 23, 25, 29, 32, 33, 35, 36, 40, 41, 52, 55-57]
- Couch in ***combining peas, spring field beans, winter field beans*** [1-5, 7, 9, 10, 12-17, 20, 22-24, 28-32, 34, 35, 37, 38, 44-46, 54, 55, 57]; ***durum wheat*** [1, 2, 4, 5, 7, 9, 16, 23, 28, 30, 31, 34, 37, 38, 44, 46, 54]; ***farm forestry*** [42]; ***forest*** [6, 8, 9, 12, 20, 23, 24, 35, 40-42, 56]; ***grassland*** [1, 4, 10, 16, 28]; ***land not intended to bear vegetation, oilseed rape for industrial use*** [53]; ***linseed*** [1, 2, 7, 9, 10, 14, 16, 17, 20, 28-30, 32, 34, 37, 38, 44, 46, 54, 55, 57]; ***mustard*** [10, 20, 30, 34, 37, 38, 46, 54]; ***spring barley, spring wheat, winter barley, winter wheat*** [1-5, 7, 9, 10, 12, 13, 15, 16, 20, 22-24, 28, 30, 31, 34, 35, 37, 38, 44-46, 49, 54]; ***spring oats, winter oats*** [1-5, 7, 9, 10, 12, 13, 15, 16, 20, 22-24, 28, 30, 31, 34, 35, 37, 38, 44-46, 49]; ***spring oilseed rape, winter oilseed***

rape [1-5, 7, 9, 10, 12-17, 20, 22-24, 28-32, 34, 35, 37, 38, 44-46, 49, 53-55, 57]; ***stubbles*** [1-5, 7, 9, 10, 12-17, 20, 22-24, 27-32, 34, 35, 37, 38, 44-46, 49, 53-55, 57]

- Creeping bent in ***combining peas, spring field beans, spring oilseed rape, stubbles, winter field beans, winter oilseed rape*** [4, 5, 7, 9, 14, 17, 23, 29, 32, 44, 55, 57]; ***cultivated land/soil*** [1]; ***durum wheat, spring barley, spring oats, spring wheat, winter barley, winter oats, winter wheat*** [4, 5, 7, 9, 23, 44]; ***linseed*** [7, 9, 14, 17, 29, 32, 44, 55, 57]
- Green cover in ***land temporarily removed from production*** [2, 5, 7, 9, 14-17, 23, 28, 29, 32, 39, 44, 53, 55, 57]
- Harvest management/desiccation in ***combining peas, spring field beans, winter field beans*** [1, 3, 10, 12, 13, 15, 16, 20, 22, 24, 28, 30, 31, 34, 35, 37, 38, 45, 46, 53]; ***durum wheat*** [1, 3, 12, 13, 15, 16, 22, 24, 28, 30, 31, 34, 35, 37, 38, 45, 46]; ***linseed*** [1, 3, 10, 12, 13, 16, 20, 22, 23, 28, 30, 34, 35, 37, 38, 44-46, 53]; ***mustard*** [1, 3, 10, 12, 13, 15, 20, 22, 27, 30, 34, 35, 37, 38, 45, 46, 53, 54]; ***oilseed rape for industrial use*** [53]; ***spring barley, winter barley*** [1, 10, 14, 16, 17, 20, 24, 27-32, 34, 37, 38, 46, 49, 53, 55, 57]; ***spring oats, winter oats*** [1, 3, 10, 12-17, 20, 22, 24, 27-32, 34, 35, 37, 38, 45, 46, 49, 55, 57]; ***spring oilseed rape, winter oilseed rape*** [1-4, 10, 12-17, 20, 22-24, 27-32, 34, 35, 37, 38, 44-46, 49, 53-55, 57]; ***spring wheat, winter wheat*** [1, 3, 10, 12-17, 20, 22, 24, 27-32, 34, 35, 37, 38, 45, 46, 49, 53, 55, 57]
- Heather in ***farm forestry*** [42]; ***forest*** [4-10, 13, 14, 16-19, 21, 23, 25, 28, 29, 32, 33, 35, 36, 41, 42, 52, 55-57]
- Japanese knotweed in ***land not intended to bear vegetation*** [53]
- Perennial dicotyledons in ***combining peas, spring field beans, spring oilseed rape, winter field beans, winter oilseed rape*** [5, 7, 9, 14, 17, 29, 32, 44, 55, 57]; ***durum wheat, spring barley, spring oats, spring wheat, winter barley, winter oats, winter wheat*** [5, 7, 9, 44]; ***forest*** [18, 21, 25, 33, 36, 52]; ***forest*** *(wiper application)* [40]; ***grassland*** *(wiper application)* [30, 34, 37, 38, 46]; ***linseed*** [7, 9, 14, 17, 29, 32, 44, 55, 57]; ***non-crop farm areas*** [7]
- Perennial grasses in ***aquatic situations*** [2, 4-10, 12-21, 23, 25, 28, 29, 32, 33, 35, 36, 40-42, 47, 48, 52, 55-57]; ***enclosed waters, land immediately adjacent to aquatic areas*** [18, 52]; ***tolerant conifers*** [10, 16, 19, 28, 40]
- Pre-harvest desiccation in ***linseed*** [7, 9]; ***poppies*** *(off-label - for morphine production)* [37]; ***spring oilseed rape, winter oilseed rape*** [5, 7, 9]
- Reeds in ***aquatic situations*** [2, 4-10, 12-21, 23, 25, 28, 29, 32, 33, 35-37, 40-42, 47, 48, 52, 55-57]; ***enclosed waters, land immediately adjacent to aquatic areas*** [18, 23, 41, 42, 52, 56]
- Rhododendrons in ***farm forestry*** [42]; ***forest*** [6, 8-10, 13, 14, 16-19, 21, 23, 25, 28, 29, 32, 33, 35, 36, 40-42, 52, 55-57]
- Rushes in ***aquatic situations*** [2, 4-10, 12-21, 23, 25, 28, 29, 32, 33, 35-37, 40-42, 47, 48, 52, 55-57]; ***enclosed waters, land immediately adjacent to aquatic areas*** [18, 23, 41, 42, 52, 56]; ***forest*** [20]; ***grassland*** [53]
- Sedges in ***aquatic situations*** [2, 4-10, 12-21, 23, 25, 28, 29, 32, 33, 35-37, 40-42, 47, 48, 52, 55-57]; ***enclosed waters, land immediately adjacent to aquatic areas*** [18, 23, 41, 42, 52, 56]
- Sucker control in ***apple orchards, cherries, damsons, pear orchards, plums*** [22, 34, 37, 38, 45, 46]
- Sward destruction in ***amenity grass*** [40]; ***grassland*** [1-5, 7, 9, 10, 12-17, 20, 22-24, 27-32, 34, 35, 37, 38, 44-46, 49, 53-55, 57]; ***rotational grassland*** [50]
- Total vegetation control in ***amenity areas*** [40, 47]; ***amenity vegetation, land not intended to bear vegetation*** [11, 19, 40]; ***fencelines, road verges*** [19, 40, 47]; ***hard surfaces*** [19, 40]; ***industrial sites*** [19, 40, 47, 48]
- Volunteer cereals in ***all edible crops, all non-edible crops*** [50]; ***combining peas*** [39]; ***cultivated land/soil*** [4, 5, 7, 9, 14, 17, 23, 29, 32, 44, 55, 57]; ***durum wheat, leeks, linseed, mustard, onions, spring barley, spring field beans, spring oats, spring oilseed rape, spring wheat, sugar beet, swedes, turnips, winter barley, winter field beans, winter oats, winter oilseed rape, winter wheat*** [30, 37-39]; ***peas*** [37, 38]; ***stubbles*** [1-5, 7, 9, 10, 12-17, 20, 22-24, 27-32, 34, 35, 37, 38, 44-46, 49, 50, 53-55, 57]; ***vining peas*** [30, 39]
- Volunteer oilseed rape in ***stubbles*** [53]
- Volunteer potatoes in ***stubbles*** [1-5, 7, 9, 10, 12, 13, 15, 16, 20, 22-24, 27, 28, 30, 31, 34, 35, 37, 38, 44-46, 49, 53, 54]
- Waterlilies in ***aquatic situations*** [2, 4-10, 12-21, 23, 25, 28, 29, 32, 33, 35, 36, 40-42, 47, 48, 52, 55-57]; ***enclosed waters*** [23, 41, 42, 56]
- Weed beet in ***sugar beet*** *(wiper application)* [3, 12, 15, 22, 24, 35, 45]

- Wild oats in ***durum wheat*** [3, 13, 15, 22, 35, 45]; ***spring barley, spring oats, spring wheat, winter barley, winter oats, winter wheat*** [3, 12, 13, 15, 22, 35, 45]
- Woody weeds in ***farm forestry*** [42]; ***forest*** [4-10, 12-14, 16-21, 23-25, 28, 29, 32, 33, 35, 36, 40-42, 52, 55-57]; ***tolerant conifers*** [10, 16, 19, 28, 40]

Specific Off-Label Approvals (SOLAs)

- ***asparagus, blackcurrants, blueberries, chestnuts, cob nuts, grapevines, hazel nuts, rhubarb, walnuts*** *(OLA 032129) Dec 2008* [4]
- ***poppies*** *(for morphine production) (OLA 031409) Dec 2008* [37]

Efficacy guidance

- For best results apply to actively growing weeds with enough leaf to absorb chemical
- A rainfree period of at least 6 h (preferably 24 h) should follow spraying
- Adjuvants are obligatory for some products and recommended for some uses with others. See labels
- Mixtures with other pesticides or fertilizers may lead to reduced control.
- Products are formulated as isopropylamine, ammonium, potassium, or trimesium salts of glyphosate and may vary in the details of efficacy claims. See individual product labels
- Some products in ready-to-use formulations for use through hand-held applicators. See label for instructions [11, 26, 51]
- When applying products through rotary atomisers the spray droplet spectra must have a minimum Volume Median Diameter (VMD) of 200 microns
- With wiper application weeds should be at least 10 cm taller than crop
- Annual weed grasses should have at least 5 cm of leaf and annual broad-leaved weeds at least 2 expanded true leaves
- Perennial grass weeds should have 4-5 new leaves and be at least 10 cm long when treated. Perennial broad-leaved weeds should be treated at or near flowering but before onset of senescence
- Volunteer potatoes and polygonums are not controlled by harvest-aid rates
- Bracken must be treated at full frond expansion
- Fruit tree suckers best treated in late spring
- Chemical thinning treatment can be applied as stump spray or stem injection
- In order to allow translocation, do not cultivate before spraying and do not apply other pesticides, lime, fertilizer or farmyard manure within 5 d of treatment
- Recommended intervals after treatment and before cultivation vary. See labels

Restrictions

- Maximum number of treatments normally 1 per crop or season on field and edible crops and unrestricted for non-crop uses. Check labels for details
- Do not treat cereals grown for seed or undersown crops
- Consult grain merchant before treating crops grown on contract or intended for malting
- Do not use treated straw as a mulch or growing medium for horticultural crops
- Do not use on grassland if the crop is to be used as animal feed or bedding
- For use in nursery stock, shrubberies, orchards, grapevines and tree nuts care must be taken to avoid contact with the trees. Do not use in orchards established less than 2 yr and keep off low-lying branches
- Certain conifers may be sprayed overall in dormant season. See label for details
- Use a tree guard when spraying in established forestry plantations
- Do not spray root suckers in orchards in late summer or autumn
- Do not use under glass or polythene as damage to crops may result
- Do not use wiper techniques in soft fruit crops
- Do not use trimesium salt products in aquatic situations or forestry
- Do not mix, store or apply in galvanised or unlined mild steel containers or spray tanks
- Do not leave diluted chemical in spray tanks for long periods and make sure that tanks are well vented

Crop-specific information

- Harvest intervals: 4 wk for blackcurrants, blueberries; 14 d for linseed, oilseed rape; 5-7 d for all other edible crops. Check label for exact details
- Latest use: 2-14 d before cultivating, drilling or planting a crop in treated land; after harvest (post-leaf fall) but before bud formation in the following season for nuts and most fruit and vegetable crops; before fruit set for grapevines. See labels for details

Following crops guidance

- Decaying remains of plants killed by spraying must be dispersed before direct drilling

Environmental safety

- Products differ in their hazard and environmental safety classification. See labels
- Do not dump surplus herbicide in water or ditch bottoms
- Maximum permitted concentration in treated water 0.2 ppm
- The Environment Agency or Local River Purification Authority must be consulted before use in or near water
- LERAP Category B [30]
- Take extreme care to avoid drift and possible damage to neighbouring crops or plants
- Treated poisonous plants must be removed before grazing or conserving
- Do not use in covered areas such as greenhouses or under polythene
- For field edge treatment direct spray away from hedge bottoms
- Exclude livestock from treated areas. Livestock may not graze or be fed the treated forage nor may it be used for hay, silage or bedding

Hazard classification and safety precautions

Hazard H03 [13, 15, 53]; H04 [1, 3-6, 10, 12, 18, 20-22, 24, 27, 30, 31, 33-36, 39, 42, 43, 45-47, 49-51, 57]; H11 [3-6, 12, 13, 15, 18, 20, 22, 24, 28-36, 39, 42, 43, 45, 46, 50, 51, 57]

Risk phrases R20 [13, 15]; R22a [53]; R36 [1, 3, 6, 12, 13, 15, 18, 20-22, 27, 33-36, 43, 45-47, 49, 51]; R38 [1, 6, 12, 21, 24, 27, 30, 31, 35, 47, 49]; R41 [4, 5, 10, 24, 31, 35, 39, 42, 57]; R50 [30, 50]; R51 [3-6, 12, 13, 15, 18, 20, 22, 24, 28, 29, 31-36, 39, 42, 43, 45, 46, 51, 57]; R52 [1, 7, 8, 25, 41, 53]; R58 [1, 3-6, 12, 13, 15, 18, 20, 22, 24, 28-36, 39, 41-43, 45, 46, 51, 53, 57]

Operator protection A [1-49, 51-57]; C [1-10, 12-29, 31-43, 45-47, 49, 51-53, 55-57]; D [2, 14, 16-19, 28, 29, 32, 41, 52, 55, 56]; F [3, 12, 15, 20-22, 33, 34, 36, 45, 46]; H [2, 3, 7-12, 14-20, 22-25, 28-30, 32, 34, 36-42, 44-46, 48, 52, 53, 55-57]; M [2-5, 7-26, 28-30, 32-42, 44-48, 51-57]; N [3-5, 12, 13, 15, 20-22, 24, 26, 33-36, 45-47, 51, 54, 57]; U02a [1, 4-10, 12, 13, 15-17, 20-29, 31-33, 35, 39, 44, 47, 49, 51, 55, 56]; U05a [1, 3-6, 10, 12, 13, 15, 18, 20-22, 24, 26, 27, 30, 31, 33-36, 39, 42, 43, 45-47, 49, 51, 53, 54, 57]; U07 [43, 53]; U08 [1, 3-10, 12, 13, 15-18, 20-29, 31-36, 39, 43-47, 49, 51, 55, 56]; U09a [26, 42, 51, 57]; U11 [3, 12, 13, 15, 18, 20, 22, 24, 31, 33-36, 39, 42, 43, 45, 46, 50, 51, 57]; U15 [3, 12, 18, 20, 22, 33, 34, 36, 43, 45, 46]; U19a [1, 3-13, 15-18, 20-29, 31-36, 39, 42-47, 49, 51, 55-57]; U20a [6-8, 16, 25-27, 30, 37, 38, 48, 53, 54]; U20b [1-5, 9-15, 17-24, 26, 28, 29, 31-36, 38-47, 49-52, 55-57]

Environmental protection E06a [54]; E13c [1, 9, 10, 12, 16, 17, 20, 21, 23, 26, 27, 38, 47, 49, 51]; E15a [2-8, 11-15, 18-20, 22, 24, 25, 28-46, 48, 50-56]; E16a, E16b [30]; E19a [2, 4, 5, 7-10, 12-17, 23, 25, 28, 29, 32, 33, 35, 47, 52, 55]; E19b [57]; E23 [56]; E31a [11]; E31b [1, 2, 4-10, 12-15, 20, 21, 23-25, 27-29, 31, 35, 48, 49, 52, 55, 57]; E31c [3, 16-19, 22, 30, 32-34, 36-38, 40, 41, 44-46, 50, 54, 56]; E32a [26, 39, 42, 51]; E32d [3, 12, 13, 15, 18, 20, 22, 30, 33, 34, 36, 39, 42, 43, 45, 46, 50, 53, 57]; E32e [28, 29, 32]; E33, E36 [43, 53]; E34 [1, 11, 20, 24, 43, 53]; E38 [3, 12, 13, 15, 18, 20, 22, 30, 33, 34, 36, 39, 42, 43, 45, 46, 50, 53]

Storage and disposal E01 [1-13, 15-22, 24-43, 45-49, 51-54, 56, 57]; E04 [1, 3-13, 15, 18-22, 24-27, 30, 31, 33-43, 45-49, 51, 53, 54, 56, 57]; E26 [1, 3-9, 11, 16, 18-26, 30, 31, 33-36, 39-46, 50, 51, 53, 54, 56]; E29a [39, 42, 53]; E30a [1-57]

Medical advice M03 [53]; M05a [57]

263 guazatine

A guanidine fungicide seed dressing for cereals

Products

1	Panoctine	Makhteshim	300 g/l	LS	11404
2	Ravine	Makhteshim	300 g/l	LS	10095

Uses

- Brown foot rot in ***spring barley*** *(seed treatment - reduction)*, ***spring oats*** *(seed treatment - reduction)*, ***winter barley*** *(seed treatment - reduction)*, ***winter oats*** *(seed treatment - reduction)* [1]
- Foot rot in ***spring barley*** *(seed treatment)*, ***winter barley*** *(seed treatment)* [1]
- Fusarium foot rot and seedling blight in ***spring barley*** *(seed treatment - reduction)*, ***spring oats*** *(seed treatment - reduction)*, ***winter barley*** *(seed treatment - reduction)*, ***winter oats*** *(seed*

treatment - reduction) [1, 2]; ***spring wheat*** *(seed treatment - reduction)*, ***winter wheat*** *(seed treatment - reduction)* [2]
- Leaf stripe in ***spring barley*** *(seed treatment)*, ***winter barley*** *(seed treatment)* [1]
- Net blotch in ***spring barley*** *(seed treatment)*, ***winter barley*** *(seed treatment)* [1]
- Pyrenophora leaf spot in ***spring oats*** *(seed treatment)*, ***winter oats*** *(seed treatment)* [1]
- Septoria seedling blight in ***spring wheat*** *(seed treatment)*, ***winter wheat*** *(seed treatment)* [2]

Efficacy guidance
- Apply with conventional seed treatment machinery

Restrictions
- Maximum number of treatments 1 per batch
- Do not treat grain with moisture content above 16% and do not allow moisture content of treated seed to exceed 16%
- Do not apply to cracked, split or sprouted seed
- Do not use treated seed as food or feed

Crop-specific information
- Latest use: pre-drilling
- After treating, bag seed immediately and keep in dry, draught-free store
- Treatment may lower germination capacity, particularly if seed grown, harvested or stored under adverse conditions

Environmental safety
- Dangerous for the environment
- Very toxic to aquatic organisms

Hazard classification and safety precautions

Hazard H03, H11
Risk phrases R20, R22a, R37, R41, R43, R50, R58
Operator protection A, C, D, H; U05a, U20b [1, 2]; U09a, U14, U15 [1]
Environmental protection E13c [1]; E15a [2]; E31a, E34, E38 [1, 2]; E33 [1] (returnable container)
Storage and disposal E01, E04, E26, E30a
Treated seed S01, S02, S04b, S05, S06a, S07
Medical advice M03 [1]; M04 [1, 2]

264 guazatine + imazalil

A fungicide seed treatment for barley and oats

Products

1	Panoctine Plus	Makhteshim	300:25 g/l	LS	11757
2	Ravine Plus	Makhteshim	300:25 g/l	LS	11832

Uses
- Brown foot rot in ***spring barley*** *(seed treatment)*, ***winter barley*** *(seed treatment)*
- Covered smut in ***spring barley*** *(seed treatment)*, ***winter barley*** *(seed treatment)*
- Leaf stripe in ***spring barley*** *(seed treatment)*, ***winter barley*** *(seed treatment)*
- Net blotch in ***spring barley*** *(seed treatment - moderate control)*, ***winter barley*** *(seed treatment - moderate control)*
- Pyrenophora leaf spot in ***spring oats*** *(seed treatment)*, ***winter oats*** *(seed treatment)*
- Seedling blight and foot rot in ***spring barley*** *(seed treatment - moderate control)*, ***winter barley*** *(seed treatment - moderate control)*
- Snow mould in ***spring barley*** *(seed treatment)*, ***winter barley*** *(seed treatment)*

Approval information
- Imazalil included in Annex I under EC Directive 91/414

Efficacy guidance
- Apply with conventional seed treatment machinery

Restrictions
- Maximum number of treatments 1 per batch
- Do not mix with other formulations
- Do not treat grain with moisture content above 16% and do not allow moisture content of treated seed to exceed 16%

- Do not apply to cracked, split or sprouted seed
- Do not store treated seed for more than 6 mth
- Treatment may lower germination capacity, particularly if seed grown, harvested or stored under adverse conditions

Crop-specific information

- Latest use: pre-drilling

Environmental safety

- Dangerous for the environment
- Very toxic to aquatic organisms
- Do not use treated seed as food or feed

Hazard classification and safety precautions

Hazard H02, H11
Risk phrases R22a, R23, R37, R41, R43, R50, R58
Operator protection A, C, D, H; U05a, U10, U11, U14, U19a, U20b
Environmental protection E15a, E31b, E34, E38
Storage and disposal E01, E04, E30a
Treated seed S02, S04b, S05, S07
Medical advice M04

265 guazatine + triticonazole

A fungicide seed treatment for wheat

Products

Premis	BASF	150:12.5 g/l	FS	11742

Uses

- Bunt in ***spring wheat*** *(seed treatment)*, ***winter wheat*** *(seed treatment)*
- Fusarium foot rot and seedling blight in ***spring wheat*** *(seed treatment)*, ***winter wheat*** *(seed treatment)*

Efficacy guidance

- Apply undiluted using conventional seed treatment machinery
- Ensure seed drill is correctly calibrated before sowing

Restrictions

- Maximum number of treatments 1 per seed batch
- Do not treat grain with moisture content above 16% and do not allow moisture content of treated seed to exceed 16%
- Do not store treated seed from one season to the next
- Do not treat cracked, split or sprouted seed
- Do not use treated seed as food or feed

Crop-specific information

- Latest use: before drilling
- After treating, bag seed immediately and keep in dry, draught-free store

Environmental safety

- Dangerous for the environment
- Toxic to aquatic organisms
- Product available in returnable containers which should not be rinsed and returned to the supplier as instructed. Used containers should be treated as if they contained pesticide

Hazard classification and safety precautions

Hazard H03, H11
Risk phrases R22a, R36, R51, R58
Operator protection A, C, D, H; U05a, U07, U08, U11, U20b
Environmental protection E15a, E32c, E32d, E33, E34, E36, E38
Storage and disposal E01, E04, E29b, E30a
Treated seed S02, S04b, S05, S06a, S07
Medical advice M03

266 hymexazol

A systemic isoxazole fungicide for pelleting sugar beet seed

Products

Tachigaren 70 WP	SumiAgro	70% w/w	WP	10496

Uses

- Black leg in ***sugar beet*** *(seed treatment)*

Efficacy guidance

- Incorporate into pelleted seed using suitable seed pelleting machinery

Restrictions

- Maximum number of treatments 1 per batch of seed
- Do not use treated seed as food or feed

Crop-specific information

- Latest use: before planting sugar beet seed

Environmental safety

- Harmful to aquatic organisms
- Treated seed harmful to game and wildlife

Hazard classification and safety precautions

Hazard H04, H07
Risk phrases R41, R52, R58
Operator protection A, C, F; U05a, U11, U20b
Environmental protection E13c, E32a
Storage and disposal E01, E04, E30a
Treated seed S01, S02, S05, S06a

267 imazalil

A systemic and protectant conazole fungicide

See also azaconazole + imazalil

Products

1	Fungaflor Smoke	Certis	15% w/w	FU	11327
2	Fungazil 100 SL	BASF	100 g/l	LS	11762
3	Magnate 100 SL	Makhteshim	100 g/l	SL	11705

Uses

- Aspergillus rot in ***food storage areas*** [1]
- Cladosporium in ***food storage areas*** [1]
- Dry rot in ***seed potatoes*** [2, 3]; ***ware potatoes*** [2]
- Gangrene in ***seed potatoes*** [2, 3]; ***ware potatoes*** [2]
- Penicillium rot in ***food storage areas*** [1]
- Powdery mildew in ***protected ornamentals***, ***protected roses*** [1]
- Silver scurf in ***seed potatoes*** [2, 3]; ***ware potatoes*** [2]
- Skin spot in ***seed potatoes*** [2, 3]; ***ware potatoes*** [2]

Approval information

- Imazalil included in Annex I under EC Directive 91/414

Efficacy guidance

- For best control of skin and wound diseases of ware potatoes treat as soon as possible after harvest, preferably within 7-10 d, before any wounds have healed [2, 3]
- Treat cucurbits before or as soon as disease appears and repeat every 10-14 d or every 7 d if infection pressure great or with susceptible cultivars [1]

Restrictions

- Maximum number of treatments 2 per batch of seed tubers [2, 3]; 1 per batch of ware tubers [2]
- Consult processor before treating potatoes for processing
- Do not treat outdoor crops of cucumbers, roses or ornamentals

- Do not spray cucurbits or ornamentals in full, bright sunshine. When spraying in the evening the spray should dry before nightfall. May cause damage if open flowers are sprayed
- Test for crop tolerance before large scale treatment. Do not use on rose cultivars Dr A.J. Verhage and Jack Frost
- With ornamentals of unknown tolerance test on a few plants in first instance

Crop-specific information
- Latest use: during storage and before chitting for seed potatoes
- HI edible crops 1 d
- Apply to clean soil-free potatoes post-harvest before putting into store, or at first grading. A further treatment may be applied in early spring before planting [2, 3]
- Apply through canopied hydraulic or spinning disc equipment prefereably diluted with up to two litres water per tonne of potatoes to obtain maximum skin cover and penetration [2, 3]
- Use on ware potatoes subject to discharges of imazalil from potato washing plants being within emission limits set by the UK monitoring authority

Environmental safety
- Dangerous for the environment
- Very toxic to aquatic organisms
- Do not empty into drains [2]
- Personal protective equipment requirements may vary for each pack size. Check label

Hazard classification and safety precautions

Hazard H03, H07 [1]; H04 [2, 3]; H11 [3]
Risk phrases R22a [1]; R36, R50 [3]; R41, R52 [2]; R58 [2, 3]
Operator protection A, C [2, 3]; D [1, 3]; H [2]; U04a, U20a [2, 3]; U05a, U19a [1-3]; U10, U20b [1]; U11 [1, 2]; U14 [2]
Environmental protection E02a, E13b, E32a [1]; E13c, E31a [3]; E15a, E19b, E31c, E32d [2]; E34 [2, 3]
Consumer protection C12 [1]
Storage and disposal E01, E04, E30a [1-3]; E26 [2, 3]
Treated seed S01, S03, S04a, S05, S06a [2, 3]; S02 [2]
Medical advice M03 [3]; M04 [1]

268 imazalil + pencycuron

A fungicide mixture for treatment of seed potatoes

Products

Monceren IM	Bayer CropScience	0.6:12.5% w/w	DS	11426

Uses
- Black scurf in ***potatoes*** *(tuber treatment)*
- Silver scurf in ***potatoes*** *(tuber treatment - reduction)*
- Stem canker in ***potatoes*** *(tuber treatment - reduction)*

Approval information
- Imazalil included in Annex I under EC Directive 91/414

Efficacy guidance
- Apply to clean seed tubers during planting (see label for suitable method) or sprinkle over tubers in chitting trays before loading into planter. It is essential to obtain an even distribution over tubers
- If seed tubers become damp from light rain distribution of product should not be affected. Tubers in the hopper should be covered if a shower interrupts planting

Restrictions
- Maximum number of treatments 1 per batch
- Operators must wear suitable respiratory equipment and gloves when handling product and when riding on planter. Wear gloves when handling treated tubers
- Do not use on tubers which have previously been treated with a dry powder seed treatment or hot water
- Do not use treated tubers for human or animal consumption

Crop-specific information

- Latest use: immediately before planting
- May be used on seed tubers previously treated with a liquid fungicide but not before 8 wk have elapsed if this contained imazalil

Environmental safety

- Harmful to aquatic organisms
- Treated seed harmful to game and wildlife

Hazard classification and safety precautions

Risk phrases R52, R58
Operator protection A, D; U20b
Environmental protection E03, E13c, E32a
Storage and disposal E30a
Treated seed S01, S02, S03, S04a, S05, S06a

269 imazalil + thiabendazole

A fungicide mixture for treatment of seed potatoes

Products

Extratect Flowable	Banks Cargill	100:300 g/l	FS	08704

Uses

- Dry rot in ***seed potatoes*** *(tuber treatment - reduction)*
- Gangrene in ***seed potatoes*** *(tuber treatment - reduction)*
- Silver scurf in ***potatoes*** *(tuber treatment - reduction)*, ***seed potatoes*** *(tuber treatment - reduction)*
- Skin spot in ***potatoes*** *(tuber treatment - reduction)*, ***seed potatoes*** *(tuber treatment - reduction)*
- Stem canker in ***potatoes*** *(tuber treatment - reduction)*

Approval information

- Imazalil and thiabendazole included in Annex I under EC Directive 91/414
- Approved for use through ULV equipment [1]

Efficacy guidance

- Apply immediately after lifting with suitable spinning disc or low volume hydraulic applicator for disease reduction during storage
- Apply within 2 mth of planting for maximum disease reduction in the progeny crop

Restrictions

- Maximum number of treatments 1 per batch
- Do not mix with any other product
- May not be used on seed tubers previously treated with a product containing imazalil
- Do not use treated seed as food or feed

Crop-specific information

- Latest use: pre-planting
- Delayed emergence noted under certain conditions but with no final effect on yield

Environmental safety

- Harmful to fish or other aquatic life. Do not contaminate surface waters or ditches with chemical or used container

Hazard classification and safety precautions

Hazard H04
Risk phrases R37, R41
Operator protection A, C, D, H; U05a, U09a, U11, U19a, U20c
Environmental protection E13c, E31a
Storage and disposal E01, E04, E30a
Treated seed S01, S02, S05, S06a

270 imazalil + triticonazole

A conazole fungicide mixture for seed treatment of barley

Products

Robust	BASF	12.5:12.5 g/l	FS	11807

Uses

- Covered smut in ***spring barley*** *(seed treatment)*, ***winter barley*** *(seed treatment)*
- Leaf stripe in ***spring barley*** *(seed treatment)*, ***winter barley*** *(seed treatment)*
- Loose smut in ***spring barley*** *(seed treatment)*, ***winter barley*** *(seed treatment)*
- Net blotch in ***spring barley*** *(seed treatment - moderate control)*, ***winter barley*** *(seed treatment - moderate control)*
- Seedling blight and foot rot in ***spring barley*** *(seed treatment - moderate control)*, ***winter barley*** *(seed treatment - moderate control)*

Approval information

- Imazalil included in Annex I under EC Directive 91/414

Efficacy guidance

- Apply undiluted through conventional seed treatment machine
- Treated seed may be stored for up to 18 mth but germination should be checked before use and sowing rate adjusted if necessary
- Keep treated seed in a dry, draught free store
- Calibrate seed drill before sowing

Restrictions

- Maximum number of treatments 1 per batch of seed
- Do not treat seed with moisture content above 16%
- Do not apply to cracked, split or sprouted seed
- Do not handle treated seed unnecessarily
- Do not use treated seed as food or feed

Crop-specific information

- Latest use: pre-drilling

Environmental safety

- Toxic to aquatic organisms
- Product available in large (1000 l) returnable containers. They should be re-circulated for 30 min before use and operated as directed on the label

Hazard classification and safety precautions

Hazard H04
Risk phrases R43, R51, R58
Operator protection A, C, H; U05a, U07, U09a, U14, U20b
Environmental protection E15a, E33, E34, E36
Storage and disposal E01, E04, E30a
Treated seed S02, S04b, S05, S07
Medical advice M03

271 imazamethabenz-methyl

A post-emergence imidazolinone grass weed herbicide for use in winter cereals

Products

Dagger	BASF	300 g/l	SC	10218

Uses

- Blackgrass in ***winter barley***, ***winter wheat***
- Charlock in ***winter barley***, ***winter wheat***
- Loose silky bent in ***winter barley***, ***winter wheat***
- Onion couch in ***winter barley***, ***winter wheat***
- Volunteer oilseed rape in ***winter barley***, ***winter wheat***
- Wild oats in ***winter barley***, ***winter wheat***

Approval information

- Imazamethabenz-methyl not supported in EU review. Timescales doe revocation of approvals nor yet decided

Efficacy guidance

- Activity is contact and residual. Grass and broad leaved weeds controlled from pre-emergence to specified post-emergence sizes
- When used alone add authorised non-ionic wetter for improved contact activity
- Best results achieved when applied to fine, firm, clod-free seedbed when soil moist
- Effects of autumn/winter treatment normally persist to control spring flushes of weeds
- Follow-up treatment with a specific blackgrass herbicide may be necessary for complete control of light infestations. For heavy infestations use a specific blackgrass herbicide
- Imazamethabenz-methyl is a member of the ALS-inhibitor group of herbicides and products should be used in a planned Resistance Management strategy. See Section 5 for more information

Restrictions

- Maximum number of treatments 2 per crop (as split dose treatment)
- Do not use on durum wheat
- Do not use on soils where surface water is likely to accumulate
- Do not use on soils with more than 10% organic matter

Crop-specific information

- Latest use: before 4th node detectable (GS 34)
- Apply from 2-fully expanded leaf stage of crop up to three nodes detectable (GS 12-33)

Following crops guidance

- Cereals, oilseed rape, maize, sweetcorn, ryegrass, peas or potatoes may be sown in the season after use. Where oilseed rape is to follow spring treatment the ground must be ploughed before sowing
- Sugar beet may be sown in the season after use provided the soil is ploughed beforehand and the pH is above 6.5. Sugar beet seed crops should not be sown in the autumn after spring use
- Where the total dose applied in the season exceeds the maximum individual dose only wheat or barley should be sown as following crops
- Where other sulfonyl ureas have also been applied during the growing season only wheat, barley, rye or triticale should be sown following crops
- In case of crop failure land may be redrilled once normal harvest would have taken place. If pendimathalin has also been applied during the growing season the land must first be ploughed to 15 cm minimum

Environmental safety

- Dangerous for the environment
- Very toxic to aquatic organisms

Hazard classification and safety precautions

Hazard H03, H11
Risk phrases R20, R50, R58
Operator protection A, C; U05a, U08, U20b
Environmental protection E15a, E31c, E32d, E38
Storage and disposal E01, E04, E29b, E30a
Medical advice M05a

272 imazaquin

An imidazolinone herbicide and plant growth regulator available only in mixtures

See also chlormequat with choline chloride + imazaquin

273 imidacloprid

A neonicotinoid insecticide for seed, soil, peat or foliar treatment

See also *beta-cyfluthrin + imidacloprid*
bitertanol + fuberidazole + imidacloprid
fuberidazole + imidacloprid + triadimenol

Products

1	Admire	Bayer CropScience	70% w/w	WG	11234
2	Gaucho	Bayer CropScience	70% w/w	WS	11281
3	Intercept 5GR	Scotts	5% w/w	GR	08126
4	Intercept 70WG	Scotts	70% w/w	WG	08585

Uses

- Aphids in ***bedding plants***, ***hardy ornamental nursery stock***, ***herbaceous perennials***, ***pot plants*** [3, 4]; ***lettuce*** *(off-label)* [2]
- Beet virus yellows vectors in ***sugar beet*** *(seed treatment)* [2]
- Cabbage aphid in ***collards*** *(off-label - seed treatment)*, ***kale*** *(off-label - seed treatment)* [2]
- Damson-hop aphid in ***hops*** [1]
- Flea beetle in ***sugar beet*** *(seed treatment)* [2]
- Glasshouse whitefly in ***bedding plants***, ***hardy ornamental nursery stock***, ***herbaceous perennials***, ***pot plants*** [3, 4]
- Mangold fly in ***sugar beet*** *(seed treatment)* [2]
- Millipedes in ***sugar beet*** *(seed treatment)* [2]
- Peach-potato aphid in ***broccoli*** *(off-label - seed treatment)*, ***brussels sprouts*** *(off-label - seed treatment)*, ***cabbages*** *(off-label - seed treatment)*, ***calabrese*** *(off-label - seed treatment)*, ***cauliflowers*** *(off-label - seed treatment)*, ***collards*** *(off-label - seed treatment)*, ***kale*** *(off-label - seed treatment)* [2]
- Pygmy beetle in ***sugar beet*** *(seed treatment)* [2]
- Sciarid flies in ***bedding plants***, ***hardy ornamental nursery stock***, ***herbaceous perennials***, ***pot plants*** [3, 4]
- Springtails in ***sugar beet*** *(seed treatment)* [2]
- Symphylids in ***sugar beet*** *(seed treatment)* [2]
- Thrips in ***protected chrysanthemums*** *(off-label)* [1]
- Tobacco whitefly in ***bedding plants***, ***hardy ornamental nursery stock***, ***herbaceous perennials***, ***pot plants*** [3, 4]
- Vine weevil in ***bedding plants***, ***hardy ornamental nursery stock***, ***herbaceous perennials***, ***pot plants*** [3, 4]

Specific Off-Label Approvals (SOLAs)

- ***broccoli***, ***brussels sprouts***, ***cabbages***, ***calabrese***, ***cauliflowers***, ***collards***, ***kale*** *(seed treatment) (OLA 023927) Dec 2008* [2]
- ***lettuce*** *(OLA 031886) Dec 2008* [2]
- ***protected chrysanthemums*** *(OLA 010584) Dec 2008* [1]

Efficacy guidance

- Treated seed should be drilled within the season of purchase [2]
- Base of hop plants should be free of weeds and debris at application [1]
- Hop bines emerging away from the main stock or adjacent to poles may require a separate application [1]
- Uptake and movement within hops requires soil moisture and good growing conditions [1]
- Control may be impaired in plantations greater than 3640 plants/ha [1]
- To minimise likelihood of resistance do not treat all the hop crop in any one yr [1] and adopt a planned programme to alternate with pesticides of different types or use other measures when using in compost [3, 4]
- When applied as drench or incorporated as granules in moist compost compound is readily absorbed and translocated to aerial parts of plant [3, 4]

Restrictions

- Maximum number of treatments 1 per batch of seed [2] or growing medium [3, 4]; 1 per yr [1]
- Product must not be used in compost that has already been treated with an imidacloprid-containing product [4]

- Product formulated for use only as a compost incorporation treatment into peat-based growing media using suitable automated equipment [3]
- For use only on container grown ornamentals [3, 4]
- The safety of seeds sown into treated compost should not be assumed. Test treat before full-scale use [3]
- Product must not be used on crops for human or animal consumption and treated compost must not be re-used for this purpose [3, 4]
- Do not use treated seed as food or feed [2]

Crop-specific information

- Latest use: before bines reach 2 m or before end 1st wk Jun for hops; before drilling for sugar beet; before sowing or planting for brassicas, bedding plants, hardy ornamental nursery stock, pot plants
- Apply to sugar beet seed as part of the normal commercial pelleting process using special treatment machinery [2]
- Apply to hops as a directed stem base spray before most bines reach a height of 2 m. If necessary treat both sides of the crown at half the normal concentration [1]
- The rate used per hop plant must be adjusted in accordance with the plant population if greater than 3640 per ha. See label for details [1]

Environmental safety

- Harmful to aquatic organisms [1]
- High risk to bees. Do not apply to crops in flower or to those in which bees are actively foraging. Do not apply when flowering weeds are present [1, 4]
- Avoid spillage or other environmental contamination when incorporating into compost [3]
- Treated seed harmful to game and wildlife [2]

Hazard classification and safety precautions

Hazard H03 [1, 4]; H04 [2]
Risk phrases R22a [1, 4]; R43 [2]; R52 [1]
Operator protection A [1-3]; C, D, H [2]; U05a [1, 4]; U20a [2]; U20b [1, 3, 4]
Environmental protection E12a, E34 [1, 4]; E15a, E32a [1-4]
Storage and disposal E01, E04 [1, 4]; E29a [3]; E30a [1-4]
Treated seed S02, S04b, S05, S06a, S07 [2]

274 imidacloprid + tebuconazole + triazoxide

A broad spectrum fungicide and insecticide seed treatment for winter barley

Products

Raxil Secur	Bayer CropScience	233:20:20 g/l	LS	11298

Uses

- Leaf stripe in ***winter barley*** *(seed treatment)*
- Loose smut in ***winter barley*** *(seed treatment)*
- Net blotch in ***winter barley*** *(seed treatment)*
- Virus vectors in ***winter barley*** *(seed treatment)*
- Wireworm in ***winter barley*** *(reduction of damage)*

Efficacy guidance

- Best applied through recommended seed treatment machines
- Evenness of seed cover improved by simultaneous application of equal volumes of product and water or dilution of product with an equal volume of water
- Drill treated seed in the same season. Use minimum 125 kg treated seed per ha. Lower drilling rates and/or early drilling affect duration of BYDV protection needed and may require follow-up aphicide treatment
- In high risk areas where aphid activity is heavy and prolonged a follow-up aphicide treatment may be required
- Protection against foliar air-borne and splash-borne diseases later in the season will require appropriate fungicide follow-up sprays

Restrictions

- Maximum number of treatments 1 per batch of seed

- Do not use on seed with more than 16% moisture content, or on sprouted, cracked or skinned seed
- Do not use treated seed as food or feed

Crop-specific information

- Latest use: before drilling
- Slightly delayed and reduced emergence may occur but this is normally outgrown
- Any delay in field emergence, for whatever reason, may be accentuated by treatment

Environmental safety

- Harmful to aquatic organisms
- Dangerous to birds, game and other wildlife. Treated seed should not be left on the soil surface. Bury spillages
- Treated seed should not be broadcast, but drilled to a depth of 4 cm in a well prepared seedbed
- If seed is left on the soil surface the field should be harrowed and rolled to ensure good incorporation

Hazard classification and safety precautions

Hazard H03
Risk phrases R22a, R43, R52, R58
Operator protection A, H; U04a, U05a, U07, U13, U20b
Environmental protection E03, E13c, E33, E34, E36
Storage and disposal E01, E04, E26, E30a
Treated seed S01, S02, S03, S04c, S05, S06a, S07, S08
Medical advice M03

275 indol-3-ylacetic acid

A plant growth regulator for promoting rooting of cuttings

Products

1	Rhizopon A Powder	Rhizopon	1% w/w	DP	09087
2	Rhizopon A Tablets	Rhizopon	50 mg a.i.	TB	09088

Uses

- Rooting of cuttings in ***ornamental specimens***

Efficacy guidance

- Apply by dipping end of prepared cuttings into powder or dissolved tablet solution
- Shake off excess powder and make planting holes to prevent powder stripping off [1]
- Consult manufacturer for details of application by spray or total immersion

Restrictions

- Maximum number of treatments 1 per cutting
- Store product in a cool, dark and dry place
- Use solutions once only. Discard after use [2]
- Use plastic, not metallic, container for solutions [2]

Crop-specific information

- Latest use: before cutting insertion

Hazard classification and safety precautions

Operator protection U19a [1]; U20a [1, 2]
Environmental protection E15a, E32a
Storage and disposal E30a

276 4-indol-3-yl-butyric acid

A plant growth regulator promoting the rooting of cuttings

Products

1	Chryzoplus Grey 0.8%	Rhizopon	0.8% w/w	DP	09094
2	Chryzopon Rose 0.1%	Rhizopon	0.1% w/w	DP	09092
3	Chryzosan White 0.6%	Rhizopon	0.6% w/w	DP	09093
4	Chryzotek Beige	Rhizopon	0.4% w/w	DP	09081
5	Chryzotop Green	Rhizopon	0.25% w/w	DP	09085

SEE SECTION 3 FOR PRODUCTS ALSO REGISTERED

Products – continued

6	Rhizopon AA Powder (0.5%)	Rhizopon	0.5% w/w	DP	09082
7	Rhizopon AA Powder (1%)	Rhizopon	1% w/w	DP	09084
8	Rhizopon AA Powder (2%)	Rhizopon	2% w/w	DP	09083
9	Rhizopon AA Tablets	Rhizopon	50 mg a.i.	TB	09086
10	Seradix 1	Certis	0.1% w/w	DP	11330
11	Seradix 2	Certis	0.3% w/w	DP	11331
12	Seradix 3	Certis	0.8% w/w	DP	11332

Uses

- Rooting of cuttings in ***ornamental specimens***

Efficacy guidance

- Dip base of cuttings into powder immediately before planting
- Powders or solutions of different concentration are required for different types of cutting. Lowest concentration for softwood, intermediate for semi-ripe, highest for hardwood
- See label for details of concentration and timing recommended for different species
- Use of planting holes recommended for powder formulations to ensure product is not removed on insertion of cutting. Cuttings should be watered in if necessary

Restrictions

- Maximum number of treatments 1 per situation
- Use of too strong a powder or solution may cause injury to cuttings
- No unused moistened powder should be returned to container

Crop-specific information

- Latest use: before cutting insertion for ornamental specimens

Hazard classification and safety precautions

Operator protection A, C [10-12]; U14, U15 [10-12]; U19a, U20a [1-12]
Environmental protection E15a, E32a
Storage and disposal E01 [10-12]; E30a [1-12]

277 4-indol-3-yl-butyric acid + 2-(1-naphthyl)acetic acid with dichlorophen

A plant growth regulator for promoting rooting of cuttings

Products

Synergol	Certis	5.0:5.0 g/l	SL	07386

Uses

- Rooting of cuttings in ***ornamental specimens***

Efficacy guidance

- Dip base of cuttings into diluted concentrate immediately before planting
- Suitable for hardwood and softwood cuttings
- See label for details of concentration and timing for different species

Hazard classification and safety precautions

Hazard H03
Risk phrases R22a
Operator protection A, C; U05a, U13, U20b
Environmental protection E15a, E32a, E32d, E34
Storage and disposal E01, E04, E30a
Medical advice M03

278 iodosulfuron-methyl-sodium

A post-emergence sulfonylurea herbicide for winter sown cereals

See also amidosulfuron + iodosulfuron-methyl-sodium

Products

Hussar	Bayer CropScience	5% w/w	WG	11205

Uses

- Chickweed in ***triticale, winter rye, winter wheat***
- Cleavers in ***triticale, winter rye, winter wheat***
- Italian ryegrass in ***triticale*** *(from seed)*, ***winter rye*** *(from seed)*, ***winter wheat*** *(from seed)*
- Mayweeds in ***triticale, winter rye, winter wheat***
- Perennial ryegrass in ***triticale*** *(from seed)*, ***winter rye*** *(from seed)*, ***winter wheat*** *(from seed)*
- Red dead-nettle in ***triticale, winter rye, winter wheat***
- Speedwells in ***triticale, winter rye, winter wheat***

Efficacy guidance

- Best results obtained from treatment in warm weather when soil is moist and the weeds are growing actively
- Weeds must be present at application to be controlled
- Weed control is slow especially under cool dry conditions
- Dry conditions resulting in moisture stress may reduce effectiveness
- Occasionally weeds may only be stunted but they will normally have little or no competitive effect on the crop
- Iodosulfuron is a member of the ALS-inhibitor group of herbicides and products should be used in a planned Resistance Management strategy. See Section 5 for more information

Restrictions

- Maximum number of treatments 1 per crop
- Must only be applied between 1 Feb in yr of harvest and specified latest time of application
- Do not apply to undersown crops, or crops to be undersown
- Do not roll or harrow within 1 wk of spraying
- Do not spray crops under stress from any cause or if the soil is compacted
- Treat broadcast crops after the plants have a well-established root system
- Do not spray if rain imminent or frost expected
- Do not apply in mixture or in sequence with any other ALS inhibitor

Crop-specific information

- Latest use: before third node detectable (GS 33)

Following crops guidance

- No restrictions apply to the sowing of cereal crops or sugar beet in the spring of the yr following treatment

Environmental safety

- Dangerous for the environment
- Very toxic to aquatic organisms
- LERAP Category B
- Take extreme care to avoid damage by drift onto broad-leaved plants outside the target area or onto ponds, waterways and ditches
- Observe carefully label instructions for sprayer cleaning

Hazard classification and safety precautions

Hazard H04, H11
Risk phrases R41, R50, R58
Operator protection A, C, H; U05a, U08, U11, U14, U15, U20b
Environmental protection E13b, E16a, E16b, E31a, E32d, E38
Storage and disposal E01, E04

279 iodosulfuron-methyl-sodium + mesosulfuron-methyl

A sulfonyl urea herbicide mixture for winter wheat

Products

1	Atlantis WG	Bayer CropScience	0.6:3.0% w/w	WG	11679
2	Greencrop Doonbeg WG	Greencrop	0.6:3.0% w/w	WG	12199
3	Standon Mimas WG	Standon	0.6:3.0% w/w	WG	12075

Uses

- Annual meadow grass in ***winter wheat***
- Blackgrass in ***winter wheat***
- Chickweed in ***winter wheat***

SEE SECTION 3 FOR PRODUCTS ALSO REGISTERED

- Italian ryegrass in ***winter wheat***
- Mayweeds in ***winter wheat***
- Perennial ryegrass in ***winter wheat***
- Rough meadow grass in ***winter wheat***
- Wild oats in ***winter wheat***

Efficacy guidance

- Optimum grass weed control obtained when all grass weeds are emerged at spraying. Activity is primarily via foliar uptake and good spray coverage of the target weeds is essential
- Translocation occurs readily within the target weeds and growth is inhibited within hours of treatment but symptoms may not be apparent for up to 4 wk, depending on weed species, timing of treatment and weather conditions
- Residual activity is important for best results and is optimised by treatment on fine moist seedbeds. Avoid application under very dry conditions
- Residual efficacy may be reduced by high soil temperatures and cloddy seedbeds
- Iodosulfuron-methyl and mesosulfuron-methyl are both members of the ALS-inhibitor group of herbicides and products should be used in a planned Resistance Management strategy. See Section 5 for more information

Restrictions

- Maximum number of treatments 1 per crop with a maximum total dose equivalent to one full dose treatment
- Do not use on crops undersown with grasses, clover or other legumes or any other broad-leaved crop
- Do not use as a stand-alone treatment for blackgrass, ryegrass or chickweed control
- Do not use as the sole means of weed control in successive crops
- Do not use in mixture or in sequence with any other ALS-inhibitor herbicide except those specified on the label
- Do not apply to crops under stress from any cause
- Do not apply when rain is imminent or during periods of frosty weather
- Specified adjuvant must be used. See label

Crop-specific information

- Latest use: flag leaf just visible (GS 39)
- Winter wheat may be treated from the two-leaf stage of the crop
- Safety to crops grown for seed not established

Following crops guidance

- In the event of crop failure sow only winter wheat in the same cropping season
- Only winter wheat or winter barley may be sown in the year of harvest of a treated crop
- Spring wheat, spring barley or sugar beet may be drilled in the following spring

Environmental safety

- Dangerous for the environment
- Very toxic to aquatic organisms
- LERAP Category B
- Take extreme care to avoid drift onto plants outside the target area or on to ponds, waterways or ditches

Hazard classification and safety precautions

Hazard H04, H11
Risk phrases R41, R50, R58
Operator protection A, C, H; U05a, U11, U20b
Environmental protection E13b, E16a, E31a [1-3]; E16b [2, 3]; E32d, E38 [1, 2]
Storage and disposal E01, E04, E26, E30a

SECTION 2

280 ioxynil

A contact acting HBN herbicide for use in turf and onions

See also *bromoxynil + diflufenican + ioxynil*
bromoxynil + ethofumesate + ioxynil
bromoxynil + ioxynil

Products

Totril	Bayer CropScience	225 g/l	EC	10026

Uses

- Annual dicotyledons in ***bulb onions***, ***carrots*** *(off-label)*, ***chives*** *(off-label)*, ***garlic***, ***leeks***, ***parsnips*** *(off-label)*, ***salad onions***, ***shallots***

Specific Off-Label Approvals (SOLAs)

- ***carrots***, ***parsnips*** *(OLA 001920) Feb 2005* [1]
- ***chives*** *(OLA 001918) Dec 2008* [1]

Approval information

- Ioxynil was reviewed in 1995 and approvals for home garden use, and most hand held applications revoked.

Efficacy guidance

- Best results on seedling to 4-leaf stage weeds in active growth during mild weather

Restrictions

- Maximum number of treatments up to 4 per crop at split doses on onions; 1 per crop on carrots, garlic, leeks, parsnips, shallots, chives
- Do not apply by hand-held equipment or at concentrations higher than those recommended

Crop-specific information

- Latest use: pre-emergence for carrots, parsnips
- HI onions, chives, shallots, garlic, leeks 14 d
- Apply to sown onion crops as soon as possible after plants have 3 true leaves or to transplanted crops when established

Environmental safety

- Dangerous for the environment [1]
- Very toxic to aquatic organisms
- Harmful to bees. Do not apply to crops in flower or to those in which bees are actively foraging. Do not apply when flowering weeds are present
- Keep livestock out of treated areas for at least 6 wk after treatment and until foliage of any poisonous weeds such as ragwort has died and become unpalatable

Hazard classification and safety precautions

Hazard H03, H08, H11
Risk phrases R22a, R22b, R36, R43, R50, R58, R63, R66, R67
Operator protection A, C; U05a, U08, U19a, U20b, U23a
Environmental protection E06a (6 wk); E12f, E31b, E32d, E34, E38
Storage and disposal E01, E04, E26, E30a
Medical advice M03, M05b

281 iprodione

A protectant dicarboximide fungicide with some eradicant activity

See also *carbendazim + iprodione*

Products

1	Amenitywise Iprodione Green	Standon	250 g/l	SC	10883
2	Chipco Green	Bayer Environ.	250 g/l	SC	11211
3	I T Iprodione	I T Agro	50% w/w	WP	08267
4	Rovral Flo	BASF	255 g/l	SC	11702
5	Rovral Green	Bayer Environ.	250 g/l	SC	09938
6	Rovral Liquid FS	BASF	500 g/l	FS	11703

Products – continued

7	Rovral WP	BASF	50% w/w	WP	11694
8	Standon Iprodione 50 WP	Standon	50% w/w	WP	10656

Uses

- Alternaria in ***brassica seed crops***, ***mustard*** *(seed treatment)*, ***stored cabbages***, ***stubble turnips*** *(seed treatment)*, ***swedes*** *(seed treatment)*, ***turnips*** *(seed treatment)* [3, 7, 8]; ***broccoli***, ***brussels sprouts***, ***cabbage seed crops***, ***calabrese***, ***cauliflower seed crops***, ***cauliflowers***, ***mustard seed crops***, ***spring oilseed rape***, ***stubble turnips***, ***winter oilseed rape***, ***winter wheat*** [4]; ***chinese cabbage*** *(seed treatment)*, ***edible brassicas*** *(seed treatment)*, ***flower seeds*** *(seed treatment)*, ***kohlrabi*** *(seed treatment)* [7]; ***flax*** *(seed treatment)* [6, 7]; ***fodder rape*** *(seed treatment)*, ***leaf brassicas*** *(seed treatment)*, ***linseed*** *(seed treatment)*, ***ornamental specimens*** *(seed treatment)* [3, 8]; ***poppies*** *(off-label - for morphine production - seed treatment)* [6]; ***spring oilseed rape*** *(seed treatment)*, ***winter oilseed rape*** *(seed treatment)* [3, 6-8]
- Black scurf in ***seed potatoes*** *(seed treatment)* [3, 6, 8]
- Black scurf and stem canker in ***potatoes*** *(seed treatment)* [7]
- Botrytis in ***cabbage seed crops***, ***cauliflower seed crops***, ***grapevines*** *(off-label)*, ***mustard seed crops***, ***spring oilseed rape***, ***winter oilseed rape***, ***winter wheat*** [4]; ***chicory*** *(off-label)*, ***combining peas*** *(off-label)*, ***dwarf beans*** *(off-label)*, ***endives*** *(off-label)*, ***fennel*** *(off-label)*, ***french beans*** *(off-label)*, ***mange-tout peas*** *(off-label)*, ***pears*** *(off-label)*, ***protected courgettes*** *(off-label)*, ***protected endives*** *(off-label)*, ***protected radicchio*** *(off-label)*, ***protected rhubarb*** *(off-label)*, ***red beet*** *(off-label)*, ***runner beans*** *(off-label)*, ***vining peas*** *(off-label)* [7]; ***cucumbers*** [3]; ***lettuce*** *(outdoor crops)*, ***outdoor tomatoes***, ***pot plants***, ***protected lettuce***, ***protected tomatoes***, ***raspberries***, ***stored cabbages*** [3, 7, 8]; ***strawberries*** [3, 4, 7, 8]
- Botrytis bunch rot in ***grapevines*** *(off-label)* [4]
- Brown patch in ***amenity grass*** [2, 5]; ***managed amenity turf*** [1, 2, 5]
- Canker in ***horseradish*** *(off-label)* [7]
- Chocolate spot in ***spring field beans***, ***winter field beans*** [4]
- Collar rot in ***onions***, ***salad onions*** [4]
- Dollar spot in ***amenity grass*** [2, 5]; ***managed amenity turf*** [1, 2, 5]
- Fusarium patch in ***amenity grass*** [2, 5]; ***managed amenity turf*** [1, 2, 5]
- Glume blotch in ***winter wheat*** [4]
- Grey mould in ***protected cucumbers*** *(off-label)* [7]
- Leaf rot in ***onions***, ***salad onions*** [4]
- Melting out in ***amenity grass*** [2, 5]; ***managed amenity turf*** [1, 2, 5]
- Net blotch in ***spring barley***, ***winter barley*** [4]
- Phoma in ***red beet*** *(off-label)* [7]
- Red thread in ***amenity grass*** [2, 5]; ***managed amenity turf*** [1, 2, 5]
- Sclerotinia in ***chicory*** *(off-label)*, ***combining peas*** *(off-label)*, ***dwarf beans*** *(off-label)*, ***french beans*** *(off-label)*, ***mange-tout peas*** *(off-label)*, ***runner beans*** *(off-label)*, ***vining peas*** *(off-label)* [7]
- Sclerotinia stem rot in ***spring oilseed rape***, ***winter oilseed rape*** [4]
- Snow mould in ***amenity grass*** [2, 5]; ***managed amenity turf*** [1, 2, 5]
- Stemphylium in ***asparagus*** *(off-label)* [7]

Specific Off-Label Approvals (SOLAs)

- ***asparagus***, ***combining peas***, ***vining peas*** *(OLA 001902) Dec 2008* [7]
- ***chicory*** *(OLA 001906) Dec 2008* [7]
- ***dwarf beans***, ***french beans***, ***mange-tout peas***, ***runner beans*** *(OLA 001904) Dec 2008* [7]
- ***endives***, ***fennel***, ***protected courgettes***, ***protected endives***, ***protected radicchio*** *(OLA 001908) Dec 2008* [7]
- ***grapevines*** *(OLA 001870) Dec 2008* [4]
- ***horseradish*** *(OLA 021254) Dec 2008* [7]
- ***pears*** *(OLA 012525) Dec 2008* [7]
- ***poppies*** *(for morphine production - seed treatment) (OLA 040682) Dec 2008* [6]
- ***protected cucumbers*** *(OLA 003070) Dec 2008* [7]
- ***protected rhubarb*** *(OLA 001912) Dec 2008* [7]
- ***red beet*** *(OLA 001910) Dec 2008* [7]

Approval information

- Iprodione included in Annex I under EC Directive 91/414

SECTION 2

Efficacy guidance

- Many diseases require a programme of 2 or more sprays at intervals of 2-4 wk. Recommendations vary with disease and crop - see label for details
- Use as a drench to control cabbage storage diseases. Spray ornamental pot plants and cucumbers to run-off
- Apply turf and amenity grass treatments after mowing to dry grass, free of dew. Do not mow again for at least 24 h
- Seed treatments on oilseed rape and linseed should be applied to dry seed prior to sowing
- Best results on seed potatoes achieved using hydraulic sprayer with solid or hollow cone jets with air or liquid pressure atomisation
- Apply to seed potatoes after harvest or after grading out and before traying out for chitting

Restrictions

- Maximum number of treatments 1 per batch for seed treatments; 1 per crop on cereals, vining peas, cabbage (as drench), chicory, borage; 2 per crop on field beans and stubble turnips; 3 per crop on brassicas (including seed crops), oilseed rape, protected winter lettuce (Oct-Feb); 4 per crop on strawberries, grapevines, salad onions, cucumbers; 5 per crop on raspberries; 6 per crop on bulb onions, tomatoes, curcurbits, peppers, aubergines, turf, nuts; 7 per crop on lettuce (Mar-Sep)
- A minimum of 3 wk must elapse between treatments on leaf brassicas
- See label for pot plants showing good tolerance. Check other species before applying on a large scale
- Do not treat oilseed rape seed that is cracked or broken or of low viability
- Do not excessively wet seed potato skins
- Do not treat oats
- Do not use treated seed as food or feed [6]

Crop-specific information

- Latest use: pre-planting for seed treatments; at planting for potatoes; before grain watery ripe (GS 69) for wheat and barley; 5 wk before forcing for chicory; 6 wk before release of pears for human consumption; 16 wk before processing or sale for red beet
- HI strawberries, protected tomatoes 1 d; outdoor tomatoes, cucurbits, aubergines 2 d; lettuce, raspberries, nuts, cress, dwarf beans, endives, fennel, French beans, runner beans, lamb's lettuce (Mar-Sep), lettuce, mange-tout peas, onions, radicchio, scarole 7 d; grapevines, protected rhubarb 14 d; brassicas, brassica seed crops, oilseed rape, mustard, field beans, peas, stubble turnips 21 d; protected lettuce (Oct-Feb) 28 d; asparagus 5 mth
- Turf and amenity grass may become temporarily yellowed if frost or hot weather follows treatment
- Personal protective equipment requirements may vary for each pack size. Check label

Environmental safety

- Dangerous for the environment
- Very toxic to aquatic organisms
- Treated brassica seed crops not to be used for human or animal consumption
- See label for guidance on disposal of spent drench liquor
- Treatment harmless to *Encarsia* or *Phytoseiulus* being used for integrated pest control

Hazard classification and safety precautions

Hazard H03, H11 [1, 2, 4-8]
Risk phrases R36, R38 [1, 2, 5]; R40, R58 [1, 2, 4-8]; R50 [1, 2, 4-6]; R51 [7, 8]
Operator protection A [1, 2, 4-6]; C [1, 2, 5]; H [6]; U04a [1, 2, 5]; U05a [1, 2, 4-6]; U08 [1, 2, 4, 5]; U19a [1, 3, 7, 8]; U20b [4]; U20c [1-3, 5-8]
Environmental protection E13c [3]; E15a [1, 2, 4-8]; E31b [1, 2, 4, 5]; E32a [3, 6-8]; E32d, E38 [2, 4-7]; E34 [1-3, 5-8]
Storage and disposal E01, E04 [1, 2, 4-6]; E26 [1, 2, 5, 6]; E30a [1-8]
Treated seed S01, S05, S06a, S07 [6-8]; S02 [7, 8]; S03, S04a [6]
Medical advice M03 [2, 5]

SEE SECTION 3 FOR PRODUCTS ALSO REGISTERED

282 iprodione + thiophanate-methyl

A protectant and systemic fungicide for oilseed rape

Products

1	Compass	BASF	167:167 g/l	SC	11740
2	Snooker	BASF	150:200 g/l	SC	11744

Uses

- Alternaria in ***carrots*** *(off-label)*, ***horseradish*** *(off-label)*, ***parsnips*** *(off-label)*, ***spring oilseed rape***, ***winter oilseed rape*** [1, 2]
- Botrytis in ***limnanthes alba (meadowfoam)*** *(off-label)* [1]
- Chocolate spot in ***spring field beans***, ***winter field beans*** [1, 2]
- Crown rot in ***carrots*** *(off-label)*, ***horseradish*** *(off-label)*, ***parsnips*** *(off-label)* [1, 2]
- Grey mould in ***spring oilseed rape***, ***winter oilseed rape*** [1, 2]
- Light leaf spot in ***spring oilseed rape***, ***winter oilseed rape*** [1, 2]
- Sclerotinia stem rot in ***spring oilseed rape***, ***winter oilseed rape*** [1, 2]
- Stem canker in ***spring oilseed rape***, ***winter oilseed rape*** [1]

Specific Off-Label Approvals (SOLAs)

- ***carrots***, ***horseradish***, ***parsnips*** *(OLA 001936) Dec 2008* [1]
- ***carrots***, ***horseradish***, ***parsnips*** *(OLA 041186) Dec 2008* [2]
- ***limnanthes alba (meadowfoam)*** *(OLA 001934) Dec 2008* [1]

Approval information

- Iprodione included in Annex I under EC Directive 91/414

Efficacy guidance

- Timing of sprays on oilseed rape varies with disease, see label for details
- For season-long control product should be applied as part of a disease control programme

Restrictions

- Maximum number of treatments (refers to total sprays containing benomyl, carbendazim and thiophanate methyl) 2 per crop for oilseed rape, field beans, meadowfoam; 3 per crop for carrots, horseradish, parsnips

Crop-specific information

- Latest use: before end of flowering for oilseed rape, field beans
- HI meadowfoam, field beans, oilseed rape 3 wk; carrots, horseradish, parsnips 28 d
- Treatment may extend duration of green leaf in winter oilseed rape

Environmental safety

- Dangerous for the environment
- Very toxic to aquatic organisms

Hazard classification and safety precautions

Hazard H03, H11
Risk phrases R36, R38 [2]; R40, R43, R50, R58, R68 [1, 2]
Operator protection A, C, H, M; U05a, U20c [1, 2]; U09a [2]
Environmental protection E15a, E31b, E32d, E38
Storage and disposal E01, E04, E30a [1, 2]; E26 [2]; E29b [1]
Medical advice M05a [1]

283 isoproturon

A residual urea herbicide for use in cereals

See also *chlorotoluron + isoproturon*
diflufenican + flurtamone + isoproturon
diflufenican + isoproturon

Products

1	Aligran	Nufarm UK	83% w/w	WG	11761
2	Alpha IPU 500	Makhteshim	500 g/l	SC	10733
3	Alpha Isoproturon 500	Makhteshim	500 g/l	SC	05882
4	Arelon 500	Nufarm UK	500 g/l	SC	11639

Products – continued

5	Bison 83 WG	Nufarm UK	83% w/w	WG	11580
6	Fieldgard	Nufarm UK	500 g/l	SC	11770
7	Isotron 500	DAPT	500 g/l	SC	10668
8	Primer	AgriGuard	500 g/l	SC	11782
9	Protugan	Makhteshim	500 g/l	SC	10043
10	Protugan 80 WDG	Makhteshim	80% w/w	WG	10430

Uses

- Annual dicotyledons in ***spring barley*** *(off-label)* [3]; ***spring wheat*** [4, 6]; ***spring wheat*** *(autumn sown)* [10]; ***triticale, winter rye*** [2, 4, 6, 9, 10]; ***winter barley, winter wheat*** [1-10]
- Annual grasses in ***spring wheat*** [4, 6]; ***spring wheat*** *(autumn sown)* [10]; ***triticale, winter rye*** [2, 4, 6, 9, 10]; ***winter barley, winter wheat*** [1-10]
- Annual meadow grass in ***spring barley*** [9]; ***spring barley*** *(off-label)* [2, 3, 5, 9, 10]
- Blackgrass in ***spring wheat*** [4, 6]; ***spring wheat*** *(autumn sown)* [10]; ***triticale, winter rye*** [2, 4, 6, 9, 10]; ***winter barley, winter wheat*** [2-4, 6-10]
- Rough meadow grass in ***spring wheat*** [4, 6]; ***spring wheat*** *(autumn sown)* [10]; ***triticale, winter rye*** [2, 4, 6, 9, 10]; ***winter barley, winter wheat*** [2-4, 6-10]
- Wild oats in ***spring wheat*** [4, 6]; ***spring wheat*** *(autumn sown)* [10]; ***triticale, winter rye*** [2, 4, 6, 9, 10]; ***winter barley, winter wheat*** [2-4, 6-10]

Specific Off-Label Approvals (SOLAs)

- ***spring barley*** *(OLA 011481) Dec 2008* [3]
- ***spring barley*** *(OLA 011488) Dec 2008* [9]
- ***spring barley*** *(OLA 012166) Dec 2008* [2]
- ***spring barley*** *(OLA 023410) Dec 2008* [10]
- ***spring barley*** *(OLA 032358) Dec 2008* [5]

Approval information

- Isoproturon included in Annex I under EC Directive 91/414
- All approvals for aerial use of isoproturon were revoked in 1995

Efficacy guidance

- May be applied in autumn or spring (but only post-emergence on wheat or barley). Best control normally achieved by early post-emergence treatment
- See label for details of rates, timings and tank mixes for different weed problems
- Apply to moist soils. Effectiveness may be reduced in seasons of above average rainfall or by prolonged dry weather
- Residual activity reduced on soils with more than 10% organic matter. Only use on such soils in spring
- Always follow WRAG guidelines for preventing and managing herbicide resistant weeds. Section 5 for more information

Restrictions

- Maximum number of treatments 1 per crop
- Maximum total dose 2.5 kg a.i./ha for any crop. The IPU stewardship programme guidelines recommend that this maximum dose should not be used after the end of Oct, and that where possible it should be reduced to 1.5 kg a.i./ha by using mixtures or sequences with other herbicides
- Do not apply pre-emergence to wheat or barley. On triticale and winter rye use pre-emergence only and on named varieties
- Do not use on durum wheats, oats, undersown cereals or crops to be undersown
- Do not apply to very wet or waterlogged soils, or when heavy rainfall is forecast
- Do not use on very cloddy soils or if frost is imminent or after onset of frosty weather
- Do not roll for 1 wk before or after treatment or harrow for 1 wk before or any time after treatment

Crop-specific information

- Latest use: not later than second node detectable stage (GS 32)
- Recommended timing of treatment varies depending on crop to be treated, method of sowing, season of application, weeds to be controlled and product being used. See label for details
- Crop damage may occur on free draining, stony or gravelly soils if heavy rain falls soon after spraying. Early sown crops may be damaged if spraying precedes or coincides with a period of rapid growth

Environmental safety

- Dangerous for the environment
- Very toxic to aquatic organisms
- Do not apply to dry, cracked or waterlogged soils where heavy rain may lead to contamination of drains by isoproturon
- Do not empty into drains

Hazard classification and safety precautions

Hazard H03, H11 [1-6, 8-10]
Risk phrases R22a, R51 [1, 5]; R40 [1-5, 8-10]; R50 [2-4, 6, 8-10]; R58 [1-6, 8-10]; R68 [6]
Operator protection A, C, H [1-10]; D [1, 5, 10]; U05a [2, 3, 6, 7, 9, 10]; U08 [1-7, 9, 10]; U14, U15 [10]; U19a [2, 3, 7, 9, 10]; U20a [3, 8]; U20b [1, 2, 4-6, 9, 10]
Environmental protection E13b [7]; E13c [3]; E15a [1, 2, 4-6, 8-10]; E19b [1, 5]; E31b [2-4, 6-9]; E32a [1, 5, 10]; E32d [4]; E32e [2, 3, 6, 9]; E34 [2, 3, 9]
Storage and disposal E01, E04 [1-3, 5, 7, 9, 10]; E26, E30a [1-10]
Treated seed S06a [7]

284 isoproturon + pendimethalin

A contact and residual herbicide for use in winter cereals

Products

1	Encore	BASF	125:250 g/l	SC	10375
2	Trump	BASF	236:236 g/l	SC	10376

Uses

- Annual dicotyledons in ***spring wheat*** *(autumn sown)* [2]; ***triticale***, ***winter barley***, ***winter rye***, ***winter wheat*** [1, 2]
- Annual grasses in ***spring wheat*** *(autumn sown)* [2]; ***triticale***, ***winter barley***, ***winter rye***, ***winter wheat*** [1, 2]
- Blackgrass in ***spring wheat*** *(autumn sown)* [2]; ***triticale***, ***winter barley***, ***winter rye***, ***winter wheat*** [1, 2]
- Wild oats in ***spring wheat*** *(autumn sown)* [2]; ***triticale***, ***winter barley***, ***winter rye***, ***winter wheat*** [1, 2]

Approval information

- Isoproturon and pendimethalin included in Annex I under EC Directive 91/414

Efficacy guidance

- May be applied in autumn or spring (but only post-emergence on wheat or barley)
- Annual grasses controlled from pre-emergence to 4-leaf stage, extended to 3-tiller stage with blackgrass by tank mixing with additional isoproturon. Best results on wild oats post-emergence in autumn
- Annual dicotyledons controlled pre-emergence and up to 8-leaf stage [1], up to 12-leaf stage [2]. See label for details
- Apply to moist soils. Best results achieved by application to fine, firm seedbed. Trash, ash or straw should have been incorporated evenly
- Contact activity is reduced by rain within 6 h of application
- Activity may be reduced on soil with more than 6% organic matter or ash. Do not use on soils with more than 10% organic matter
- Always follow WRAG guidelines for preventing and managing herbicide resistant weeds. Section 5 for more information

Restrictions

- Maximum number of treatments 1 per crop. Maximum total dose of isoproturon 2.5 kg a.i./ha per crop
- Do not apply pre-emergence to wheat or barley. On winter rye and triticale use pre-emergence only on named cultivars
- Do not use on durum wheat or spring cereals
- Do not use pre-emergence on winter rye or triticale drilled after 30 Nov [2]
- Do not use on crops suffering stress from disease, drought, waterlogging, poor seedbed conditions or other causes or apply post-emergence when frost imminent
- Do not apply to very wet or waterlogged soils, or when heavy rain is forecast

- Do not undersow treated crops
- Do not roll or harrow after application

Crop-specific information

- Latest use: pre-emergence for winter rye and triticale; before leaf sheath erect (GS 30) [2] or before 1st node detectable (GS 31) [1] for winter wheat and barley
- Apply before first node detectable (GS 31) [1], to before leaf sheath erect (GS 30) [2]
- May be used on autumn sown spring wheat [2]

Following crops guidance

- In the event of autumn crop failure spring wheat, spring barley, maize, potatoes, beans or peas may be grown following ploughing to at least 150 mm

Environmental safety

- Dangerous for the environment
- Very toxic to aquatic organisms
- Do not apply to dry, cracked or waterlogged soils where heavy rain may lead to contamination of drains by isoproturon

Hazard classification and safety precautions

Hazard H03, H11
Risk phrases R40, R50, R58 [1, 2]; R68 [2]
Operator protection A, C, H; U05a, U08, U13, U19a, U20b
Environmental protection E15a, E31b, E32d, E38 [1, 2]; E34 [1]
Storage and disposal E01, E04, E26, E30a
Medical advice M05a

285 isoproturon + simazine

A contact and residual herbicide for winter wheat and barley

Products

1	Alpha Protugan Plus	Makhteshim	375:60 g/l	SC	08799
2	Harlequin 500 SC	Makhteshim	450:50 g/l	SC	09779

Uses

- Annual dicotyledons in ***winter barley, winter wheat***
- Annual grasses in ***winter barley, winter wheat***
- Blackgrass in ***winter barley, winter wheat***

Approval information

- Isoproturon included in Annex I under EC Directive 91/414

Efficacy guidance

- Annual grasses controlled up to early tillering, annual dicotyledons to 50 mm
- Apply to fine, firm, even seedbed. Any trash or burnt straw should be dispersed in preparing seedbed
- Apply to moist soils. Weed control may be reduced in excessively wet autumns or if prolonged dry weather follows application to dry soil
- Always follow WRAG guidelines for preventing and managing herbicide resistant weeds. Section 5 for more information

Restrictions

- Maximum number of treatments 1 per crop. Maximum total dose of isoproturon 2.5 kg a.i./ha per crop
- Do not use on durum wheat, undersown crops or those due to be undersown
- Do not use on sand or on soils with more than 10% organic matter
- On stony or gravelly soils there is risk of crop damage, especially with heavy rain soon after application
- Do not apply when frost or heavy rain is forecast or to crops severely checked by frost, waterlogging, pest or disease attack.
- Do not harrow for 7 d before treatment or roll for 7 d before or after treatment in spring
- Do not harrow after application

Crop-specific information

- Latest use: end of tillering [1]; leaf sheath erect (GS 30) [2]

- May be applied in autumn or spring (but only post-emergence of the crop)
- Apply from 2-leaf stage of crop to before leaf sheath erect (GS 12-30)
- With direct drilled crops soil surface should be broken by surface cultivation and seed covered by 12-25 mm soil
- Early sown crops may be prone to damage if spraying precedes or coincides with period of rapid growth in autumn

Following crops guidance

- In the event of crop failure land should be inverted by mouldboard ploughing to at least 150 mm and harrowed before drilling/planting another crop

Environmental safety

- Dangerous for the environment
- Very toxic to aquatic organisms
- Do not apply to dry, cracked or waterlogged soils where heavy rain may lead to contamination of drains by isoproturon
- LERAP Category B

Hazard classification and safety precautions

Hazard H03, H11 [1]; H04 [2]
Risk phrases R38 [2]; R40, R50, R58 [1]
Operator protection A, C, H [1, 2]; D, M [2]; U05a, U08, U19a, U20b [1, 2]; U14, U15 [1]
Environmental protection E13b [2]; E15a, E32e [1]; E16a, E16b, E31b [1, 2]
Storage and disposal E01, E04, E26, E30a [1, 2]; E29a [2]

286 isoproturon + trifluralin

A residual early post-emergence herbicide for cereals

Products

Autumn Kite	Nufarm UK	300:200 g/l	EC	11549

Uses

- Annual dicotyledons in ***winter barley, winter wheat***
- Annual grasses in ***winter barley, winter wheat***
- Blackgrass in ***winter barley, winter wheat***

Efficacy guidance

- Provides contact and residual control. Best results achieved by application to dicotyledons pre-emergence or up to 2-leaf stage, to blackgrass up to 2-3 tillers
- Apply post-emergence when leaves dry, weeds are actively growing and rain not expected for 2 h
- Effectiveness may be reduced by prolonged dry or sunny weather after application
- Do not use on soils with more than 10% organic matter
- If used in conjunction with minimum cultivation ensure that all trash and burnt straw is removed, buried or dispersed before spraying

Restrictions

- Maximum number of treatments 1 per crop. Maximum total dose of isoproturon 2.5 kg a.i./ha per crop
- Apply post-emergence to 4-leaf and 2 tillers stage and to broadcast crops after 3-leaf stage (GS 13)
- Do not use on durum wheat or crops to be undersown
- Do not use on Sands and do not incorporate in soil. On very stony, gravelly or other free draining soils crops may be damaged if heavy rain falls soon after treatment
- Do not spray when heavy rain is forecast or to crops stressed by frost, waterlogging, deficiency or pest attack
- Do not roll after treatment until following spring
- In case of crop failure only sow carrots, peas or sunflowers within 5 mth and plough to at least 15 cm. Do not sow sugar beet in spring following treatment
- Do not harrow treated crops

Crop-specific information

- Before 6 leaf stage (GS 16)

Environmental safety

- Dangerous for the environment

- Very toxic to aquatic organisms
- Keep in original container, tightly closed, in a safe place, under lock and key
- Do not spray where soils are cracked, to avoid run-off through drains

Hazard classification and safety precautions

Hazard H03, H11
Risk phrases R22b, R36, R40, R43, R50, R58, R66, R67
Operator protection A, C; U05a, U08, U11, U13, U19a, U20b
Environmental protection E15a, E31b, E32d, E34
Storage and disposal E01, E04, E26, E29b, E30a
Medical advice M05b

287 isoxaben

A soil-acting amide herbicide for use in grass and fruit

Products

Flexidor 125	Landseer	125 g/l	SC	10946

Uses

- Annual dicotyledons in ***amenity vegetation***, ***apple orchards***, ***blackberries***, ***blackcurrants***, ***cherries***, ***container-grown woody ornamentals***, ***forestry transplants***, ***gooseberries***, ***grapevines***, ***hardy ornamental nursery stock***, ***hops***, ***pear orchards***, ***plums***, ***raspberries***, ***strawberries***

Efficacy guidance

- When used alone apply pre-weed emergence
- Effectiveness is reduced in dry conditions. Weed seeds germinating at depth are not controlled
- Activity reduced on soils with more than 10% organic matter. Do not use on peaty soils
- Various tank mixtures are recommended for early post-weed emergence treatment (especially for grass weeds). See label for details

Restrictions

- Maximum number of treatments 1 per crop for all edible crops; 2 per yr on amenity vegetation and non-edible crops

Crop-specific information

- Latest use: 3 true leaves for cucurbits; before 1 Apr in yr of harvest for other edible crops

Following crops guidance

- See label for details of crops which may be sown in the event of failure of a treated crop

Environmental safety

- Keep all livestock out of treated areas for at least 50 d

Hazard classification and safety precautions

Operator protection A, C; U05a, U20b
Environmental protection E06a (50 d); E15a, E32a
Storage and disposal E01, E26, E30a

288 isoxaben + trifluralin

A residual herbicide mixture for amenity and horticultural use

Products

1	Axit GR	Fargro	0.5:2.0% w/w	GR	08892
2	Premiere	Dow	0.5:2.0 %w/w	GR	10866

Uses

- Annual dicotyledons in ***amenity trees and shrubs***, ***amenity vegetation*** [2]; ***container-grown ornamentals*** *(outdoor stock only)*, ***hardy ornamental nursery stock*** *(outdoor stock only)* [1]
- Annual meadow grass in ***amenity trees and shrubs***, ***amenity vegetation*** [2]; ***container-grown ornamentals*** *(outdoor stock only)*, ***hardy ornamental nursery stock*** *(outdoor stock only)* [1]
- Fat hen in ***amenity trees and shrubs***, ***amenity vegetation*** [2]; ***container-grown ornamentals*** *(outdoor stock only)*, ***hardy ornamental nursery stock*** *(outdoor stock only)* [1]
- Groundsel in ***amenity trees and shrubs***, ***amenity vegetation*** [2]

- Hairy bittercress in ***amenity trees and shrubs***, ***amenity vegetation*** [2]; ***container-grown ornamentals*** *(outdoor stock only)*, ***hardy ornamental nursery stock*** *(outdoor stock only)* [1]
- Shepherd's purse in ***hardy ornamental nursery stock*** *(outdoor stock only)* [1]
- Sowthistle in ***amenity trees and shrubs***, ***amenity vegetation*** [2]; ***container-grown ornamentals*** *(outdoor stock only)*, ***hardy ornamental nursery stock*** *(outdoor stock only)* [1]

Efficacy guidance

- Best results obtained when granules applied to firm, moist soil, free from clods. 20-30 mm irrigation or rainfall required within 3 d after treatment
- Weed control reduced under dry soil conditions
- Hoeing after application will reduce weed control
- Treatment should be made before mulching and before weed emergence
- Uniform cover with granules optimises results. Improved uniformity of cover may be achieved by spreading the granules twice at half dose at right angles
- Control may be reduced on soils with high organic matter or where organic manure has been applied
- Perennial weeds regenerating from underground rootstocks or stolons are not controlled

Restrictions

- Maximum number of treatments 1 per yr
- Products contain up to 1% crystalline silica. The MEL for respirable silica dust is 0.4 mg/cu m
- Do not use on soils with more than 10% organic matter (except in container grown hardy nursery stock)
- Do not mix in the soil or compost
- Do not use on ornamentals grown under protection
- Do not apply to plants with wet foliage or where granules can lodge in the foliage
- Do not apply to heathers or *Potentilla spp.*

Crop-specific information

- Ensure soil has settled and no cracks are present before application to transplanted trees and shrubs [2]
- May be used on Light, Medium and Heavy soils, but on Light soils early growth may be reduced under adverse conditions
- For newly planted containerised stock first irrigate and allow compost to settle and foliage and stems to dry before application [1]

Following crops guidance

- Land must be mouldboard ploughed to at least 20 cm before drilling or planting succeeding plants except well rooted forestry trees and ornamentals
- Do not use in the 12 mth prior to lifting field grown nursery stock if edible crops are to be grown as a following crop
- After an autumn application cereals or grass crops should not be sown until the following autumn
- Crops most sensitive to treatment residues are oilseed rape, stubble turnips, other brassicae, sugar beet, fodder beet, herbage seed, grass leys and amenity grass

Environmental safety

- Harmful to fish or other aquatic life. Do not contaminate surface waters or ditches with chemical or used container
- Do not allow direct applications of granules from ground based/vehicle mounted applicators to fall within 6 m, or from hand-held applicators to within 2 m, of surface waters or ditches. Direct applications away from water

Hazard classification and safety precautions

Operator protection A; U20b
Environmental protection E13c, E16f, E32a, E34
Storage and disposal E01, E30a

289 kresoxim-methyl

A protectant strobilurin fungicide for apples

See also *epoxiconazole + fenpropimorph + kresoxim-methyl*
epoxiconazole + kresoxim-methyl
epoxiconazole + kresoxim-methyl + pyraclostrobin
fenpropimorph + kresoxim-methyl

Products

Stroby WG	BASF	50% w/w	WG	08653

Uses
- Black spot in ***roses***
- Powdery mildew in ***apples*** *(reduction)*, ***blackcurrants***, ***protected strawberries***, ***roses***, ***strawberries***
- Scab in ***apples***

Approval information
- Kresoxim-methyl included in Annex I under EC Directive 91/414
- Approved for use in ULV systems

Efficacy guidance
- Activity is protectant. Best results achieved from treatments prior to disease development. See label for timing details on each crop. Treatments should be repeated at 10-14 d intervals but note limitations below
- To minimise the likelihood of development of resistance to strobilurin fungicides these products should be used in a planned Resistance Management strategy. See Section 5 for more information
- Product may be applied in ultra low volumes (ULV) but disease control may be reduced

Restrictions
- Maximum number of treatments 4 per yr on apples; 3 per yr on other crops. See notes in Efficacy about limitations on consecutive treatments
- Consult before using on crops intended for processing

Crop-specific information
- HI 14 d for blackcurrants, protected strawberries, strawberries; 35 d for apples
- On apples do not spray product more than twice consecutively and separate each block of two consecutive treatments with at least two applications from a different cross-resistance group. For all other crops do not apply consecutively and use a maximum of once in every three fungicide sprays
- Product should not be used as final spray of the season on apples

Environmental safety
- Dangerous for the environment
- Very toxic to aquatic organisms
- Broadcast air-assisted LERAP (5 m)
- Harmless to ladybirds and predatory mites
- Harmless to honey bees and may be applied during flowering. Nevertheless local beekeepers should be notified when treatment of orchards in flower is to occur

Hazard classification and safety precautions

Hazard H03, H11
Risk phrases R40, R50, R58
Operator protection U05a, U20b
Environmental protection E15a, E31c, E32d, E38; E17b (5 m)
Storage and disposal E01, E04, E29b, E30a
Medical advice M05a

290 lambda-cyhalothrin

A quick-acting contact and ingested pyrethroid insecticide

Products

1	Hallmark with Zeon Technology	Syngenta	100 g/l	CS	10480
2	Stealth	Syngenta	2.5% w/w	WG	11514

Uses

- Aphids in ***chestnuts*** *(off-label)*, ***cob nuts*** *(off-label)*, ***durum wheat***, ***hazel nuts*** *(off-label)*, ***potatoes***, ***short rotation coppice willow***, ***spring barley***, ***spring oats***, ***spring wheat***, ***walnuts*** *(off-label)*, ***winter barley***, ***winter oats***, ***winter wheat*** [1]
- Barley yellow dwarf virus vectors in ***durum wheat***, ***winter barley***, ***winter oats***, ***winter wheat*** [1, 2]; ***spring barley***, ***spring wheat*** [1]
- Beet leaf miner in ***sugar beet*** [1]
- Beet virus yellows vectors in ***spring oilseed rape*** [1]; ***winter oilseed rape*** [1, 2]
- Beetles in ***short rotation coppice willow*** [1]
- Cabbage stem flea beetle in ***spring oilseed rape*** [1]; ***winter oilseed rape*** [1, 2]
- Cabbage stem weevil in ***radishes*** *(off-label)* [1]
- Carrot fly in ***carrots*** *(off-label)*, ***celeriac*** *(off-label)*, ***celery*** *(off-label)*, ***fennel*** *(off-label)*, ***parsnips*** *(off-label)* [1]
- Caterpillars in ***broccoli***, ***brussels sprouts***, ***cabbages***, ***calabrese***, ***cauliflowers***, ***chestnuts*** *(off-label)*, ***cob nuts*** *(off-label)*, ***hazel nuts*** *(off-label)*, ***walnuts*** *(off-label)* [1]
- Cutworms in ***carrots***, ***celeriac*** *(off-label)*, ***chicory*** *(off-label - for forcing)*, ***fennel*** *(off-label)*, ***lettuce***, ***red beet*** *(off-label)*, ***sugar beet*** [1]
- Flea beetle in ***poppies*** *(off-label - for morphine production)*, ***spring oilseed rape***, ***sugar beet*** [1]; ***winter oilseed rape*** [1, 2]
- Frit fly in ***sweetcorn*** *(off-label)* [1]
- Pea and bean weevil in ***peas***, ***spring field beans***, ***winter field beans*** [1]
- Pea aphid in ***peas*** [1]
- Pea moth in ***peas*** [1]
- Pear sucker in ***pears*** [1]
- Pod midge in ***spring oilseed rape***, ***winter oilseed rape*** [1]
- Pollen beetle in ***poppies*** *(off-label - for morphine production)*, ***spring oilseed rape***, ***winter oilseed rape*** [1]
- Sawflies in ***short rotation coppice willow*** [1]
- Seed weevil in ***spring oilseed rape***, ***winter oilseed rape*** [1]
- Silver Y moth in ***celery*** *(off-label)*, ***dwarf beans*** *(off-label)*, ***navy beans*** *(off-label)*, ***red beet*** *(off-label)*, ***runner beans*** *(off-label)* [1]
- Thrips in ***bulb onions*** *(off-label)*, ***leeks*** *(off-label)*, ***salad onions*** *(off-label)* [1]
- Whitefly in ***broccoli***, ***brussels sprouts***, ***cabbages***, ***calabrese***, ***cauliflowers*** [1]
- Yellow cereal fly in ***winter wheat*** [1]

Specific Off-Label Approvals (SOLAs)

- ***bulb onions***, ***leeks***, ***salad onions*** *(OLA 032571) Dec 2008* [1]
- ***carrots***, ***parsnips*** *(OLA 011642) Dec 2008* [1]
- ***celeriac***, ***radishes*** *(OLA 011430) Dec 2008* [1]
- ***celery*** *(OLA 011644) Dec 2008* [1]
- ***chestnuts***, ***cob nuts***, ***hazel nuts***, ***walnuts*** *(OLA 011646) Dec 2008* [1]
- ***chicory*** *(for forcing) (OLA 011649) Dec 2008* [1]
- ***dwarf beans***, ***navy beans***, ***runner beans*** *(OLA 011645) Dec 2008* [1]
- ***fennel*** *(OLA 011648) Dec 2008* [1]
- ***poppies*** *(for morphine production) (OLA 030768) Dec 2008* [1]
- ***red beet*** *(OLA 011643) Dec 2008* [1]
- ***sweetcorn*** *(OLA 011647) Dec 2008* [1]

Approval information

- Lambda-cyhalothrin included in Annex I under EC Directive 91/414

Efficacy guidance

- Best results normally obtained from treatment when pest attack first seen. See label for detailed recommendations on each crop

- Timing for control of barley yellow dwarf virus vectors depends on specialist assessment of the level of risk in the area
- Repeat applications recommended in some crops where prolonged attack occurs, up to maximum total dose. See label for details
- Where strains of aphids resistant to lambda-cyhalothrin occur control is unlikely to be satisfactory
- Addition of wetter recommended for control of certain pests in brassicas and oilseed rape
- Use of sufficient water volume to ensure thorough crop penetration recommended for optimum results
- Use of drop-legged sprayer gives improved results in crops such as Brussels sprouts

Restrictions

- Maximum number of applications or maximum total dose per crop varies - see labels
- Do not apply to a cereal crop if any product containing a pyrethroid insecticide or dimethoate has been applied to the crop after the start of ear emergence (GS 51)
- Do not spray cereals in the spring/summer (ie after 1 Apr) within 6 m of edge of crop

Crop-specific information

- Latest use on cereals and oilseed rape depends on product used. See labels
- HI 3 d for lettuce, radishes, red beet; 7 d for celery, chicory, dwarf beans, fennel, navy beans, runner beans, sweetcorn; 14 d for carrots, celeriac, nuts, parsnips; 6 wk for spring oilseed rape; 8 wk for sugar beet
- If using mixtures, add lambda-cyhalothrin product to the tank last

Environmental safety

- Dangerous for the environment
- Very toxic to aquatic organisms
- High risk to non-target insects or other arthropods. Do not spray within 6 m of the field boundary [2]
- LERAP Category A
- Broadcast air-assisted LERAP (38 m in pears) [1]

Hazard classification and safety precautions

Hazard H03, H11 [1, 2]; H04 [2]
Risk phrases R20, R22a, R43, R50, R58 [1, 2]; R36, R38 [2]
Operator protection A, C, H, J, K, M [1, 2]; L [1]; U02a, U05a, U08, U14, U20b [1, 2]; U11 [2]
Environmental protection E15a, E16c, E16d, E32d, E34, E38 [1, 2]; E17b [1] (pears 38 m); E22a, E31c [2]; E31b [1]
Storage and disposal E01, E04, E30a [1, 2]; E26 [2]
Medical advice M03

291 lambda-cyhalothrin + pirimicarb

An insecticide mixture combining translaminar, contact, fumigant and stomach activity

Products

Dovetail	Syngenta	5:100 g/l	EC	10479

Uses

- Aphids in ***broccoli, brussels sprouts, cabbages, calabrese, carrots, cauliflowers, durum wheat, lettuce, peas, potatoes, spring barley, spring oats, spring rye, spring wheat, sugar beet, triticale, winter barley, winter oats, winter rye, winter wheat***
- Beet virus yellows vectors in ***spring oilseed rape, winter oilseed rape***
- Cabbage stem flea beetle in ***spring oilseed rape, winter oilseed rape***
- Caterpillars in ***broccoli, brussels sprouts, cabbages, calabrese, cauliflowers***
- Cutworms in ***carrots, lettuce, potatoes, sugar beet***
- Flea beetle in ***spring oilseed rape, sugar beet, winter oilseed rape***
- Leaf miner in ***sugar beet***
- Mealy aphid in ***broccoli, brussels sprouts, cabbages, calabrese, cauliflowers, spring oilseed rape, winter oilseed rape***
- Pea and bean weevil in ***peas, spring field beans, winter field beans***
- Pea midge in ***peas***
- Pea moth in ***peas***
- Pod midge in ***spring oilseed rape, winter oilseed rape***

- Pollen beetle in ***broccoli, brussels sprouts, cabbages, calabrese, cauliflowers, spring oilseed rape, winter oilseed rape***
- Seed weevil in ***spring oilseed rape, winter oilseed rape***
- Whitefly in ***broccoli, brussels sprouts, cabbages, calabrese, cauliflowers***

Approval information

- Lambda-cyhalothrin included in Annex I under EC Directive 91/414

Efficacy guidance

- Best results obtained from treatment when pest attack first seen or after warning issued
- Repeat applications recommended in some crops where prolonged attacks occur, up to maximum total dose
- Addition of Agral recommended for certain uses in brassicas and oilseed rape. See label
- Use of drop-leg sprayers improves efficacy in crops such as Brussels sprouts
- Control unlikely to be satisfactory if aphids resistant to lambda-cyhalothrin or pirimicarb present

Restrictions

- Maximum total dose equivalent to two full dose treatments on carrots, lettuce, peas, field beans; three full dose treatments on brassicas, oilseed rape, cereals; four full dose treatments on sugar beet; eight full dose treatments on potatoes
- Must not be applied to cereals if any product containing a pyrethroid insecticide or dimethoate has been sprayed after the start of ear emergence (GS 51)

Crop-specific information

- Latest use: before late milky ripe (GS 77) for cereals; before end of flowering for oilseed rape
- HI sugar beet 8 wk; carrots 14 d; brassicas, lettuce, potatoes, peas, field beans 3 d

Environmental safety

- Dangerous for the environment
- Very toxic to aquatic organisms
- Do not spray cereals after 1 Apr within 6 m of the edge of the crop
- Harmful to livestock. Keep all livestock out of treated areas for at least 7 d following treatment
- Keep in original container, tightly closed, in a safe place, under lock and key
- LERAP Category A

Hazard classification and safety precautions

Hazard H03, H08, H11
Risk phrases R20, R22a, R22b, R37, R38, R41, R43, R50, R58
Operator protection A, C, E, H, J, K, M; U02a, U05a, U08, U11, U19a, U20b
Environmental protection E06c (7 d); E15a, E16c, E16d, E31c, E32d, E34, E38
Storage and disposal E01, E04, E30b
Medical advice M02, M03, M05b

292 lenacil

A residual, soil-acting uracil herbicide for beet and horticultural crops

See also ethofumesate + lenacil + phenmedipham

Products

1	Agricola Lenacil FL	Interfarm	440 g/l	SC	09481
2	Agriguard Lenacil	AgriGuard	80% w/w	WP	10488
3	Fernpath Lenzo Flo	AgriGuard	440 g/l	SC	11919
4	Venzar Flowable	DuPont	440 g/l	SC	06907

Uses

- Annual dicotyledons in ***farm woodland*** *(off-label)* [4]; ***fodder beet, mangels, sugar beet*** [1-4]; ***red beet*** [2-4]
- Annual meadow grass in ***farm woodland*** *(off-label)* [4]; ***fodder beet, mangels, sugar beet*** [1-4]; ***red beet*** [2-4]

Specific Off-Label Approvals (SOLAs)

- ***farm woodland*** *(OLA 971282) Dec 2008* [4]

Efficacy guidance

- On beet crops may be used pre- or post-emergence, alone or in mixture to broaden weed spectrum

FOR FULL CONDITIONS OF USE ALWAYS READ THE PRODUCT LABEL

- Apply overall or as band spray to beet crops pre-drilling incorporated, pre- or post-emergence
- All labels have limitations on soil types that may be treated. Residual activity reduced on soils with high OM content
- Best results achieved on a firm moist tilth. Continuing presence of moisture from rain or irrigation gives improved residual control of later germinating weeds. Effectiveness may be reduced by dry conditions

Restrictions

- Maximum number of treatments in beet crops 1 pre-emergence + 3 post-emergence per crop; 2 per yr on established woody ornamentals and roses
- See label for soil type restrictions
- Do not apply other residual herbicides within 3 mth before or after spraying
- Do not treat crops under stress from drought, low temperatures, nutrient deficiency, pest or disease attack, or waterlogging

Crop-specific information

- Latest use: pre-emergence for red beet, fodder beet, spinach, spinach beet, mangels; before leaves meet over rows when used on these crops post-emergence
- Heavy rain after application may cause damage especially if followed by very hot weather
- Reduction in beet stand may occur where crop emergence or vigour is impaired by soil capping or pest attack

Following crops guidance

- Succeeding crops should not be planted or sown for at least 4 mth (6 mth on organic soils) after treatment following ploughing to at least 150 mm

Environmental safety

- Dangerous for the environment
- Very toxic to aquatic organisms

Hazard classification and safety precautions

Hazard H04 [1-3]; H11 [1, 2, 4]
Risk phrases R36, R38 [1-3]; R37 [2, 3]; R50, R58 [1, 4]
Operator protection A, C; U05a, U08, U19a [1-4]; U14, U15 [1-3]; U20a [1, 2, 4]; U20b [3]
Environmental protection E15a [1-4]; E31a [3, 4]; E31b, E34 [1, 2]; E32a, E32d, E38 [4]; E32e [1]
Storage and disposal E01, E04, E30a [1-4]; E26 [1, 3]

293 lenacil + phenmedipham

A contact and residual herbicide for use in sugar beet

Products

Agricola Lens	Interfarm	66.6:95 g/l	SC	10257

Uses

- Annual dicotyledons in ***fodder beet***, ***mangels***, ***sugar beet***

Efficacy guidance

- Apply at any stage of crop when weeds at cotyledon stage
- Germinating weeds controlled by root uptake for several weeks
- Best results achieved under warm, moist conditions on a fine, firm seedbed
- Treatment may be repeated up to a maximum of 3 applications on later weed flushes
- Residual activity may be reduced on highly organic soils or under very dry conditions
- Rain falling 1 h after application does not reduce activity

Restrictions

- Maximum number of treatments 3 per crop
- Do not apply when temperature above or likely to exceed 21°C on day of spraying or under conditions of high light intensity
- Do not spray any crop under stress from drought, waterlogging, cold, wind damage or any other cause

Crop-specific information

- Latest use: before crop leaves meet across rows
- Heavy rain after application may reduce stand of crop particularly in very hot weather. In severe cases yield may be reduced

Following crops guidance

- Do not sow or plant any crop within 4 mth of treatment. In case of crop failure only sow or plant beet crops, strawberries or other tolerant horticultural crop within 4 mth

Environmental safety

- Dangerous for the environment
- Toxic to aquatic organisms

Hazard classification and safety precautions

Hazard H03, H11
Risk phrases R21, R22a, R22b, R36, R37, R40, R41, R43, R51, R58
Operator protection A, C, H; U02a, U05a, U08, U11, U14, U19a, U20b
Environmental protection E15a, E31b, E34
Storage and disposal E01, E04, E30a
Medical advice M04, M05b

294 lindane

An organochlorine insecticide all approvals for which were revoked in 2001 following non-inclusion of the substance in Annex I of EC Directive 91/414

295 linuron

A contact and residual urea herbicide for various field crops

Products

1	Afalon	Makhteshim	450 g/l	SC	11665
2	Alpha Linuron 50 SC	Makhteshim	500 g/l	SC	06967
3	Lincon 50	DAPT	500 g/l	SC	10639
4	Linuron 500	Nufarm UK	480 g/l	SC	11590
5	UPL Linuron 45% Flowable	United Phosphorus	450 g/l	SC	07435

Uses

- Annual dicotyledons in ***bulb onions*** *(off-label)*, ***celeriac*** *(off-label)*, ***garlic*** *(off-label)*, ***leeks*** *(off-label)* [2]; ***carrots, parsley, potatoes*** [1-5]; ***celery*** [1]; ***parsnips*** [1-3, 5]; ***spring barley, spring wheat*** [3-5]; ***spring oats*** [2, 3]
- Annual meadow grass in ***carrots, parsley, potatoes*** [1-5]; ***celery*** [1]; ***parsnips*** [1-3, 5]; ***spring barley, spring wheat*** [2-5]; ***spring oats*** [3]
- Black bindweed in ***carrots, parsley, potatoes*** [1-5]; ***celery*** [1]; ***parsnips*** [1-3, 5]; ***spring barley, spring wheat*** [2-5]; ***spring oats*** [2, 3]
- Chickweed in ***carrots, parsley, potatoes*** [1-5]; ***celery*** [1]; ***parsnips*** [1-3, 5]; ***spring barley, spring wheat*** [2-5]; ***spring oats*** [2, 3]
- Corn marigold in ***carrots, parsley, potatoes, spring barley, spring wheat*** [2-4]; ***parsnips, spring oats*** [2, 3]
- Fat hen in ***carrots, parsley, potatoes*** [1-5]; ***celery*** [1]; ***parsnips*** [1-3, 5]; ***spring barley, spring wheat*** [2-5]; ***spring oats*** [2, 3]
- Redshank in ***carrots, parsley, potatoes*** [1-5]; ***celery*** [1]; ***parsnips*** [1-3, 5]; ***spring barley, spring wheat*** [2-5]; ***spring oats*** [2, 3]

Specific Off-Label Approvals (SOLAs)

- ***bulb onions, celeriac, garlic, leeks*** *(OLA 021748) Dec 2008* [2]

Approval information

- Linuron included in Annex I under EC Directive 91/414

Efficacy guidance

- Many weeds controlled pre-emergence or post-emergence to 2-3 leaf stage, some (annual meadow grass, mayweed) only susceptible pre-emergence. See label for details
- Best results achieved by application to firm, moist soil of fine tilth
- Little residual effect on soil with more than 10% organic matter

Restrictions

- Maximum number of treatments 1 per crop on most crops or 1 pre-em + 1 post-em for carrots, parsley, parsnips

- Maximum total dose equivalent to one full dose treatment
- Apply only pre-crop emergence [5]
- Do not use on undersown cereals or crops grown on Sands or Very Light soils or soils heavier than Sandy Clay Loam or with more than 10% organic matter
- Do not apply to emerged crops of carrots, parsnips or parsley under stress
- Do not apply by hand-held sprayers

Crop-specific information

- Latest use: pre-emergence for cereals and linseed; pre- or up to 40% emergence for potatoes (products differ, see label for details).
- HI onions, garlic 8 wk; celeriac 12 wk; leeks 16 wk
- Drill spring cereals at least 3 cm deep and apply pre-emergence of crop or weeds
- Apply to potatoes well earthed up to a rounded ridge pre-crop emergence and do not cultivate after spraying
- Apply to carrots or parsley at any time after drilling on organic soils, and within 4 d of drilling on other soils. Apply post-emergence as soon as weeds appear but after first rough leaf stage.
- Recommendations for parsnips, parsley and celery vary. See label for details

Following crops guidance

- Potatoes, carrots and parsnips may be planted at any time after application. Lettuce should not be grown within 12 mth of treatment. Transplanted brassicas may be grown from 3 mth after treatment

Environmental safety

- Dangerous for the environment [4]
- Very toxic to aquatic organisms
- Keep livestock out of treated areas for at least 5 mth [4]
- LERAP Category B

Hazard classification and safety precautions

Hazard H03, H11 [1, 2, 4, 5]
Risk phrases R22a, R40 [1, 2, 5]; R48, R50, R58 [1, 2, 4, 5]; R68 [4]
Operator protection A, H [1-5]; C [2-5]; U05a, U20b [1, 2, 4, 5]; U08 [1-5]; U19a [2-5]; U23a [1]
Environmental protection E07b [4] (5 mth); E13b, E16b [3]; E15a [1, 2, 4, 5]; E16a [1-5]; E31b, E32d [1, 4]; E32a [2, 3, 5]; E32e [1, 2, 5]; E38 [4]
Storage and disposal E01 [1, 2, 4, 5]; E04 [1, 2, 5]; E26, E30a [1-5]; E29b [1]
Treated seed S06a [3]

296 linuron + trifluralin

A residual pre-emergence herbicide for use in winter cereals

Products

1	Arizona	Makhteshim	120:240 g/l	EC	11722
2	Uranus	Makhteshim	120:240 g/l	EC	11637

Uses

- Annual dicotyledons in ***triticale***, ***winter barley***, ***winter wheat***
- Annual grasses in ***triticale***, ***winter barley***, ***winter wheat***

Approval information

- Linuron included in Annex I under EC Directive 91/414

Efficacy guidance

- Effective against weeds germinating near soil surface. Best results achieved by application to fine, firm, moist seedbed free of clods, crop residues or established weeds
- Effectiveness reduced by long dry periods after application or on waterlogged soil
- Autumn application residual effects normally last until spring but further herbicide treatment may be needed on thin or backward crops
- Results in loose seedbeds on lighter soils improved by rolling after drilling

Restrictions

- Maximum number of treatments 1 per crop
- Do not treat durum wheat or undersown crops

- Do not use on soils classed as Sands. Do not harrow after treatment
- Do not use on peaty soils or where organic matter exceeds 10%

Crop-specific information

- Latest use: pre-emergence of crop
- Apply without incorporation as soon as possible after drilling and before crop emergence, within 3 d on early drilled crops
- Crop seed must be well covered to a minimum depth of 30 mm

Following crops guidance

- In the event of failure of a treated crop, only peas, carrots or sunflowers may be sown within 5 mth of application. Grass and cereals should not be planted until the following autumn, and sugar beet not for one yr
- Before drilling or planting a following crop mouldboard plough to at least 150 mm

Environmental safety

- Dangerous for the environment
- Very toxic to aquatic organisms
- LERAP Category B

Hazard classification and safety precautions

Hazard H03, H08, H11
Risk phrases R22a, R36, R40, R43, R48, R50, R58 [1, 2]; R22b [2]
Operator protection A, C, H; U05a, U08, U11, U13, U19a, U20b
Environmental protection E15a, E16a, E16b, E31b [1, 2]; E32d [2]; E32e [1]
Storage and disposal E01, E04, E26, E30a
Medical advice M05b

297 magnesium phosphide

A phosphine generating compound used to control insect pests in stored commodities

Products

Degesch Plates	Rentokil	56% w/w	GE	07603

Uses

- Insect pests in ***stored grain***

Efficacy guidance

- Product acts as fumigant by releasing poisonous hydrogen phosphide gas on contact with moisture in the air
- Place plates on the floor or wall of the building or on the surface of the commodity. Exposure time varies depending on temperature and pest. See label

Restrictions

- Magnesium phosphide is subject to the Poisons Rules 1982 and the Poisons Act 1972. See notes in Section 5
- Only to be used by professional operators trained in the use of magnesium phosphide and familiar with the precautionary measures to be observed. See label for full precautions

Environmental safety

- Highly flammable
- Prevent access to buildings under fumigation by livestock, pets and other non-target mammals and birds
- Dangerous to fish or other aquatic life. Do not contaminate surface waters or ditches with chemical or used container
- Keep in original container, tightly closed, in a safe place, under lock and key
- Do not allow plates or their spent residues to come into contact with food other than raw cereal grains
- Remove used plates after treatment. Do not bulk spent plates and residues: spontaneous ignition could result
- Keep livestock out of treated areas

Hazard classification and safety precautions

Hazard H01, H07
Risk phrases R21, R26, R28

Operator protection A, D, H; U01, U05b, U07, U13, U19a, U20a
Environmental protection E02a (4 h min); E02b, E13b, E32b, E34
Storage and disposal E01, E04, E26, E29a, E30b
Medical advice M04

298 malathion

A broad-spectrum contact organophosphorus insecticide and acaricide

See also bifenthrin + malathion

Products

Fyfanon 440	Headland	440 g/l	EW	11134

Uses
- Aphids in ***carnations, chrysanthemums, dahlias, gladioli, outdoor cyclamen, protected carnations, protected chrysanthemums, protected cyclamen, protected dahlias, protected gladioli, protected roses, protected tulips, roses, tulips***

Efficacy guidance
- Spray when pest first seen and repeat as necessary, usually at 7-14 d intervals
- Number and timing of sprays vary with crop and pest. See label for details
- Repeat spray routinely for scale insect and whitefly control in glasshouses
- Where aphids resistant to malathion occur control is unlikely to be satisfactory

Restrictions
- Malathion is an anticholinesterase organophosphorus compound. Do not use if under medical advice not to work with such compounds
- Do not use on antirrhinums, crassula, ferns, fuchsias, gerberas, petunias, pileas, sweet peas or zinnias as damage may occur
- Do not apply by tractor mounted/drawn or broadcast air assisted sprayers

Environmental safety
- Dangerous for the environment
- Very toxic to aquatic organisms
- Dangerous to bees. Do not apply to crops in flower or to those in which bees are actively foraging. Do not apply when flowering weeds are present

Hazard classification and safety precautions
Hazard H11
Risk phrases R50, R58
Operator protection A, H, M, P; U08, U19a, U20b
Environmental protection E12d, E15a, E22b, E31a, E34, E38
Storage and disposal E26, E30a
Medical advice M01

299 maleic hydrazide

A pyridazinone plant growth regulator suppressing sprout and bud growth

See also dicamba + maleic hydrazide + MCPA

Products

1	Fazor	Dow	60% w/w	SG	05558
2	Mazide 25	Vitax	250 g/l	SL	02067
3	Regulox K	Bayer Environ.	250 g/l	SL	09937
4	Rouge	Nufarm UK	60% w/w	WB	11090
5	Royal MH 180	Nufarm UK	180 g/l	SL	10982
6	Royal MH 180	Certis	180 g/l	SL	11062
7	Source II	Chiltern	60% w/w	WB	08314

Uses
- Growth retardation in ***amenity grass, hedges, roadside grass*** [2]

- Growth suppression in ***amenity grass*** [3, 5, 6]; ***areas around farm buildings***, ***golf courses***, ***road verges*** [5]; ***carrots*** *(off-label)*, ***parsnips*** *(off-label)* [1]; ***grass near water*** [3, 5]; ***industrial sites***, ***roadside grass***, ***waste ground*** [3]; ***managed amenity turf*** [6]
- Sprout suppression in ***bulb onions*** [1, 2, 4, 6, 7]; ***ware potatoes*** [1, 7]
- Sucker inhibition in ***amenity trees and shrubs*** [2, 3]
- Volunteer suppression in ***ware potatoes*** [1, 4, 7]

Specific Off-Label Approvals (SOLAs)

- ***carrots***, ***parsnips*** *(OLA 012159) Dec 2008* [1]

Approval information

- Maleic hydrazide included in Annex I under EC Directive 91/414
- Approved for use on grass near water [5] . See notes in Section 5 on use of herbicides in or near water

Efficacy guidance

- Apply to grass at any time of yr when growth active, best when growth starting in Apr-May and repeated when growth recommences
- Uniform coverage and dry weather necessary for effective results
- Accurate timing essential for good results on potatoes but rain or irrigation within 24 h may reduce effectiveness on onions and potatoes
- Mow 2-3 d before and 5-10 d after spraying for best results. Need for mowing reduced for up to 6 wk
- When used for suppression of volunteer potatoes treatment may also give some suppression of sprouting in store but separate treatment will be necessary if sprouting occurs
- To control suckers wet trunks thoroughly, especially pruned and basal bud areas [2]

Restrictions

- Maximum number of treatments 2 per yr on amenity grass, land not intended for cropping and land adjacent to aquatic areas; 1 per crop on onions, potatoes and on or around tree trunks
- Do not apply in drought or when crops are suffering from pest, disease or herbicide damage. Do not treat fine turf or grass seeded less than 8 mth previously
- Do not treat potatoes within 3 wk of applying a haulm desiccant or if temperatures above 26°C
- Consult processor before use on potato crops for processing
- Do not apply by knapsack sprayer [1]

Crop-specific information

- Latest use: 3 wk before haulm destruction for potatoes; before 50% necking for onions
- HI onions 4-7 d; potatoes, carrots, parsnips 3 wk
- Apply to onions at 10% necking and not later than 50% necking stage when the tops are still green
- Only treat onions in good condition and properly cured, and do not treat more than 2 wk before maturing. Treated onions may be stored until Mar but must then be removed to avoid browning
- Apply to second early or maincrop potatoes at least 3 wk before haulm destruction
- Only treat potatoes of good keeping quality; not on seed, first earlies or crops grown under polythene
- Spray hawthorn hedges in full leaf, privet 7 d after cutting, in Apr-May [2]
- May be applied to grass along water courses but not to water surface [3, 5]

Environmental safety

- Only apply to grass not to be used for grazing
- Do not use treated water for irrigation purposes within 3 wk of treatment or until concentration in water falls below 0.02 ppm [3, 5, 6]
- Maximum permitted concentration in water 2 ppm
- Do not dump surplus product in water or ditch bottoms
- Avoid drift onto nearby vegetables, flowers or other garden plants

Hazard classification and safety precautions

Hazard H04 [4, 7]

Risk phrases R41 [4, 7]

Operator protection A [1, 4, 7]; C [4, 7]; H [1]; U02a, U05a, U13, U14, U15, U19a, U20a, U22a [4, 7]; U08 [1]; U20b [1, 5]; U20c [2, 3, 6]

Environmental protection E15a [1-7]; E19a [3, 5, 6]; E21 [3, 5, 6] (3 wk); E31a [2, 5, 6]; E31c [3]; E32a [1]

Storage and disposal E01, E04 [4, 7]; E26 [3, 4, 7]; E29a [3]; E30a [1-7]

300 mancozeb

A protective dithiocarbamate fungicide for potatoes and other crops

See also benalaxyl + mancozeb
chlorothalonil + mancozeb
cymoxanil + mancozeb
dimethomorph + mancozeb
fenamidone + mancozeb

Products

1	Dithane 945	Interfarm	80% w/w	WP	09897
2	Dithane 945	Dow	80% w/w	WP	10985
3	Dithane NT	Dow	75% w/w	WG	10986
4	Dithane NT Dry Flowable	Interfarm	75% w/w	WG	09898
5	Dithane Superflo	Dow	455 g/l	SC	10799
6	Karamate Dry Flo Newtec	Landseer	75% w/w	WG	09759
7	Manzate 75 WG	Headland	75% w/w	WG	11046
8	Micene 80	Sipcam	80% w/w	WP	09112
9	Micene DF	Sipcam	77% w/w	WG	09957
10	Penncozeb WDG	Nufarm UK	77% w/w	WG	11622
11	Quell Flo	Interfarm	455 g/l	SC	09894

Uses

- Black spot in ***roses*** [6]
- Blight in ***potatoes*** [1-5, 7-11]
- Brown rust in ***spring barley***, ***winter barley*** [8-10]; ***spring wheat***, ***winter wheat*** [1-5, 9-11]
- Currant leaf spot in ***blackcurrants***, ***gooseberries*** [6]
- Downy mildew in ***bulb onions*** *(off-label)*, ***garlic*** *(off-label)* [2, 9]; ***lettuce***, ***protected lettuce*** [6]; ***poppies*** *(off-label - for morphine production)* [4]; ***winter oilseed rape*** [1-5, 11]
- Fire in ***tulips*** [6]
- Fungus diseases in ***grapevines*** *(off-label)* [6]
- Net blotch in ***spring barley***, ***winter barley*** [8-10]
- Ray blight in ***chrysanthemums*** [6]
- Rhynchosporium in ***spring barley*** [9, 10]; ***winter barley*** [1-3, 9, 10]
- Rust in ***carnations***, ***geraniums***, ***roses*** [6]
- Scab in ***apples*** [6, 8-10]; ***pears*** [6]
- Septoria diseases in ***spring wheat***, ***winter wheat*** [9, 10]
- Septoria leaf spot in ***spring wheat***, ***winter wheat*** [1-5, 8, 11]
- Sooty moulds in ***spring barley***, ***winter barley*** [9, 10]; ***spring wheat***, ***winter wheat*** [1-5, 9-11]
- White mould in ***daffodils*** *(off-label - for galanthamine production)* [2]
- Yellow rust in ***spring barley***, ***spring wheat***, ***winter barley*** [9, 10]; ***winter wheat*** [1-5, 9-11]

Specific Off-Label Approvals (SOLAs)

- ***bulb onions***, ***garlic*** *(OLA 020158) Jun 2006* [9]
- ***bulb onions***, ***garlic*** *(OLA 021430) Jun 2006* [2]
- ***daffodils*** *(for galanthamine production) (OLA 041519) Dec 2008* [2]
- ***grapevines*** *(OLA 021426) Dec 2008* [6]
- ***poppies*** *(for morphine production) (OLA 040162) Dec 2008* [4]

Approval information

- Approved for aerial application on potatoes [1-5, 8-11]. See notes in Section 5

Efficacy guidance

- Mancozeb is a protectant fungicide and will give moderate control, suppression or reduction of the cereal diseases listed if treated before they are established but in many cases mixture with carbendazim is essential to achieve satisfactory results. See labels for details
- May be recommended for suppression or control of mildew in cereals depending on product and tank mix. See label for details

Restrictions

- Maximum number of treatments varies with crop and product used - check labels for details
- Check labels for minimum interval that must elapse between treatments

- On protected lettuce only 2 post-planting applications of mancozeb or of any combination of products containing EBDC fungicide (mancozeb, maneb, thiram, zineb) either as a spray or a dust are permitted within 2 wk of planting out and none thereafter.
- Avoid treating wet cereal crops or those suffering from drought or other stress
- Keep dry formulations away from fire and sparks
- Use dry formulations immediately. Do not store

Crop-specific information

- Latest use: before early milk stage (GS 73) for cereals; before 6 true leaf stage and before 31 Dec for winter oilseed rape.
- HI potatoes 7 d; outdoor lettuce 14 d; protected lettuce 21 d; apples, blackcurrants, bulb onions, garlic, gooseberries, pears 4 wk; grapevines 9 wk
- Apply to potatoes before haulm meets across rows (usually mid-Jun) or at earlier blight warning, and repeat every 7-14 d depending on conditions and product used (see label)
- May be used on potatoes up to desiccation of haulm
- On oilseed rape apply as soon as disease develops between cotyledon and 5-leaf stage (GS 1,0-1,5)
- Apply to cereals from 4-leaf stage to before early milk stage (GS 71). Recommendations vary, see labels for details
- Treat winter oilseed rape before 6 true leaf stage (GS 1,6) and before 31 Dec

Environmental safety

- Dangerous for the environment
- Very toxic to aquatic organisms
- Do not empty into drains

Hazard classification and safety precautions

Hazard H03 [10]; H04 [1-9, 11]; H11 [1-7, 10, 11]
Risk phrases R36 [7-9]; R37 [1-6, 8-11]; R38 [8, 9]; R42 [10]; R43 [1-6, 10, 11]; R50, R58 [1-7, 10, 11]
Operator protection A [1-11]; C [7-9]; D [2-4, 6, 8, 9]; U05a [1-11]; U08 [5, 7-9, 11]; U11 [7]; U13 [8, 9]; U14 [1-6, 11]; U19a [7-10]; U20a [1, 8]; U20b [2-7, 9, 11]
Environmental protection E13c [8, 9]; E15a [1-7, 10, 11]; E19b [10]; E31a [7]; E32a [1-6, 8-11]; E32d, E38 [1-7, 11]; E34 [5, 7, 11]
Storage and disposal E01, E04, E30a [1-11]; E26 [3-9, 11]; E29a [7]

301 mancozeb + metalaxyl-M

A systemic and protectant fungicide mixture

Products

Fubol Gold WG	Syngenta	64:4% w/w	WG	10184

Uses

- Blight in ***potatoes***
- Downy mildew in ***rhubarb*** *(off-label)*, ***salad onions*** *(off-label)*
- Fungus diseases in ***chives*** *(off-label)*, ***herbs (see appendix 6)*** *(off-label)*, ***lettuce*** *(off-label)*, ***parsley*** *(off-label)*, ***protected chives*** *(off-label)*, ***protected herbs (see appendix 6)*** *(off-label)*, ***protected lettuce*** *(off-label)*, ***protected parsley*** *(off-label)*
- Phytophthora fruit rot in ***apples*** *(off-label - applied to orchard floor)*
- White blister in ***cabbages*** *(off-label)*, ***protected radishes*** *(off-label)*, ***radishes*** *(off-label)*

Specific Off-Label Approvals (SOLAs)

- ***apples*** *(applied to orchard floor) (OLA 011610) Dec 2008* [1]
- ***cabbages, protected radishes, radishes*** *(OLA 011610) Dec 2008* [1]
- ***chives, herbs (see appendix 6), lettuce, parsley, protected chives, protected herbs (see appendix 6), protected lettuce, protected parsley*** *(OLA 032142) Dec 2008* [1]
- ***rhubarb*** *(OLA 022044) Aug 2006* [1]
- ***salad onions*** *(OLA 032324) Dec 2008* [1]

Approval information

- Metalaxyl-M included in Annex I under EC Directive 91/414

Efficacy guidance

- Commence potato blight programme before risk of infection occurs as crops begin to meet along the rows and repeat every 10-14 d according to blight risk. Do not exceed a 14 d interval between sprays
- If infection risk conditions occur earlier than the above growth stage commence spraying potatoes immediately
- Complete the potato blight programme using a protectant fungicide starting no later than 10 d after the last phenylamide spray. At least 2 such sprays should be applied
- Monitor sprayed potato crops and stop using any phenylamide product if active blight is identified. Switch to a tin-based product programme within 7 d and continue up to haulm destruction or harvest
- To minimise the likelihood of development of resistance these products should be used in a planned Resistance Management strategy. See Section 5 for more information

Restrictions

- Maximum number of treatments or maximum total dose varies with crop. See labels

Crop-specific information

- Latest use: before end of active potato haulm growth or before end Aug, whichever is earlier; before transplanting protected herbs; before dormancy in yr of harvest for rhubarb
- HI 14 d for outdoor herbs; 21 d for protected herbs; 28 d for apples
- After treating early potatoes destroy and remove any remaining haulm after harvest to minimise blight pressure on neighbouring maincrop potatoes

Environmental safety

- Dangerous for the environment
- Very toxic to aquatic organisms
- Do not harvest crops for human consumption for at least 7 d after final application

Hazard classification and safety precautions

Hazard H04, H11
Risk phrases R37, R43, R50, R58
Operator protection A; U05a, U08, U20b
Environmental protection E15a, E32a, E32d, E34, E38
Consumer protection C02 (7 d)
Storage and disposal E01, E04, E30a

302 mancozeb + propamocarb hydrochloride

A systemic and contact protectant fungicide for potato blight control

Products

Tattoo	Bayer CropScience	301.6:248 g/l	SC	07293

Uses

- Blight in ***potatoes***

Efficacy guidance

- Commence treatment as soon as there is risk of blight infection
- In the absence of a warning treatment should start before the crop meets within the row
- Repeat treatments every 10-14 d depending on degree of infection risk
- Irrigated crops should be regarded as at high risk and treated every 10 d
- Do not spray when rainfall is imminent and apply only to dry foliage
- To complete the spray programme after the end of Aug, make subsequent treatments up to haulm destruction with a protectant fungicide, preferably fentin based
- To minimise the likelihood of development of resistance these products should be used in a planned Resistance Management strategy. See Section 5 for more information

Restrictions

- Maximum number of treatments 5 per crop
- Do not use once blight has become readily visible

Crop-specific information

- HI potatoes 14 d

Environmental safety

- Dangerous for the environment
- Very toxic to aquatic organisms

Hazard classification and safety precautions

Hazard H04, H11
Risk phrases R43, R50, R58
Operator protection A, H; U05a, U08, U14, U19a, U20a
Environmental protection E13c, E31b, E32d, E38
Consumer protection C02 (14 d)
Storage and disposal E01, E04, E26, E30a

303 mancozeb + zoxamide

A protectant fungicide mixture for potatoes

Products

1	Electis 75 WG	Dow	66.7:8.3% w/w	WG	11013
2	Roxam 75 WG	Dow	66.7:8.3% w/w	WG	11017
3	Unikat 75 WG	Landseer	66.7:8.3% w/w	WG	10567

Uses

- Blight in ***potatoes*** [1, 2]
- Downy mildew in ***grapevines*** [3]

Efficacy guidance

- Apply as protectant spray on potatoes immediately risk of blight in district or as crops begin to meet along the rows and repeat every 7-14 d according to blight risk [1, 2]
- Do not use if potato blight present in crop. Products are not curative [1, 2]
- Spray irrigated potato crops as soon as possible after irrigation once the crop leaves are dry [1, 2]
- In the absence of an official warning commence treatment of grapevines at the 5-6 leaf stage and repeat at a minimum interval of 10 d [3]
- Increase spray volume with growth of grapevines but avoid excessive application [3]

Restrictions

- Maximum number of treatments 6 per yr for grapevines; 10 per crop for potatoes
- Do not use on grapevines for plant propagation [3]
- Consult processors before use on crops for juice production [3]

Crop-specific information

- HI 7 d for potatoes [1, 2]; 56 d for grapevines [3]

Environmental safety

- Dangerous for the environment
- Very toxic to aquatic organisms
- LERAP Category B
- Broadcast air-assisted LERAP (20 m) [3]
- Keep away from fire and sparks

Hazard classification and safety precautions

Hazard H04, H11
Risk phrases R37, R43, R50, R58
Operator protection A, H; U05a, U14, U20b
Environmental protection E15a, E16b, E32a, E32d, E38 [1-3]; E16a [1, 2]; E17b [3] (20 m)
Storage and disposal E01, E04, E26, E30a

304 MCPA

A translocated phenoxyacetic herbicide for cereals and grassland

See also *2,4-D + dichlorprop-P + MCPA + mecoprop-P*
2,4-D + MCPA
2,4-DB + MCPA
bentazone + MCPA + MCPB
clopyralid + 2,4-D + MCPA
clopyralid + diflufenican + MCPA
clopyralid + fluroxypyr + MCPA
dicamba + dichlorprop-P + ferrous sulphate + MCPA
dicamba + dichlorprop-P + MCPA
dicamba + maleic hydrazide + MCPA
dicamba + MCPA + mecoprop-P
dichlorprop-P + MCPA
dichlorprop-P + MCPA + mecoprop-P

Products

1	Agricorn 500 II	FCC	500 g/l	SL	09155
2	Agritox 50	Nufarm UK	500 g/l	SL	07400
3	Agritox Dry	Nufarm UK	80% w/w	SG	10554
4	Agroxone	Headland	485 g/l	SL	09947
5	Campbell's MCPA 50	United Phosphorus	500 g/l	SL	00381
6	Circium II	Nufarm UK	500 g/l	SL	11801
7	Headland Spear	Headland	500 g/l	SL	07115
8	HY-MCPA	Agrichem	500 g/l	SL	06293
9	Tasker 75	Headland	750 g/l	SL	10544

Uses

- Annual dicotyledons in ***amenity grass*** [1-4, 6, 7]; ***established grassland, spring barley, spring oats, spring wheat, winter barley, winter oats, winter wheat*** [1-5, 7-9]; ***grass seed crops*** [1-4, 9]; ***land not intended to bear vegetation*** [2, 4]; ***linseed*** [1-4, 8, 9]; ***managed amenity turf*** [1-4, 7]; ***newly sown grass*** [4]; ***rotational grassland*** [9]; ***spring rye*** [1-3]; ***undersown barley*** *(red clover or grass)*, ***undersown wheat*** *(red clover or grass)* [4, 9]; ***undersown spring cereals, undersown winter cereals*** [1, 8]; ***winter rye*** [1-3, 5, 7]
- Buttercups in ***amenity grass, managed amenity turf*** [3, 6-8]; ***established grassland*** [5, 7, 8]
- Charlock in ***established grassland*** [1-4, 9]; ***grass seed crops*** [9]; ***linseed, rotational grassland*** [8, 9]; ***newly sown grass*** [4]; ***spring barley, spring oats, spring wheat, winter barley, winter oats, winter wheat*** [1-5, 7-9]; ***spring rye*** [1-3]; ***undersown barley*** *(red clover or grass)*, ***undersown wheat*** *(red clover or grass)* [4, 9]; ***winter rye*** [1-3, 5, 7]
- Dandelions in ***amenity grass*** [3]; ***established grassland*** [5, 8]; ***land not intended to bear vegetation*** [6]; ***managed amenity turf*** [3, 6]
- Docks in ***established grassland*** [5, 7, 8]
- Fat hen in ***amenity grass, hard surfaces, managed amenity turf, natural surfaces not intended to bear vegetation, permeable surfaces overlying soil*** [8]; ***established grassland*** [1-4, 9]; ***grass seed crops*** [9]; ***linseed, rotational grassland*** [8, 9]; ***newly sown grass*** [4]; ***spring barley, spring oats, spring wheat, winter barley, winter oats, winter wheat*** [1-5, 7-9]; ***spring rye*** [1-3]; ***undersown barley*** *(red clover or grass)*, ***undersown wheat*** *(red clover or grass)* [4, 9]; ***winter rye*** [1-3, 5, 7]
- Hemp-nettle in ***established grassland*** [1-4, 9]; ***grass seed crops, rotational grassland*** [9]; ***linseed*** [8, 9]; ***newly sown grass*** [4]; ***spring barley, spring oats, spring wheat, winter barley, winter oats, winter wheat*** [1-5, 7-9]; ***spring rye*** [1-3]; ***undersown barley*** *(red clover or grass)*, ***undersown wheat*** *(red clover or grass)* [4, 9]; ***winter rye*** [1-3, 5, 7]
- Perennial dicotyledons in ***amenity grass*** [1-4, 6, 7]; ***established grassland, spring barley, spring oats, spring wheat, winter barley, winter oats, winter wheat*** [1-5, 7-9]; ***grass seed crops, linseed, rotational grassland*** [9]; ***land not intended to bear vegetation*** [2, 4]; ***managed amenity turf*** [1-4, 7]; ***spring rye*** [1-3]; ***undersown barley*** *(red clover or grass)*, ***undersown wheat*** *(red clover or grass)* [4, 9]; ***winter rye*** [1-3, 5, 7]
- Plantains in ***amenity grass, established grassland, grass seed crops, managed amenity turf*** [8]
- Ragwort in ***established grassland*** [7]

- Rushes in ***established grassland*** [7]
- Stinging nettle in ***hard surfaces, natural surfaces not intended to bear vegetation, permeable surfaces overlying soil*** [8]; ***industrial sites*** [7]; ***land not intended to bear vegetation*** [6, 7]
- Thistles in ***established grassland*** [5, 7, 8]; ***hard surfaces, natural surfaces not intended to bear vegetation, permeable surfaces overlying soil*** [8]; ***industrial sites*** [7]; ***land not intended to bear vegetation*** [6, 7]
- Wild radish in ***established grassland*** [1-4, 9]; ***grass seed crops, rotational grassland*** [9]; ***linseed*** [8, 9]; ***newly sown grass*** [4]; ***spring barley, spring oats, spring wheat, winter barley, winter oats, winter wheat*** [1-5, 7-9]; ***spring rye*** [1-3]; ***undersown barley*** *(red clover or grass)*, ***undersown wheat*** *(red clover or grass)* [4, 9]; ***winter rye*** [1-3, 5, 7]

Efficacy guidance

- Best results achieved by application to weeds in seedling to young plant stage under good growing conditions when crop growing actively
- Spray perennial weeds in grassland before flowering. Most susceptible growth stage varies between species. See label for details
- Do not spray during cold weather, drought, if rain or frost expected or if crop wet

Restrictions

- Maximum number of treatments normally 1 per crop or yr except grass (2 per yr) for some products. See label
- Do not treat grass within 3 mth of germination and preferably not in the first yr of a direct sown ley or after reseeding
- Do not use on cereals before undersowing
- Do not roll, harrow or graze for a few days before or after spraying; see label
- Do not use on grassland where clovers are an important part of the sward
- Do not use on any crop suffering from stress or herbicide damage
- Avoid spray drift onto nearby susceptible crops

Crop-specific information

- Latest use: before 1st node detectable (GS 31) for cereals; 4-6 wk before heading for grass seed crops; before crop 15-25 cm high for linseed
- Apply to winter cereals in spring from fully tillered, leaf sheath erect stage to before first node detectable (GS 31)
- Apply to spring barley and wheat from 5-leaves unfolded (GS 15), to oats from 1-leaf unfolded (GS 11) to before first node detectable (GS 31)
- Apply to cereals undersown with grass after grass has 2-3 leaves unfolded
- Recommendations for crops undersown with legumes vary. Red clover may withstand low doses after 2-trifoliate leaf stage, especially if shielded by taller weeds but white clover is more sensitive. See label for details
- Apply to grass seed crops from 2-3 leaf stage to 5 wk before head emergence
- Temporary wilting may occur on linseed but without long term effects

Following crops guidance

- Do not direct drill brassicas or legumes within 6 wk of spraying grassland

Environmental safety

- Harmful to aquatic organisms
- MCPA is active at low concentrations. Take extreme care to avoid drift onto neighbouring crops, especially beet crops, brassicas, most market garden crops including lettuce and tomatoes under glass, pears and vines
- Harmful to fish or other aquatic life. Do not contaminate surface waters or ditches with chemical or used container [1]
- Keep livestock out of treated areas until foliage of poisonous weeds such as ragwort has died and become unpalatable

Hazard classification and safety precautions

Hazard H03
Risk phrases R20, R21 [4, 5, 8]; R22a [4, 5, 7-9]; R41 [1-3, 5-9]; R52 [1, 7]
Operator protection A, C [1-8]; U05a, U08, U11 [1-9]; U14 [8]; U15 [7, 8]; U20a [1, 5]; U20b [2-4, 6-9]
Environmental protection E07a, E15a, E34 [1-9]; E31a [1-5, 8, 9]; E31b [6, 7]; E38 [7]
Storage and disposal E01, E04, E26 [1-9]; E30a [1-5, 7-9]
Medical advice M03 [1-9]; M05a [5]

305 MCPA + MCPB

A translocated herbicide for undersown cereals, grassland and various legumes

Products

1	Bellmac Plus	United Phosphorus	38:262 g/l	SL	07521
2	Impetus	Headland	25:275 g/l	SL	11021
3	Trifolex-Tra	BASF	34:216 g/l	SL	10396
4	Tropotox Plus	Nufarm UK	37.5:262.5 g/l	SL	11142

Uses

- Annual dicotyledons in ***all cereals, clover seed crops, direct-sown seedling clovers, dredge corn containing peas, established leys*** [4]; ***combining peas, red clover, spring barley, spring oats, spring wheat, vining peas, white clover, winter barley, winter oats, winter wheat*** [2]; ***established grassland, peas*** [3]; ***permanent pasture*** [2, 4]; ***rotational grassland*** [1-3]; ***sainfoin*** [1, 4]; ***undersown spring cereals, undersown winter cereals*** [1-4]
- Perennial dicotyledons in ***all cereals, clover seed crops, direct-sown seedling clovers, dredge corn containing peas, established leys*** [4]; ***combining peas, red clover, spring barley, spring oats, spring wheat, vining peas, white clover, winter barley, winter oats, winter wheat*** [2]; ***established grassland, peas*** [3]; ***permanent pasture*** [2, 4]; ***rotational grassland*** [1-3]; ***sainfoin*** [1, 4]; ***undersown spring cereals, undersown winter cereals*** [1-4]

Efficacy guidance

- Best results achieved by application to weeds in seedling to young plant stage under good growing conditions when crop growing actively. Spray perennials when adequate leaf surface before flowering. Retreatment often needed in following year
- Spray leys and sainfoin before crop provides cover for weeds
- Rain, cold or drought may reduce effectiveness

Restrictions

- Maximum number of treatments 1 per crop or yr
- Do not spray clovers for seed
- Do not spray peas [1]
- Do not roll or harrow for a few days before or after spraying

Crop-specific information

- Latest use: before first node detectable (GS 31) for cereals; before flower buds visible for peas
- Apply to cereals from 2-expanded leaf stage to before jointing (GS 12-30) and, where undersown, after 1-trifoliate leaf stage of clover
- Apply to direct-sown seedling clover after 1-trifoliate leaf stage
- Apply to mature white clover for fodder at any stage. Do not spray red clover after flower stalk has begun to form
- Apply to sainfoin after first trifoliate leaf stage

Environmental safety

- Harmful to aquatic organisms
- Keep livestock out of treated areas until foliage of any poisonous weeds such as ragwort has died and become unpalatable
- Take extreme care to avoid drift onto neighbouring crops, especially beet crops, brassicas, most market garden crops including lettuce and tomatoes under glass, pears and vines

Hazard classification and safety precautions

Hazard H03
Risk phrases R20, R21 [2]; R22a [1-4]; R38, R58 [1]; R41, R52 [1, 2]
Operator protection A, C [2]; U05a [2-4]; U08, U20b [1-4]; U11, U14, U15 [1, 2]; U19a [3]
Environmental protection E07a, E31b, E34 [1-4]; E13c [1]; E15a [2-4]
Storage and disposal E01 [2-4]; E04 [2, 3]; E26, E30a [1-4]
Medical advice M03 [2]; M05a [1, 2, 4]

SEE SECTION 3 FOR PRODUCTS ALSO REGISTERED

306 MCPA + mecoprop-P

A translocated selective herbicide for amenity grass

Products

Greenmaster Extra	Scotts	0.49:0.29 % w/w	GR	11563

Uses

- Annual dicotyledons in ***managed amenity turf***
- Perennial dicotyledons in ***managed amenity turf***

Efficacy guidance

- Apply from Apr to Sep, when weeds growing actively and have large leaf area available for chemical absorption

Restrictions

- Maximum total dose 105 g/sq m per yr. The total amount of mecoprop-P applied in a single yr must not exceed the maximum total dose approved for any single product for use on turf
- Avoid contact with cultivated plants
- Do not use first 4 mowings as compost or mulch unless composted for 6 mth
- Do not treat newly sown or turfed areas for at least 6 mth
- Do not reseed bare patches for 8 wk after treatment
- Do not apply when heavy rain expected or during prolonged drought. Irrigate after 1-2 d unless rain has fallen
- Do not mow within 2-3 d of treatment
- Treat areas planted with bulbs only after the foliage has died down
- Avoid walking on treated areas until it has rained or irrigation has been applied

Crop-specific information

- Granules contain NPK fertilizer to encourage grass growth

Environmental safety

- Take extreme care to avoid drift onto neighbouring crops, especially beet crops, brassicas, most market garden crops including lettuce and tomatoes under glass, pears and vines
- Harmful to fish or other aquatic life. Do not contaminate surface waters or ditches with chemical or used container
- Keep livestock out of treated areas for at least 2 wk and until foliage of any poisonous weeds such as ragwort has died and become unpalatable

Hazard classification and safety precautions

Operator protection A, C, H, M; U20a
Environmental protection E07a, E13c, E32a
Storage and disposal E30a

307 MCPB

A translocated phenoxybutyric herbicide

See also *bentazone + MCPA + MCPB*
bentazone + MCPB
MCPA + MCPB

Products

1	Bellmac Straight	United Phosphorus	400 g/l	SL	07522
2	Butoxone	Headland	400 g/l	SL	10501
3	Tropotox	Nufarm UK	400 g/l	SL	11141

Uses

- Annual dicotyledons in ***blackberries, combining peas, gooseberries, loganberries, permanent pasture, raspberries, spring barley, spring oats, spring wheat, vining peas, white clover*** *(seed crops)*, ***winter barley, winter oats, winter wheat*** [2]; ***blackcurrants*** [1-3]; ***clover seed crops, peas*** [1, 3]; ***rotational grassland, undersown spring cereals, undersown winter cereals*** [1, 2]
- Perennial dicotyledons in ***blackberries, combining peas, gooseberries, loganberries, permanent pasture, raspberries, spring barley, spring oats, spring wheat, vining peas,***

white clover *(seed crops)*, ***winter barley***, ***winter oats***, ***winter wheat*** [2]; ***blackcurrants*** [1-3]; ***clover seed crops***, ***peas*** [1, 3]; ***rotational grassland***, ***undersown spring cereals***, ***undersown winter cereals*** [1, 2]

Efficacy guidance

- Best results achieved by spraying young seedling weeds in good growing conditions
- Best results on perennials by spraying before flowering
- Effectiveness may be reduced by rain within 12 h, by very cold or dry conditions

Restrictions

- Maximum number of treatments 1 per crop or yr. One label allows two treatments on blackcurrants [3]
- Do not roll or harrow for 7-10 d before or after treatment (check label)

Crop-specific information

- Latest use: first node detectable stage (GS 31) for cereals; before flower buds appear in terminal leaf (GS 201) for peas; before flower buds form for clover; before weeds damaged by frost for cane and bush fruit
- Apply to undersown cereals from 2-leaves unfolded to first node detectable (GS 12-31), and after first trifoliate leaf stage of clover
- Red clover seedlings may be temporarily damaged but later growth is normal
- Apply to white clover seed crops in Mar to early Apr, not after mid-May, and allow 3 wk before cutting and closing up for seed
- Apply to peas from 3-6 leaf stage but before flower bud detectable (GS 103-201). Consult PGRO (see Appendix 2) or label for information on susceptibility of cultivars. Do not treat peas grown for seed [2]
- Do not use on leguminous crops not mentioned on the label
- Apply to cane and bush fruit after harvest and after shoot growth ceased but before weeds are damaged by frost, usually in late Aug or Sep; direct spray onto weeds as far as possible

Environmental safety

- Harmful to aquatic organisms
- Keep livestock out of treated areas until foliage of any poisonous weeds such as ragwort has died and become unpalatable
- Take extreme care to avoid drift onto neighbouring sensitive crops

Hazard classification and safety precautions

Hazard H03
Risk phrases R22a [1-3]; R38, R41 [1, 2]; R52, R58 [1]
Operator protection A, C; U05a, U08, U20b [1-3]; U11, U14, U15 [1, 2]; U19a [3]
Environmental protection E07a, E31b, E34 [1-3]; E13c [1, 2]; E15a [3]
Storage and disposal E01, E04, E30a [1-3]; E26 [1, 2]
Medical advice M03, M05a

308 mecoprop-P

A translocated phenoxypropionic herbicide for cereals and grassland

See also *2,4-D + dichlorprop-P + MCPA + mecoprop-P*
2,4-D + mecoprop-P
carfentrazone-ethyl + mecoprop-P
dicamba + MCPA + mecoprop-P
dicamba + mecoprop-P
dichlorprop-P + MCPA + mecoprop-P
fluroxypyr + mecoprop-P
MCPA + mecoprop-P

Products

1	Clenecorn Super	FCC	600 g/l	SL	09818
2	Clovotox	Bayer Environ.	142.5 g/l	SL	09928
3	Compitox Plus	Nufarm UK	600 g/l	SL	10077
4	Duplosan KV	Nufarm UK	600 g/l	SL	12073
5	Isomec	Nufarm UK	600 g/l	SL	11156
6	Landgold Mecoprop-P 600	Landgold	600 g/l	SL	09014
7	Optica	Headland	600 g/l	SL	09963

SEE SECTION 3 FOR PRODUCTS ALSO REGISTERED

Uses

- Annual dicotyledons in ***established grassland*** [6, 7]; ***grass seed crops, young leys*** [1, 3-5, 7]; ***grassland*** [1, 3-5]; ***managed amenity turf*** [2, 3, 5, 7]; ***rotational grassland*** [6]; ***spring barley, spring oats, spring wheat, winter barley, winter oats, winter wheat*** [1, 3-7]; ***undersown spring cereals, undersown winter cereals*** [1]
- Chickweed in ***established grassland*** [6, 7]; ***grass seed crops, young leys*** [1, 3-5, 7]; ***grassland*** [1, 3-5]; ***managed amenity turf*** [2, 3, 5, 7]; ***rotational grassland*** [6]; ***spring barley, spring oats, spring wheat, winter barley, winter oats, winter wheat*** [1, 3-7]; ***undersown spring cereals, undersown winter cereals*** [1]
- Cleavers in ***established grassland*** [6, 7]; ***grass seed crops, young leys*** [1, 3-5, 7]; ***grassland*** [1, 3-5]; ***managed amenity turf*** [3, 5, 7]; ***rotational grassland*** [6]; ***spring barley, spring oats, spring wheat, winter barley, winter oats, winter wheat*** [1, 3-7]; ***undersown spring cereals, undersown winter cereals*** [1]
- Clovers in ***managed amenity turf*** [2]
- Perennial dicotyledons in ***established grassland, rotational grassland*** [6]; ***grass seed crops, grassland, young leys*** [3-5]; ***managed amenity turf*** [2, 3, 5]; ***spring barley, spring oats, spring wheat, winter barley, winter oats, winter wheat*** [3-6]

Approval information

- Mecoprop-P included in Annex I under EC Directive 91/414

Efficacy guidance

- Best results achieved by application to seedling weeds which have not been frost hardened, when soil warm and moist and expected to remain so for several days

Restrictions

- Maximum number of treatments normally 1 per crop for spring cereals and newly sown grass; 2 per crop for winter cereals and established grass. Check labels for details
- The total amount of mecoprop-P applied in a single yr must not exceed the maximum total dose approved for any single product for the crop/situation
- Do not spray cereals undersown with clovers or legumes or to be undersown with legumes or grasses
- Do not spray grass seed crops within 5 wk of seed head emergence
- Do not spray crops suffering from herbicide damage or physical stress
- Do not spray during cold weather, periods of drought, if rain or frost expected or if crop wet
- Do not roll or harrow for 7 d before or after treatment

Crop-specific information

- Latest use: generally before 1st node detectable (GS 31) for spring cereals and before 3rd node detectable (GS 33) for winter cereals, but individual labels vary; 5 wk before emergence of seed head for grass seed crops
- Spray winter cereals from 1 leaf stage in autumn up to and including first node detectable in spring (GS 10-31) or up to second node detectable (GS 32) if necessary. Apply to spring cereals from first fully expanded leaf stage (GS 11) but before first node detectable (GS 31)
- Spray cereals undersown with grass after grass starts to tiller
- Spray newly sown grass leys when grasses have at least 3 fully expanded leaves and have begun to tiller. Any clovers will be damaged

Environmental safety

- Harmful to aquatic organisms
- Keep livestock out of treated areas for at least 2 wk and until foliage of any poisonous weed, such as ragwort, has died and become unpalatable
- Take extreme care to avoid drift onto neighbouring crops, especially beet crops, brassicas, most market garden crops including lettuce and tomatoes under glass, pears and vines

Hazard classification and safety precautions

Hazard H03 [1, 3-7]; H04 [2]

Risk phrases R21 [6]; R22a [1, 3-7]; R36 [2, 6]; R38 [4, 6, 7]; R41 [1, 3-5, 7]; R43 [1, 3, 5]; R52 [7]

Operator protection A, C [1-7]; B [1]; H [2, 4, 7]; M [1, 4, 7]; U05a, U08, U20b [1-7]; U11 [1, 3-5]; U14 [1, 3, 5]; U15 [7]

Environmental protection E07a [1-7]; E13c [1, 2, 6, 7]; E15a [3-5]; E31a [3, 6]; E31b [1, 5, 7]; E31c [2, 4]; E34 [1, 3-7]; E38 [7]

Storage and disposal E01, E04, E26 [1-7]; E29b [1, 5]; E30a [1-6]

Medical advice M03 [1, 3-7]; M05a [4]

309 mepanipyrim

An anilinopyrimidine fungicide for use on strawberries

Products

1	Frupica	Certis	50% w/w	WP	11178
2	Frupica SC	Certis	450 g/l	SC	12067

Uses
- Botrytis in ***protected strawberries***, ***strawberries***

Approval information
- Mepanipyrim included in Annex I under Directive 91/414

Efficacy guidance
- Product is protectant and should be applied as a preventative spray when conditions favourable for Botrytis development occur
- To maintain Botrytis control use as part of a programme with other fungicides that control the disease
- To minimise the possibility of development of resistance adopt resistance management procedures by using products from different chemical groups as part of a mixed spray programme

Restrictions
- Maximum number of treatments 2 per crop (including other anilinopyrimidine products)
- Consult processor before use on crops for processing
- Use spray mixture immediately after preparation

Crop-specific information
- HI 3 d

Environmental safety
- Dangerous for the environment
- Very toxic to aquatic organisms
- LERAP Category B

Hazard classification and safety precautions

Hazard H11
Risk phrases R50, R58
Operator protection A, H; U05a [1]; U20c [1, 2]
Environmental protection E15a [1]; E16a, E16b, E32a [1, 2]; E31c, E32e, E34 [2]
Storage and disposal E01, E04, E26
Medical advice M04 [1]

310 mepiquat chloride

A quaternary ammonium plant growth regulator available only in mixtures

See also *2-chloroethylphosphonic acid + mepiquat chloride*
chlormequat + 2-chloroethylphosphonic acid + mepiquat chloride
chlormequat + mepiquat chloride

311 mesosulfuron-methyl

A sulfonyl urea herbicide for cereals available in mixtures

See also *iodosulfuron-methyl-sodium + mesosulfuron-methyl*

312 metalaxyl-M

A phenylamide systemic fungicide

See also *chlorothalonil + metalaxyl-M*
cymoxanil + fludioxonil + metalaxyl-M
mancozeb + metalaxyl-M

Products

SL 567A	Syngenta	480 g/l	EC	10594

Uses

- Cavity spot in ***carrots***, ***parsnips*** *(off-label)*
- Crown rot in ***water lilies*** *(off-label)*
- Downy mildew in ***grapevines*** *(off-label)*, ***hops*** *(off-label)*, ***leaf spinach*** *(off-label)*, ***protected cucumbers*** *(off-label)*, ***protected spinach*** *(off-label)*, ***salad onions*** *(off-label)*, ***spinach beet*** *(off-label)*
- Phytophthora in ***asparagus*** *(off-label)*
- Phytophthora root rot in ***raspberries*** *(off-label)*
- Pythium in ***watercress*** *(off-label)*
- Root malformation disorder in ***red beet*** *(off-label)*
- Storage rots in ***apples*** *(off-label)*, ***cabbages*** *(off-label)*, ***pears*** *(off-label)*
- White blister in ***horseradish*** *(off-label)*

Specific Off-Label Approvals (SOLAs)

- ***apples***, ***cabbages***, ***pears*** *(OLA 040723) Apr 2007* [1]
- ***asparagus*** *(OLA 040611) Sep 2012* [1]
- ***grapevines*** *(OLA 040949) Sep 2012* [1]
- ***hops*** *(OLA 040718) Mar 2005* [1]
- ***horseradish*** *(OLA 040722) Sep 2012* [1]
- ***leaf spinach***, ***protected spinach***, ***spinach beet*** *(OLA 040948) Sep 2012* [1]
- ***parsnips*** *(OLA 040613) Sep 2012* [1]
- ***protected cucumbers*** *(OLA 041136) Sep 2012* [1]
- ***raspberries***, ***salad onions*** *(OLA 040720) Apr 2007* [1]
- ***red beet*** *(OLA 040835) Sep 2012* [1]
- ***water lilies*** *(OLA 040721) Sep 2012* [1]
- ***watercress*** *(OLA 040719) Apr 2007* [1]

Approval information

- Metalaxyl-M included in Annex I under EC Directive 91/414

Efficacy guidance

- Best results achieved when applied to damp soil
- Efficacy may be reduced in prolonged dry weather. Irrigation immediately before or after application may be necessary
- Results may not be satisfactory on soils with high organic matter content
- Control of cavity spot on carrots overwintered in the ground or lifted in winter may be lower than expected

Restrictions

- Maximum number of treatments or maximum total dose varies with crop and nature of treatment. See label for details
- Do not use where carrots have been grown on the same site for either of the previous two yrs
- Consult before use on crops intended for processing

Crop-specific information

- Latest use: 4 wk before removal of apples or pears from store; 6 wk after drilling for carrots; 10 wk before sale or supply (after rinsing) of water lilies
- HI raspberries 3 mth; horseradish 8 wk; grapevines 30 d; red beet 28 d; hops, leaf spinach, protected spinach, salad onions, spinach beet 14 d; asparagus 7 d; protected cucumbers 2 d

Environmental safety

- Dangerous for the environment
- Harmful to aquatic organisms

Hazard classification and safety precautions

Hazard H03, H11

Risk phrases R22a, R36, R43, R52, R58

Operator protection A, C, H; U02a, U04a, U05a, U10, U20b

Environmental protection E15a, E31c, E32d, E34, E38

Storage and disposal E01, E04, E30a

Medical advice M03

313 metaldehyde

A molluscicide bait for controlling slugs and snails

Products

	Product	Company	Concentration	Formulation	Reg. No.
1	Allure	Chiltern	1.5% w/w	PT	11089
2	Appeal	Chiltern	1.5% w/w	PT	12022
3	Attract	Chiltern	1.5% w/w	PT	12023
4	Brits	Doff Portland	6% w/w	PT	11792
5	Chiltern Blues	Chiltern	6% w/w	PT	10071
6	Chiltern Hundreds	Chiltern	3% w/w	PT	10072
7	Dixie 6	Greencrop	6% w/w	PT	11790
8	Doff Horticultural Slug Killer Blue Mini Pellets	Doff Portland	3% w/w	PT	11463
9	Escar-Go 6	Chiltern	6% w/w	PT	06076
10	Hardy	Chiltern	6% w/w	PT	06948
11	Luxan 9363	Luxan	6% w/w	PT	07359
12	Luxan 9363 Red	Luxan	6% w/w	PT	11480
13	Luxan Deal	Luxan	6% w/w	PT	11101
14	Luxan Metaldehyde	Luxan	6% w/w	PT	06564
15	Luxan Trigger	Luxan	6% w/w	PT	10419
16	Metarex Amba	De Sangosse	5% w/w	PT	11157
17	Metarex Green	De Sangosse	5% w/w	PT	10113
18	Metarex RG	De Sangosse	5% w/w	PT	10115
19	Molotov	Chiltern	3% w/w	PT	08295
20	Regel	De Sangosse	5% w/w	PT	10114
21	Rowent	Chiltern	6% w/w	PT	12024

Uses

- Slugs in ***all edible crops*** [1-21]; ***all non-edible crops*** [1-14, 16-21]; ***amenity grass, managed amenity turf*** [11, 12, 14, 15]; ***cultivated land/soil*** [1-3, 8, 21]; ***natural surfaces not intended to bear vegetation*** [4, 7, 11, 12, 14]; ***ornamental plant production*** [16-18, 20]; ***protected crops*** [1-3, 16-18, 20, 21]
- Snails in ***all edible crops*** [1-21]; ***all non-edible crops*** [1-14, 16-21]; ***amenity grass, managed amenity turf*** [11, 12, 14, 15]; ***cultivated land/soil*** [1-3, 8, 21]; ***natural surfaces not intended to bear vegetation*** [4, 7, 11, 12, 14]; ***ornamental plant production*** [16-18, 20]; ***protected crops*** [1-3, 16-18, 20, 21]

Approval information

- Approved for aerial application on all edible crops, all non-edible crops [4, 7, 8, 13]; on natural surfaces not intended to be cultivated [4, 7]; on bare soil [8]. See notes in Section 5

Efficacy guidance

- Apply pellets by hand, fiddle drill, fertilizer distributor, by air (check label) or in admixture with seed. See labels for rates and timing.
- Best results achieved from an even spread of granules applied during mild, damp weather when slugs and snails most active. May be applied in standing crops
- Varieties of oilseed rape low in glucosinolates can be more acceptable to slugs than "single low" varieties and control may not be as good
- To prevent slug build up apply at end of season to brassicas and other leafy crops
- To reduce tuber damage in potatoes apply twice in Jul and Aug

Restrictions

- Do not apply when rain imminent or water glasshouse crops within 4 d of application
- Take care to avoid lodging of pellets in the foliage when making late applications to edible crops

Environmental safety

- Dangerous to game, wild birds and animals

SEE SECTION 3 FOR PRODUCTS ALSO REGISTERED

- Some products contain proprietary cat and dog deterrent
- Keep poultry out of treated areas for at least 7 d

Hazard classification and safety precautions

Operator protection A, H [1-4, 6, 7, 11-21]; J [4, 7, 13]; U05a [1-3, 5, 6, 9-15, 19, 21]; U15, U20a [16-18, 20]; U20b [5, 6, 9, 10, 19, 21]; U20c [1-4, 7, 8, 11-15]

Environmental protection E05b [1-6] (7 d); E05b [7]; E05b [8-12] (7 d); E05b [13]; E05b [14-21] (7 d); E10a, E15a, E32a [1-21]; E32d [16-18, 20]

Storage and disposal E01, E04 [1-3, 5, 6, 9-21]; E26 [11-15]; E29a [1-6, 8-10, 16-21]; E30a [1-3, 5-7, 9-21]

Treated seed S04a [16-18, 20]

Medical advice M04 [11-15]; M05a [16-18, 20]

314 metamitron

A contact and residual triazinone herbicide for use in beet crops

See also *chloridazon + chlorpropham + metamitron*
chloridazon + metamitron
chlorpropham + metamitron

Products

1	Alpha Metamitron	Makhteshim	70% w/w	WG	11081
2	Barclay Seismic	Barclay	70% w/w	WG	11378
3	Bettix 70 WG	United Phosphorus	70% w/w	WG	11154
4	Bettix Flo	United Phosphorus	700 g/l	SC	11959
5	Fernpath Haptol	AgriGuard	70% w/w	WG	11951
6	Fernpath Haptol Flo	AgriGuard	700 g/l	SC	11481
7	Goldbeet	Makhteshim	90% w/w	WG	11538
8	Goltix 90	Makhteshim	90% w/w	WG	11578
9	Goltix Flowable	Makhteshim	700 g/l	SC	11540
10	Goltix WG	Makhteshim	70% w/w	WG	11539
11	Landgold Metamitron	Landgold	70% w/w	WG	06287
12	Marquise	Makhteshim	70% w/w	WG	08738
13	MM 70	Nufarm UK	70% w/w	WG	11582
14	MM 70 Flo	Nufarm UK	700 g/l	SC	11197
15	Skater	Makhteshim	700 g/l	SC	11159
16	Target SC	Unicrop	700 g/l	SC	10861
17	Target SC	Unicrop	700 g/l	SC	11070
18	Volcan	Sipcam	70% w/w	WG	09295

Uses

- Annual dicotyledons in ***asparagus*** *(off-label)*, ***forest*** *(off-label)* [12, 15]; ***fodder beet***, ***mangels*** [1-12, 14-18]; ***red beet*** [1-12, 14, 16-18]; ***sugar beet*** [1-18]
- Annual grasses in ***asparagus*** *(off-label)*, ***forest*** *(off-label)* [12, 15]; ***fodder beet***, ***mangels*** [1-12, 14-18]; ***red beet*** [1-12, 14, 16-18]; ***sugar beet*** [1-18]
- Annual meadow grass in ***fodder beet***, ***mangels*** [1-12, 14-18]; ***red beet*** [1-12, 14, 16-18]; ***sugar beet*** [1-18]
- Fat hen in ***fodder beet***, ***mangels*** [1-12, 14-18]; ***red beet*** [1-12, 14, 16-18]; ***sugar beet*** [1-18]
- Groundsel in ***chives*** *(off-label)*, ***herbs (see appendix 6)*** *(off-label)*, ***parsley*** *(off-label)* [12, 15]

Specific Off-Label Approvals (SOLAs)

- ***asparagus***, ***chives***, ***forest***, ***herbs (see appendix 6)***, ***parsley*** *(OLA 030737) Dec 2008* [12]
- ***asparagus***, ***chives***, ***forest***, ***herbs (see appendix 6)***, ***parsley*** *(OLA 030738) Dec 2008* [15]

Efficacy guidance

- May be used pre-emergence alone or post-emergence in tank mixture or with an authorised adjuvant oil
- Low dose programme (LDP). Apply a series of low-dose post-weed emergence sprays, including adjuvant oil, timing each treatment according to weed emergence and size. See label for details and for recommended tank mixes and sequential treatments. On mineral soils the LDP should be preceded by pre-drilling or pre-emergence treatment
- Traditional application. Apply either pre-drilling before final cultivation with incorporation to 8-10 cm, or pre-crop emergence at or soon after drilling into firm, moist seedbed to emerged weeds from cotyledon to first true leaf stage

- On emerged weeds at or beyond 2-leaf stage addition of adjuvant oil advised
- Up to 3 post-emergence sprays may be used on soils with over 10% organic matter
- For control of wild oats and certain other weeds, tank mixes with other herbicides or sequential treatments are recommended. See label for details

Restrictions

- Maximum total dose equivalent to three full dose treatments for most products; 4 lower maximum dose treatments [12]
- Using traditional method post-crop emergence on mineral soils do not apply before first true leaves have reached 1 cm long

Crop-specific information

- Latest use: before crop foliage meets across rows for beet crops; 12 wk after transplanting asparagus
- HI herbs 6 wk
- Crop tolerance may be reduced by stress caused by growing conditions, effects of pests, disease or other pesticides, nutrient deficiency etc

Following crops guidance

- Only sugar beet, fodder beet or mangels may be drilled within 4 mth after treatment. Winter cereals may be sown in same season after ploughing, provided 16 wk passed since last treatment

Environmental safety

- Dangerous for the environment
- Very toxic to aquatic organisms
- Do not empty into drains

Hazard classification and safety precautions

Hazard H03 [1-4, 6-14]; H04 [5]; H11 [1-17]
Risk phrases R22a [1-4, 6-14]; R41 [5, 6]; R43 [1, 5, 6, 12]; R50 [1, 4, 7-9, 11, 12, 14]; R51 [2, 3, 5, 10, 13, 15-17]; R58 [1-5, 7-17]
Operator protection A [1, 2, 4-6, 9, 10, 12, 14, 16, 17]; C [2, 5]; H [1, 2, 5, 12]; U05a [1, 5, 6, 12, 14]; U08, U14 [1, 6, 12, 14]; U19a [1-3, 6, 10, 12-14]; U20a [1-3, 5-8, 10-14]; U20b [4, 9]; U20c [15-18]
Environmental protection E15a [1-18]; E19b [4, 7-9, 15-17]; E31c [15-17]; E32a [1-14, 18]; E32d [2-4, 7-10, 13, 15-17]; E32e [1, 12, 14]; E34 [4, 6, 9, 14]; E38 [2, 3, 10, 13]
Storage and disposal E01 [1, 4-9, 12, 14]; E04 [1, 4-6, 9, 12, 14]; E26 [6, 14]; E30a [1-18]
Medical advice M03 [4, 6, 9]; M05a [4, 7-9, 11]

315 metam-sodium

A methyl isothiocyanate producing sterilant for glasshouse, nursery and outdoor soils

Products

Sistan 51	Unicrop	510 g/l	SL	10046

Uses

- Nematodes in ***glasshouse soils, nursery soils, outdoor soils, potting soils***
- Soil pests in ***glasshouse soils, nursery soils, outdoor soils, potting soils***
- Soil-borne diseases in ***glasshouse soils, nursery soils, outdoor soils, potting soils***
- Weed seeds in ***glasshouse soils, nursery soils, outdoor soils, potting soils***

Efficacy guidance

- Metam-sodium is a partial soil sterilant and acts by breaking down in contact with soil to release methyl isothiocyanate (MIT)
- Apply to glasshouse soils as a drench, or inject undiluted to 20 cm at 30 cm intervals and seal immediately, or apply to surface and rotavate
- May also be used by mixing into potting soils
- Apply when soil temperatures exceed 7°C, preferably above 10°C, between 1 Apr and 31 Oct. Soil must be of fine tilth, free from debris and with 'potting moisture' content. If soil is too dry postpone treatment and water soil

Restrictions

- No plants must be present during treatment
- Crops must not be planted until a cress germination test has been completed satisfactorily

- Do not treat glasshouses within 2 m of growing crops. Fumes are damaging to all plants
- Avoid using in equipment incorporating natural rubber parts

Crop-specific information

- Latest use: pre-planting of crop

Following crops guidance

- Do not plant until soil is entirely free of fumes

Environmental safety

- Keep unprotected persons, livestock and pets out of treated areas for at least 24 h following treatment
- When diluted breakdown commences almost immediately. Only quantities for immediate use should be made up
- Divert or block drains which could carry solution under untreated glasshouses
- After treatment allow sufficient time (several weeks) for residues to dissipate and aerate soil by forking. Time varies with soil and season. Soils with high clay or organic matter content will retain gas longer than lighter soils

Hazard classification and safety precautions

Hazard H03, H04
Risk phrases R22a, R37, R38, R43
Operator protection A, C, H, M; U02a, U04a, U05a, U08, U19a, U20b
Environmental protection E02a (24 h); E15a, E31a, E34
Storage and disposal E01, E04, E30a
Medical advice M03

316 metazachlor

A residual anilide herbicide for use in brassicas, nurseries and forestry

Products

1	Agriguard Metazachlor	AgriGuard	500 g/l	SC	10417
2	Alpha Metazachlor 50 SC	Makhteshim	500 g/l	SC	10669
3	Butisan S	BASF	500 g/l	SC	11733
4	Gharda Bonanza	Nufarm UK	500 g/l	SC	11571
5	Greencrop Monogram	Greencrop	500 g/l	SC	12048
6	Landgold Metazachlor 50	Landgold	500 g/l	SC	09726
7	Me2 Booty	Me2	500 g/l	SC	10659
8	Me2 Booty 2	Me2	500 g/l	SC	12010
9	Standon Metazachlor 50	Standon	500 g/l	SC	05581
10	Standon Metazachlor 500	Standon	500 g/l	SC	12012
11	Sultan 50 SC	Makhteshim	500 g/l	SC	10418

Uses

- Annual dicotyledons in ***broccoli*** [2-11]; ***brussels sprouts, cabbages, cauliflowers, spring oilseed rape, swedes, turnips, winter oilseed rape*** [1-11]; ***calabrese*** [1-4, 6-11]; ***farm woodland*** [1-3, 7, 8, 10, 11]; ***forest, nursery fruit trees and bushes, ornamental trees, shrubs*** [1-3, 5, 7, 8, 10, 11]
- Annual grasses in ***farm woodland*** [1-3, 7, 8, 10, 11]; ***forest, ornamental trees, shrubs*** [1-3, 5, 7, 8, 10, 11]
- Annual meadow grass in ***broccoli, brussels sprouts, cabbages, cauliflowers, spring oilseed rape, swedes, turnips, winter oilseed rape*** [1-11]; ***calabrese*** [1-4, 6-11]; ***nursery fruit trees and bushes, ornamental trees, shrubs*** [1-3, 5, 7, 8, 10, 11]
- Blackgrass in ***broccoli*** [2-11]; ***brussels sprouts, cabbages, cauliflowers, spring oilseed rape, swedes, turnips, winter oilseed rape*** [1-11]; ***calabrese*** [1-4, 6-11]; ***nursery fruit trees and bushes, ornamental trees, shrubs*** [1-3, 5, 7, 8, 10, 11]

Efficacy guidance

- Activity is dependent on root uptake. For pre-emergence use apply to firm, moist, clod-free seedbed
- Some weeds (chickweed, mayweed, blackgrass etc) susceptible up to 2- or 4-leaf stage. Moderate control of cleavers achieved provided weeds not emerged and adequate soil moisture present

- Split pre- and post-emergence treatments recommended for certain weeds in winter oilseed rape on light and/or stony soils
- Effectiveness is reduced on soils with more than 10% organic matter
- Various tank-mixtures recommended to broaden spectrum. See label for details
- Always follow WRAG guidelines for preventing and managing herbicide resistant weeds. Section 5 for more information

Restrictions

- Maximum number of treatments 1 per crop for spring oilseed rape, swedes, turnips and brassicas; 2 per crop for winter oilseed rape and honesty (split dose treatment); 3 per yr for ornamentals, nursery stock, nursery fruit trees, forestry and farm forestry
- Do not use on sand, very light or poorly drained soils
- Do not treat protected crops or spray overall on ornamentals with soft foliage
- Do not spray crops suffering from wilting, pest or disease
- Do not spray broadcast crops or if a period of heavy rain forecast
- When used on nursery fruit trees any fruit harvested within 1 yr of treatment must be destroyed

Crop-specific information

- Latest use: pre-emergence for swedes and turnips; before 10 leaf stage for spring oilseed rape; before end of Jan for winter oilseed rape; before 6 pairs of true leaves visible for honesty
- HI brassicas 6 wk
- On winter oilseed rape may be applied pre-emergence from drilling until seed chits, post-emergence after fully expanded cotyledon stage (GS 1,0) or by split dose technique depending on soil and weeds. See label for details
- On spring oilseed rape, swedes, turnips metazachlor recommended as a pre-emergence sequential treatment following trifluralin
- On spring oilseed rape may also be used pre-weed-emergence from cotyledon to 10-leaf stage of crop (GS 1,0-1,10)
- With pre-emergence treatment ensure seed covered by 15 mm of well consolidated soil. Harrow across slits of direct-drilled crops
- Ensure brassica transplants have roots well covered and are well established. Direct drilled brassicas should not be treated before 3 leaf stage
- In ornamentals and hardy nursery stock apply after plants established and hardened off as a directed spray or, on some subjects, as an overall spray. See label for list of tolerant subjects. Do not treat plants in containers

Following crops guidance

- Any crop can follow normally harvested treated winter oilseed rape. See label for details of crops which may be planted after spring treatment and in event of crop failure

Environmental safety

- Dangerous for the environment
- Very toxic to aquatic organisms
- Keep livestock out of treated areas until foliage of any poisonous weeds such as ragwort has died and become unpalatable
- Keep livestock out of treated areas of swede and turnip for at least 5 wk following treatment

Hazard classification and safety precautions

Hazard H03 [1-3, 5-11]; H04 [4]; H11 [1-11]

Risk phrases R22a [1-3, 5-11]; R38 [2, 3, 5-9, 11]; R43, R50 [1-11]; R58 [3, 4, 6-9]

Operator protection A, C [1-11]; H, M [1-3, 5-11]; U05a, U08, U19a, U20b [1-11]; U14 [2, 3, 11]; U15 [2, 11]

Environmental protection E06a [1-11] (5 wk for swedes, turnips); E07a, E15a [1-11]; E31b [1, 4-10]; E31c [2, 3, 11]; E32d, E38 [3]; E32e [2, 11]; E34 [1-3, 5-11]

Storage and disposal E01, E04, E30a [1-11]; E26 [1, 2, 4-11]

Medical advice M03 [1-3, 5-11]; M05a [3]

317 metazachlor + quinmerac

A residual herbicide mixture for oilseed rape

Products

1	Katamaran	BASF	375:125 g/l	SC	11732
2	Novall	BASF	400:100 g/l	SC	12031
3	Standon Metazachlor-Q	Standon	375:125 g/l	SC	09676

Uses

- Annual dicotyledons in ***winter oilseed rape***
- Annual meadow grass in ***winter oilseed rape***
- Blackgrass in ***winter oilseed rape***
- Cleavers in ***winter oilseed rape***
- Poppies in ***winter oilseed rape***

Efficacy guidance

- Activity is dependent on root uptake. Pre-emergence treatments should be applied to firm moist seedbeds. Applications to dry soil do not become effective until after rain has fallen
- Maximum activity achieved from treatment before weed emergence for some species
- Weed control may be reduced if excessive rain falls shortly after application especially on light soils
- May be used on all soil types except Sands, Very Light Soils, and soils containing more than 10% organic matter. Crop vigour and/or plant stand may be reduced on brashy and stony soils

Restrictions

- Maximum total dose equivalent to one full dose treatment
- Damage may occur in waterlogged conditions. Do not use on poorly drained soils
- Do not treat stressed crops. In frosty conditions transient scorch may occur

Crop-specific information

- Latest use: end Jan in yr of harvest
- To ensure crop safety it is essential that crop seed is well covered with soil to 15 mm. Loose or puffy seedbeds must be consolidated before treatment. Do not use on broadcast crops
- Crop vigour and possibly plant stand may be reduced if excessive rain falls shortly after treatment especially on light soils

Following crops guidance

- In the event of crop failure after use, wheat or barley may be sown in the autumn after ploughing to 15 cm. Spring cereals or brassicas may be planted after ploughing in the spring

Environmental safety

- Dangerous for the environment
- Very toxic to aquatic organisms
- Keep livestock out of treated areas until foliage of any poisonous weeds such as ragwort has died and become unpalatable
- To reduce risk of movement to water do not apply to dry soil or if heavy rain is forecast. On clay soils create a fine consolidated seedbed
- LERAP Category B

Hazard classification and safety precautions

Hazard H04, H11
Risk phrases R43, R50, R58
Operator protection A [1-3]; C, H [2]; U04a [2]; U05a, U08, U14, U19a, U20b [1-3]
Environmental protection E07a, E15a, E16a [1-3]; E16b, E31b [3]; E31c, E32d, E38 [1, 2]
Storage and disposal E01, E04, E30a [1-3]; E26 [3]; E29b [1, 2]
Medical advice M05a [1, 2]

318 metconazole

A conazole fungicide

Products

1	Caramba	BASF	60 g/l	SL	10213
2	Sunorg Pro	BASF	90 g/l	SL	11112

Uses

- Alternaria in ***spring oilseed rape***, ***winter oilseed rape*** [1, 2]
- Brown rust in ***spring barley***, ***winter barley***, ***winter wheat*** [1]
- Fusarium ear blight in ***winter wheat*** *(reduction)* [1]
- Net blotch in ***spring barley*** *(reduction)*, ***winter barley*** *(reduction)* [1]
- Phoma in ***spring oilseed rape*** *(foliar disease - reduction)*, ***winter oilseed rape*** *(foliar disease - reduction)* [1]; ***spring oilseed rape*** *(reduction)*, ***winter oilseed rape*** *(reduction)* [2]
- Powdery mildew in ***spring barley*** *(moderate control)*, ***winter barley***, ***winter wheat*** *(moderate control)* [1]
- Rhynchosporium in ***spring barley***, ***winter barley*** [1]
- Sclerotinia stem rot in ***spring oilseed rape*** *(reduction)*, ***winter oilseed rape*** *(reduction)* [1, 2]
- Septoria leaf spot in ***winter wheat*** [1]
- Yellow rust in ***winter wheat*** [1]

Efficacy guidance

- Best results from application to healthy, vigorous crops when disease starts to develop
- Treatment for leaf blotch should be made after GS 33 and when weather favouring development of the disease has occurred
- Treat mildew infections before 3% infection on any green leaf. A specific mildewicide will improve control of established infections
- Treat yellow rust before 1% infection on any leaf or as preventive treatment after GS 39
- For brown rust spray susceptible varieties before any of top 3 leaves has more than 2% infection
- Sclerotinia in oilseed rape should be treated at petal fall. A tank mixture with carbendazim may be needed for fully effective control

Restrictions

- Maximum total dose equivalent to two full dose treatments
- Do not apply to oilseed rape crops that are damaged or stressed from previous treatments, adverse weather, nutrient deficiency or pest attack. Spring application may lead to reduction of crop height
- The addition of adjuvants is neither advised nor necessary and can lead to enhanced growth regulatory effects on stressed crops of oilseed rape
- Ensure sprayer is free from residues of previous treatments that may harm the crop, especially oilseed rape. Use of a detergent cleaner is advised before and after use
- Do not apply with pyrethroid insecticides on oilseed rape at flowering [2]

Crop-specific information

- Latest use: up to and including milky ripe stage (GS 71) for cereals; 10% pods at final size for oilseed rape

Following crops guidance

- Only cereals, oilseed rape, sugar beet, linseed, maize, clover, beans, peas, carrots, potatoes or onions may be sown as following crops after treatment

Environmental safety

- Dangerous for the environment
- Very toxic to aquatic organisms
- Risk to certain non-target insects or other arthropods
- LERAP Category B but avoid treatment close to field boundary, even if permitted by LERAP assessment, to reduce effects on non-target insects or other arthropods

Hazard classification and safety precautions

Hazard H03, H08 [1]; H04 [2]; H11 [1, 2]
Risk phrases R22b, R38, R41, R43, R50 [1]; R36, R51 [2]; R58 [1, 2]
Operator protection A, C [1, 2]; H [1]; U02a, U05a, U20c [1, 2]; U07, U11, U14 [1]
Environmental protection E15a, E31a, E32d, E34, E38 [1, 2]; E16a, E16b [1]
Storage and disposal E01, E04, E26, E30a
Medical advice M05b [1]

SEE SECTION 3 FOR PRODUCTS ALSO REGISTERED

319 methiocarb

A stomach acting carbamate molluscicide and insecticide

Products

1	Decoy Wetex	Bayer CropScience	2% w/w	PT	11266
2	Exit Wetex	Bayer CropScience	3% w/w	PT	11276
3	Huron	Bayer CropScience	3% w/w	PT	11288
4	Karan	Bayer CropScience	3% w/w	PT	11289
5	Lupus	Bayer CropScience	3% w/w	PT	11291
6	New Draza	Bayer CropScience	3% w/w	PT	11294
7	Rivet	Bayer CropScience	3% w/w	PT	11300

Uses

- Cutworms in ***sugar beet*** *(reduction)*
- Leatherjackets in ***all cereals*** *(reduction)*, ***potatoes*** *(reduction)*, ***ryegrass*** *(seed admixture)*, ***sugar beet*** *(reduction)*
- Millipedes in ***sugar beet*** *(reduction)*
- Slugs in ***all cereals***, ***all non-edible crops*** *(outdoor only)*, ***brussels sprouts***, ***cabbages***, ***cauliflowers***, ***leaf spinach***, ***lettuce***, ***maize***, ***potatoes***, ***ryegrass*** *(seed admixture)*, ***spring oilseed rape***, ***strawberries***, ***sugar beet***, ***sunflowers***, ***winter oilseed rape***
- Strawberry seed beetle in ***strawberries***

Efficacy guidance

- Use as a surface, overall application when pests active (normally mild, damp weather), pre-drilling or post-emergence. May also be used on cereals or ryegrass in admixture with seed at time of drilling
- Also reduces populations of cutworms and millipedes
- Best on potatoes in late Jul to Aug
- Apply to strawberries before strawing down to prevent seed beetles contaminating crop
- See label for details of suitable application equipment

Restrictions

- This product contains an anticholinesterase carbamate compound. Do not use if under medical advice not to work with such compounds
- Maximum number of treatments 3 per crop for potatoes; 2 per crop for cereals, Brussels sprouts, cauliflowers, maize, oilseed rape, sunflowers; 1 per crop for cabbages, lettuce, leaf spinach, sugar beet; 1 per yr for strawberries
- Do not treat any protected crops
- Do not allow pellets to lodge in edible crops

Crop-specific information

- Latest use: before first node detectable for cereals, maize; before three visibly extended internodes for oilseed rape, sunflowers
- HI 7 d for spinach, strawberrries; 14 d for Brussels sprouts, cabbages, cauliflowers, lettuce; 18 d for potatoes; 6 mth for sugar beet

Environmental safety

- Harmful to aquatic organisms
- Dangerous to game, wild birds and animals
- Risk to certain non-target insects or other arthropods
- Avoid surface broadcasting application within 6 m of field boundary to reduce effects on non-target species
- Admixed seed should be drilled and not broadcast, and may not be applied from the air

Hazard classification and safety precautions

Hazard H03
Risk phrases R22a, R52, R58
Operator protection A, H, J; U05a, U20b
Environmental protection E05b [1-7] (7 d); E10a, E13b, E32a, E34 [1-7]; E22b [2-7]; E22c [1]
Storage and disposal E01, E04, E30a
Medical advice M02, M03

320 methoxyfenozide

A moulting accelerating diacylhydrazine insecticide

Products

Runner	Bayer CropScience	240 g/l	SC	11470

Uses

- Codling moth in ***apples, pears***
- Tortrix moths in ***apples, pears***
- Winter moth in ***apples, pears***

Efficacy guidance

- To achieve best results uniform coverage of the foliage and full spray penetration of the leaf canopy is important, particularly when spraying post-blossom
- For maximum effectiveness on winter moth and tortrix spray pre-blossom when first signs of active larvae are seen, followed by a further spray in June if larvae of the summer generation are present
- For codling moth spray post-blossom to coincide with early to peak egg deposition. Follow-up treatments will normally be needed
- Methoxyfenozide is a moulting accelerating compound (MAC) and may be used in an anti-resistance strategy with other top fruit insecticides (including chitin biosynthesis inhibitors and juvenile hormones) which have a different mode of action
- To reduce further the likelihood of resistance development use at full recommended dose in sufficient water volume to achieve required spray penetration

Restrictions

- Maximum number of treatments 3 per yr but no more than two should be sprayed consecutively

Crop-specific information

- HI 14 d for apples, pears

Environmental safety

- LERAP Category B
- Broadcast air-assisted LERAP (5 m)

Hazard classification and safety precautions

Operator protection U20b
Environmental protection E16b, E17b, E32a
Storage and disposal E26, E30a

321 1-methylcyclopropene

An inhibitor of ethylene production for use in stored apples

Products

SmartFresh	Landseer	3.3% w/w	SP	11799

Uses

- Ethylene inhibition in ***apples*** *(post-harvest use)*
- Scald in ***apples*** *(post-harvest use)*

Efficacy guidance

- Best results obtained from treatment of fruit in good condition and of proper quality for long-term storage
- Effects may be reduced in fruit that is in poor condition or ripe prior to storage or harvested late
- Product acts by releasing vapour into store when mixed with water
- Apply as soon as possible after harvest
- Treatment controls superficial scald and maintains fruit firmness and acid content for 3-6 mth in normal air, and 6-9 mth in controlled atmosphere storage
- Ethylene production recommences after removal from storage

Restrictions

- Maximum number of treatments 1 per batch of apples
- Must only be used by suitably trained and competent persons in fumigation operations
- Consult processors before treatment of fruit destined for processing or cider making

- Do not apply in mixture with other products
- Ventilate all areas thoroughly with all refrigeration fans operating at maximum power for at least 15 min before re-entry

Crop-specific information
- Latest use: 7d after harvest of apples
- Product tested on Granny Smith, Gala, Jonagold, Bramley and Cox. Consult distributor or supplier before treating other varieties

Environmental safety
- Unprotected persons must be kept out of treated stores during the 24 h treatment period
- Prior to application ensure that the store can be properly and promptly sealed

Hazard classification and safety precautions

Operator protection U05a
Environmental protection E02a (24 h); E15a, E33, E34
Consumer protection C12
Storage and disposal E01, E04, E29a

322 metoxuron

A contact and residual urea herbicide for carrots

Products

Dosaflo	Syngenta	500 g/l	SC	09351

Uses
- Annual dicotyledons in ***carrots***
- Mayweeds in ***carrots***

Approval information
- Products containing this active ingredient have been granted derogations for specified 'Essential Uses' for use until 31 December 2007. Sale and supply must cease by 30 June 2007 but growers have no guarantee that the products will continue to be available until then.
  For more information see 'The Review Programme' under 'Pesticide Legislation' in Section 5

Efficacy guidance
- Best results achieved by application to weeds, especially mayweeds, in seedling to young plant stage
- Product will suppress growth of annual meadow-grass and wild oats between emergence and the 3 leaf stage of growth

Restrictions
- Maximum number of treatments not specified
- Carrots should be drilled to an even depth of not less than 1.5 cms otherwise some plants may be treated at cotyledon stage and their growth checked
- Do not spray when soil very dry or wet, when heavy rain is imminent, or when shade temperature exceeds 25°C
- Do not spray carrots on soils with more than 80% sand or less than 1% organic matter
- Do not apply during prolonged frosty weather, when temperatures are below freezing, or when frost is imminent
- Do not spray crops checked by pests, wind, frost or waterlogging until recovered
- This product must not be used on any crops other than those listed, including any extrapolations that would normally be permissible under the Long Term Arrangements for Extension of Use (see Section 5)

Crop-specific information
- Latest use: not specified for carrots

Following crops guidance
- No crop should be sown within 6 wk of treatment

Environmental safety
- Dangerous for the environment
- Very toxic to aquatic organisms

Hazard classification and safety precautions

Hazard H11

Risk phrases R50, R58

Operator protection U05a, U08, U19a, U20b

Environmental protection E15a, E31c, E32d, E38

Storage and disposal E01, E04, E26, E30a

323 metrafenone

A benzophenone protectant and curative fungicide for cereals

Products

Flexity	BASF	300 g/l	SC	11775

Uses

- Eyespot in ***spring wheat*** *(reduction)*, ***winter wheat*** *(reduction)*
- Powdery mildew in ***spring barley***, ***spring wheat***, ***winter barley***, ***winter wheat***

Efficacy guidance

- Best results obtained from treatment at the start of foliar disease attack
- Activity against mildew in wheat is mainly protectant with moderate curative control in the latent phase; activity is entirely protectant in barley
- Useful reduction of eyespot in wheat is obtained if treatment applied at GS 30-32
- Should be used as part of a resistance management strategy that includes mixtures or sequences effective against mildew and non-chemical methods

Restrictions

- Maximum number of treatments 2 per crop
- Do not use sequential treatments containing metrafenone

Crop-specific information

- Latest use: beginning of flowering (GS 31) for wheat, barley

Following crops guidance

- Cereals, oilseed rape, sugar beet, linseed, maize, clover, field beans, peas, turnips, carrots, cauliflowers, onions, lettuce or potatoes may follow a treated cereal crop

Environmental safety

- Dangerous for the environment
- Toxic to aquatic organisms

Hazard classification and safety precautions

Hazard H04, H11

Risk phrases R43, R51, R58

Operator protection A, H; U05a, U20b

Environmental protection E15a, E31c, E34

Storage and disposal E01, E04, E26, E30a

Medical advice M03

324 metribuzin

A contact and residual triazinone herbicide for use in potatoes

See also flufenacet + metribuzin

Products

1	Ag-Chem Metribuzin	Ag-Chem	70% w/w	WG	12007
2	Agriguard Metribuzin	AgriGuard	70% w/w	WG	09853
3	Citation 70	United Phosphorus	70% w/w	WG	09370
4	Citation 70	United Phosphorus	70% w/w	WG	11929
5	Inter-Metribuzin WG	I T Agro	70% w/w	WG	09801
6	Lexone 70DF	DuPont	70% w/w	WG	04991
7	Python	Makhteshim	70% w/w	WG	11166
8	Sencorex WG	Bayer CropScience	70% w/w	WG	11304
9	Shotput	Makhteshim	70% w/w	WG	11960
10	Tuberon	Unicrop	480 g/l	SC	11061

SEE SECTION 3 FOR PRODUCTS ALSO REGISTERED

Uses

- Annual and perennial weeds in ***asparagus*** *(off-label)* [8]
- Annual dicotyledons in ***asparagus*** *(off-label)*, ***carrots*** *(off-label)*, ***parsnips*** *(off-label)* [7]; ***early potatoes***, ***maincrop potatoes*** [1-10]
- Annual grasses in ***asparagus*** *(off-label)*, ***carrots*** *(off-label)*, ***parsnips*** *(off-label)* [7]; ***early potatoes***, ***maincrop potatoes*** [1-10]
- Perennial dicotyledons in ***asparagus*** *(off-label)*, ***carrots*** *(off-label)*, ***parsnips*** *(off-label)* [7]
- Volunteer oilseed rape in ***early potatoes***, ***maincrop potatoes*** [1-9]

Specific Off-Label Approvals (SOLAs)

- ***asparagus***, ***carrots***, ***parsnips*** *(OLA 030368) Dec 2008* [7]
- ***asparagus*** *(OLA 031888) Dec 2008* [8]

Efficacy guidance

- Best results achieved on weeds at cotyledon to 1-leaf stage
- Water dispersible granule formulations may be applied pre- or post-emergence of crop [3, 5-8]
- Suspension concentrate product may be applied pre-emergence only [10]
- Apply to moist soil with well-rounded ridges and few clods
- Activity reduced by dry conditions and on soils with high organic matter content
- On fen and moss soils pre-planting incorporation to 10-15 cm gives increased activity. Incorporate thoroughly and evenly
- With named maincrop and second early varieties on soils with more than 10% organic matter shallow pre- or post-planting incorporation may be used. See label for details
- Effective control using a programme of reduced doses is made possible by using a spray of smaller droplets, thus improving retention. See label for details [3, 5-8]

Restrictions

- Maximum number of treatments 3 per crop on potatoes subject to maximum permitted dose for water dispersible granule formulations [3, 5-8]; 2 per crop on carrots; 1 per crop on asparagus. See labels for details
- Maximum number of treatments 2 per crop on potatoes for suspension concentrate product [10]
- Apply pre-emergence only on named first earlies, pre- or post-emergence on named second earlies. On named maincrop varieties apply pre-emergence (except for certain varieties on Sands or Very Light soils) or post-emergence before longest shoots reach 15 cm. See labels for details [3, 5-8]
- Do not treat any variety on Light soils [10]
- Do not cultivate after treatment
- Some varieties may be sensitive to post-emergence treatment if crop under stress [3, 5-8]

Crop-specific information

- Latest use: pre-emergence for early potatoes; before most advanced shoots reach 15 cm for maincrop potatoes [3, 5-8]; whilst crop still dormant in yr of harvest for asparagus
- Latest use for suspension concentrate product: pre-emergence for all crops [10]
- HI carrots, parsnips 4 wk
- On stony or gravelly soils there is risk of crop damage, especially if heavy rain falls soon after application
- When days are hot and sunny delay spraying until evening

Following crops guidance

- Ryegrass, cereals or winter beans may be sown in same season provided at least 16 wk elapsed after treatment and ground ploughed to 15 cm and thoroughly cultivated as soon as possible after harvest and no later than end Dec
- In W Cornwall on soil with more than 5% organic matter early potatoes treated as recommended may be followed by summer planted brassica crops provided the soil has been ploughed, spring rainfall has been normal and at least 14 wk have elapsed since treatment
- Do not grow any vegetable brassicas on silt soils in Lincs, and lettuces or radishes anywhere in UK on land treated the previous yr. Other crops may be sown normally in spring of next yr

Environmental safety

- Dangerous for the environment [1-10]
- Very toxic to aquatic organisms
- Do not empty into drains
- LERAP Category B

Hazard classification and safety precautions

Hazard H03, H11

Risk phrases R22a [1-10]; R43 [7, 9]; R50, R58 [1, 5-10]

Operator protection A [7, 9, 10]; H [7, 9]; U04a [10]; U05a [7, 9, 10]; U08, U13, U19a [1-10]; U14 [7, 9]; U20a [1, 6, 8]; U20b [2-5, 7, 9, 10]

Environmental protection E13b, E38 [1, 8]; E15a [2-7, 9, 10]; E16a, E16b [1-10]; E19b [7, 9]; E31c, E34 [10]; E32a [1-9]; E32d [1, 7-9]

Storage and disposal E01, E04 [5, 7, 9, 10]; E26 [2-4, 10]; E29a [2-5]; E30a [1-10]

Medical advice M03 [10]; M05a [7, 9]

325 metsulfuron-methyl

A contact and residual sulfonylurea herbicide for use in cereals and land temporarily removed from production

See also carfentrazone-ethyl + metsulfuron-methyl

Products

1	Ally	DuPont	20% w/w	WG	02977
2	Landgold Metsulfuron	Landgold	20% w/w	WG	06280

Uses

- Annual dicotyledons in ***linseed, triticale*** [1]; ***spring barley, spring oats, spring wheat, winter barley, winter oats, winter wheat*** [1, 2]
- Chickweed in ***linseed, triticale*** [1]; ***spring barley, spring oats, spring wheat, winter barley, winter oats, winter wheat*** [1, 2]
- Green cover in ***land temporarily removed from production*** [1]
- Mayweeds in ***linseed, triticale*** [1]; ***spring barley, spring oats, spring wheat, winter barley, winter oats, winter wheat*** [1, 2]

Approval information

- Metsulfuron-methyl included in Annex I under EC Directive 91/414

Efficacy guidance

- Best results achieved on small, actively growing weeds up to 6-true leaf stage. Good spray cover is important
- Commonly used in tank-mixture on wheat and barley with other cereal herbicides to improve control of resistant dicotyledons (cleavers, fumitory, ivy-leaved speedwell), larger weeds and grasses. See label for recommended mixtures
- When tank mixing, always add metsulfuron-methyl product to tank first
- Metsulfuron-methyl is a member of the ALS-inhibitor group of herbicides and products should be used in a planned Resistance Management strategy. See Section 5 for more information

Restrictions

- Maximum number of treatments 1 per crop
- Product must only be used after 1 Feb
- Do not use on cereal crops undersown with grass or legumes
- Do not use on any crop suffering stress from drought, waterlogging, frost, deficiency, pest or disease attack or apply within 7 d of rolling
- Do not spray a cereal crop in tank mixture, or in sequence, with a product containing any other sulfonylurea except as directed on the label
- Spraying equipment should not be drained or flushed onto land planted, or to be planted, with trees or crops other than cereals and should be thoroughly cleansed after use - see label for instructions

Crop-specific information

- Latest use: before flag leaf sheath extending stage for cereals (GS 41)
- Apply to wheat and oats from 2-leaf (GS 12), to barley and triticale from 3-leaf stage (GS 13) until flag-leaf fully emerged (GS 39). Do not spray Igri barley before leaf sheath erect stage (GS 30)
- Recommendations for oats, triticale and linseed apply to product alone

Following crops guidance

- Only cereals, oilseed rape, field beans or grass may be sown in same calendar year after treating cereals with the product alone. Other restrictions apply to tank mixtures. See label for details

SEE SECTION 3 FOR PRODUCTS ALSO REGISTERED

- In the event of crop failure sow only wheat within 3 mth after treatment
- Only cereals should be planted within 16 mth of applying to a linseed crop

Environmental safety

- Dangerous for the environment
- Very toxic to aquatic organisms
- Extremely dangerous to aquatic higher plants. Do not contaminate surface waters or ditches with chemical or used container
- Take extreme care to avoid damage by drift onto broad-leaved plants outside the target area, onto surface waters or ditches or onto land intended for cropping
- LERAP Category B
- A range of broad leaved species will be fully or partially controlled when used in land temporarily removed from production, hence product may be suitable where wild flower borders or other forms of conservation headland are being developed
- Before use on land temporarily removed from production as part of grant-aided scheme, ensure compliance with the management rules
- Green cover on land temporarily removed from production must not be grazed by livestock or harvested for human or animal consumption or used for animal bedding

Hazard classification and safety precautions

Hazard H11
Risk phrases R50, R58
Operator protection U08, U19a, U20a
Environmental protection E15a, E16a, E16b, E32a
Storage and disposal E30a

326 metsulfuron-methyl + thifensulfuron-methyl

A contact residual and translocated sulfonylurea herbicide mixture for use in cereals

Products

1	Finish PX	DuPont	8.6:42.9% w/w	WG	10989
2	Harmony M	DuPont	7:68% w/w	WG	03990

Uses

- Annual dicotyledons in ***spring barley***, ***spring wheat***, ***winter wheat*** [1, 2]; ***winter barley***, ***winter oats*** [1]
- Chickweed in ***spring barley***, ***spring wheat***, ***winter barley***, ***winter oats***, ***winter wheat*** [1]
- Cleavers in ***spring barley***, ***spring wheat***, ***winter wheat*** [2]
- Field pansy in ***spring barley***, ***spring wheat***, ***winter wheat*** [1, 2]; ***winter barley***, ***winter oats*** [1]
- Knotgrass in ***spring barley***, ***spring wheat***, ***winter barley***, ***winter oats***, ***winter wheat*** [1]
- Mayweeds in ***spring barley***, ***spring wheat***, ***winter barley***, ***winter oats***, ***winter wheat*** [1]
- Polygonums in ***spring barley***, ***spring wheat***, ***winter wheat*** [2]
- Speedwells in ***spring barley***, ***spring wheat***, ***winter wheat*** [1, 2]; ***winter barley***, ***winter oats*** [1]

Approval information

- Metsulfuron-methyl and thifensulfuron-methyl included in Annex I under EC Directive 91/414

Efficacy guidance

- Best results by application to small, actively growing weeds up to 6-true leaf stage
- Effectiveness may be reduced if heavy rain occurs within 4 h of application or if soil conditions very dry
- Tank-mixture with mecoprop, mecoprop-P or reduced rate of fluroxypyr improves control of cleavers and other problem weeds
- Metsulfuron-methyl and thifensulfuron-methyl are members of the ALS-inhibitor group of herbicides and products should be used in a planned Resistance Management strategy. See Section 5 for more information

Restrictions

- Maximum number of treatments 1 per crop
- Products may be used only once on any crop and must be used after 1 Feb, and after crop has three leaves
- Do not use on any crop suffering stress from drought, waterlogging, frost, deficiency, pest or disease attack or any other cause

- Do not use on crops undersown with grasses, clover or legumes, or any other broad leaved crop
- Do not spray in tank mixture, or in sequence, with a product containing any other sulfonylurea herbicide except as directed on the label
- Do not apply within 7 d of rolling
- Restrictions apply to certain tank mixtures. See label

Crop-specific information

- Latest use: before flag leaf sheath extending (GS 39) for barley and wheat; before second node detectable (GS 32) for oats
- When tank mixing, always add metsulfuron-methyl/thifensulfuron-methyl product to tank first and fully disperse before adding second product
- Apply in spring to crops from 3-leaf stage to flag-leaf fully emerged (GS 13-39)

Following crops guidance

- Only cereals, oilseed rape, field beans or grass may be sown in same calendar year after treatment
- Additional constraints apply after use of certain tank mixtures. See label
- In the event of crop failure sow only winter wheat within 3 mth after treatment and after ploughing and cultivating to a depth of at least 15 cm

Environmental safety

- Dangerous for the environment
- Very toxic to aquatic organisms
- Extremely dangerous to aquatic higher plants. Do not contaminate surface waters or ditches with chemical or used container
- LERAP Category B
- Take extreme care to avoid damage by drift onto broad-leaved plants outside the target area, or onto ponds, waterways or ditches
- Spraying equipment should not be drained or flushed onto land planted, or to be planted, with trees or crops other than cereals and should be thoroughly cleansed after use - see label for instructions

Hazard classification and safety precautions

Hazard H11
Risk phrases R50, R58
Operator protection U08, U19a [1, 2]; U20a [2]; U20b [1]
Environmental protection E14a [2]; E15a [1]; E16a, E16b, E32a, E32d, E38 [1, 2]
Storage and disposal E30a

327 metsulfuron-methyl + tribenuron-methyl

A sulfonylurea herbicide mixture for winter wheat

Products

1	Biplay PX	DuPont	13:26.1% w/w	WG	10990
2	DP 911 PX	DuPont	13:26.1% w/w	WG	10992
3	DP 911 WSB	DuPont	13:26.1% w/w	WB	09867

Uses

- Annual dicotyledons in ***spring barley, spring wheat, triticale, winter barley, winter oats, winter wheat***
- Chickweed in ***spring barley, spring wheat, triticale, winter barley, winter oats, winter wheat***
- Field pansy in ***spring barley, spring wheat, triticale, winter barley, winter oats, winter wheat***
- Hemp-nettle in ***spring barley, spring wheat, triticale, winter barley, winter oats, winter wheat***
- Mayweeds in ***spring barley, spring wheat, triticale, winter barley, winter oats, winter wheat***
- Red dead-nettle in ***spring barley, spring wheat, triticale, winter barley, winter oats, winter wheat***

Approval information

- Metsulfuron-methyl included in Annex I under EC Directive 91/414

Efficacy guidance

- Best results obtained when applied to small actively growing weeds
- Product acts by foliar and root uptake. Good spray cover essential but performance may be reduced when soil conditions are very dry and residual effects may be reduced by heavy rain

SEE SECTION 3 FOR PRODUCTS ALSO REGISTERED

- Weed growth inhibited within hours of treatment and many show marked colour changes as they die back
- Place water soluble bags whole, directly into spray tank [3]
- Metsulfuron-methyl and tribenuron-methyl are members of the ALS-inhibitor group of herbicides and products should be used in a planned Resistance Management strategy. See Section 5 for more information

Restrictions

- Maximum number of treatments 1 per crop
- Product must only be used after 1 Feb and after crop has three leaves
- Do not apply to a crop suffering from drought, water-logging, low temperatures, pest or disease attack, nutrient deficiency, soil compaction or any other stress
- Do not use on crops undersown with grasses, clover or other legumes
- Do not apply within 7 d of rolling
- Do not apply in mixture, or in sequence, with a product containing any other sulfonylurea or 'ALS inhibiting' herbicide except as directed on the product label

Crop-specific information

- Latest use: before flag leaf sheath extending

Following crops guidance

- Only cereals, field beans or oilseed rape may be sown in the same calendar yr as harvest of a treated crop
- In the event of failure of a treated crop only winter wheat may sown within 3 mth of treatmernt, and only after ploughing and cultivation to 15 cm min

Environmental safety

- Dangerous for the environment
- Very toxic to aquatic organisms
- LERAP Category B
- Some non-target crops are highly sensitive. Take extreme care to avoid drift outside the target area, or onto ponds, waterways or ditches
- Spraying equipment should be thoroughly cleaned in accordance with manufacturer's instructions

Hazard classification and safety precautions

Hazard H04, H11
Risk phrases R42, R50, R58
Operator protection A; U05a, U08, U19a, U20b
Environmental protection E15a, E16a, E16b, E32a
Storage and disposal E01, E04, E30a

328 myclobutanil

A systemic, protectant and curative conazole fungicide

Products

1	Systhane 20EW	Landseer	200 g/l	EW	09396
2	Systhane 6 W	Dow	6% w/w	WP	10808

Uses

- American gooseberry mildew in ***blackcurrants***, ***gooseberries*** [1]
- Black spot in ***ornamental plant production*** [1]; ***roses*** [1, 2]
- Blossom wilt in ***cherries*** *(off-label)*, ***mirabelles*** *(off-label)* [1]
- Mildew in ***roses*** [2]
- Plum rust in ***plums*** *(off-label)* [1]
- Powdery mildew in ***apples***, ***grapevines*** *(off-label)*, ***hops*** *(off-label)*, ***ornamental plant production***, ***pears***, ***protected blackberries*** *(off-label)*, ***protected raspberries*** *(off-label)*, ***protected rubus hybrids*** *(off-label)*, ***roses***, ***strawberries*** [1]
- Rust in ***ornamental plant production*** [1]; ***roses*** [1, 2]
- Scab in ***apples***, ***pears*** [1]

Specific Off-Label Approvals (SOLAs)

- ***cherries***, ***mirabelles*** *(OLA 991535) Dec 2008* [1]
- ***hops*** *(OLA 021412) Dec 2008* [1]

FOR FULL CONDITIONS OF USE ALWAYS READ THE PRODUCT LABEL

- ***plums*** *(OLA 012459) Dec 2008* [1]
- ***protected blackberries, protected raspberries, protected rubus hybrids*** *(OLA 023195) Dec 2008* [1]

Efficacy guidance

- Best results achieved when used as part of routine preventive spray programme from bud burst to end of flowering in apples and pears and from just before the signs of mildew infection in blackcurrants and gooseberries [1]
- In strawberries commence spraying at, or just prior to, first flower. Post-harvest sprays may be required on mildew-susceptible varieties where mildew is present and likely to be damaging [1]
- Spray at 7-14 d intervals depending on disease pressure and dose applied
- For improved scab control in post-blossom period tank-mix with mancozeb or captan
- Apply alone from mid-Jun for control of secondary mildew on apples and pears
- On roses spray at first signs of disease and repeat every 2 wk. In high risk areas spray when leaves emerge in spring, repeat 1 wk later and then continue normal programme

Restrictions

- Maximum total dose equivalent to ten full dose treatments in apples and pears; six full dose treatments in blackcurrants, gooseberries, strawberries; 5 full dose treatments in plums; 3.5 full dose treatments in cherries, mirabelles

Crop-specific information

- HI 28 d (grapevines); 21 d (cherries, mirabelles); 14 d (apples, pears, blackcurrants, gooseberries, hops); 3 d (plums, protected blackberries, protected raspberries, protected Rubus hybrids, strawberries)

Environmental safety

- Dangerous for the environment
- Toxic to aquatic organisms

Hazard classification and safety precautions

Hazard H03, H11 [1]
Risk phrases R22b, R51, R58, R63 [1]
Operator protection A, H, J [1]; U08, U15 [1]; U20a [1, 2]
Environmental protection E13c, E32a [2]; E15a, E31c, E32d [1]
Storage and disposal E01, E04, E26 [1]; E30a [2]
Medical advice M05b [1]

329 2-(1-naphthyl)acetic acid

A plant growth regulator to promote rooting of cuttings

See also 4-indol-3-yl-butyric acid + 2-(1-naphthyl)acetic acid with dichlorophen

Products

1	Rhizopon B Powder (0.1%)	Rhizopon	0.1% w/w	SP	09089
2	Rhizopon B Powder (0.2%)	Rhizopon	0.2% w/w	SP	09090
3	Rhizopon B Tablets	Rhizopon	25 mg a.i.	TB	09091

Uses

- Rooting of cuttings in ***ornamental specimens***

Efficacy guidance

- Dip moistened base of cuttings into powder immediately before planting [1, 2]
- See label for details of concentrations recommended for promotion of rooting in cuttings of different species [3]
- Dip prepared cuttings in solution for 4-24 h depending on species [3]

Restrictions

- Maximum number of treatments 1 per cutting

Crop-specific information

- Latest use: before cutting insertion for ornamental specimens

Hazard classification and safety precautions

Operator protection U08, U19a, U20a

Environmental protection E15a, E32a, E34

Storage and disposal E30a

330 (2-naphthyloxy)acetic acid

A plant growth regulator for setting tomato fruit

Products

Betapal Concentrate	Vitax	16 g/l	SL	00234

Uses

- Increasing fruit set in ***outdoor tomatoes, protected tomatoes***

Efficacy guidance

- Apply with any fine sprayer or syringe when first half dozen flowers are open

Restrictions

- Do not spray growing head of tomato plants

Crop-specific information

- Spray actual trusses only once as they develop when about half the flowers are open
- Take care not to spray the growing head of the plant

Hazard classification and safety precautions

Operator protection U20c

Environmental protection E15a, E31a

Storage and disposal E30a

331 napropamide

A soil applied amide herbicide for oilseed rape, fruit and woody ornamentals

Products

1	AC 650	United Phosphorus	450 g/l	SC	11102
2	Devrinol	United Phosphorus	450 g/l	SC	09374

Uses

- Annual dicotyledons in ***apple orchards, blackcurrants, forest nurseries, gooseberries, ornamental plant production, pear orchards, plums, raspberries, strawberries*** [2]; ***winter oilseed rape*** [1, 2]
- Annual grasses in ***apple orchards, blackcurrants, forest nurseries, gooseberries, ornamental plant production, pear orchards, plums, raspberries, strawberries*** [2]; ***winter oilseed rape*** [1, 2]
- Cleavers in ***apple orchards, blackcurrants, forest nurseries, gooseberries, ornamental plant production, pear orchards, plums, raspberries, strawberries, winter oilseed rape*** [2]
- Groundsel in ***apple orchards, blackcurrants, forest nurseries, gooseberries, ornamental plant production, pear orchards, plums, raspberries, strawberries, winter oilseed rape*** [2]

Efficacy guidance

- Best results obtained from treatment pre-emergence of weeds but product may be used in conjunction with contact herbicide such as paraquat for control of emerged weeds. Otherwise remove existing weeds before application
- Product broken down by sunlight, so application during summer not recommended. Most crops recommended for treatment between Nov and end-Feb
- Apply to winter oilseed rape as pre-drilling treatment in tank-mixture with trifluralin and incorporate within 30 min. Post-emergence use of specific grass weedkilller recommended where volunteer cereals are a serious problem
- Increase water volume to ensure adequate dampening of compost when treating containerised nursery stock with dense leaf canopy

Restrictions

- Maximum number of treatments 1 per crop or yr
- Do not use on Sands

- Do not use on soils with more than 10% organic matter
- Do not treat ornamentals in containers of less than 1 litre
- Some phytotoxicity seen on yellow and golden varieties of conifers and container grown alpines. On any ornamental variety treat only a small number of plants in first season

Crop-specific information

- Latest use: at drilling for winter oilseed rape; before end Feb for apples, pears, strawberries, bush and cane fruit; before end of Apr for field and container grown ornamental trees and shrubs
- Where minimal cultivation used to establish oilseed rape, tank-mixture may be applied directly to stubble and mixed into top 25 mm as part of surface cultivations
- Apply up to 14 d prior to drilling winter oilseed rape
- Apply to strawberries established for at least one season or to maiden crops as long as planted carefully and no roots exposed, between 1 Nov and end Feb. Do not treat runners of poor vigour or with shallow roots, or runner beds
- Newly planted ornamentals should have no roots exposed. Do not treat stock of poor vigour or with shallow roots. Treatments made in Mar and Apr must be followed by 25 mm irrigation within 24 h
- Bush and cane fruit must be established for at least 10 mth before spraying and treated between 1 Nov and end Feb

Following crops guidance

- After use in fruit or ornamentals no crop can be drilled within 7 mth of treatment. Leaf, flowerhead, root and fodder brassica crops may be drilled after 7 mth; potatoes, maize, peas or dwarf beans after 9 mth; autumn sown wheat or grass after 18 mth; any crop after 2 yr
- After use in oilseed rape only oilseed rape, swedes, fodder turnips, brassicas or potatoes should be sown within 12 mth of application
- Soil should be mould-board ploughed to a depth of at least 200 mm before drilling or planting any following crop

Environmental safety

- Dangerous for the environment
- Toxic to aquatic organisms

Hazard classification and safety precautions

Hazard H04, H11
Risk phrases R36, R38, R51, R58
Operator protection A, C; U05a, U11
Environmental protection E32e, E34, E38
Storage and disposal E26, E30a

332 natural plant extracts

A contact insecticide that works by physical action

Products

Majestik	Certis	SL	-

Uses

- Aphids in ***fruit crops***, ***ornamental plant production***, ***protected fruit***, ***protected ornamentals***, ***protected vegetables***, ***vegetables***
- Leafhoppers in ***fruit crops***, ***ornamental plant production***, ***protected fruit***, ***protected ornamentals***, ***protected vegetables***, ***vegetables***
- Mites in ***fruit crops***, ***ornamental plant production***, ***protected fruit***, ***protected ornamentals***, ***protected vegetables***, ***vegetables***
- Thrips in ***fruit crops***, ***ornamental plant production***, ***protected fruit***, ***protected ornamentals***, ***protected vegetables***, ***vegetables***
- Whitefly in ***fruit crops***, ***ornamental plant production***, ***protected fruit***, ***protected ornamentals***, ***protected vegetables***, ***vegetables***

Approval information

- Product not controlled by Control of Pesticides Regulations/Plant Protection Products Regulations because it acts by physical means only

Efficacy guidance

- Product acts by coating target pests and inhibiting respiration or movement by physical action
- Effective on all growth stages of mites, aphids, thrips, whitefly and leafhoppers. Treat as soon as insects are seen and repeat as necessary
- Ensure thorough spray coverage of plant, paying special attention to growing points and the underside of leaves
- Consult manufacturer for guidance on efficacy before treating outdoor crops for the first time
- Product may be used in conjunction with biological control agents. Spray 24 h before they are introduced

Restrictions

- No limit on the number of treatments and no minimum interval between applications
- Before large scale use on a new crop, treat a few plants to check for crop safety
- Do not treat ornamental crops when in flower
- Avoid treating fruit or vegetables close to harvest as product may leave a sticky deposit

Crop-specific information

- HI zero. Product is not taken into plant

Environmental safety

- If bees are being used for pollination ensure hives are closed before treatment and do not re-open before spray has completely dried on the plants

Hazard classification and safety precautions

Operator protection U05a, U09a, U19a, U20c

333 nicosulfuron

A sulfonylurea herbicide for maize

Products

Samson	Syngenta	40 g/l	SC	10433

Uses

- Annual dicotyledons in ***maize***
- Annual meadow grass in ***maize***
- Chickweed in ***maize***
- Couch in ***maize***
- Grass weeds in ***sweetcorn*** *(off-label)*
- Mayweeds in ***maize***
- Ryegrass in ***maize***
- Shepherd's purse in ***maize***
- Volunteer oilseed rape in ***maize***

Specific Off-Label Approvals (SOLAs)

- ***sweetcorn*** *(OLA 040840) Dec 2008* [1]

Efficacy guidance

- Product should be applied post-emergence between 2 and 8 crop leaf stage to emerged weeds from the 2-leaf stage
- Product acts mainly by foliar activity. Ensure good spray cover of the weeds
- Always follow WRAG guidelines for preventing and managing herbicide resistant weeds, especially when using in continuously grown maize. Section 5 for more information

Restrictions

- Maximum total dose equivalent to one full dose treatment
- Do not use if an organophosphorus soil insecticide has been used or if an organophosphorus foliar insecticide will be used on the same crop
- Do not mix with foliar or liquid fertilisers or specified herbicides. See label for details
- Do not apply in mixture, or sequence, with any other sulfonyl-urea containing product
- Do not treat crops under stress
- Do not apply if rainfall is forecast to occur within 6 h of application

Crop-specific information

- Latest Use: up to and including 8 true leaves for maize

- Some transient yellowing may be seen from 1-2 wk after treatment
- Only healthy maize crops growing in good field conditions should be treated

Following crops guidance

- Winter wheat (not undersown) may be sown 4 mth after treatment; maize may be sown in the spring following treatment; all other crops may be sown from the next autumn

Environmental safety

- LERAP Category B

Hazard classification and safety precautions

Hazard H04
Risk phrases R38
Operator protection A, H; U02a, U05a, U10, U14, U20b
Environmental protection E15a, E16a, E16b, E31c
Storage and disposal E01, E04
Medical advice M05a

334 nicotine

A general purpose, non-persistent, contact, alkaloid insecticide

Products

1	Nico Soap	United Phosphorus	75 g/l	SL	07517
2	Nicotine 40% Shreds	Dow	40% w/w	FU	05725
3	No-Fid	Certis	75 g/l	LI	11183
4	Stalwart	United Phosphorus	75 g/l	SL	11877
5	XL-All Insecticide	Vitax	70 g/l	SL	02369
6	XL-All Nicotine 95%	Vitax	950 g/l	SL	07402

Uses

- Aphids in ***all soft fruit, protected asparagus, protected celery, protected marrows, vegetables*** [5]; ***all top fruit, flower bulbs, leaf brassicas, navy beans, nursery stock, ornamental specimens, parsley, peas, protected carnations, protected chrysanthemums, protected crops, protected flowers, protected melons, protected roses*** [5, 6]; ***apples, artichokes, asparagus, broccoli, brussels sprouts, cabbages, calabrese, cauliflowers, celeriac, chicory, chinese cabbage, chives, combining peas, courgettes, cress, cucumbers, dwarf beans, edible podded peas, endives, fennel, frise, garlic, kale, lamb's lettuce, marrows, ornamental plant production, outdoor tomatoes, peppers, potatoes, pumpkins, radicchio, radishes, shallots, spring field beans, squashes, sweetcorn, vining peas, winter field beans*** [1, 3, 4]; ***aubergines, protected herbs (see appendix 6), protected potatoes, protected salad onions*** [2]; ***broad beans, carrots, celery, leaf spinach, lettuce, parsnips, red beet, runner beans, swedes, turnips*** [1, 3-6]; ***bush fruit, protected figs, protected grapevines, protected peaches*** [6]; ***leeks, onions*** [1, 3-5]; ***protected courgettes, protected peppers*** [2, 5]; ***protected cucumbers, protected lettuce*** [2, 5, 6]; ***protected ornamentals, protected tomatoes*** [1-6]; ***strawberries*** [1, 3, 4, 6]
- Capsids in ***all soft fruit, vegetables*** [5]; ***all top fruit, flower bulbs, protected crops*** [5, 6]; ***apples*** [1, 3, 4]; ***bush fruit*** [6]
- Caterpillars in ***artichokes, asparagus, broad beans, broccoli, brussels sprouts, cabbages, calabrese, carrots, cauliflowers, celeriac, celery, chicory, chinese cabbage, chives, combining peas, courgettes, cress, cucumbers, dwarf beans, edible podded peas, endives, fennel, frise, garlic, kale, lamb's lettuce, leaf spinach, leeks, lettuce, marrows, onions, ornamental plant production, outdoor tomatoes, parsnips, peppers, potatoes, protected ornamentals, protected tomatoes, pumpkins, radicchio, radishes, red beet, runner beans, shallots, spring field beans, squashes, swedes, sweetcorn, turnips, vining peas, winter field beans*** [1, 3, 4]
- Glasshouse whitefly in ***aubergines, protected courgettes, protected cucumbers, protected herbs (see appendix 6), protected lettuce, protected ornamentals, protected peppers, protected potatoes, protected salad onions, protected tomatoes*** [2]
- Green leafhopper in ***aubergines, protected courgettes, protected cucumbers, protected herbs (see appendix 6), protected lettuce, protected ornamentals, protected peppers, protected potatoes, protected salad onions, protected tomatoes*** [2]
- Insect pests in ***french beans*** *(off-label)* [5]

SEE SECTION 3 FOR PRODUCTS ALSO REGISTERED

- Leaf miner in ***artichokes, asparagus, broad beans, broccoli, brussels sprouts, cabbages, calabrese, carrots, cauliflowers, celeriac, celery, chicory, chinese cabbage, chives, combining peas, courgettes, cress, cucumbers, dwarf beans, edible podded peas, endives, fennel, frise, garlic, kale, lamb's lettuce, leaf spinach, leeks, lettuce, marrows, onions, ornamental plant production, outdoor tomatoes, parsnips, peppers, potatoes, protected ornamentals, protected tomatoes, pumpkins, radicchio, radishes, red beet, runner beans, shallots, spring field beans, squashes, swedes, sweetcorn, turnips, vining peas, winter field beans*** [1, 3, 4]; ***flower bulbs, leaf brassicas, protected crops*** [5, 6]
- Leafhoppers in ***all soft fruit, vegetables*** [5]; ***bush fruit*** [6]; ***flower bulbs, protected crops*** [5, 6]
- Peach-potato aphid in ***winter oilseed rape*** *(off-label)* [1]
- Potato virus vectors in ***chitting potatoes*** [2]
- Sawflies in ***all soft fruit, vegetables*** [5]; ***all top fruit, flower bulbs, protected crops*** [5, 6]; ***bush fruit*** [6]; ***gooseberries*** [1, 3, 4]
- Thrips in ***aubergines, protected courgettes, protected cucumbers, protected herbs (see appendix 6), protected lettuce, protected ornamentals, protected peppers, protected potatoes, protected salad onions, protected tomatoes*** [2]; ***flower bulbs, protected crops*** [5, 6]; ***vegetables*** [5]

Specific Off-Label Approvals (SOLAs)

- ***french beans*** *(OLA 920080) Dec 2008* [5]
- ***winter oilseed rape*** *(OLA 012480) Dec 2008* [1]

Efficacy guidance

- Apply as foliar spray, taking care to cover undersides of leaves and repeat as necessary or dip young plants, cuttings or strawberry runners before planting out
- Best results achieved by spraying at air temperatures above 16°C but do not treat in bright sunlight or windy weather [2]
- Fumigate glasshouse crops at temperatures of at least 16°C. See label for details of recommended fumigation procedure [2]
- Potatoes may be fumigated in chitting houses to control virus-spreading aphids [2]
- May be used in integrated control systems to give partial control of organophosphorus-, organochlorine- and pyrethroid-resistant whitefly

Restrictions

- Nicotine is subject to the Poisons Rules 1982 and the Poisons Act 1972. See notes in Section 5 [2, 6]
- Maximum number of treatments not specified
- On plants of unknown sensitivity test first on a small scale
- Keep unprotected persons out of treated glasshouses for at least 12 h
- Wear overall, hood, rubber gloves and respirator if entering glasshouse within 12 h of fumigating [2]

Crop-specific information

- HI 2 d for most crops - check labels

Environmental safety

- Dangerous for the environment
- Toxic to aquatic organisms
- Dangerous to livestock. Keep all livestock out of treated areas/away from treated water for at least 12 h
- Dangerous to game, wild birds and animals
- Harmful to bees. Do not apply to crops in flower or to those in which bees are actively foraging. Do not apply when flowering weeds are present
- Keep in original container, tightly closed, in a safe place, under lock and key

Hazard classification and safety precautions

Hazard H01 [1, 3, 4, 6]; H02 [5]; H03, H07 [2]; H11 [6]

Risk phrases R20, R21 [2]; R22a [2, 3, 5]; R24 [5]; R25, R27 [1, 3, 4, 6]; R36 [1, 3, 4]; R37 [3]; R38 [1, 4]; R51 [6]; R52 [1, 3-5]; R58 [1, 3-6]

Operator protection A [1-6]; C [1, 3, 4, 6]; D, J [2]; H [2, 6]; K, M [6]; U02a, U04a, U05a, U10, U13 [1, 3-6]; U05b [2]; U08 [5, 6]; U09a [1, 3, 4]; U19a, U20a [1-6]

Environmental protection E02a, E06b [1-5] (12 h); E02a, E06b [6]; E10a, E12f [1-6]; E13b [1-4]; E15a, E31a [5, 6]; E31b [1, 3, 4]; E32a [2]; E34 [2, 5, 6]

Consumer protection C02 [2] (24 h); C02 [5] (2 d); C02 [6]

Storage and disposal E01, E04 [1, 2, 4-6]; E26 [1, 3, 4]; E30a [1, 3-5]; E30b [2, 6]
Medical advice M03 [2]; M04 [1, 3-6]

335 oxadiazon

A residual and contact oxadiazolone herbicide for fruit and ornamentals

See also carbetamide + diflufenican + oxadiazon

Products

1	Ronstar 2G	Certis	2% w/w	GR	10011
2	Ronstar Liquid	Certis	250 g/l	EC	11215

Uses

- Annual dicotyledons in ***apple orchards, blackcurrants, gooseberries, grapevines, hops, pear orchards, raspberries, woody ornamentals*** [2]; ***container-grown ornamentals*** [1]
- Annual grasses in ***apple orchards, blackcurrants, gooseberries, grapevines, hops, pear orchards, raspberries, woody ornamentals*** [2]; ***container-grown ornamentals*** [1]
- Bindweeds in ***apple orchards, blackcurrants, gooseberries, grapevines, hops, pear orchards, raspberries, woody ornamentals*** [2]
- Cleavers in ***apple orchards, blackcurrants, gooseberries, grapevines, hops, pear orchards, raspberries, woody ornamentals*** [2]
- Knotgrass in ***apple orchards, blackcurrants, gooseberries, grapevines, hops, pear orchards, raspberries, woody ornamentals*** [2]

Efficacy guidance

- Rain or overhead watering is needed soon after application for effective results
- Pre-emergence activity reduced on soils with more than 10% organic matter and, in these conditions, post-emergence treatment is more effective [2]
- Best results on bindweed when first shoots are 10-15 cm long

Restrictions

- See label for list of ornamental species which may be treated with granules. Treat small numbers of other species to check safety. Do not treat hydrangea, spiraea or genista
- Do not cultivate after treatment [2]
- Do not treat container stock under glass or use on plants rooted in media with high sand or non-organic content. Do not apply to plants with wet foliage
- Do not use more than 8 l/ha in any 12 mth period [2]

Crop-specific information

- Latest use: Jul for apples, grapevines, hops, pears; Jun for raspberries, woody ornamentals
- Apply spray to apples and pears from Jan to Jul, avoiding young growth
- Treat bush fruit from Jan to bud-break, avoiding bushes, grapevines in Feb/Mar before start of new growth or in Jun/Jul avoiding foliage
- Treat hops cropped for at least 2 yr in Feb or in Jun/Jul after deleafing
- Treat woody ornamentals from Jan to Jun, avoiding young growth. Do not spray container stock overall

Environmental safety

- Dangerous for the environment
- Very toxic to aquatic organisms

Hazard classification and safety precautions

Hazard H03, H08, H11 [2]
Risk phrases R22b, R36, R38, R41, R50, R67 [2]; R52 [1]; R58 [1, 2]
Operator protection A, C; U05a, U14, U15, U20b [1, 2]; U08 [2]; U09a, U19a [1]
Environmental protection E15a, E32d [1, 2]; E31a, E38 [2]; E32a [1]
Storage and disposal E01, E04, E30a
Medical advice M05b [2]

336 oxamyl

A soil-applied, systemic oxime carbamate nematicide and insecticide

Products

Vydate 10G	DuPont	10% w/w	GR	02322

Uses

- American serpentine leaf miner in ***aubergines*** *(off-label)*, ***ornamental plant production*** *(off-label)*, ***protected broad beans*** *(off-label)*, ***protected ornamentals*** *(off-label)*, ***protected peppers*** *(off-label)*, ***protected soya beans*** *(off-label)*, ***protected tomatoes*** *(off-label)*
- Aphids in ***potatoes***, ***sugar beet***
- Docking disorder vectors in ***sugar beet***
- Free-living nematodes in ***potatoes***
- Leaf miner in ***aubergines*** *(off-label)*, ***ornamental plant production*** *(off-label)*, ***protected broad beans*** *(off-label)*, ***protected ornamentals*** *(off-label)*, ***protected peppers*** *(off-label)*, ***protected soya beans*** *(off-label)*, ***protected tomatoes*** *(off-label)*
- Mangold fly in ***sugar beet***
- Millipedes in ***sugar beet***
- Potato cyst nematode in ***potatoes***
- Pygmy beetle in ***sugar beet***
- South American leaf miner in ***aubergines*** *(off-label)*, ***ornamental plant production*** *(off-label)*, ***protected broad beans*** *(off-label)*, ***protected ornamentals*** *(off-label)*, ***protected peppers*** *(off-label)*, ***protected soya beans*** *(off-label)*, ***protected tomatoes*** *(off-label)*
- Spraing vectors in ***potatoes***
- Stem nematodes in ***carrots*** *(off-label)*, ***garlic*** *(off-label)*, ***parsnips*** *(off-label)*, ***protected garlic*** *(off-label)*, ***protected onions*** *(off-label)*

Specific Off-Label Approvals (SOLAs)

- ***aubergines***, ***ornamental plant production***, ***protected broad beans***, ***protected ornamentals***, ***protected peppers***, ***protected soya beans***, ***protected tomatoes*** *(OLA 930020) Dec 2008* [1]
- ***carrots***, ***parsnips*** *(OLA 040617) Dec 2008* [1]
- ***garlic*** *(OLA 920163) Dec 2008* [1]
- ***protected garlic***, ***protected onions*** *(OLA 940925) Dec 2008* [1]

Approval information

- Approvals for oxamyl allowed to continue, subject to label amendments, following the UK Review programme of cholinesterase compounds

Efficacy guidance

- Apply granules with suitable applicator before drilling or planting. See label for details of recommended machines
- In potatoes incorporate thoroughly to 10 cm and plant within 3-4 d
- In sugar beet apply in seed furrow at drilling

Restrictions

- Oxamyl is subject to the Poisons Rules 1982 and the Poisons Act 1972. See notes in Section 5
- This product contains an anticholinesterase carbamate compound. Do not use if under medical advice not to work with such compounds
- Maximum number of treatments 1 per crop or yr
- Keep in original container, tightly closed, in a safe place, under lock and key
- Wear protective gloves if handling treated compost or soil within 2 wk after treatment
- Allow at least 12 h, followed by at least 1 h ventilation, before entry of unprotected persons into treated glasshouses

Crop-specific information

- Latest use: at drilling/planting for vegetables; before drilling/planting for potatoes and peas.
- HI tomatoes, ornamentals 2 wk

Environmental safety

- Dangerous for the environment
- Toxic to aquatic organisms
- Dangerous to game, wild birds and animals. Bury spillages
- Keep in original container, tightly closed, in a safe place, under lock and key

Hazard classification and safety precautions

Hazard H02, H11
Risk phrases R23, R25, R51, R58
Operator protection A, B, H, K, M [1]; C [1] (or D); U02a, U04a, U05a, U09a, U13, U19a, U20a
Environmental protection E10a, E15a, E32a, E32d, E34, E38
Consumer protection C02
Storage and disposal E01, E04, E30b
Medical advice M02, M04

337 oxycarboxin

A protectant, eradicant and systemic carboxamide fungicide for use on ornamentals

Products

Plantvax 75	Fargro	75% w/w	WP	01601

Uses

- Rust in ***ornamental specimens***, ***protected ornamentals***

Approval information

- Products containing this active ingredient have been granted derogations for specified 'Essential Uses' for use until 31 December 2007. Sale and supply must cease by 30 June 2007 but growers have no guarantee that the products will continue to be available until then.
  For more information see 'The Review Programme' under 'Pesticide Legislation' in Section 5

Efficacy guidance

- Apply as a routine to established carnation beds from Sep to Feb at 7-10 d intervals. On rust-susceptible species or if rust already established add authorised wetting agent and repeat every 10-14 d until infection pressure ceases
- On stock plants or cuttings of carnations or geraniums apply after striking and repeat 3 times at 10-14 d intervals. Remove and burn any infected cuttings
- Can be used as a drench on carnations grown in peat bags

Restrictions

- This product must not be used on any crops other than those listed, including any extrapolations that would normally be permissible under the Long Term Arrangements for Extension of Use (see Section 1)

Crop-specific information

- Spray product alone and preferably not within 2 d of other sprays
- Can cause marginal leaf scorch when taken up by roots. Avoid spraying to run-off
- Product leaves visible deposit on plants so avoid treating when blooms are open

Environmental safety

- Harmful to aquatic organisms

Hazard classification and safety precautions

Hazard H03
Risk phrases R22a, R38, R52, R58
Operator protection A, C; U02a, U05a, U09a, U19a, U20b
Environmental protection E13b, E32a, E38
Storage and disposal E01, E04, E30a
Medical advice M03

338 paclobutrazol

A conazole plant growth regulator for ornamentals and fruit

Products

1	Bonzi	Syngenta Bioline	4 g/l	SC	10517
2	Cultar	Syngenta	250 g/l	SC	10523

Uses

- Controlling vigour in ***apples***, ***pears*** [2]
- Growth regulation in ***cherries*** *(off-label)*, ***plums*** *(off-label)* [2]

- Improving colour in ***poinsettias*** [1]
- Increasing flowering in ***azaleas, bedding plants, begonias, kalanchoes, lilies, roses, tulips*** [1]
- Increasing fruit set in ***apples, pears*** [2]
- Stem shortening in ***azaleas, bedding plants, begonias, kalanchoes, lilies, poinsettias, roses, tulips*** [1]

Specific Off-Label Approvals (SOLAs)

- ***cherries, plums*** *(OLA 031235) Dec 2008* [2]

Efficacy guidance

- Chemical is active via both foliage and root uptake. For best results apply in dull weather when relative humidity not high
- Apply as spray to produce compact pot plants and to improve bract colour of poinsettias [1]
- Apply as compost drench to reduce flower stem length of potted tulips and Mid-Century hybrid lilies [1]
- Timing is critical and varies with species. See label for details [1]

Restrictions

- Maximum number of treatments 1 per specimen for some species [1]
- Maximum total dose equivalent to 3 full dose treatments for pears and 4 full dose treatments for apples [2]
- Some varietal restrictions apply in top fruit (see label for details) [2]
- Do not use on trees of low vigour or under stress [2]
- Do not use on trees from green cluster to 2 wk after full petal fall [2]
- Do not use in underplanted orchards or those recently interplanted [2]

Crop-specific information

- HI apples, pears 14 d [2]
- Apply to apple and pear trees under good growing conditions as pre-blossom spray (apples only) and post-blossom at 7-14 d intervals [2]
- Timing and dose of orchard treatments vary with species and cultivar. See label [2]

Following crops guidance

- Chemical has residual soil activity which can affect growth of following crops. Withhold treatment from orchards due for grubbing to allow the following intervals between the last treatment and planting the next crop: apples, pears 1 yr; beans, peas, onions 2 yr; stone fruit 3 yr; cereals, grass, oilseed rape, carrots 4 yr; market brassicas 6 yr; potatoes and other crops 7 yr [2]

Environmental safety

- Harmful to aquatic organisms
- Do not use on food crops [1]
- Permissible in organic systems
- Keep livestock out of treated areas for at least 2 yr following treatment

Hazard classification and safety precautions

Hazard H04 [2]
Risk phrases R36, R52, R58 [2]
Operator protection A, H, M [2]; U05a [2]; U08 [1]; U20a [1, 2]
Environmental protection E06a [2] (2 yr); E13c, E31b [1]; E15a [1, 2]; E31c, E32d, E34, E38 [2]
Consumer protection C01 [1]; C02 [2] (2 wk)
Storage and disposal E01, E04 [2]; E26 [1]; E30a [1, 2]
Medical advice M03 [2]

339 paraffin oil (commodity substance)

An agent for the control of birds by egg treatment

Products

paraffin oil	various	OL

Uses

- Birds in ***miscellaneous pest control situations*** *(egg treatment)*

Approval information
- Approval for the use of paraffin oil as a commodity substance was granted on 25 April 1995 by Ministers under regulation 5 of the Control of Pesticides Regulations 1986
- Only to be used where a licence has been approved in accordance with Section 16(1) of the Wildlife and Countryside Act (1981)

Efficacy guidance
- Egg treatment should be undertaken as soon as clutch is complete
- Eggs should be treated by complete immersion in liquid paraffin

Restrictions
- Use to control eggs of birds covered by licences issued by the Agriculture and Environment Departments under Section 16(1) of the Wildlife and Countryside Act (1981)
- Treat eggs once only

Hazard classification and safety precautions

Operator protection A, C

340 paraquat

A non-selective, non-residual contact bipyridilium herbicide

See also *diquat + paraquat*
diuron + paraquat

Products

1	Agriguard Paraquat	AgriGuard	200 g/l	SL	09951
2	Dextrone X	Nomix Enviro	200 g/l	SL	00687
3	Fernpath Graminite	AgriGuard	200 g/l	SL	11646
4	Gramoxone 100	Syngenta	200 g/l	SL	10526

Uses
- Annual dicotyledons in ***all top fruit***, ***aubergines*** *(off-label - inter-row directed treatment)*, ***chives*** *(off-label)*, ***forest nursery beds***, ***grapevines***, ***grassland*** *(sward destruction/direct drilling)*, ***hardy ornamentals***, ***herbs (see appendix 6)*** *(off-label)*, ***lettuce*** *(off-label - inter-row directed treatment)*, ***lucerne*** *(off-label)*, ***parsley*** *(off-label)*, ***protected celery*** *(off-label - inter-row directed treatment)*, ***protected courgettes*** *(off-label - inter-row directed treatment)*, ***protected cucumbers*** *(off-label - inter-row directed treatment)*, ***protected gherkins*** *(off-label - inter-row directed treatment)*, ***protected lettuce*** *(off-label - inter-row directed treatment)*, ***protected marrows*** *(off-label - inter-row directed treatment)*, ***protected melons*** *(off-label - inter-row directed treatment)*, ***protected onions*** *(off-label - inter-row directed treatment)*, ***protected ornamentals*** *(off-label - inter-row directed treatment)*, ***protected peppers*** *(off-label - inter-row directed treatment)*, ***protected pumpkins*** *(off-label - inter-row directed treatment)*, ***protected squashes*** *(off-label - inter-row directed treatment)*, ***protected tomatoes*** *(off-label - inter-row directed treatment)*, ***rhubarb*** *(off-label)*, ***row crops*** *(stale seedbed/inter-row)* [4]; ***blackcurrants***, ***cultivated land/soil*** *(minimum cultivation)*, ***gooseberries***, ***hops***, ***potatoes***, ***raspberries***, ***stubbles***, ***sugar beet*** [1, 3, 4]; ***flower bulbs***, ***forest***, ***forestry transplants***, ***non-crop areas*** [1-4]; ***forest nursery beds*** *(stale seedbed)* [1-3]; ***orchards***, ***redcurrants***, ***row crops***, ***whitecurrants*** [1, 3]; ***woody ornamentals*** [2]
- Annual grasses in ***all top fruit***, ***aubergines*** *(off-label - inter-row directed treatment)*, ***chives*** *(off-label)*, ***grapevines***, ***hardy ornamentals***, ***herbs (see appendix 6)*** *(off-label)*, ***lettuce*** *(off-label - inter-row directed treatment)*, ***lucerne*** *(off-label)*, ***parsley*** *(off-label)*, ***protected celery*** *(off-label - inter-row directed treatment)*, ***protected courgettes*** *(off-label - inter-row directed treatment)*, ***protected cucumbers*** *(off-label - inter-row directed treatment)*, ***protected gherkins*** *(off-label - inter-row directed treatment)*, ***protected lettuce*** *(off-label - inter-row directed treatment)*, ***protected marrows*** *(off-label - inter-row directed treatment)*, ***protected melons*** *(off-label - inter-row directed treatment)*, ***protected onions*** *(off-label - inter-row directed treatment)*, ***protected ornamentals*** *(off-label - inter-row directed treatment)*, ***protected peppers*** *(off-label - inter-row directed treatment)*, ***protected pumpkins*** *(off-label - inter-row directed treatment)*, ***protected squashes*** *(off-label - inter-row directed treatment)*, ***protected tomatoes*** *(off-label - inter-row directed treatment)*, ***rhubarb*** *(off-label)*, ***row crops*** *(stale seedbed/inter-row)* [4]; ***blackcurrants***, ***cultivated land/soil*** *(minimum cultivation)*, ***gooseberries***, ***grassland*** *(sward destruction/direct drilling)*, ***hops***, ***potatoes***, ***raspberries***, ***stubbles***, ***sugar beet*** [1, 3, 4]; ***flower bulbs***, ***forest***, ***forest***

SEE SECTION 3 FOR PRODUCTS ALSO REGISTERED

nursery beds (stale seedbed), ***forestry transplants***, ***non-crop areas*** [1-4]; ***orchards***, ***redcurrants***, ***row crops***, ***whitecurrants*** [1, 3]; ***woody ornamentals*** [2]
- Barren brome in ***field crops*** *(stubble treatment)* [1, 3]; ***stubbles*** [4]
- Chemical stripping in ***hops*** [1, 3, 4]
- Creeping bent in ***all top fruit***, ***hardy ornamentals*** [4]; ***blackcurrants***, ***cultivated land/soil*** *(minimum cultivation)*, ***gooseberries***, ***grassland*** *(sward destruction/direct drilling)*, ***hops***, ***raspberries***, ***stubbles*** [1, 3, 4]; ***flower bulbs***, ***forest***, ***forestry transplants***, ***non-crop areas*** [1-4]; ***orchards***, ***redcurrants***, ***whitecurrants*** [1, 3]; ***woody ornamentals*** [2]
- Firebreak desiccation in ***forest*** [1-4]
- Green cover in ***field margins*** [4]; ***land temporarily removed from production*** [1, 3, 4]
- Perennial ryegrass in ***all top fruit*** [4]; ***blackcurrants***, ***gooseberries***, ***grassland*** *(sward destruction/direct drilling)*, ***hops***, ***raspberries*** [1, 3, 4]; ***non-crop areas*** [1-4]; ***orchards***, ***redcurrants***, ***whitecurrants*** [1, 3]
- Rough meadow grass in ***all top fruit*** [4]; ***blackcurrants***, ***gooseberries***, ***grassland*** *(sward destruction/direct drilling)*, ***hops***, ***raspberries*** [1, 3, 4]; ***non-crop areas*** [1-4]; ***orchards***, ***redcurrants***, ***whitecurrants*** [1, 3]
- Runner desiccation in ***strawberries*** [1, 3, 4]
- Volunteer cereals in ***cultivated land/soil*** *(minimum cultivation)*, ***potatoes***, ***stubbles***, ***sugar beet*** [1, 3, 4]; ***row crops*** [1, 3]; ***row crops*** *(stale seedbed/inter-row)* [4]
- Wild oats in ***stubbles*** [1, 3, 4]

Specific Off-Label Approvals (SOLAs)
- ***aubergines***, ***lettuce***, ***protected celery***, ***protected courgettes***, ***protected cucumbers***, ***protected gherkins***, ***protected lettuce***, ***protected marrows***, ***protected melons***, ***protected onions***, ***protected ornamentals***, ***protected peppers***, ***protected pumpkins***, ***protected squashes***, ***protected tomatoes*** *(inter-row directed treatment) (OLA 020225) Dec 2008* [4]
- ***chives***, ***herbs (see appendix 6)***, ***parsley*** *(OLA 021322) Apr 2005* [4]
- ***lucerne***, ***rhubarb*** *(OLA 020225) Dec 2008* [4]

Approval information
- Paraquat included in Annex I under EC Directive 91/414

Efficacy guidance
- Apply to green weeds preferably less than 15 cm high. Spray is rainfast after 10 min
- Addition of wetter recommended with lower application rates
- Spray in autumn to suppress couch when shoots have 2 leaves, repeat as necessary and plough after last treatment
- For direct-drilling land should be free of perennial weeds and protection against slugs should be provided. Allow 7-10 d before drilling into sprayed grass
- Best time for use in top fruit is Nov-Apr
- For forestry fire-break use apply in Jul-Aug and fire 7-10 d later
- For interrow use in row crops apply with guarded, no-drift sprayer
- Use as a carefully directed spray in blackcurrants, gooseberries, grapevines and other fruit crops, in raspberries only when dormant
- For stawberry runner control use guarded sprayer, not when flowers or fruit present
- Apply to bulbs pre-emergence (at least 3 d pre-emergence on sandy soils) or at end of season, provided no attached foliage and bulbs well covered (not on sandy soils)
- In forestry seedbed apply up to 3 d before seedling emergence

Restrictions
- Paraquat is subject to the Poisons Rules 1982 and the Poisons Act 1972. See notes in Section 5
- Maximum number of treamtments not stated for many situations but normally 1 per crop or yr; 3 per yr on leafy herbs
- Products in this profile are for professional use only
- Do not put in a food or drinks container
- Do not use on straw or other artificial growing media
- Do not use on potatoes under hot, dry conditions or on hops under drought conditions
- If using around glasshouses ensure vents and doors closed
- Only use clean water for mixing up spray
- Allow at least 4 h before cultivating, leave overnight if possible
- Observe label restrictions for maximum permitted concentration when spraying at low volume

Crop-specific information

- Latest use: before shoots 15 cm high and before 10% of shoots emerged for early potatoes, before 40% of shoots emerged for maincrop potatoes [1, 3, 4]; before 31 Mar for lucerne; see labels for other crops and situations
- HI mint, leafy herbs 8 wk
- Apply to lucerne in late Feb/early Mar when crop dormant

Following crops guidance

- Chemical is inactivated on contact with soil. Crops can be sown or planted soon after spraying on most soils but 3 d should be allowed on sandy or immature peat soils

Environmental safety

- Dangerous for the environment
- Very toxic to aquatic organisms
- Paraquat may be harmful to hares; stubbles must be sprayed early in the day [1, 3, 4]
- Harmful to fish or other aquatic life. Do not contaminate surface waters or ditches with chemical or used container [2]
- Keep in original container, tightly closed, in a safe place, under lock and key
- Harmful to livestock
- Keep all livestock out of treated areas for at least 24 h
- Do not contaminate surface waters or ditches with chemical or used container

Hazard classification and safety precautions

Hazard H02, H11
Risk phrases R20 [1, 3]; R21, R36, R37, R38 [1-4]; R22a [4]; R25 [1-3]; R48, R50, R58 [2, 4]
Operator protection A, C; U02a [1-3]; U04a, U20a [1, 3]; U04b, U20b [2, 4]; U04c [4]; U05a, U09a, U19a [1-4]; U08 [3]
Environmental protection E06c [1-4] (24 h); E11, E15a, E34 [1-4]; E31a [1, 3]; E31c [4]; E32a [2]; E32d, E38 [2, 4]
Storage and disposal E01, E04, E30b [1-4]; E26 [1, 3]
Medical advice M04 [1-4]; M05c [3, 4]

341 penconazole

A protectant conazole fungicide with antisporulant activity

Products

Topas	Syngenta	100 g/l	EC	09717

Uses

- Powdery mildew in ***apples***, ***blackcurrants***, ***hops***, ***ornamental trees***
- Rust in ***roses***
- Scab in ***ornamental trees***

Efficacy guidance

- Use as a protectant fungicide by treating at the earliest signs of disease
- Treat crops every 10-14 d (every 7-10 d in warm, humid weather) at first sign of infection or as a protective spray ensuring complete coverage. See label for details of timing
- Increase dose and volume with growth of hops but do not exceed 2000 l/ha. Little or no activity will be seen on established powdery mildew
- Antisporulant activity reduces development of secondary mildew in apples

Restrictions

- Maximum number of treatments 10 per yr for apples, 6 per yr for hops, 4 per yr for blackcurrants
- Check for varietal susceptibility in roses. Some defoliation may occur after repeat applications on Dearest

Crop-specific information

- HI apples, hops 14 d; blackcurrants 4 wk

Environmental safety

- Dangerous for the environment
- Toxic to aquatic organisms

Hazard classification and safety precautions

Hazard H04, H11

SEE SECTION 3 FOR PRODUCTS ALSO REGISTERED

Risk phrases R36, R51, R58
Operator protection A, C; U02a, U05a, U08, U20b
Environmental protection E15a, E31c, E32d, E38
Storage and disposal E01, E04, E26, E30a
Medical advice M05b

342 pencycuron

A non-systemic urea fungicide for use on seed potatoes

See also imazalil + pencycuron

Products

1	Agriguard Pencycuron	AgriGuard	12.5% w/w	DP	10445
2	Me2 Penny	Me2	12.5% w/w	DS	10369
3	Monceren DS	Bayer CropScience	12.5% w/w	DS	11292
4	Standon Pencycuron DP	Standon	12.5% w/w	DS	08774

Uses

- Black scurf in ***potatoes*** *(tuber treatment)*
- Stem canker in ***potatoes*** *(tuber treatment)*

Efficacy guidance

- Provides control of tuber-borne disease and gives some reduction of stem canker
- Apply to seed tubers in chitting trays, to bulk bins immediately before planting or in hopper at planting
- Apply in accordance with detailed guidelines in manufacturer's literature
- If rain interrupts planting cover tubers in hopper

Restrictions

- Maximum number of treatments 1 per batch of seed potatoes
- Treated tubers must be used only as seed and not for human or animal consumption
- Do not use on tubers previously treated with a dry powder seed treatment or hot water
- Use of suitable dust mask is mandatory when applying dust, filling the hopper or riding on planter

Crop-specific information

- Latest use: at planting

Environmental safety

- Do not use treated seed as food or feed
- Treated seed harmful to game and wildlife

Hazard classification and safety precautions

Operator protection A, F; U20b
Environmental protection E03, E15a, E32a
Storage and disposal E29a [1, 2, 4]; E30a [1-4]
Treated seed S01, S02, S03, S04a, S05, S06a

343 pendimethalin

A residual dinitroaniline herbicide for cereals and other crops

See also bentazone + pendimethalin
cyanazine + pendimethalin
flufenacet + pendimethalin
isoproturon + pendimethalin

Products

1	Agriguard Pendimethalin	AgriGuard	400 g/l	SC	10864
2	Bema	Makhteshim	40% w/w	WG	11756
3	Blazer	Headland	330 g/l	EC	11217
4	Claymore	BASF	400 g/l	SC	11779
5	Greencrop Estuary	Greencrop	400 g/l	SC	12095
6	Inter Pendimethalin	I T Agro	330 g/l	EC	09841
7	Landgold Pendimethalin 400	Landgold	400 g/l	SC	10288

Products – continued

8	Plinth	Interfarm	330 g/l	EC	11094
9	Stomp 400 SC	BASF	400 g/l	SC	11777

Uses

- Annual dicotyledons in ***apple orchards, blackcurrants, cherries, gooseberries, hops, pear orchards, plums, strawberries, sunflowers*** [1-6, 9]; ***blackberries, broccoli, brussels sprouts, cabbages, calabrese, cauliflowers, forage maize, leeks, loganberries, raspberries*** [1-5, 9]; ***bulb onions*** [4, 9]; ***carrots, parsley, parsnips*** [1-5, 8, 9]; ***combining peas, durum wheat, potatoes, spring barley, triticale, winter barley, winter rye, winter wheat*** [1-9]; ***evening primrose*** *(off-label)*, ***farm forestry*** *(off-label)*, ***herbs (see appendix 6)*** *(off-label)*, ***maize*** *(off-label)*, ***onion sets*** *(off-label)*, ***runner beans*** *(off-label)*, ***sweetcorn*** *(off-label)*, ***winter field beans*** *(off-label)* [2]; ***onions*** [1-3, 5]; ***rubus hybrids*** [1, 2, 4, 5, 9]; ***tayberries*** [3]
- Annual grasses in ***apple orchards, blackcurrants, cherries, gooseberries, hops, pear orchards, plums, strawberries, sunflowers*** [6]; ***combining peas, durum wheat, potatoes, spring barley, triticale, winter barley, winter rye, winter wheat*** [6, 7]
- Annual meadow grass in ***apple orchards, blackberries, blackcurrants, broccoli, brussels sprouts, cabbages, calabrese, cauliflowers, cherries, forage maize, gooseberries, hops, leeks, loganberries, pear orchards, plums, raspberries, strawberries, sunflowers*** [1-5, 9]; ***bulb onions*** [4, 9]; ***carrots, combining peas, durum wheat, parsley, parsnips, potatoes, spring barley, triticale, winter barley, winter rye, winter wheat*** [1-5, 8, 9]; ***onions*** [1-3, 5]; ***rubus hybrids*** [1, 2, 4, 5, 9]; ***tayberries*** [3]
- Blackgrass in ***apple orchards, blackberries, blackcurrants, broccoli, brussels sprouts, cabbages, calabrese, cauliflowers, cherries, gooseberries, hops, leeks, loganberries, pear orchards, plums, raspberries, strawberries, sunflowers*** [1-5, 9]; ***bulb onions*** [4, 9]; ***carrots, combining peas, durum wheat, parsley, parsnips, spring barley, triticale, winter barley, winter rye, winter wheat*** [1-5, 8, 9]; ***onions*** [1-3, 5]; ***rubus hybrids*** [1, 2, 4, 5, 9]; ***tayberries*** [3]
- Field pansy in ***daffodils*** *(off-label - for galanthamine production)* [9]
- Knotgrass in ***daffodils*** *(off-label - for galanthamine production)* [9]
- Rough meadow grass in ***apple orchards, blackberries, blackcurrants, broccoli, brussels sprouts, cabbages, calabrese, cauliflowers, cherries, forage maize, gooseberries, hops, leeks, loganberries, pear orchards, plums, raspberries, strawberries, sunflowers*** [1-5, 9]; ***bulb onions*** [4, 9]; ***carrots, combining peas, durum wheat, parsley, parsnips, potatoes, spring barley, triticale, winter barley, winter rye, winter wheat*** [1-5, 8, 9]; ***onions*** [1-3, 5]; ***rubus hybrids*** [1, 2, 4, 5, 9]; ***tayberries*** [3]
- Speedwells in ***daffodils*** *(off-label - for galanthamine production)* [9]
- Volunteer oilseed rape in ***apple orchards, blackberries, blackcurrants, broccoli, brussels sprouts, cabbages, calabrese, cauliflowers, cherries, forage maize, gooseberries, hops, leeks, loganberries, pear orchards, plums, raspberries, strawberries, sunflowers*** [1-5, 9]; ***bulb onions*** [4, 9]; ***carrots, combining peas, durum wheat, parsley, parsnips, potatoes, spring barley, triticale, winter barley, winter rye, winter wheat*** [1-5, 8, 9]; ***daffodils*** *(off-label - for galanthamine production)* [9]; ***onions*** [1-3, 5]; ***rubus hybrids*** [1, 2, 4, 5, 9]; ***tayberries*** [3]
- Wild oats in ***durum wheat, triticale, winter barley, winter rye, winter wheat*** [1-5, 8, 9]

Specific Off-Label Approvals (SOLAs)

- ***daffodils*** *(for galanthamine production) (OLA 041515) Dec 2008* [9]
- ***evening primrose*** *(OLA 041057) Dec 2008* [2]
- ***farm forestry*** *(OLA 041059) Dec 2008* [2]
- ***herbs (see appendix 6), onion sets, runner beans, sweetcorn, winter field beans*** *(OLA 041058) Dec 2008* [2]
- ***maize*** *(OLA 041060) Dec 2008* [2]

Approval information

- Pendimethalin included in Annex I under EC Directive 91/414

Efficacy guidance

- Apply as soon as possible after drilling. Weeds are controlled as they germinate and emerged weeds will not be controlled by use of the product alone
- For effective blackgrass control apply not more than 2 d after final cultivation and before weed seeds germinate
- Tank mixes with approved formulations of isoproturon or chlorotoluron recommended for improved pre- and post-emergence control of blackgrass in cereals

- Tank mixture with atrazine recommended for weed control in fodder maize
- Best results by application to fine firm, moist, clod-free seedbeds when rain follows treatment. Effectiveness reduced by prolonged dry weather after treatment
- Effectiveness reduced on soils with more than 6% organic matter. Do not use where organic matter exceeds 10%
- Any trash, ash or straw should be incorporated evenly during seedbed preparation
- Do not disturb soil after treatment
- On peas drilled after end Mar (mid-Apr in Scotland) tank-mix with cyanazine
- Apply to potatoes as soon as possible after planting and ridging in tank-mix with cyanazine or metribuzin but note that the use of cyanazine or metribuzin will restrict the varieties that may be treated

Restrictions

- Maximum number of treatments 1 per crop or yr
- Maximum total dose equivalent to one full dose treatment on most crops; 2 full dose treatments on leeks
- May be applied pre-emergence of cereal crops sown before 30 Nov provided seed covered by at least 32 mm soil, or post-emergence to early tillering stage (GS 23)
- Do not undersow treated crops
- Do not use on crops suffering stress due to disease, drought, waterlogging, poor seedbed conditions or chemical treatment or on soils where water may accumulate

Crop-specific information

- Latest use: pre-emergence for spring barley, carrots, lettuce, fodder maize, parsnips, parsley, sage, peas, runner beans, potatoes, onions and leeks; before transplanting for brassicas; before leaf sheaths erect for winter cereals; before bud burst for blackcurrants, gooseberries, cane fruit, hops; before flower trusses emerge for strawberries; 14 d after transplanting for leaf herbs
- Do not use on spring barley after end Mar (mid-Apr in Scotland on some labels) because dry conditions likely. Do not apply to dry seedbeds in spring unless rain imminent
- Apply to combining peas as soon as possible after sowing. Do not spray if plumule less than 13 mm below soil surface
- Apply to potatoes up to 7 d before first shoot emerges
- Apply to drilled crops as soon as possible after drilling but before crop and weed emergence
- Apply in top fruit, bush fruit and hops from autumn to early spring when crop dormant
- In cane fruit apply to weed free soil from autumn to early spring, immediately after planting new crops and after cutting out canes in established crops
- Apply in strawberries from autumn to early spring (not before Oct on newly planted bed). Do not apply pre-planting or during flower initiation period (post-harvest to mid-Sep)
- Apply pre-emergence in drilled onions or leeks, not on Sands, Very Light, organic or peaty soils or when heavy rain forecast
- Apply to brassicas after final plant-bed cultivation but before transplanting. Avoid unnecessary soil disturbance after application and take care not to introduce treated soil into the root zone when transplanting. Follow transplanting with specified post-planting treatments - see label
- Do not use on protected crops or in greenhouses

Following crops guidance

- Before ryegrass is drilled after a very dry season plough or cultivate to at least 15 cm. If treated spring crops are to be followed by crops other than cereals, plough or cultivate to at least 15 cm
- In the event of crop failure land must be ploughed or thoroughly cultivated to at least 15 cm. See label for minimum intervals that should elapse between treatment and sowing a range of replacement crops

Environmental safety

- Dangerous for the environment
- Very toxic to aquatic organisms
- Do not empty into drains [2]
- LERAP Category B [3]

Hazard classification and safety precautions

Hazard H03 [3, 6, 8]; H04 [2, 4, 9]; H08 [3]; H11 [2-9]
Risk phrases R20, R21, R38 [6]; R22b, R51 [3, 8]; R36 [6, 8]; R43 [2, 6]; R50 [2, 4-7, 9]; R58 [2-9]
Operator protection A [2, 5, 6, 8]; C [6, 8]; H [2, 8]; M [8]; U02a, U11 [6]; U05a [3, 4, 6, 9]; U08, U19a [1-7, 9]; U13, U20b [1, 2, 4-7, 9]; U14, U15 [8]

Environmental protection E13b [1]; E15a [2-9]; E16a [3]; E19b, E32a [2]; E31b [1, 3-9]; E32d [2-4, 9]; E32e [8]; E34 [3, 6, 8]; E38 [3, 4, 9]

Storage and disposal E01 [2-4, 6, 8, 9]; E04 [3, 4, 6, 9]; E26 [1, 3-7, 9]; E30a [1-9]

Medical advice M03 [3]; M05a [2]; M05b [3, 8]

344 pendimethalin + picolinafen

A post-emergence broad-spectrum herbicide mixture for winter cereals

Products

1	Pico Stomp	BASF	320:16 g/l	SC	11987
2	PicoMax	BASF	320:16 g/l	SC	10739
3	Picona	BASF	320:16 g/l	SC	10715
4	PicoPro	BASF	320:16 g/l	SC	10740

Uses

- Annual dicotyledons in ***winter barley, winter wheat***
- Annual meadow grass in ***winter barley, winter wheat***
- Chickweed in ***winter barley, winter wheat***
- Cleavers in ***winter barley, winter wheat***
- Rough meadow grass in ***winter barley, winter wheat***
- Speedwells in ***winter barley, winter wheat***

Approval information

- Pendimethalin and picolinafen included in Annex I under EC Directive 91/414

Efficacy guidance

- Best results obtained on crops growing in a fine, firm tilth and when rain falls within 7 d of application
- Loose or cloddy seed beds must be consolidated prior to application otherwise reduced weed control may occur
- Residual weed control may be reduced on soils with more than 6% organic matter or under prolonged dry conditions
- Always follow WRAG guidelines for preventing and managing herbicide resistant weeds. Section 5 for more information

Restrictions

- Maximum number of treatments 1 per crop
- Do not treat undersown cereals or those to be undersown
- Do not roll emerged crops before treatment nor autumn treated crops until the following spring
- Do not use on stony or gravelly soils, on soils that are waterlogged or prone to waterlogging, or on soils with more than 10% organic matter
- Do not treat crops under stress from any cause

Crop-specific information

- Latest use: before pseudo-stem erect stage (GS 30)
- Transient bleaching may occur after treatment but it does not lead to yield loss

Following crops guidance

- In the event of failure of a treated crop plough to at least 15 cm and allow at least 8 wk from the time of treatment before re-drilling either spring wheat or spring barley
- After a normally harvested treated crop there are no restrictions on following crops except that, in a very dry season, plough or cultivate to at least 15 cm before drilling rye-grass

Environmental safety

- Dangerous for the environment
- Very toxic to aquatic organisms
- Product binds strongly to soil minimising likelihood of movement into groundwater

Hazard classification and safety precautions

Hazard H04 [1]; H11 [1-4]

Risk phrases R43 [1]; R50, R58 [1-4]

Operator protection A, H; U02a, U05a, U20c

Environmental protection E15a, E31c, E32d, E34, E38

SEE SECTION 3 FOR PRODUCTS ALSO REGISTERED

Storage and disposal E01, E04, E26, E30a
Medical advice M03

345 pentanochlor

A contact anilide herbicide

See also *chlorpropham + pentanochlor*

Products

1	Croptex Bronze	Certis	400 g/l	EC	11329
2	Solan 40	Nufarm UK	400 g/l	EC	11897

Uses

- Annual dicotyledons in ***anemones***, ***carnations***, ***celeriac***, ***foxgloves***, ***freesias***, ***larkspur***, ***roses***, ***sweet williams***, ***wallflowers*** [2]; ***carrots***, ***celery***, ***chrysanthemums***, ***nursery stock***, ***parsley***, ***parsnips***, ***sweet peas*** [1, 2]
- Annual meadow grass in ***anemones***, ***carnations***, ***celeriac***, ***foxgloves***, ***freesias***, ***larkspur***, ***roses***, ***sweet williams***, ***wallflowers*** [2]; ***carrots***, ***celery***, ***chrysanthemums***, ***nursery stock***, ***parsley***, ***parsnips***, ***sweet peas*** [1, 2]

Approval information

- Products containing this active ingredient have been granted derogations for specified 'Essential Uses' for use until 31 December 2007. Sale and supply must cease by 30 June 2007 but growers have no guarantee that the products will continue to be available until then.
  For more information see 'The Review Programme' under 'Pesticide Legislation' in Section 5

Efficacy guidance

- Pentanochlor acts mainly through contact activity with some residual effect in moist soils. Best results achieved on young weed seedlings under warm, moist conditions
- Weeds most susceptible in cotyledon to 2-3 leaf stage although some important species controlled at later stages
- Effectiveness reduced by very cold weather or drought

Restrictions

- Maximum number of treatments 2 per crop for edible crops and sweet peas; 1 per yr for other horticultural ornamental crops
- Not recommended for flower crops on light sandy soils
- Do not treat crops suffering from stress, drought, mineral deficiency, pest attack or damage from previous herbicide applications
- Do not harvest crops for human or animal consumption for at least 28 d after last application
- These products must not be used on any crops other than those listed, including any extrapolations that would normally be permissible under the Long Term Arrangements for Extension of Use (see Section 5), except on herbs and other ornamentals

Crop-specific information

- HI 28 d for all edible crops
- Apply pre-emergence in anemones, freesias, foxgloves, larkspur

Following crops guidance

- In the event of failure of a treated crop only recommended crops should be re-planted
- Any crop may be planted 4-5 wk after application following mould-board ploughing or repeated cultivation

Environmental safety

- Dangerous to fish or other aquatic life. Do not contaminate surface waters or ditches with chemical or used container

Hazard classification and safety precautions

Hazard H03 [2]; H04 [1]
Risk phrases R21, R22a, R22b, R37, R40 [2]; R36 [1, 2]; R38 [1]
Operator protection A, C; U05a, U08, U11, U19a [1, 2]; U14, U15, U20b [1]; U20a [2]
Environmental protection E13b [1]; E15a [2]; E31b [1, 2]
Consumer protection C02 [1] (28 d)
Storage and disposal E01, E04, E30a
Medical advice M05b [2]

346 petroleum oil

An insecticidal and acaricidal hydrocarbon oil

Products

Certis Spraying Oil	Certis	710 g/l	EC	-

Uses

- Mealybugs in ***bush fruit, cane fruit, fruit trees, grapevines, hops, nursery stock, protected cucumbers, protected pot plants, protected tomatoes***
- Red spider mites in ***bush fruit, cane fruit, fruit trees, grapevines, hops, nursery stock, protected cucumbers, protected pot plants, protected tomatoes***
- Scale insects in ***bush fruit, cane fruit, fruit trees, grapevines, hops, nursery stock, protected cucumbers, protected pot plants, protected tomatoes***

Approval information

- Product not controlled by Control of Pesticides Regulations because it acts by physical means only

Efficacy guidance

- Spray at 1% (0.5% on tender foliage) to wet plants thoroughly, particularly the underside of leaves, and repeat as necessary
- Petroleum oil acts by blocking pest breathing pores and making leaf surfaces inhospitable to pests seeking to attack or attach to the leaf surface
- On outdoor crops apply when dormant or on plants with a known tolerance
- On plants of unknown sensitivity test first on a small scale
- Mixtures with certain pesticides may damage crop plants. If mixing, spray a few plants to test for tolerance before treating larger areas

Restrictions

- Do not mix with sulphur or iprodione products or use such mixtures within 28 d of treatment
- Do not treat protected crops in bright sunshine unless the glass is well shaded
- Test species tolerance before large scale treatment

Crop-specific information

- HI zero
- Treat grapevines before flowering

Hazard classification and safety precautions

Hazard H03
Risk phrases R22a, R36, R37, R38
Operator protection A, C; U05a, U08, U11, U19a
Environmental protection E31a, E32d, E34
Storage and disposal E01, E04, E30a
Medical advice M03

347 phenmedipham

A contact carbamate herbicide for beet crops and strawberries

See also *desmedipham + ethofumesate + phenmedipham*
desmedipham + phenmedipham
ethofumesate + lenacil + phenmedipham
ethofumesate + phenmedipham
lenacil + phenmedipham

Products

1	Beetup	United Phosphorus	114 g/l	EC	07520
2	Betanal Flow	Bayer CropScience	160 g/l	SE	10997
3	Cleancrop Phenmedipham	United Phosphorus	114 g/l	EC	11586
4	Crotale	Sipcam	471 g/l	SC	11593
5	Dancer FL	Sipcam	160 g/l	SC	10048
6	Herbasan Flow	Nufarm UK	160 g/l	SC	11448
7	Mandolin Flow	Nufarm UK	160 g/l	SE	11162
8	Tripart Beta	Tripart	118 g/l	EC	03111

Uses

- Annual dicotyledons in ***fodder beet, mangels, sugar beet*** [1-8]; ***red beet*** [1-7]; ***strawberries*** [1, 3]

Efficacy guidance

- Best results achieved by application to young seedling weeds, preferably cotyledon stage, under good growing conditions when low doses are effective
- 2-3 repeat applications at 7-10 d intervals using a low dose are recommended on mineral soils, 3-5 applications may be needed on organic soils
- Addition of adjuvant oil may improve effectiveness on some weeds
- Various tank-mixtures with other beet herbicides recommended. See label for details
- Use of certain pre-emergence herbicides is recommended in combination with post-emergence treatment. See label for details

Restrictions

- Maximum number of treatments and maximum total dose varies with crop and product used. See label for details
- At high temperatures (above 21°C) reduce rate and spray after 5 pm
- Do not apply immediately after frost or if frost expected
- Do not spray wet foliage or if rain imminent
- Do not spray crops stressed by wind damage, nutrient deficiency, pest or disease attack etc. Do not roll or harrow for 7 d before or after treatment
- Do not use on strawberries under cloches or polythene tunnels
- Do not use after 31 Jul in yr of harvest [2, 4-7]

Crop-specific information

- Latest use: before crop leaves meet between rows for beet crops; before flowering for strawberries
- Apply to beet crops at any stage as low dose/low volume spray or from fully developed cotyledon stage with full rate. Apply to red beet after fully developed cotyledon stage
- Apply to strawberries at any time when weeds in susceptible stage, except in period from start of flowering to picking

Environmental safety

- Dangerous for the environment
- Very toxic to aquatic organisms

Hazard classification and safety precautions

Hazard H03 [1, 3, 8]; H04 [4]; H11 [1-3, 6-8]

Risk phrases R21, R40, R51 [1, 3]; R22a, R36, R37 [1, 3, 8]; R38 [8]; R43 [4]; R50 [2, 6, 7]; R58 [1-3, 6, 7]

Operator protection A [1, 2, 4, 7, 8]; C [1, 8]; H [4]; P [8]; U05a [1, 3, 4, 8]; U08 [1, 3, 8]; U09a [5]; U13 [1, 3]; U19a [1, 3, 5, 8]; U20b [1-8]

Environmental protection E13b, E31c [5]; E13c [2, 4]; E15a [1, 3, 6-8]; E31b [1-4, 6-8]; E32d, E38 [2, 6, 7]; E34 [1, 3, 8]

Storage and disposal E01, E04 [1, 3, 4, 8]; E26 [1-3, 5, 7, 8]; E30a [1-8]

Medical advice M03 [1, 3, 8]

348 d-phenothrin

A non-systemic pyrethroid insecticide available only in mixtures

349 d-phenothrin + tetramethrin

A pyrethroid insecticide mixture for control of flying insects

Products

1	Killgerm ULV 500	Killgerm	36.8:18.4 g/l	UL	H4647
2	Sorex Super Fly Spray	Sorex	0.10:0.25% w/w	AE	H6297

Uses

- Flies in ***agricultural premises*** [1, 2]
- Grain storage mite in ***grain stores*** [1]

- Mosquitoes in ***agricultural premises*** [1, 2]
- Wasps in ***agricultural premises*** [1, 2]

Approval information

- Product approved for ULV application. See label for details [1]

Efficacy guidance

- Close doors and windows and spray in all directions for 3-5 sec. Keep room closed for at least 10 min
- Product synergised with piperonyl butoxide [2]

Restrictions

- For use only by profesional operators
- Do not use space sprays containing pyrethrins or pyrethroid more than once per week in intensive or controlled environment animal houses in order to avoid development of resistance. If necessary, use a different control method or product [1]

Crop-specific information

- May be used in the presence of poultry and livestock

Environmental safety

- Dangerous for the environment
- Toxic to aquatic organisms
- Do not apply directly to livestock/poultry
- Remove exposed milk and collect eggs before application. Protect milk machinery and containers from contamination

Hazard classification and safety precautions

Hazard H03, H11 [1]; H06 [2]
Risk phrases R22b, R51, R58 [1]
Operator protection A, C, D, E, H [1]; U02b, U09b, U20a, U20b [1]; U14, U19a [1, 2]; U20c [2]
Environmental protection E02a, E13a [2]; E05a [1, 2]; E15a, E31a, E32d, E38 [1]
Consumer protection C04, C05, C10 [2]; C06, C09, C11, C12 [1]; C07, C08 [1, 2]
Storage and disposal E01, E30a [1, 2]; E26, E27 [1]
Medical advice M03 [2]; M05b [1]

350 picloram

A persistent, translocated pyridine carboxylic acid herbicide for non-crop areas

See also 2,4-D + picloram
clopyralid + picloram

Products

Tordon 22K	Nomix Enviro	240 g/l	SL	05790

Uses

- Annual dicotyledons in ***land not intended for cropping***
- Bracken in ***land not intended for cropping***
- Japanese knotweed in ***land not intended for cropping***
- Perennial dicotyledons in ***land not intended for cropping***
- Woody weeds in ***land not intended for cropping***

Efficacy guidance

- May be applied at any time of year. Best results achieved by application as foliage spray in late winter to early spring
- For bracken control apply 2-4 wk before frond emergence
- Clovers are highly sensitive and eliminated at very low doses
- Persists in soil for up to 2 yr

Restrictions

- Maximum number of treatments 1 per yr on land not intended for cropping
- Do not apply around desirable trees or shrubs where roots may absorb chemical
- Do not apply on slopes where chemical may be leached onto areas of desirable plants

Environmental safety

- Harmful to fish or other aquatic life. Do not contaminate surface waters or ditches with chemical or used container
- Keep livestock out of treated areas for at least 2 wk and until foliage of any poisonous weeds such as ragwort has died and become unpalatable

Hazard classification and safety precautions

Hazard H04
Risk phrases R43
Operator protection A, H; U05a, U08, U14, U20b
Environmental protection E07a, E13c, E31b
Storage and disposal E01, E04, E30a

351 picolinafen

An aryloxypicolinamide herbicide for cereals available only in mixtures

See also pendimethalin + picolinafen

352 picoxystrobin

A broad spectrum strobilurin fungicide for cereals

See also cyprodinil + picoxystrobin

Products

Acanto	Syngenta	250 g/l	SC	10978

Uses

- Brown rust in ***spring barley***, ***spring wheat***, ***winter barley***, ***winter wheat***
- Crown rust in ***spring oats***, ***winter oats***
- Eyespot in ***spring wheat*** *(reduction)*, ***winter wheat*** *(reduction)*
- Late ear diseases in ***spring wheat***, ***winter wheat***
- Net blotch in ***spring barley***, ***winter barley***
- Powdery mildew in ***spring barley***, ***spring oats***, ***winter barley***, ***winter oats***
- Rhynchosporium in ***spring barley***, ***winter barley***
- Septoria diseases in ***spring wheat***, ***winter wheat***
- Yellow rust in ***spring wheat***, ***winter wheat***

Efficacy guidance

- Best results achieved from protectant treatments made before disease establishes in the crop
- Use of tank mixtures is recommended where disease has become established at the time of treatment
- Persistence of yellow rust control is less than for other foliar diseases but can be improved by use of an appropriate tank mixture
- Eyespot control is not reliable and an appropriate tank mix should be used where crops are at risk
- Picoxystrobin is a member of the QoI cross resistance group. Product should be used preventatively and not relied on for its curative potential
- Use product as part of an Integrated Crop Management strategy incorporating other methods of control, including where appropriate other fungicides with a different mode of action. Do not apply more than two foliar applications of QoI containing products to any cereal crop
- There is a significant risk of widespread resistance occurring in *Septoria tritici* populations in UK. Failure to follow resistance management action may result in reduced levels of disease control
- On cereal crops product must always be used in mixture with another product, recommended for control of the same target disease, that contains a fungicide from a different cross resistance group and is applied at a dose that will give robust control
- Strains of barley powdery mildew resistant to QoIs are common in the UK

Restrictions

- Maximum number of treatments 2 per crop per yr

Crop-specific information

- Latest use: grain watery ripe (GS 71)

Environmental safety

- Dangerous for the environment
- Very toxic to aquatic organisms

Hazard classification and safety precautions

Hazard H11
Risk phrases R50, R58
Operator protection U05a, U09a, U14, U15, U20a
Environmental protection E15a, E31c, E32d, E38
Storage and disposal E01, E04, E26, E30a

353 pirimicarb

A carbamate insecticide for aphid control

See also *lambda-cyhalothrin + pirimicarb*

Products

1	Agriguard Pirimicarb	AgriGuard	50% w/w	WG	09620
2	Aphox	Syngenta	50% w/w	WG	10515
3	Greencrop Glenroe	Greencrop	50% w/w	WG	09903
4	Landgold Pirimicarb 50	Landgold	50% w/w	WG	09018
5	Phantom	Syngenta	50% w/w	WG	11954
6	Standon Pirimicarb 50	Standon	50% w/w	WG	08878

Uses

- Aphids in ***apples***, ***broad beans***, ***broccoli***, ***brussels sprouts***, ***cabbages***, ***calabrese***, ***cauliflowers***, ***durum wheat***, ***pears***, ***peas***, ***potatoes***, ***spring barley***, ***spring field beans***, ***spring oats***, ***spring rye***, ***spring wheat***, ***strawberries***, ***sugar beet***, ***swedes***, ***triticale***, ***turnips***, ***winter barley***, ***winter field beans***, ***winter oats***, ***winter rye***, ***winter wheat*** [1-6]; ***blackcurrants***, ***carrots***, ***chinese cabbage***, ***collards***, ***dwarf beans***, ***gooseberries***, ***maize***, ***parsnips***, ***raspberries***, ***redcurrants***, ***runner beans***, ***spring oilseed rape***, ***sweetcorn***, ***winter oilseed rape*** [1-3, 5, 6]; ***carnations***, ***celeriac*** *(off-label)*, ***celery*** *(off-label)*, ***chicory*** *(off-label - for forcing)*, ***chrysanthemums***, ***cinerarias***, ***courgettes*** *(off-label)*, ***cyclamen***, ***endives*** *(off-label)*, ***fennel*** *(off-label)*, ***gherkins*** *(off-label)*, ***honesty*** *(off-label)*, ***horseradish*** *(off-label)*, ***leaf spinach*** *(off-label)*, ***marrows*** *(off-label)*, ***parsley*** *(off-label)*, ***parsley root*** *(off-label)*, ***protected courgettes*** *(off-label)*, ***protected endives*** *(off-label)*, ***protected gherkins*** *(off-label)*, ***protected radishes*** *(off-label)*, ***protected spinach*** *(off-label)*, ***protected spinach beet*** *(off-label)*, ***radishes*** *(off-label)*, ***red beet*** *(off-label)*, ***roses***, ***salad brassicas***, ***salad brassicas*** *(off-label - for baby leaf production)*, ***spinach beet*** *(off-label)*, ***sweetcorn*** *(off-label)* [2]; ***cherries*** [1-3, 5]; ***cucumbers***, ***forest nurseries***, ***lettuce*** *(outdoor crops)*, ***outdoor tomatoes***, ***peppers***, ***protected carnations***, ***protected chrysanthemums***, ***protected cinerarias***, ***protected cyclamen***, ***protected lettuce***, ***protected roses***, ***protected tomatoes*** [2, 3, 5]; ***grassland*** [1]; ***kale*** [1, 2, 4-6]; ***ornamental specimens*** [3, 5]
- Blackfly in ***cherries*** [6]
- Carrot willow aphid in ***celery*** [1]
- Leaf curling plum aphid in ***plums*** *(off-label)* [2]
- Mealy plum aphid in ***plums*** *(off-label)* [2]
- Thrips in ***endives*** *(off-label)*, ***parsley*** *(off-label)*, ***protected endives*** *(off-label)* [2]

Specific Off-Label Approvals (SOLAs)

- ***celeriac***, ***fennel***, ***red beet*** *(OLA 011292) Dec 2008* [2]
- ***celery*** *(OLA 040680) Dec 2008* [2]
- ***chicory*** *(for forcing) (OLA 011283) Dec 2008* [2]
- ***courgettes***, ***gherkins***, ***horseradish***, ***leaf spinach***, ***marrows***, ***parsley root***, ***protected radishes***, ***protected spinach***, ***protected spinach beet***, ***radishes***, ***spinach beet***, ***sweetcorn*** *(OLA 011298) Dec 2008* [2]
- ***endives***, ***parsley***, ***protected endives*** *(OLA 011286) Dec 2008* [2]
- ***honesty*** *(OLA 011289) Dec 2008* [2]
- ***plums*** *(OLA 012687) Dec 2008* [2]
- ***protected courgettes***, ***protected gherkins*** *(OLA 011301) Dec 2008* [2]
- ***salad brassicas*** *(for baby leaf production) (OLA 011295) Dec 2008* [2]

Approval information

- Pirimicarb has been included in a comprehensive review of anticholinesterase compounds and considered by the ACP. As a result approvals for some SOLAs, and for uses on grassland and celery were revoked and approvals expire in Dec 2003
- Approved for aerial application on cereals [1-3, 5]. See notes in Section 5

Efficacy guidance

- Chemical has contact, fumigant and translaminar activity
- Best results achieved under warm, calm conditions when plants not wilting and spray does not dry too rapidly. Little vapour activity at temperatures below 15°C
- Apply as soon as aphids seen or warning issued and repeat as necessary
- Addition of non-ionic wetter recommended for use on brassicas
- On cucumbers and tomatoes a root drench is preferable to spraying when using predators in an integrated control programme
- Where aphids resistant to pirimicarb occur control is unlikely to be satisfactory

Restrictions

- This product contains an anticholinesterase carbamate compound. Do not use if under medical advice not to work with such compounds
- Maximum number of treatments not specified in some cases but normally 2-6 depending on crop
- When treating ornamentals check safety by treating a small number of plants first
- Spray equipment must only be used where the operator's normal working position is within a closed cab on a tractor or self-propelled sprayer when making air-assisted applications to apples or other top fruit

Crop-specific information

- Latest use in accordance with harvest intervals below
- HI oilseed rape, cereals, maize, sweetcorn, lettuce under glass 14 d (sweetcorn off-label 3 d); grassland, chicory, plums 7 d; cucumbers, tomatoes and peppers under glass 2 d; protected courgettes and gherkins 24 h; other edible crops 3 d; flowers and ornamentals zero

Environmental safety

- Dangerous for the environment
- Very toxic to aquatic organisms
- Chemical has little effect on bees, ladybirds and other insects and is suitable for use in integrated control programmes on apples and pears
- Keep all livestock out of treated areas for at least 7 d. Bury or remove spillages

Hazard classification and safety precautions

Hazard H02, H11 [2-6]; H03 [1]
Risk phrases R20 [2, 5]; R22a [1]; R25, R36, R50, R58 [2-6]
Operator protection A, C, D, H, J, M; U05a, U08, U19a [1-6]; U20a [3, 5]; U20b [1, 2, 4, 6]
Environmental protection E06c [1-6] (7 d); E13b [1]; E15a [2-6]; E32a, E34 [1-6]; E32d, E38 [2, 5]
Storage and disposal E01, E04, E30a [1-6]; E26, E29a [3]
Medical advice M02 [1-6]; M03 [1]; M04 [2-6]

354 pirimiphos-methyl

A contact, fumigant and translaminar organophosphorus insecticide

Products

1	Actellic D	Syngenta	250 g/l	EC	10509
2	Actellic Smoke Generator No. 20	Syngenta	22.5% w/w	FU	10540
3	Fumite Pirimiphos Methyl Smoke	Certis	22.5% w/w	FU	00941

Uses

- Ants in ***protected cucumbers, protected ornamentals, protected peppers, protected tomatoes*** [3]
- Aphids in ***protected cucumbers, protected ornamentals, protected peppers, protected tomatoes*** [3]
- Capsids in ***protected cucumbers, protected ornamentals, protected peppers, protected tomatoes*** [3]

- Earwigs in ***protected cucumbers, protected ornamentals, protected peppers, protected tomatoes*** [3]
- Flour beetle in ***stored grain*** [1]
- Flour moth in ***stored grain*** [1]
- Grain beetle in ***stored grain*** [1]
- Grain storage mite in ***stored grain*** [1]
- Grain storage pests in ***grain stores*** [1, 2]
- Grain weevil in ***stored grain*** [1]
- Leaf miner in ***protected cucumbers, protected ornamentals, protected peppers, protected tomatoes*** [3]
- Red spider mites in ***protected cucumbers, protected ornamentals, protected peppers, protected tomatoes*** [3]
- Sawflies in ***protected cucumbers, protected ornamentals, protected peppers, protected tomatoes*** [3]
- Thrips in ***protected cucumbers, protected ornamentals, protected peppers, protected tomatoes*** [3]
- Warehouse moth in ***stored grain*** [1]
- Whitefly in ***protected cucumbers, protected ornamentals, protected peppers, protected tomatoes*** [3]

Efficacy guidance

- Chemical acts rapidly and has short persistence in plants but persists for long periods on inert surfaces
- Best results for protection of stored grain achieved by cleaning store thoroughly before use and employing a combination of pre-harvest and grain/seed treatments
- Best results for admixture treatment obtained when grain stored at 15% moisture or less. Dry and cool moist grain coming into store but then treat as soon as possible, ideally as it is loaded
- Surface admixture can be highly effective on localised surface infestations but should not be relied on for long-term control unless application can be made to the full depth of the infestation
- For control of whitefly and other glasshouse pests apply as smoke or by thermal fogging. See label for details of techniques and suitable machines [3]
- Where insect pests resistant to pirimiphos-methyl occur control is unlikely to be satisfactory

Restrictions

- This product contains an organophosphorus anticholinesterase compound. Do not use if under medical advice not to work with such compounds
- Maximum number of treatments 2 per grain store; 1 per batch of stored grain
- Do not mix with grain store disinfectants because of risk of chemical interaction [1]
- Do not apply surface admixture or complete admixture using hand held equipment [1]
- Do not apply treated seed from the air [1]
- Do not fumigate glasshouses in bright sunshine or when foliage is wet or roots dry
- Do not fumigate young seedlings or plants being hardened off
- Do not fog open flowers of ornamentals without first consulting firm
- Do not fog mushrooms when wet as slight spotting may occur

Crop-specific information

- Latest use: before storing grain
- HI for edible crops zero [3]
- Disinfect empty grain stores by spraying surfaces and/or fumigation and treat grain by full or surface admixture. Treat well before harvest in late spring or early summer and repeat 6 wk later or just before harvest if heavily infested. See label for details of treatment and suitable application machinery
- Treatment volumes on structural surfaces should be adjusted according to surface porosity. See label for guidelines
- Treat inaccessible areas with smoke generating product used in conjunction with spray treatment of remainder of store

Environmental safety

- Dangerous for the environment
- Toxic to aquatic organisms
- Ventilate fumigated or fogged spaces thoroughly before re-entry

SEE SECTION 3 FOR PRODUCTS ALSO REGISTERED

- Unprotected persons must be kept out of fumigated areas within 3 h of ignition and for 4 h after treatment
- Do not harvest for human or animal consumption for at least 5 d after last application
- Extremely dangerous to bees. Do not apply to crops in flower or to those in which bees are actively foraging. Do not apply when flowering weeds are present [3]
- Wildlife must be excluded from buildings during treatment [1]
- Grain treated by admixture as specified may be consumed by humans and livestock

Hazard classification and safety precautions

Hazard H03, H11 [1-3]; H07 [2, 3]; H08 [1]

Risk phrases R20, R51, R58 [1-3]; R22b, R38 [1]; R36, R37 [1, 3]

Operator protection A, C [1]; D, H [1, 2]; U05a [1-3]; U10, U20b [2]; U14, U19a [2, 3]; U20a [3]

Environmental protection E02a [2, 3] (4 h); E12c [3]; E15a [1-3]; E31c, E32d, E38 [1]; E32a, E34 [2, 3]

Consumer protection C02 [3] (5 d); C11 [2]; C12 [2, 3]

Storage and disposal E01, E04, E30a

Treated seed S05 [1]

Medical advice M01 [1-3]; M03 [2]; M05b [1]

355 prochloraz

A broad-spectrum protectant and eradicant conazole fungicide

See also fluquinconazole + prochloraz

Products

	Product	Company	Content	Form	Reg. no.
1	Alpha Prochloraz 40 EC	Makhteshim	400 g/l	EC	11002
2	Barclay Eyetak 40	Barclay	400 g/l	EC	11352
3	Mirage 40 EC	Makhteshim	400 g/l	EC	06770
4	Panache 40	DAPT	400 g/l	EC	10632
5	Poraz	BASF	450 g/l	EC	11701
6	Prelude 20LF	Agrichem	200 g/l	LS	04371
7	Prospero	Makhteshim	400 g/l	EC	11931
8	Scotts Octave	Scotts	46% w/w	WP	09275

Uses

- Alternaria in ***spring oilseed rape*** [1, 3]; ***winter oilseed rape*** [1, 3, 5, 7]
- Eyespot in ***spring wheat*** [1-4]; ***winter barley, winter wheat*** [1-5, 7]; ***winter rye*** [1, 3, 5, 7]
- Fungus diseases in ***container-grown ornamentals, hardy ornamentals, woody ornamentals*** [8]
- Glume blotch in ***spring wheat*** [1, 3]; ***winter wheat*** [1, 3, 5, 7]
- Grey mould in ***spring oilseed rape*** [1, 3]; ***winter oilseed rape*** [1, 3, 5, 7]
- Light leaf spot in ***spring oilseed rape*** [1, 3, 4]; ***winter oilseed rape*** [1, 3-5, 7]
- Net blotch in ***spring barley, winter barley*** [1, 3, 5, 7]
- Phoma in ***spring oilseed rape*** [1, 3, 4]; ***winter oilseed rape*** [1, 3-5, 7]
- Powdery mildew in ***spring barley, spring wheat, winter barley, winter wheat*** [5, 7]; ***winter rye*** [1, 3, 5, 7]
- Rhynchosporium in ***spring barley, winter barley, winter rye*** [1, 3, 5, 7]
- Ring spot in ***chives*** *(off-label)*, ***herbs (see appendix 6)*** *(off-label)*, ***lettuce*** *(off-label)*, ***parsley*** *(off-label)*, ***protected chives*** *(off-label)*, ***protected herbs (see appendix 6)*** *(off-label)*, ***protected lettuce*** *(off-label)*, ***protected parsley*** *(off-label)* [3]
- Sclerotinia stem rot in ***spring oilseed rape*** [1, 3]; ***winter oilseed rape*** [1, 3, 5, 7]
- Seed-borne diseases in ***flax*** *(seed treatment)*, ***linseed*** *(seed treatment)* [6]
- Septoria leaf spot in ***spring wheat*** [1, 3]; ***winter rye*** [5, 7]; ***winter wheat*** [1, 3, 5, 7]
- White leaf spot in ***spring oilseed rape*** [1, 3]; ***winter oilseed rape*** [1, 3, 5, 7]

Specific Off-Label Approvals (SOLAs)

- ***chives, herbs (see appendix 6), lettuce, parsley, protected chives, protected herbs (see appendix 6), protected lettuce, protected parsley*** *(OLA 032623) Dec 2008* [3]

Efficacy guidance

- Spray cereals at first signs of disease. Protection of winter crops through season usually requires at least 2 treatments. See label for details of rates and timing. Treatment active against strains of eyespot resistant to benzimidazole fungicides

- Tank mixes with other fungicides recommended to improve control of rusts in wheat and barley. See label for details
- A period of at least 3 h without rain should follow spraying
- Can be used through most seed treatment machines if good even seed coverage is obtained. Check drill calibration before drilling treated seed. Use inert seed flow agent supplied by manufacturer [6]
- Use methylated spirits, rather than water, to clean residual material from seed treatment machinery, then carry out final rinse with water and detergent [6]
- Apply as drench against soil diseases, as a spray against aerial diseases, as a dip for cuttings, or as a drench at propagation. Spray applications may be repeated at 10-14 d intervals. Under mist propagation use 7 d intervals [8]

Restrictions

- Maximum number of treatments 1 per batch for flax, linseed [6]; varies with dose and crop for other uses - see labels for details
- Do not treat linseed varieties Linda, Bolas, Karen, Laura, Mikael, Norlin, Moonraker, Abbey, Agriace [6]
- Do not use treated seed as food or feed [6]

Crop-specific information

- Latest use: milky ripe stage (GS 77) for cereals; before drilling flax or linseed [6]
- HI 6 wk for oilseed rape and cereals; 21 d for herbs
- Application with other fungicides may cause cereal crop scorch [1-5, 7]

Environmental safety

- Dangerous for the environment
- Very toxic to aquatic organisms
- Treated seed harmful to game and wildlife. Product contains red dye for easy identification of treated seed [6]

Hazard classification and safety precautions

Hazard H03 [5, 6]; H04 [1-4, 7, 8]; H08 [2, 4]; H11 [1-3, 5-8]
Risk phrases R22a [5, 6]; R36 [1-8]; R38 [1-5, 7, 8]; R43 [2, 4]; R50 [2, 5, 8]; R51 [1, 3, 6, 7]; R58 [1-3, 5-8]
Operator protection A [1-8]; C [1-6, 8]; U05a, U20b [1-8]; U08 [1-7]; U09a [8]; U11 [1, 3, 6, 7]; U14 [1, 3, 5, 7]; U15 [1, 3, 7]; U19a [2, 8]
Environmental protection E13b [4]; E15a [1-3, 5, 7, 8]; E31a [6]; E31b [1-5, 7, 8]; E32d [5, 6, 8]; E34 [5]; E38 [5, 6]
Storage and disposal E01, E30a [1-8]; E04 [1-5, 7, 8]; E26 [1-4, 6, 7]
Treated seed S01, S02, S03, S04b, S05, S06a, S07 [6]
Medical advice M03, M05a [5]; M05b [1, 3, 7]

356 prochloraz + propiconazole

A broad spectrum fungicide mixture for wheat, barley and oilseed rape

Products

1	Bumper P	Makhteshim	400:90 g/l	EC	08548
2	Greencrop Twinstar	Greencrop	400:90 g/l	EC	09516

Uses

- Eyespot in ***spring barley***, ***spring wheat*** [1]; ***winter barley***, ***winter wheat*** [1, 2]
- Light leaf spot in ***spring oilseed rape***, ***winter oilseed rape*** [1]
- Phoma in ***spring oilseed rape***, ***winter oilseed rape*** [1]
- Rhynchosporium in ***spring barley*** [1]; ***winter barley*** [1, 2]
- Septoria leaf spot in ***spring wheat*** [1]; ***winter wheat*** [1, 2]

Approval information

- Propiconazole included in Annex I under EC Directive 91/414

Efficacy guidance

- Best results obtained from treatment when disease is active but not well established
- Treat wheat normally from flag leaf ligule just visible stage (GS 39) but earlier if there is a high risk of Septoria

SEE SECTION 3 FOR PRODUCTS ALSO REGISTERED

- Treat barley from the first node detectable stage (GS 31)
- If disease pressure persists a second application may be necessary

Restrictions

- Maximum number of treatments 2 per crop

Crop-specific information

- Latest use: before grain watery ripe (GS 71) for cereals; before most seeds green for winter oilseed rape

Environmental safety

- Dangerous for the environment
- Very toxic to aquatic organisms

Hazard classification and safety precautions

Hazard H04, H11
Risk phrases R36, R58 [1, 2]; R50 [2]; R51 [1]
Operator protection A, C; U05a, U08, U20b [1, 2]; U11, U15 [1]
Environmental protection E15a, E31b
Storage and disposal E01, E04, E26, E30a
Medical advice M05a [1]

357 prochloraz + tebuconazole

A broad spectrum systemic fungicide mixture for cereals and oilseed rape

Products

Agate	Makhteshim	267:133 g/l	EC	12016

Uses

- Brown rust in ***spring barley, spring rye, spring wheat, winter barley, winter rye, winter wheat***
- Eyespot in ***spring barley, spring rye, spring wheat, winter barley, winter rye, winter wheat***
- Glume blotch in ***spring wheat, winter wheat***
- Late ear diseases in ***spring wheat, winter wheat***
- Net blotch in ***spring barley, winter barley***
- Powdery mildew in ***spring barley, spring rye, spring wheat, winter barley, winter rye, winter wheat***
- Rhynchosporium in ***spring barley, spring rye, winter barley, winter rye***
- Sclerotinia stem rot in ***spring oilseed rape, winter oilseed rape***
- Septoria leaf spot in ***spring wheat, winter wheat***
- Yellow rust in ***spring barley, spring rye, spring wheat, winter barley, winter rye, winter wheat***

Efficacy guidance

- Best results achieved from applications at an early stage of disease development before infection spreads to new crop growth
- Adequate protection of winter cereals will usually require a programme of at least two treatments
- Optimum application timing is normally when disease first seen but varies with main target disease - see label
- To minimise possibility of development of resistance repeated applications should not be made against the same pathogen on cereals. See notes on Resistance Management in Section 5

Restrictions

- Maximum total dose equivalent to two full dose treatments

Crop-specific information

- Latest use: before grain milky ripe (GS 73) for cereals; most seeds green for oilseed rape
- HI 6 wk
- Occasionally transient leaf speckling may occur after treating wheat. Yield responses should not be affected

Environmental safety

- Dangerous for the environment
- Very toxic to aquatic organisms

Hazard classification and safety precautions

Hazard H03, H11

Risk phrases R22a, R36, R43, R50, R58

Operator protection A, C, H; U05a, U09b, U14, U20a

Environmental protection E13b, E31c, E32d, E34, E38

Storage and disposal E01, E04, E26, E30a

Medical advice M03

358 prochloraz + triticonazole

A broad spectrum fungicide seed treatment mixture for winter cereals

Products

Kinto	BASF	60:20 g/l	FS	12038

Uses

- Bunt in ***winter wheat*** *(seed treatment)*
- Covered smut in ***winter barley*** *(seed treatment)*
- Fusarium foot rot and seedling blight in ***winter barley*** *(seed treatment)*, ***winter wheat*** *(seed treatment)*
- Leaf stripe in ***winter barley*** *(seed treatment)*
- Loose smut in ***winter barley*** *(seed treatment)*, ***winter wheat*** *(seed treatment)*
- Septoria seedling blight in ***winter wheat*** *(seed treatment)*

Efficacy guidance

- Apply undiluted through conventional seed treatment machine
- Calibrate seed drill with treated seed before sowing
- Bag treated seed immediately and keep in a dry, draught free store
- Ensure good even coverage of seed to achieve best results
- Treatment may cause some slight delay in crop emergence

Restrictions

- Maximum number of treatments 1 per batch of seed
- Do not treat seed with moisture content above 16% and do not allow moisture content of treated seed to exceed 16%
- Do not treat cracked, split or sprouted seed
- Do not handle treated seed unnecessarily
- Do not use treated seed as food or feed

Crop-specific information

- Latest use: before drilling wheat or barley

Environmental safety

- Dangerous for the environment
- Toxic to aquatic organisms

Hazard classification and safety precautions

Hazard H11

Risk phrases R51, R58

Operator protection A, H; U05a, U07, U20b

Environmental protection E15a, E32a, E32d, E34, E38

Storage and disposal E01, E04, E29b, E30a

Treated seed S01, S02, S03, S04a, S04b, S05, S07

359 prometryn

A contact and residual triazine herbicide

Products

1	Alpha Prometryne 50 WP	Makhteshim	50% w/w	WP	04871
2	Gesagard	Syngenta	50% w/w	WP	08410
3	Pennine 50	DAPT	50% w/w	WP	10638

SEE SECTION 3 FOR PRODUCTS ALSO REGISTERED

Uses

- Annual dicotyledons in ***bulb onions*** *(off-label)*, ***chives*** *(off-label)*, ***leeks*** *(off-label)*, ***salad onions*** *(off-label)*, ***transplanted leeks*** *(off-label)* [2]; ***carrots***, ***celery***, ***parsley***, ***protected parsley*** [1-3]; ***herbs (see appendix 6)*** *(off-label)*, ***protected chives*** *(off-label)*, ***protected herbs (see appendix 6)*** *(off-label)* [1, 2]; ***parsnips*** [3]; ***protected celery***, ***transplanted leeks*** [1, 3]
- Annual grasses in ***bulb onions*** *(off-label)*, ***leeks*** *(off-label)*, ***salad onions*** *(off-label)*, ***transplanted leeks*** *(off-label)* [2]; ***carrots***, ***celery***, ***parsley***, ***protected parsley*** [1-3]; ***chives*** *(off-label)*, ***herbs (see appendix 6)*** *(off-label)*, ***protected chives*** *(off-label)*, ***protected herbs (see appendix 6)*** *(off-label)* [1, 2]; ***parsnips*** [3]; ***protected celery***, ***transplanted leeks*** [1, 3]

Specific Off-Label Approvals (SOLAs)

- ***bulb onions***, ***salad onions*** *(OLA 011658) Dec 2008* [2]
- ***chives***, ***herbs (see appendix 6)***, ***protected chives***, ***protected herbs (see appendix 6)*** *(OLA 023394) Dec 2007* [2]
- ***chives***, ***herbs (see appendix 6)***, ***protected chives***, ***protected herbs (see appendix 6)*** *(OLA 031794) Dec 2007* [1]
- ***leeks***, ***transplanted leeks*** *(OLA 011659) Dec 2008* [2]

Approval information

- Products containing this active ingredient have been granted derogations for specified 'Essential Uses' for use until 31 December 2007. Sale and supply must cease by 30 June 2007 but growers have no guarantee that the products will continue to be available until then.
  For more information see 'The Review Programme' under 'Pesticide Legislation' in Section 5

Efficacy guidance

- Best results achieved by application to young seedling weeds up to 5 cm high (cotyledon stage for knotgrass, mayweed and corn marigold) on fine, moist seedbed when rain falls afterwards. Do not use on very cloddy soils
- On organic soils only contact action effective and repeat application may be needed (certain crops only)
- Excessive rain after treatment may check crop

Restrictions

- Maximum number of treatments 1 per crop for transplanted leeks; 1 per crop for transplanted celery
- Maximum total dose equivalent to one full dose treatment on all crops
- This product must not be used on any crops other than those listed, including any extrapolations that would normally be permissible under the Long Term Arrangements for Extension of Use (see Section 5)

Crop-specific information

- HI onions 10 wk; carrots, parsley, celery, leeks, herbs 6 wk
- Apply to carrots, celery, parsley post-emergence after 2-rough leaf stage or after transplants established
- Apply to transplanted leeks or celery after transplants established. Do not use on drilled leeks

Following crops guidance

- In the event of crop failure within 8 wk of treatment only peas, potatoes, carrots, parsley or celery may be planted. After 8 wk the land should be cultivated or ploughed to 150 mm min before planting the next crop

Environmental safety

- Dangerous for the environment
- Toxic to aquatic organisms

Hazard classification and safety precautions

Hazard H11 [1, 2]
Risk phrases R50 [1]; R51 [2]; R58 [1, 2]
Operator protection U05a [1, 2]; U14, U15 [1]; U20b [1-3]
Environmental protection E13c [3]; E15a [1, 2]; E32a [1-3]; E32d, E38 [2]; E32e [1]
Consumer protection C02 [1, 3] (6 wk)
Storage and disposal E01, E04 [1, 2]; E30a [1-3]

360 propachlor

A pre-emergence chloroacetanilide herbicide for various horticultural crops

See also *chloridazon + propachlor*
chlorthal-dimethyl + propachlor

Products

1	Alpha Propachlor 50 SC	Makhteshim	500 g/l	SC	04873
2	Brasson	Monsanto	480 g/l	SC	10560
3	Ramrod Flowable	Monsanto	480 g/l	SC	10314
4	Tripart Sentinel 2	Tripart	480 g/l	SC	05140

Uses

- Annual dicotyledons in ***blackcurrants*** *(off-label)*, ***blueberries*** *(off-label)*, ***chinese cabbage*** *(off-label)*, ***chives*** *(off-label)*, ***gooseberries*** *(off-label)*, ***herbs (see appendix 6)*** *(off-label)*, ***kohlrabi*** *(off-label)*, ***lettuce*** *(off-label)*, ***parsley*** *(off-label)*, ***radicchio*** *(off-label)*, ***redcurrants*** *(off-label)*, ***strawberries*** *(off-label)*, ***whitecurrants*** *(off-label)* [1, 3]; ***broccoli, brussels sprouts, cabbages, calabrese, cauliflowers, kale, leeks, onions, spring oilseed rape, swedes, turnips*** [1-4]; ***brown mustard, sage, white mustard*** [2, 3]; ***cress*** *(off-label)*, ***frise*** *(off-label - under crop covers)*, ***lamb's lettuce*** *(off-label - under crop covers)*, ***lettuce*** *(off-label - under crop covers)*, ***radicchio*** *(off-label - under crop covers)*, ***ribes hybrids*** *(off-label)*, ***salad brassicas*** *(off-label - for baby leaf production)*, ***scarole*** *(off-label - under crop covers)* [3]; ***fodder rape, ornamental plant production*** [1-3]; ***mustard, rubus hybrids*** *(off-label)*, ***salad onions, strawberries, winter oilseed rape*** [1]
- Annual grasses in ***blackcurrants*** *(off-label)*, ***blueberries*** *(off-label)*, ***chinese cabbage*** *(off-label)*, ***chives*** *(off-label)*, ***gooseberries*** *(off-label)*, ***herbs (see appendix 6)*** *(off-label)*, ***kohlrabi*** *(off-label)*, ***lettuce*** *(off-label)*, ***parsley*** *(off-label)*, ***radicchio*** *(off-label)*, ***redcurrants*** *(off-label)*, ***strawberries*** *(off-label)*, ***whitecurrants*** *(off-label)* [1, 3]; ***broccoli, brussels sprouts, cabbages, calabrese, cauliflowers, kale, leeks, onions, spring oilseed rape, swedes, turnips*** [1-4]; ***brown mustard, sage, white mustard*** [2, 3]; ***cress*** *(off-label)*, ***frise*** *(off-label - under crop covers)*, ***lamb's lettuce*** *(off-label - under crop covers)*, ***lettuce*** *(off-label - under crop covers)*, ***radicchio*** *(off-label - under crop covers)*, ***ribes hybrids*** *(off-label)*, ***scarole*** *(off-label - under crop covers)* [3]; ***fodder rape, ornamental plant production*** [1-3]; ***mustard, rubus hybrids*** *(off-label)*, ***salad onions, strawberries, winter oilseed rape*** [1]

Specific Off-Label Approvals (SOLAs)

- ***blackcurrants, blueberries, chinese cabbage, chives, gooseberries, herbs (see appendix 6), kohlrabi, lettuce, parsley, radicchio, redcurrants, rubus hybrids, strawberries, whitecurrants*** *(OLA 002424) Dec 2008* [1]
- ***blackcurrants, blueberries, chinese cabbage, chives, cress, gooseberries, herbs (see appendix 6), kohlrabi, lettuce, parsley, radicchio, redcurrants, ribes hybrids, strawberries, whitecurrants*** *(OLA 021159) Dec 2008* [3]
- ***frise, lamb's lettuce, lettuce, radicchio, scarole*** *(under crop covers) (OLA 021159) Dec 2008* [3]
- ***salad brassicas*** *(for baby leaf production) (OLA 030436) Dec 2008* [3]

Efficacy guidance

- Controls germinating (not emerged) weeds for 6-8 wk
- Best results achieved by application to fine, firm, moist seedbed free of established weeds in spring, summer and early autumn
- Use higher rate on soils with more than 10% organic matter
- Recommended as tank-mix with glyphosate or chlorthal-dimethyl on onions, leeks, swedes, turnips. See label for details

Restrictions

- Maximum number of treatments 1 or 2 per crop, varies with product and crop - see label for details
- Maximum total dose equivalent to two full dose treatments on brassicas and leeks; one full dose treatment on most other crops [1]
- Do not use after products that de-wax the leaves of plants, especially brassica crops
- Do not use on crops under glass or polythene
- Do not use under extremely wet, dry or other adverse growth conditions

Crop-specific information

- Latest use: ranges from before crop or weed emergence to 3-4 leaf stage of brasssica crops and pre-blossom for strawberries. Consult label for details
- HI salad brassicas 28 d; other salad vegetables, herbs 6 wk
- Apply in brassicas from drilling to time seed chits or after 3-4 true leaf stage but before weed emergence, in swedes and turnips pre-emergence only
- Apply in transplanted brassicas within 48 h of planting in warm weather. Plants must be hardened off and special care needed with block sown or modular propagated plants
- Apply in onions and leeks pre-emergence or from post-crook to young plant stage
- Apply to newly planted strawberries soon after transplanting, to weed-free soil in established crops in early spring before new weeds emerge
- Granules may be applied to onion, leek and brassica nurseries and to most flower crops after bedding out and hardening off
- Do not use pre-emergence of drilled wallflower seed

Following crops guidance

- In the event of crop failure only replant recommended crops in treated soil

Environmental safety

- Dangerous for the environment
- Very toxic to aquatic organisms

Hazard classification and safety precautions

Hazard H03, H11
Risk phrases R21, R36 [1, 4]; R22a, R38, R43 [1-4]; R37 [4]; R50, R58 [1-3]
Operator protection A, C; U02a, U04a, U05a, U08, U13, U19a [1-4]; U10, U11 [1]; U14 [2, 3]; U20a [4]; U20b [1-3]
Environmental protection E15a, E34 [1-4]; E31b [1, 4]; E31c, E32d [2, 3]; E32e [1]
Storage and disposal E01, E04, E26, E30a
Medical advice M03 [1-3]; M04 [1]

361 propamocarb hydrochloride

A translocated protectant carbamate fungicide

See also *chlorothalonil + propamocarb hydrochloride*
fenamidone + propamocarb hydrochloride
mancozeb + propamocarb hydrochloride

Products

1	Filex	Scotts	722 g/l	SL	07631
2	Flash	AgriGuard	722 g/l	SL	11989
3	Proplant	Fargro	722 g/l	SL	08572

Uses

- Blight in ***outdoor tomatoes***, ***protected tomatoes*** [2]
- Botrytis in ***chives*** *(off-label)*, ***herbs (see appendix 6)*** *(off-label)*, ***lettuce*** *(off-label)*, ***parsley*** *(off-label)*, ***protected herbs (see appendix 6)*** *(off-label)* [1, 3]; ***cress*** *(off-label)* [1]; ***protected chives*** *(off-label)*, ***protected lettuce*** *(off-label)*, ***protected parsley*** *(off-label)*, ***protected radishes*** *(off-label)* [3]
- Damping off in ***inert substrate cucumbers*** *(off-label)*, ***inert substrate peppers*** *(off-label)*, ***inert substrate tomatoes*** *(off-label)*, ***nft peppers*** *(off-label)*, ***nft tomatoes*** *(off-label)* [1, 3]; ***protected peppers*** *(off-label)*, ***protected tomatoes*** *(off-label)* [1]
- Downy mildew in ***broccoli***, ***brussels sprouts***, ***cabbages***, ***calabrese***, ***cauliflowers***, ***chinese cabbage*** [1-3]; ***chives*** *(off-label)*, ***herbs (see appendix 6)*** *(off-label)*, ***lettuce*** *(off-label)*, ***parsley*** *(off-label)*, ***protected herbs (see appendix 6)*** *(off-label)*, ***protected radishes*** *(off-label)*, ***radishes*** *(off-label)* [1, 3]; ***cress*** *(off-label)*, ***watercress*** *(off-label - under protection)* [1]; ***protected chives*** *(off-label)*, ***protected lettuce*** *(off-label)*, ***protected parsley*** *(off-label)*, ***watercress*** *(off-label - during propagation)* [3]
- Phytophthora in ***aubergines***, ***bedding plants***, ***broccoli***, ***brussels sprouts***, ***cabbages***, ***calabrese***, ***cauliflowers***, ***chinese cabbage***, ***cucumbers***, ***leeks***, ***nursery stock***, ***onions***, ***peppers***, ***pot plants***, ***tulips*** [1-3]; ***container-grown ornamentals***, ***ornamental specimens***, ***watercress*** *(off-label - under protection)* [1]; ***flower bulbs***, ***outdoor tomatoes***, ***protected tomatoes***,

rockwool aubergines*, *rockwool cucumbers*, *rockwool peppers*, *rockwool tomatoes [2, 3]; ***watercress*** *(off-label - during propagation)* [3]

- Pythium in ***aubergines*, *bedding plants*, *broccoli*, *brussels sprouts*, *cabbages*, *calabrese*, *cauliflowers*, *chinese cabbage*, *cucumbers*, *flower bulbs*, *leeks*, *nursery stock*, *onions*, *outdoor tomatoes*, *peppers*, *pot plants*, *protected tomatoes*, *rockwool aubergines*, *rockwool cucumbers*, *rockwool peppers*, *rockwool tomatoes*, *tulips*** [1-3]; ***ornamental specimens*, *watercress*** *(off-label - under protection)* [1]; ***watercress*** *(off-label - during propagation)* [3]
- Root malformation disorder in ***red beet*** *(off-label)* [1, 3]
- Root rot in ***inert substrate cucumbers*** *(off-label)*, ***inert substrate peppers*** *(off-label)*, ***inert substrate tomatoes*** *(off-label)*, ***nft peppers*** *(off-label)*, ***nft tomatoes*** *(off-label)* [3]
- White blister in ***horseradish*** *(off-label)*, ***protected horseradish*** *(off-label)* [1, 3]

Specific Off-Label Approvals (SOLAs)

- ***chives*, *herbs (see appendix 6)*, *horseradish*, *lettuce*, *parsley*, *protected chives*, *protected herbs (see appendix 6)*, *protected horseradish*, *protected lettuce*, *protected parsley*, *protected radishes*, *radishes*** *(OLA 040625) Dec 2008* [3]
- ***chives*, *cress*, *herbs (see appendix 6)*, *lettuce*, *parsley*, *protected herbs (see appendix 6)*** *(OLA 040626) Dec 2008* [1]
- ***horseradish*, *protected horseradish*** *(OLA 992036) Dec 2008* [1]
- ***inert substrate cucumbers*, *inert substrate peppers*, *inert substrate tomatoes*, *nft peppers*, *nft tomatoes*** *(OLA 002527) Dec 2008* [3]
- ***inert substrate cucumbers*, *inert substrate peppers*, *inert substrate tomatoes*, *nft peppers*, *nft tomatoes*, *protected peppers*, *protected tomatoes*** *(OLA 992032) Dec 2008* [1]
- ***protected radishes*, *radishes*** *(OLA 992030) Dec 2008* [1]
- ***red beet*** *(OLA 031247) Dec 2008* [1]
- ***red beet*** *(OLA 031453) Jul 2008* [3]
- ***watercress*** *(during propagation) (OLA 040625) Dec 2008* [3]
- ***watercress*** *(under protection) (OLA 010439) Dec 2008* [1]

Efficacy guidance

- Chemical is absorbed through roots and translocated throughout plant
- Incorporate in compost before use or drench moist compost or soil before sowing, pricking out, striking cuttings or potting up
- Drench treatment can be repeated at 3-6 wk intervals
- Concentrated solution is corrosive to all metals other than stainless steel
- May also be applied in trickle irrigation systems
- To prevent root rot in tulip bulbs apply as dip for 20 min but full protection only achieved by soil drench treatment as well

Restrictions

- Maximum number of treatments 4 per crop for cucumbers, tomatoes, peppers, aubergines; 1 per crop for listed brassicas; 1 compost incorporation and/or 1 drench treatment for leeks, onions; 1 dip and 1 soil drench for tulips
- When applied over established seedlings rinse off foliage with water and do not apply under hot, dry conditions
- On plants of unknown tolerance test first on a small scale
- Do not apply in a recirculating system
- Store away from seeds and fertilizers

Crop-specific information

- Latest use: before transplanting for brassicas, herbs; before crop emergence for watercress
- HI 2 d for inert substrate and NFT vegetable crops; 14 d for cucumbers, tomatoes, peppers, aubergines, horseradish, radishes, herbs and protected herbs; 4 wk for calabrese, cauliflower, sprouting broccoli, Brussels sprouts, Chinese cabbage. watercress; 19 wk for leeks, onions

Hazard classification and safety precautions

Operator protection A, B, C, H, K, M; U05a [2, 3]; U08, U20b [1-3]
Environmental protection E15a, E32a
Storage and disposal E01, E04, E26, E29a [2, 3]; E30a [1-3]

SEE SECTION 3 FOR PRODUCTS ALSO REGISTERED

362 propaquizafop

A phenoxy alkanoic acid foliar acting grass herbicide

Products

1	Bulldog	Makhteshim	100 g/l	EC	11723
2	Cleancrop GYR	Makhteshim	100 g/l	EC	10646
3	Falcon	Makhteshim	100 g/l	EC	10585
4	Greencrop Satchmo	Greencrop	100 g/l	EC	10748
5	Landgold PQF 100	Landgold	100 g/l	EC	10763
6	Raptor	Makhteshim	100 g/l	EC	11092
7	Shogun	Makhteshim	100 g/l	EC	10584
8	Standon Propaquizafop	Standon	100 g/l	EC	11536

Uses

- Annual grasses in ***bulb onions, carrots, combining peas, early potatoes, farm forestry, fodder beet, linseed, maincrop potatoes, parsnips, spring field beans, spring oilseed rape, sugar beet, swedes, turnips, winter field beans, winter oilseed rape*** [1-8]; ***cut logs, forest, forest nurseries*** [1-3, 6, 7]; ***poppies*** *(off-label - for morphine production)* [1, 6]
- Annual meadow grass in ***chives*** *(off-label)*, ***parsley*** *(off-label)* [3]; ***herbs (see appendix 6)*** *(off-label)*, ***leaf spinach*** *(off-label)*, ***leeks*** *(off-label)*, ***red beet*** *(off-label)* [1, 3]
- Couch in ***chives*** *(off-label)*, ***parsley*** *(off-label)* [3]; ***herbs (see appendix 6)*** *(off-label)*, ***leaf spinach*** *(off-label)*, ***leeks*** *(off-label)*, ***red beet*** *(off-label)* [1, 3]
- Perennial grasses in ***bulb onions, carrots, combining peas, early potatoes, farm forestry, fodder beet, linseed, maincrop potatoes, parsnips, spring field beans, spring oilseed rape, sugar beet, swedes, turnips, winter field beans, winter oilseed rape*** [1-8]; ***celery*** *(off-label)* [3]; ***cut logs, forest, forest nurseries*** [1-3, 6, 7]
- Volunteer cereals in ***celery*** *(off-label)*, ***chives*** *(off-label)*, ***parsley*** *(off-label)* [3]; ***herbs (see appendix 6)*** *(off-label)*, ***leaf spinach*** *(off-label)*, ***leeks*** *(off-label)*, ***red beet*** *(off-label)* [1, 3]; ***poppies*** *(off-label - for morphine production)* [1, 6]
- Wild oats in ***celery*** *(off-label)* [3]

Specific Off-Label Approvals (SOLAs)

- ***celery*** *(OLA 021502) Dec 2008* [3]
- ***chives, herbs (see appendix 6), leaf spinach, leeks, parsley, red beet*** *(OLA 020148) Dec 2008* [3]
- ***herbs (see appendix 6), leaf spinach, leeks, red beet*** *(OLA 032022) Dec 2008* [1]
- ***poppies*** *(for morphine production) (OLA 030770) Dec 2008* [1, 6]

Efficacy guidance

- Apply to emerged weeds when they are growing actively with adequate soil moisture
- Activity is slower under cool conditions
- Broad-leaved weeds and any weeds germinating after treatment are not controlled
- Annual meadow grass up to 3 leaves checked at low doses and severely checked at highest dose
- Spray barley cover crops when risk of wind blow has passed and before there is serious competition with the crop
- Various tank mixtures and sequences recommended for broader spectrum weed control in oilseed rape, peas and sugar beet. See label for details
- Severe couch infestations may require a second application at reduced dose when regrowth has 3-4 leaves unfolded
- Products contain surfactants. Tank mixing with adjuvants not required or recommended
- Always follow WRAG guidelines for preventing and managing herbicide resistant weeds. Section 5 for more information

Restrictions

- Maximum total dose varies according to crop treated. See label for details
- See label for list of tolerant tree species
- Do not treat seed potatoes

Crop-specific information

- Latest use: before crop flower buds visible for winter oilseed rape, linseed, field beans; before 8 fully expanded leaf stage for spring oilseed rape, mustard; before weeds are covered by the crop for potatoes, sugar beet, fodder beet; when flower buds visible for peas

- HI field beans 2 wk; herbs 3 wk; celery, leeks, red beet, onions, carrots, potatoes, parsnips 4 wk; peas, early potatoes 7 wk; sugar beet, fodder beet, maincrop potatoes, swedes, turnips 8 wk
- Application in high temperatures and/or low soil moisture content may cause chlorotic spotting especially on combining peas and field beans
- Overlaps at the highest dose can cause damage from early applications to carrots and parsnips

Following crops guidance

- In the event of a failed treated crop an interval of 2 wk must elapse between the last application and redrilling with winter wheat or winter barley. 4 wk must elapse before sowing oilseed rape, peas or field beans, and 16 wk before sowing ryegrass or oats

Environmental safety

- Dangerous for the environment
- Toxic to aquatic organisms
- Risk to certain non-target insects or other arthropods. See directions for use

Hazard classification and safety precautions

Hazard H04, H11
Risk phrases R36, R38, R51, R58
Operator protection A, C [3-5, 8]; U02a, U05a, U08, U19a, U20b [1-8]; U11 [3, 5, 8]; U13 [1-3, 6, 7]; U14, U15 [1-3, 5-8]; U23a [4]
Environmental protection E15a, E22b, E31b [1-8]; E32e, E34 [1-3, 6, 7]; E33 [1, 3, 6, 7] (returnable container)
Storage and disposal E01, E04, E26, E30a
Medical advice M05b

363 propham

A pre-sowing carbamate herbicide, all approvals for which were revoked in 1997

364 propiconazole

A systemic, curative and protectant conazole fungicide

See also *cyproconazole + propiconazole*
prochloraz + propiconazole

Products

Bumper 250 EC	Makhteshim	250 g/l	EC	09039

Uses

- Brown rust in ***spring barley, spring wheat, winter barley, winter rye, winter wheat***
- Crown rust in ***grass for ensiling, grass seed crops, spring oats, winter oats***
- Drechslera leaf spot in ***grass for ensiling, grass seed crops***
- Leaf blotch in ***garlic*** *(off-label)*, ***onions*** *(off-label)*, ***shallots*** *(off-label)*
- Light leaf spot in ***spring oilseed rape*** *(reduction)*, ***winter oilseed rape*** *(reduction)*
- Mildew in ***grass for ensiling, grass seed crops***
- Powdery mildew in ***spring barley, spring oats, spring wheat, winter barley, winter oats, winter rye, winter wheat***
- Ramularia leaf spots in ***sugar beet*** *(reduction)*
- Rhynchosporium in ***grass for ensiling, grass seed crops, spring barley, winter barley, winter rye***
- Rust in ***garlic*** *(off-label)*, ***sugar beet***
- Septoria in ***spring wheat, winter rye, winter wheat***
- Sooty moulds in ***spring wheat, winter wheat***
- White blister in ***honesty*** *(off-label)*
- White rust in ***chrysanthemums*** *(off-label)*, ***protected chrysanthemums*** *(off-label)*
- Yellow rust in ***spring barley, spring wheat, winter barley, winter wheat***

Specific Off-Label Approvals (SOLAs)

- ***chrysanthemums, garlic, honesty, onions, protected chrysanthemums, shallots*** *(OLA 012142) Dec 2008* [1]

Approval information

- Propiconazole included in Annex I under EC Directive 91/414

Efficacy guidance

- Best results achieved by applying at early stage of disease. Recommended spray programmes vary with crop, disease, season, soil type and product. See label for details

Restrictions

- Maximum number of treatments 4 per crop for wheat (up to 3 in yr of harvest); 3 per crop for barley, oats, rye, triticale (up to 2 in yr of harvest), leeks; 2 per crop or yr for oilseed rape, sugar beet, grass seed crops, honesty; 1 per yr on grass for ensiling
- Propiconazole may produce an allergic reaction
- On oilseed rape do not apply during flowering. Apply first treatment before flowering, the second afterwards
- Grass seed crops must be treated in yr of harvest
- A minimum interval of 14 d must elapse between treatments on leeks and 21 d on sugar beet
- Avoid spraying crops under stress, eg during cold weather or periods of frost

Crop-specific information

- Latest use: before 4 pairs of true leaves for honesty
- HI cereals, grass seed crops 35 d; oilseed rape, sugar beet, garlic, onions, shallots, grass for ensiling 28 d

Environmental safety

- Dangerous for the environment
- Toxic to aquatic organisms

Hazard classification and safety precautions

Hazard H11
Risk phrases R51, R58
Operator protection A, C; U02a, U05a, U08, U19a, U20b
Environmental protection E15a, E31b, E34
Storage and disposal E01, E04, E26, E30a
Treated seed S06a
Medical advice M05a

365 propiconazole + trifloxystrobin

A mixture of strobilurin and conazole fungicides for cereals

Products

Rombus	Bayer CropScience	125:125 g/l	EC	11301

Uses

- Brown rust in ***spring barley, winter barley, winter wheat***
- Net blotch in ***spring barley, winter barley***
- Powdery mildew in ***spring barley, winter barley***
- Rhynchosporium in ***spring barley, winter barley***
- Septoria in ***winter wheat***
- Yellow rust in ***spring barley, winter barley, winter wheat***

Approval information

- Propiconazole included in Annex I under EC Directive 91/414

Efficacy guidance

- Best results obtained from treatment at early stages of disease development
- Trifloxystrobin is a member of the QoI cross resistance group. Product should be used preventatively and not relied on for its curative potential
- Use product as part of an Integrated Crop Management strategy incorporating other methods of control, including where appropriate other fungicides with a different mode of action. Do not apply more than two foliar applications of QoI containing products to any cereal crop
- There is a significant risk of widespread resistance occurring in *Septoria tritici* populations in UK. Failure to follow resistance management action may result in reduced levels of disease control

Restrictions

- Maximum number of treatments 2 per crop and maximum total dose of propiconazole (including other products containing propiconazole) equivalent to two full dose treatments

Crop-specific information

- HI 35 d for all crops

Environmental safety

- Dangerous for the environment
- Very toxic to aquatic organisms

Hazard classification and safety precautions

Hazard H04, H11
Risk phrases R36, R50, R58
Operator protection A, C; U02a, U09a, U19a, U20b
Environmental protection E13b, E15a, E31b, E32d, E38
Consumer protection C02 (35 d)
Storage and disposal E01, E04, E30a

366 propoxycarbazone-sodium

A sulfonylaminocarbonyltriazolinone residual grass weed herbicide for winter wheat

Products

1	Attribut	Bayer CropScience	70% w/w	SG	11221
2	Ethos	Bayer CropScience	70% w/w	SG	11223

Uses

- Blackgrass in ***winter wheat***
- Couch in ***winter wheat***

Efficacy guidance

- Activity by root and foliar absorption but depends on presence of sufficient soil moisture to ensure root uptake by weeds. Control is enhanced by use of an adjuvant to encourage foliar uptake
- Best results obtained from treatments applied when weed grasses are growing actively. Symptoms may not become apparent for 3-4 wk after application
- Ensure good even spray coverage
- To achieve best control of couch and reduction of infestation in subsequent crop treat between 2 true leaves and first node stage of the weed. A split dose programme may improve control
- Blackgrass should be treated after using specialist blackgrass herbicides with a different mode of action
- Proxycarbazone-sodium is a member of the ALS-inhibitor group of herbicides and products should be used in a planned Resistance Management strategy. See Section 5 for more information

Restrictions

- Maximum total dose equivalent to one full dose treatment
- Do not use in a programme with other aceto-lactase synthesis (ALS) inhibitors
- Do not apply when temperature near or below freezing
- Avoid treatment under dry soil conditions

Crop-specific information

- Latest use: flag leaf just visible (GS 37)

Following crops guidance

- Only winter wheat, field beans or winter barley may be sown in the autumn following spring treatment
- Any crop may be grown in the spring on land treated during the previous calendar yr

Environmental safety

- Dangerous to fish or other aquatic life. Do not contaminate surface waters or ditches with chemical or used container
- LERAP Category B
- Take extreme care to avoid drift onto adjacent plants or land as this could result in severe damage

Hazard classification and safety precautions

Operator protection A; U05a, U20b

Environmental protection E13b, E16a, E16b, E32a

Storage and disposal E01, E04, E30a

367 propyzamide

A residual amide herbicide for use in a wide range of crops

Products

1	Agriguard Propyzamide 50 WP	AgriGuard	50% w/w	WP	10187
2	Agriguard Propyzamide Flo	AgriGuard	400 g/l	SC	10435
3	Bulwark Flo	Interfarm	400 g/l	SC	10161
4	Flomide	Interfarm	400 g/l	SC	11171
5	Greencrop Saffron FL	Greencrop	400 g/l	SC	10244
6	Kerb 50 W	Dow	50% w/w	WP	10771
7	Kerb 80 EDF	Dow	80% w/w	WG	10761
8	Kerb Flo	Dow	400 g/l	SC	10768
9	Kerb Granules	SumiAgro	4% w/w	GR	08917
10	Kerb Pro Flo	Dow	400 g/l	SC	10760
11	Kerb Pro Granules	SumiAgro Amenity	4% w/w	GR	09600
12	Menace 80 EDF	Dow	80% w/w	WG	10766
13	Mithras 80 EDF	Dow	80% w/w	WG	10767
14	Precis	Dow	400 g/l	SC	10762
15	Propose 50 W	Interfarm	50% w/w	WP	11170
16	Quaver Flo	Dow	400 g/l	SC	10769
17	Standon Propyzamide 400 SC	Standon	400 g/l	SC	10255
18	Standon Propyzamide 50	Standon	50% w/w	WP	10814

Uses

- Annual dicotyledons in ***apple orchards, blackberries, blackcurrants, clover seed crops, gooseberries, loganberries, lucerne, plums, redcurrants, strawberries, sugar beet seed crops, winter field beans, winter oilseed rape*** [1, 2, 4-8, 12-18]; ***brassica seed crops, lettuce*** *(outdoor crops)*, ***rhubarb*** *(outdoor)* [7, 12, 13]; ***camomile*** *(off-label)*, ***fenugreek*** *(off-label)*, ***radicchio*** *(off-label)*, ***sage*** *(off-label)*, ***tarragon*** *(off-label)* [6]; ***fodder rape seed crops, kale seed crops, lettuce, rhubarb, turnip seed crops*** [1, 2, 4-6, 8, 14-18]; ***forest, trees and shrubs*** [1, 3, 6, 9-11, 15]; ***ornamental plant production, raspberries*** [18]; ***pear orchards*** [1, 2, 4-8, 12, 14-18]; ***pears*** [13]; ***raspberries*** *(England only)* [1, 2, 4-8, 12-17]; ***roses*** [1, 6, 15, 18]; ***woody ornamentals*** [1, 3, 6, 7, 9-13, 15]
- Annual grasses in ***apple orchards, blackberries, blackcurrants, clover seed crops, gooseberries, loganberries, lucerne, plums, redcurrants, strawberries, sugar beet seed crops, winter field beans, winter oilseed rape*** [1, 2, 4-8, 12-18]; ***brassica seed crops, lettuce*** *(outdoor crops)*, ***rhubarb*** *(outdoor)* [7, 12, 13]; ***camomile*** *(off-label)*, ***cherries*** *(off-label)*, ***chicory*** *(off-label)*, ***courgettes*** *(off-label)*, ***evening primrose*** *(off-label)*, ***fenugreek*** *(off-label)*, ***grapevines*** *(off-label)*, ***honesty*** *(off-label)*, ***hops*** *(off-label)*, ***marrows*** *(off-label)*, ***mirabelles*** *(off-label)*, ***protected chives*** *(off-label)*, ***protected courgettes*** *(off-label)*, ***protected herbs (see appendix 6)*** *(off-label)*, ***protected lamb's lettuce*** *(off-label)*, ***protected lettuce*** *(off-label)*, ***protected marrows*** *(off-label)*, ***protected parsley*** *(off-label)*, ***protected pumpkins*** *(off-label)*, ***protected radicchio*** *(off-label)*, ***protected scarole*** *(off-label)*, ***protected squashes*** *(off-label)*, ***pumpkins*** *(off-label)*, ***radicchio*** *(off-label)*, ***sage*** *(off-label)*, ***squashes*** *(off-label)*, ***tarragon*** *(off-label)* [6]; ***fodder rape seed crops, kale seed crops, lettuce, rhubarb, turnip seed crops*** [1, 2, 4-6, 8, 14-18]; ***forest, trees and shrubs*** [1, 3, 6, 9-11, 15]; ***ornamental plant production, raspberries*** [18]; ***pear orchards*** [1, 2, 4-8, 12, 14-18]; ***pears*** [13]; ***raspberries*** *(England only)* [1, 2, 4-8, 12-17]; ***roses*** [1, 6, 15, 18]; ***woody ornamentals*** [1, 3, 6, 7, 9-13, 15]
- Horsetails in ***forest*** [2, 3, 7, 10, 12, 13]; ***woody ornamentals*** [3, 10]
- Perennial grasses in ***apple orchards, blackberries, blackcurrants, gooseberries, loganberries, plums, redcurrants*** [1, 2, 4-8, 12-18]; ***camomile*** *(off-label)*, ***cherries*** *(off-label)*, ***chicory*** *(off-label)*, ***courgettes*** *(off-label)*, ***evening primrose*** *(off-label)*, ***fenugreek*** *(off-label)*, ***grapevines*** *(off-label)*, ***honesty*** *(off-label)*, ***hops*** *(off-label)*, ***marrows*** *(off-label)*, ***mirabelles*** *(off-label)*, ***protected chives*** *(off-label)*, ***protected courgettes*** *(off-label)*, ***protected herbs (see appendix 6)*** *(off-label)*, ***protected lamb's lettuce*** *(off-label)*, ***protected lettuce*** *(off-label)*, ***protected marrows*** *(off-label)*, ***protected parsley*** *(off-label)*, ***protected pumpkins*** *(off-label)*, ***protected radicchio***

(off-label), ***protected scarole*** *(off-label)*, ***protected squashes*** *(off-label)*, ***pumpkins*** *(off-label)*, ***radicchio*** *(off-label)*, ***sage*** *(off-label)*, ***squashes*** *(off-label)*, ***tarragon*** *(off-label)* [6]; ***clover seed crops***, ***fodder rape seed crops***, ***kale seed crops***, ***lettuce***, ***lucerne***, ***rhubarb***, ***strawberries***, ***sugar beet seed crops***, ***turnip seed crops***, ***winter field beans***, ***winter oilseed rape*** [1, 2, 4-6, 8, 14-18]; ***forest*** [1-3, 6, 7, 9-13, 15]; ***ornamental plant production***, ***raspberries*** [18]; ***pear orchards*** [1, 2, 4-8, 12, 14-18]; ***pears*** [13]; ***raspberries*** *(England only)* [1, 2, 4-8, 12-17]; ***rhubarb*** *(outdoor)* [7, 12, 13]; ***roses*** [1, 6, 15, 18]; ***trees and shrubs*** [1, 3, 6, 9-11, 15]; ***woody ornamentals*** [1, 3, 6, 7, 9-13, 15]
- Sedges in ***forest*** [2, 3, 7, 10, 12, 13]; ***woody ornamentals*** [3, 10]
- Volunteer cereals in ***sugar beet seed crops***, ***winter field beans***, ***winter oilseed rape*** [7, 12, 13]
- Wild oats in ***sugar beet seed crops***, ***winter field beans***, ***winter oilseed rape*** [7, 12, 13]

Specific Off-Label Approvals (SOLAs)

- ***camomile***, ***cherries***, ***chicory***, ***courgettes***, ***evening primrose***, ***fenugreek***, ***grapevines***, ***honesty***, ***hops***, ***marrows***, ***mirabelles***, ***protected chives***, ***protected courgettes***, ***protected herbs (see appendix 6)***, ***protected lamb's lettuce***, ***protected lettuce***, ***protected marrows***, ***protected parsley***, ***protected pumpkins***, ***protected radicchio***, ***protected scarole***, ***protected squashes***, ***pumpkins***, ***radicchio***, ***sage***, ***squashes***, ***tarragon*** *(OLA 021416) Dec 2008* [6]

Approval information

- Propyzamide included in Annex I under EC Directive 91/414
- Some products may be applied through CDA equipment. See labels for details

Efficacy guidance

- Active via root uptake. Weeds controlled from germination to young seedling stage, some species (including many grasses) also when established
- Best results achieved by winter application to fine, firm, moist soil. Rain is required after application if soil dry
- Uptake is slow and and may take up to 12 wk
- Excessive organic debris or ploughed-up turf may reduce efficacy
- For heavy couch infestations a repeat application may be needed in following winter
- Always follow WRAG guidelines for preventing and managing herbicide resistant weeds. Section 5 for more information

Restrictions

- Maximum number of treatments 1 per crop or yr
- Maximum total dose equivalent to one full dose treatment for all crops
- Do not treat protected crops
- Apply to listed edible crops only between 1 Oct and the date specified as the latest time of application (except lettuce)
- Do not apply in windy weather and avoid drift onto non-target crops
- Do not use on soils with more than 10% organic matter except in forestry

Crop-specific information

- Latest use: labels vary but normally before 31 Dec for rhubarb, lucerne, strawberries and winter field beans; before 31 Jan for other crops; pre-emergence for chicory
- HI normally 6 wk for edible crops and 3 mth for protected herbs but labels vary in detail
- Apply as soon as possible after 3-true leaf stage of oilseed rape (GS 1,3) and seed brassicas, after 4-leaf stage of sugar beet for seed, within 7 d after sowing but before emergence for field beans, after perennial crops established for at least 1 season, strawberries after 1 yr
- Only apply to strawberries on heavy soils. Do not use on matted row crops
- Only apply to field beans on medium and heavy soils
- Only apply to established lucerne not less than 7 d after last cut
- In lettuce lightly incorporate in top 25 mm pre-drilling or irrigate on dry soil
- See label for lists of ornamental and forest species which may be treated

Following crops guidance

- Following an application between 1 Apr and 31 Jul at any dose the following minimum intervals must be observed before sowing the next crop: lettuce 0 wk; broad beans, chicory, clover, field beans, lucerne, radishes, peas 5 wk; brassicas, celery, leeks, oilseed rape, onions, parsley, parsnips 10 wk
- Following an application between 1 Aug and 31 Mar at any dose the following minimum intervals must be observed before sowing the next crop: lettuce 0 wk; broad beans, chicory, clover, field

beans, lucerne, radishes, peas 10 wk; brassicas, celery, leeks, oilseed rape, onions, parsley, parsnips 25 wk or after 15 Jun, whichever occurs sooner

- Cereals or grasses or other crops not listed may be sown 30 wk after treatment up to 840 g ai/ha between 1 Aug and 31 Mar or 40 wk after treatment at higher doses at any time and after mouldboard ploughing to at least 15 cm
- A period at least 9 mth must elapse between applications of propyzamide to the same land

Environmental safety

- Dangerous for the environment
- Very toxic to aquatic organisms

Hazard classification and safety precautions

Hazard H03, H11 [1-6, 8, 9, 12, 14, 15, 17, 18]

Risk phrases R40, R58 [1-6, 8, 9, 12, 14, 15, 17, 18]; R50 [3-5, 8, 9, 12, 14, 17, 18]; R51 [1, 2, 6, 15]

Operator protection A, H [3, 15]; U20a [9, 11]; U20c [1-8, 10, 12-18]

Environmental protection E15a [1-18]; E31a [9]; E31b [5, 17]; E32a [1-4, 6-8, 10-16, 18]; E32d, E38 [3, 4, 6, 8, 9, 12, 14, 15]; E34 [1, 2, 11]

Consumer protection C02 [1, 2, 4-6] (6 wk); C02 [7]; C02 [8] (6 wk); C02 [11]; C02 [12] (6 wk); C02 [13]; C02 [14-18] (6 wk)

Storage and disposal E01 [2, 5, 11, 17, 18]; E26 [1-8, 10, 12-18]; E29a [2]; E30a [1-18]

368 pymetrozine

A novel azomethine insecticide

Products

1	Chess WG	Syngenta Bioline	50% w/w	WG	10651
2	Plenum WG	Syngenta	50% w/w	WG	10652

Uses

- Aphids in ***ornamental specimens***, ***protected cucumbers***, ***protected ornamentals*** [1]; ***potatoes***, ***sweetcorn*** *(off-label)* [2]; ***protected blackberries*** *(off-label)*, ***protected blackcurrants*** *(off-label)*, ***protected gooseberries*** *(off-label)*, ***protected loganberries*** *(off-label)*, ***protected raspberries*** *(off-label)*, ***protected redcurrants*** *(off-label)*, ***protected ribes hybrids*** *(off-label)*, ***protected rubus hybrids*** *(off-label)*, ***protected strawberries*** *(off-label)* [1, 2]
- Currant lettuce aphid in ***lettuce*** *(off-label)*, ***protected chives*** *(off-label)*, ***protected herbs (see appendix 6)*** *(off-label)*, ***protected lettuce*** *(off-label)*, ***protected parsley*** *(off-label)*, ***protected salad brassicas*** *(off-label - for baby leaf production)* [2]
- Glasshouse whitefly in ***protected tomatoes*** *(off-label)* [1]
- Peach-potato aphid in ***broccoli*** *(off-label)*, ***brussels sprouts*** *(off-label)*, ***cabbages*** *(off-label)*, ***calabrese*** *(off-label)*, ***cauliflowers*** *(off-label)*, ***collards*** *(off-label)*, ***kale*** *(off-label)* [2]; ***lettuce*** *(off-label)*, ***protected chives*** *(off-label)*, ***protected herbs (see appendix 6)*** *(off-label)*, ***protected lettuce*** *(off-label)*, ***protected parsley*** *(off-label)*, ***protected salad brassicas*** *(off-label - for baby leaf production)* [1, 2]; ***protected aubergines*** *(off-label)*, ***protected endives*** *(off-label)*, ***protected peppers*** *(off-label)* [1]
- Pre-harvest desiccation in ***hops*** *(off-label)* [2]
- Tobacco whitefly in ***protected tomatoes*** *(off-label)* [1]

Specific Off-Label Approvals (SOLAs)

- ***broccoli***, ***brussels sprouts***, ***cabbages***, ***calabrese***, ***cauliflowers*** *(OLA 031478) Oct 2011* [2]
- ***collards***, ***kale*** *(OLA 041317) Oct 2011* [2]
- ***hops*** *(OLA 031423) Oct 2011* [2]
- ***lettuce***, ***protected chives***, ***protected endives***, ***protected herbs (see appendix 6)***, ***protected lettuce***, ***protected parsley*** *(OLA 030843) Oct 2011* [1]
- ***lettuce***, ***protected chives***, ***protected herbs (see appendix 6)***, ***protected lettuce***, ***protected parsley*** *(OLA 030844) Oct 2011* [2]
- ***protected aubergines***, ***protected peppers***, ***protected tomatoes*** *(OLA 030845) Oct 2011* [1]
- ***protected blackberries***, ***protected blackcurrants***, ***protected gooseberries***, ***protected loganberries***, ***protected raspberries***, ***protected redcurrants***, ***protected ribes hybrids***, ***protected rubus hybrids***, ***protected strawberries*** *(OLA 031072) Oct 2011* [1]

- ***protected blackberries, protected blackcurrants, protected gooseberries, protected loganberries, protected raspberries, protected redcurrants, protected ribes hybrids, protected rubus hybrids, protected strawberries*** *(OLA 031073) Oct 2011* [2]
- ***protected salad brassicas*** *(for baby leaf production) (OLA 030843) Oct 2011* [1]
- ***protected salad brassicas*** *(for baby leaf production) (OLA 030844) Oct 2011* [2]
- ***protected tomatoes*** *(OLA 030170) Dec 2008* [1]
- ***sweetcorn*** *(OLA 041318) Oct 2011* [2]

Approval information
- Pymetrozine included in Annex I under EC Directive 91/414

Efficacy guidance
- Pymetrozine moves systemically in the plant and acts by preventing feeding leading to death by starvation in 1-4 d. There is no immediate knockdown
- Aphids controlled include those resistant to organophosphorus and carbamate insecticides
- To prevent development of resistance do not use continuously or as the sole method of control
- Best results achieved by starting spraying as soon as aphids seen in crop and repeating as necessary.
- To limit spread of persistent viruses such as potato leaf roll virus, apply from 90% crop emergence

Restrictions
- Maximum total dose equivalent to two full dose treatments on potatoes; three full dose treatments on edible crops; four full dose treatments on ornamentals
- Consult processors before use on potatoes for processing
- Check tolerance of ornamental species before large scale use. See label for list of species known to have been treated without damage. Visible spray deposits may be seen on leaves of some species [1]

Crop-specific information
- HI aubergines, peppers, cucumbers 3 d; potatoes, collards, kale, lettuce outdoor herbs 7 d, brassicas, sweetcorn, protected herbs 14 d; protected cane fruit, bush fruit, soft fruit 12 wk

Environmental safety
- High risk to bees (outdoor use only). Do not apply to crops in flower or to those in which bees are actively foraging. Do not apply when flowering weeds are present
- Avoid spraying within 6 m of field boundaries to reduce effects on non-target insects or arthropods

Hazard classification and safety precautions

Hazard H03 [2]
Risk phrases R40 [2]
Operator protection A; U05a, U20c
Environmental protection E12a (outdoor use only); E15a, E32a
Storage and disposal E01, E04 [1]; E30a [1, 2]

369 pyraclostrobin

A protectant and curative strobilurin fungicide for cereals

See also *boscalid + pyraclostrobin*
epoxiconazole + fenpropimorph + pyraclostrobin
epoxiconazole + kresoxim-methyl + pyraclostrobin
epoxiconazole + pyraclostrobin

Products

1	Comet	BASF	250 g/l	EC	10875
2	Insignia	Vitax	20% w/w	WG	11865
3	Tucana	BASF	250 g/l	EC	10899
4	Vivid	BASF	250 g/l	EC	10898

Uses
- Brown rust in ***spring barley, spring wheat, winter barley, winter wheat*** [1, 3, 4]
- Crown rust in ***spring oats, winter oats*** [1, 3, 4]
- Dollar spot in ***managed amenity turf*** *(reduction)* [2]
- Fusarium patch in ***managed amenity turf*** [2]

- Net blotch in ***spring barley*, *winter barley*** [1, 3, 4]
- Red thread in ***managed amenity turf*** [2]
- Rhynchosporium in ***spring barley*** *(moderate)*, ***winter barley*** *(moderate)* [1, 3, 4]
- Septoria diseases in ***spring wheat*, *winter wheat*** [1, 3, 4]
- Yellow rust in ***spring barley*, *spring wheat*, *winter barley*, *winter wheat*** [1, 3, 4]

Efficacy guidance

- For best results apply at the start of disease attack. This is especially for the control of Fusarium Patch [2] as severe damage to turf can occur once it is established
- Regular turf aeration, appropriate scarification and judicious use of nitrogenous fertiliser will assist the control of Fusarium Patch [2]
- Best results on Septoria glume blotch achieved when used as a protective treatment and against Septoria leaf blotch when treated in the latent phase [1, 3, 4]
- Yield response may be obtained in the absence of visual disease symptoms [1, 3, 4]
- Pyraclostrobin is a member of the QoI cross resistance group. Product should be used preventatively and not relied on for its curative potential
- Use product as part of an Integrated Crop Management strategy incorporating other methods of control, including where appropriate other fungicides with a different mode of action. Do not apply more than two foliar applications of QoI containing products to any cereal crop or to grass
- There is a significant risk of widespread resistance occurring in *Septoria tritici* populations in UK. Failure to follow resistance management action may result in reduced levels of disease control [1, 3, 4]
- On cereal crops product must always be used in mixture with another product, recommended for control of the same target disease, that contains a fungicide from a different cross resistance group and is applied at a dose that will give robust control [1, 3, 4]

Restrictions

- Maximum total dose on cereals or turf equivalent to two full dose treatments
- Do not apply during drought conditions or to frozen turf [2]

Crop-specific information

- Latest use: before grain watery ripe (GS 71) for wheat; up to and including emergence of ear just complete (GS 59) for barley and oats
- Avoid applying to turf immediately after cutting or 48 hr before mowing [2]

Environmental safety

- Dangerous for the environment
- Very toxic to aquatic organisms
- LERAP Category B

Hazard classification and safety precautions

Hazard H03, H11
Risk phrases R20, R50 [1-4]; R22a, R38, R58 [1, 3, 4]
Operator protection A [1-4]; D [2]; U05a, U14, U20b [1, 3, 4]
Environmental protection E15a, E16a, E16b, E31c, E32d, E38 [1-4]; E34 [1, 3, 4]
Storage and disposal E01, E04, E26 [1, 3, 4]; E29b [2]; E30a [1-4]
Medical advice M03 [1, 3, 4]

370 pyrethrins

A non-persistent, contact acting insecticide extracted from Pyrethrum

Products

1	Dairy Fly Spray	B H & B	0.75 g/l	AL	H5579
2	Killgerm ULV 400	Killgerm	30 g/l	UL	H4838
3	Turbair Super Flydown	SumiAgro	2.5 g/l	RH	H7225

Uses

- Flies in ***dairies*, *farm buildings*** [1, 2]; ***livestock houses*, *poultry houses*** [1-3]

Approval information

- Products formulated for ULV application [1, 2]. May be applied through fogging machine or sprayer [1]. See label for details

Efficacy guidance

- For fly control close doors and windows and spray or apply fog as appropriate

FOR FULL CONDITIONS OF USE ALWAYS READ THE PRODUCT LABEL

- For best results outdoors spray during early morning or late afternoon and evening when conditions are still
- Ensure spray is directed to all areas where flies are seen to be numerous
- To avoid possibility of development of resistance do not spray more frequently than once per week

Restrictions

- For use only by professional operators
- Do not allow spray to contact open food products or food preparing equipment or utensils
- Remove exposed milk and collect eggs before application
- Do not treat plants
- Do not use space sprays containing pyrethrins or pyrethroid more than once per week in intensive or controlled environment animal houses in order to avoid development of resistance. If necessary, use a different control method or product [2]

Crop-specific information

- Product formulated for use through Turbair Flydowner machines [3]

Environmental safety

- Dangerous for the environment
- Toxic to aquatic organisms
- Do not apply directly to livestock
- Exclude all persons and animals during treatment [1]

Hazard classification and safety precautions

Hazard H03 [1-3]; H04 [1]; H11 [2]
Risk phrases R22a, R36, R38 [1]; R22b, R58 [2, 3]; R51 [2]; R66 [3]
Operator protection A [1-3]; B [1]; C [2, 3]; D, H [2]; E [1, 2]; P [3]; U02b, U14 [2, 3]; U05a [1, 3]; U09a [1]; U09b, U20a [2]; U11, U15 [3]; U19a, U20b [1-3]
Environmental protection E02a [3] (until surfaces dry); E05a [1, 2]; E13c, E32a [1]; E15a, E31a, E32d, E38 [2]; E32e, E34 [3]
Consumer protection C04, C10 [1]; C06, C11 [1-3]; C07, C08, C09 [1, 2]; C12 [2]
Storage and disposal E01, E30a [1-3]; E04 [1, 3]; E26 [2]; E27 [2, 3]
Medical advice M05b [2, 3]

371 pyridate

A contact pyridazine herbicide for cereals, maize and brassicas

Products

Lentagran WP	Syngenta	45% w/w	WB	08478

Uses

- Annual dicotyledons in ***spring oilseed rape*** *(off-label)*, ***winter oilseed rape*** *(off-label)*
- Black nightshade in ***brussels sprouts***, ***cabbages***, ***maize***, ***onions***
- Cleavers in ***brussels sprouts***, ***cabbages***, ***maize***, ***onions***, ***spring oilseed rape*** *(off-label)*, ***winter oilseed rape*** *(off-label)*
- Fat hen in ***brussels sprouts***, ***cabbages***, ***maize***, ***onions***

Specific Off-Label Approvals (SOLAs)

- ***spring oilseed rape***, ***winter oilseed rape*** *(OLA 011663) Dec 2008* [1]

Approval information

- Pyridate included in Annex I under EC Directive 91/414

Efficacy guidance

- Best results achieved by application to actively growing weeds at 6-8 leaf stage when temperatures are above 8°C before crop foliage forms canopy

Restrictions

- Maximum number of treatments 1 per crop
- Do not apply in mixture with or within 14 d of any other product which may result in dewaxing of crop foliage
- Apply to maize after first-leaf stage. Do not use on cv. Meritos, Sunrise or Tainon 236
- Do not use on crops suffering stress from frost, drought, disease or pest attack

Crop-specific information

- Latest use: before 7 leaf stage for maize; before 5 true leaves for onions; before flower buds visible for oilseed rape
- HI 6 wk for Brussels sprouts, cabbages
- Apply to cabbages and Brussels sprouts after 4 fully expanded leaf stage. Allow 2 wk after transplanting before treating

Environmental safety

- Dangerous for the environment
- Toxic to aquatic organisms

Hazard classification and safety precautions

Hazard H04, H11
Risk phrases R43, R51, R58
Operator protection A, C; U05a, U14, U22a
Environmental protection E15a, E31c, E32d, E38
Storage and disposal E01, E04, E30a

372 pyrimethanil

An anilinopyrimidine fungicide for apples and strawberries

Products

Scala	BASF	400 g/l	SC	11695

Uses

- Botrytis in ***aubergines*** *(off-label)*, ***bilberries*** *(off-label)*, ***blackcurrants*** *(off-label)*, ***cranberries*** *(off-label)*, ***gooseberries*** *(off-label)*, ***grapevines*** *(off-label)*, ***outdoor tomatoes*** *(off-label)*, ***protected blackberries*** *(off-label)*, ***protected boysenberries*** *(off-label)*, ***protected lettuce*** *(off-label)*, ***protected loganberries*** *(off-label)*, ***protected marionberries*** *(off-label)*, ***protected raspberries*** *(off-label)*, ***protected sunberries*** *(off-label)*, ***protected tayberries*** *(off-label)*, ***protected tomatoes*** *(off-label)*, ***redcurrants*** *(off-label)*, ***strawberries***, ***whitecurrants*** *(off-label)*
- Scab in ***apples***

Specific Off-Label Approvals (SOLAs)

- ***aubergines***, ***outdoor tomatoes***, ***protected tomatoes*** *(OLA 021321) Dec 2008* [1]
- ***bilberries***, ***blackcurrants***, ***cranberries***, ***gooseberries***, ***redcurrants***, ***whitecurrants*** *(OLA 001883) Dec 2008* [1]
- ***grapevines*** *(OLA 021414) Dec 2008* [1]
- ***protected blackberries***, ***protected boysenberries***, ***protected loganberries***, ***protected marionberries***, ***protected raspberries***, ***protected sunberries***, ***protected tayberries*** *(OLA 023411) Dec 2008* [1]
- ***protected lettuce*** *(OLA 001590) Dec 2008* [1]

Efficacy guidance

- On apples a programme of sprays will give early season control of scab. Season long control can be achieved by continuing programme with other approved fungicides
- In strawberries product should be used as part of a programme of disease control treatments which should alternate with other materials to prevent or limit development of less sensitive strains of grey mould

Restrictions

- Maximum number of treatments 5 per yr for apples; 4 per yr for protected crops; 3 per yr for grapevines; 2 per yr for cane fruit, bush fruit, strawberries, tomatoes
- Product does not taint apples. Processors should be consulted before use on strawberries

Crop-specific information

- Latest use: before end of flowering for apples
- HI protected cane fruit, strawberries 1 d; aubergines, tomatoes 3 d; protected lettuce 14 d; bush fruit, grapevines 21 d
- All varieties of apples and strawberries may be treated

- Treat apples from bud burst at 10-14 d intervals
- In strawberries start treatments at white bud to give maximum protection of flowers against grey mould and treat every 7-10 d. Product should not be used more than once in a 3 or 4 spray programme

Environmental safety

- Harmful to aquatic organisms
- Product has negligible effect on hoverflies and lacewings. Limited evidence indicates some margin of safety to *Typhlodromus pyri*
- Broadcast air-assisted LERAP (20 m)

Hazard classification and safety precautions

Risk phrases R52, R58
Operator protection U08, U20b
Environmental protection E15a, E31b, E32d; E17b (20 m)
Storage and disposal E01, E04, E26, E29b, E30a

SECTION 2

373 quinmerac

A residual herbicide available only in mixtures

See also *chloridazon + quinmerac*
metazachlor + quinmerac

374 quinoxyfen

A systemic protectant fungicide for cereals

See also *fenpropimorph + quinoxyfen*

Products

1	Apres	Dow	500 g/l	SC	08881
2	Erysto	Dow	250 g/l	SC	08697
3	Fortress	Dow	500 g/l	SC	08279
4	Me2 After	Me2	500 g/l	SC	10463

Uses

- Powdery mildew in ***durum wheat***, ***spring oats***, ***spring rye***, ***sugar beet***, ***triticale***, ***winter oats***, ***winter rye*** [1-3]; ***spring barley***, ***spring wheat***, ***winter barley***, ***winter wheat*** [1-4]

Approval information

- Quinoxyfen included in Annex I under Directive 91/414

Efficacy guidance

- For best results treat at early stage of disease development before infection spreads to new crop growth. Further treatment may be necessary if disease pressure remains high
- Product not curative and will not control latent or established disease infections
- For broad spectrum control in cereals use in tank mixtures - see label
- Product rainfast after 1 h
- Systemic activity may be reduced in severe drought

Restrictions

- Maximum total dose equivalent to two full dose treatments in cereals; see labels for split dose recommendation in sugar beet
- On cereals apply only in the spring from mid-tillering stage (GS 25)

Crop-specific information

- Latest use: first awns visible stage (GS 49) for cereals
- HI sugar beet 28 d

Environmental safety

- Dangerous for the environment
- Very toxic to aquatic organisms

Hazard classification and safety precautions

Hazard H04, H11

SEE SECTION 3 FOR PRODUCTS ALSO REGISTERED

Risk phrases R43, R50, R58
Operator protection A, C, H; U05a, U14
Environmental protection E15a, E16a, E16b, E32d, E34, E38
Storage and disposal E01, E04

375 quizalofop-P-ethyl

An aryl phenoxypropionic acid post-emergence grass herbicide

Products

1	CoPilot	Interfarm	100 g/l	EC	08042
2	Sceptre	Bayer CropScience	250 g/l	SC	08043

Uses

- Annual grasses in ***combining peas, fodder beet, linseed, mangels, red beet, spring field beans, spring oilseed rape, sugar beet, vining peas, winter field beans, winter oilseed rape***
- Couch in ***combining peas, fodder beet, linseed, mangels, red beet, spring field beans, spring oilseed rape, sugar beet, vining peas, winter field beans, winter oilseed rape***
- Perennial grasses in ***combining peas, fodder beet, linseed, mangels, red beet, spring field beans, spring oilseed rape, sugar beet, vining peas, winter field beans, winter oilseed rape***
- Volunteer cereals in ***combining peas, fodder beet, linseed, mangels, red beet, spring field beans, spring oilseed rape, sugar beet, vining peas, winter field beans, winter oilseed rape***

Efficacy guidance

- Best results achieved by application to emerged weeds growing actively in warm conditions with adequate soil moisture. Use split treatment to extend period of control in oilseed rape
- Weed control may be reduced under conditions such as drought that limit uptake and translocation
- Annual meadow-grass is not controlled
- Various spray programmes and tank-mixtures recommended to control mixed dicotyledon/grass weed populations. See label for details
- For effective couch control do not hoe beet crops within 21 d after spraying
- At least 2 h without rain should follow application otherwise results may be reduced
- Must be used with recommended authorised adjuvant [2]

Restrictions

- Maximum number of treatments 1 on peas and beans, 2 as split dose on oilseed rape, 2 for couch control in all other recommended crops. See labels for maximum total dose of individual products
- Do not spray crops under stress from any cause or in frosty weather
- May cause taint in peas. Consult processor before use on peas or beans for processing
- An interval of at least 3 d must elapse between treatment and use of another herbicide on beet crops, 14 d on oilseed rape, 21 d on linseed and other recommended crops

Crop-specific information

- HI 16 wk for beet crops; 11 wk for oilseed rape, linseed; 8 wk for field beans; 5 wk for peas
- In some situations treatment can cause yellow patches on foliage of peas, especially vining varieties. Symptome usually rapidly and completely outgrown

Following crops guidance

- In the event of crop failure broad-leaved crops may be resown at any time, cereals, onions, leeks or maize may be sown after 2-6 wk depending on dose used. See label for details. Onions, leeks and maize are not recommended to follow a failed treated crop

Environmental safety

- Dangerous for the environment
- Toxic to aquatic organisms

Hazard classification and safety precautions

Hazard H03, H08 [1]; H11 [1, 2]
Risk phrases R22b, R37, R41, R66, R67 [1]; R51, R58 [1, 2]
Operator protection A, C; U05a, U11, U20b [1]; U09a, U20a [2]
Environmental protection E13b, E32d, E38 [1, 2]; E31b [1]; E31c, E34 [2]
Storage and disposal E01, E04 [1]; E26, E30a [1, 2]
Medical advice M05b [1]

376 resmethrin

A contact acting pyrethroid insecticide available only in mixtures

377 resmethrin + tetramethrin

A contact acting pyrethroid insecticide mixture

Products

Sorex Wasp Nest Destroyer	Sorex	0.10:0.05% w/w	AE	H6294

Uses
- Wasps in ***miscellaneous pest control situations***

Efficacy guidance
- Product acts by lowering temperature in and around nest and producing a stupefying vapour in addition to its contact insecticidal effect. Spray only on surfaces
- Apply by directing jet at nest from up to 3 m away and approach nest gradually until jet can be directed into entrance hole. Continue spraying until nest saturated

Restrictions
- For use only by trained operators and owners of commercial and agricultural premises

Environmental safety
- Harmful to caged birds and pets
- Dangerous to bees. Do not apply to crops in flower or to those in which bees are actively foraging. Do not apply when flowering weeds are present
- Extremely dangerous to fish or other aquatic life. Remove fish tanks before spraying

Hazard classification and safety precautions

Hazard H03
Risk phrases R20
Operator protection A, C, H; U14, U19a, U20a
Environmental protection E10b, E12d, E13a, E15a
Storage and disposal E01, E04, E30a

378 rimsulfuron

A selective systemic sulfonylurea herbicide

Products

1	Me2 Aducksbackside	Me2	25% w/w	SG	10065
2	Standon Rimsulfuron	Standon	25% w/w	SG	09955
3	Tarot	Makhteshim	25% w/w	SG	11896
4	Titus	Makhteshim	25% w/w	SG	11895

Uses
- Annual dicotyledons in ***forage maize, potatoes***
- Volunteer oilseed rape in ***forage maize, potatoes***

Efficacy guidance
- Product should be used with a suitable adjuvant or a suitable herbicide tank-mix partner. See label for details
- Product acts by foliar action. Best results obtained from good spray cover of small actively growing weeds. Effectiveness is reduced in very dry conditions
- Split application provides control over a longer period and improves control of fat hen and polygonums
- Weed spectrum can be broadened by tank mixture with other herbicides. See label for details
- Susceptible weeds cease growth immediately and symptoms can be seen 10 d later
- Rimsulfuron is a member of the ALS-inhibitor group of herbicides

Restrictions
- Maximum number of treatments 2 per crop
- Do not treat maize previously treated with organophosphorus insecticides
- Avoid high light intensity (full sunlight) and high temperatures on the day of spraying

- Do not treat during periods of substantial diurnal temperature fluctuation or when frost anticipated
- Do not apply to any crop stressed by drought, water-logging, low temperatures, pest or disease attack, nutrient or lime deficiency

Crop-specific information

- Latest use: before most advanced potato plants are 25 cm high; before 4-collar stage of fodder maize
- All varieties of ware potatoes and fodder maize may be treated, but variety restrictions of any tank-mix partner must be observed

Following crops guidance

- Only winter wheat should follow a treated crop in the same calendar yr
- Only barley, wheat or maize should be sown in the spring of the yr following treatment
- In the second autumn after treatment any crop except brassicas or oilseed rape may be drilled

Environmental safety

- Dangerous for the environment
- Toxic to aquatic organisms
- Herbicide is very active. Take particular care to avoid drift onto plants outside the target area
- Spraying equipment should not be drained or flushed onto land planted, or to be planted, with trees or crops other than potatoes or forage maize and should be thoroughly cleansed after use - see label for instructions

Hazard classification and safety precautions

Hazard H11 [1, 2]
Risk phrases R51, R58 [1, 2]
Operator protection U08, U19a [1-4]; U20a [3, 4]; U20b [1, 2]
Environmental protection E13a [3, 4]; E15a [1, 2]; E32a, E34 [1-4]
Storage and disposal E01, E30a

379 rotenone

A natural, contact insecticide of low persistence

Products

FS Liquid Derris	Ford Smith	50 g/l	EC	01213

Uses

- Aphids in ***all soft fruit***, ***all top fruit***, ***miscellaneous flowers***, ***ornamental specimens***, ***protected crops***, ***vegetables***
- Raspberry beetle in ***cane fruit***
- Sawflies in ***gooseberries***
- Slug sawfly in ***pears***, ***roses***

Efficacy guidance

- Apply as high volume spray when pest first seen and repeat as necessary
- Spray raspberries at first pink fruit, loganberries when most of blossom over, blackberries as first blossoms open. Repeat 2 wk later if necessary
- Spray to obtain thorough coverage, especially on undersurfaces of leaves

Crop-specific information

- HI 1 d

Environmental safety

- Flammable
- Dangerous to fish or other aquatic life. Do not contaminate surface waters or ditches with chemical or used container

Hazard classification and safety precautions

Hazard H08
Operator protection U20a
Environmental protection E13b, E31a, E34
Consumer protection C02
Storage and disposal E30a

380 silthiofam

A benzamide fungicide seed dressing for cereals

Products

Latitude	Monsanto	125 g/l	FS	10695

Uses

- Take-all in ***spring wheat*** *(seed treatment)*, ***winter barley*** *(seed treatment)*, ***winter wheat*** *(seed treatment)*

Efficacy guidance

- Apply using approved seed treatment equipment which has been accurately calibrated
- Apply simultaneously (i.e. not in mixture) with a standard seed treatment
- Drill treated seed in the season of purchase. The viability of treated seed and fungicide activity may be reduced by physical storage
- Drill treated seed at 2.5-4 cm into a well prepared firm seedbed
- As precaution against possible development of disease resistance do not treat more than three consecutive susceptible cereal crops in any one rotation

Restrictions

- Maximum number of treatments one per batch of seed
- Do not use on seed with more than 16% moisture content, or on sprouted, cracked or skinned seed
- Test germination of all seed batches before treatment

Crop-specific information

- Latest use: immediately prior to drilling

Environmental safety

- Harmful to fish or other aquatic life. Do not contaminate surface waters or ditches with chemical or used container

Hazard classification and safety precautions

Operator protection A, H; U20b
Environmental protection E03, E15a, E31a, E34
Storage and disposal E26, E30a
Treated seed S01, S02, S04b, S05, S06a, S07

381 simazine

A soil-acting triazine herbicide with restricted permitted uses

See also *isoproturon + simazine*

Products

1	Alpha Simazine 50 SC	Makhteshim	500 g/l	SC	04801
2	Gesatop	Syngenta	500 g/l	SC	08412
3	Sipcam Simazine Flowable	Sipcam	500 g/l	SC	07622

Uses

- Annual dicotyledons in ***asparagus***, ***nursery stock***, ***rhubarb***, ***trees and shrubs***, ***woody ornamentals*** [3]; ***asparagus*** *(off-label)*, ***rhubarb*** *(off-label)*, ***runner beans*** *(off-label)* [1, 2]; ***broad beans***, ***spring field beans***, ***winter field beans*** [1-3]; ***ornamental plant production***, ***roses*** [2, 3]; ***ornamental plant production*** *(for resale only)* [1]
- Annual grasses in ***asparagus***, ***nursery stock***, ***rhubarb***, ***trees and shrubs***, ***woody ornamentals*** [3]; ***asparagus*** *(off-label)*, ***rhubarb*** *(off-label)*, ***runner beans*** *(off-label)* [1, 2]; ***broad beans***, ***spring field beans***, ***winter field beans*** [1-3]; ***ornamental plant production***, ***roses*** [2, 3]; ***ornamental plant production*** *(for resale only)* [1]

Specific Off-Label Approvals (SOLAs)

- ***asparagus***, ***rhubarb*** *(OLA 011660) Dec 2008* [2]
- ***asparagus***, ***rhubarb***, ***runner beans*** *(OLA 021255) Dec 2008* [1]
- ***runner beans*** *(OLA 011662) Dec 2008* [2]

Approval information

- Approvals for sale, supply and use of simazine on non-crop land were revoked in 1992 and 1993
- Products containing this active ingredient have been granted derogations for specified 'Essential Uses' for use until 31 December 2007. Sale and supply must cease by 30 June 2007 but growers have no guarantee that the products will continue to be available until then.
  For more information see 'The Review Programme' under 'Pesticide Legislation' in Section 5

Efficacy guidance

- Active via root uptake. Best results achieved by application to fine, firm, moist soil, free of established weeds, when rain falls after treatment
- Do not use on highly organic soils as efficacy will be severely impaired
- Following repeated use of simazine or other triazine herbicides resistant strains of groundsel and some other annual weeds may develop

Restrictions

- Use must be restricted to one product containing simazine or atrazine applied either as a single application at the maximum approved dose or (subject to any existing maximum permitted number of treatments) to several applications at lower doses up to the maximum approved dose for a single application
- Do not spray beans on sandy or gravelly soils or cultivate after treatment. Do not treat varieties Beryl, Feligreen or Rowena
- Allow at least 7 mth before drilling or planting other crops, longer if weather dry
- Do not sow oats in autumn following spring application in maize

Crop-specific information

- Latest use: before emergence of spears for asparagus; 7 d after drilling for sweetcorn; before end of Feb (autumn planted), 10 d after drilling for spring planted field and broad beans; before end of Mar for top, bush and cane fruit; after harvest, before end of Nov for strawberries; after harvest, before 1 May for hops.
- HI mint and sage 4 mth; rhubarb 3 mth; herbs 14 wk; asparagus 5 wk [1, 2]
- Apply in top, bush and cane fruit and woody ornamentals established at least 12 mth in Feb-Mar. Roses may be sprayed immediately after planting. See label for lists of resistant and susceptible species
- Apply to strawberries in Jul-Dec, not in spring. Do not treat spring-planted crops established less than 6 mth, or winter-planted crops less than 9 mth. Do not spray varieties Huxley Giant, Madame Montal or Regina
- Apply to hops overall in Feb-Apr before weeds emerge. Newly planted sets must be covered with minimum of 50 mm soil
- Apply to forest nursery seedbed in second yr or to transplant lines after plants 5 cm tall. Do not treat Norway spruce (Christmas trees)
- On Sands, stony or gravelly soils there is risk of crop damage, especially with heavy rain

Environmental safety

- Dangerous for the environment
- Very toxic to aquatic organisms
- LERAP Category B
- Use must be restricted to one product containing atrazine or simazine, and either to a single application at the maximum approved rate or (subject to any existing maximum permitted number of treatments) to several applications at lower doses up to the maximum approved rate for a single application
- To reduce soil run-off on gradients especially in orchard and forest plantations, users are advised to plant grass strips or leave 6 m wide strips between treated areas and surface waters

Hazard classification and safety precautions

Hazard H03, H11 [1, 2]
Risk phrases R40, R50, R58 [1, 2]; R43 [2]
Operator protection A, C, H [1-3]; B [3]; D, M [1, 3]; U05a, U14 [1, 2]; U15 [1]; U20c [1-3]; U23a [2]
Environmental protection E13b, E31a [3]; E15a [1, 2]; E16a [1-3]; E16b [1, 3]; E31b, E32e [1]; E31c, E32d, E38 [2]
Storage and disposal E01, E04 [1]; E26 [2, 3]; E30a [1-3]
Medical advice M05a [2]

382 sodium chlorate

A non-selective inorganic herbicide for total vegetation control

Products

1	Atlacide Soluble Powder	Nomix Enviro	58% w/w	SP	00125
2	Doff Sodium Chlorate Weedkiller	Doff Portland	53% w/w	SP	06049

Uses
- Total vegetation control in ***land not intended to bear vegetation*** [1]; ***non-crop areas, paths and drives*** [2]

Efficacy guidance
- Active through foliar and root uptake
- Apply as overall spray at any time during growing season. Best results obtained from application to moist soil in spring or early summer

Restrictions
- Maximum number of treatments 2 per yr for non-crop areas, paths and drives [2]
- Treated ground must not be replanted for at least 6 mth after treatment [2]
- Do not apply before heavy rain

Environmental safety
- Dangerous for the environment
- Contact with combustible material may cause fire
- Harmful to aquatic organisms
- Keep livestock out of treated areas for at least two weeks following treatment and until poisonous weeds, such as ragwort, have died down and become unpalatable
- Clothing, paper, plant debris etc become highly inflammable when dry if contaminated with sodium chlorate
- Fire risk has been reduced by inclusion of fire depressant but product should not be used in areas of exceptionally high fire risk eg oil installations, timber yards
- Wash clothing thoroughly after use
- If clothes become contaminated do not stand near an open fire
- Must not be sold to persons under 18
- Keep livestock out of treated areas until foliage of any poisonous weeds such as ragwort has died and become unpalatable

Hazard classification and safety precautions

Hazard H03, H09 [1, 2]; H11 [1]
Risk phrases R08, R22a [1, 2]; R52 [1]
Operator protection A, H [2]; L [1]; M; U02a [1]; U05a, U20b [1, 2]
Environmental protection E07a, E32d [1]; E13c [2]; E15a, E32a, E34 [1, 2]
Storage and disposal E01, E04, E30a [1, 2]; E29a [2]
Medical advice M03 [1, 2]; M05a [2]

383 sodium chloride (commodity substance)

An inorganic salt for use as a sugar beet herbicide

Products

sodium chloride	various	-	SL

Uses
- Polygonums in ***sugar beet***
- Volunteer potatoes in ***sugar beet***

Approval information
- Approval for the use of sodium chloride as a commodity substance was granted on 11 September 1996 by Ministers under regulation 5 of the Control of Pesticides Regulations 1986

Efficacy guidance
- Best results obtained from good spray cover treatment in hot, humid weather
- Use as a follow-up treatment where earlier sprays have failed to control large volunteer potatoes or polygonums

SEE SECTION 3 FOR PRODUCTS ALSO REGISTERED

- Acts by contact scorch and has no direct effect on daughter potato tubers
- Saturated spray solution should contain 0.1% w/w non-ionic wetter and be applied at 1000 l/ha
- Spray between three true leaf stage of crops and end Jul
- Crop scorch may occur after treatment

384 sodium hypochlorite (commodity substance)

An inorganic horticultural bactericide for use in mushrooms

Products

sodium hypochlorite	various	100% w/v	ZZ

Uses

- Bacterial blotch in ***mushrooms***

Approval information

- Approval for the use of sodium hypochlorite as a commodity substance was granted on 5 December 1996 by Ministers under regulation 5 of the Control of Pesticides Regulations 1986

Restrictions

- Maximum concentration 315 mg/litre of water
- Mixing and loading must only take place in a ventilated area
- Must only be used by suitably trained and competent operators

Crop-specific information

- HI 1 d

Environmental safety

- Harmful to fish or other aquatic life. Do not contaminate surface waters or ditches with chemical or used container

Hazard classification and safety precautions

Operator protection A, C, H
Environmental protection E13c

385 sodium monochloroacetate

A contact herbicide for various horticultural crops

Products

1	Atlas Somon	Nufarm UK	96% w/w	SP	07727
2	Croptex Steel	Certis	95% w/w	SP	02418

Uses

- Annual dicotyledons in ***broccoli*** *(off-label)*, ***calabrese*** *(off-label)*, ***cauliflowers*** *(off-label)* [2]; ***brussels sprouts***, ***bulb onions***, ***cabbages***, ***kale***, ***leeks*** [1, 2]; ***salad onions*** [1]
- Basal defoliation in ***hops*** *(off-label)* [2]
- Sucker control in ***blackberries*** *(off-label)*, ***loganberries*** *(off-label)*, ***raspberries*** *(off-label)*, ***rubus hybrids*** *(off-label)* [2]
- Sucker inhibition in ***blackberries*** *(off-label)*, ***loganberries*** *(off-label)*, ***raspberries*** *(off-label)*, ***rubus hybrids*** *(off-label)* [1]

Specific Off-Label Approvals (SOLAs)

- ***blackberries***, ***loganberries***, ***raspberries***, ***rubus hybrids*** *(OLA 041460) Dec 2007* [1]
- ***blackberries***, ***loganberries***, ***raspberries***, ***rubus hybrids*** *(OLA 930115) Dec 2008* [2]
- ***broccoli***, ***calabrese***, ***cauliflowers*** *(OLA 992602) Dec 2008* [2]
- ***hops*** *(OLA 930357) Dec 2008* [2]

Approval information

- Products containing this active ingredient have been granted derogations for specified 'Essential Uses' for use until 31 December 2007. Sale and supply must cease by 30 June 2007 but growers have no guarantee that the products will continue to be available until then.
  For more information see 'The Review Programme' under 'Pesticide Legislation' in Section 5

Efficacy guidance

- Sodium monochloroacetate acts by contact action. Best results achieved by application to emerged weed seedlings up to young plant stage in warm humid weather when weeds are growing actively
- Effectiveness reduced by rain within 12 h

Restrictions

- Maximum number of treatments 1 per crop; 2 per yr for sucker control in cane fruit
- Do not spray cabbage that has begun to heart
- Do not treat transplanted onions or onion sets
- Safety on brassicas, onions and leeks depends on presence of adequate leaf wax, check by crystal violet wax test. Do not add wetters, pesticides or nutrients to spray
- These products must not be used on any crops other than those listed, including any extrapolations that would normally be permissible under the Long Term Arrangements for Extension of Use (see Section 5)
- Do not spray if frost likely or if temperature likely to exceed 27°C
- Do not apply any other pesticide within 7 d after use

Crop-specific information

- Latest use: before 4 true leaf stage for onions and leeks.
- HI 21 d for cabbage, kale, Brussels sprouts; 28 d for cane fruit; 14 wk for hops [2]
- Apply to brassicas from 2-4 leaf stage or after recovery from transplanting
- Apply to onions and leeks after crook stage but before 4-leaf stage
- Apply to cane fruit crops established for at least 1 yr as a directed spray
- Apply for sucker control in raspberries when canes 10-20 cm high with addition of Wayfarer adjuvant [2]

Environmental safety

- Corrosive. Causes burns
- Dangerous for the environment [2]
- Very toxic to aquatic organisms. May cause long-term adverse effects in the aquatic environment [2]
- Do not apply directly to livestock/poultry [1]
- Keep livestock, especially poultry, out of treated areas for at least 2 wk
- Harmful to bees. Do not apply to crops in flower or to those in which bees are actively foraging. Do not apply when flowering weeds are present

Hazard classification and safety precautions

Hazard H02, H11 [1, 2]; H05 [2]
Risk phrases R20, R21, R22a [2]; R25, R34, R41, R50, R58 [1, 2]; R38 [1]
Operator protection A, C, D, E; U05a, U10, U11, U19a, U20b
Environmental protection E05b [1] (2wk); E05b [2] (2 wk); E06a [1, 2] (2 wk); E12f, E34 [1, 2]; E15a, E31a, E32d, E38 [1]; E32a [2]
Storage and disposal E01, E04, E30a
Medical advice M04

386 spinosad

A selective insecticide derived from naturally occurring soil fungi (naturalyte)

Products

1	Conserve	Fargro	120 g/l	SC	11011
2	Conserve	Fargro	120 g/l	SC	12058

Uses

- Western flower thrips in ***protected cucumbers***, ***protected ornamentals***

Efficacy guidance

- Product enters insects by contact from a treated surface or ingestion of treated plant material therefore good spray coverage is essential
- Some plants, for example Fuchsia flowers, can provide effective refuges from spray deposits and control may be reduced
- Apply when nymphs or adults are first seen

SEE SECTION 3 FOR PRODUCTS ALSO REGISTERED

- Monitor pest development carefully to see whether further applications are necessary. A 2 or 3 spray programme at 5-7 d intervals may be needed when conditions favour rapid pest development
- To reduce possibility of development of resistance, adopt resistance management measures. See label and Section 5

Restrictions
- Maximum number of treatments 3 per crop for protected cucumbers or 6 per structure per yr for protected ornamentals
- Establish whether any incoming plants have been treated and apply no more than 3 consecutive sprays. Maximum number of treatments 6 per structure per yr
- Avoid application in bright sunlight or into open flowers

Crop-specific information
- HI 3 d for protected cucumbers
- Test for tolerance on a small number of ornamentals or cucumbers before large scale treatment
- Some spotting of african violet flowers may occur

Environmental safety
- Dangerous for the environment
- Toxic to aquatic organisms
- Whenever possible use an Integrated Pest Management system
- Product has low impact on many insect and mite predators but is harmful to adults of most parasitic wasps. Most beneficials may be introduced to treated plants when spray deposits are dry but an interval of 2 wk should elapse before introduction of parasitic wasps. See label for details
- Treatment may cause temporary reduction in abundance of insect and mite predators if present at application
- Exposure to direct spray is harmful to bumble bees but dry spray deposits are harmless

Hazard classification and safety precautions

Hazard H11
Risk phrases R51, R58
Operator protection A, H; U05a
Environmental protection E15a, E31b, E32d, E34, E38
Storage and disposal E01, E04, E26

387 spiromesifen

A ketoenol insecticide for protected tomatoes

Products

Oberon	Bayer CropScience	240 g/l	SC	11819

Uses
- Whitefly in ***inert substrate tomatoes***, ***nft tomatoes***

Efficacy guidance
- Apply to run off and ensure that all sides of the crop are covered
- Apply when infestation first observed and repeat 7-10 d later if necessary
- Always follow guidelines for preventing and managing insect resistance. Section 5 for more information

Restrictions
- Maximum number of treatments 2 per cropping cycle
- Do not use on cherry tomatoes or on any outdoor crops
- Product should be used in rotation with compounds with different modes of action

Crop-specific information
- HI 3 d for tomatoes

Environmental safety
- Dangerous for the environment
- Very toxic to aquatic organisms

Hazard classification and safety precautions

Hazard H04, H11
Risk phrases R43, R50, R58

Operator protection A, H; U05a, U14, U20b
Environmental protection E15a, E32a, E32d, E34, E38
Storage and disposal E01, E04, E26, E30a

388 spiroxamine

A spiroketal amine fungicide for cereals

Products

1	Torch	Bayer CropScience	500 g/l	EW	11258
2	Zenon	Bayer CropScience	800 g/l	EC	11232

Uses

- Brown rust in ***spring barley***, ***spring rye***, ***spring wheat***, ***winter barley***, ***winter rye***, ***winter wheat*** [1, 2]
- Powdery mildew in ***spring barley***, ***spring rye***, ***spring wheat***, ***winter barley***, ***winter rye***, ***winter wheat*** [1, 2]
- Rhynchosporium in ***spring barley*** *(reduction)*, ***winter barley*** *(reduction)* [1, 2]; ***spring rye*** *(reduction)*, ***winter rye*** *(reduction)* [1]
- Yellow rust in ***spring barley***, ***spring rye***, ***spring wheat***, ***winter barley***, ***winter rye***, ***winter wheat*** [1, 2]

Approval information

- Spiroxamine included in Annex I under EC Directive 91/414

Efficacy guidance

- For best results treat at an early stage of disease development before infection spreads to new growth
- Active against net blotch when applied at first signs of disease but better control achieved with mixtures. See label for details
- To reduce the risk of development of resistance avoid repeat treatments on diseases such as powdery mildew. If necessary tank mix or alternate with other non-morpholine fungicides

Restrictions

- Maximum total dose equivalent to two full dose treatments

Crop-specific information

- Latest use: before caryopsis watery ripe (GS 71) for spring wheat, winter rye, winter wheat; ear emergence complete (GS 59) for spring barley, winter barley

Environmental safety

- Dangerous for the environment
- Very toxic to aquatic organisms
- LERAP Category B

Hazard classification and safety precautions

Hazard H03, H11
Risk phrases R20, R22a, R38, R41, R43, R50, R58
Operator protection A, C, H; U05a, U11, U13, U14, U15, U20a
Environmental protection E13b, E16a, E31b, E32d, E34, E38
Storage and disposal E01, E04, E26, E30a
Medical advice M03 [1, 2]; M05a [2]

389 spiroxamine + tebuconazole

A broad spectrum systemic fungicide mixture for cereals

Products

1	Beam	Bayer CropScience	250:133 g/l	EW	11255
2	Sage	Bayer CropScience	250:133 g/l	EW	11303

Uses

- Brown rust in ***spring barley***, ***spring rye***, ***spring wheat***, ***winter barley***, ***winter rye***, ***winter wheat***
- Fusarium ear blight in ***spring wheat***, ***winter wheat***

- Glume blotch in ***spring wheat, winter wheat***
- Net blotch in ***spring barley, winter barley***
- Powdery mildew in ***spring barley, spring rye, spring wheat, winter barley, winter rye, winter wheat***
- Rhynchosporium in ***spring barley, winter barley***
- Septoria leaf spot in ***spring wheat, winter wheat***
- Sooty moulds in ***spring wheat, winter wheat***
- Yellow rust in ***spring barley, spring rye, spring wheat, winter barley, winter rye, winter wheat***

Approval information

- Spiroxamine included in Annex I under EC Directive 91/414

Efficacy guidance

- Best results achieved from treatment at early stage of disease development before infection spreads to new growth
- To protect flag leaf and ear from Septoria apply from flag leaf emergence to ear fully emerged (GS 37-59). Earlier treatment may be necessary where there is high disease risk
- Control of rusts, powdery mildew, leaf blotch and net blotch may require second treatment 2-3 wk later

Restrictions

- Maximum total dose equivalent to two full dose treatments
- Do not use on durum wheat

Crop-specific information

- Latest use: before caryopsis watery ripe (GS 71) for rye, wheat; ear emergence complete (GS 59) for barley
- Some transient leaf speckling may occur on wheat but this has not been shown to reduce yield reponse or disease control

Environmental safety

- Dangerous for the environment
- Very toxic to aquatic organisms
- LERAP Category B

Hazard classification and safety precautions

Hazard H03 [1, 2]; H11 [2]
Risk phrases R20, R22a, R38, R41, R43, R50, R58
Operator protection A, C, H; U05a, U11, U13, U14, U15, U20b
Environmental protection E13b, E16a, E16b, E31b, E32d, E34, E38
Storage and disposal E01, E04, E26, E30a
Medical advice M03

390 strychnine hydrochloride (commodity substance)

A vertebrate control agent for destruction of moles underground

Products

strychnine	various	-	ZZ

Uses

- Moles in ***grassland*** *(areas of restricted public access)*, ***miscellaneous pest control situations*** *(areas of restricted public access)*

Approval information

- Approval for the use of strychnine hydrochloride as a commodity substance was granted on 19 June 1997 by Ministers under regulation 5 of the Control of Pesticides Regulations 1986

Efficacy guidance

- Only for use as poison bait against moles on commercial agricultural/horticultural land where public access restricted, on grassland associated with aircraft landing strips, horse paddocks, race and golf courses and other areas specifically approved by Defra in England and the appropriate authorities in Wales and Scotland

Restrictions

- Strychnine is subject to the Poisons Rules 1982 and the Poisons Act 1972. See notes in Section 5

- Must only be supplied to holders of a permit issued by Defra in England or the appropriate authorities in Wales and Scotland. Permits may only be issued to persons who satisfy the appropriate authority that they are trained and competent in its use
- Only to be supplied in original sealed pack in units up to 2 g. Quantities of more than 8 g may be supplied only to providers of a commercial service
- Other restrictions apply, see the PSD website at: www.pesticides.gov.uk/approvals.asp?id=310

Environmental safety

- Store in original container under lock and key and only on premises under control of a permit holder or of a named individual who satisfies the designated requirements of competence
- A written COSHH assessment must be made before use
- Must be prepared for application with great care so that there is no contamination of the ground surface
- Any prepared bait remaining at the end of the day must be buried
- Operators must be provided with a Chemicals (Hazard Information and Supply) (CHIP) Safety Data Sheet, obtainable from the supplier, before commencing work

Hazard classification and safety precautions

Operator protection A

391 sulfosulfuron

A sulfonylurea herbicide for grass and broad-leaved weed control in winter wheat

Products

1	Ag-Chem Prefect	Ag-Chem	80% w/w	WG	11985
2	Landgold Sulfosulfuron	Landgold	80% w/w	WG	10908
3	Monitor	Monsanto	80% w/w	WG	10495
4	Safeguard	AgriGuard	80% w/w	WG	12020

Uses

- Brome grasses in ***winter wheat***
- Chickweed in ***winter wheat***
- Cleavers in ***winter wheat***
- Loose silky bent in ***winter wheat***
- Mayweeds in ***winter wheat***
- Onion couch in ***winter wheat***

Approval information

- Sulfosulfuron included in Annex I under EC Directive 91/414

Efficacy guidance

- For best results treat in early spring when annual weeds are small and growing actively. Avoid treatment when weeds are dormant for any reason
- Best control of onion couch is achieved when the weed has more than two leaves. Effects on bulbils or on growth in the following yr have not been examined
- An extended period of dry weather before or after treatment may result in reduced control
- The addition of a recommended surfactant is essential for full activity
- Specific follow-up treatments may be needed for complete control of some weeds
- Use only where competitively damaging weed populations have emerged otherwise yield may be reduced
- Sulfosulfuron is a member of the ALS-inhibitor group of herbicides and products should be used in a planned Resistance Management strategy. See Section 5 for more information

Restrictions

- Maximum total dose equivalent to one treatment at full dose
- Do not treat crops under stress
- Apply only after 1 Feb and from 3 expanded leaf stage
- Do not treat durum wheat or any undersown wheat crop
- Do not use in mixture or in sequence with any other sulfonyl urea herbicide on the same crop

Crop-specific information

- Latest use: flag leaf ligule just visible (GS 39)

Following crops guidance

- In the autumn following treatment winter wheat, winter rye, winter oats, triticale, winter oilseed rape, winter peas or winter field beans may be sown on any soil, and winter barley on soils with less than 60% sand
- In the next spring following application crops of wheat, barley, oats, maize, peas, beans, linseed, oilseed rape, potatoes or grass may be sown
- Sugar beet or any other crop not mentioned above must not be drilled until the second spring following application

Environmental safety

- Dangerous for the environment
- Very toxic to aquatic organisms
- LERAP Category B
- Take extreme care to avoid drift onto broad leaved plants or other crops, or onto ponds, waterways or ditches.
- Follow detailed label instructions for cleaning the sprayer to avoid damage to sensitive crops during subsequent use

Hazard classification and safety precautions

Hazard H11
Risk phrases R50 [1-4]; R58 [4]
Operator protection U08 [2]; U20b [1-4]
Environmental protection E13a [4]; E15a [1-3]; E16a, E16b, E34 [1-4]; E31b [2, 4]; E32a, E32d, E38 [1, 3]
Storage and disposal E04 [4]; E30a [1-4]

392 sulphur

A broad-spectrum inorganic protectant fungicide, foliar feed and acaricide

Products

1	Cosavet DF	Headland	80% w/w	WG	11477
2	Headland Sulphur	Headland	800 g/l	SC	03714
3	Kumulus DF	BASF	80% w/w	SG	04707
4	Luxan Micro-Sulphur	Luxan	80% w/w	WG	06565
5	Solfa WG	Nufarm UK	80% w/w	WG	11602
6	Sulphur Flowable	United Phosphorus	800 g/l	SC	07526
7	Thiovit Jet	Syngenta	80% w/w	WG	10928
8	Venus	Headland	80% w/w	WG	11856

Uses

- American gooseberry mildew in ***gooseberries*** [5]
- Foliar feed in ***grassland*** [4, 5, 7]; ***spring barley, spring wheat, winter barley, winter wheat*** [1, 3, 4, 7, 8]; ***spring oilseed rape*** [1, 4, 5, 7, 8]; ***sugar beet, swedes*** [4]; ***winter oilseed rape*** [1, 3-5, 7, 8]
- Gall mite in ***blackcurrants*** [1, 3, 5, 8]
- Powdery mildew in ***apples, strawberries*** [1-3, 5, 6, 8]; ***aubergines*** *(off-label)*, ***parsnips*** *(off-label)*, ***protected chives*** *(off-label)*, ***protected cucumbers*** *(off-label)*, ***protected herbs (see appendix 6)*** *(off-label)*, ***protected parsley*** *(off-label)*, ***protected peppers*** *(off-label)*, ***protected tomatoes*** *(off-label)* [7]; ***gooseberries*** [1, 3, 8]; ***grapevines*** [1, 3, 5, 8]; ***hops*** [1-3, 5-8]; ***pears*** [2, 6]; ***spring barley, spring wheat, winter barley, winter wheat*** [2, 4-6]; ***spring oats, winter oats*** [2, 5]; ***spring oilseed rape, spring rye, winter oilseed rape, winter rye*** [2]; ***sugar beet*** [1-8]; ***swedes*** [2, 5-7]; ***turnips*** [5]
- Scab in ***apples*** [1-3, 6, 8]; ***pears*** [2, 6]

Specific Off-Label Approvals (SOLAs)

- ***aubergines, protected chives, protected cucumbers, protected herbs (see appendix 6), protected parsley, protected peppers, protected tomatoes*** *(OLA 023652) Dec 2008* [7]
- ***parsnips*** *(OLA 023654) Dec 2008* [7]

Efficacy guidance

- Apply when disease first appears and repeat 2-3 wk later. Details of application rates and timing vary with crop, disease and product. See label for information

- Sulphur acts as foliar feed as well as fungicide and with some crops product labels vary in whether treatment recommended for disease control or growth promotion
- In grassland best results obtained at least 2 wk before cutting for hay or silage, 3 wk before grazing
- Treatment unlikely to be effective if disease already established in crop

Restrictions

- Maximum number of treatments normally 2 per crop for grassland, sugar beet, parsnips, swedes, protected herbs; 3 per yr for blackcurrants, gooseberries; 4 per crop on apples, hops; variable on cereals. See labels
- Do not use on sulphur-shy apples (Beauty of Bath, Belle de Boskoop, Cox's Orange Pippin, Lanes Prince Albert, Lord Derby, Newton Wonder, Rival, Stirling Castle) or pears (Doyenne du Comice)
- Do not use on gooseberry cultivars Careless, Early Sulphur, Golden Drop, Leveller, Lord Derby, Roaring Lion, or Yellow Rough
- Do not use on apples or gooseberries when young, under stress or if frost imminent
- Do not use on fruit for processing, on grapevines during flowering or near harvest on grapes for wine-making
- Do not use on hops at or after burr stage
- Do not spray top or soft fruit with oil or within 30 d of an oil-containing spray

Crop-specific information

- Latest use: before burr stage for hops; before end Sep for parsnips, swedes; before 1st wk of August [2] or end Sep (other products) for sugar beet; fruit swell for gooseberries
- HI cutting grass for hay or silage 2 wk; grazing grassland 3 wk

Environmental safety

- Sulphur products are attractive to livestock and must be kept out of their reach
- Do not empty into drains

Hazard classification and safety precautions

Hazard H04 [5]
Risk phrases R37 [5]; R43 [7]
Operator protection A, H [7]; U02a, U08 [1, 7, 8]; U05a [5, 7]; U14 [1, 6-8]; U19a [5]; U20a [2]; U20b [6]; U20c [1, 3-5, 7, 8]
Environmental protection E15a [1-8]; E19b [5]; E31b [2, 6]; E32a [1, 3, 4, 7, 8]; E34 [2, 4, 6]
Storage and disposal E01 [4-7]; E04 [5-7]; E26 [6]; E30a [1-8]

393 sulphuric acid (commodity substance)

A strong acid used as an agricultural desiccant

Products

sulphuric acid	various	77% w/w	SL

Uses

- Haulm destruction in ***potatoes***
- Pre-harvest desiccation in ***bulbs/corms***, ***peas***

Approval information

- Approval for the use of sulphuric acid as a commodity substance was granted on 23 November 1995 by Ministers under regulation 5 of the Control of Pesticides Regulations 1986

Efficacy guidance

- Apply with suitable equipment between 1 Mar and 15 Nov

Restrictions

- Sulphuric acid is subject to the Poisons Rules 1982 and the Poisons Act 1972. See notes in Section 5
- Must only be used by suitably trained operators competent in use of equipment for applying sulphuric acid
- Not to be applied using hand-held or pedestrian controlled applicators
- A written COSHH assessment must be made before use. Operators should observe OES set out in HSE guidance note EH40/90 or subsequent issues
- Operators must have liquid suitable for eye irrigation immediately available at all times throughout spraying operation

- Maximum number of treatments 3 per crop for potatoes; 1 per crop for peas; 1 per yr for bulbs and corms
- Only 'sulphur burnt' sulphuric acid to be used

Crop-specific information

- Latest use: 15 Nov for bulbs, corms, peas, potatoes

Environmental safety

- Spray must not be deposited within 1 m of public footpaths
- Written notice of any intended spraying must be given to owners of neighbouring land and readable warning notices posted beforehand and left in place for 96 h afterwards
- Unprotected persons must be kept out of treated areas for at least 96 h after treatment
- Do not apply to crops in which bees are actively foraging. Do not apply when flowering weeds are present

Hazard classification and safety precautions

Hazard H05

Risk phrases R34

Operator protection B, C, D, H, K, M

394 tau-fluvalinate

A contact pyrethroid insecticide for cereals and oilseed rape

Products

1	Greencrop Malin	Greencrop	240 g/l	EW	11787
2	Klartan	Makhteshim	240 g/l	EW	11074
3	Mavrik	Makhteshim	240 g/l	EW	10612

Uses

- Aphids in ***spring barley***, ***spring oilseed rape***, ***spring wheat***, ***winter barley***, ***winter oilseed rape***, ***winter wheat***
- Barley yellow dwarf virus vectors in ***winter barley***, ***winter wheat***
- Pollen beetle in ***spring oilseed rape***, ***winter oilseed rape***

Efficacy guidance

- For BYDV control on winter cereals follow local warnings or spray high risk crops in mid-Oct and make repeat application in late autumn/early winter if aphid activity persists
- For summer aphid control on cereals spray once when aphids present on two thirds of ears and increasing
- On oilseed rape treat peach potato aphids in autumn in response to local warning and repeat if necessary
- Best control of pollen beetle in oilseed rape obtained from treatment at green to yellow bud stage and repeat if necessary
- Good spray cover of target essential for best results

Restrictions

- Maximum total dose equivalent to two full dose treatments on oilseed rape. See label for dose rates on cereals
- A minimum of 14 d must elapse between applications to cereals

Crop-specific information

- Latest use: before caryopsis watery ripe (GS 71) for barley; before flowering for oilseed rape; before kernel medium milk (GS 75) for wheat

Environmental safety

- Dangerous for the environment
- Very toxic to aquatic organisms
- High risk to non-target insects or other arthropods. Do not spray within 6 m of the field boundary [1-3]
- LERAP Category A
- Avoid spraying oilseed rape within 6 m of field boundary to reduce effects on certain non-target species or other arthropods
- Must not be applied to cereals if any product containing a pyrethroid insecticide or dimethoate has been sprayed after the start of ear emergence (GS 51)

Hazard classification and safety precautions

Hazard H04 [1]; H11 [1-3]

Risk phrases R36, R38, R51 [1]; R50 [2, 3]; R58 [1-3]

Operator protection A, C, H; U05a, U10, U11, U19a, U20a

Environmental protection E15a, E16c, E16d, E22a, E31c

Storage and disposal E01, E04, E26 [1-3]; E30a [2, 3]; E30b [1]

395 tebuconazole

A systemic conazole fungicide for cereals and other field crops

See also *imidacloprid + tebuconazole + triazoxide*
prochloraz + tebuconazole
spiroxamine + tebuconazole

Products

1	Barclay Busker	Barclay	250 g/l	EC	11345
2	Folicur	Bayer CropScience	250 g/l	EW	11278
3	Greencrop Tabloid	Greencrop	250 g/l	EW	11969
4	Me2 Tebuconazole	Me2	250 g/l	EW	09751
5	Standon Tebuconazole	Standon	250 g/l	EC	09056

Uses

- Alternaria in ***cabbages*** [1-3]; ***carrots***, ***horseradish*** [2, 3]; ***spring oilseed rape***, ***winter oilseed rape*** [1-5]
- Botrytis in ***daffodils*** *(off-label - for galanthamine production)* [2]; ***linseed*** *(reduction)* [1-5]
- Brown rust in ***spring barley***, ***spring rye***, ***spring wheat***, ***winter barley***, ***winter rye***, ***winter wheat*** [1-5]
- Chocolate spot in ***spring field beans***, ***winter field beans*** [1-5]
- Crown rust in ***spring oats*** *(reduction)*, ***winter oats*** *(reduction)* [2, 3]
- Fungus diseases in ***kohlrabi*** *(off-label)* [2]
- Fusarium ear blight in ***spring wheat***, ***winter wheat*** [2, 3]
- Glume blotch in ***spring wheat***, ***winter wheat*** [2, 3]
- Light leaf spot in ***cabbages*** [2-5]; ***spring oilseed rape***, ***winter oilseed rape*** [1-5]
- Lodging control in ***spring oilseed rape***, ***winter oilseed rape*** [2, 3]
- Net blotch in ***spring barley***, ***winter barley*** [1-5]
- Phoma in ***spring oilseed rape***, ***winter oilseed rape*** [2, 3]
- Powdery mildew in ***cabbages***, ***spring barley***, ***spring oats***, ***spring rye***, ***spring wheat***, ***swedes***, ***turnips***, ***winter barley***, ***winter oats***, ***winter rye***, ***winter wheat*** [1-5]; ***carrots***, ***linseed*** [2, 3]; ***chives*** *(off-label)*, ***herbs (see appendix 6)*** *(off-label)*, ***parsley*** *(off-label)*, ***parsnips*** [2]
- Rhynchosporium in ***spring barley***, ***spring rye***, ***winter barley***, ***winter rye*** [1-5]
- Ring spot in ***broccoli*** *(off-label)*, ***calabrese*** *(off-label)*, ***cauliflowers*** *(off-label)*, ***chinese cabbage*** *(off-label)*, ***collards*** *(off-label)*, ***kale*** *(off-label)*, ***leaf brassicas*** *(off-label)* [2]; ***cabbages*** [1-5]; ***spring oilseed rape*** *(reduction)*, ***winter oilseed rape*** *(reduction)* [2, 3]
- Rust in ***broad beans*** *(off-label)*, ***chives*** *(off-label)*, ***dwarf beans*** *(off-label)*, ***french beans*** *(off-label)*, ***herbs (see appendix 6)*** *(off-label)*, ***parsley*** *(off-label)*, ***runner beans*** *(off-label)* [2]; ***leeks*** [2-4]; ***spring field beans***, ***winter field beans*** [1-5]
- Sclerotinia in ***carrots*** [2, 3]
- Sclerotinia stem rot in ***spring oilseed rape***, ***winter oilseed rape*** [1-5]
- Septoria diseases in ***spring wheat***, ***winter wheat*** [1, 4, 5]
- Septoria leaf spot in ***spring wheat***, ***winter wheat*** [2, 3]
- Sooty moulds in ***spring wheat***, ***winter wheat*** [2, 3]
- Stem canker in ***spring oilseed rape***, ***winter oilseed rape*** [2, 3]
- White rot in ***bulb onions*** *(off-label)*, ***garlic*** *(off-label)*, ***salad onions*** *(off-label)*, ***shallots*** *(off-label)* [2]
- Yellow rust in ***spring barley***, ***spring rye***, ***spring wheat***, ***winter barley***, ***winter rye***, ***winter wheat*** [1-5]

Specific Off-Label Approvals (SOLAs)

- ***broad beans*** *(OLA 031878) Dec 2008* [2]
- ***broccoli***, ***calabrese***, ***cauliflowers***, ***chinese cabbage*** *(OLA 031874) Dec 2008* [2]
- ***bulb onions*** *(OLA 031877) Dec 2008* [2]

- ***bulb onions**, **garlic**, **shallots*** *(OLA 031879) Dec 2008* [2]
- ***chives**, **herbs (see appendix 6)**, **parsley*** *(OLA 031873) Dec 2008* [2]
- ***collards**, **kale*** *(OLA 031872) Dec 2008* [2]
- ***daffodils*** *(for galanthamine production) (OLA 041516) Dec 2008* [2]
- ***dwarf beans**, **french beans**, **runner beans*** *(OLA 031880) Dec 2008* [2]
- ***kohlrabi*** *(OLA 031881) Dec 2008* [2]
- ***leaf brassicas*** *(OLA 031876) Dec 2008* [2]
- ***salad onions*** *(OLA 031875) Jun 2007* [2]

Efficacy guidance

- For best results apply at an early stage of disease development before infection spreads to new crop growth
- To protect flag leaf and ear from Septoria diseases apply from flag leaf emergence to ear fully emerged (GS 37-59). Earlier application may be necessary where there is a high risk of infection
- Improved control of established mildew can be obtained by tank mixture with fenpropimorph
- For light leaf spot control in oilseed rape apply in autumn/winter with a follow-up spray in spring/ summer if required
- For control of most other diseases spray at first signs of infection with a follow-up spray 2-4 wk later if necessary. See label for details
- For disease control in cabbages a 3-spray programme at 21-28 d intervals will give good control

Restrictions

- Maximum total dose equivalent to 1 full dose treatment on linseed, green beans, runner beans; 2 full dose treatments on wheat, barley, rye, field beans, broad beans, swedes, turnips, onions, garlic, shallots; 2.25 full dose treatments on cabbages, herbs; 2.5 full dose treatments on oilseed rape; 3 full dose treatments on leeks, horseradish, parsnips, carrots
- Do not treat durum wheat
- Apply only to listed oat varieties (see label)
- Do not apply before swedes and turnips have a root diameter of 2.5 cm, or before heart formation in cabbages

Crop-specific information

- Latest use: before grain milky-ripe for cereals (GS 71); when most seed green-brown mottled for oilseed rape (GS 6,3); before brown capsule for linseed
- HI field beans, swedes, turnips 35 d; market brassicas, kale, carrots, parsnips, garlic, salad onions, salad brassicas, shallots 21 d; leeks, herbs 14 d; broad beans, green beans, runner beans 7 d
- Some transient leaf speckling on wheat or leaf reddening/scorch on oats may occur but this has not been shown to reduce yield response to disease control

Environmental safety

- Dangerous for the environment
- Toxic to aquatic organisms

Hazard classification and safety precautions

Hazard H03, H11
Risk phrases R20 [2]; R22a, R41, R51, R58 [1-5]; R38 [1, 3-5]
Operator protection A, C, H; U05a, U11, U20a [1-5]; U19a [1, 3-5]
Environmental protection E15a, E31b [1-5]; E32d, E38 [2]; E34 [2, 3]
Storage and disposal E01, E04, E26, E30a
Medical advice M03

396 tebuconazole + triadimenol

A broad spectrum systemic fungicide for cereals

Products

1	Silvacur	Bayer CropScience	250:125 g/l	EC	11309
2	Veto F	Bayer CropScience	225:75 g/l	EC	11317

Uses

- Brown rust in ***spring barley**, **spring rye**, **spring wheat**, **winter barley**, **winter rye**, **winter wheat***
- Crown rust in ***spring oats**, **winter oats***
- Fusarium ear blight in ***spring wheat**, **winter wheat***

- Glume blotch in ***spring wheat***, ***winter wheat***
- Net blotch in ***spring barley***, ***winter barley***
- Powdery mildew in ***spring barley***, ***spring oats***, ***spring rye***, ***spring wheat***, ***winter barley***, ***winter oats***, ***winter rye***, ***winter wheat***
- Rhynchosporium in ***spring barley***, ***winter barley***
- Septoria leaf spot in ***spring wheat***, ***winter wheat***
- Sooty moulds in ***spring wheat***, ***winter wheat***
- Yellow rust in ***spring barley***, ***spring rye***, ***spring wheat***, ***winter barley***, ***winter rye***, ***winter wheat***

Efficacy guidance

- For best results apply at an early stage of disease development before infection spreads to new crop growth
- To protect flag leaf and ear from Septoria diseases apply from flag leaf emergence to ear fully emerged (GS 37-59). Earlier application may be necessary where there is a high risk of infection
- For control of rust, powdery mildew, leaf and net blotch apply at first signs of disease with a second application 2-3 wk later if necessary

Restrictions

- Maximum total dose equivalent to two full dose treatments
- Do not use on durum wheat
- Use only on listed varieties of oats. See label [1]

Crop-specific information

- Latest use: before grain milky-ripe (GS 71)
- Some transient leaf speckling may occur on wheat or leaf reddening/scorch on oats but this has not been shown to reduce yield response or disease control

Environmental safety

- Harmful to aquatic organisms

Hazard classification and safety precautions

Hazard H03 [1]
Risk phrases R20, R36 [1]; R52, R58 [1, 2]
Operator protection A, C, H; U05a, U09b, U20a
Environmental protection E13c [1, 2]; E31b, E34 [1]; E31c [2]
Storage and disposal E01, E04, E26, E30a
Medical advice M03 [2]

397 tebuconazole + triazoxide

A triazole and benzotriazine fungicide seed treatment for use in barley

Products

Raxil S	Bayer CropScience	20:20 g/l	LS	11296

Uses

- Leaf stripe in ***spring barley*** *(seed treatment)*, ***winter barley*** *(seed treatment)*
- Loose smut in ***spring barley*** *(seed treatment)*, ***winter barley*** *(seed treatment)*
- Net blotch in ***spring barley*** *(seed treatment)*, ***winter barley*** *(seed treatment)*

Efficacy guidance

- Best applied through recommended seed treatment machines
- Evenness of seed cover improved by simultaneous application of equal volumes of product and water or dilution of product with an equal volume of water
- Drill treated seed in the same season

Restrictions

- Maximum number of treatments 1 per batch of seed
- Do not use on seed with more than 16% moisture content, or on sprouted, cracked or skinned seed
- If product has been left to stand for more than 2 mth full agitation may be necessary to re-suspend it before use
- Diluted product must be used immediately
- Do not use treated seed as food or feed

SEE SECTION 3 FOR PRODUCTS ALSO REGISTERED

Crop-specific information

- Latest use: before drilling
- Slightly delayed and reduced emergence may occur but this is normally outgrown
- Any delay in field emergence, for whatever reason, may be accentuated by treatment

Environmental safety

- Harmful to aquatic organisms
- Treated seed harmful to game and wildlife
- Product also supplied in returnable containers. See label for guidance on handling, storage, protective clothing and precautions
- Seed should be drilled to a depth of 4 cm into a well-prepared seed bed
- If seed is present on the soil surface, or if spills have occurred, the field should be harrowed and rolled if conditions are appropriate to ensure good incorporation

Hazard classification and safety precautions

Risk phrases R52, R58
Operator protection A, H; U07, U20b
Environmental protection E03, E13c, E33, E34, E36
Storage and disposal E26, E30a
Treated seed S01, S02, S03, S04b, S05, S06a, S07

398 tebufenpyrad

A pyrazole mitochondrial electron transport inhibitor (METI) aphicide and acaricide

Products

Masai	BASF	20% w/w	WB	10223

Uses

- Damson-hop aphid in ***hops***, ***plums*** *(off-label)*
- Fruit tree red spider mite in ***blackberries*** *(off-label)*, ***raspberries*** *(off-label)*
- Gall mite in ***bilberries*** *(off-label)*, ***blackcurrants*** *(off-label)*, ***blueberries*** *(off-label)*, ***cranberries*** *(off-label)*, ***gooseberries*** *(off-label)*, ***redcurrants*** *(off-label)*, ***whitecurrants*** *(off-label)*
- Red spider mites in ***apples***, ***pears***
- Two-spotted spider mite in ***blackberries*** *(off-label)*, ***hops***, ***protected roses***, ***raspberries*** *(off-label)*, ***strawberries***

Specific Off-Label Approvals (SOLAs)

- ***bilberries***, ***blackcurrants***, ***blueberries***, ***cranberries***, ***gooseberries***, ***redcurrants***, ***whitecurrants*** *(OLA 031741) Aug 2007* [1]
- ***blackberries***, ***raspberries*** *(OLA 041498) Aug 2006* [1]
- ***plums*** *(OLA 031742) Dec 2008* [1]

Efficacy guidance

- Acts on eggs (except winter eggs) and all motile stages of spider mites up to adults
- Treat spider mites from 80% egg hatch but before mites become established
- For effective control total spray cover of the crop is required
- Product can be used in a programme to give season-long control of damson/hop aphids coupled with mite control
- Where aphids resistant to tebufenpyrad occur in hops control is unlikely to be satisfactory and repeat treatments may result in lower levels of control. Where possible use different active ingredients in a programme

Restrictions

- Maximum total dose equivalent to one full dose treatment on apples, pears, strawberries; 3 full dose treatments on hops
- Other mitochondrial electron transport inhibitor (METI) acaricides should not be applied to the same crop in the same calendar yr either separately or in mixture
- Do not treat apples before 90% petal fall
- Small-scale testing of rose varieties to establish tolerance recommended before use
- Inner liner of container must not be removed

Crop-specific information

- Latest use: end of burr stage for hops

- HI strawberries 3 d; apples, bilberries, blackcurrants, blueberries, cranberries, gooseberries, pears, redcurrants, whitecurrants 7d; blackberries, plums, raspberries 21 d
- Product has no effect on fruit quality or finish

Environmental safety

- Dangerous for the environment
- Very toxic to aquatic organisms
- High risk to bees. Do not apply to crops in flower or to those in which bees are actively foraging. Do not apply when flowering weeds are present
- LERAP Category B
- Broadcast air-assisted LERAP (18 m)

Hazard classification and safety precautions

Hazard H03, H11
Risk phrases R20, R22a, R37, R50, R58
Operator protection A; U02a, U05a, U09a, U13, U14, U20b
Environmental protection E12a, E13b, E15a, E16a, E16b, E32a, E32d, E38; E17b (18 m)
Storage and disposal E01, E04, E30a
Medical advice M05a

399 teflubenzuron

A benzoylurea insecticide for use on ornamentals

Products

Nemolt	Fargro	150 g/l	SC	10226

Uses

- Browntail moth in ***ornamental plant production***
- Caterpillars in ***ornamental plant production***
- Whitefly in ***ornamental plant production***

Efficacy guidance

- Product acts as larval stomach poison interfering with moulting process leading to cessation of feeding and larval death
- Apply as soon as first stage larvae seen. This will often coincide the peak of moth flight

Restrictions

- Maximum number of treatments 3 per yr
- Test specific varieties before carrying out extensive treatments

Environmental safety

- Dangerous for the environment
- Very toxic to aquatic organisms
- Limited evidence suggests some margin of safety to *Encarsia formosa*. Effects on other parasites and predators not fully tested

Hazard classification and safety precautions

Hazard H11
Risk phrases R50, R58
Operator protection A, C; U05a, U20c
Environmental protection E15a, E32a, E32d, E38
Storage and disposal E01, E04, E26, E30a

400 tefluthrin

A soil acting pyrethroid insecticide seed treatment

Products

1	Evict	Syngenta	100 g/l	CF	11735
2	Force ST	Bayer CropScience	200 g/l	CF	11671
3	Force ST	Syngenta	200 g/l	CF	11752

Uses

- Bean seed fly in ***bulb onions*** *(off-label - seed treatment)*, ***leeks*** *(off-label - seed treatment)*, ***salad onions*** *(off-label - seed treatment)* [2]
- Carrot fly in ***carrots*** *(off-label - seed treatment)*, ***parsnips*** *(off-label - seed treatment)* [2]
- Millipedes in ***fodder beet*** *(seed treatment)*, ***sugar beet*** *(seed treatment)* [2, 3]
- Onion fly in ***bulb onions*** *(off-label - seed treatment)*, ***leeks*** *(off-label - seed treatment)*, ***salad onions*** *(off-label - seed treatment)* [2]
- Pygmy beetle in ***fodder beet*** *(seed treatment)*, ***sugar beet*** *(seed treatment)* [2, 3]
- Springtails in ***fodder beet*** *(seed treatment)*, ***sugar beet*** *(seed treatment)* [2, 3]
- Symphylids in ***fodder beet*** *(seed treatment)*, ***sugar beet*** *(seed treatment)* [2, 3]
- Wheat bulb fly in ***spring barley*** *(seed treatment)*, ***spring oats*** *(seed treatment)*, ***spring wheat*** *(seed treatment)*, ***winter barley*** *(seed treatment)*, ***winter oats*** *(seed treatment)*, ***winter wheat*** *(seed treatment)* [1]
- Wireworm in ***spring barley*** *(seed treatment)*, ***spring oats*** *(seed treatment)*, ***spring wheat*** *(seed treatment)*, ***winter barley*** *(seed treatment)*, ***winter oats*** *(seed treatment)*, ***winter wheat*** *(seed treatment)* [1]

Specific Off-Label Approvals (SOLAs)

- ***bulb onions*, *leeks*, *salad onions*** *(seed treatment) (OLA 001748) Dec 2008* [2]
- ***carrots*, *parsnips*** *(seed treatment) (OLA 000873) Dec 2008* [2]

Efficacy guidance

- Apply during process of pelleting beet seed. Consult manufacturer for details of specialist equipment required [2, 3]
- Apply to cereal seed through a suitable liquid seed treater calibrated to achieve even coverage [1]
- Micro-capsule formulation allows slow release to provide a protection zone around treated seed during establishment
- Product is non-systemic and may not protect against wireworm attack at the soil surface [1]
- Where egg counts indicate a high risk of wheat bulb fly attack, a follow-up spray treatment may be needed [1]
- Where wireworm populations exceed 1.25 million per ha a suitable spray treatment must also be applied
- Factors that adversely affect crop establishment may reduce the level of pest control [1]

Restrictions

- Maximum number of treatments 1 per batch of seed
- Sow treated seed as soon as possible. Do not store treated seed from one drilling season to next
- Must be co-applied with a suitable fungicide seed treatment to protect against seed and soil-borne disease [1]
- Do not use on seed that is sprouted, carcked, damage or over 16% moisture content [1]
- If used in areas where soil erosion by wind or water likely preventative measures must be taken to prevent this happening
- Can cause a transient tingling or numbing sensation to exposed skin. Avoid skin contact with product, treated seed and dust throughout all operations in the seed treatment plant and at drilling

Crop-specific information

- Latest use: before drilling seed
- Treated seed must be drilled within the season of treatment [1, 3] or stored in cool, dry, ventilated conditions for drilling one season later [2]

Environmental safety

- Dangerous for the environment
- Very toxic to aquatic organisms
- Keep treated seed secure from people, domestic stock/pets and wildlife at all times during storage and use
- Treated seed harmful to game and wild life. Bury spillages
- In the event of seed spillage clean up as much as possible into the related seed sack and bury the remainder completely
- Do not apply treated seed from the air
- Keep livestock out of areas drilled with treated seed for at least 80 d

Hazard classification and safety precautions

Hazard H03, H11

Risk phrases R20, R43, R58 [1-3]; R50 [2, 3]; R51 [1]

Operator protection A, D, H [1-3]; C [1]; E [2, 3]; U02a, U04a, U08 [2, 3]; U05a, U20b [1-3]; U07 [3]; U14 [1]

Environmental protection E06a [2, 3] (80 d); E13a [1, 2]; E15a [1, 3]; E32a, E32d, E38 [1-3]; E34 [3]

Storage and disposal E01, E04 [1-3]; E26, E30a [2, 3]

Treated seed S01, S03 [1, 2]; S02, S04b, S05, S07 [1-3]; S08 [1]

Medical advice M05b [2, 3]

401 tepraloxydim

A systemic post-emergence herbicide for control of annual grass weeds

Products

1	Aramo	BASF	50 g/l	EC	10280
2	Landgold Tepraloxydim	Landgold	50 g/l	EC	11808
3	Standon Tepraloxydim	Standon	50 g/l	EC	11119

Uses

- Annual grasses in ***bulb onions, cabbages, carrots, cauliflowers, combining peas, fodder beet, land temporarily removed from production, leeks, linseed, spring field beans, sugar beet, vining peas, winter field beans, winter oilseed rape*** [1, 2]; ***flax, linseed for industrial use, winter oilseed rape for industrial use*** [1]
- Annual meadow grass in ***bulb onions, cabbages, carrots, cauliflowers, combining peas, fodder beet, leeks, linseed, spring field beans, spring oilseed rape, sugar beet, vining peas, winter field beans, winter oilseed rape*** [3]; ***golf courses*** *(off-label)*, ***salad onions*** *(off-label)* [1]
- Blackgrass in ***bulb onions, cabbages, carrots, cauliflowers, combining peas, fodder beet, leeks, linseed, spring field beans, spring oilseed rape, sugar beet, vining peas, winter field beans, winter oilseed rape*** [3]
- Green cover in ***land temporarily removed from production*** [3]
- Perennial grasses in ***bulb onions, cabbages, carrots, cauliflowers, combining peas, fodder beet, land temporarily removed from production, leeks, linseed, spring field beans, sugar beet, vining peas, winter field beans, winter oilseed rape*** [1, 2]; ***flax, linseed for industrial use, winter oilseed rape for industrial use*** [1]
- Volunteer barley in ***bulb onions, cabbages, carrots, cauliflowers, combining peas, fodder beet, leeks, linseed, spring field beans, spring oilseed rape, sugar beet, vining peas, winter field beans, winter oilseed rape*** [3]
- Volunteer cereals in ***bulb onions, cabbages, carrots, cauliflowers, combining peas, fodder beet, land temporarily removed from production, leeks, linseed, spring field beans, sugar beet, vining peas, winter field beans, winter oilseed rape*** [1, 2]; ***flax, linseed for industrial use, winter oilseed rape for industrial use*** [1]
- Volunteer wheat in ***bulb onions, cabbages, carrots, cauliflowers, combining peas, fodder beet, leeks, linseed, spring field beans, spring oilseed rape, sugar beet, vining peas, winter field beans, winter oilseed rape*** [3]

Specific Off-Label Approvals (SOLAs)

- ***golf courses*** *(OLA 040612) May 2005* [1]
- ***salad onions*** *(OLA 032474) May 2005* [1]

Efficacy guidance

- Best results obtained from applications when weeds small and have not begun to compete with crop
- Only emerged weeds are controlled
- Cool conditions slow down activity and very dry conditions reduce activity by interfering with uptake and translocation
- Foliar death of susceptible weeds evident after 3-4 wks in warm conditions
- Reduced doses must not be used on resistant grass weed populations

Restrictions

- Maximum total dose equivalent to one full dose treatment on all crops
- Consult processors or contract agents before treatment of crops intended for processing or for seed
- Do not apply to crops damaged or stressed by factors such as previous herbicide treatments, pest or disease attack

- Do not spray if rain or frost expected, or if foliage is wet
- Do not treat oilseed rape with very low vigour and poor yield potential. Overlapping on oilseed rape may cause damage and reduce yields
- On peas a satisfactory crystal violet leaf wax test must be carried out before treatment. Winter varieties may be treated only in the spring
- For sugar beet, fodder beet, linseed and green cover on land temporarily removed from production, applications are prohibited between 1 Nov and 31 Mar
- For field beans, peas, leeks, bulb onions, carrots, cabbages and cauliflowers applications are prohibited between 1 Nov and 1 Mar

Crop-specific information

- Latest use : before end Nov or before crop has 9 true leaves, whichever is first for winter oilseed rape; before crop has 6 true leaves for spring oilseed rape [3]; before flower buds visible for flax [1], linseed; before head formation for cabbages, cauliflowers
- HI 3 wk for carrots; 4 wk for bulb onions, leeks; 5 wk for peas; 8 wk for fodder beet, field beans, sugar beet
- Treatment prohibited between 1 Nov and 31 Mar on golf courses [1]

Following crops guidance

- In the event of failure of a treated crop, wheat or barley may be drilled after 2 wk following normal seedbed cultivations, maize or Italian rye-grass may be planted after 8 wk following cultivation to 20 cm
- Graminaceous crops other than those mentioned above should not follow a treated crop in the rotation.
- Broad leaved crops may be planted at any time following a failed or normally harvested treated crop

Environmental safety

- Dangerous for the environment
- Toxic to aquatic organisms

Hazard classification and safety precautions

Hazard H03, H11
Risk phrases R22b, R38, R51, R58, R63, R66
Operator protection A; U05a, U08, U20b [1-3]; U23a [1, 2]
Environmental protection E15a [1-3]; E31b [2, 3]; E31c, E32d, E38 [1]
Storage and disposal E01, E04, E29b, E30a [1-3]; E26 [2, 3]; E29a [2]
Medical advice M05b

402 terbuthylazine

A triazine herbicide available only in mixtures

403 terbuthylazine + terbutryn

A pre-emergence herbicide for peas, beans and lupins

Products

1	Batallion	Makhteshim	150:350 g/l	SC	08305
2	Opogard	Syngenta	150:350 g/l	SC	08427

Uses

- Annual dicotyledons in ***broad beans***, ***combining peas***, ***spring field beans***, ***vining peas***, ***winter field beans*** [1, 2]; ***edible podded peas*** [1]; ***lupins*** [2]
- Annual grasses in ***broad beans***, ***combining peas***, ***spring field beans***, ***vining peas***, ***winter field beans*** [1, 2]; ***edible podded peas*** [1]; ***lupins*** [2]

Approval information

- Products containing terbuthylazine and terbutryn have been granted derogations for specified 'Essential Uses' for use until 31 December 2007. Sale and supply must cease by 30 June 2007 but growers have no guarantee that the products will continue to be available until then.
For more information see 'The Review Programme' under 'Pesticide Legislation' in Section 5

Efficacy guidance

- Active via root uptake and with foliar activity on cotyledon stage weeds
- Best results achieved by application to fine, firm, moist seedbed, preferably at weed emergence, when rain falls after spraying
- Effectiveness may be reduced by excessive rain, drought or cold
- Residual control lasts for up to 8 wk on mineral soils. Effectiveness reduced on highly organic soils and subsequent use of post-emergence treatment recommended
- Do not use on soils which are very cloddy or have more than 10% organic matter otherwise weed control will be reduced

Restrictions

- Maximum number of treatments 1 per crop
- Vedette and Printana peas may be damaged by treatment. Do not use on forage peas
- Do not treat peas, beans or lupins on soils lighter than loamy fine sand, or lupins on silty clay soil
- Do not cultivate after treatment

Crop-specific information

- Latest use: 3 d before crop emergence
- Apply as soon as possible after drilling
- Crop seed must be covered by at least 25 mm of settled soil
- Heavy rain after application may cause damage to peas on light soils

Following crops guidance

- Treated land should be mould-board ploughed to 150 mm before any succeeding crop is planted
- A period of 12 wk (14 wk in dry conditions) since application must elapse before planting any succeeding crop

Environmental safety

- Dangerous for the environment
- Very toxic to aquatic organisms

Hazard classification and safety precautions

Hazard H03 [1]; H04 [2]; H11 [1, 2]
Risk phrases R22a [1]; R43 [2]; R50, R58 [1, 2]
Operator protection U02a, U08, U14, U15, U19a [1]; U05a, U20a [1, 2]
Environmental protection E15a [1, 2]; E31b, E32e, E34 [1]; E31c, E32d, E38 [2]
Storage and disposal E01, E04, E30a
Medical advice M05a [1]

404 terbutryn

A residual triazine for control of aquatic algae

See also *fomesafen + terbutryn*
terbuthylazine + terbutryn

Products

Clarosan	Scotts	1% w/w	GR	H7692

Uses

- Algae in ***areas of water***

Approval information

- Approved for aquatic weed control [1]. See notes in Section 5 on use of herbicides in or near water
- Certain products containing terbutryn have been granted derogations for specified 'Essential Uses' for use until 31 December 2007. Sale and supply must cease by 30 June 2007 but growers have no guarantee that the products will continue to be available until then.
  For more information see 'The Review Programme' under 'Pesticide Legislation' in Section 5

Efficacy guidance

- For best results apply when algal growth active and before heavy infestations have built up. Control is more rapid in warm weather
- Apply granules evenly across either from the bank using a machine with a long throw or through suitable equipment mounted on a boat

- Granular formulation permits early application to avoid deoxygenation caused by dead or dying vegetation
- Use only in static or sluggishly moving water. The flow in moving water should be stopped for at least 7 d after treatment otherwise weed control may be reduced. Treat small ponds before start of new growth, usually before end Apr
- Algal growth ceases immediately but visible signs may not be obvious for 2-4 wk after treatment. Re-growth should not occur for at least 3-4 mth
- Effectiveness reduced in water with peaty bottom

Restrictions

- This product must not be used for any purpose other than that listed, including any extrapolations that would normally be permissible under the Long Term Arrangements for Extension of Use (see Section 5)
- Product must only be used in static or sluggishly moving water where the maximum water flow is less than 1 metre per 3 minutes
- Do not use in trout farms or areas where fish are intensively reared and dissolved oxygen levels are critical

Crop-specific information

- Latest use: Aug for areas of water

Environmental safety

- Dangerous for the environment
- Toxic to aquatic organisms
- Do not dump surplus herbicide in water or ditch bottoms
- Do not use treated water for irrigation purposes within 7 d of treatment
- If dense weed growth in watercourses to be controlled without de-oxygenation treat in sections of about 400 m at a time with intervals of at least 14 d
- If the water body to be treated is subject to the Water Resources Act and the Control of Pollution Act, users must consult the approriate water regulatory body before use

Hazard classification and safety precautions

Hazard H11
Risk phrases R51, R58
Operator protection U20b
Environmental protection E19a, E32e; E21 (7 d)
Storage and disposal E29a, E30a

405 tetraconazole

A systemic, protectant and curative triazole fungicide for cereals

See also chlorothalonil + tetraconazole

Products

Juggler	Sipcam	100 g/l	EC	09391

Uses

- Brown rust in ***spring barley***, ***spring wheat***, ***winter barley***, ***winter wheat***
- Crown rust in ***spring oats***, ***winter oats***
- Powdery mildew in ***spring barley***, ***spring wheat***, ***winter barley***, ***winter wheat***
- Septoria leaf spot in ***spring wheat***, ***winter wheat***
- Sooty moulds in ***spring wheat***, ***winter wheat***
- Yellow rust in ***spring barley***, ***spring wheat***, ***winter barley***, ***winter wheat***

Efficacy guidance

- Apply at any time from the tillering stage up to the latest times indicated for each crop
- Best results achieved from applications before disease becomes established
- If disease pressure persists a second treatment may be required to prevent late season attacks
- Best control of ear diseases in wheat obtained from application at or after the ear completely emerged stage

Restrictions

- Maximum total dose equivalent to three full dose applications on wheat and two full dose applications on other cereals

Crop-specific information

- Latest use: before end of ear emergence (GS 59) for oats; before end of flowering (GS 69) for barley; before grain watery ripe (GS 71) for wheat

Environmental safety

- Harmful to fish or other aquatic life. Do not contaminate surface waters or ditches with chemical or used container
- LERAP Category B

Hazard classification and safety precautions

Hazard H03, H04
Risk phrases R22a, R36, R38
Operator protection A, C, H; U05a, U19a, U20a
Environmental protection E13c, E16a, E16b, E31b
Storage and disposal E01, E04, E26, E30a
Medical advice M03

406 tetramethrin

A contact acting pyrethroid insecticide

See also *d-phenothrin + tetramethrin*
resmethrin + tetramethrin

Products

Product	Company	Concentration	Formulation	MAFF No.
Killgerm Py-Kill W	Killgerm	10.2% w/v	EC	H4632

Uses

- Flies in ***agricultural premises***, ***livestock houses***

Efficacy guidance

- Dilute in accordance with directions and apply as space or surface spray

Restrictions

- For use only by professional operators
- Maximum number of treatments 1 per wk in agricultural premises and livestock houses
- Do not apply directly on food or livestock
- Remove exposed milk before application. Protect milk machinery and containers from contamination
- Do not use space sprays containing pyrethrins or pyrethroid more than once per wk in intensive or controlled environment animal houses in order to avoid development of resistance. If necessary, use a different control method or product

Environmental safety

- Dangerous for the environment [1]
- Toxic to aquatic organisms

Hazard classification and safety precautions

Hazard H03, H08, H11
Risk phrases R22b, R41, R51, R58
Operator protection A, H; U02b, U09a, U19a, U20b
Environmental protection E15a, E32d
Consumer protection C05, C06, C07, C08, C09, C10, C11
Storage and disposal E01, E30a
Medical advice M05b

407 thiabendazole

A systemic, curative and protectant benzimidazole (MBC) fungicide

See also *imazalil + thiabendazole*

Products

	Product	Company	Concentration	Formulation	MAFF No.
1	Hykeep	Agrichem	2% ww	DS	06744
2	Storite Clear Liquid	Syngenta	220 g/l	SL	09503

Products – continued

3 Storite Excel	Syngenta	500 g/l	SC	09542
4 Storite Flowable	Banks Cargill	450 g/l	FS	08703

Uses

- Dry rot in ***seed potatoes*** *(off-label)* [3]; ***seed potatoes*** *(tuber treatment - post-harvest)* [2, 4]; ***ware potatoes*** *(tuber treatment - post-harvest)* [1-4]
- Dutch elm disease in ***elm trees*** [2]
- Fusarium basal rot in ***narcissi*** *(bulb dip or spray)* [2]
- Gangrene in ***seed potatoes*** *(tuber treatment - post-harvest)* [2, 4]; ***ware potatoes*** *(tuber treatment - post-harvest)* [1-4]
- Silver scurf in ***seed potatoes*** *(tuber treatment - post-harvest)* [2, 4]; ***ware potatoes*** *(tuber treatment - post-harvest)* [1-4]
- Skin spot in ***seed potatoes*** *(tuber treatment - post-harvest)* [2, 4]; ***ware potatoes*** *(tuber treatment - post-harvest)* [1-4]

Specific Off-Label Approvals (SOLAs)

- ***seed potatoes*** *(OLA 041322) Dec 2008* [3]

Approval information

- Thiabendazole included in Annex I under EC Directive 91/414

Restrictions

- Maximum number of treatments 1 per batch for seed potato treatments; 2 per yr for bulbs; 1 per yr on elm trees
- Treated seed potatoes must not be used for food or feed
- Do not mix with any other product
- Injection of trees for dutch elm disease control must only be carried out by trained arborists or others trained in injection techniques and identification of the disease [2]

Crop-specific information

- Latest use: before planting for seed potatoes; 21 d before removal from store for sale, processing or consumption for ware potatoes
- Apply to potatoes as soon as possible after harvest using suitable equipment and always within 2 wk of lifting provided the skins are set. See label for details
- Potatoes should only be treated by systems that provide an accurate dose to tubers not carrying excessive quantities of soil
- Use as a post-lifting or hot water treatment to reduce basal and neck rot in narcissus bulbs. Ensure bulbs are clean [2]
- Some reduction in flower number may occur in the first spring following the double treatment [2]

Environmental safety

- Dangerous for the environment
- Toxic to aquatic organisms

Hazard classification and safety precautions

Hazard H04 [3]; H11 [2, 3]
Risk phrases R43 [3]; R51 [2, 3]; R52 [1]; R58 [1-3]
Operator protection A, C, D, H [1-3]; B, K [2]; M [1, 2]; U05a [2, 3]; U14 [3]; U19a, U20c [1-4]
Environmental protection E13c [1, 4]; E15a, E32d, E38 [2, 3]; E31a [3, 4]; E31c [2]; E32a [1]
Storage and disposal E01, E30a [1-4]; E04 [2, 3]; E26 [2-4]; E29a [1]
Treated seed S03, S04a, S05 [2]

408 thiabendazole + thiram

A fungicide seed dressing mixture for field and vegetable crops

Products

Hy-TL	Agrichem	225:300 g/l	FS	06246

Uses

- Ascochyta in ***broad beans*** *(seed treatment)*, ***peas*** *(seed treatment)*, ***spring field beans*** *(seed treatment)*, ***winter field beans*** *(seed treatment)*
- Damping off in ***broad beans*** *(seed treatment)*, ***peas*** *(seed treatment)*, ***spring field beans*** *(seed treatment)*, ***winter field beans*** *(seed treatment)*

- Fusarium in ***bulb onions*** *(off-label - seed treatment)*
- Neck rot in ***bulb onions*** *(off-label - seed treatment)*

Specific Off-Label Approvals (SOLAs)

- ***bulb onions*** *(seed treatment) (OLA 021299) Dec 2008* [1]

Approval information

- Thiabendazole included in Annex I under EC Directive 91/414

Efficacy guidance

- Dress seed as near to sowing as possible
- Dilution may be needed with particularly absorbent types of seed. If diluted material used, seed may require drying before storage

Restrictions

- Maximum number of treatments 1 per batch
- Seed to be treated should be of satisfactory quality and moisture content
- Do not use treated seed as food or feed

Crop-specific information

- Latest use: before drilling

Environmental safety

- Dangerous for the environment
- Treated seed harmful to game and wildlife

Hazard classification and safety precautions

Hazard H03, H11
Risk phrases R20, R22a, R36, R37, R43, R68
Operator protection A, C, D, H, M; U05a, U08, U14, U15, U20a
Environmental protection E03, E31c, E34
Storage and disposal E01, E04, E26, E30a
Treated seed S01, S02, S03, S04a, S04b, S05, S06a
Medical advice M03, M04

409 thiacloprid

A chloronicotinyl insecticide for use in apples

Products

Calypso	Bayer CropScience	480 g/l	SC	11257

Uses

- Aphids in ***cherries*** *(off-label - under temporary covers)*, ***mirabelles*** *(off-label - under temporary covers)*, ***protected herbs (see appendix 6)*** *(off-label)*, ***protected lettuce*** *(off-label)*, ***protected salad brassicas*** *(off-label - for baby leaf production)*
- Capsids in ***strawberries*** *(off-label)*
- Common green capsid in ***blackberries*** *(off-label)*, ***raspberries*** *(off-label)*, ***rubus hybrids*** *(off-label)*
- Damson-hop aphid in ***plums*** *(off-label)*
- Palm thrips in ***protected aubergines*** *(off-label)*, ***protected courgettes*** *(off-label)*, ***protected cucumbers*** *(off-label)*, ***protected ornamentals*** *(off-label)*, ***protected peppers*** *(off-label)*, ***protected tomatoes*** *(off-label)*
- Rosy apple aphid in ***apples***
- South American leaf miner in ***protected aubergines*** *(off-label)*, ***protected courgettes*** *(off-label)*, ***protected cucumbers*** *(off-label)*, ***protected ornamentals*** *(off-label)*, ***protected peppers*** *(off-label)*, ***protected tomatoes*** *(off-label)*
- Tarnished plant bug in ***blackberries*** *(off-label)*, ***protected strawberries*** *(off-label)*, ***raspberries*** *(off-label)*, ***rubus hybrids*** *(off-label)*
- Tobacco whitefly in ***protected aubergines*** *(off-label)*, ***protected courgettes*** *(off-label)*, ***protected cucumbers*** *(off-label)*, ***protected ornamentals*** *(off-label)*, ***protected peppers*** *(off-label)*, ***protected tomatoes*** *(off-label)*

- Western flower thrips in ***protected aubergines*** *(off-label)*, ***protected courgettes*** *(off-label)*, ***protected cucumbers*** *(off-label)*, ***protected ornamentals*** *(off-label)*, ***protected peppers*** *(off-label)*, ***protected tomatoes*** *(off-label)*
- Woolly currant scale in ***bilberries*** *(off-label)*, ***blackcurrants*** *(off-label)*, ***blueberries*** *(off-label)*, ***cranberries*** *(off-label)*, ***gooseberries*** *(off-label)*, ***redcurrants*** *(off-label)*, ***whitecurrants*** *(off-label)*

Specific Off-Label Approvals (SOLAs)

- ***bilberries***, ***blackcurrants***, ***blueberries***, ***cranberries***, ***gooseberries***, ***redcurrants***, ***whitecurrants*** *(OLA 041493) Dec 2005* [1]
- ***blackberries***, ***raspberries***, ***rubus hybrids*** *(OLA 041494) Dec 2008* [1]
- ***cherries***, ***mirabelles*** *(under temporary covers) (OLA 041496) Dec 2005* [1]
- ***plums*** *(OLA 041495) Dec 2005* [1]
- ***protected aubergines***, ***protected courgettes***, ***protected cucumbers***, ***protected ornamentals***, ***protected peppers***, ***protected tomatoes*** *(OLA 040128) Dec 2005* [1]
- ***protected herbs (see appendix 6)***, ***protected lettuce*** *(OLA 032726) Dec 2005* [1]
- ***protected salad brassicas*** *(for baby leaf production) (OLA 032726) Dec 2005* [1]
- ***protected strawberries*** *(OLA 041497) Dec 2005* [1]
- ***strawberries*** *(OLA 032727) Dec 2005* [1]

Efficacy guidance

- Best results obtained from a programme of sprays commencing pre-blossom at the first sign of aphids
- Minimise the possibility of development of resistance by alternating insecticides with different modes of action in the programme
- In dense canopies and on larger trees increase water volume to ensure full coverage

Restrictions

- Maximum number of treatments 3 per yr on plums; 2 per yr on all other crops

Crop-specific information

- HI apples 14 d; protected vegetables 3 d

Environmental safety

- Harmful to aquatic organisms
- Risk to certain non-target insects or other arthropods. See directions for use
- Broadcast air-assisted LERAP (30 m)

Hazard classification and safety precautions

Hazard H03
Risk phrases R20, R22a, R40, R43, R52, R58
Operator protection A, H; U05a, U14, U20b
Environmental protection E13c, E22b, E31b, E34; E17b (30 m)
Storage and disposal E01, E04, E30a
Medical advice M03, M05b

410 thifensulfuron-methyl

A translocated sulfonylurea herbicide

See also *flupyrsulfuron-methyl + thifensulfuron-methyl*
metsulfuron-methyl + thifensulfuron-methyl

Products

1	DUK 118	Headland	75% w/w	WG	11923
2	Prospect	Nufarm UK	75% w/w	WG	09389

Uses

- Annual dicotyledons in ***spring barley***, ***spring wheat***, ***winter barley***, ***winter wheat*** [1]
- Charlock in ***spring barley***, ***spring wheat***, ***winter barley***, ***winter wheat*** [1]
- Chickweed in ***spring barley***, ***spring wheat***, ***winter barley***, ***winter wheat*** [1]
- Docks in ***established grassland***, ***land temporarily removed from production*** *(in green cover)* [2]
- Knotgrass in ***spring barley***, ***spring wheat***, ***winter barley***, ***winter wheat*** [1]
- Mayweeds in ***spring barley***, ***spring wheat***, ***winter barley***, ***winter wheat*** [1]

Approval information

- Thifensulfuron-methyl included in Annex I under EC Directive 91/414

Efficacy guidance

- Best results achieved from application to small emerged weeds when growing actively
- Ensure good spray coverage and apply to dry foliage
- Susceptible weeds stop growing almost immediately but symptoms may not be visible for about 2 wk
- Product sold in twin pack with mecoprop-P to provide option for improved control of cleavers [1]
- Only broad-leaved docks (*Rumex obtusifolius*) are controlled; curled docks (*Rumex crispus*) are resistant [2]
- Docks with developing or mature seed heads should be topped and the regrowth treated later [2]
- Established docks with large tap roots may require follow-up treatment [2]
- High populations of docks in grassland will require further treatment in following yr [2]
- Thifensulfuron-methyl is a member of the ALS-inhibitor group of herbicides and products should be used in a planned Resistance Management strategy. See Section 5 for more information

Restrictions

- Maximum number of treatments 1 per crop for cereals or one per yr for grassland and cereals. Must only be applied from 1 Feb in yr of harvest
- Do not apply to cereal crops undersown with grasses, clover or other legumes, or any other broad-leaved crop [1]
- Do not treat new leys in year of sowing [2]
- Do not treat where nutrient imbalances, drought, waterlogging, low temperatures, lime deficiency, pest or disease attack have reduced crop or sward vigour
- Do not roll or harrow within 7 d of spraying
- Do not apply to any cereal crop in sequence or mixture with another sulfonyl urea or 'ALS inhibiting' herbicide except as permitted on the label [1]

Crop-specific information

- Latest use: before 1 Aug on grass; flag leaf fully emerged stage (GS 39) on cereals
- On grass apply 7-10 d before grazing and do not graze for 7 d afterwards [2]
- Product may cause a check to both sward and clover which is usually outgrown [2]

Following crops guidance

- Only grass or cereals may be sown within 4 wk of application to grassland or setaside, or in the event of failure of any treated crop
- No restrictions apply after normal harvest of a treated cereal crop

Environmental safety

- Dangerous for the environment
- Very toxic to aquatic organisms
- LERAP Category B
- Keep livestock out of treated areas for at least 7 d following treatment
- Take extreme care to avoid drift onto broad-leaved plants outside the target area or onto surface waters or ditches, or land intended for cropping
- Spraying equipment should not be drained or flushed onto land planted, or to be planted, with trees or crops other than cereals and should be thoroughly cleansed after use - see label for instructions

Hazard classification and safety precautions

Hazard H11
Risk phrases R50, R58
Operator protection U08, U19a, U20b
Environmental protection E06a [2] (7 d); E15a, E16a, E16b, E32a, E32d, E38 [1, 2]
Storage and disposal E30a

411 thifensulfuron-methyl + tribenuron-methyl

A mixture of two sulfonylurea herbicides for cereals

Products

Calibre	DuPont	50:25% w/w	WG	07795

SEE SECTION 3 FOR PRODUCTS ALSO REGISTERED

Uses

- Annual dicotyledons in ***spring barley, spring wheat, winter barley, winter wheat***
- Charlock in ***spring barley, spring wheat, winter barley, winter wheat***
- Chickweed in ***spring barley, spring wheat, winter barley, winter wheat***
- Mayweeds in ***spring barley, spring wheat, winter barley, winter wheat***

Approval information

- Thifensulfuron-methyl included in Annex I under EC Directive 91/414

Efficacy guidance

- Apply after 1 Feb when weeds are small and actively growing
- Ensure good spray cover of the weeds
- Ensure that weeds present are those that are susceptible. See label
- Follow label mixing instructions
- Effectiveness reduced by rain within 4 h of treatment
- Various tank mixtures recommended to broaden weed control spectrum. Other mixtures are specifically excluded. See label
- Thifensulfuron-methyl and tribenuron-methyl are members of the ALS-inhibitor group of herbicides and products should be used in a planned Resistance Management strategy. See Section 5 for more information

Restrictions

- Maximum number of treatments 1 per crop
- Do not apply to cereals undersown with grass, clover or other legumes
- Do not apply within 7 d of rolling
- Do not apply in sequence or in tank mixture with a product containing any other sulfonyl urea except as directed on label
- Do not apply to any crop suffering from stress or not actively growing

Crop-specific information

- Latest use: before flag leaf ligule first visible (GS 39)
- May be sprayed on all varieties of wheat and barley from 3 leaf stage (GS 13) up to and including flag leaf fully emerged (GS 39)
- Igri winter barley may suffer damage during period of rapid growth and must not be treated before leaf sheath erect stage (GS 30)

Following crops guidance

- Only cereals, field beans or oilseed rape may be sown in the same calendar year as harvest of a treated crop
- In the event of failure of a treated crop sow only a cereal crop within 3 mth of product application

Environmental safety

- Dangerous for the environment
- Very toxic to aquatic organisms
- LERAP Category B
- Spraying equipment should not be drained or flushed onto land planted, or to be planted, with trees or crops other than cereals and should be thoroughly cleansed after use - see label for instructions
- Take particular care to avoid damage by drift onto broad-leaved plants outside the target area or onto surface waters or ditches

Hazard classification and safety precautions

Hazard H11
Risk phrases R50, R58
Operator protection U08, U19a, U20a
Environmental protection E15a, E16a, E16b, E32a, E32d, E38
Storage and disposal E30a

412 thiodicarb

A carbamate insecticide with molluscicide uses

Products

Me2 Exodus	Me2	4% w/w	RB	09786

Uses

- Slugs in ***durum wheat, potatoes, spring barley, spring oats, spring oilseed rape, spring wheat, triticale, winter barley, winter oats, winter oilseed rape, winter wheat***

Efficacy guidance

- Apply bait as broadcast treatment or admixed with seed
- Additional broadcast treatments may be needed after drilling admixed seed
- Sow admixed seed as soon as possible after treatment

Restrictions

- Product contains an anticholinesterase carbamate compound. Do not use if under medical advice not to work with such compounds
- Maximum number of treatments 1 per crop (admixture), 3 per crop (broadcast)
- Do not treat grain with more than 16% moisture content or allow treated seed to rise above this level
- May only be applied to oilseed rape and potatoes as broadcast application
- Must not be admixed with cereal seed already admixed with any other product containing thiodicarb
- Do not use treated seed as food or feed

Crop-specific information

- Latest use: before drilling when admixed with barley, durum wheat, oats, triticale, wheat; before second node detectable (GS 32) for barley, durum wheat, oats, triticale, wheat; before stem extension (GS 2,0) for oilseed rape
- HI potatoes, Brussels sprouts 3 wk

Environmental safety

- Dangerous for the environment
- Toxic to aquatic organisms
- Harmful to game, wild birds and animals

Hazard classification and safety precautions

Hazard H03, H11
Risk phrases R22a, R36, R43, R51
Operator protection A, D, H; U05a, U13, U20b
Environmental protection E10b, E15a, E32a, E34
Storage and disposal E01, E04, E30a
Treated seed S01, S02, S05, S07
Medical advice M02, M03

413 thiophanate-methyl

A carbendazim precursor fungicide with protectant and curative activity

See also iprodione + thiophanate-methyl

Products

1	Mildothane Turf Liquid	Bayer Environ.	500 g/l	SC	09935
2	Snare	Headland Amenity	500 g/l	SC	11873

Uses

- Dollar spot in ***managed amenity turf***
- Fusarium patch in ***managed amenity turf***
- Red thread in ***managed amenity turf***
- Wormcast formation in ***managed amenity turf***

Approval information

- Following implementation of Directive 98/82/EC, approval for use of thiophanate-methyl on several crops was revoked in 1999

Efficacy guidance

- Treat turf at the first sign of disease and repeat at monthly intervals as necessary
- For earthworm cast suppression apply to moist turf at first sign of casting activity and repeat as necessary

- Apply to turf during period of active growth and do not mow for 48 h
- Control of Red Thread and Dollar Spot will be helped by application of a nitrogenous fertilizer 7 d before fungicide treatment

Restrictions

- On mown turf apply after cutting and at least 48 hr before the next cut
- Do not use when earthworms are inactive during periods of drought or when ground is frozen

Environmental safety

- Dangerous for the environment
- Very toxic to aquatic organisms

Hazard classification and safety precautions

Hazard H03, H11
Risk phrases R43, R50, R58, R68
Operator protection A, C, H, M; U05a, U14, U20c
Environmental protection E15a, E31a, E32d, E34, E38
Storage and disposal E01, E04, E26, E30a
Medical advice M03

414 thiram

A protectant dithiocarbamate fungicide

See also *carboxin + thiram*
thiabendazole + thiram

Products

1	Agrichem Flowable Thiram	Agrichem	600 g/l	LS	10784
2	Thiraflo	Crompton	480 g/l	FS	11526
3	Thyram Plus	Agrichem	600 g/l	LS	10785
4	Unicrop Thianosan DG	Unicrop	80% w/w	WG	05454

Uses

- Botrytis in ***chrysanthemums***, ***freesias***, ***lettuce*** *(outdoor crops)*, ***ornamental specimens*** *(except Hydrangea)*, ***outdoor tomatoes***, ***protected lettuce***, ***protected tomatoes***, ***raspberries***, ***strawberries*** [4]
- Botrytis fruit rot in ***apples*** [4]
- Cane spot in ***raspberries*** [4]
- Damping off in ***broad beans*** *(seed treatment)*, ***dwarf beans*** *(seed treatment)*, ***grass seed*** *(seed treatment)*, ***maize*** *(seed treatment)*, ***runner beans*** *(seed treatment)*, ***spring field beans*** *(seed treatment)*, ***spring oilseed rape*** *(seed treatment)*, ***winter field beans*** *(seed treatment)*, ***winter oilseed rape*** *(seed treatment)* [1-3]; ***cabbages*** *(seed treatment)*, ***carrots*** *(seed treatment)*, ***cauliflowers*** *(seed treatment)*, ***leeks*** *(seed treatment)*, ***lettuce*** *(seed treatment)*, ***onions*** *(seed treatment)*, ***peas*** *(seed treatment)*, ***radishes*** *(seed treatment)*, ***salad onions*** *(seed treatment)*, ***turnips*** *(seed treatment)* [1, 3]; ***combining peas*** *(seed treatment)*, ***vining peas*** *(seed treatment)* [2]; ***poppies*** *(off-label - for morphine production - seed treatment)* [1]
- Downy mildew in ***protected lettuce*** [4]
- Fire in ***tulips*** [4]
- Gloeosporium in ***apples*** [4]
- Pythium in ***poppies*** *(off-label - for morphine production - seed treatment)* [1]
- Rust in ***blackcurrants***, ***carnations***, ***chrysanthemums*** [4]
- Scab in ***apples***, ***pears*** [4]
- Seed-borne diseases in ***carrots*** *(seed soak)*, ***celery*** *(seed soak)*, ***fodder beet*** *(seed soak)*, ***mangels*** *(seed soak)*, ***parsley*** *(seed soak)*, ***red beet*** *(seed soak)*, ***sugar beet*** *(seed soak)* [1]
- Spur blight in ***raspberries*** [4]

Specific Off-Label Approvals (SOLAs)

- ***poppies*** *(for morphine production - seed treatment) (OLA 040681) Dec 2008* [1]

Efficacy guidance

- Spray before onset of disease and repeat every 7-14 d. Spray interval varies with crop and disease. See label for details [4]

- Apply as seed treatment for protection against damping off. May be applied through most types of seed treatment machinery, from automated continuous flow machines to smaller batch treating apparatus [1, 3]
- Co-application of 175 ml water per 100 kg of seed likely to improve evenness of seed coverage [1-3]
- Do not spray when rain imminent [4]
- For use on tulips, chrysanthemums and carnations add non-ionic wetter [4]

Restrictions

- Maximum number of treatments 3 per crop for protected winter lettuce (thiram based products only [4]; 2 per crop if sequence of thiram and other EBDC fungicides used) [4]; 2 per crop for protected summer lettuce [4]; 1 per batch of seed for seed treatments
- Do not apply to hydrangeas [4]
- Notify processor before dusting or spraying crops for processing [4]
- Do not dip roots of forestry transplants [4]
- Do not treat seed of tomatoes, peppers or aubergines [1, 3]

Crop-specific information

- Latest use: pre-drilling for broad beans, peas, dwarf beans, grass seed, maize, poppies, runner beans, field beans, oilseed rape, vining peas; 21 d after planting out or 21 d before harvest, whichever is earlier, for protected winter lettuce; 14 d after planting out or 21 d before harvest, whichever is earlier, for protected summer lettuce; before drilling for soya beans [1]
- HI protected lettuce 21 d; outdoor lettuce 14 d; apples, pears, blackcurrants, raspberries, strawberries, tomatoes 7 d
- Seed to be treated should be of satisfactory quality and moisture content [1-3]
- Follow label instructions for treating small quantities of seed [1, 3]

Environmental safety

- Dangerous for the environment
- Very toxic to aquatic organisms
- Do not use treated seed as food or feed [2]
- Treated seed harmful to game and wildlife [2]
- A red dye is available from manufacturer to colour treated seed. See label for details [1, 3]
- Additional precautions apply if using small volume returnable container- see labels [1, 3]

Hazard classification and safety precautions

Hazard H03 [1-3]; H04 [4]; H11 [2]
Risk phrases R20, R43, R68 [1, 3]; R22a [1-3]; R36, R37 [1, 3, 4]; R38 [4]; R48, R50, R58 [2]
Operator protection A; U02a, U14, U15 [1, 3]; U05a, U20b [1-4]; U07 [1, 3] (SVR only); U08 [2-4]; U09a [1]; U19a [4]
Environmental protection E13b, E32a [4]; E15a, E31c, E34 [1, 3]; E31a, E38 [2]; E33, E36 [1, 3] (SVR only)
Storage and disposal E01, E04, E30a [1-4]; E26 [1-3]
Treated seed S01, S02, S03, S04b, S05, S06a, S07 [1-3]; S04a [1, 3]
Medical advice M03 [2, 3]; M04 [1, 3]

415 tolclofos-methyl

A protectant organophosphorus fungicide for soil-borne diseases

Products

1	Basilex	Scotts	50% w/w	WP	07494
2	Me2 Cindy	Me2	500 g/l	FS	10634
3	Rizolex	Certis	10% w/w	DS	09673
4	Rizolex Flowable	Certis	500 g/l	FS	11399

Uses

- Black scurf and stem canker in ***potatoes*** *(tuber treatment)* [2-4]
- Bottom rot in ***protected lettuce*** [1]
- Damping off in ***seedlings of ornamentals*** [1]
- Damping off and wirestem in ***brussels sprouts***, ***cabbages***, ***calabrese***, ***cauliflowers***, ***chinese cabbage*** [1]
- Foot rot in ***ornamental specimens***, ***seedlings of ornamentals*** [1]

SEE SECTION 3 FOR PRODUCTS ALSO REGISTERED

- Rhizoctonia in ***potatoes*** *(off-label - chitted seed treatment)* [4]; ***protected celery*** *(off-label)*, ***protected radishes*** *(off-label)*, ***swedes*** *(off-label - with fleece covers)*, ***swedes*** *(off-label - without covers)*, ***turnips*** *(off-label - with fleece covers)*, ***turnips*** *(off-label - without covers)* [1]
- Root rot in ***ornamental specimens***, ***seedlings of ornamentals*** [1]

Specific Off-Label Approvals (SOLAs)

- ***potatoes*** *(chitted seed treatment) (OLA 041323) Dec 2008* [4]
- ***protected celery*** *(OLA 992009) Dec 2008* [1]
- ***protected radishes*** *(OLA 002006) Dec 2008* [1]
- ***swedes***, ***turnips*** *(with fleece covers) (OLA 023749) Dec 2006* [1]
- ***swedes***, ***turnips*** *(without covers) (OLA 023749) Dec 2006* [1]

Approval information

- Tolclofos-methyl has been included in a comprehensive review of anticholinesterase compounds and considered by the ACP. As a result approvals have been allowed to continue subject to the imposition of additional operator protection requirements.
- May be applied by misting equipment mounted over roller table. See label for details [3, 4]

Efficacy guidance

- Apply dust to seed potatoes during hopper loading [3]
- Apply flowable formulation to clean tubers with suitable misting equipment over a roller table. Spray as potatoes taken into store (first earlies) or as taken out of store (second earlies, maincrop, crops for seed) pre-chitting [2, 4]
- Do not mix flowable formulation with any other product [2, 4]
- To control Rhizoctonia in vegetables and ornamentals apply as drench before sowing, pricking out or planting [1]
- On established seedlings and pot plants apply as drench and rinse off foliage [1]

Restrictions

- Tolclofos-methyl is an atypical organophosphorus compound which has weak anticholinesterase activity. Do not use if under medical advice not to work with such compounds
- Maximum number of treatments 1 per batch for seed potatoes; 1 per crop for lettuce, brassicas, celery, radishes, swedes, turnips; 1 at each stage of growth (ie sowing, pricking out, potting) to a maximum of 3, for ornamentals
- Only to be used with automatic planters [3]
- Not recommended for use on seed potatoes where hot water treatment used or to be used [3]
- Do not apply as overhead drench to vegetables or ornamentals when hot and sunny [1]
- Do not use on heathers [1]
- Must not be used via hand-held equipment
- Application to protected crops must only be made where the operator is outside the structure at the time of treatment

Crop-specific information

- Latest use: at planting of seed potatoes [2-4]; before planting celery [1]; before transplanting for lettuce, brassicas, ornamental specimens [1]; before 2 true lvs for swedes, turnips [1]
- HI potatoes 12 wk [4]

Environmental safety

- Dangerous for the environment [1-4]
- Very toxic to aquatic organisms
- LERAP Category B [1]
- Treated tubers to be used as seed only, not for food or feed [2, 3]

Hazard classification and safety precautions

Hazard H04, H11

Risk phrases R36, R37, R38 [1-4]; R50 [3]; R51, R58 [1, 2, 4]

Operator protection A, D, H [1-4]; C [2, 4]; E [2-4]; J, M [3]; U05a, U11 [3]; U16 [2, 4]; U19a, U20b [1-4]

Environmental protection E15a [2-4]; E16a [1]; E31a [2, 4]; E32a [1, 3]; E32d [3]

Storage and disposal E01, E04 [3]; E29b [2-4]; E30a [1-4]

Medical advice M01 [1]; M04 [2-4]

416 tolylfluanid

A multi-site protectant fungicide

See also fenhexamid + tolylfluanid

Products

Elvaron Multi	Bayer CropScience	50.5% w/w	WG	11422

Uses

- Botrytis in ***blackberries***, ***blackcurrants***, ***gooseberries***, ***loganberries***, ***raspberries***, ***redcurrants***, ***strawberries***, ***whitecurrants***
- Red spider mites in ***apples*** *(reduction)*, ***pears*** *(reduction)*
- Rust mite in ***apples*** *(reduction)*, ***pears*** *(reduction)*
- Scab in ***apples***, ***pears***

Efficacy guidance

- Best results in soft fruit achieved by good spray coverage of flowers and young fruitlets
- To achieve required coverage of flowers and young fruitlets spray using a hand lance, drop arm attachments or a special strawberry boom with leaf rolling bar
- Best results using high volume directed over the crop rows

Restrictions

- Maximum total dose equivalent to 2.5 full dose treatments on currants, gooseberries; 4 full dose treatments on blackberries, loganberries, raspberries, strawberries; 8 full dose treatments on apples, pears

Crop-specific information

- HI apples, pears 7 d; blackberries, loganberries, raspberries, strawberries 14 d; currants, gooseberries 21 d

Environmental safety

- Dangerous for the environment
- Very toxic to aquatic organisms
- Risk to non-target insects or other arthropods
- LERAP Category B
- Broadcast air-assisted LERAP
- Use best available application technique that minimises off-target drift to reduce effects on non-target insects or other arthropods

Hazard classification and safety precautions

Hazard H04, H11
Risk phrases R36, R43, R50
Operator protection A, H, M; U02a, U05a, U14, U20b
Environmental protection E13b, E16a, E16b, E17b, E22c, E32a, E32d, E38
Storage and disposal E01, E04, E26, E30a

417 tralkoxydim

A foliar applied oxime herbicide for grass weed control in cereals.

Products

1	Grasp	Syngenta	250 g/l	SC	10441
2	Greencrop Gweedore	Greencrop	250 g/l	SC	09882
3	Landgold Tralkoxydim	Landgold	250 g/l	SC	08604
4	Standon Tralkoxydim	Standon	250 g/l	SC	09579

Uses

- Awned canary grass in ***spring barley*** *(qualified minor use)*, ***spring wheat*** *(qualified minor use)* [1]
- Blackgrass in ***durum wheat***, ***spring barley***, ***spring wheat***, ***triticale***, ***winter barley***, ***winter rye***, ***winter wheat*** [1-4]; ***spring barley*** *(autumn sown)*, ***spring wheat*** *(autumn sown)* [1]
- Rough meadow grass in ***durum wheat***, ***spring barley*** *(autumn sown)*, ***spring wheat***, ***spring wheat*** *(autumn sown)*, ***triticale***, ***winter barley***, ***winter rye***, ***winter wheat*** [1]
- Ryegrass in ***durum wheat***, ***spring barley***, ***spring barley*** *(autumn sown)*, ***spring wheat***, ***spring wheat*** *(autumn sown)*, ***triticale***, ***winter barley***, ***winter rye***, ***winter wheat*** [1]

SEE SECTION 3 FOR PRODUCTS ALSO REGISTERED

- Wild oats in ***durum wheat*, *spring barley*, *spring wheat*, *triticale*, *winter barley*, *winter rye*, *winter wheat*** [1-4]; ***spring barley*** *(autumn sown)*, ***spring wheat*** *(autumn sown)* [1]
- Yorkshire fog in ***durum wheat*** *(qualified minor use)*, ***spring barley*** *(autumn sown - qualified minor use)*, ***spring wheat*** *(autumn sown - qualified minor use)*, ***triticale*** *(qualified minor use)*, ***winter barley*** *(qualified minor use)*, ***winter rye*** *(qualified minor use)*, ***winter wheat*** *(qualified minor use)* [1]

Efficacy guidance

- Product leaf-absorbed and translocated rapidly to growing points. Best results achieved when weeds growing actively in competitive crops under warm humid conditions with adequate soil moisture
- Activity not dependent on soil type. Weeds germinating after application will not be controlled
- Best control of wild oats obtained from 2 leaf to 1st node detectable stage of weeds, and of blackgrass up to 3 tillers
- Always follow WRAG guidelines for preventing and managing herbicide resistant weeds. Section 5 for more information

Restrictions

- Maximum number of treatments 1 or 2 per crop. See label
- Authorised adjuvant must always be added. See label
- Do not spray undersown crops or crops to be undersown
- Do not spray when foliage wet or covered in ice or crop otherwise under stress
- Do not spray if a protracted period of cold weather forecast
- Do not spray crops under stress from chemical treatment, grazing, pest attack, mineral deficiency or low fertility
- Do not roll or harrow within 1 wk of spraying
- Do not tank-mix with phenoxy hormone or sulfonylurea herbicides or apply these chemicals within 15 d before and 5 d after treatment

Crop-specific information

- Latest use: before booting (GS 41)
- Apply to winter cereals from 2 leaves unfolded up to and including flag leaf sheath extending (GS 12-41). If necessary winter cereals may be sprayed twice: once in autumn and once in spring
- Apply to spring cereals from end of tillering up to and including flag leaf sheath extending (GS 29-41)

Environmental safety

- Harmful to aquatic organisms. May cause long-term adverse effects to the aquatic environment [1]

Hazard classification and safety precautions

Hazard H03 [1]; H04 [2-4]
Risk phrases R36 [2-4]; R43, R52, R58 [1-4]
Operator protection A, C, H; U02a, U04a, U05a, U08, U20b [1-4]; U14, U19a [2, 3]
Environmental protection E15a [1-4]; E31b [2-4]; E31c, E32d, E38 [1]; E34 [4]
Storage and disposal E01, E04, E30a [1-4]; E26 [2-4]
Medical advice M03 [4]

418 tri-allate

A soil-acting thiocarbamate herbicide for grass weed control

Products

Avadex Excel 15G	Gowan	15% w/w	GR	10299

Uses

- Blackgrass in ***combining peas*, *durum wheat*, *fodder beet*, *lucerne*, *mangels*, *red beet*, *red clover*, *sainfoin*, *spring barley*, *spring field beans*, *sugar beet*, *triticale*, *vetches*, *vining peas*, *white clover*, *winter barley*, *winter field beans*, *winter wheat***

- Meadow grasses in ***combining peas, durum wheat, fodder beet, lucerne, mangels, red beet, red clover, sainfoin, spring barley, spring field beans, sugar beet, triticale, vetches, vining peas, white clover, winter barley, winter field beans, winter wheat***
- Wild oats in ***combining peas, durum wheat, fodder beet, lucerne, mangels, red beet, red clover, sainfoin, spring barley, spring field beans, sugar beet, triticale, vetches, vining peas, white clover, winter barley, winter field beans, winter wheat***

Approval information
- Approved for aerial application on wheat, barley, rye, triticale, field beans, peas, fodder beet, sugar beet, red beet, lucerne, sainfoin, vetches, clover [1]. See notes in Section 5

Efficacy guidance
- Incorporate or apply to surface pre-emergence (post-emergence application possible in winter cereals up to 2-leaf stage of wild oats)
- Do not use on soils with more than 10% organic matter
- Wild oats controlled up to 2-leaf stage
- If applied to dry soil, rainfall needed for full effectiveness, especially with granules on surface. Do not use if top 5-8 cm bone dry
- Do not apply with spinning disc granule applicator; see label for suitable types
- Do not apply to cloddy seedbeds
- Use sequential treatments to improve control of barren brome and annual dicotyledons (see label for details)

Restrictions
- Maximum number of treatments 1 per crop
- Consolidate loose, puffy seedbeds before drilling to avoid chemical contact with seed
- Drill cereals well below treated layer of soil (see label for safe drilling depths)
- Do not use on direct-drilled crops or undersow grasses into treated crops
- Do not sow oats or grasses within 1 yr of treatment

Crop-specific information
- Latest use: pre-drilling for beet crops; before crop emergence for field beans, spring barley, peas, forage legumes; before first node detectable stage (GS 31) for winter wheat, winter barley, durum wheat, triticale, winter rye

Environmental safety
- Irritating to eyes and skin
- May cause sensitization by skin contact
- Harmful to fish or other aquatic life. Do not contaminate surface waters or ditches with chemical or used container

Hazard classification and safety precautions

Hazard H04
Risk phrases R36, R38, R43
Operator protection A, C, H; U02a, U05a, U20a
Environmental protection E13c, E32a
Storage and disposal E01, E04, E30a

419 triazamate

A carbamoyl triazole insecticide for apples and specified field crops

Products

Aztec	BASF	140 g/l	EW	10211

Uses
- Aphids in ***brussels sprouts, sugar beet*** *(including Myzus persicae)*
- Green aphid in ***apples***
- Pea aphid in ***combining peas, vining peas***
- Rosy apple aphid in ***apples***

Efficacy guidance
- Treat as soon as aphids seen in crop or (on sugar beet) immediately after official warnings are issued. Treat vining peas when 15% of plants infested and combining peas when 20% of plants infested

- Products are systemic and act through contact and stomach action. Ensure foliage well covered with spray by using higher volumes in dense crops
- Controls aphids resistant to other chemical groups
- To reduce risk of development of resistance consider use of products with alternative modes of action in intensive pest control programmes. If tank mixtures used, do not reduce dose of either component

Restrictions

- Maximum number of treatments 1 per crop on peas; 2 per crop on apples, 3 per crop on sugar beet; 4 per crop on Brussels sprouts
- Maximum total dose on brassicas equivalent to one full dose treatment on peas, 2 on apples, 3 on sugar beet, 4 on Brussels sprouts
- Product must be used with recommended adjuvant, except on apples. Consult manufacturer
- Consult processor before use on crops for processing

Crop-specific information

- HI apples, Brussels sprouts, sugar beet 28 d; peas 21 d
- Label warns that damage may result unless all recommendations carefully followed
- Treatment may cause increased russetting of apples

Environmental safety

- Dangerous for the environment
- Very toxic to aquatic organisms
- Product not harmful to bees when used as directed but spraying in late evening/early morning or in dull weather recommended to avoid unnecessary stress on foraging bees. Do not apply to apples during flowering
- Do not store above 40°C

Hazard classification and safety precautions

Hazard H03, H11
Risk phrases R20, R22a, R43, R50, R58
Operator protection A, H; U02a, U05a, U09a, U13, U14, U19a, U20b
Environmental protection E15a, E32a, E32d, E34, E38
Storage and disposal E01, E04, E26, E30a
Medical advice M03, M05a, M06

420 triazoxide

A benzotriazine fungicide available only in mixtures

See also *imidacloprid + tebuconazole + triazoxide*
tebuconazole + triazoxide

421 tribenuron-methyl

A foliar acting sulfonylurea herbicide with some root activity for use in cereals

See also *thifensulfuron-methyl + tribenuron-methyl*

Products

1	Landgold Tribenuron 75	Landgold	75% w/w	WG	11203
2	Quantum 75 DF	DuPont	75% w/w	WG	09340

Uses

- Annual dicotyledons in ***durum wheat***, ***spring barley***, ***spring oats***, ***spring wheat***, ***triticale***, ***winter barley***, ***winter oats***, ***winter rye***, ***winter wheat***
- Charlock in ***durum wheat***, ***spring barley***, ***spring oats***, ***spring wheat***, ***triticale***, ***winter barley***, ***winter oats***, ***winter rye***, ***winter wheat***
- Chickweed in ***durum wheat***, ***spring barley***, ***spring oats***, ***spring wheat***, ***triticale***, ***winter barley***, ***winter oats***, ***winter rye***, ***winter wheat***
- Mayweeds in ***durum wheat***, ***spring barley***, ***spring oats***, ***spring wheat***, ***triticale***, ***winter barley***, ***winter oats***, ***winter rye***, ***winter wheat***

Efficacy guidance

- Best control achieved when weeds small and actively growing
- Good spray cover must be achieved since larger weeds often become less susceptible
- Susceptible weeds cease growth almost immediately after treatment and symptoms can be seen in about 2 wk
- Products can be used on all soil types
- Weed control may be reduced when conditions very dry
- See label for details of technique to be used for dissolving tablets in spray tank
- When tank mixing ensure product fully dispersed before adding other products
- Tribenuron-methyl is a member of the ALS-inhibitor group of herbicides and products should be used in a planned Resistance Management strategy. See Section 5 for more information

Restrictions

- Maximum number of treatments 1 per crop
- Do not spray in tank mixture, or in sequence, with a product containing any other sulfonylurea except as directed on label
- Do not apply to crops undersown with grass, clover or other broad-leaved crops
- Do not apply to any crop suffering stress from any cause or not actively growing
- Do not apply within 7 d of rolling

Crop-specific information

- Latest use: up to and including flag leaf ligule/collar just visible (GS 39)
- Apply in autumn or in spring (after 1 Feb) from 3 leaf stage of crop up to and including flag leaf fully emerged (GS 13-39)

Following crops guidance

- In the event of crop failure sow only a cereal within 3 mth of application. See label for other restrictions on subsequent cropping

Environmental safety

- Dangerous for the environment
- Very toxic to aquatic organisms
- Special care must be taken to avoid damage by drift onto nearby broad-leaved crops, surface waters or ditches
- Spraying equipment should not be drained or flushed onto land planted, or to be planted, with trees or crops other than cereals and should be thoroughly cleansed after use - see label for instructions

Hazard classification and safety precautions

Hazard H04, H11
Risk phrases R43, R50, R58
Operator protection A; U05a, U08, U20b [1, 2]; U19a [1]
Environmental protection E15a, E32a [1, 2]; E32d, E38 [2]
Storage and disposal E01, E04, E30a

422 triclopyr

An aryloxyalkanoic acid herbicide for perennial and woody weed control

See also *2,4-D + dicamba + triclopyr*
clopyralid + fluroxypyr + triclopyr
clopyralid + triclopyr
fluroxypyr + triclopyr

Products

1	Garlon 2	Syngenta	240 g/l	EC	10672
2	Garlon 4	Dow	480 g/l	EC	05090
3	Nomix Garlon 4	Nomix Enviro	480 g/l	EC	12081
4	Timbrel	Dow	480 g/l	EC	05815

Uses

- Brambles in ***established grassland, land not intended for cropping, non-crop areas*** *(directed treatment)* [1]; ***forest, land not intended to bear vegetation*** [2-4]

SEE SECTION 3 FOR PRODUCTS ALSO REGISTERED

- Broom in ***established grassland, land not intended for cropping, non-crop areas*** *(directed treatment)* [1]; ***forest, land not intended to bear vegetation*** [2-4]
- Brush clearance in ***industrial sites*** [2-4]
- Docks in ***established grassland, land not intended for cropping, non-crop areas*** [1]; ***forest, land not intended to bear vegetation*** [2-4]
- Gorse in ***established grassland, land not intended for cropping, non-crop areas*** *(directed treatment)* [1]; ***forest, land not intended to bear vegetation*** [2-4]
- Hard rush in ***established grassland, land not intended for cropping, non-crop areas*** [1]
- Perennial dicotyledons in ***established grassland, land not intended for cropping, non-crop areas*** [1]; ***forest, industrial sites, land not intended to bear vegetation*** [2-4]
- Rhododendrons in ***forest, land not intended to bear vegetation*** [2-4]
- Scrub clearance in ***industrial sites*** [2-4]; ***land not intended for cropping, non-crop areas*** *(directed treatment)* [1]
- Stinging nettle in ***established grassland, land not intended for cropping, non-crop areas*** [1]; ***forest, land not intended to bear vegetation*** [2-4]
- Woody weeds in ***established grassland, land not intended for cropping, non-crop areas*** *(directed treatment)* [1]; ***forest, industrial sites, land not intended to bear vegetation*** [2-4]

Efficacy guidance

- Apply in grassland as spot treatment or overall foliage spray when weeds in active growth in spring or summer. Details of dose and timing vary with species. See label
- Apply to woody weeds as summer foliage, winter shoot, basal bark, cut stump or tree injection treatment
- Apply foliage spray in water when leaves fully expanded but not senescent
- Do not spray in drought, in very hot or cold conditions
- Control may be reduced if rain falls within 2 h of application
- Control of rhododendron can be variable. If higher than 1.8 m cut stump treatment recommended. A follow-up shoot treatment may be required
- See label for maximum concentrations when applying in oil, water or via watering can

Restrictions

- Maximum number of treatments 1 per yr on non-crop land (as directed spray); 2 per yr on established grassland (including land not intended for cropping) and forestry
- Do not apply to grass leys less than 1 yr old
- Do not drill kale, swedes, turnips, grass or mixtures containing clover within 6 wk of treatment. Allow at least 6 wk before planting trees
- Not to be used on food crops
- Do not apply through hand held rotary atomisers
- Not to be applied in or near water

Crop-specific information

- Latest use: 6 wk before replanting; 7 d before grazing
- HI grass 6 wk
- Uses on land not intended for cropping include grassland of no agricultural interest such as roadside verges, railway and motorway embankments
- Apply winter shoot, basal bark or cut stump sprays in paraffin or diesel oil. Dose and timing vary with species. See label for details
- Inject undiluted or 1:1 dilution into cuts spaced every 7.5 cm round trunk
- Clover will be killed or severely checked by application in grassland

Environmental safety

- Dangerous for the environment
- Very toxic to aquatic organisms
- LERAP Category B
- Do not allow spray to drift onto agricultural or horticultural crops, amenity plantings, gardens, ponds, lakes or water courses. Vapour drift may occur under hot conditions
- Keep livestock out of treated areas for at least 7 d and until foliage of any poisonous weeds such as buttercups or ragwort has died and become unpalatable

Hazard classification and safety precautions

Hazard H03, H11 [1-4]; H08 [1]

Risk phrases R22a, R50 [2-4]; R22b, R38, R43, R58 [1-4]; R51 [1]

Operator protection A, C, H, M; U02a, U05a, U08, U14, U20b [1-4]; U19a [1]

Environmental protection E07a (7 d); E15a, E16a, E16b, E23, E31b, E32d, E34, E38
Consumer protection C01
Storage and disposal E01, E04, E30a
Medical advice M05b

423 trifloxystrobin

A protectant strobilurin fungicide for cereals

See also *cyproconazole + trifloxystrobin*

Products

1	Me2 Buy	Me2	125 g/l	EC	10655
2	Swift SC	Bayer CropScience	500 g/l	SC	11227
3	Twist	Bayer CropScience	125 g/l	EC	11230

Uses
- Brown rust in ***spring barley***, ***winter barley***, ***winter wheat*** [1-3]
- Net blotch in ***spring barley***, ***winter barley*** [1-3]
- Powdery mildew in ***spring barley***, ***winter barley*** [1]
- Rhynchosporium in ***spring barley***, ***winter barley*** [1-3]
- Septoria diseases in ***winter wheat*** [1-3]

Approval information
- Trifloxystrobin included in Annex I under EC Directive 91/414

Efficacy guidance
- Should be used protectively before disease is established in crop. Further treatment may be necessary if disease attack prolonged
- Trifloxystrobin is a member of the QoI cross resistance group. Product should be used preventatively and not relied on for its curative potential
- Use product as part of an Integrated Crop Management strategy incorporating other methods of control, including where appropriate other fungicides with a different mode of action. Do not apply more than two foliar applications of QoI containing products to any cereal crop
- There is a significant risk of widespread resistance occurring in *Septoria tritici* populations in UK. Failure to follow resistance management action may result in reduced levels of disease control
- On cereal crops product must always be used in mixture with another product, recommended for control of the same target disease, that contains a fungicide from a different cross resistance group and is applied at a dose that will give robust control
- Strains of barley powdery mildew resistant to QoIs are common in the UK

Restrictions
- Maximum number of treatments 2 per crop per yr

Crop-specific information
- HI barley, wheat 35 d

Environmental safety
- Dangerous for the environment
- Very toxic to aquatic organisms
- Do not harvest for human or animal consumption for at least 35 days after last application

Hazard classification and safety precautions

Hazard H04 [1, 3]; H11 [1-3]
Risk phrases R36, R43 [1, 3]; R50, R58 [1-3]
Operator protection A, H [1-3]; C [1, 3]; U02a, U05a, U09a, U19a [2]; U14 [1, 3]; U20b [1-3]
Environmental protection E13b, E32d, E38 [1-3]; E31b, E32a [1, 3]; E31c [2]
Consumer protection C02 (35 d)
Storage and disposal E01, E04, E30a [1-3]; E26, E29a [1, 3]

SEE SECTION 3 FOR PRODUCTS ALSO REGISTERED

424 trifluralin

A soil-incorporated dinitroaniline herbicide for use in various crops

See also *clodinafop-propargyl + trifluralin*
diflufenican + trifluralin
isoxaben + trifluralin

Products

1	Alpha Trifluralin 48 EC	Makhteshim	480 g/l	EC	07406
2	Tandril 48	DAPT	480 g/l	EC	10643
3	Treflan	Dow	480 g/l	EC	05817
4	Triflurex 48 EC	Makhteshim	480 g/l	EC	07947
5	Trimaran	Nufarm UK	480 g/l	EC	11400

Uses

- Annual dicotyledons in ***broad beans, broccoli, brussels sprouts, cabbages, carrots, cauliflowers, french beans, kale, lettuce, parsnips, raspberries, runner beans, strawberries, swedes, turnips, winter barley, winter wheat*** [1-5]; ***calabrese, spring linseed*** [3-5]; ***celeriac*** *(off-label)*, ***evening primrose*** *(off-label)*, ***ornamental plant production*** *(off-label)*, ***parsley, sugar beet*** [3]; ***chives*** *(off-label)*, ***combining peas*** *(off-label)*, ***herbs (see appendix 6)*** *(off-label)*, ***kohlrabi*** *(off-label)*, ***linseed*** *(off-label)*, ***nursery fruit trees and bushes*** *(off-label)*, ***ornamental specimens*** *(off-label)*, ***parsley*** *(off-label)*, ***radish seed crops*** *(off-label)*, ***soya beans*** *(off-label)*, ***sunflowers*** *(off-label)* [1, 3]; ***linseed*** [1]; ***mustard*** [1-3]; ***navy beans*** [3, 5]; ***spring field beans, winter field beans*** [1, 2]; ***spring oilseed rape, winter oilseed rape*** [1-4]; ***winter linseed*** [3, 4]
- Annual grasses in ***broad beans, broccoli, brussels sprouts, cabbages, carrots, cauliflowers, french beans, kale, lettuce, parsnips, raspberries, runner beans, strawberries, swedes, turnips, winter barley, winter wheat*** [1-5]; ***calabrese, spring linseed*** [3-5]; ***celeriac*** *(off-label)*, ***evening primrose*** *(off-label)*, ***ornamental plant production*** *(off-label)*, ***parsley, sugar beet*** [3]; ***chives*** *(off-label)*, ***combining peas*** *(off-label)*, ***herbs (see appendix 6)*** *(off-label)*, ***kohlrabi*** *(off-label)*, ***linseed*** *(off-label)*, ***nursery fruit trees and bushes*** *(off-label)*, ***ornamental specimens*** *(off-label)*, ***parsley*** *(off-label)*, ***radish seed crops*** *(off-label)*, ***soya beans*** *(off-label)*, ***sunflowers*** *(off-label)* [1, 3]; ***linseed*** [1]; ***mustard*** [1-3]; ***navy beans*** [3, 5]; ***spring field beans, winter field beans*** [1, 2]; ***spring oilseed rape, winter oilseed rape*** [1-4]; ***winter linseed*** [3, 4]
- Fat hen in ***chives*** *(off-label)*, ***combining peas*** *(off-label)*, ***herbs (see appendix 6)*** *(off-label)*, ***kohlrabi*** *(off-label)*, ***linseed*** *(off-label)*, ***nursery fruit trees and bushes*** *(off-label)*, ***ornamental specimens*** *(off-label)*, ***parsley*** *(off-label)*, ***radish seed crops*** *(off-label)*, ***soya beans*** *(off-label)*, ***sunflowers*** *(off-label)* [1]

Specific Off-Label Approvals (SOLAs)

- ***celeriac, combining peas, linseed, ornamental specimens, sunflowers*** *(OLA 981560) Dec 2008* [3]
- ***chives, combining peas, herbs (see appendix 6), kohlrabi, linseed, nursery fruit trees and bushes, ornamental specimens, parsley, radish seed crops, soya beans, sunflowers*** *(OLA 012763) Dec 2008* [1]
- ***chives, herbs (see appendix 6), nursery fruit trees and bushes, ornamental plant production, parsley, radish seed crops, soya beans*** *(OLA 930074) Dec 2008* [3]
- ***evening primrose*** *(OLA 921080) Dec 2008* [3]
- ***kohlrabi*** *(OLA 982341) Dec 2008* [3]

Efficacy guidance

- Acts on germinating weeds and requires soil incorporation to 5 cm (10 cm for crops to be grown on ridges) within 30 min of spraying. See label for details of suitable application equipment
- Apply and incorporate at any time during 2 wk before sowing or planting
- Best results achieved by application to fine, firm seedbed, free of clods, crop residues and established weeds
- In winter cereals normally applied as surface treatment without incorporation in tank-mixture with other herbicides to increase spectrum of control. See label for details
- Follow-up herbicide treatment recommended with some crops. See label for details

Restrictions

- Maximum number of treatments 1 per crop or per yr
- Do not apply to brassica plant raising beds

- Minimum interval between application and drilling or planting may be up to 12 mth. See label for details
- Do not use on sand, fen soil or soils with more than 10% organic matter

Crop-specific information

- Latest use: varies between products; check labels. Normally before 4 leaves unfolded (GS 13) for cereals; up to 6, 8 or 10 leaves for sugar beet; pre-sowing/planting for other crops
- Transplants should be hardened off prior to transplanting
- Apply in sugar beet after plants 10 cm high with 4-8 leaves and harrow into soil
- Apply in cereals after drilling up to and including 3 leaf stage (GS 13)

Environmental safety

- Dangerous for the environment
- Very toxic to aquatic organisms
- Keep livestock out of treated areas for at least two weeks following treatment

Hazard classification and safety precautions

Hazard H03, H08 [1-5]; H04 [2]; H11 [1, 3-5]

Risk phrases R20, R22a, R66 [5]; R22b [1-5]; R36 [1, 2, 4, 5]; R37, R43, R67 [3, 5]; R38 [2]; R50, R58 [1, 3-5]

Operator protection A, C [1, 2, 4]; U05a [1, 2, 4, 5]; U08, U13, U20b [1-5]; U11 [1, 4, 5]; U14 [3, 5]; U19a [3]

Environmental protection E07a [5]; E13c [2]; E15a, E34 [1, 4, 5]; E31b [1-5]; E32d [1, 3, 4]; E38 [3]

Storage and disposal E01 [1, 2, 4, 5]; E04, E29a [1, 2, 4]; E26, E27 [2, 5]; E28 [3]; E30a [1-5]

Medical advice M05b

425 triflusulfuron-methyl

A sulfonyl urea herbicide for sugar beet

Products

Debut	DuPont	50% w/w	WG	07804

Uses

- Annual dicotyledons in ***fodder beet***, ***sugar beet***
- Charlock in ***red beet*** *(off-label)*
- Cleavers in ***red beet*** *(off-label)*
- Flixweed in ***red beet*** *(off-label)*
- Fool's parsley in ***red beet*** *(off-label)*
- Nipplewort in ***red beet*** *(off-label)*
- Ox-eye daisy in ***red beet*** *(off-label)*

Specific Off-Label Approvals (SOLAs)

- ***red beet*** *(OLA 023629) Dec 2008* [1]

Efficacy guidance

- Product should be used with a recommended adjuvant or a suitable herbicide tank-mix partner - see label for details
- Product acts by foliar action. Best results obtained from good spray cover of small actively growing weeds
- Susceptible weeds cease growth immediately and symptoms can be seen 5-10 d later
- Best results achieved from a programme of up to 4 treatments starting when first weeds have emerged with subsequent applications every 5-14 d when new weed flushes at cotyledon stage
- Weed spectrum can be broadened by tank mixture with other herbicides. See label for details
- Product may be applied overall or via band sprayer
- Triflusulfuron-methyl is a member of the ALS-inhibitor group of herbicides

Restrictions

- Maximum number of treatments 4 per crop
- Do not apply to any crop stressed by drought, water-logging, low temperatures, pest or disease attack, nutrient or lime deficiency

Crop-specific information

- Latest use: before crop leaves meet between rows

SEE SECTION 3 FOR PRODUCTS ALSO REGISTERED

- HI 4 wk for red beet
- All varieties of sugar beet and fodder beet may be treated from early cotyledon stage until the leaves begin to meet between the rows

Following crops guidance

- Only winter cereals should follow a treated crop in the same calendar yr. Any crop may be sown in the next calendar yr
- After failure of a treated crop, sow only spring barley, linseed or sugar beet within 4 mth of spraying unless prohibited by tank-mix partner

Environmental safety

- Dangerous for the environment
- Very toxic to aquatic organisms
- LERAP Category B
- Triflusulfuron-methyl is very active. Take particular care to avoid drift onto plants outside the target area
- Spraying equipment should not be drained or flushed onto land planted, or to be planted, with trees or crops other than sugar beet and should be thoroughly cleansed after use - see label for instructions

Hazard classification and safety precautions

Hazard H04, H11
Risk phrases R43, R50, R58
Operator protection A; U08, U19a, U20a
Environmental protection E15a, E16a, E32a, E32d, E38
Storage and disposal E26, E30a

426 trinexapac-ethyl

A novel cyclohexanecarboxylate plant growth regulator for cereals, turf and amenity grassland

Products

1	Moddus	Syngenta	250 g/l	EC	08801
2	Shortcut	Scotts	25% w/w	WB	09254

Uses

- Growth retardation in ***amenity grass***, ***amenity turf*** [2]
- Lodging control in ***durum wheat***, ***ryegrass seed crops***, ***spring barley***, ***spring oats***, ***spring rye***, ***triticale***, ***winter barley***, ***winter oats***, ***winter rye***, ***winter wheat*** [1]

Efficacy guidance

- Best results on turf achieved from application to actively growing weed free turf grass that is adequately fertilized and watered and is not under stress [2]
- Product rainfast after 12 h [2]

Restrictions

- Maximum total dose equivalent to one full dose on cereals, ryegrass seed crops [1]
- Maximum number of treatments on turf 5 per yr [2]
- Do not apply if rain or frost expected or if crop wet
- Only use on crops at risk of lodging [1]
- Turf under stress when treated may show signs of damage. Do not apply within 12 h of mowing [2]
- Not recommended for closely mown fine turf [2]
- Do not treat newly sown turf [2]
- Not to be used on food crops [2]
- Do not compost or mulch clippings [2]

Crop-specific information

- Latest use: before 2nd node detectable (GS 32) for oats, ryegrass seed crops; before 3rd node detectable (GS 33) for durum wheat, spring barley, triticale, rye; before flag leaf sheath extending (GS 41) for winter barley, winter wheat
- On wheat apply as single treatment between leaf sheath erect stage (GS 30) and flag leaf fully emerged (GS 39) [1]

- On barley, rye, triticale and durum wheat apply as single treatment between leaf sheath erect stage (GS 30) and second node detectable (GS 32), or on winter barley at higher dose between flag leaf just visible (GS 37) and flag leaf fully emerged (GS 39) [1]
- On oats and ryegrass seed crops apply between leaf sheath erect stage (GS 30) and first node detectable stage (GS 32) [1]
- Treatment may cause ears to remain erect through to harvest [1]

Environmental safety

- Dangerous for the environment
- Toxic to aquatic organisms
- Avoid drift outside target area

Hazard classification and safety precautions

Hazard H03, H11 [1]
Risk phrases R43, R51 [1]; R58 [1, 2]
Operator protection A [1, 2]; C [1]; U05a [1, 2]; U15, U20c [1]; U20b [2]
Environmental protection E13c, E31a, E32d, E38
Consumer protection C01 [2]
Storage and disposal E01, E04, E30a [1, 2]; E26 [2]

427 triticonazole

A conazole fungicide available only in mixtures

See also *guazatine + triticonazole*
imazalil + triticonazole
prochloraz + triticonazole

428 Verticillium lecanii

A fungal parasite of aphids and whitefly

Products

1	Mycotal	Koppert	16.1% w/w	WP	04782
2	Vertalec	Koppert	2.5% w/w	WP	04781

Uses

- Aphids in ***aubergines***, ***protected beans***, ***protected chrysanthemums***, ***protected cucumbers***, ***protected lettuce***, ***protected ornamentals***, ***protected peppers***, ***protected roses***, ***protected tomatoes*** [2]
- Whitefly in ***aubergines***, ***protected beans***, ***protected cucumbers***, ***protected lettuce***, ***protected ornamentals***, ***protected peppers***, ***protected tomatoes*** [1]

Efficacy guidance

- *Verticillium lecanii* is a pathogenic fungus that infects the target pests and destroys them
- Apply spore powder as spray as part of biological control programme keeping the spray liquid well agitated
- Pre-soak the product for 2-4 h before application to rehydrate the spores and assist in dispersion
- Treat before infestations build to high levels and repeat as directed on the label
- Spray during late afternoon and early evening directing spray onto underside of leaves and to growing points
- Best results require minimum 80% relative humidity and 18°C within the crop canopy
- Product highly infective to many aphid species except the chrysanthemum aphid. Follow specific label directions for this pest [2]

Restrictions

- Never use in tank mixture
- Do not use a fungicide within 3 d of treatment. Pesticides containing captan, chlorothalonil, fenarimol, dichlofluanid, imazalil, maneb, prochloraz, quinomethionate, thiram or tolylfluanid may not be used on the same crop
- Keep in a refrigerated store at 2-6°C but do not freeze

SEE SECTION 3 FOR PRODUCTS ALSO REGISTERED

Environmental safety

- Products have negligible effects on commercially available natural predators or parasites but consult manufacturer before using with a particular biological control agent for the first time

Hazard classification and safety precautions

Operator protection U19a, U20b
Environmental protection E15a, E32a
Storage and disposal E30a

429 vinclozolin

A protectant dicarboximide fungicide

Products

1	Ronilan FL	BASF	500 g/l	SC	02960
2	Standon Vinclozolin	Standon	500 g/l	SC	07836

Uses

- Alternaria in ***spring oilseed rape***, ***winter oilseed rape*** [1, 2]
- Ascochyta in ***combining peas*** [1, 2]
- Blossom wilt in ***apples*** [1]
- Botrytis in ***combining peas***, ***dwarf beans***, ***navy beans***, ***runner beans***, ***spring oilseed rape***, ***vining peas***, ***winter oilseed rape*** [1, 2]; ***daffodils*** *(off-label - for galanthamine production)* [1]
- Chocolate spot in ***broad beans***, ***spring field beans***, ***winter field beans*** [1, 2]
- Mycosphaerella in ***combining peas*** [1, 2]
- Sclerotinia stem rot in ***spring oilseed rape***, ***winter oilseed rape*** [1, 2]

Specific Off-Label Approvals (SOLAs)

- ***daffodils*** *(for galanthamine production) (OLA 041517) Dec 2008* [1]

Efficacy guidance

- Timing of sprays varies with crop and disease. See label for details
- May be used at reduced rate on peas and field beans in tank-mix with chlorothalonil
- Where dicarboximide resistant strains have developed product may not be effective

Restrictions

- Maximum number of treatments 2 per crop or per yr
- Do not treat mange-tout varieties of peas
- Do not spray if crop wet or if rain or frost expected
- Operator must use a vehicle fitted with a cab and forced air filtration unit with a pesticide filter complying with HSE Guidance Note PM 74 or to an equally effective standard

Crop-specific information

- Before end of petal fall for apples
- HI beans, peas 2 wk; daffodils 4 wk; oilseed rape 7 wk

Environmental safety

- Dangerous for the environment
- Toxic to aquatic organisms
- Product presents a minimal hazard to bees when used as directed but consider informing local bee-keepers if intending to spray crops in flower

Hazard classification and safety precautions

Hazard H02 [1, 2]; H11 [2]
Risk phrases R40, R43, R58, R60, R61 [1, 2]; R51 [2]; R52 [1]
Operator protection A, C, H, K [1, 2]; M [1]; U05a, U19a [1, 2]; U08, U14, U20b [1]; U09a, U20a [2]
Environmental protection E15a, E31c [1, 2]; E31b, E34 [2]
Storage and disposal E01, E04, E30a [1, 2]; E26 [2]
Medical advice M04

430 warfarin

A hydroxycoumarin rodenticide

Products

1	Grey Squirrel Liquid Concentrate	Killgerm	0.5% w/w	CB	06455
2	Sakarat Ready-to-Use (Cut Wheat Base)	Killgerm	0.025% w/w	RB	H6807
3	Sakarat Ready-to-Use (Whole Wheat)	Killgerm	0.025% w/w	RB	H6808
4	Sakarat X	Killgerm	0.05% w/w	RB	H6809
5	Sewarin Extra	Killgerm	0.05% w/w	RB	H6810
6	Sewarin P	Killgerm	0.025% w/w	RB	H6811
7	Sewercide Cut Wheat Rat Bait	Killgerm	0.05% w/w	RB	H6805
8	Sewercide Whole Wheat Rat Bait	Killgerm	0.05% w/w	RB	H6806
9	Sorex Warfarin 250 ppm Rat Bait	Sorex	0.025% w/w	AB	H6812
10	Sorex Warfarin 500 ppm Rat Bait	Sorex	0.05% w/w	AB	H6813
11	Sorex Warfarin Sewer Bait	Sorex	0.05% w/w	AB	H6814
12	Warfarin 0.5% Concentrate	B H & B	0.5% w/w	CB	H6815
13	Warfarin Ready Mixed Bait	B H & B	0.025% w/w	RB	H6816

Uses

- Grey squirrels in ***agricultural premises***, ***forest***, ***industrial sites*** [1]
- Mice in ***agricultural premises*** [9-11]
- Rats in ***agricultural premises*** [2-13]

Approval information

- Product not approved for use in N Ireland [1]

Efficacy guidance

- For rodent control place ready-to-use or prepared baits at many points wherever rats active. Out of doors shelter bait from weather
- Inspect baits frequently and replace or top up as long as evidence of feeding. Do not underbait
- For grey squirrel control mix with whole wheat and leave to stand for 2-3 h before use
- Use bait in specially constructed hoppers and inspect every 2-3 d. Replace as necessary

Restrictions

- For use only by local authorities, professional operators providing a pest control service and persons occupying industrial, agricultural or horticultural premises
- For use only between 15 Mar and 15 Aug for tree protection [1]
- Must not be used outdoors at all in Scotland, nor in areas of England or Wales where pine martens occur naturally [1]

Environmental safety

- Prevent access to baits by children and animals, especially cats, dogs and pigs
- Rodent bodies must be searched for and burned or buried, not placed in refuse bins or rubbish tips. Remains of bait and containers must be removed after treatment and burned or buried
- Bait must not be used where food, feed or water could become contaminated
- The use of warfarin to control grey squirrels is illegal unless the provisions of the Grey Squirrels Order 1973 are observed. See label for list of counties in which bait may not be used [1]

Hazard classification and safety precautions

Operator protection A, C, D, E, H [2-8]; U13 [1-13]; U20a [1-6]; U20b [7-13]

Environmental protection E31a [12, 13]; E32a [1-11]

Storage and disposal E30a

Vertebrate/rodent control products V01a, V03a [1, 4, 7-10, 12, 13]; V01b, V03b, V04b [2, 3, 5, 6]; V02 [1-10, 12, 13]; V04a [1, 4, 7-13]; V05 [1]

Medical advice M03 [2, 3, 5, 6]; M05b [9-11]

SEE SECTION 3 FOR PRODUCTS ALSO REGISTERED

431 zeta-cypermethrin

A contact and stomach acting pyrethroid insecticide

Products

1	Fury 10 EW	Belchim	100 g/l	EW	11608
2	Minuet EW	Belchim	100 g/l	EW	11610

Uses

- Aphids in ***spring barley***, ***spring oats***, ***spring wheat***, ***winter barley***, ***winter oats***, ***winter wheat***
- Barley yellow dwarf virus vectors in ***spring barley***, ***spring wheat***, ***winter barley***, ***winter wheat***
- Cabbage stem flea beetle in ***spring oilseed rape***, ***winter oilseed rape***
- Cutworms in ***potatoes***, ***sugar beet***
- Flax flea beetle in ***linseed***
- Flea beetle in ***spring oilseed rape***, ***winter oilseed rape***
- Large flax flea beetle in ***linseed***
- Pea and bean weevil in ***combining peas***, ***spring field beans***, ***vining peas***, ***winter field beans***
- Pea aphid in ***combining peas***, ***vining peas***
- Pea moth in ***combining peas***, ***vining peas***
- Pod midge in ***spring oilseed rape***, ***winter oilseed rape***
- Pollen beetle in ***spring oilseed rape***, ***winter oilseed rape***
- Rape winter stem weevil in ***spring oilseed rape***, ***winter oilseed rape***
- Seed weevil in ***spring oilseed rape***, ***winter oilseed rape***

Approval information

- Following implementation of Directive 98/82/EC, approval for use of zeta-cypermethrin on numerous crops was revoked in 1999

Efficacy guidance

- On winter cereals spray when aphids first found in the autumn for BYDV control. A second spray may be required on late drilled crops or in mild conditions
- For summer aphids on cereals spray when treatment threshold reached
- For listed pests in other crops spray when feeding damage first seen or when treatment threshold reached. Under high infestation pressure a second treatment may be necessary
- Best results for pod midge and seed weevil control in oilseed rape obtained from treatment after pod set but before 80% petal fall
- Pea moth treatments should be applied according to ADAS/PGRO warnings or when economic thresholds reached as indicated by pheromone traps
- Treatments for cutworms should be made at egg hatch and repeated no sooner than 10 d later

Restrictions

- Maximum number of treatments 2 per crop
- Consult processors before use on crops for processing

Crop-specific information

- Latest use: before end of flowering for oilseed rape; before flowering completed (GS 69) for cereals
- HI potatoes, field beans 14 d; sugar beet 60 d

Environmental safety

- Dangerous for the environment
- Very toxic to aquatic organisms
- High risk to non-target insects or other arthropods. Do not spray within 6 m of the field boundary
- Risk to certain non-target insects or other arthropods. See directions for use
- LERAP Category A

Hazard classification and safety precautions

Hazard H03, H11
Risk phrases R20, R22a, R43, R50, R58
Operator protection A, C, H; U05a, U08, U14, U15, U19a, U20b
Environmental protection E15a, E16c, E16d, E22a, E31b, E32d, E34, E38
Storage and disposal E01, E04, E30a
Medical advice M05a

432 zinc phosphide

A phosphine generating rodenticide

Products

ZP Rodent Pellets	Antec Biosentry	2.0% w/w	GB	07814

Uses

- Mice in ***farm buildings***
- Rats in ***farm buildings***

Efficacy guidance

- Use to prepare baits either by dry or wet baiting as directed

Restrictions

- Zinc phosphide is subject to the Poisons Rules 1982 and the Poisons Act 1972. See notes in Section 5
- For use only by local authorities, professional operators providing a pest control service and persons occupying industrial, agricultural or horticultural premises

Environmental safety

- Spontaneously inflammable in contact with acid
- Keep in original container, tightly closed, in a safe place, under lock and key
- Wear respirator if mixing baits in a confined space
- Prevent access to baits by children and domestic animals, especially cats, dogs and pigs
- Search for and burn or bury all rodent bodies. Do not place in refuse bins or on rubbish tips
- Do not prepare or use baits where food or water could be contaminated
- Remove all remains of baits and bait containers after treatment and burn or bury
- Wash out all mixing equipment thoroughly at the end of every operation

Hazard classification and safety precautions

Operator protection A; U13, U20a
Environmental protection E15a, E32a
Storage and disposal E30b
Vertebrate/rodent control products V01a, V02, V03b, V04a
Medical advice M03

433 ziram

A dithiocarbamate bird and animal repellent

Products

AAprotect	Unicrop	32% w/w	PA	03784

Uses

- Birds in ***all top fruit, field crops, forest, ornamental specimens***
- Deer in ***all top fruit, field crops, forest, ornamental specimens***
- Hares in ***all top fruit, field crops, forest, ornamental specimens***
- Rabbits in ***all top fruit, field crops, forest, ornamental specimens***

Efficacy guidance

- Apply undiluted to main stems up to knee height to protect against browsing animals at any time of yr or spray 1:1 dilution on stems and branches in dormant season
- Use dilute spray on fully dormant fruit buds to protect against bullfinches
- Only apply to dry stems, branches or buds
- Use of diluted spray can give limited protection to field crops in areas of high risk during establishment period

Restrictions

- Maximum number of treatments 3 per season as spray for all top fruit, field crops, forest and ornamental specimens; 1 per season as paste treatment on trees and ornamental specimens
- Do not spray elongating shoots or buds about to open
- Do not apply concentrated spray to foliage, fruit buds or field crops

Crop-specific information

- HI edible crops 8 wk

SEE SECTION 3 FOR PRODUCTS ALSO REGISTERED

Environmental safety

- Harmful to fish or other aquatic life. Do not contaminate surface waters or ditches with chemical or used container

Hazard classification and safety precautions

Hazard H04
Risk phrases R36, R37, R38
Operator protection A, C; U05a, U08, U19a, U20b
Environmental protection E13c, E31c
Consumer protection C02 (8 wk)
Storage and disposal E01, E04, E30a

434 zoxamide

A substituted benzamide available only in mixtures

See also mancozeb + zoxamide

SECTION 3
PRODUCTS ALSO REGISTERED

Products also Registered

Products listed in the table below have not been notified for inclusion in Section 2 of this edition of the *Guide*. However they have extant approval until 31 December 2008 unless an earlier expiry date is shown. These products may legally be stored and used in accordance with their label until their approval expires, but they may not still be available for purchase.

Product	Marketing Co.	Reg. No.	Expiry Date
acibenzolar-S-methyl			
Bion	Syngenta	09803	
aldicarb			
Agriguard Aldicarb	AgriGuard	10481	31-12-07
Aventis Temik 10G	Aventis	09749	
Marnoch Carbadil 10	Marnoch	10338	31-12-07
Standon Aldicarb 10G	Standon	11024	31-12-07
alpha-cypermethrin			
Alert	BASF	10272	
Alpha C 6 ED	Techneat	11838	
Alphaguard 100 EC	Nufarm UK	12003	
Alphatop 100 EC	Interfarm	12102	
Bestseller 100 EC	Chimac-Agriphar	10461	
Cleancrop Acymet	United Agri	10497	
Cleancrop Scope	United Agri	10273	
Contest	Cyanamid	09024	
Fastac	BASF	10220	
Fastac	Cyanamid	07008	
Fastac Dry	BASF	10221	
I T Alpha-cypermethrin	I T Agro	08274	
aluminium ammonium sulphate			
Guardsman B	Chiltern	05494	
Guardsman M	Chiltern	05495	
Rezist	Barrettine	08576	
Scoot	Garotta	07388	
Stay Off	Vitax	02019	
aluminium phosphide			
Detia Gas Ex-P	Igrox	09802	
Detia Gas-Ex-B	Detia Degesch	06927	
Detia Gas-Ex-T	Igrox	03792	
Phostoxin I	Rentokil	05694	
aluminium sulphate			
Growing Success Slug Killer	Growing Success	04386	

Product	Marketing Co.	Reg. No.	Expiry Date
amidosulfuron			
Aventis Eagle	Aventis	09765	
Barclay Cleave	Barclay	09489	
Barclay Cleave	Barclay	11340	
Druid	Aventis	08714	30-06-05
Pursuit 50	Bayer CropScience	08716	
2-aminobutane			
Hortichem 2-Aminobutane	Certis	06147	31-12-07
amitraz			
Bye Bye 20 EC	Chimac-Agriphar	09346	
Cleancrop Bye Bye 20EC	United Agri	11530	
Ovasyn	Bayer CropScience	09190	
amitrole			
Aminotriazole Technical	Nufarm UK	10855	
Loft	Marks	06030	
MSS Aminotriazole Technical	Mirfield	04645	
Weedazol Pro	Nufarm UK	11995	
Weedazol-TL	Nufarm UK	11968	
Weedazol-TL	Marks	02349	
Weedazol-TL	Bayer	02979	30-06-05
amitrole + 2,4-D + diuron			
Trik	Mirfield	07853	31-10-05
asulam			
Asulox	RP Agric.	06124	30-04-05
Cleancrop Asulam	United Agri	10465	
atrazine			
Alpha Atrazine 50 WP	Makhteshim	04793	
Atlas Atrazine	Nufarm UK	07702	
Atrazine 90 WG	Sipcam	09310	31-08-05
DAPT Atrazine 50 SC	DAPT	08031	
MSS Atrazine 50 FL	Nufarm UK	01398	
MSS Atrazine 80 WP	Mirfield	04360	
Unicrop Atrazine 50	Unicrop	02645	
Unicrop Atrazine FL	Unicrop	08045	
Unicrop Flowable Atrazine	Unicrop	05446	
azamethiphos			
Alfacron 10 WP	Novartis A H	09439	30-09-05
aziprotryne			
Brasoran 50 WP	Novartis	08394	
azoxystrobin			
Amistar	Zeneca	08517	

Product	Marketing Co.	Reg. No.	Expiry Date
Barclay ZX	Barclay	09570	
Barclay ZX	Barclay	11336	
Clayton Stobik	Clayton	09440	01-07-08
Marnoch Binoxy	Marnoch	10622	01-07-08
Olympus	Syngenta	08520	01-10-05
Olympus	Syngenta	10541	01-07-08
Ortiva	Syngenta	09843	02-10-05
Ortiva	Syngenta	10542	01-07-08
Priori	Syngenta	08516	01-07-08
Priori	Syngenta	10543	01-07-08
azoxystrobin + fenpropimorph			
Amistar Pro	Zeneca	08871	
Aspect	Syngenta	10516	
benalaxyl + mancozeb			
Galben M	Sipcam	05092	
benazolin + bromoxynil + ioxynil			
Asset	Aventis	07243	
benazolin + clopyralid			
Benazalox	Aventis	07246	
benfuracarb			
Oncol 10G	Nufarm UK	12139	
bentazone			
Agrotech Bentazone	Agrotech-Trading	12143	01-08-05
Basagran	BASF	00188	
I T Bentazone	I T Agro	08677	
I T Bentazone 48	I T Agro	09283	
Standon Bentazone	Standon	09204	
bentazone + dichlorprop-P			
Quitt SL	BASF	08108	
bentazone + MCPA + MCPB			
Headland Archer	Headland	08814	
bentazone + pendimethalin			
Impuls	BASF	09720	30-09-05
beta-cyfluthrin + clothianidin			
Poncho Beta	Bayer CropScience	12076	27-04-08
beta-cyfluthrin + imidacloprid			
Chinook	Bayer CropScience	11421	
Chinook	Bayer	11151	
Chinook	Bayer	10660	
Chinook Blue	Bayer	10912	

Product	Marketing Co.	Reg. No.	Expiry Date
bifenox + chlorotoluron			
Dicurane Duo 446 SC	Syngenta	08404	
bifenox + isoproturon			
Banco	Feinchemie	10606	31-08-05
Banco	Makhteshim	11191	
RP 4169	Feinchemie	10605	31-08-05
RP 4169	Makhteshim	11190	
bifenox + MCPA + mecoprop-P			
Sirocco	Feinchemie	10620	31-08-05
Sirocco	Makhteshim	11192	
Sirocco	Aventis Environ.	09939	
bifenthrin + flutriafol			
Roseclear Gun !	Scotts	09923	
bitertanol + fuberidazole			
Sibutol	Bayer	07238	31-05-05
Sibutol CF	Bayer CropScience	11306	
Sibutol CF	Bayer	08174	31-05-05
Sibutol New Formula	Bayer CropScience	11307	
Sibutol New Formula	Bayer	08270	31-05-05
UK 743	Bayer CropScience	11315	
UK 743	Bayer	10707	
bitertanol + fuberidazole + imidacloprid			
Sibutol Secur	Bayer	09131	31-05-05
bitertanol + fuberidazole + imidacloprid + triadimenol			
Cereline Secur	Bayer CropScience	11261	
Cereline Secur	Bayer	09511	31-10-05
bitertanol + fuberidazole + triadimenol			
Cereline	Bayer CropScience	11260	
Cereline	Bayer	07239	31-10-05
boscalid			
Me2 Succotash	Me2	12147	24-11-06
bromacil + diuron			
Borocil K	Aventis Environ.	09924	
Borocil K	RP Amenity	05183	
bromoxynil			
Barclay Mutiny	Barclay	11136	
bromoxynil + clopyralid			
Vindex	Dow	05470	

Product	Marketing Co.	Reg. No.	Expiry Date
bromoxynil + clopyralid + fluroxypyr + ioxynil			
Crusader S	Dow	05174	
bromoxynil + ethofumesate + ioxynil			
Stefes Leyclene	Bayer CropScience	10086	31-10-05
bromoxynil + fluroxypyr			
Sickle	Dow	05187	
Tomahawk Plus	Makhteshim	09836	
bromoxynil + fluroxypyr + ioxynil			
Advance	Dow	05173	
Treble	Barclay	10577	
bromoxynil + ioxynil			
Deloxil	Bayer CropScience	07405	
Deloxil	Bayer CropScience	09987	31-10-05
Deloxil	Bayer CropScience	07313	
Oxytril CM	RP Agric.	08667	
Percept	United Phosphorus	05481	
Status	Bayer CropScience	08668	31-10-05
Status	Bayer CropScience	10019	
Stellox	Nufarm UK	11551	
Stellox 380 EC	Syngenta	08451	
bromoxynil + ioxynil + mecoprop-P			
Swipe P	Syngenta	08452	
bromoxynil + ioxynil + triasulfuron			
Teal	Nufarm UK	11940	
Teal	Syngenta	08453	
bromoxynil + ioxynil + trifluralin			
Masterspray	SumiAgro	02971	
Masterspray	SumiAgro	09603	
bromoxynil + terbuthylazine			
Alpha Bromotril PT	Makhteshim	09435	
Cleancrop Amaize	United Agri	11990	
Templar	Makhteshim	10254	
bromuconazole			
Granit	Aventis	09995	
Granit	RP Agric.	08268	
bupirimate			
Nimrod	Syngenta	10530	
bupirimate + triforine			
Nimrod-T	Miracle	07865	

Product	Marketing Co.	Reg. No.	Expiry Date
buprofezin			
Applaud	Certis	10930	31-01-06
Applaud	Syngenta	10678	
captan			
Alpha Captan 50 WP	Makhteshim	04797	
PP Captan 80 WG	Tomen	08971	28-02-05
PP Captan 83	Tomen	08768	
captan + penconazole			
Topas C	Syngenta	08459	
carbendazim			
Barclay Shelter	Barclay	09000	
BASF Turf Systemic Fungicide	BASF	05774	
Bavistin	BASF	00217	
Bavistin DF	BASF	03848	
Bavistin FL	BASF	00218	
Carbate Flowable	SumiAgro	08957	
Clayton Am-Carb	Clayton	11906	
Clayton Chizm FL	Clayton	11050	
Cleancrop Curve	United Agri	11774	
Delsene 50 Flo	Nufarm UK	09469	30-11-05
Derosal Liquid	Bayer CropScience	07315	
Derosal WDG	Aventis	07316	
Goldazim 500 SC	Chimac-Agriphar	10675	
Headland Addstem	Headland	06755	
Headland Addstem DF	Headland	08904	
Headland Regain	Headland	08675	
I.T. Carbendazim	I T Agro	12155	
Mascot Systemic	Rigby Taylor	08776	
Mascot Systemic	Scotts	09132	
MSS Mircarb	Mirfield	08788	
Occidor 500 SC	Chimac-Agriphar	10937	
Quadrangle Hinge	Quadrangle	08070	
S.M.I Carbendazim	Sub-Micron	12040	
Stefes Derosal Liquid	Bayer CropScience	10079	31-10-05
Stefes Derosal WDG	Bayer CropScience	10078	31-10-05
Supercarb	SumiAgro	01560	
Supercarb	SumiAgro	09610	
Top Farm Carbendazim 435	Top Farm	05307	
Turf Systemic Fungicide	SumiAgro	10821	
Turf Systemic Fungicide	SumiAgro Amenity	09349	
Turfclear WDG	Scotts	07490	
carbendazim + chlorothalonil			
Bravocarb	Syngenta	10520	

Product	Marketing Co.	Reg. No.	Expiry Date
Bravocarb	Zeneca	09105	
Greenshield	Scotts	07988	
carbendazim + flusilazole			
Landgold Flusilazole MBC	Landgold	08528	
Standon Flusilazole Plus	Standon	07403	
carbendazim + iprodione			
Calidan	BASF	11687	
Calidan	Aventis	09980	
Calidan	RP Agric.	06536	
Vitesse	Bayer Environ.	12037	
Vitesse	RP Amenity	06537	
carbendazim + maneb			
Multi-W FL	SumiAgro	09447	
Protector	Procam	09448	
Tripart 147	Tripart	07978	
carbendazim + maneb + sulphur			
Bolda FL	Atlas	07653	
carbendazim + metalaxyl			
Ridomil mbc 60 WP	Syngenta	08437	
carbendazim + prochloraz			
Novak	BASF	11693	
Novak	Aventis	08020	31-01-05
Sportak Alpha HF	BASF	11698	
Sportak Alpha HF	Aventis	07225	
carbendazim + propiconazole			
Hispor 45 WP	Syngenta	08418	
Sparkle 45 WP	Syngenta	08450	
carbendazim + tebuconazole			
Bayer UK 413	Bayer CropScience	11243	
Bayer UK 413	Bayer	08277	
Tricur	Bayer CropScience	11314	
Tricur	Bayer	10281	30-06-05
Tricur	Bayer	10377	
carbendazim + vinclozolin			
Konker	BASF	03988	
carbetamide			
Carbetamex	Feinchemie	10265	
carbetamide + diflufenican + oxadiazon			
Buffalo G	Bayer Environ.	11088	

Product	Marketing Co.	Reg. No.	Expiry Date
carbosulfan			
Marshal 10G	Belchim	11682	
Marshal 10G	FMC	09804	
carboxin + imazalil + thiabendazole			
Vitaflo Extra	Crompton	07048	
carfentrazone-ethyl			
Aurora 40 WG	Belchim	11614	
Platform	Belchim	11615	
Shark	Belchim	11616	31-03-06
Spotlight Plus	Belchim	12100	17-08-07
carfentrazone-ethyl + flupyrsulfuron-methyl			
Lexus Class	DuPont	10809	31-03-05
Lexus Class WSB	DuPont	08637	
carfentrazone-ethyl + isoproturon			
Affinity	FMC	10723	
Affinity	FMC	09745	
carfentrazone-ethyl + mecoprop-P			
Platform S	FMC	10726	
carfentrazone-ethyl + thifensulfuron-methyl			
Harmony Express	DuPont	09467	
chloridazon			
Cleancrop AKAZON	United Agri	11903	
Gladiator DF	Tripart	06342	
Lidazone 65 WG	Globachem	10847	
Luxan Booster WG	Luxan	11709	
Luxan Chloridazon	Luxan	06304	
PG Chloridazon 430 SC	PG-Crop Protection	11517	
Portman Weedmaster	Portman	06018	
Pyramin FL	BASF	01661	
Pyramin FL	BASF	11628	
Questar	BASF	11629	
Questar	BASF	07955	
Starter Flowable	Truchem	03421	
Stefes Chloridazon	Bayer CropScience	10110	
Takron	BASF	06237	30-04-05
Tripart Gladiator	Tripart	00986	
Weedmaster SC	Portman	08793	
chloridazon + chlorpropham + metamitron			
Newtron	Nufarm UK	10943	31-08-05

Product	Marketing Co.	Reg. No.	Expiry Date
chloridazon + ethofumesate			
Magnum	BASF	08635	
Spectron	Aventis	07284	
chloridazon + lenacil			
Advizor	DuPont	06571	
Advizor S	Griffin	10948	
chloridazon + metamitron			
Volcan Combi FL	Sipcam	11442	
chloridazon + quinmerac			
Fiesta T	BASF	10260	30-06-05
chlormequat			
Agriguard Chlormequat 700	AgriGuard	09782	
Agriguard Chlormequat 760	AgriGuard	10290	
Alpha Chlormequat 460	Makhteshim	04804	
Alpha Pentagan	Makhteshim	04794	
Alpha Pentagan Extra	Makhteshim	04796	
Ashlade 460 CCC	Ashlade	06474	
Ashlade 700 CCC	Ashlade	06473	
Ashlade Brevis	Nufarm UK	08119	30-11-05
Atlas 3C:645 Chlormequat	Nufarm UK	07700	
Atlas 3C:645 Chlormequat	Nufarm UK	05710	
Atlas Chlormequat 46	Nufarm UK	07704	
Atlas Chlormequat 460:46	Nufarm UK	06258	
Atlas Chlormequat 700	Nufarm UK	10706	
Atlas Quintacel	Nufarm UK	07706	30-06-05
Atlas Terbine	Nufarm UK	06523	
Atlas Terbine	Nufarm UK	07709	
Atlas Tricol	Nufarm UK	07707	
Atlas Tricol	Nufarm UK	07190	
Barclay Holdup	Barclay	06799	
Barclay Holdup	Barclay	11365	
Barclay Holdup 600	Barclay	11373	
Barclay Holdup 600	Barclay	08794	
Barclay Holdup 640	Barclay	08795	
Barclay Holdup 640	Barclay	11374	
Barclay Liffey	Barclay	09856	
Barclay Liffey	Barclay	11366	
Barclay Lucan	Barclay	09855	
Barclay Lucan	Barclay	11367	
Barleyquat B	Mandops	06001	
BASF 3C Chlormequat 600	BASF	04077	
BASF 3C Chlormequat 750	BASF	06878	
Belcocel	Taminco	11881	
Belcocel	Griffin	08248	

Product	Marketing Co.	Reg. No.	Expiry Date
Bettaquat B	Mandops	06004	
Chlormequat 46	Nufarm UK	11504	
Ciba Chlormequat 460	Ciba Specialty	09525	
Ciba Chlormequat 5C 460:320	Ciba Specialty	09527	
Ciba Chlormequat 730	Ciba Specialty	09526	
Clayton CCC 750	Clayton	07952	
Clayton Manquat	Clayton	09916	
Clayton Standup	Clayton	11760	
Cleancrop Chlormequat 700	United Agri	10143	
Cropsafe 5C Chlormequat	Certis	07897	
Greencrop Cong 750	Greencrop	09383	
Hive	Nufarm UK	10953	31-10-05
Hyquat 70	Agrichem	03364	
Intracrop Balance	Intracrop	08037	
Intracrop MCCC	Intracrop	08506	
Larke	Nufarm UK	10045	30-11-05
Larke	Nufarm UK	11453	
Mandops Chlormequat 460	Mandops	06090	
Mirquat	Nufarm UK	11406	
MSS Chlormequat 40	Mirfield	01401	
MSS Chlormequat 460	Mirfield	03935	
MSS Chlormequat 60	Mirfield	03936	
MSS Chlormequat 70	Mirfield	03937	
MSS Mirquat	Mirfield	08166	
Portman Chlormequat 400	Portman	01523	
Portman Chlormequat 460	Portman	02549	
Portman Chlormequat 700	Portman	03465	
Quadrangle Chlormequat 700	Quadrangle	03401	
Renown	Bayer CropScience	10103	
Sigma PCT	Nufarm UK	10705	30-09-05
Stabilan 460	Nufarm UK	09304	
Stabilan 5C	Nufarm UK	08144	
Stabilan 640	Nufarm UK	09401	
Stabilan 670	Nufarm UK	09402	
Stabilan 700	Nufarm UK	10579	31-10-05
Stabilan 750	Nufarm UK	09303	
Stay Up	Nufarm UK	09444	
Stefes CCC	Bayer CropScience	10104	31-10-05
Stefes CCC 640	Bayer CropScience	10111	31-10-05
Stefes CCC 700	Bayer CropScience	10105	31-10-05
Stefes CCC 720	Bayer CropScience	10108	31-10-05
Supaquat 720	Portman	09381	
Terbine	Nufarm UK	11407	
Tripart 5C	Tripart	04726	
Tripart Brevis	Tripart	03754	
Tripart Brevis 2	Tripart	06612	
Tripart Chlormequat 460	Tripart	03685	

Product	Marketing Co.	Reg. No.	Expiry Date
Uplift	United Phosphorus	07527	
Whyte Chlormequat 700	Nufarm UK	09641	
chlormequat + 2-chloroethylphosphonic acid			
Barclay Banshee XL	Barclay	09201	
Barclay Banshee XL	Barclay	11339	
Nomad	Bayer CropScience	10004	
Terpal C	BASF	07062	
chlormequat + 2-chloroethylphosphonic acid + imazaquin			
Satellite	BASF	10395	
Satellite	Cyanamid	08969	
chlormequat with choline chloride			
Agriguard 5C Chlormequat 460	AgriGuard	09851	
Ashlade 5C	Ashlade	06227	
Ashlade 700 5C	Ashlade	07046	
Atlas 460:46	Atlas	07705	31-12-05
Atlas 5C Chlormequat	Nufarm UK	03084	
Atlas 5C Chlormequat	Atlas	07701	31-12-05
Atlas 5C Quintacel	Nufarm UK	11130	
Barclay Take 5	Barclay	08524	
Fernpath Tangent	AgriGuard	10341	
MSS Mircell	Mirfield	06939	
New 5C Cycocel	BASF	01483	
Portman Supaquat	Portman	03466	
chlormequat with choline chloride + imazaquin			
Meteor	Cyanamid	06505	
Upright	BASF	10404	
2-chloroethylphosphonic acid			
Aventis Cerone	Bayer CropScience	09748	31-10-05
Aventis Cerone	Aventis	09972	
Barclay Coolmore	Barclay	07917	
Barclay Coolmore	Barclay	11349	
Cerone	RP Agric.	06185	
Charger	Bayer CropScience	08827	31-10-05
Charger	Aventis	09986	
Ethrel C	Hortichem	06995	31-10-05
EXPO 3149D	Bayer CropScience	08828	31-10-05
EXPO 3149D	Bayer CropScience	09989	
Pan Stiffen	Pan Agriculture	11644	
2-chloroethylphosphonic acid + mepiquat chloride			
Barclay Banshee	Barclay	08175	
CleanCrop Fonic M	United Agri	09553	
Cleancrop Fonic M	United Agri	09533	

Product	Marketing Co.	Reg. No.	Expiry Date
Marnoch Phonet	Marnoch	09979	
Terpal	Clayton	07626	
chlorothalonil			
Atlas Cropguard	Nufarm UK	09123	30-06-05
Barclay Corrib 500	Barclay	11350	
Barclay Corrib 500	Barclay	08981	
BB Chlorothalonil	Brown Butlin	03320	
Bombardier	Unicrop	02675	
Bravo 500	BASF	05637	
Bravo 500	Syngenta	09059	31-01-06
Bravo 720	Syngenta	10519	
Bravo 720	Zeneca	09104	
Clayton Turret	Clayton	09400	
Cleancrop Chlorothalonil 720	United Agri	10102	
Cleancrop Rover	United Agri	10500	
Cleancrop Rover 2	United Agri	11965	
Clortosip	Sipcam	06126	
Cracker	Tronsan	12149	
Cropguard	Nufarm UK	11835	
Delphi	Sipcam	11937	
ISK 375	Zeneca	09103	
Jupital	Syngenta	09109	
Jupital	Syngenta	10528	
Jupital DG	Syngenta	10529	
Jupital DG	Syngenta	09181	
Mace	Nufarm UK	11957	
Mainstay	Quadrangle	05625	
Miros DF	Sipcam	04966	
Mycoguard	Chiltern	08115	31-01-05
Mycoguard	Gharda	10945	
Mycoguard	Nufarm UK	12008	
Nuturf Chlorothalonil	Nufarm UK	11974	
Repulse	Syngenta	06705	
Repulse	Syngenta	10535	28-02-05
Repulse	Certis	07641	
Standon Chlorothalonil 500	Standon	08597	
Tripart Faber	Tripart	04549	
Ultrafaber	Tripart	05627	
Visclor 500 SC	Sipcam	09404	
Visclor 75 DF	Sipcam	09361	
chlorothalonil + cymoxanil			
Gex 44	Griffin	10168	
chlorothalonil + cyproconazole			
Bravo Xtra	Syngenta	11824	
Cleancrop Cyprothal	United Agri	09580	

Product	Marketing Co.	Reg. No.	Expiry Date
Octolan	Syngenta	11675	
Octolan	Novartis	08480	
SAN 703	Syngenta	11676	
chlorothalonil + fluquinconazole			
Vista CT	Aventis	09368	
chlorothalonil + flutriafol			
Prospa	Headland	11548	
chlorothalonil + mancozeb			
Guru	Interfarm	10056	
Sipcam Flo	Sipcam	07601	
chlorothalonil + metalaxyl			
Folio	Syngenta	08547	
chlorothalonil + propamocarb hydrochloride			
Aventis Merlin	Aventis	09719	
Pan Wizard	Pan Agriculture	11953	
Tattoo C	Bayer CropScience	07623	
chlorothalonil + propiconazole			
Sambarin 312.5 SC	Novartis	08439	
chlorothalonil + tetraconazole			
Eminent Star	Isagro	10447	
chlorothalonil + vinclozolin			
Curalan CL	BASF	07174	
chlorotoluron			
Atol	Nufarm UK	10235	
Clayton Chloron	Clayton	08148	
Clayton Chloron 500 FL	Clayton	10791	
Dicurane	Novartis	08403	
Luxan Chlorotoluron 500 Flowable	Luxan	09165	
MSS Chlorotoluron 500	Nufarm UK	07871	
NWA CTU 500	Nufarm UK	11140	
NWA CTU 500	Nufarm UK	10173	
Stefes Toluron	Bayer CropScience	10106	31-10-05
Talisman	FCC	03109	
Tolugan 700	Makhteshim	07874	
Top Farm Toluron 500	Top Farm	05986	
chlorotoluron + pendimethalin			
Totem	BASF	10485	
chlorpropham			
Aceto Chlorpropham 50M	Aceto	08929	
Atlas CIPC 40	Nufarm UK	07710	

Product	Marketing Co.	Reg. No.	Expiry Date
Croptex Pewter	Certis	02507	
Luxan Gro Stop 100	Luxan	12052	
Luxan Gro-Stop 300 EC	Luxan	08602	
Luxan Gro-Stop Basis	Luxan	08601	
MSS CIPC 30M	Nufarm UK	10064	
MSS CIPC 40 EC	Whyte Agrochemicals	01403	
MSS CIPC 50 LF	Whyte Agrochemicals	03285	
MSS CIPC 50 M	Whyte Agrochemicals	01404	31-07-05
Sprout Nip	Aceto	11786	
Spud-Nic 100 Briquette	Aceto	11633	28-02-05
Standon CIPC 300 HN	Standon	09187	
Triherbicide CIPC	Cerexagri	06874	
Triherbicide CIPC	Atofina	06426	
Whyte CIPC 40EC	Whyte Agrochemicals	11952	
chlorpropham + fenuron			
Atlas Red	Nufarm UK	07724	
Croptex Chrome	Certis	02415	31-12-07
chlorpropham + fenuron + propham			
Atlas Herbon Pabrac	Atlas	03997	
Atlas Herbon Pabrac	Atlas	07714	
chlorpropham + metamitron			
Buckler	Nufarm UK	11032	
chlorpyrifos			
Agriguard Chlorpyrifos	AgriGuard	10626	
Barclay Clinch II	Barclay	11346	
Barclay Clinch II	Barclay	08596	
Choir	Nufarm UK	09778	
Cleancrop Pychlorex	United Agri	11681	
Crossfire 480	Dow	08141	
Dispatch	Dow	08139	
Dursban 4	Dow	07815	
Lorsban 480	Dow	08076	
Lorsban WG	Dow	10139	
suSCon Indigo	Fargro	09902	
chlorthal-dimethyl			
Dacthal W75	AMVAC	10289	
Dacthal W75	Certis	05500	
Dacthal W75	United Agri	10617	
Dacthal W-75	ISK Biosciences	05556	
Dacthal W-75	Certis	10623	28-02-06
cinidon-ethyl			
Lotus	BASF	09231	31-03-05

Product	Marketing Co.	Reg. No.	Expiry Date
clodinafop-propargyl			
Unite A	Bayer CropScience	07686	
clodinafop-propargyl + diflufenican			
Amazon	Aventis	10266	
Amazon	Novartis	08128	
clodinafop-propargyl + trifluralin			
Reserve	Syngenta	12077	
clofentezine			
Apollo 50 SC	Aventis	07242	
clomazone			
Centium 360 CS	FMC	10720	
Cirrus	Belchim	11751	
clopyralid			
Agriguard Clopyralid	AgriGuard	10625	31-08-05
Barclay Karaoke	Barclay	08185	
Barclay Karaoke	Barclay	11357	
Cliophar	Chimac-Agriphar	09430	
Dow Shield	Dow	05578	28-02-05
Glopyr 200 SL	Globachem	10979	
I T Clopyralid	I T Agro	09808	
Lontrel 200	Dow	11558	
clopyralid + fluroxypyr + ioxynil			
Hotspur	Dow	05185	
clopyralid + fluroxypyr + MCPA			
Greenor	Rigby Taylor	07848	28-02-05
clopyralid + fluroxypyr + triclopyr			
Pastor	Dow	07440	31-10-05
clopyralid + ioxynil			
Escort	Dow	05466	
clopyralid + propyzamide			
Matrikerb	Dow	10806	
Matrikerb	SumiAgro	01308	
Matrikerb	Rohm & Haas	02443	
Matrikerb	Dow	09604	
clothianidin			
Bayer UK 978	Bayer CropScience	11201	
copper ammonium carbonate			
Croptex Fungex	Certis	02888	31-03-05

Product	Marketing Co.	Reg. No.	Expiry Date
copper hydroxide			
Spin Out	DuPont	12069	
copper oxychloride + maneb + sulphur			
Ashlade SMC Flowable	Ashlade	06494	
copper oxychloride + metalaxyl			
Ridomil Plus	Syngenta	08353	
cyanazine			
Barclay Canter	Barclay	07530	
Cleancrop Vectro	United Agri	10475	31-12-07
Fortrol	Feinchemie	10666	
Fortrol	Cyanamid	07009	
cyanazine + pendimethalin			
Bullet	Feinchemie	10654	31-08-05
Bullet	Cyanamid	08049	31-12-07
cyazofamid			
Ranman Twinpack	BASF	11396	
Ranman TwinPack	BASF	10628	
cycloxydim			
Marnoch Clodim	Marnoch	09852	31-12-05
Stratos	BASF	06891	
cyfluthrin			
Baythroid	Makhteshim	11663	
cymoxanil			
Curzate 60DF	DuPont	11515	
Scribe	Syngenta	11967	30-06-05
Scribe 60WG	Syngenta	11971	
cymoxanil + mancozeb			
Besiege	DuPont	08086	
Clayton Krypton	Clayton	09398	
Cleancrop Xanilite	United Agri	10050	
Curzate M WG	DuPont	11901	
Curzate M68 WSB	DuPont	08073	
Fytospore 68	DuPont	09827	
Marnoch Mancym	Marnoch	10130	
Matilda	Nufarm UK	12006	
Rhapsody	DuPont	11958	
Rhythm	Interfarm	09636	
Rhythm CM	Interfarm	11825	
Standon Cymoxanil Extra	Standon	09442	
Systol M	Quadrangle	03480	

Product	Marketing Co.	Reg. No.	Expiry Date
Systol M	DuPont	08085	
Zetanil	Sipcam	11993	
cypermethrin			
Afrisect 10	Mitchell Cotts	10849	
Afrisect 10	Stefes	09114	
Afrisect 10	Chimac-Agriphar	12027	
Agriguard Cypermethrin	AgriGuard	12134	
Arrivo	FMC	10731	
Brazil	Nufarm UK	11135	
Cleancrop Pyrimet	United Agri	10619	
Cyperguard 100 EC	Nufarm UK	10779	31-12-05
Cyperguard 100 EC	Gharda	10775	
Cyperkill 10	Chimac-Agriphar	12028	
Cyperkill 10	Chiltern	04119	
Cyperkill 25	Chiltern	09038	
Cyperkill 25	Chimac-Agriphar	12029	
Cyperkill 5	Chimac-Agriphar	12026	
Cypertox	FCC	05122	
Mcc 25 Ec	Chimac-Agriphar	12030	
MCC 25 EC	Chiltern	09115	
Permasect C	Nufarm UK	09200	30-06-05
cyproconazole			
Alto 240 EC	Bayer CropScience	11236	
Alto 240 EC	Bayer	10462	
Alto 240 EC	Novartis	08354	
Caddy 240 EC	Bayer	10631	30-06-05
Fort	Bayer	11027	31-03-06
Fort	Bayer CropScience	11606	31-07-05
cyproconazole + mancozeb			
Alto Eco	Syngenta	08466	
cyproconazole + propiconazole			
Alto Extra	Syngenta	10850	
cyproconazole + quinoxyfen			
Choice	Dow	10680	
DOE 1762	Dow	08962	
cyproconazole + trifloxystrobin			
Fencer	Bayer CropScience	11277	
Fencer	Bayer	10887	
Sphere	Bayer	10586	
cyprodinil			
Barclay Amtrak	Barclay	09562	
Barclay Amtrak	Barclay	11338	

Product	Marketing Co.	Reg. No.	Expiry Date
Cleancrop Cyprodinil	United Agri	09668	
Standon Cyprodinil	Standon	09345	
Unix	Syngenta	08764	31-01-06
2,4-D			
Agrichem 2,4-D	Agrichem	04098	
Agricorn 2,4-D	FCC	07349	01-04-05
Agricorn D	FCC	00056	01-04-05
Atlas 2,4-D	Atlas	07699	
Barclay Haybob II	Barclay	08532	
Depitox	Nufarm UK	10746	31-07-05
Dupont 24-D	DuPont	11737	
Easel	Nufarm UK	09878	
Easel	Nufarm UK	11146	
Forester	Vitax	00914	
GroWell 2,4-D Amine	GroWell	11395	
GroWell 2,4-D Amine	Marks	08750	
Headland Staff 500	Headland	12087	
Herboxone 60	Marks	09693	
HY-D	Agrichem	06278	
Luxan 2,4-D	Luxan	09379	
Marks 2,4-D-A	Marks	01282	01-04-05
Maton	Headland	10366	
MSS 2,4-D Amine	Nufarm UK	11139	
MSS 2,4-D Amine	Nufarm UK	10183	
MSS 2,4-D Ester	Nufarm UK	01393	
Nomix 2,4-D Herbicide	Nomix Enviro	06394	
Palormone D	Unicrop	01534	01-04-05
Ragox	Nufarm UK	11145	
Ragox	Nufarm UK	10407	
Ritefeed 2,4-D Amine	Ritefeed	05309	01-04-05
Silvapron D	BP	01935	01-04-05
Syford	Vitax	02026	
2,4-D + dicamba			
Green-Up Weedfree Spot Weedkiller for Lawns	Vitax	06321	
Lawn Builder Plus Weed Control	Scotts	08499	
2,4-D + dicamba + triclopyr			
Cleancrop Broadshot	United Agri	11664	
2,4-D + MCPA			
Agroxone Combi	Headland	10277	31-03-05
Agroxone Combi	Headland	11025	
2,4-D + mecoprop-P			
Mascot Selective-P	Rigby Taylor	06105	01-04-05

Product	Marketing Co.	Reg. No.	Expiry Date
daminozide			
B-Nine	Crompton	11465	
B-Nine	Certis	11057	30-11-05
B-Nine	Certis	07844	30-04-05
B-Nine	Certis	04468	30-11-05
Dazide Enhance	Fine	11943	
Dazide WSG	Fine	11753	
dazomet			
Basamid	BASF	00192	
Basamid	Certis	07204	
2,4-DB			
Butoxone DB	Marks	10482	
2,4-DB + linuron + MCPA			
Alistell	Syngenta	06515	30-04-05
Alistell	Syngenta	10510	
2,4-DB + MCPA			
Butoxone DB Extra	Marks	12152	
Butoxone DB Extra	Marks	10455	
Headland Cedar	Headland	12180	
MSS 2,4-DB + MCPA	Nufarm UK	01392	
deltamethrin			
Agriguard Deltamethrin	AgriGuard	10770	
Agrotech Deltamethrin	Agrotech-Trading	12165	
Aventis Decis	Aventis	09710	
Decis Micro	Bayer CropScience	08618	
Deleet	Rentokil	09312	
Delta-M 2.5 EC	O Endres	11334	
Deltamex 2.5 EC	O Endres	10291	30-09-05
Deltaplan	Chimac-Agriphar	11678	
deltamethrin + pirimicarb			
Best	Bayer CropScience	08988	
Evidence	Aventis	06934	30-06-05
Patriot EC	Aventis	08990	30-06-05
desmedipham + ethofumesate + phenmedipham			
Aventis Betanal Progress OF	Aventis	09722	
Betanal Progress OF	Aventis	07629	
desmedipham + phenmedipham			
Betanal Compact	Aventis	07247	
dicamba			
Cadence	Syngenta	08796	

Product	Marketing Co.	Reg. No.	Expiry Date
Cadence	Barclay	09578	
SAN 845H	Syngenta	08797	
dicamba + dichlorprop-P + MCPA			
Intrepid	Miracle	07819	
dicamba + MCPA			
Banvel M	Syngenta	08469	
dicamba + MCPA + mecoprop			
Tribute	Nomix Enviro	06921	
dicamba + MCPA + mecoprop-P			
ALS Premier Selective Plus	Amenity Land	08940	
Docklene Super	Bayer CropScience	10728	
Herrisol New	Bayer CropScience	11445	
Herrisol New	Bayer	07166	
Mascot Super Selective-P	Rigby Taylor	06106	
Mircam Plus	Nufarm UK	11004	31-01-06
Mircam Plus	Nufarm UK	09884	28-02-05
Mircam Super	Nufarm UK	11836	
MSS Mircam Plus	Mirfield	01416	
Outrun	SumiAgro	10426	
Pasturol D	FCC	07033	
Pasturol P	FCC	09099	
Premier Amenity Selective	Amenity Land	10098	
Quadrangle Quadban	Nufarm UK	07090	
Super Selective Plus	Nufarm UK	11666	
Super Selective Plus	Rigby Taylor	11928	
Triad	Headland	11944	
Trireme	Nufarm UK	10425	31-01-06
UPL Amenity Grassland Herbicide	United Phosphorus	11713	
dicamba + mecoprop-P			
Condox	Syngenta	10522	
Headland Saxon	Headland	10436	31-08-05
Headland Swift	Headland	11945	
Headland Swift	Headland	10437	
Optica Forte	Marks	11913	
Optica Forte	Marks	07432	
Prompt	Headland	11535	31-08-05
dicamba + triasulfuron			
Banvel T	Syngenta	08470	
dichlobenil			
Casoron G	Crompton	09022	
Casoron G	Rigby Taylor	09326	
Casoron G	Zeneca	08065	

Product	Marketing Co.	Reg. No.	Expiry Date
Casoron G	Miracle	07926	
Casoron G4	Crompton	09215	
Casoron G4	Miracle	07927	
Casoron G-SR	Miracle	07925	
Dichlobenil Granules	Certis	11871	
Scotts Dichlo G Macro	Scotts	12011	
Scotts Dichlo G Micro	Scotts	11857	
Sierraron G	Scotts	09263	
Sierraron G	Scotts	09675	
Sierraron G4	Scotts	10491	
Standon Dichlobenil 6G	Standon	08874	
Viking Granules	Nomix Enviro	11859	
dichlorophen			
Halophen RE 49	Superfog	10863	31-01-06
n2n Mosskiller	Nomix Enviro	11823	
Nomix-Chipman Mosskiller	Nomix Enviro	06271	
Ritefeed Dichlorophen	Ritefeed	05265	
dichlorophen + ferrous sulphate			
Aitken's Lawn Sand Plus	Aitken	04542	
1,3-dichloropropene			
Telone 2000	Dow	05748	
dichlorprop + MCPA			
Redipon Extra	United Phosphorus	07518	
dichlorprop-P			
Headland Link	Headland	10603	31-05-05
Optica DP	Marks	07818	
Optica DP	Marks	11067	
dichlorprop-P + ferrous sulphate + MCPA			
Vitagrow Granular Feed, Weed & Mosskiller	Sinclair	10971	
dichlorprop-P + ioxynil			
Duet	DuPont	11553	
Mextrol DP	Nufarm UK	11529	
dichlorprop-P + MCPA			
Optica Duo	Marks	10343	
dichlorprop-P + MCPA + mecoprop-P			
Optica Trio	Marks	09747	
dicloran			
Fumite Dicloran Smoke	Certis	09291	

Product	Marketing Co.	Reg. No.	Expiry Date
difenacoum			
Neokil	Sorex	05564	
Neosorexa	Sorex	07756	
Neosorexa Ratpacks	Sorex	04653	
difenoconazole			
Landgold Difenoconazole	Landgold	09964	
Plover	Syngenta	08429	30-04-05
difenzoquat			
Avenge 2	BASF	10210	
Avenge 2	Cyanamid	03241	
Match	Cyanamid	07186	
diflubenzuron			
Dimilin 25-WP	Hortichem	08902	
Dimilin Flo	Certis	10618	30-04-05
Dimilin WP	Syngenta	08870	
diflufenican			
EXP 4005	Bayer CropScience	10582	
diflufenican + flufenacet			
Regatta	Bayer CropScience	12054	
Zephyr	Bayer CropScience	12053	
diflufenican + flurtamone + isoproturon			
EXP 31502A	Bayer CropScience	09991	31-01-05
diflufenican + glyphosate			
Pistol	Bayer Environ.	12173	
diflufenican + isoproturon			
Clayton Fenican 550	Clayton	11160	
DFF + IPU WDG	Aventis	10204	
Gavel	Bayer CropScience	09992	31-10-05
Panther WDG	Aventis	10650	
Shire	Bayer CropScience	08117	
Tolkan Turbo	Bayer CropScience	10024	30-06-05
Unite B	Bayer CropScience	07686	
diflufenican + terbuthylazine			
Bolero	Syngenta	08392	
dimethoate			
Danadim	Cheminova	09583	28-02-06
Sector	Cheminova	10492	
dimethomorph + mancozeb			
Invader	BASF	11861	31-10-05
Invader	BASF	10390	31-05-05

Product	Marketing Co.	Reg. No.	Expiry Date
Invader	Cyanamid	06989	
Invader WDG	BASF	10349	
Saracen	BASF	12005	
Saracen	BASF	11860	30-11-05
Saracen	BASF	10392	
Saracen WDG	BASF	10393	
dimoxystrobin			
BAS 505 04f	BASF	11926	27-01-07
diphenylamine			
No-Scald DPA	Cerexagri	08312	
diquat			
Clayton Diquat	Clayton	10730	
Cleancrop Diquat	United Agri	09687	
Midstream	Scotts	09267	
Midstream	Miracle	07739	
Reglone	Zeneca	09646	
Standon Diquat	Standon	10674	
Waterloo	Alfa	10839	31-08-05
diquat + paraquat			
Clayton Paradigm	Clayton	12001	
PDQ	Zeneca	10241	
dithianon			
Barclay Cluster	Barclay	08792	
Barclay Cluster	Barclay	11347	
Dithianon Flowable	Cyanamid	07007	
dithianon + penconazole			
Topas D 275 SC	Syngenta	08460	
diuron			
Atlas Diuron	Nufarm UK	08214	
Chipko Diuron 80	Nomix Enviro	00497	
Chipman Diuron 80	Nomix Enviro	08054	
Chipman Diuron Flowable	Nomix Enviro	05701	
Diuron 50 FL	Staveley	02814	
Diuron 80% WP	Staveley	00730	
Freeway	RP Amenity	06047	
Karmex	DuPont	12068	
Karmex	Nufarm UK	11647	
Karmex	Headland	11045	
Mascot Diuron Flow	Rigby Taylor	10889	30-09-05
Mascot Diuron Flow	Rigby Taylor	11212	
MSS Diuron 50 FL	Nufarm UK	07160	
MSS Diuron 500 FL	Mirfield	08171	31-01-06

Product	Marketing Co.	Reg. No.	Expiry Date
MSS Diuron 80 WP	Nufarm UK	09787	31-01-06
n2n Diuron 80	Nomix Enviro	11821	
n2n Diuron Flowable	Nomix Enviro	11820	
n2n Diuron Flowable	Nomix Enviro	11822	
Nomix Chipman Flowable Diuron	Nomix Enviro	10837	
Rescind	RP Amenity	08036	
Sanuron	Headland	09236	
UPL Diuron 80	United Phosphorus	07619	
diuron + glyphosate			
Cleanguard	Bayer Environ.	12055	
NX 2075	Nomix-Chipman	11104	
NX 3083	Nomix-Chipman	08712	
NX 3083	Nomix-Chipman	11100	
Total	Nomix-Chipman	07932	
Total	Nomix-Chipman	11099	
Touche	Nomix-Chipman	06797	01-01-05
Touche	Nomix-Chipman	11097	
Trymark	Nomix Enviro	11867	
dodeca-8,10-dienyl acetate			
Exosex CM	Exosect	12103	
dodine			
Barclay Dodex	Barclay	09055	
Barclay Dodex	Barclay	11351	
Dodifun 400 SC	Hermoo	11657	
Syllit 400 SC	Chimac-Agriphar	11079	
endosulfan			
Thiodan 20 EC	Aventis	07335	
epoxiconazole			
Clayton Oust	Clayton	11588	
Cleancrop EPX	United Agri	12135	
Cleancrop EPX	United Agri	10381	
Epic	BASF	08320	31-08-06
Opus	BASF	08319	31-07-06
epoxiconazole + fenpropimorph			
Barclay Riverdance	Barclay	09658	
Barclay Riverdance	Barclay	11341	
Eclipse	BASF	07361	30-06-05
Landgold Epoxiconazole FM	Landgold	08806	
Opus Team	BASF	07362	31-07-05
epoxiconazole + fenpropimorph + kresoxim-methyl			
BAS 493F	BASF	10438	31-01-05
Cleancrop Chant	United Agri	11746	

Product	Marketing Co.	Reg. No.	Expiry Date
Cleancrop Chant	United Agri	10466	31-01-05
Mantra	BASF	08886	
Mastiff	BASF	10857	31-01-05
epoxiconazole + kresoxim-methyl			
Barclay Avalon	Barclay	09466	
Barclay Avalon	Barclay	11337	
Clayton Gantry	Clayton	09482	
Cleancrop Kresoxazole	United Agri	09698	
Landgold Strobilurin KE	Landgold	09908	
Landmark	BASF	08889	
Marnoch Kempo	Marnoch	09918	
epoxiconazole + kresoxim-methyl + quinoxyfen			
BAS 541 00F	BASF	10924	
TPF003	Dow	10916	
epoxiconazole + pyraclostrobin			
Opera	BASF	12167	
esfenvalerate			
Clayton Estate	Clayton	11155	
Greencrop Cajole	Greencrop	11402	
Sumi-Alpha	Cyanamid	07207	
ethofumesate			
Atlas Thor	Atlas	07732	
Barclay Keeper 500 Flow	Barclay	11843	
Barclay Keeper EC	Barclay	11887	
Cleancrop EC 200	Barclay	11886	
Cleancrop SC 500	Barclay	11844	
I T Ethofumesate	I T Agro	11108	
Kubist	Griffin	09454	31-08-05
Kubist 2	Nufarm UK	11126	
Landgold Ethofumesate 200	Landgold	08980	
Linesman 200 EC	Barclay	11885	
Linesman 500 SC	Barclay	11845	
MSS Thor	Nufarm UK	08817	
Nortron	Bayer CropScience	07266	
Salute	United Phosphorus	07660	
Standon Ethofumesate 200	Standon	08726	
Stefes Fumat 2	Bayer CropScience	10084	
ethofumesate + metamitron			
Torero	Feinchemie	10250	
ethofumesate + metamitron + phenmedipham			
Bayer UK 540	Bayer CropScience	11244	28-02-06
Bayer UK 540	Bayer	07399	28-02-06

Product	Marketing Co.	Reg. No.	Expiry Date
Betanal Trio WG	Aventis	07537	
MAUK 540	Makhteshim	11545	
ethofumesate + phenmedipham			
Barclay Goalpost	Barclay	09192	
Barclay Goalpost	Barclay	11372	
Betanal Tandem	Aventis	07254	
Betosip Combi	Sipcam	08630	
Magic Tandem	Bayer CropScience	11814	
Powertwin	Feinchemie	10442	
Thunder	Feinchemie	10958	31-08-05
Twin	Feinchemie	09875	
ethoprophos			
Aventis Mocap 10G	Aventis	09973	
Mocap 10G	RP Agric.	06773	
etridiazole			
Aaterra WP	Zeneca	06625	
Terrazole 35 WP	Crompton	09800	
fatty acids			
Safers Insecticidal Soap	Safer	07197	
fatty acids + sulphur			
Nature's Answer Pest and Disease Control	Scotts	11022	31-05-05
fenarimol			
Rimidin	Dow	07938	
Rubigan	Dow	05489	
fenazaquin			
Matador 200 SC	Margarita	11058	
Matador 200 SC	Dow	07960	
fenbuconazole			
Indar 5EW	Whelehan	09644	
Kruga 5EC	Interfarm	09874	31-01-05
Kruga 5EC	Interfarm	09863	
Reward 5EC	Interfarm	09862	
Surpass 5EC	Interfarm	09861	
fenbuconazole + fenpropidin			
Accolade	SumiAgro	09950	
fenbuconazole + prochloraz			
Mirage Extra	Makhteshim	10558	
fenbuconazole + propiconazole			
Graphic	Interfarm	10987	

Product	Marketing Co.	Reg. No.	Expiry Date
Graphic	Rohm & Haas	09907	
Graphic	Dow	10800	
fenbutatin oxide			
Torque	BASF	07148	
Torque	BASF	10399	
fenhexamid			
Lattice	Bayer CropScience	11224	
Lattice	Bayer	09198	31-05-11
Teldor	Bayer	08955	30-06-05
fenhexamid + tolylfluanid			
Talat	Bayer	09655	
fenoxaprop-P-ethyl			
Aventis Cheetah Super	Aventis	09730	
fenoxaprop-P-ethyl + isoproturon			
Pinion	Griffin	10287	31-03-05
Puma X	Aventis	08779	
fenpropathrin			
Meothrin	Cyanamid	07206	
fenpropidin			
Cleancrop Fulmar	United Agri	12033	
Landgold Fenpropidin 750	Landgold	08973	
Mallard	Syngenta	08662	
Patrol	Syngenta	10531	
Patrol	Zeneca	08661	
fenpropidin + prochloraz			
SL 552A	Syngenta	08673	
Sponsor	BASF	11902	
Sponsor	Aventis	08674	
fenpropidin + propiconazole			
Prophet	Syngenta	08433	
Sheen	Syngenta	08442	
Zulu	Syngenta	08464	
fenpropidin + propiconazole + tebuconazole			
Bayer UK 593	Bayer CropScience	11416	
Bayer UK 593	Bayer	08285	
Gladio	Novartis	08413	
fenpropidin + tebuconazole			
Monicle	Bayer CropScience	11428	
Monicle	Bayer	07375	
SL 556 500 EC	Syngenta	08449	

Product	Marketing Co.	Reg. No.	Expiry Date
fenpropimorph			
BAS 421F	BASF	06127	
Clayton Spigot	Clayton	11560	
Cleancrop Fenpro	United Agri	09885	
Cleancrop Fenpropimorph	United Agri	09445	
Keetak	BASF	06950	
Marnoch Phorm	Marnoch	10773	31-05-05
fenpropimorph + flusilazole			
BAS 48500F	BASF	06784	
Colstar	DuPont	12175	
fenpropimorph + kresoxim-methyl			
Ensign	BASF	08362	
Landgold Strobilurin KF	Landgold	09196	
Splice	BASF	11918	
fenpropimorph + metrafenone			
Flexity TP	BASF	11855	
fenpropimorph + pyraclostrobin			
BAS 528 00f	BASF	11444	
fenpropimorph + quinoxyfen			
EF-1288	Dow	08241	
fenpyroximate			
NNI 850 5SC	Nihon Nohyaku	09887	
Sequel	Hortichem	09886	31-10-05
fentin acetate + maneb			
Brestan 60 SP	Aventis	07305	
fentin hydroxide			
Barclay Fentin Flow	Barclay	07914	
Barclay Fentin Flow 532	Barclay	09434	
Super-Tin 4L	Griffin	09028	
Super-Tin 4L	Griffin	09559	
Super-Tin 80 WP	Griffin	09560	
ferrous sulphate			
Aitken's Lawn Sand	Aitken	05253	
Fisons Greenmaster Autumn	Fisons	03211	28-02-06
Fisons Greenmaster Mosskiller	Fisons	00881	28-02-06
Greenmaster Autumn	Scotts	07508	28-02-06
Greenmaster Mosskiller	Scotts	07509	28-02-06
Maxicrop Moss Killer & Conditioner	Maxicrop	04635	
Moss Control Plus Lawn Fertilizer	Miracle	07912	
No More Moss Lawn Feed	Wolf	10754	

Product	Marketing Co.	Reg. No.	Expiry Date
Pentagon Prestige Lawn Sand	Sinclair	10456	
Vitagrow Lawn Sand	Vitagrow	05097	
fipronil			
Regent 1 GR	BASF	11743	
Regent 1 GR	Bayer CropScience	11078	
florasulam			
Primus 25 SC	Dow	10175	
florasulam + fluroxypyr			
Starane Vantage	Dow	10922	25-02-07
fluazifop-P-butyl			
Fusilade 250 EW	Zeneca	06531	
Fusilade 250 EW	Syngenta	10525	
PP 007	Syngenta	06533	
PP 007	Syngenta	10533	
Wizzard	Syngenta	06521	
Wizzard	Syngenta	10539	
fluazinam			
Barclay Cobbler	Barclay	08349	
Barclay Cobbler	Barclay	11348	
Legacy	ISK Biosciences	09966	
Shirlan	Zeneca	07091	
Shirlan Programme	Syngenta	08761	
Shirlan Programme	Syngenta	10574	
Standon Fluazinam 500	Standon	08670	
Top Farm Fluazinam	Top Farm	07683	
fluazinam + metalaxyl-M			
Emrald Flumet	Emrald	12000	
Epok	ISK Biosciences	10429	31-10-05
fludioxonil			
Beret Gold	BASF	12013	
flufenacet + metribuzin			
Artist	Bayer	10658	30-06-05
flufenacet + pendimethalin			
210	BASF	10683	
Cleancrop Hector	United Agri	12163	
Emrald Berg	Emrald	12086	
Greencrop Duet 603	Greencrop	12171	
Landgold Bedrock	Teliton	12145	
fluoroglycofen-ethyl + isoproturon			
Competitor	Bayer CropScience	07310	

Product	Marketing Co.	Reg. No.	Expiry Date
fluoxastrobin			
Bayer UK 831	Bayer CropScience	12091	16-08-08
flupyrsulfuron-methyl			
Bullion	DuPont	12050	06-08-05
DPX-KE459 WSB	DuPont	08540	
KE 459 DF	DuPont	09835	
KE 459 PX	DuPont	11189	06-08-05
Lexus 50 DF	DuPont	09026	
Lexus 50 PX	DuPont	11188	06-08-05
flupyrsulfuron-methyl + metsulfuron-methyl			
Lexus XPE	DuPont	08541	
Standon Flupyrsulfuron MM	Standon	09098	
flupyrsulfuron-methyl + picolinafen			
BUK 960	BASF	11152	01-03-07
Buk 960 01H	BASF	11662	03-03-07
DP 944	DuPont	11120	01-03-07
DP 945	DuPont	11447	
flupyrsulfuron-methyl + thifensulfuron-methyl			
Lexus Millenium WSB	DuPont	09207	
flupyrsulfuron-methyl + tribenuron-methyl			
Excalibur	DuPont	12047	
fluquinconazole			
Diablo	BASF	11688	
Diablo	Bayer CropScience	09914	
Flamenco	Aventis	09913	30-06-05
Jockey F	Aventis	10074	30-06-05
Jockey Flexi	BASF	11691	
Jockey Flexi	Bayer CropScience	10073	
fluquinconazole + prochloraz			
Aventis Foil	Aventis	09709	
Baron	BASF	11686	
Baron	Bayer CropScience	10904	
Foil	Aventis	10905	30-06-05
Jockey	Aventis	10076	30-06-05
Jockey Plus	BASF	11692	
Jockey Plus	Bayer CropScience	10075	
fluroxypyr			
Barclay Hurler	Barclay	08791	
BAS 91092H	BASF	11044	
Clayton Fluroxypyr	Clayton	06356	
Emrald Chopper	Emrald	11716	
Gala	Dow	12019	30-11-10

Product	Marketing Co.	Reg. No.	Expiry Date
Gala	Dow	09793	
Gex 353 B	Griffin	10185	30-11-05
Tomahawk 2000	Makhteshim	09307	
Vizzler	Interfarm	11409	01-12-05
fluroxypyr + MCPA + mecoprop-P			
EM1658/01	SumiAgro	10892	
fluroxypyr + metosulam			
EF 1166	Bayer CropScience	11381	
EF1166	Dow	07966	
fluroxypyr + thifensulfuron-methyl + tribenuron-methyl			
GEX 353	DuPont	11474	
fluroxypyr + tribenuron-methyl			
DP953 PX	DuPont	11321	
fluroxypyr + triclopyr			
Doxstar	Dow	06050	31-10-05
Evade	SumiAgro Amenity	08071	
flusilazole			
Genie	DuPont	08238	
Sanction	DuPont	08237	
flutolanil			
EXP 10066A	Bayer CropScience	11584	
NNF-136	Nihon Nohyaku	11585	
flutriafol			
Impact	Headland	11521	
fosetyl-aluminium			
Aliette	Certis	02484	
Aliette	Hortichem	05648	
Aliette 80 WG	Certis	09156	30-06-05
fuberidazole + imazalil + triadimenol			
Baytan IM	Bayer CropScience	11418	
Baytan IM	Bayer	00226	
fuberidazole + imidacloprid + triadimenol			
Baytan Secur	Bayer	09510	31-01-05
fuberidazole + triadimenol			
Baytan	Bayer CropScience	11250	
Baytan	Bayer	00225	
Baytan CF	Bayer CropScience	11251	
Baytan CF	Bayer	08193	

Product	Marketing Co.	Reg. No.	Expiry Date
gibberellins			
Berelex	Nufarm UK	08903	
GIBB Plus	Globachem	11875	
Gistar	Hermoo	11397	
Regulex	Sumitomo	10147	
glufosinate-ammonium			
Aventis Challenge	Aventis	09721	
Aventis Harvest	Aventis	09810	
Dash	Nomix Enviro	05177	
Headland Sword	Headland	07676	
Kaspar	Certis	10615	30-06-05
Nomix Touchweed	Nomix Enviro	09596	
glyphosate			
Acrion	Bayer Environ.	09965	
Acrion	Bayer Environ.	11718	
Agriguard Glyphosate 180	AgriGuard	09806	01-01-05
AgriGuard Glyphosate 360	AgriGuard	09184	
Apache	Syngenta	06748	01-01-05
Apache	Syngenta	10514	
Azural	Monsanto	09582	01-01-05
Barbarian	Barclay	07625	
Barclay Barbarian	Barclay	09865	31-03-06
Barclay Cleanup	Barclay	09179	
Barclay Dart	Barclay	05129	
Barclay Gallup	Barclay	05161	
Barclay Gallup 360	Barclay	09127	28-02-06
Barclay Gallup Biograde 360	Barclay	09840	30-04-05
Barclay Gallup Biograde 450	Dalgety	10151	31-03-06
Barclay Gallup Biograde Amenity	Barclay	10203	
Barclay Gallup Hi Aktiv	Barclay	10457	31-10-05
Barclay Garryowen	Barclay	09869	
Barclay Garryowen	Barclay	11364	
Biactive 270	Monsanto	10031	
Biactive 270	Monsanto	10301	
Bioglyce	Austrital	11459	
Bioglyce	Applyworld	10054	30-11-05
Cardel Egret	Cardel	09703	
Cardel Glyphosate	Cardel	09581	01-01-05
Cardel Glyphosate	Cardel	10253	01-01-05
Clarion	Syngenta	10521	
Clarion	Syngenta	08272	01-01-05
Clayton Glyphosate	Clayton	06608	
Clayton Rhizeup	Clayton	08920	
Clayton Swath	Clayton	06715	
CleanCrop Egret	United Agri	10689	

Product	Marketing Co.	Reg. No.	Expiry Date
Cleancrop Hoedown	Loveland	11194	
Cleancrop Hoedown	United Agri	10852	
Clean-Up-360	Top Farm	05076	
Do-Away	NCH	10112	31-03-06
Do-Away	Barclay	11380	
Economix	Nomix Enviro	08008	
Fernpath Opsen	AgriGuard	10687	31-01-05
Fernpath Statis	AgriGuard	10663	01-01-05
First Line	Linemark	11386	
Glister	Sinon EU	11413	
Gly-480	Cardel	11528	31-08-05
Glyfonex 480	Danagri	07464	01-01-05
Glyfos 480	Headland	10996	
Glyfos 480	Cheminova	08014	31-03-05
Glyfos Proactive	Nomix-Chipman	11976	
Glyfosate-360	Top Farm	05319	
Glymark	Nomix Enviro	11868	
Glyper	PBI	07968	
Glyphosate 120B	Monsanto	10302	01-01-05
Glyphosate 120B	Monsanto	08292	01-01-05
Glyphosate 360	Austrital	11555	
Glyphosate 360	Danagri	09233	28-02-06
Glyphosate 360A	Danagri	09234	01-01-05
Glyphosate 360B	Danagri	09235	01-01-05
Glyphosate Biactive	Monsanto	10303	01-01-05
Gorgon	Portman	09182	01-01-05
Helosate	Helm	06499	
Initial Line	Linemark	11384	
Marnoch Glyphosate	Marnoch	09744	
Mogul	Monsanto	07076	01-01-05
Mogul	Monsanto	10306	01-01-05
Mogul	Monsanto	12045	02-06-07
MON 240	Monsanto	10307	01-01-05
MON 240	Monsanto	04538	
MON 52276	Monsanto	10310	01-01-05
MON 52276	Monsanto	06949	28-02-05
Monty	Quadrangle	07796	01-01-05
MSS Glyfield	Nufarm UK	08009	31-07-05
MSS Glyfield	Nufarm UK	11117	
Muster	Syngenta	06685	01-01-05
Muster LA	Zeneca Prof.	05762	01-01-05
New Total Weedkiller	Headland	11207	
Nomix G	Monsanto	08781	
Nomix Nova	Nomix Enviro	11829	
Nomix Nova	Nomix Enviro	08647	
Nomix Revenge	Nomix Enviro	11827	
Nomix Revenge	Nomix Enviro	09483	

Product	Marketing Co.	Reg. No.	Expiry Date
Nomix Supernova	Nomix Enviro	09473	
NX 2031	Nomix Enviro	11105	
NX 2033	Nomix Enviro	11106	
NX 8049	Nomix Enviro	10099	
PMG	AgriGuard	10153	01-01-05
Poise	Unicrop	08276	
Pontil 360	Mastra	11433	
Portman Glider	Portman	04695	
Portman Glyphosate	Portman	05891	
Portman Glyphosate 360	Portman	04699	
Preline	Linemark	10977	
Preline Plus	Linemark	11385	
Reliance	Monsanto	09954	01-01-05
Rival	Monsanto	10316	
Rival	Monsanto	09220	
Roundup	Monsanto	10317	
Roundup	Monsanto	01828	
Roundup 2000	Monsanto	08069	
Roundup Amenity	Monsanto	08721	
Roundup Biactive	Monsanto	06941	28-02-05
Roundup Biactive Dry	Monsanto	06942	
Roundup Biactive Dry	Monsanto	10321	
Roundup CF	Monsanto	10323	01-01-05
Roundup CF	Monsanto	09547	01-01-05
Roundup Express	Monsanto	10318	
Roundup Four 80	Monsanto	03176	
Roundup Four 80	Monsanto	10325	
Roundup Greenscape	Monsanto	10599	
Roundup GT	Monsanto	10327	
Roundup GT	Monsanto	08068	
Roundup MX	Monsanto	09631	01-01-05
Roundup MX	Monsanto	10328	01-01-05
Roundup Pro	Monsanto	04146	
Roundup Pro	Monsanto	10329	
Roundup Pro Biactive	Monsanto	06954	
Roundup Rapide	Monsanto	10331	
Roundup Rapide	Monsanto	08067	30-09-05
Samurai	Monsanto	09952	
Scorpion	Monsanto	09953	01-01-05
Smart 360	Mastra	09476	
Spasor	RP Amenity	07211	
Stampede	Zeneca	06327	
Stampede	Syngenta	10536	
Sting	Monsanto	10439	01-01-05
Sting	Monsanto	02789	01-01-05
Sting CT	Monsanto	04754	
Sting CT	Monsanto	10336	01-01-05

Product	Marketing Co.	Reg. No.	Expiry Date
Stride	Syngenta	10537	
Stride	Syngenta	08282	01-01-05
Touchdown	Zeneca	06326	
Touchdown LA	Scotts	09270	
Typhoon 360	Feinchemie	09792	31-07-05
Unistar Glyfosate 360	Unistar	05928	
Unistar Glyphosate 360	Unistar	06332	
guazatine			
Panoctine	Makhteshim	10094	
guazatine + triticonazole			
Premis	Aventis	10177	28-02-05
hymexazol			
Tachigaren 70 WP	Sumitomo	02649	
imazalil			
Fungaflor Smoke	Brinkman	06009	
Fungazil 100 SL	Aventis	10270	
Sphinx	BASF	11764	
Sphinx	Aventis	10561	
Sphinx	Aventis	07607	
Stryper	Crompton	08310	
imazalil + pencycuron			
Monceren IM	Bayer	06259	30-06-05
Monceren IM Flowable	Bayer CropScience	11427	
Monceren IM Flowable	Bayer	06731	
imazalil + triticonazole			
Robust	Aventis	10176	31-03-05
imazamethabenz-methyl			
Assert	BASF	08656	31-01-05
Assert	BASF	10209	
Dagger	Cyanamid	03737	
imazapyr			
Arsenal	BASF	10208	
Arsenal	Cyanamid	04064	
Arsenal	Nomix Enviro	05537	
Arsenal 50	Nomix Enviro	05567	
imidacloprid			
Admire	Bayer	07481	30-06-05
Bayer UK 368	Bayer	08125	31-12-05
Bayer UK 397	Bayer CropScience	11242	
Bayer UK 397	Bayer	08930	
Gaucho	Bayer	06590	30-06-05

Product	Marketing Co.	Reg. No.	Expiry Date
Gaucho FS	Bayer	08496	
Gaucho FS	Bayer CropScience	11282	
Levington Professional Plus Intercept	BASF	08569	
Rentokil Desyst	Rentokil	09446	
imidacloprid + tebuconazole + triazoxide			
Raxil Secur	Bayer	08966	31-01-05
indol-3-ylacetic acid			
Rhizopon A Powder	Fargro	07131	
Rhizopon A Tablets	Fargro	07132	
4-indol-3-yl-butyric acid			
Chryzoplus Grey	Fargro	07984	
Chryzopon Rose	Fargro	07982	
Chryzosan White	Fargro	07983	
Chryzotek Beige	Fargro	07125	
Chryzotop Green	Fargro	07129	
Rhizopon AA Powder (0.5%)	Fargro	07126	
Rhizopon AA Powder (1%)	Fargro	07127	
Rhizopon AA Powder (2%)	Fargro	07128	
Rhizopon AA Tablets	Fargro	07130	
Seradix 1	Certis	10422	
Seradix 2	Certis	10423	
Seradix 3	Certis	10424	
iodosulfuron-methyl-sodium + mesosulfuron-methyl			
Nemesis	Bayer CropScience	11927	31-01-06
Pacifica	Bayer CropScience	12049	08-02-08
ioxynil			
Actrilawn 10	RP Amenity	05247	
Totril	RP Agric.	06116	
iprodione			
Aventis Rovral Flo	Aventis	09974	
CDA Rovral	RP Amenity	04679	
Cleancrop Gavotte SC	Willmot Pertwee	11869	30-06-05
Cleancrop Gavotte WP	Willmot Pertwee	11870	30-06-05
I.T. Iprodione	I T Agro	11892	30-06-05
I.T. Iprodione Flow	I T Agro	11893	30-06-05
Ipromex 50% WP	O Endres	11704	
Rovral Flo	Aventis	10013	30-06-05
Rovral Green	RP Amenity	05702	
Rovral Liquid FS	Aventis	10551	30-06-05
Rovral WP	Aventis	10015	30-06-05
Rovral WP	RP Agric.	06091	

Product	Marketing Co.	Reg. No.	Expiry Date
iprodione + thiophanate-methyl			
Aventis Compass	Aventis	10040	
Compass	Aventis	10041	30-06-05
Snooker	Aventis	10018	30-06-05
isoproturon			
Agriguard IPU	AgriGuard	10818	
Alpha Isoproturon 650	Makhteshim	07034	
Arelon 2	Nufarm UK	11670	
Arelon 500	Aventis	08100	
Arelon 500	DuPont	11110	30-06-05
Arelon 700	Nufarm UK	11542	
Atum	Bayer CropScience	10897	
Atum WDG	RP Agric.	07778	
Atum WDG	Bayer CropScience	09971	
Auger	RP Agric.	06581	
Auger	Headland	10896	
Bison	Gharda	08699	
Bison 83 WG	Nufarm UK	10063	28-02-06
Bison 83 WG	Gharda	10062	
Clayton Siptu 50 FL	Clayton	08751	
Cordelia 2	Griffin	10238	30-04-06
Cordelia 2	Nufarm UK	11591	
Cordelia 2	DuPont	11111	30-06-05
DAPT Isoproturon 500 FL	DAPT	08092	
Emrald Wotsit	Emrald	12060	
Fieldgard	Nufarm UK	11541	
Fieldgard	Nufarm UK	10710	31-03-06
Griffin IPU 700	Griffin	10409	31-03-06
Isoguard	Nufarm UK	10829	
Isoguard	Gharda	10817	
Isoguard 83 WG	Gharda	10061	
Isoguard SVR	BASF	10402	
Isoproturon 500	Griffin	10261	31-03-06
Isoproturon 500	Nufarm UK	11603	
Mysen	Portman	07695	
Mysen WDG	Portman	09141	
Portman Isotop SC	Portman	07663	
Stefes IPU	Bayer CropScience	10107	31-10-05
Tolkan	Bayer CropScience	07365	31-10-05
Tolkan	Bayer CropScience	10022	
Tolkan Liquid	Bayer CropScience	11994	
Tolkan Liquid	Aventis	10895	30-04-05
isoproturon + isoxaben			
Ipso	Dow	05736	31-01-05

Product	Marketing Co.	Reg. No.	Expiry Date
isoproturon + pendimethalin			
Artillery	BASF	10372	
Artillery	BASF	07768	
Artillery SC	BASF	08504	
Artillery SC	BASF	10385	
Encore	Cyanamid	04737	
Encore SC	BASF	10386	
Jolt	Cyanamid	05488	28-02-05
Jolt	BASF	10373	
Jolt SC	BASF	10387	
Jolt SC	BASF	08503	
Orient	BASF	10374	
Orient	BASF	07798	
Orient SC	BASF	10388	
Orient SC	BASF	08501	
Trump	Cyanamid	03687	
Trump SC	BASF	10389	
isoproturon + simazine			
Cleancrop Nevada	United Agri	11771	
isoproturon + trifluralin			
Autumn Kite	Griffin	10294	
isoxaben			
Agriguard Isoxaben	AgriGuard	11652	
Flexidor	Dow	05121	
Flexidor 125	Landseer	05104	
Gallery 125	Rigby Taylor	06889	
Knot Out	Vitax	05163	
isoxaben + methabenzthiazuron			
Glytex	Bayer CropScience	11424	
Glytex	Bayer	04230	
isoxaben + terbuthylazine			
Skirmish	Novartis	08444	
isoxaben + trifluralin			
Premiere Granules	Dow	07987	
lambda-cyhalothrin			
Hallmark with Zeon technology	Zeneca	09809	
Zeus	Interfarm	11852	
lambda-cyhalothrin + pirimicarb			
Dovetail	Zeneca	07973	
lenacil			
Clayton Lenacil 80 W	Clayton	09488	

Product	Marketing Co.	Reg. No.	Expiry Date
Clayton Lenaflo	Clayton	11031	
Cleancrop Lenflow	United Agri	10059	
Lenazar Flowable	Hermoo	11068	
Lenazar WP	Hermoo	10792	
Venzar 80 WP	DuPont	09981	
lenacil + phenmedipham			
DUK 880	DuPont	04121	
lenacil + triflusulfuron-methyl			
Safari Lite WSB	DuPont	12169	
linuron			
Afalon	Aventis	08186	
Alpha Linuron 50 WP	Makhteshim	04870	
Ashlade Linuron FL	Nufarm UK	06221	
DAPT Linuron 50 SC	DAPT	08058	
Linurex 50 SC	Makhteshim	07950	
Linuron Flowable	SumiAgro	09602	
MSS Linuron 500	Nufarm UK	08893	
linuron + trifluralin			
Chandor	Dow	05631	31-01-05
Linnet	SumiAgro	09601	
Neminfest	Montedison	02546	
Neminfest	Sipcam	07219	
magnesium phosphide			
Degesch Fumigation Pellets	Rentokil	11436	
Detia Gas-Ex-B Forte	Detia Degesch	10661	
malathion			
Fyfanon	Headland	11014	
Malathion 60	United Phosphorus	08018	
maleic hydrazide			
Bos MH 180	Crompton	06502	
Cleancrop Malahide	United Agri	11066	
Fazor	Crompton	05461	
Royal MH 180	Mirfield	07840	
Royal MH 180	Crompton	07043	
mancozeb			
Agrichem Mancozeb 80	Agrichem	06354	
Agrizeb	Chimac-Agriphar	10980	
Ashlade Mancozeb FL	Nufarm UK	06226	
Barclay Manzeb 455	Barclay	07990	
Barclay Manzeb 455	Barclay	11358	
Cleancrop Mancozeb	United Agri	11193	
Deny DF	Interfarm	10144	

Product	Marketing Co.	Reg. No.	Expiry Date
Deny DF	Rohm & Haas	10797	
Deny WP	Interfarm	10798	
Deny WP	Rohm & Haas	10140	
Dequiman MZ	Elf	06870	
Dithane 945	SumiAgro	09889	
Dithane 945	PBI	00719	
Dithane Dry Flowable Newtec	SumiAgro	09892	31-01-05
Dithane NT Dry Flowable	Interfarm	09868	
Dithane Superflo	Rohm & Haas	10142	
Headland Kor Flo	Headland	08019	
Headland Quell	Headland	08317	
Headland Zebra WP	Headland	07441	
Karamate Dry Flo	Landseer	09259	
Karamate N	Dow	10802	31-01-05
Karamate N	Rohm & Haas	01125	
Kor Flo	Interfarm	09895	
Kor NT Dry Flowable	Interfarm	09893	
Laminator DG	Interfarm	11073	
Laminator FL	Interfarm	11072	
Laminator WP	Interfarm	11071	
Manconex	Griffin	09555	
Mancozeb 80	Dow	10804	
Mancozeb 80	Rohm & Haas	09896	
Manex II	Agrichem	07637	
Manfil	Indofil	11093	
Manzate 200 PI	Griffin	09480	
Manzate 75 WG	DuPont	12070	
Manzate 75 WG	Griffin	10547	30-04-05
Micene 80	Sipcam	08560	
Mortar Flo	Griffin	09592	31-01-05
Nemispor	Sipcam	07348	
Opie 80 WP	SumiAgro	08301	
Opie 80 WP	SumiAgro	09890	31-01-05
Penncozeb WDG	Nufarm UK	09690	31-07-05
Penncozeb WDG	Nufarm UK	11065	
Restraint DF Newtec	Griffin	09755	31-01-05
Sabero Mancozeb 80% WP	Sabero	11879	
Stefes Restraint	Bayer CropScience	08945	
Tariff 75 WG Newtec	SumiAgro	09891	31-01-05
Unicrop Mancozeb	Unicrop	05467	
mancozeb + metalaxyl			
Fubol 75 WP	Novartis	08409	
Osprey 58 WP	Novartis	08428	
Ridomil MZ 75	Syngenta	07640	
Ridomil MZ 75WP	Novartis	08438	31-01-06

Product	Marketing Co.	Reg. No.	Expiry Date
mancozeb + metalaxyl-M			
Fubol Gold WP	Syngenta	08812	
mancozeb + propamocarb hydrochloride			
Aventis Tattoo	Aventis	09811	
mancozeb + zoxamide			
Electis 75 WG	Interfarm	10565	
RH 7281/Mancozeb 75 WG	Rohm & Haas	10564	30-09-05
RH 7281/Mancozeb 75 WG	Dow	10807	
Roxam 75 WG	Interfarm	10566	
Unikat 75 WG	Dow	11018	
maneb			
Agrichem Maneb 80	Agrichem	05474	
Ashlade Maneb Flowable	Ashlade	06477	
Luxan Maneb 80	Luxan	06570	
Maneb 80	Dow	10805	31-01-05
Manex	Griffin	09554	
Manex	Agrichem	07935	
Trimangol 80	Cerexagri	06070	
Trimangol 80	Cerexagri	06871	
Trimangol WDG	Cerexagri	06992	
Trimanzone	Intracrop	09584	
Unicrop Maneb 80	Unicrop	06926	
Unicrop Maneb FL	Unicrop	08025	
X-Spor SC	United Phosphorus	08077	
maneb + zinc			
Manex	Chiltern	05731	31-01-06
MCPA			
Agrichem MCPA-50	Agrichem	04097	
Agricorn 500	FCC	00055	
Agroxone 40	Marks	10264	
Agroxone 50	Mirfield	08345	
Agroxone 75	Marks	09208	
Atlas MCPA	Atlas	07717	
Atlas MCPA	Nufarm UK	03055	
Barclay Meadowman	Barclay	07639	
Barclay Meadowman II	Barclay	08525	
BASF MCPA Amine 50	BASF	00209	31-01-06
BH MCPA 75	RP Amenity	05395	
Circium II	Nufarm UK	11143	31-03-05
Circium II	Whyte Agrochem.	10293	
Dupont MCPA	DuPont	11738	
Empal	Unicrop	00795	
FCC Agricorn 50M	FCC	09032	
Luxan MCPA 500	Luxan	07470	

Product	Marketing Co.	Reg. No.	Expiry Date
Marks MCPA P30	Marks	01292	
Marks MCPA S25	Marks	01294	
Marks MCPA SP	Marks	01293	
MCPA 25%	Nufarm UK	02766	
MCPA 25%	Nufarm UK	07998	
MCPA 500	Nufarm UK	08655	
MSS MCPA 50	Nufarm UK	11144	
MSS MCPA 50	Nufarm UK	10240	
Nufarm MCPA 750	Nufarm UK	11768	
Nufarm Mcpa Amine 50	Nufarm UK	12046	
Nufarm MCPA DMA 480	Nufarm UK	09686	
Nufarm MCPA DMA 500	Nufarm UK	09685	
Quadrangle MCPA 50	Quadrangle	10493	
Tripart MCPA 50	Tripart	02206	
MCPA + MCPB			
Butoxone Plus	Headland	11080	
Butoxone Plus	Headland	10499	
MSS MCPB + MCPA	Nufarm UK	01413	
Trifolex-Tra	Cyanamid	07147	
Tropotox Plus	Nufarm UK	10701	
MCPA + mecoprop			
Greenmaster Extra	Scotts	07594	31-03-06
MCPA + mecoprop-P			
Cleanrun Pro	Scotts	12083	
No More Weeds Lawn Feed	Wolf	10747	
Optica Combi	Marks	10118	
MCPB			
Tropotox	Nufarm UK	10697	
mecoprop			
Clovotox	RP Amenity	05354	
mecoprop-P			
Astix	Aventis	06174	
Astix K	Bayer CropScience	06904	
BAS 03729H	Nufarm UK	05912	
Duplosan 500	Nufarm UK	12108	
Duplosan 500	BASF	07889	
Duplosan KV	BASF	09431	31-01-06
Duplosan KV 500	BASF	08027	
Duplosan KV 500	Nufarm UK	12107	
Duplosan New System CMPP	BASF	04481	
Duplosan New System CMPP	Nufarm UK	12106	
Dupont Mecoprop-p	DuPont	11739	
Headland Charge	Headland	10981	

Product	Marketing Co.	Reg. No.	Expiry Date
Isomec	Nufarm UK	09881	31-07-05
Landgold Mecoprop-p	Landgold	06052	
MSS Mirprop	Nufarm UK	05911	
MSS Optica	Mirfield	04973	
Optica	Marks	05814	
Optica 50	Marks	08283	
mecoprop-P + metribuzin			
Centra	Bayer CropScience	11420	
Centra	Bayer	08112	
Optica Plus	Marks	09325	
mecoprop-P + metsulfuron-methyl			
Headland Neptune	Headland	10230	
mecoprop-P + triasulfuron			
Raven	Syngenta	08434	
mefluidide			
Check Turf II	Certified Laboratories	04463	
Embark	Gordon International	04810	
Embark Lite	Intracrop	08749	26-10-05
Gro-Tard II	Chemsearch	04462	
mepanipyrim			
Frupica	Certis	10944	27-01-05
mesosulfuron-methyl			
AEF 6012-33H	Bayer CropScience	11218	17-09-05
metalaxyl			
Polycote Universal	Syngenta	08431	31-01-06
metalaxyl + thiabendazole			
Apron T 69 WS	Syngenta	08387	
Polycote Select	Germains	09718	31-01-06
metalaxyl + thiabendazole + thiram			
Apron Combi FS	Novartis	08386	
metalaxyl + thiram			
Favour 600 SC	Syngenta	08405	
metaldehyde			
Antares	CDP-Clartex	11654	
Aristo	De Sangosse	11797	
Aristo	De Sangosse	11484	
Aristo M	De Sangosse	11599	
Bristol Blues	Doff Portland	11492	
Bristol Blues	Doff Portland	11798	
Brits	Doff Portland	11435	

Product	Marketing Co.	Reg. No.	Expiry Date
Clartex	SumiAgro	09213	31-03-05
Clartex	CDP-Clartex	10942	
Clean Crop Hyde	United Agri	12025	
Cleancrop Jekyll	United Agri	11724	
Condor	Doff Portland	11791	
Condor	Doff Portland	11490	
Dixie 6	Greencrop	11488	
Dixie 6 M	De Sangosse	11597	
EM 1617/01	SumiAgro	09344	
Entice	CDP-Clartex	11096	
ESP	SumiAgro	09428	31-05-05
ESP	CDP-Clartex	10938	
FP 107	SumiAgro	09060	
Helimax	De Sangosse	07350	
Helimax	De Sangosse	11720	
Lincoln VI	Doff Portland	11491	28-02-05
Lincoln VI	Doff Portland	11793	
Lynx	De Sangosse	11767	
Lynx	De Sangosse	11482	
Lynx M	De Sangosse	11598	
Metasec M	De Sangosse	11600	
Mifaslug	FCC	10292	
Murphy Slugits	Scotts	09576	
Optimol	CDP-Clartex	10939	
Optimol	SumiAgro	09061	
Optimol XL	CDP-Clartex	10940	
Optimol XL	SumiAgro	10502	
Pastel M	De Sangosse	11736	
Pesta	De Sangosse	11483	
Pesta	De Sangosse	11796	
Pesta M	De Sangosse	11596	
Slug Pellets	SumiAgro	10150	
Slug Pellets	CDP-Clartex	10941	
Super Six	Doff Portland	11489	28-02-05
Super Six	Doff Portland	11794	
Super-flor 6% Metaldehyde Slug Killer Mini Pellets	CMI	11487	28-02-05
Super-Flor 6% Metaldehyde Slug Killer Mini Pellets	De Sangosse	11789	
Unicrop 6% Mini Slug Pellets	Unicrop	11795	
Unicrop 6% Mini Slug Pellets	Unicrop	11486	28-02-05
metamitron			
Agriguard Metamitron	AgriGuard	09859	
Barclay Seismic	Barclay	11377	
Barclay Seismic	Barclay	09323	
Barclay Seismic XL	Barclay	09328	

Product	Marketing Co.	Reg. No.	Expiry Date
Bettix 70 WG	United Phosphorus	11019	
Clayton Mitrex	Clayton	11921	
Cobra	Interfarm	10578	
Goltix 90	Bayer CropScience	11283	28-02-06
Goltix 90	Bayer	08654	30-06-05
Goltix Flowable	Bayer CropScience	11284	
Goltix Flowable	Bayer	08986	30-06-05
Goltix WG	Bayer CropScience	11285	28-02-06
Goltix WG	Bayer	02430	30-06-05
Homer	Bayer CropScience	11287	28-02-06
Homer	Griffin	09834	
Homer	Bayer	10867	28-02-06
Homer	Makhteshim	11544	
Lektan	Bayer	06111	28-02-06
Lektan	Makhteshim	11543	
Lektan	Bayer CropScience	11290	28-02-06
Mitron 70 WG	Hermoo	11516	
Mitron 90 WG	Hermoo	10888	
Mitron SC	Hermoo	11643	
MM70	Gharda	09490	
Skater	Feinchemie	09790	
Standon Metamitron	Standon	07885	
metam-sodium			
Discovery	United Phosphorus	10416	
Fumetham	Chemical Nutrition	10047	
Metam 510	UCB	09796	
Metham Sodium 400	United Phosphorus	08051	
Sistan	Unicrop	01957	
Sistan 38	Unicrop	08646	
Vapam	United Agri	09194	
metazachlor			
Barclay Metaza	Barclay	09169	
Barclay Metaza	Barclay	11359	
Butisan S	BASF	00357	
Clayton Metazachlor	Clayton	09688	
Clayton Metazachlor 50 SC	Clayton	11719	
Clean Crop MTZ 500	United Agri	09222	
Fuego	Makhteshim	11177	
Fuego	Feinchemie	10296	
Fuego 50	Makhteshim	11473	
Landgold Metazachlor 50 Sc	Teliton	12133	
Marnoch Metazachlor	Marnoch	09761	
Rapsan 500 SC	Q-Chem	12088	
Rapsan 500 SC	Globachem	11468	

Product	Marketing Co.	Reg. No.	Expiry Date
metazachlor + quinmerac			
Clayton Mazarac	Clayton	11161	
Katamaran	BASF	09049	
metconazole			
Caramba	Cyanamid	09864	
Clayton Tunik	Clayton	10545	
Pan Metconazole	Pan Agriculture	10473	
methabenzthiazuron			
Clayton Benson	Clayton	08687	
Tribunil	Bayer CropScience	11312	
Tribunil	Bayer	02169	
Tribunil WG	Bayer	03260	
Tribunil WG	Bayer CropScience	11313	
methiocarb			
Barclay Poacher	Barclay	09031	
Bayer UK 808	Bayer	09513	31-01-05
Bayer UK 808	Bayer CropScience	11246	31-01-05
Bayer UK 809	Bayer	09514	
Bayer UK 809	Bayer CropScience	11247	
Bayer UK 892	Bayer	09540	31-01-05
Bayer UK 892	Bayer CropScience	11248	31-01-05
Bayer UK 935	Bayer CropScience	11249	28-02-05
Bayer UK 935	Bayer	09541	28-02-05
Club	Bayer CropScience	11263	31-01-05
Club	Syngenta	07176	31-01-05
Decoy	Bayer	06535	
Decoy	Bayer CropScience	11264	
Decoy Plus	Bayer CropScience	11265	31-01-05
Decoy Plus	Bayer	07615	31-01-05
Draza	Bayer	00765	
Draza	Bayer CropScience	11268	31-01-06
Draza 2	Bayer CropScience	11269	
Draza 2	Bayer	04748	31-01-05
Draza Plus	Bayer CropScience	11270	31-01-05
Draza Plus	Bayer	06553	31-01-05
Draza ST	Bayer CropScience	11320	31-01-05
Draza ST	Bayer	05315	
Draza Wetex	Bayer	09704	31-01-05
Draza Wetex	Bayer CropScience	11271	
Elvitox	Bayer	06738	31-01-05
Elvitox	Bayer CropScience	11273	31-01-05
Epox	Bayer CropScience	11274	31-01-05
Epox	Bayer	06737	31-01-05
Exit	Bayer	07632	
Exit	Bayer CropScience	11275	31-01-05

Product	Marketing Co.	Reg. No.	Expiry Date
Exit Wetex	Bayer	10149	31-01-05
Huron	Bayer	10148	31-01-05
Karan	Bayer	09637	31-01-05
Lupus	Bayer	09638	31-01-05
Rescur	Bayer	07942	31-01-05
Rescur	Bayer CropScience	11299	31-01-05
Rivet	Bayer	09512	31-01-05
methoxyfenozide			
RH-2485 240 SC	Bayer CropScience	11322	21-10-05
methyl bromide			
Bromomethane 100%	Brian Jones	09244	
Mebrom 100	Mebrom	04869	
Methyl Bromide	Rentokil	09316	
Methyl Bromide 100%	Bromine & Chem.	01336	
Rentokil Methyl Bromide	Rentokil	05646	
Sobrom BM 100	Brian Jones	04381	31-12-05
methyl bromide with amyl acetate			
Fumyl-O-Gas	Brian Jones	04833	31-12-05
methyl bromide with chloropicrin			
Methyl Bromide 98	Bromine & Chem.	01335	
Sobrom BM 98	Brian Jones	04189	31-12-05
metosulam			
EF 1077	Bayer CropScience	11479	
EF 1077	Dow	07965	31-12-05
metrafenone			
Attenzo	BASF	11917	17-08-06
metribuzin			
Bayer UK 093	Bayer CropScience	11240	
Bayer UK 093	Bayer	07903	30-06-05
Cleancrop Metribuzin	United Agri	09621	
Cleancrop Solmet	United Agri	10865	
Lexone 2	DuPont	12051	
Python	Feinchemie	09791	
Sencorex WG	Bayer	03755	30-06-05
Tuberon	Unicrop	10613	30-04-05
metsulfuron-methyl			
Agform MTS 20	Agform	10685	
Agriguard Metsulfuron	AgriGuard	09569	
Ally PX	DuPont	10844	
Ally SX	DuPont	12059	12-07-07
Ally WSB	DuPont	06588	
Barclay Flumen	Barclay	08752	

Product	Marketing Co.	Reg. No.	Expiry Date
Barclay Flumen	Barclay	11355	
Clayton Metsulfuron	Clayton	06734	
Jubilee (WSB)	DuPont	06082	31-03-05
Jubilee 20 DF	DuPont	06136	
Lorate 20 DF	DuPont	06135	
Simba 20 DF	Griffin	09437	
Standon Metsulfuron	Standon	05670	
metsulfuron-methyl + thifensulfuron-methyl			
Concert	DuPont	10984	
DP 928	DuPont	09632	
DP 928 PX	DuPont	10991	
myclobutanil			
Systhane 20 EW	Whelehan	09397	
Systhane 6 Flo	Whelehan	06551	
Systhane 6 Flo	Landseer	07334	
Systhane 6 W	Rohm & Haas	04570	
Systhane 6 W	Dow	09611	
Systhane 6W	SumiAgro	04571	
2-(1-naphthyl)acetic acid			
Rhizopon B Powder (0.1%)	Fargro	07133	
Rhizopon B Powder (0.2%)	Fargro	07134	
Rhizopon B Tablets	Fargro	07135	
Tipoff	Unicrop	05878	
napropamide			
Banweed	United Phosphorus	09376	
Devrinol	United Phosphorus	09375	
nicotine			
No-FID	Certis	07959	
oxadiazon			
Ronstar Liquid	Certis	08974	30-06-05
paclobutrazol			
Bonzi	Syngenta	06640	
Cultar	Zeneca	06649	
paraquat			
Barclay Total	Barclay	11122	
Barclay Total	Barclay	11361	
Clayton Paragon	Clayton	11210	
CleanCrop Parachute	United Agri	11876	
Gramoxone 100	Zeneca	06674	
Speedway Liquid	Scotts	07744	

Product	Marketing Co.	Reg. No.	Expiry Date
penconazole			
Topenco 100 EC	Globachem	11972	
pencycuron			
Luxan Pencycuron	Luxan	12144	
Luxan Pencycuron 250 SC	Luxan	12142	
Marnoch Penor D	Marnoch	09871	
Monceren DS	Bayer	04160	30-06-05
Monceren Flowable	Bayer CropScience	11425	
Monceren Flowable	Bayer	04907	
Pan Pencycuron P	Pan Agriculture	10494	
Tubercare 12.5 DS	PG-Crop Protection	11147	
pendimethalin			
Barclay Tremor	Barclay	09188	
Barclay Tremor	Barclay	11342	
Blazer	Cheminova	10258	
Claymore	BASF	10214	30-09-05
Clayton Pendalin	Clayton	09708	
Cleancrop Stomp	United Agri	10696	
Cleancrop Stomp	United Agri	11778	
Ipimethalin	I Pi Ci	10664	
Parapet	Dow	10700	
PDM 330 EC	BASF	11084	
Plinth	SumiAgro	10665	30-06-05
Sidewinder	Dow	10699	
Sovereign	BASF	11781	
Sovereign	Novartis	08533	
Sovereign	BASF	10228	
Standon Pendimethalin 400 SC	Standon	10729	
Stomp 400 SC	BASF	10229	28-02-05
Stomp 400 SC	Cyanamid	04183	
pendimethalin + picolinafen			
Stomp Pico	BASF	10738	
pendimethalin + simazine			
Deuce	BASF	06746	
Deuce	BASF	10383	
Merit	BASF	10384	10-09-05
Merit	Cyanamid	04976	
Peniophora gigantea			
PG Suspension	Ecological	08975	28-02-05
PG Suspension	Forest Research	11772	
pentanochlor			
Atlas Solan 40	Nufarm UK	07726	31-12-07
Croptex Bronze	Certis	04087	31-12-07

Product	Marketing Co.	Reg. No.	Expiry Date
permethrin			
Permit	SumiAgro	01577	
Permit	SumiAgro	09609	
phenmedipham			
Agriguard Phenmedipham	AgriGuard	12148	
Atlas Protrum K	Nufarm UK	03089	
Atlas Protrum K	Atlas	07723	
Barclay Punter XL	Barclay	08047	
Beetup Flo	United Phosphorus	11916	
Beetup Flo 160	United Phosphorus	10382	
Betanal E	Aventis	07248	
Betanal Flo	Aventis	08898	
Betosip	Sipcam	06787	
Betosip 114	Sipcam	05910	
Dancer Flow	Sipcam	11405	
Herbasan	Nufarm UK	10734	30-06-05
Hickson Phenmedipham	Griffin	09961	
Hickson Phenmedipham	Nufarm UK	11124	
Kemifam E	Bayer CropScience	08978	31-10-05
Kemifam E	Kemira	06104	
Luxan Phenmedipham	Luxan	06933	
Mandolin	Nufarm UK	11127	
Mandolin	Griffin	09456	
Mandolin Flo	Nufarm UK	11125	
Mandolin Flo	Griffin	10368	
MSS Protrum G	Nufarm UK	08342	
Protrum G	Nufarm UK	11133	
Stefes Forte 2	Stefes	08204	
Stefes Medipham 2	Stefes	08203	
picloram			
Tordon 22K	Dow	05083	
picolinafen			
AC 900001	BASF	10714	30-09-12
pirimicarb			
Aphox	Zeneca	06633	
Barclay Pirimisect	Barclay	11360	
Barclay Pirimisect	Barclay	09057	
Clayton Pirimicarb 50 SG	Clayton	09221	
Cleancrop Miricide	United Agri	11776	
Pirimate	Portman	09568	
pirimiphos-methyl			
Actellic Smoke Generator No 20	Zeneca	06627	
Actellic Smoke Generator No. 10	Syngenta	10448	

Product	Marketing Co.	Reg. No.	Expiry Date
prochloraz			
Barclay Eyetak 40	Barclay	07843	
Barclay Eyetak 450	Barclay	09484	
Mirage 45 SC	Makhteshim	10894	
Octave	BASF	11741	
Octave	Aventis	07267	
Poraz	Aventis	10474	30-06-05
Prelude 20 LF	BASF	11904	
Prelude 20 LF	Bayer CropScience	07269	
Sporgon 50WP	Sylvan	08802	
Sportak 45 EW	BASF	11696	
Sportak 45 EW	Aventis	07996	30-06-05
Sportak 45 HF	BASF	11697	
Sportak 45 HF	Bayer CropScience	07287	
Stefes Poraz	Stefes	07528	
prochloraz + propiconazole			
Bumper Excell	Makhteshim	11015	
prochloraz + tebuconazole			
Agate	Bayer CropScience	11235	
Agate	Bayer	08826	30-06-05
prochloraz + triticonazole			
Premis Pro	BASF	12056	
prometryn			
Alpha Prometryne 80 WP	Makhteshim	04795	31-12-07
prometryn + terbutryn			
P-Weed	SumiAgro	08574	31-12-07
P-Weed	SumiAgro	09608	31-12-07
propachlor			
Alpha Propachlor 65 WP	Makhteshim	04807	
Atlas Propachlor	Atlas	06462	
Portman Brasson	Portman	08158	
Portman Propachlor 50 FL	Portman	06892	
Portman Propachlor SC	Portman	08159	
Ramrod 20 Granular	Monsanto	10313	
Ramrod 20 Granular	Monsanto	01687	
Ramrod Flowable	Monsanto	01688	
Ramrod Granular	Hortichem	05806	
Tripart Sentinel	AgriGuard	03250	
propamocarb hydrochloride			
Previcur N	Bayer CropScience	08575	
Propeller	PG-Crop Protection	11811	

Product	Marketing Co.	Reg. No.	Expiry Date
propaquizafop			
Agil	Makhteshim	11048	
Barclay Rebel	Barclay	11648	
Emrald Eyetort	Emrald	12034	
Falcon	Novartis	09384	
Headland Rocket	Headland	10998	
propiconazole			
Barclay Bolt	Barclay	08341	
Barclay Bolt	Barclay	11344	
Controller Fungicide	Ritefeed	10340	
Mantis	Syngenta	08423	
Radar	Syngenta	09168	
Tilt	Syngenta	08456	
Tumbleblite	Scotts	07567	
propiconazole + tebuconazole			
Cogito	Syngenta	11833	
Cogito	Syngenta	08397	
Endeavour	Bayer CropScience	11423	
Endeavour	Bayer	07385	
propiconazole + trifloxystrobin			
Rombus	Bayer	10464	
propineb			
Antracol	Bayer CropScience	11237	
Antracol	Bayer	00104	
propyzamide			
Barclay Piza 400FL	Barclay	11495	
Base 50 W	Interfarm	10202	31-01-05
Base 50 W	Dow	10765	31-01-05
Bulwark Flo	Interfarm	10755	
Clayton Propel	Clayton	09783	
Clayton Propel 80 WG	Clayton	11216	
Flomide	Alfa	11103	
Kerb 50 W	Interfarm	10200	31-01-05
Kerb 50 W	SumiAgro	10166	
Kerb 50W	SumiAgro	01133	
Kerb 80 EDF	Interfarm	10201	31-01-05
Kerb 80 EDF	SumiAgro	10163	31-01-05
Kerb Flo	SumiAgro	10157	31-01-05
Kerb Flo	Interfarm	10160	31-01-05
Kerb Granules	Dow	10803	
Kerb Granules	Rohm & Haas	01136	
Kerb Granules	SumiAgro	01135	
Kerb Pro Flo	SumiAgro Amenity	10158	31-01-05
Kerb Pro Granules	SumiAgro	08698	

Product	Marketing Co.	Reg. No.	Expiry Date
Menace 80 EDF	Interfarm	10165	31-01-05
Mithras 80 EDF	Interfarm	10164	31-01-05
Precis	Interfarm	10737	31-01-05
Propose	Alfa	10890	31-08-05
Propose	Interfarm	11195	
Quaver Flo	Interfarm	10162	31-01-05
Rapier	MTM Agrochem.	05314	
Redeem Flo	Dow	10764	
Redeem Flo	Interfarm	10236	31-01-05
Top Farm Propyzamide 500	Top Farm	05484	
prothioconazole			
Proline	Bayer CropScience	12084	16-08-08
Redigo	Bayer CropScience	12085	16-08-08
pymetrozine			
Chess	Syngenta Bioline	09817	
Plenum	Syngenta	09816	
pyraclostrobin			
BASF Insignia	BASF	11900	01-12-06
pyrethrins			
Advanced Bug Killer Concentrate	Growing Success	12062	
Natures Answer Natural Pest Control	Scotts	11030	
Py Spray Garden Insect Killer	Vitax	06085	
pyrethrins + resmethrin			
Pynosect 30 Water Miscible	Mitchell Cotts	01653	
pyrimethanil			
Scala	Aventis	07806	30-06-05
Standon Pyrimethanil	Standon	10576	
quinomethionate			
Morestan	Hortichem	01376	
quinoxyfen			
Clean Crop QFN	United Agri	11966	20-12-05
quizalofop-P-ethyl			
EXP31922D	Bayer CropScience	12166	
Pilot D	Aventis	08041	
Poseidon	Nissan	11040	
rimsulfuron			
Landgold Rimsulfuron	Landgold	08959	
Titus	DuPont	07908	30-06-05
rotenone			
Devcol Liquid Derris	Nehra	06063	

Product	Marketing Co.	Reg. No.	Expiry Date
simazine			
Alpha Simazine 50 WP	Makhteshim	04879	
Alpha Simazine 80 WP	Makhteshim	04800	
Atlas Simazine	Nufarm UK	05610	
Atlas Simazine	Nufarm UK	07725	
MSS Simazine 50 FL	Mirfield	01418	
Simazine 500	Makhteshim	12004	10-09-05
Simazine 90WG	Sipcam	10641	
Unicrop Flowable Simazine	Unicrop	05447	
Unicrop Simazine 50	Unicrop	02646	
Unicrop Simazine FL	Unicrop	08032	
sodium chlorate			
Deosan Chlorate Weedkiller (Fire Suppressed)	JohnsonDiversey	08521	
Gem Sodium Chlorate Weedkiller	Joseph Metcalf	04276	
Growing Success Sodium Chlorate Weedkiller	Growing Success	10787	
Sodium Chlorate	Marlow Chemical	06294	
Strathclyde Sodium Chlorate Weedclear	Strathclyde	07420	
TWK Total Weedkiller	Reabrook	06393	28-02-06
sodium monochloroacetate			
Atlas Somon	Nufarm UK	03045	31-12-07
spiroxamine			
Accrue	Bayer CropScience	11219	
Accrue	Bayer	08335	01-09-09
Bayer UK 477	Bayer CropScience	11222	
Bayer UK 477	Bayer	08330	01-09-09
Neon	Bayer CropScience	11226	01-09-09
Neon	Bayer	08337	30-06-05
Talvin	Bayer CropScience	11228	
Talvin	Bayer	09253	01-09-09
Torch	Bayer	08336	30-06-05
Torch Extra	Bayer CropScience	12140	01-09-09
Zenon	Bayer	09193	30-06-05
spiroxamine + tebuconazole			
Array	Bayer CropScience	11238	
Array	Bayer	09352	
Bayer UK 552	Bayer CropScience	11245	
Bayer UK 552	Bayer	10410	
Beam	Bayer	10411	30-06-05
Bronze	Bayer CropScience	11419	
Bronze	Bayer	10412	
Draco	Bayer CropScience	11267	
Draco	Bayer	09353	

Product	Marketing Co.	Reg. No.	Expiry Date
Folicur Star	Bayer CropScience	11531	
Sage	Bayer	10413	30-06-05
sulfuryl difluoride			
Profume	Dow	12035	06-07-07
sulphonated cod liver oil			
Scuttle	Fine	06232	
sulphur			
Ashlade Sulphur FL	Nufarm UK	06478	
Headland Venus	Headland	10611	
Headland Venus	Headland	09572	
Kumulus S	BASF	01170	
Microthiol Special	Cerexagri	06268	
MSS Sulphur 80	Mirfield	05752	
MSS Sulphur 80 WP	Mirfield	03225	31-01-05
Solfa	Nufarm UK	11390	31-03-06
Solfa	Nufarm UK	06959	31-10-05
Thiomex	Omex	11176	
Thiovit	Syngenta	08493	
Tripart Imber	Tripart	04050	
tar acids			
Armillatox	Armillatox	06234	
tar oils			
Jeyes Fluid	Jeyes	04606	
tebuconazole			
Aurigen	Syngenta	12043	
Barclay Busker	Barclay	08994	
Bayer UK 226	Bayer CropScience	11241	
Bayer UK 226	Bayer	09504	
Clayton Tebucon	Clayton	08707	
Folicur	Bayer	08691	30-06-05
Gainer	Bayer CropScience	11279	
Gainer	Bayer	08692	
Halt	Bayer CropScience	11286	
Halt	Bayer	08693	
Raxil	Bayer CropScience	11295	
Raxil	Bayer	06460	
UK 200	Bayer CropScience	11316	
UK 200	Bayer	08246	
tebuconazole + triadimenol			
Garnet	Bayer CropScience	11280	
Garnet	Bayer	06391	
Ruby	Bayer CropScience	11302	

Product	Marketing Co.	Reg. No.	Expiry Date
Ruby	Bayer	06389	
Silvacur	Bayer	06387	30-06-05
Silvacur 300	Bayer	08056	
Silvacur 300	Bayer CropScience	11310	
Veto F	Bayer	08057	30-06-05
tebuconazole + triazoxide			
Raxil S	Bayer	06974	30-06-05
Raxil S CF	Bayer CropScience	11297	
Raxil S CF	Bayer	08192	
tebufenpyrad			
Masai	Cyanamid	07452	
Masai G	BASF	07453	31-03-05
Masai G	BASF	10224	
tebutam			
Comodor 600	Agrichem	08398	
teflubenzuron			
Nemolt	Fargro	07012	
tefluthrin			
Evict	Bayer CropScience	11673	
Force ST	Bayer	09713	31-05-05
terbutryn			
Alpha Terbutryn 50 SC	Makhteshim	04809	
Clarosan	Scotts	09394	30-09-05
Prebane	Novartis	08432	
tetraconazole			
Digit	Monsanto	09409	
Digit	Monsanto	10503	
Domark	Isagro	10452	
Emerald	Isagro	10449	
Eminent	Monsanto	09410	
Eminent	Monsanto	10504	
MON 10 EC	Monsanto	09347	
MON 10 EC	Monsanto	10505	
Omen	Sipcam	09413	
Tetra 5634	Monsanto	09411	
Tetra 5634	Monsanto	10507	
thiabendazole			
Tecto Flowable Turf Fungicide	Vitax	06273	
Tecto Superflowable Turf Fungicide	Vitax	09258	

Product	Marketing Co.	Reg. No.	Expiry Date
thiabendazole + thiram			
Hy-Vic	Agrichem	06247	
sHYlin	Agrichem	10030	
thiacloprid			
Calypso	Bayer	10342	30-06-05
thifensulfuron-methyl			
DUK 118	DuPont	04596	
Harmony 75DF	DuPont	11884	
Harmony PX	DuPont	10845	
Harmony SX	DuPont	12181	31-10-07
thifensulfuron-methyl + tribenuron-methyl			
DUK 110	DuPont	09189	
Gex 353 A	Griffin	10185	30-11-05
thiodicarb			
Genesis	Sipcam	11846	
Genesis	Bayer CropScience	11712	
Genesis	RP Agric.	06168	
Genesis ST	Sipcam	11853	
Genesis ST	Bayer CropScience	11710	
Judge	Sipcam	11847	
Judge	Bayer CropScience	11711	
Toro	Sipcam	12138	
thiophanate-methyl			
Cercobin Liquid	Certis	11941	
Cercobin Liquid	Bayer CropScience	06188	31-10-05
Cercobin Liquid	Aventis	09984	
Mildothane Liquid	Certis	11942	
Mildothane Liquid	Aventis	10002	
Mildothane Liquid	Hortichem	06211	
thiram			
Hy-Flo	Agrichem	04637	
Robinson's Thiram 60	Agrichem	04638	
Thiraflo	Crompton	11225	
Thiraflo	Uniroyal	09496	31-10-05
tolclofos-methyl			
Rizolex 50 WP	Bayer CropScience	07272	
Rizolex Flowable	Certis	09358	30-11-05
tolylfluanid			
Elvaron Multi	Bayer	10080	30-06-05

Product	Marketing Co.	Reg. No.	Expiry Date
tralkoxydim			
Grasp	Zeneca	06675	
Standon Tralkoxydim	Standon	08326	
triadimefon			
Bayleton	Bayer	00221	
Bayleton 5	Bayer	00222	
triadimenol			
Bayfidan	Bayer CropScience	11417	
Bayfidan	Bayer	02672	
Hi-Shot	BASF	06508	
Hi-Shot	BASF	10397	
Spinnaker	BASF	10398	
Spinnaker	Cyanamid	07023	
tri-allate			
Avadex 15G	Monsanto	10297	
Avadex BW Granular	Gowan	12104	
Avadex BW Granular	Monsanto	10298	
Avadex BW Granular	Monsanto	00174	
Avadex Excel 15G	Gowan	12109	
triasulfuron			
Lo-Gran	Novartis	08421	
triazamate			
Aztec	Cyanamid	07817	
tribenuron-methyl			
Quantum	DuPont	06270	
Quantum PX	DuPont	10843	
triclopyr			
Agriguard Triclopyr	AgriGuard	10679	
Altix 240 EC	Chimac-Agriphar	10648	
Chipman Garlon 4	Nomix Enviro	06016	
Cleancrop Triptic 48 EC	United Agri	11568	
Cleancrop Unival	United Agri	11388	
Garlon 2	Dow	05682	
Garlon 2	Zeneca	06616	
n2n Garlon 4	Nomix Enviro	11806	
Tricle	PG-Crop Protection	11862	
Triptic 48 EC	United Phosphorus	09294	
Unival	Chimac-Agriphar	10636	31-10-05
trifloxystrobin			
Aprix	Bayer CropScience	11220	
Flint	Bayer CropScience	11259	

Product	Marketing Co.	Reg. No.	Expiry Date
Twist 500 SC	Bayer CropScience	11231	
Zest SC	Bayer CropScience	11233	
trifluralin			
Ashlade Trifluralin	Nufarm UK	08303	
Ashlade Trimaran	Nufarm UK	06228	
Atlas Trifluralin	Nufarm UK	08498	
DAPT Trifluralin 48 EC	DAPT	07906	
Digermin	Sipcam	07221	
Digermin	Montedison	00701	
FCC Trigard	FCC	09030	
Ipifluor	I Pi Ci	04692	
MSS Trifluralin 48 EC	Nufarm UK	10749	31-12-05
MSS Trifluralin 48 EC	Nufarm UK	07753	
Portman Trifluralin	Portman	05751	
Triflur	Nufarm UK	10750	
Trifsan	Dow	09237	
Trigard	SumiAgro	09699	
Trilogy	United Phosphorus	08996	
Triplen	Sipcam	05897	
Tristar	SumiAgro	09612	
Whyte Trifluralin	Nufarm UK	09286	
triflusulfuron-methyl			
Debut WSB	DuPont	07809	
Landgold TFS 50	Landgold	08941	
Standon Triflusulfuron	Standon	09487	
triforine			
Saprol	Cyanamid	07016	
trinexapac-ethyl			
Cleancrop Alatrin	United Agri	11805	
Primo Maxx	Scotts	11878	
vinclozolin			
Barclay Flotilla	Barclay	07905	
Barclay Flotilla	Barclay	11354	
Landgold Vinclozolin DG	Landgold	06500	
Landgold Vinclozolin SC	Landgold	06459	
Marnoch Vinol	Marnoch	09815	
Ronilan DF	BASF	04456	

Product	Marketing Co.	Reg. No.	Expiry Date
zeta-cypermethrin			
Fury 10 EW	FMC	10268	31-01-05
Fury 10 EW	FMC	10718	
Minuet EW	FMC	10269	31-01-05
Minuet EW	FMC	10716	
zineb-ethylene thiuram disulphide adduct			
Polyram DF	BASF	08234	

SECTION 4
ADJUVANTS

Adjuvants

Adjuvants are not themselves classed as pesticides and there is considerable misunderstanding over the extent to which they are legally controlled under the Food and Environment Protectiion Act. An adjuvant is a substance other than water which enhances the effectiveness of a pesticide with which it is mixed. Consent C(i)5 under the Control of Pesticides Regulations allows that an adjuvant can be used with a pesticide only if that adjuvant is authorised and on a list published from time to time in *The Pesticides Register*. An authorised adjuvant has an *adjuvant number* and may have specific requirements about the circumstances in which it may be used.

Adjuvant product labels must be consulted for full details of authorised use, but the table below provides a summary of the label information to indicate the area of use of the adjuvant. Label precautions refer to the keys given in Appendix 4, and may include warnings about products harmful or dangerous to fish. The table includes all adjuvants notified by suppliers as available in 2004.

Product	Supplier	Adj. No.	Type
Abacus	Loveland	A0434	adjuvant/wetter
Contains	505.2 g/l esterified rapeseed oil, 300 g/l alkylphenyl hydroxypolyoxyethylene and 60 g/l natural fatty acids		
Use with	All approved pesticides on edible and non-edible crops when used at half their approved dose or less. Also with approved pesticides on specified crops, up to specified growth stages, at up to their full approved dose		
Protective clothing	A, C, H		
Precautions	R36, R58, U05a, U11, U14, U19a, U20b, E15a, E19b, E31a, E32d, E34, E01, E04, E26, H04		
Activator 90	Loveland	A0337	non-ionic surfactant/spreader/wetter
Contains	750 g/l alkylphenyl hydroxypolyoxyethylene and 150 g/l natural fatty acids		
Use with	Any spray for which additional wetter is approved and recommended		
Protective clothing	A, C, H		
Precautions	R36, R58, U02a, U04a, U05a, U10, U11, U20b, E15a, E19b, E31a, E01, E04, E26, H04		
Admix-P	Loveland	A0301	wetter
Contains	80% w/w polyalkylene oxide modified heptamethyltrisiloxane and a maximum of 20 % w/w allyloxypolyethylene glycol methyl ether		
Use with	A wide range of pesticides applied as corm, tuber, onion and other bulb treatments in seed production and in seed potato treatment		
Protective clothing	A, C, H		
Precautions	R20, R21, R22a, R36, R43, R48, R51, R58, U11, U15, U19a, E15a, E19b, E31a, E32d, E01, E04, E26, H03, H11		
Agral	Syngenta	A0421	non-ionic surfactant/spreader/wetter
Contains	948 g/l alky phenol ethylene oxide		
Use with	Any spray for which additional wetter is approved and recommended		
Protective clothing	A, C		
Precautions	R36, R38, U05a, U08, U20c, E15a, E31a, E34, E01, E04, E26, E30a, H04, H08		

SECTION 4

Product	Supplier	Adj. No.	Type
Amber	Interagro	A0367	vegetable oil
Contains	95% w/w methylated rapeseed oil		
Use with	Sugar beet herbicides, oilseed rape herbicides, cereal graminicides and a wide range of other pesticides that have a label recommendation for use with authorised adjuvant oils on specified crops. See label for details		
Precautions	U05a, U20b, E15a, E31a, E01, E04, E30a		
Arma	Interagro	A0306	penetrant
Contains	500 g/l alkoxylated fatty amine + 500 g/l polyoxyethylene monolaurate		
Use with	Cereal growth regulators, cereal herbicides, cereal fungicides, oilseed rape fungicides and a wide range of other pesticides on specified crops		
Precautions	R51, R58, U05a, E15a, E31a, E34, E37, E01, E04, E26, E30a, H11		
BackRow	Interagro	A0472	mineral oil
Contains	60% w/w refined paraffinic petroleum oil		
Use with	Pre-emergence herbicides. Refer to label or contact supplier for further details		
Protective clothing	A		
Precautions	R22b, R38, U02a, U05a, U08, U20b, E15a, E31a, E01, E04, E26, E30a, M05b, H03		
Bandrift Plus	Ciba Specialty	-	drift retardant
Contains	A non-ionic polyamide dispersed in oil		
Use with	A wide range of pesticides. See label for restrictions on use with wettable powders, suspension concentrates and water dispersible granules, and other usage limitations		
Protective clothing	A, C, M		
Precautions	U05a, U08, U20a, E13c, E34, E01, E04, E30a		
Banka	Interagro	A0245	spreader/wetter
Contains	29.2% w/w alkyl pyrrolidones		
Use with	Potato fungicides and a wide range of other pesticides on specified crops		
Protective clothing	A, C		
Precautions	R38, R41, R52, R58, U02a, U05a, U11, U19a, U20b, E15a, E34, E37, E01, E04, E26, E30a, H04		
Barclay Actol	Barclay	A0126	mineral oil
Contains	99% highly refined paraffinic oil		
Use with	Any approved pesticide for which the addition of a spraying oil is recommended		
Precautions	U19a, U20b, E13c, E31b, E30a		
Barclay Clinger	Barclay	A0198	sticker/wetter
Contains	96% w/v poly-1-p-menthene		
Use with	Barclay Gallup, Barclay Gallup Amenity and a wide range of approved fungicides and insecticides. See label for details		
Precautions	U19a, U20b, E13c, E31a, E30a		

Product	Supplier	Adj. No.	Type
Barramundi	Interagro	A0376	mineral oil
Contains	95% w/w mineral oil		
Use with	Sugar beet herbicides, oilseed rape herbicides, cereal graminicides and a wide range of other pesticides that have a label recommendation for use with authorised adjuvant oils on specified crops. Refer to label or contact supplier for further details		
Protective clothing	A		
Precautions	R22b, R38, U02a, U05a, U08, U20b, E15a, E31a, E01, E04, E26, E30a, M05b, H03		
Bio Syl	Intracrop	A0385	spreader/sticker/wetter
Contains	1% w/w polyoxyethylen-alpha-methyl-omega-[3-(1,3,3,3-tetramethyl-3-trimethylsiloxy)-disiloxanyl] propylether and 32.67% w/w ethylene oxide condensate		
Use with	Recommended rates of approved pesticides on a range of specified fruit and vegetable crops and on managed amenity turf. See label for details		
Protective clothing	A, C		
Precautions	R36, R38, U05a, U08, U20c, E15a, E31a, E01, E04, H04		
Bioduo	Intracrop	A0299	wetter
Contains	85% alkyl polyglycol ether and fatty acids		
Use with	A wide range of pesticides used in grassland, agriculture and horticulture and with pesticides used in non-crop situations		
Protective clothing	A, C		
Precautions	R22a, R36, R38, U05a, U08, U20c, E13c, E31a, E34, E01, E04, E30a, M03, H03, H04, H08		
Biofilm	Intracrop	A0359	sticker/wetter
Contains	96% poly-1-p-menthene		
Use with	All approved fungicides and insecticides on edible crops and all approved formulations of glyphosate used pre-harvest on wheat, barley, oilseed rape, stubble, and in non-crop situations and grassland destruction		
Precautions	U19a, U20c, E13c, E32a, E30a		
BioPower	Bayer CropScience	A0454	wetter
Contains	6.7% w/w 3,6-dioxaeicosylsulphate sodium salt and 20.2% w/w 3,6-dioxaoctadecylsulphate sodium salt		
Use with	Atlantis and all other approved cereal herbicides		
Protective clothing	A, C		
Precautions	R36, R38, U02a, U05a, U08, U13, U19a, U20b, E13c, E31a, E34, E01, E04, E26, E30a, H04		
Bond	Loveland	A0184	drift retardant/extender/sticker/wetter
Contains	450 g/l synthetic latex and 100 g/l alkylphenylhydroxy-polyoxyethylene		
Use with	Blight sprays on potatoes and other pesticides on a wide range of crops. See label or contact supplier for details		
Protective clothing	A, C, H		
Precautions	R36, R38, U11, U14, U19a, E15a, E19b, E31a, E32d, E01, E04, E26, H04		

SECTION 4

Product	Supplier	Adj. No.	Type
Byo-Flex	Greenaway	A0481	sticker/wetter
Contains	95 g/l vegetable oil		
Use with	GLY 490 (MAPP 11064)		
Precautions	U05a, U08, U20b, E31a, E37, E04, E26		
C-Cure	Interagro	A0467	mineral oil
Contains	60% w/w refined mineral oil		
Use with	Pre-emergence herbicides		
Protective clothing	A		
Precautions	R22b, R38, U02a, U05a, U08, U20b, E15a, E31a, E01, E04, E26, E30a, M05b, H03		
Celect	Fargro	A0334	extender/sticker/wetter
Contains	95% emulsified vegetable oil		
Use with	approved pesticides in horticulture (but see label for certain active ingredient exclusions)		
Protective clothing	A, C		
Precautions	U19a, U20b, E31b, E30a		
Celect	Microcide	A0413	extender/sticker/wetter
Contains	95% emulsified vegetable oil		
Use with	Approved pesticides in horticulture (but see label for certain active ingredient exclusions)		
Protective clothing	A, C		
Precautions	U19a, U20b, E31b, E30a		
Ceres Platinum	Interagro	A0445	penetrant/spreader
Contains	500 g/l alkoxylated fatty amine + 500 g/l polyoxyethylene monolaurate		
Use with	Cereal growth regulators, cereal herbicides, cereal fungicides, oilseed rape fungicides and a wide range of other pesticides on specified crops		
Precautions	R51, R58, U05a, E15a, E31a, E34, E01, E04, E26, E30a, H11		
Cerround	Helena	A0510	adjuvant/vegetable oil
Contains	80% w/w methylated soybean oil and 15% w/w polyalkylene oxide modified heptamethyl trisiloxane		
Use with	Approved pesticides on non-edible crops, cereals and stubbles of all edible crops, grassland (destruction), and pesticides approved for use on growing edible crops when used at half recommended dose or less. On specified crops product may be used at a maximum spray concentration of 0.5% with approved pesticides at their full approved rate up to the growth stages shown in the label		
Protective clothing	A, H		
Precautions	R38, R41, U02a, U05a, U08, U11, U19a, U20b, E13c, E34, E37, E01, E04, E26, E30a, M03, H04		
Codacide Oil	Microcide	A0011	vegetable oil
Contains	95% emulsified vegetable oil		
Use with	All approved pesticides and tank mixes. See label for details		
Protective clothing	A, C		
Precautions	U20b, E31b, E30a		

Product	Supplier	Adj. No.	Type
Companion PCT12	Ciba Specialty	A0482	spreader/sticker/wetter
Contains	25% w/w polyacryalmide		
Use with	Approved herbicides on cereals and any pesticide on non-food crops. See label for restrictions on use with wettable powders, suspension concentrates and water dispersible granules, and other usage limitations		
Precautions	U05a, U08, E13c, E01, E04, E30a		
Contact Plus	Interagro	A0418	mineral oil
Contains	95% w/w mineral oil		
Use with	Sugar beet herbicides, oilseed rape herbicides, cereal graminicides and a wide range of other pesticides that have a label recommendation for use with authorised adjuvant oils on specified crops. Refer to label or contact supplier for further details		
Protective clothing	A, C		
Precautions	R22b, R38, U02a, U05a, U08, U20b, E15a, E31a, E01, E04, E30a, M05b, H03		
Designer	Loveland	A0322	drift retardant/extender/sticker/wetter
Contains	8.44% w/w organosilicone wetter and 50% w/w synthetic latex		
Use with	A wide range of fungicides, insecticides and trace elements for cereals and specified agricultural and horticultural crops		
Protective clothing	A, C, H		
Precautions	R36, R38, R52, U02a, U11, U14, U15, U19a, E15a, E31a, E37, E01, E04, E26, E30a, H04		
Drill	Loveland	A0409	adjuvant/wetter
Contains	589.4 g/l esterified rapeseed oil, 225 g/l alkylphenyl hydroxypolyoxyethylene and 45 g/l natural fatty acids		
Use with	All approved pesticides on edible and non-edible crops when used at half their approved dose or less. Also with approved pesticides on specified crops, up to specified growth stages, at up to their full approved dose		
Protective clothing	A, C, H		
Precautions	R36, R58, U05a, U11, U14, U19a, U20b, E15a, E19b, E31a, E32d, E34, E01, E04, E26, H04		
Du Pont Adjuvant	DuPont	A0119	non-ionic surfactant/wetter
Contains	900 g/l ethylene oxide condensate		
Use with	Any spray for which a non-ionic wetter is recommended		
Protective clothing	A, C		
Precautions	R36, R38, U05a, U08, U20c, E13c, E31a, E01, E04, E30a, H04, H08		
Esterol	Interagro	A0330	vegetable oil
Contains	95% w/w methylated rapeseed oil		
Use with	Plant protection products that have a label recommendation for use with adjuvant oils. Refer to label or contact supplier for further details		
Precautions	U05a, U20b, E15a, E31a, E01, E04, E30a		

Product	Supplier	Adj. No.	Type
Felix	Intracrop	A0178	spreader/wetter
Contains	60% ethylene oxide condensate		
Use with	Mecoprop, 2,4-D in cereals and amenity turf, and a range of grass weedkillers in agriculture. See label for details		
Protective clothing	A, C		
Precautions	R36, R38, U05a, U08, U20a, E15a, E31a, E01, E04, E30a, H04, H08		
FMC Oil	Belchim	A0435	adjuvant/vegetable oil
Contains	842 g/l esterified rapeseed oil		
Use with	Spotlight 24 EC (MAPP 10702 and 11617)		
Precautions	E13c, E37		
Frigate	Unicrop	A0325	sticker/wetter
Contains	800 g/l tallow amine ethoxylate		
Use with	Roundup formulations		
Protective clothing	A, C		
Precautions	R22a, R23, R41, R51, R58, R67, U05a, U11, U19a, E32d, E38, E01, E04, M04, H02, H08, H11		
Galion	Intracrop	A0162	spreader/wetter
Contains	60% polyoxyalkylene glycol		
Use with	Herbicides in grassland and amenity turf and with grass weedkillers in cereals, oilseed rape, sugar beet and other crops		
Protective clothing	A, C		
Precautions	R36, R38, U05a, U08, U20b, E15a, E31a, E01, E04, E30a, H04, H08		
Gex 1664	Headland	A0366	spreader/wetter
Contains	900 g/l ethylene oxide concentrate		
Use with	All approved pesticides on edible and non-edible crops when used at half their approved dose or less. Also with approved pesticides on specified crops, up to specified growth stages, at up to their full approved dose		
Protective clothing	A, C		
Precautions	R36, R58, U05a, U08, U11, U12, U15, U20b, E13c, E31b, E34, E37, E38, E01, E04, E30b, H04		
Gly-Flex	Greenaway	A0370	spreader/wetter
Contains	140 g/l refined mineral oil		
Use with	All approved formulations of 360 g/l glyphosate		
Protective clothing	A, C, D, H, M		
Precautions	R36, R38, U08, U20b, E13c, E15a, E31a, E37, E04, E26, H04		
Green Gold	Intracrop	A0250	spreader/wetter
Contains	95% refined rapeseed oil		
Use with	All pesticides which have a recommendation for the addition of a wetter/ spreader		
Precautions	U08, U20b, E13c, E31b, E34, E30a		

Product	Supplier	Adj. No.	Type
Greencrop Astra	Greencrop	A0417	adjuvant/penetrant/spreader
Contains	500 g/l alkoxylated fatty amine and 500 g/l polyoxylene monolaurate		
Use with	Approved pesticides on a range of specified arable crops, and in forestry, managed amenity turf, stubbles and non-crop areas and situations		
Precautions	U05a, E13c, E31b, E34, E01, E04, E26, E30a		
Greencrop Dolmen	Greencrop	A0419	adjuvant/spreader
Contains	64% w/w polyalkeneoxide modified heptamethyltrisiloxane		
Use with	Approved pesticides on a range of specified arable crops, and in forestry, managed amenity turf, stubbles and non-crop areas and situations		
Protective clothing	A, C, H		
Precautions	R21, R22a, R38, R41, R43, U02a, U05a, U08, U11, E13c, E31b, E34, E01, E04, E30a, M03, H03, H04		
Grip	Loveland	A0211	drift retardant/extender/sticker/wetter
Contains	450 g/l synthetic latex and 100 g/l alkylphenyl hydroxypolyoxyethylene		
Use with	Blight sprays on potatoes and other pesticides on a wide range of crops. See label or contact supplier for details		
Protective clothing	A, C, H		
Precautions	R36, R38, U11, U14, U19a, U20b, E15a, E19b, E31a, E32d, E01, E04, E26, M03, H04		
Guide	Loveland	A0208	acidifier/drift retardant/penetrant
Contains	350 g/l modified soya lecithin, 100 g/l alkylphenyl hydroxypolyoxyethylene and 350 g/l propionic acid		
Use with	All plant growth regulators for cereals and oilseed rape, dimethoate and chlorpyrifos for wheat bulb fly control. For other uses see label or contact supplier for details		
Protective clothing	A, C, H		
Precautions	R36, R38, U11, U14, U15, U19a, E15a, E19b, E31a, E32d, E01, E04, E26, H04		
Headland Fortune	Headland	A0277	penetrant/spreader/vegetable oil/wetter
Contains	75% w/w mixed methylated fatty acid esters of seed oil and N-butanol		
Use with	Herbicides and fungicides in a wide range of crops. See label for details		
Protective clothing	A, C		
Precautions	R43, U02a, U05a, U14, U20a, E15a, E31b, E34, E38, E01, E04, E26, E30a, H04		
Headland Guard 2000	Headland	A0369	extender/sticker/wetter
Contains	52% w/w synthetic latex solution		
Use with	All approved pesticides that recommend the use of a sticking, wetting or extending agent; also with all approved pesticides for use on crops not destined for human or animal consumption; also with all approved pesticides on beans, peas, edible podded peas, oilseed rape, linseed, sugar beet, cereals, maize, Brussels sprouts, potatoes, cauliflowers		
Precautions	U05a, U08, U19a, U20b, E13b, E31b, E34, E37, E01, E04, E26		

Product	Supplier	Adj. No.	Type
Headland Guard Pro	Headland Amenity	A0423	extender/sticker
Contains	52% synthetic latex solution		
Use with	All approved pesticides, micronutrients and sea-weed based plant growth stimulants in amenity grass, managed amenity turf and amenity vegetation		
Precautions	U05a, U08, U19a, U20b, E13b, E31b, E34, E37, E01, E04, E26		
Headland Inflo XL	Headland Amenity	A0329	wetter
Contains	85% w/w polyalkylene oxide modified heptamethyl siloxane		
Use with	Any pesticide for use on amenity grassland or managed amenity turf (except any product applied in or near water) where the use of a wetting, spreading and penetrating surfactant is recommended to improve foliar coverage, and with any approved pesticide on crops not intended for human or animal consumption		
Protective clothing	A, C		
Precautions	R21, R22a, R38, R41, R43, R51, R58, U02a, U05a, U08, U14, U19a, U20b, E15a, E31b, E34, E38, E01, E04, E26, E30a, M03, H03, H11		
Headland Intake	Headland	A0074	penetrant
Contains	450 g/l propionic acid		
Use with	All approved pesticides on any crop not intended for human or animal consumption, and with all approved pesticides on beans, peas, edible podded peas, oilseed rape, linseed, sugar beet, cereals (except triazole fungicides), maize, Brussels sprouts, potatoes, cauliflowers. See label for detailed advice on timing on these crops		
Protective clothing	A, C		
Precautions	R34, U02a, U05a, U10, U11, U14, U15, U19a, E31b, E34, E01, E04, E26, E30b, M04, H05		
Headland Rhino	Headland	A0328	wetter
Contains	85% w/w polyalkylene oxide modified heptamethyl siloxane		
Use with	Any herbicide, systemic fungicide, systemic insecticide or plant growth regulator (except any product applied in or near water) where the use of a wetting, spreading and penetrating surfactant is recommended to improve foliar coverage		
Protective clothing	A, C		
Precautions	R21, R22a, R38, R41, R43, R51, R58, U02a, U05a, U08, U11, U14, U19a, U20b, E13c, E31b, E34, E01, E04, E26, E30a, M03, H03, H11		
Hyspray	Fine	A0020	cationic surfactant
Contains	800 g/l polyethoxylated tallow amine		
Use with	All approved formulations of glyphosate		
Protective clothing	A, C		
Precautions	R22a, R36, R38, U02a, U05a, U08, U13, U19a, U20a, E13b, E31b, E34, E01, E04, E30a, H03, H04, H08		

Product	Supplier	Adj. No.	Type
Intracrop Bla-Tex	Intracrop	A0173	extender/sticker
Contains	450 g/l synthetic latex		
Use with	Any approved pesticide where the addition of a sticker/extender is required and recommended		
Protective clothing	A, C		
Precautions	R36, R38, U02a, U05a, U08, U19a, U20a, E13c, E31a, E01, E04, E30a, M03, H04		
Intracrop Neotex	Intracrop	A0460	extender/sticker
Contains	42.5% w/w styrene butadiene co-polymer and 20% w/w alkylphenol ethylene oxide condensate		
Use with	Recommended rates of approved pesticides up to specified growth stages of a wide range of agricultural arable and horticultural crops, and for non-crop uses. Use on edible crops beyond specified growth stages, and in grass, should only be with half recommended rates of the pesticide or less. See label for details of growth stage restrictions. In addition may be used with all potato blight fungicides at their recommended rates of use up to the latest recommended timing of the fungicide		
Precautions	U08, U20b, E13c, E31a, E34, E37, E01, E26, E30a		
Intracrop Predict	Intracrop	A0503	adjuvant/vegetable oil
Contains	91% w/w methylated rapeseed oil		
Use with	All approved pesticides on non-edible crops, and all pesticides approved for use on growing edible crops when used at half recommended dose or less. On specified crops product may be used at a maximum spray concentration of 1% with approved pesticides at their full approved rate up to the growth stages shown in the label		
Precautions	U19a, U20b, E31b, E26, E30a		
Intracrop Questor	Intracrop	A0495	activator/non-ionic surfactant/spreader
Contains	75% polyoxyethylene polypropoxypropanol and 15% alkyl polyglycol ether		
Use with	All approved pesticides on non-edible crops and pesticides used in non-crop production, and all pesticides approved for use on growing edible crops when used at half recommended dose or less. On specified crops product may be used at a maximum spray concentration of 0.3% with approved pesticides at their full approved rate up to the growth stages shown in the label		
Protective clothing	A, C		
Precautions	R36, R38, U04a, U05a, U08, U19a, E13b, E31b, E01, E04, E30a, M03, M05a, H04		
Intracrop Retainer	Intracrop	A0508	adjuvant/vegetable oil
Contains	60% w/w methylated rapeseed oil		
Use with	All approved pesticides on non-edible crops, and all pesticides approved for use on growing edible crops when used at half recommended dose or less. On specified crops product may be used at a maximum spray concentration of 1% with approved pesticides at their full approved rate up to the growth stages shown in the label		
Precautions	U19a, U20b, E31b, E26, E30a		

Product	Supplier	Adj. No.	Type
Intracrop Rigger	Intracrop	A0479	adjuvant/vegetable oil
Contains	91.7% w/w methylated rapeseed oil		
Use with	All approved pesticides on non-edible crops, and all pesticides approved for use on growing edible crops when used at half recommended dose or less. On specified crops product may be used at a maximum spray concentration of 1.78% with approved pesticides at their full approved rate up to the growth stages shown in the label		
Precautions	U19a, U20b, E31b, E26, E30a		
Intracrop Saturn	Intracrop	A0494	activator/non-ionic surfactant/spreader
Contains	75% polyoxyethylene polypropoxypropanol and 15% alkyl polyglycol ether		
Use with	All approved pesticides on non-edible crops and pesticides used in non-crop production, and all pesticides approved for use on growing edible crops when used at half recommended dose or less. On specified crops product may be used at a maximum spray concentration of 0.3% with approved pesticides at their full approved rate up to the growth stages shown in the label		
Protective clothing	A, C		
Precautions	R36, R38, U04a, U05a, U08, U19a, E13b, E31b, E01, E04, E30a, M03, M05a, H04		
Intracrop Status	Intracrop	A0506	adjuvant/vegetable oil
Contains	91% w/w methylated rapeseed oil		
Use with	All approved pesticides on non-edible crops, and all pesticides approved for use on growing edible crops when used at half recommended dose or less. On specified crops product may be used at a maximum spray concentration of 1.0% with approved pesticides at their full approved rate up to the growth stages shown in the label		
Precautions	U19a, U20b, E31b, E26, E30a		
Intracrop Super Rapeze MSO	Intracrop	A0491	adjuvant/vegetable oil
Contains	91.7% w/w methylated rapeseed oil		
Use with	All approved pesticides on non-edible crops, and all pesticides approved for use on growing edible crops when used at half recommended dose or less. On specified crops product may be used at a maximum spray concentration of 1.78% with approved pesticides at their full approved rate up to the growth stages shown in the label		
Precautions	U19a, U20b, E31b, E26, E30a		
Katalyst	Interagro	A0450	penetrant/water conditioner
Contains	90% w/w alkoxylated fatty amine		
Use with	Glyphosate and a wide range of other pesticides on specified crops. Refer to label or contact supplier for further details		
Protective clothing	A, C		
Precautions	R22a, R36, R38, R50, R58, U02a, U05a, U08, U19a, U20b, E15a, E31a, E34, E37, E01, E04, E26, E30a, M03, H03, H11		

Product	Supplier	Adj. No.	Type
Kinetic	Helena	A0252	spreader/wetter
Contains	80% w/w polyoxypropylene-polyoxyethylene glycol 20% w/w polyalkylene oxide modified heptamethyl trisiloxane		
Use with	Approved pesticides on non-edible crops and cereals and stubbles of all edible crops when used at full recommended dose, and with pesticides approved for use on growing edible crops when used at half recommended dose or less. On specified crops product may be used at a maximum spray concentration of 0.2% with approved pesticides at their full approved rate up to the growth stages shown in the label		
Protective clothing	A, H		
Precautions	R38, R41, U02a, U05a, U08, U19a, U20b, E13c, E31c, E34, E37, E01, E04, E26, E30a, M03, H04		
Klipper	Amega	A0260	spreader/wetter
Contains	600 g/l ethylene oxide condensate		
Use with	Approved salt formulations of mecoprop alone or in mixtures with 2,4-D on managed amenity turf		
Protective clothing	A, C		
Precautions	R36, R38, U05a, U08, U20c, E15a, E31a, E01, E04, E30a, H04, H08		
Leaf-Koat	Helena	A0511	spreader/wetter
Contains	80% w/w polyoxypropylene-polyoxyethylene glycol 20% w/w polyalkylene oxide modified heptamethyl trisiloxane		
Use with	Approved pesticides on non-edible crops and cereals and stubbles of all edible crops when used at full recommended dose, and with pesticides approved for use on growing edible crops when used at half recommended dose or less. On specified crops product may be used at a maximum spray concentration of 0.2% with approved pesticides at their full approved rate up to the growth stages shown in the label		
Protective clothing	A, H		
Precautions	R38, R41, U02a, U05a, U08, U19a, U20b, E13c, E31c, E34, E37, E01, E04, E26, E30a, M03, H04		
LI-700	Loveland	A0176	acidifier/drift retardant/penetrant
Contains	350 g/l modified soya lecithin, 100 g/l alkylphenyl hydroxypolyoxyethylene and 350 g/l propionic acid		
Use with	All plant growth regulators for cereals and oilseed rape, dimethoate and chlorpyrifos for wheat bulb fly control. For other uses see label or contact supplier for details		
Protective clothing	A, C, H		
Precautions	R36, R38, U11, U14, U15, U19a, E15a, E19b, E31a, E32d, E01, E04, E26, H04		
Libsorb	Ciba Specialty	A0438	spreader/wetter
Contains	alkyl alcohol ethoxylate		
Use with	Any spray for which additional wetter is approved and recommended		
Protective clothing	A, C		
Precautions	R36, R38, U05a, U08, U20a, E13c, E31a, E01, E04, E30a, H04		

Product	Supplier	Adj. No.	Type
Logic	Microcide	A0288	vegetable oil
Contains	95% emulsified vegetable oil		
Use with	All approved pesticides on edible and non-edible crops for ground or aerial application		
Protective clothing	A, C		
Precautions	U19a, U20b, E31b, E30a		
Logic Oil	Microcide	A0293	vegetable oil
Contains	95% emulsified vegetable oil		
Use with	All approved pesticides on edible and non-edible crops for ground or aerial application		
Protective clothing	A, C		
Precautions	U19a, U20b, E31b, E30a		
Luxan Non-Ionic Wetter	Luxan	A0139	non-ionic surfactant/spreader/wetter
Contains	900 g/l alkyl phenol ethylene oxide condensate		
Use with	Any spray for which additional wetter/spreader is approved and recommended		
Protective clothing	A, C		
Precautions	R36, R38, U05a, U08, U20c, E13c, E31b, E34, E01, E04, E30a, H04, H08		
Mangard	Mandops	A0132	adjuvant/anti-transpirant
Contains	96% w/w di-1-p-menthene		
Use with	Glyphosate and a wide range of other herbicides, fungicides and insecticides. See label for details		
Precautions	U20c, E15a, E31a, E30a		
Mixture B	Amega	A0161	non-ionic surfactant/spreader/wetter
Contains	500 g/l nonyl phenol ethylene oxide condensate and 500 g/l primary alcohol ethylene oxide condensate		
Use with	Approved pesticides applied in amenity and forestry situations and on a range of specified arable and vegetable crops		
Protective clothing	A, C		
Precautions	R21, R22a, R36, R38, U05a, U08, U20b, E13c, E31a, E34, E37, E01, E04, E30a, M03, H03, H04		
Nettle	Interagro	A0466	mineral oil
Contains	60% w/w mineral oil		
Use with	Cereal graminicides and a wide range of other pesticides that have a label recommendation for use with authorised adjuvant oils on specified crops. Refer to label or contact supplier for further details		
Protective clothing	A		
Precautions	R22b, R38, U02a, U05a, U08, U20b, E15a, E31a, E01, E04, E26, E30a, M05b, H03		

Product	Supplier	Adj. No.	Type
Newman Cropspray 11E	Loveland	A0195	mineral oil
Contains	99% highly refined mineral oil		
Use with	Approved herbicides for which the addition of an adjuvant oil is recommended		
Protective clothing	A, C		
Precautions	R22a, U10, U19a, U20a, E15a, E19b, E31a, E34, E37, E01, E04, E26, M05b, H03		
Newman's T-80	Loveland	A0192	spreader/wetter
Contains	78% w/w polyoxyethylene tallow amine		
Use with	Glyphosate		
Protective clothing	A, C, H		
Precautions	R22a, R37, R38, R41, R50, R58, R67, U05a, U11, U14, U19a, U20a, E15a, E19b, E31a, E32d, E34, E37, E01, E04, E26, M03, M04, H03, H08, H11		
Nion	Amega	A0415	spreader/wetter
Contains	90% (900 g/l) ethylene oxide condensate		
Use with	Recommended rates of approved pesticides up to specified growth stages of a wide range of agricultural arable and horticultural crops. Use on edible crops beyond specified growth stages should only be with half recommended rates of the pesticide or less. See label for details of growth stage restrictions.		
Protective clothing	A, C		
Precautions	R36, R38, U05a, U08, U19a, U20b, E13c, E31a, E37, E01, E04, E30a, H04		
Nu Film P	Intracrop	A0039	sticker/wetter
Contains	96% poly-1-p-menthene		
Use with	Glyphosate and many other pesticides and growth regulators for which a protectant is recommended. Do not use in mixture with adjuvant oils or surfactants		
Precautions	U19a, U20b, E13c, E31a, E30a		
Nufarm Adjuvant Oil	Nufarm UK	A0474	sticker/wetter
Contains	95% highly refined mineral oil and surfactants		
Use with	Recommended rates of approved pesticides on a range of specified crops, and with half rates of pesticides in grassland and beet crops. See label or contact supplier for details		
Protective clothing	A, C		
Precautions	R36, U05a, U08, U11, U20b, E13c, E31b, E37, E01, E04, E26, E30a, H04		

Product	Supplier	Adj. No.	Type
Nufarm Cropoil	Nufarm UK	A0447	mineral oil
Contains	99% highly refined mineral oil		
Use with	Approved pesticides for use on certain specified crops (cereals, combinable break crops, field and dwarf beans, peas, oilseed rape, brassicas, potatoes, carrots, parsnips, sugar beet, fodder beet, mangels, red beet, maize, sweetcorn, onions, leeks, horticultural crops, forestry, amenity, grassland)		
Protective clothing	A, C		
Precautions	R43, U19a, U20b, E13c, E31b, E26, E30a, H04		
Nufarm Cropoil Gold	Nufarm UK	A0446	adjuvant/vegetable oil
Contains	97.5% w/w methylated rapeseed oil		
Use with	A wide range of pesticides used in agriculture, horticulture, forestry and amenity situations. See label for details of crops and dose rates.		
Protective clothing	A, H		
Precautions	R43, U05a, U15, U19a, U20b, E13c, E31b, E01, E04, E26, E30a, H04		
Output	Syngenta	A0429	mineral oil/surfactant
Contains	60% mineral oil and 40% surfactants		
Use with	Grasp		
Protective clothing	A, C		
Precautions	R38, U05a, U08, U19a, U20a, E13c, E31b, E01, E04, E30a, H04		
Pan Oasis	Pan Agriculture	A0411	vegetable oil
Contains	95% w/w methylated rapeseed oil		
Use with	A wide range of pesticides that have a label recommendation for use with authorised adjuvant oils. Contact distributor for further details		
Protective clothing	A		
Precautions	R36, U02a, U05a, U08, U20b, E15a, E31a, E01, E04, E26, E30a, H04		
Pan Panorama	Pan Agriculture	A0412	mineral oil
Contains	95% w/w mineral oil		
Use with	A wide range of pesticides that have a label recommendation for use with adjuvant oils. Contact distributor for details		
Protective clothing	A		
Precautions	R22b, R38, U02a, U05a, U08, U20b, E13c, E31a, E01, E04, E26, E30a, M05b, H03		
Phase II	Loveland	A0331	vegetable oil
Contains	95% w/w esterified rapeseed oil		
Use with	Pesticides approved for use in sugar beet, oilseed rape, cereals (for grass weed control), and other specified agricultural and horticultural crops		
Protective clothing	A, C, H		
Precautions	U19a, U20b, E15a, E19b, E31a, E32d, E34, E01, E04, E26		

Product	Supplier	Adj. No.	Type
Planet	Intracrop	A0224	non-ionic surfactant/spreader/wetter
Contains	85% alkyl polyglycol ether and fatty acid		
Use with	Any spray for which additional wetter is recommended		
Precautions	R22a, R36, U05a, U08, U19a, U20b, E13c, E31a, E32a, E01, E30a, H03, H04, H08		
Prima	Loveland	A0310	mineral oil
Contains	99% highly refined mineral oil		
Use with	All pesticides on all crops when used up to 50% of maximum approved dose for that use (mixtures with Roundup must only be used for treatment of stubbles); all approved pesticides at full dose on listed crops (see label). For other uses see label or contact supplier for details		
Protective clothing	A, C		
Precautions	R22a, U10, U19a, E15a, E19b, E31a, E32d, E34, E37, E01, E04, E26, M05b, H03		
Profit Oil	Microcide	A0294	extender/sticker/wetter
Contains	95% rape/soya oil		
Use with	All approved pesticides		
Protective clothing	A, C		
Precautions	U19a, U20b, E31b, E30a		
QuikSilver	Interagro	A0471	non-ionic surfactant/spreader/wetter
Contains	75% w/w nonylphenol ethylene oxide condensate + fatty acids		
Use with	Pesticides that have a label recommendation for the use of an authorised non-ionic spreader or wetter		
Protective clothing	A, C		
Precautions	R22a, R36, R38, R51, R58, U02a, U05a, U08, U20a, E15a, E34, E01, E04, E26, E30a, M03, H03, H11		
Ranger	Loveland	A0296	acidifier/drift retardant/penetrant
Contains	350 g/l modified soya lecithin, 100 g/l alkylphenyl hydroxypolyoxyethylene and 350 g/l propionic acid		
Use with	All plant growth regulators for cereals and oilseed rape, dimethoate and chlorpyrifos for wheat bulb fly control. For other uses see label or contact supplier for details		
Protective clothing	A, C, H		
Precautions	R36, R38, U11, U14, U15, U19a, E15a, E19b, E31a, E32d, E01, E04, E26, H04		
Rapide	Intracrop	A0116	penetrant/surfactant
Contains	40% propionic acid		
Use with	A wide range of pesticides and growth regulators, especially chlormequat		
Protective clothing	A, C		
Precautions	R34, R36, R38, U02a, U05a, U08, U19a, U20a, E31a, E01, E04, E30a, H04, H05		

Product	Supplier	Adj. No.	Type
Reward Oil	Microcide	A0295	extender/sticker/wetter
Contains	95% rape/soya oil		
Use with	All approved pesticides		
Protective clothing	A, C		
Precautions	U19a, U20b, E31b, E30a		
Ryda	Interagro	A0168	cationic surfactant/wetter
Contains	800 g/l polyethoxylated tallow amine		
Use with	Glyphosate		
Protective clothing	A, C		
Precautions	R22a, R36, R38, R41, R51, R58, U02a, U05a, U08, U13, U19a, U20a, E15a, E31b, E34, E37, E01, E04, E26, E30a, M03, H03, H08, H11		
Saracen	Interagro	A0368	vegetable oil
Contains	95% w/w methylated vegetable oil		
Use with	Cereal graminicides and a wide range of other pesticides that have a recommendation for use with authorised adjuvant oils on specified crops. Refer to label or contact supplier for further details		
Precautions	U05a, U20b, E15a, E31a, E01, E04, E30a		
SAS 90	Intracrop	A0311	spreader/wetter
Contains	100% polyoxyethylen-alpha-methyl-omega-[3-(1,3,3,3-tetramethyl-3-trimethylsiloxy)-disiloxanyl] propylether		
Use with	A wide range of pesticides on specified crops in agriculture and horticulture, and on amenity vegetation and land not intended for cropping		
Precautions	U02a, U08, U19a, U20b, E13c, E31a, E34, E30a		
Scout	Loveland	A0207	acidifier/drift retardant/non-ionic surfactant/penetrant
Contains	350 g/l modified soya lecithin, 100 g/l alkylphenyl hydroxypolyoxyethylene and 350 g/l propionic acid		
Use with	All plant growth regulators for cereals and oilseed rape, dimethoate and chlorpyrifos for wheat bulb fly control. For other uses see label or contact supplier for details		
Protective clothing	A, C, H		
Precautions	R36, R38, U11, U14, U15, U19a, E15a, E19b, E31a, E32d, E01, E04, E26, H04		
Signal	Intracrop	A0308	extender/sticker
Contains	450 g/l synthetic latex		
Use with	All approved pesticides where addition of a sticker/extender is required and recommended		
Protective clothing	A, C		
Precautions	R36, R38, U02a, U05a, U08, U19a, U20b, E13c, E31a, E01, E04, E30a, M03, H04		

Product	Supplier	Adj. No.	Type
Siltex	Intracrop	A0398	spreader/sticker/wetter
Contains	27% w/w polyoxyethylen-alpha-methyl-omega-[3-(1,3,3,3-tetramethyl-3-trimethylsiloxy)-disiloxanyl] propylether and 7.3% w/w cocoiminodipropionate		
Use with	Approved pesticides used at full dose on a range of specified fruit and vegetable crops, and on managed amenity turf		
Protective clothing	A, C		
Precautions	U05a, U08, U20c, E13c, E31a, E37, E01, E04, E30a		
Silwet L-77	Loveland	A0193	drift retardant/spreader/wetter
Contains	80% w/w polyalkylene oxide modified heptamethyltrisiloxane and a maximum of 20 % w/w allyloxypolyethylene glycol methyl ether		
Use with	All approved fungicides on winter and spring sown cereals; all approved pesticides applied at 50% or less of their full approved dose. A wide range of other uses. See label or contact supplier for details		
Protective clothing	A, C, H		
Precautions	R20, R21, R22a, R36, R43, R48, R51, R58, U11, U15, U19a, E15a, E19b, E31a, E32d, E01, E04, E26, H03, H11		
Slippa	Interagro	A0206	spreader/wetter
Contains	655 g/l polyalkyleneoxide modified heptamethyltrisiloxane		
Use with	Cereal fungicides and a wide range of other pesticides and trace elements on specified crops		
Protective clothing	A, C, H		
Precautions	R20, R38, R41, R43, R48, R51, R58, U02a, U05a, U08, U11, U19a, U20a, E15a, E31a, E34, E37, E01, E04, E26, E30a, M03, H03, H11		
SM99	Loveland	A0134	mineral oil
Contains	99% w/w highly refined paraffinic oil		
Use with	Approved herbicides for which the addition of an adjuvant oil is recommended		
Protective clothing	A, C		
Precautions	R22a, U10, U19a, E15a, E19b, E31a, E32d, E34, E37, E01, E04, E26, M05b, H03		
Solar	Intracrop	A0225	activator/non-ionic surfactant/spreader
Contains	75% polypropoxypropanol and 15% alkyl polyglycol ether		
Use with	Foliar applied plant growth regulators		
Protective clothing	A, C		
Precautions	R36, R38, U04a, U05a, U08, U20a, E13c, E31b, E32a, E34, E01, E04, E30a, M03, H04		
Spartan	Interagro	A0375	penetrant/water conditioner
Contains	500 g/l alkoxylated fatty amine + 400 g/l polyoxyethylene monolaurate		
Use with	Cereal growth regulators, cereal herbicides, oilseed rape fungicides and a wide range of other pesticides on specified crops. Refer to label or contact supplier for further details		
Protective clothing	A, C		
Precautions	R22a, R36, R38, R51, R58, U02a, U05a, U19a, U20b, E15a, E31a, E34, E37, E01, E04, E26, E30a, M03, H03, H11		

Product	Supplier	Adj. No.	Type
Sprayfast	Mandops	A0131	extender/sticker/wetter
Contains	334 g/l di-1-p-menthene and nonyl phenol ethylene oxide condensate		
Use with	Glyphosate and other pesticides, growth regulators or nutrients for which a coating agent is approved and recommended		
Precautions	U20b, E13c, E31a, E30a		
Sprayfix	Loveland	A0297	drift retardant/extender/sticker/wetter
Contains	450 g/l synthetic latex and 100 g/l alkylphenyl hydroxypolyoxyethylene		
Use with	Blight sprays on potatoes and other pesticides on a wide range of crops. See label or contact supplier for details		
Protective clothing	A, C, H		
Precautions	R36, R38, U11, U14, U19a, E15a, E19b, E31a, E32d, E01, E04, E26, E30a, H04		
Spraygard	Mandops	A0394	anti-transpirant/extender/sticker/wetter
Contains	400 g/l di-1-p-menthene		
Use with	Approved pesticides on all edible crops (except herbicides on peas) and as an anti-transpirant on vegetables before transplanting, evergreens, deciduous trees, shrubs, bushes and on turf		
Precautions	U20b, E13c, E31a, E37, E30a, M03		
Spraymac	Loveland	A0298	acidifier/non-ionic surfactant
Contains	350 g/l propionic acid and 100 g/l alkylphenyl hydroxypolyoxyethylene		
Use with	All plant growth regulators for cereals and oilseed rape and any spray for which the addition of a wetter/spreader is recommended. See label or contact supplier for details		
Protective clothing	A, C, H		
Precautions	R34, U04a, U10, U11, U14, U19a, U20b, E15a, E19b, E31a, E32d, E34, E01, E04, E26, M04, H05		
Sprayprover	Fine	A0238	mineral oil
Contains	95% highly refined mineral oil		
Use with	Any spray for which the addition of a mineral oil is approved and recommended		
Protective clothing	A, C		
Precautions	R36, R38, U05a, U08, U19a, U20a, E13c, E31b, E01, E04, E30a, H04		
Spread and Seal	Mandops	A0449	extender/sticker/wetter
Contains	334 g/l d-1-p-menthene		
Use with	All pesticides on all edible crops when used at up to 50% of their approved dose, and all approved forulations of glyphosate when used on cereals. Must not be used in cobination with herbicides applied to pea crops		
Precautions	U20c, E13c, E15a, E31a, E30a		
Stamina	Interagro	A0202	penetrant
Contains	100% w/w alkoxylated fatty amine		
Use with	Glyphosate and a wide range of other pesticides on specified crops		
Protective clothing	A, C		
Precautions	R22a, R38, R50, R58, U02a, U05a, U20a, E15a, E31a, E34, E01, E04, E26, E30a, M03, H03, H11		

Product	Supplier	Adj. No.	Type
Stika	Loveland	A0452	extender/sticker/wetter
Contains	225 g/l synthetic latex and 100 g/l alkylphenyl hydroxypolyoxyethylene		
Use with	Approved pesticides on non-edible crops where the use of a sticking, wetting or extending agent is recommended, and pesticides approved for use on growing edible crops when used at half recommended dose or less. On specified crops product may be used at a maximum spray concentration of 0.14% with approved pesticides at their full approved rate up to the growth stages shown in the label		
Protective clothing	A, C, H		
Precautions	U11, U14, U19a, E15a, E19b, E31a, E32d, E01, E04, E26		
Super Nova	Interagro	A0364	penetrant
Contains	500 g/l alkoxylated fatty amine + 500 g/l polyoxyethylene monolaurate		
Use with	Cereal growth regulators, cereal herbicides, cereal fungicides, oilseed rape fungicides and a wide range of other pesticides on specified crops		
Precautions	R51, R58, U05a, E15a, E31a, E34, E01, E04, E26, E30a, H11		
Sward	Amega	A0215	spreader/wetter
Contains	15.2% w/w polyalkylene oxide modified heptomethyltrisiloxane		
Use with	Specified pesticides in amenity situations		
Protective clothing	A, C		
Precautions	R41, R43, U05a, U08, U20c, E13c, E31a, E01, E04, E30a, H04		
Toil	Interagro	A0248	vegetable oil
Contains	95% w/w methylated rapeseed oil		
Use with	Sugar beet herbicides, oilseed rape herbicides, cereal graminicides and a wide range of other pesticides that have a label recommendation for use with authorised adjuvant oils on specified crops. Refer to label or contact supplier for further details		
Precautions	U05a, U20b, E15a, E31a, E01, E04, E26, E30a		
Torpedo	Loveland	A0336	penetrant/wetter
Contains	400 g/l alkoxylated tallow amine, 375 g/l alkylphenylhydroxypolyoxyethylene and 75 g/l natural fatty acids		
Use with	Sulfonylurea herbicides and their approved tank mixes, cereal fungicides, oilseed rape fungicides and a wide range of other pesticides on specified crops		
Protective clothing	A, C, H		
Precautions	R22a, R36, R38, R51, R58, U02a, U05a, U19a, U20b, E15a, E31a, E34, E37, E01, E04, E26, E30a, M03, H03, H11		
Torpedo II	Loveland	A0393	penetrant/wetter
Contains	215.5 g/litre alkoxylated tallow amine, 375 g/litre alkylphenyl hydroxypolyoxyethylene and 75 g/litre natural fatty acids		
Use with	All pesticides on edible and non-edible crops when used at half or less than the full approved dose and with products up to their full approved dose on a wide range of specified fruit and vegetable crops.		
Protective clothing	A, C, H		
Precautions	R38, R41, R52, R58, U05a, U10, U11, U19a, E15a, E19b, E31a, E32d, E34, E37, E01, E04, E26, M03, H04, H08, H11		

Product	Supplier	Adj. No.	Type
Transcend	Helena	A0333	adjuvant/vegetable oil
Contains	80% w/w methylated soybean oil and 15% w/w polyalkylene oxide modified heptamethyl trisiloxane		
Use with	Approved pesticides on non-edible crops, cereals and stubbles of all edible crops, grassland (destruction), and pesticides approved for use on growing edible crops when used at half recommended dose or less. On specified crops product may be used at a maximum spray concentration of 0.5% with approved pesticides at their full approved rate up to the growth stages shown in the label		
Protective clothing	A, H		
Precautions	R38, R41, U02a, U05a, U08, U11, U19a, U20b, E13c, E34, E37, E01, E04, E26, E30a, M03, H04		
Try-Flex	Greenaway	A0489	sticker/wetter
Contains	95% refined rapeseed oil		
Use with	Gly-480 (MAPP 09837) and Freeway (MAPP 11129) on natural Surfaces not intended to bear vegetation, permeable surfaces overlying soil, hard surfaces and forest		
Protective clothing	A, C, D, H, M		
Precautions	U08, U20b, E13c, E31a, E37, E04, E26		
Validate	Loveland	A0500	adjuvant/wetter
Contains	50% w/w lecithin, 25% w/w esterified vegetable oil and 25% w/w alcohol ethoxylate		
Use with	Approved pesticides on non-edible crops where the addition of a wetter/spreader or adjuvant oil is recommended on the pesticide label, and pesticides approved for use on growing edible crops when used at half recommended dose or less. On specified crops product may be used at a maximum spray concentration of 0.5% with approved pesticides at their full approved rate up to the growth stages shown in the label		
Protective clothing	A, C, H		
Precautions	R52, R58, U05a, U11, U14, U19a, U20b, E15a, E19b, E31a, E32d, E34, E01, E04, E26		
Wetta Plus	Interagro	A0480	non-ionic surfactant/spreader/wetter
Contains	75% w/w nonylphenol ethylene oxide condensate + fatty acids		
Use with	Pesticides that recommend the use of an authorised non-ionic spreader or wetter		
Protective clothing	A, C		
Precautions	R22a, R36, R38, R51, R58, U02a, U05a, U08, U20a, E15a, E34, E01, E04, E26, E30a, M03, H03, H11		

SECTION 5
USEFUL INFORMATION

Pesticide Legislation

Anyone who advertises, sells, supplies, stores or uses a pesticide is bound by legislation, including those who use pesticides in their own homes, gardens or allotments. There are numerous UK statutory controls, but the major legal instruments are outlined below.

The Food and Environment Protection Act 1985 (FEPA) and Control of Pesticides Regulations 1986 (COPR)

FEPA introduced statutory powers to control pesticides, with the aims of protecting human beings, creatures and plants, safeguarding the environment, ensuring safe, effective and humane methods of controlling pests, and making pesticide information available to the public.

Control of pesticides is achieved by COPR. These Regulations lay down the Approvals required before any pesticide may be sold, stored, supplied, advertised or used, subject to conditions which are contained in a series of Schedules. Schedule 1 relates to advertisement; Schedule 2 to sale, supply and storage; Schedule 3 to use; and Schedule 4 to aerial application of pesticides. The Schedules may be changed at any time, following Parliamentary approval. Details are given on the websites of the Pesticides Safety Directorate (PSD, www.pesticides.gov.uk) and the Health and Safety Executive (HSE, www.hse.gov.uk).

The controls currently in force include the following.

- Only approved products may be sold, supplied, stored, advertised or used.
- No advertisement may contain any claim for safety beyond that which is permitted in the approved label text.
- Only products specifically approved for the purpose may be applied from the air.
- A recognised Storeman's Certificate of Competence is required by anyone who stores for sale or supply pesticides approved for agricultural use.
- A recognised Certificate of Competence is required by anyone who gives advice when selling or supplying pesticides approved for agricultural use.
- Users of pesticides must comply with the Conditions of Approval relating to use.
- A recognised Certificate of Competence is required for all contractors and for persons born after 31 December 1964 applying pesticides approved for agricultural use (unless working under direct supervision of a certificate holder). Proposals are now in place that every user of agricultural pesticides will have to hold a Certificate of Competence regardless of age or supervision. A suitable transition period will allow this requirement to be enacted.
- Only those adjuvants authorised by PSD may be used.
- Regarding tank-mixes, 'no person shall combine or mix for use two or more pesticides which are anti-cholinesterase compounds unless such a mixture is expressly permitted by the conditions of approval given in relation to at least one of those pesticides', and 'no person shall combine or mix for use two or more pesticides unless all the conditions of Approval given in relation to each of those pesticides can be complied with'.

The Plant Protection Products Directive

European Council Directive 91/414/EEC is intended to harmonise national arrangements for the authorisation of plant protection products within the European Community. It became effective in the UK on 25 July 1993. Under the provisions of the Directive, individual Member States are responsible for authorisation within their own territory of products containing active substances that appear in a list agreed at Community level. This list, known as Annex I, is being created over a period of time by review of existing active ingredients (to ensure they meet present safety standards) and authorisation of new ones. About 60 active substances had achieved Annex I listing by the time of printing this edition.

Individual Member States are required to amend their national arrangements and legislation in order to meet the requirements of Directive 91/414/EEC. In the UK this has been achieved by a series of Plant Protection Products Regulations (PPPR) under which, over a period of time, all agricultural and horticultural pesticides will come to be regulated. Meanwhile, existing product approvals are being maintained under COPR, and new ones are granted for products containing active ingredients that were already on the market by 25 July 1993. Products containing new active substances that were not on the market at this date are being granted provisional approval in advance of Annex I listing of their active ingredients. As active ingredients are placed on Annex I of the Directive, products containing only those active ingredients will be regulated solely under PPPR. Active ingredients in this *Guide* that have been included in Annex I are identified in the 'Approval' section of the profile.

The Directive also provides for a system of 'mutual recognition' of products registered in other Member States. Annex I listing, and a relevant approval in the Member State on which the mutual recognition is to be based, are essential prerequisites. Authorisation may be subject to conditions set to take account of differences such as climate and agricultural practice.

The Review Programme

The process of reviewing existing active ingredients is taking considerably longer than originally anticipated. The programme is designed to ensure that all available plant protection products are supported by up-to-date information on safety and efficacy. At the start of the programme it was envisaged that all active substances that were on the market on 25 July 1993 would have been reviewed by 25 July 2003, but the process is now not scheduled for completion until December 2008. Approximately 850 substances on the EU market are being handled in four phases but, because of the cost of providing this information, many are not being supported and their approvals are being revoked. By the end of 2003 about half the original list had been lost. For the remainder, the complex packages of data have to be evaluated, and further losses are likely following scientific risk assessments.

A few substances are temporarily reprieved by derogations for 'Essential Uses' granted by the European Commission (see below). However, even allowing for this temporary relief, the effect on the horticulture industry will be serious, with fewer products, especially herbicides, available and little prospect of new developments for vegetable crops coming forward.

Derogations for 'Essential Uses' Permitted until 31 December 2007

Because of the scale of the likely loss of compounds and products, and the impact this would have on some sectors, requests have been made to the European Commission by Member States to allow the use of a few active substances after 2003 for key uses where no alternative exists. UK requests were coordinated by BCPC and the Horticulture Development Council (HDC). Against this background, the Commission has made provision for certain 'Essential Uses' to be maintained until December 2007, in order to provide time for alternatives to be researched and developed. In the UK this includes 11 herbicides for vegetables, two fungicides, two insecticides, two disinfectants, one acaricide, one nematicide and one fumigant (see list below). These 'Essential Uses' are only for the crops specified and apply only to SOLAs or products already registered for the use. Extrapolations to other crops are not permitted.

It is possible that there will be further losses of products as a result of product rationalisation by companies, or failure of some active substances to achieve Annex I listing. Even where active substances are supported in EC Review rounds 2 and 3, not all uses will be supported, and growers need to be aware of these so that requests for SOLAs can be made.

Table 1.1 Active substances and products permitted, for the following 'Essential Uses' only, until 31 Dec 2007 (unless approval expiry precedes this date)

Activity	Product	Active substance	Approved Essential Use(s)	Off-label Essential Use(s)
Acaricides	Bye Bye 20EC (09346)	amitraz	pear trees after harvest	
	Cleancrop Bye Bye 20EC (11530)	amitraz	pear trees after harvest	
	Mitac HF (07358)	amitraz	pear trees after harvest	
Fumigants	Certis 2-Aminobutane (11182)	2-aminobutane	potato (seed) stored	
	Hortichem 2-Aminobutane (06147)	2-aminobutane	potato (seed) stored	
Fungicides	Nectec Paste* (08510)	azaconazole + imazalil	ornamental plant production	
	Plantvax 75 (01601)	oxycarboxin	ornamentals (outdoor and protected)	
Herbicides	Aconite 50 (10642)	atrazine	sweetcorn	
	Alpha Atrazine 50 SC	atrazine	farm forestry, forest, sweetcorn	
	Alpha Atrazine 50 WP	atrazine	sweetcorn	
	Alpha Prometryne 50 WP (04871)	prometryn	carrots, celery, leeks (transplanted), parsley, parsnips (LTAEU)	outdoor and protected herbs
	Alpha Prometryne 80 WP (04795)	prometryn	carrots, celery, leeks (transplanted), parsnips (LTAEU)	
	Alpha Simazine 50 SC (04801)	simazine	broad beans, field beans, ornamental plant production	asparagus, rhubarb, runner beans
	Alpha Simazine 50 WP (04879)	simazine	asparagus, broad beans, field beans, ornamental shrubs, ornamental trees, rhubarb	
	Alpha Simazine 80 WP (04800)	simazine	asparagus, broad beans, field beans, ornamental shrubs, ornamental trees, rhubarb	
	Atlas Brown* (07703)	pentanochlor + chlorpropham	carrots, celeriac, celery, Chrysanthemums, herbs (LTAEU), *Narcissus*, ornamentals (other) (LTAEU), parsley, parsnips, tulips	

Activity	Product	Active substance	Approved Essential Use(s)	Off-label Essential Use(s)
Herbicides *cont.*	Atlas Solan 40 (07726)	pentanochlor	Anemone de Caen, carnations, carrots, celeriac, celery, chrysanthemums, foxgloves, freesias, herbs (LTAEU), larkspur, nursery stock (including conifers), ornamentals (other) (LTAEU), parsley, parsnips, roses, sweet peas, sweet williams, wallflowers	
	Atlas Somon (03045 and 07727)	sodium monochloro-acetate	Brussels sprouts, cabbages, collards (LTAEU), kale, leeks, onions	raspberries, loganberries, hybrid *Rubus* spp.
	Atrazol (07598)	atrazine	coniferous, forestry, sweetcorn	
	Barclay Canter (07530)	cyanazine	narcissi (LTAEU), winter oilseed rape, onions, peas	
	Batallion* (08305)	terbutryn + terbuthylazine	broad beans, combining peas, field beans, lupins (LTAEU), edible podded peas, vining peas	
	Bullet* (10654 exp 31 8 05, and 11204)	cyanazine + pendimethalin	spring field beans, combining peas	
	CleanCrop Vectro (10475)	cyanazine	bulb onions, broad beans, combining peas, leeks (LTAEU), winter oilseed rape, ornamental plant production (narcissi only), salad onions, vining peas	
	Croptex Bronze (04087)	pentanochlor	carrots, chrysanthemums, nursery stock, ornamentals (other) (LTAEU), parsnips, sweet peas	
	Croptex Bronze (11329)	pentanochlor	carrots, celery, chrysanthemums, herbs (LTAEU), nursery stock, ornamentals (other) (LTAEU), parsley, parsnips, sweet peas	
	Croptex Chrome* (02415 and 11180)	fenuron + chlorpropham	spinach	runner beans (including crops grown under covers)

Activity	Product	Active substance	Approved Essential Use(s)	Off-label Essential Use(s)
Herbicides *cont.*	Croptex Steel (02418)	sodium monochloro-acetate	Brussels sprouts, bulb onions, cabbages, collards (LTAEU), kale, leeks	cauliflowers, broccoli, calabrese, hops, raspberries, blackberries, loganberries, hybrid *Rubus* spp
	DG90 (11200)	atrazine	sweetcorn	
	Dosaflo (09351)	metoxuron	carrots, parsnips (LTAEU)	
	Flex (08885)	fomesafen	dwarf French beans, runner beans (LTAEU)	outdoor soybeans
	Fortrol (10666 exp 31 7 05)	cyanazine	combining peas, bulb onions, ornamental plant production (propagating material) (narcissi only), winter oilseed rape, vining peas	broccoli, cabbages, calabrese, cauliflowers, collards, kale
	Fortrol (11174)	cyanazine	combining peas, bulb onions, broad beans, leeks (LTAEU), ornamental plant production (propagating material) (narcissi only), winter oilseed rape, salad onions, vining peas	broccoli, cabbages, calabrese, cauliflowers, collards, kale, farm forestry, forestry
	Gesagard (08410)	prometryn	carrots, celery (outdoor), parsley, parsnips (LTAEU)	outdoor and protected herbs, direct drilled and transplanted leeks and onions
	Gesaprim (08411)	atrazine	sweetcorn	
	Gesatop (08412)	simazine	broad beans, field beans, ornamental plant production, roses, tic beans	asparagus, runner beans, rhubarb
	Greencrop Amaize (11973)	atrazine	farm forestry, forest, sweetcorn	
	IT Cyanazine (10179)	cyanazine	bulb onions, broad beans, combining peas, winter oilseed rape, ornamental plant production (propagating material) (narcissi only), salad onions, vining peas	
	Opogard* (08427)	terbutryn + terbuthylazine	broad beans, field beans, lupins, peas	
	Pennine 50 (10638)	prometryn	carrots, celery, leeks (transplanted), parsley, parsnips, herbs (LTAEU)	

Activity	Product	Active substance	Approved Essential Use(s)	Off-label Essential Use(s)
Herbicides *cont.*	Reflex T* (08884)	fomesafen + terbutryn	broad beans (spring sown), spring field beans, lupins (LTAEU), peas (spring sown)	
	Simazine 90 WG (10641)	simazine	asparagus, broad beans, field beans, ornamental plant production	
	Sinbar (01956)	terbacil		protected and outdoor herbs
	Sipcam Simazine Flowable (07622)	simazine	asparagus, broad beans, field beans, ornamental plant production, rhubarb, roses (field grown nursery stock), shrubs (field grown nursery stock), trees (field grown nursery stock)	
	Solan 40 (11897)	pentanochlor	Anemone de Caen, carnations, carrots, celeriac, celery, chrysanthemums, foxgloves, freesias, herbs (LTAEU), larkspur, nursery stock (including conifers), ornamental (other) (LTAEU), parsley, parsnips, roses, sweet peas, sweet williams, wallflowers	
	Unicrop Atrazine 50 (02645)	atrazine	forestry (coniferous), sweetcorn	
	Unicrop Atrazine FL (08045)	atrazine	sweetcorn	
	Unicrop Flowable Atrazine (05446)	atrazine	forestry (coniferous), sweetcorn	
	Unicrop Flowable Simazine (05447 exp 31 12 05)	simazine	asparagus, broad beans, field beans, rhubarb, roses (field grown nursery stock), shrubs	
	Unicrop Simazine 50 (02646)	simazine	asparagus, broad beans, field beans, rhubarb, roses, shrubs, trees (field grown nursery stock)	
	Unicrop Simazine FL (08032)	simazine	broad beans, field beans	
Insecticides	Meothrin (07206 and 10400)	fenpropathrin	blackcurrants	
	Pynosect 30 Water Miscible* (01653)	resmethrin + pyrethrins		mushrooms

Activity	Product	Active substance	Approved Essential Use(s)	Off-label Essential Use(s)
Nematicides	Temik 10 G (10021)	aldicarb	potatoes, carrots, parsnips, onions	
Disinfectants	Enforcer (09288)	dichlorophen	managed amenity turf, hard surfaces, ornamentals, glasshouses, crop standing areas	
	Jeyes Fluid (04606)	tar acids	surfaces in greenhouses, containers, boxes etc	

*Product in mixture with a supported active substance.
SOLA, specific off-label approval; LTAEU, long-term arrangements for extension of use.

The Control of Substances Hazardous to Health Regulations 1988 (COSHH)

The COSHH regulations, which came into force on 1 October 1989, were made under the Health and Safety at Work Act 1974, and are also important as a means of regulating the use of pesticides. The regulations cover virtually all substances hazardous to health, including those pesticides classed as Very toxic, Toxic, Harmful, Irritant or Corrosive; other chemicals used in farming or industry; and substances with occupational exposure limits. They also cover harmful micro-organisms, dusts and any other material, mixture, or compound used at work which can harm people's health.

The original Regulations, together with all subsequent amendments, have been consolidated into a single set of regulations: The Control of Substances Hazardous to Health Regulations 1994 (COSHH 1994).

The basic principle underlying the COSHH regulations is that the risks associated with the use of any substance hazardous to health must be assessed before it is used, and the appropriate measures taken to control the risk. The emphasis is changed from that pertaining under the Poisonous Substances in Agriculture Regulations 1984 (now repealed), whereby the principal method of ensuring safety was the use of protective clothing, to preventing or controlling exposure to hazardous substances by a combination of measures. In order of preference, the measures should be:

1 substitution with a less hazardous chemical or product
2 technical or engineering controls (e.g. the use of closed handling systems, etc.)
3 operational controls (e.g. operators located in cabs fitted with air-filtration systems, etc.)
4 use of personal protective equipment, which includes protective clothing.

Consideration must be given to whether it is necessary to use a pesticide at all in a given situation, and if so, the product posing the least risk to humans, animals and the environment must be selected. Where other measures do not provide adequate control of exposure and the use of personal protective equipment is necessary, the items stipulated on the product label must be used as a minimum. It is essential that equipment is properly maintained and the correct procedures adopted. Where necessary, the exposure of workers must be monitored, health checks carried out, and employees must be instructed and trained in precautionary techniques. Adequate records of all operations involving pesticide application must be made and retained for at least 3 years.

Certificates of Competence – the roles of BASIS and the NPTC

COPR, COSHH and other legislation places certain obligations on those who handle and use pesticides. Minimum standards are laid down for the transport, storage and use of pesticides, and the law requires those who act as storekeepers, sellers and advisors to hold recognised Certificates of Competence.

BASIS is an independent registration scheme for the pesticide industry, recognised under COPR. It is responsible for organising training courses and examinations to enable staff to obtain a Certificate of Competence.

In addition, BASIS undertakes annual assessments of pesticide supply stores, enabling distributors, contractors and seedsmen to meet their obligations under the Code of Practice for Suppliers of Pesticides. Further information can be obtained from BASIS (www.basis-reg.co.uk; see Appendix 2).

Certain spray operators also require Certificates of Competence under the Control of Pesticides Regulations. These certificates are awarded by NPTC to candidates who pass an assessment carried by an approved NPTC or Scottish Skills Testing Service Assessor in one or more of a series of competence modules. Holders are required to produce their certificate on demand for inspection to any person authorised to enforce COPR. Further information can be obtained from NPTC (www.nptc.org.uk; see Appendix 2).

Maximum Residue Levels

A small number of pesticides are liable to leave residues in foodstuffs, even when used correctly. Where residues can occur, statutory limits, known as Maximum Residue Levels (MRLs), have been established. MRLs provide a check that products have been used as directed; they are not safety limits. However, they do take account of consumer safety because they are set at levels that ensure normal dietary intake of residues presents no risk to health. Wide safety margins are built in, and eating food containing residues above the MRL does not automatically imply a risk to health. Nevertheless, it is an offence to put into circulation any produce where the MRL is exceeded.

The surrounding legislation is complex. MRLs may be specified by several different bodies. The UK has set statutory MRLs since 1988. The European Union intends eventually to introduce MRLs for all pesticide/commodity combinations. These are being introduced initially by a series of priority lists, but will subsequently be covered by the review programme under Directive 91/414. However, in cases where no information is available, EU Directive 2000/42/EC requires many MRLs to be set at the limit of determination (LOD). As a result of this, certain approvals are being withdrawn where such use would leave residues above the MRL set in the Directive.

MRLs apply to imported as well as to home-produced foodstuffs. Details of those that have been set have been published in *The Pesticides (Maximum Residue Levels in Crops, Food and Feeding Stuffs) Regulations 1994*, and successive amendments to these Regulations. These Statutory Instruments are available from The Stationery Office.

MRLs are set for many chemicals not currently marketed in Britain. Because of this, and the ever-changing information on MRLs, no quantitative information is given in this *Guide*. Instead, users of the sister CD-ROM publication *The e-UK Pesticide Guide 2005* may gain direct weblink access to the comprehensive and authoritative MRL database compiled and maintained by Leatherhead Food International. This online database sets out in table form those levels specified by UK Regulations, EC Directives, and the Codex Alimentarius for each commodity.

The e-UK Pesticide Guide 2005 edition may be purchased from BCPC Publication Sales, 01420 593200, or online via www.bcpc.org/bookshop

Approval (On-label and Off-label)

Only officially approved pesticides may be marketed and used in the UK. Approvals are granted by UK Government Ministers in response to applications that are supported by satisfactory data on safety, efficacy and, where relevant, humaneness. The Pesticides Safety Directorate (PSD, www.pesticides.gov.uk) is the UK Government Agency for regulating agricultural pesticides and plant protection products. The Health and Safety Executive (HSE, www.hse.gov.uk) fulfils the same role for other pesticides. The main focus of the regulatory process is the protection of human health and the environment.

Approvals are normally granted only in relation to individual products and for specified uses. It is an offence to use non-approved products or to use approved products in a manner that does not comply with the statutory conditions of use, except where the crop or situation is the subject of an off-label extension of use (see below).

Statutory Conditions of Use

Statutory conditions have been laid down for the use of individual products and may include:

- field of use (e.g. agriculture, horticulture etc.)
- crop or situations for which treatment is permitted
- maximum individual dose
- maximum number of treatments or maximum total dose
- maximum area or quantity which may be treated
- latest time of application or harvest interval
- operator protection or training requirements
- any other specific restrictions relating to particular pesticides.

All products must display these statutory conditions of use in a 'statutory box' on the label.

Types of Approval

There are three categories of approval that may be granted, under the Control of Pesticides Regulations, to products containing active ingredients that were already on the market by 25 July 1993:

- **Full** (granted for an unstipulated period)
- **Provisional** (granted for a stipulated period)
- **Experimental Permit** (granted for the purposes of testing and developing new products, formulations or uses). Products with only an Experimental Permit may not be advertised or sold and do not appear in this *Guide*.

Products containing new active substances, and those containing older ingredients once they have been listed in Annex I to Directive 91/414 (see previous section), are granted approval under the Plant Protection Products Regulations. Again, there are three categories, similar to those listed above:

- **Standard approval** (granted for a period not exceeding 10 years). Only applicable to products whose active substances are listed in Annex I.
- **Provisional approval** (granted for a period not exceeding three years, but renewable). Applicable where Annex I listing of the active substance is awaited.
- **Approval for research and development** (to enable field experiments or tests to be carried out with active substances not otherwise approved).

The official lists of approved products, including all the above categories except those for experimental purposes, are shown on the websites of PSD (www.pesticides.gov.uk) and HSE (www.hse.gov.uk).
Details of new approvals, amendments to existing approvals and off-label approvals (see below) are published regularly on the above websites.

Withdrawal of Approval

Product approvals may be reviewed, amended, suspended or revoked at any time. Revocation may occur for various reasons, such as commercial withdrawal or failure by the approval holder to meet data requirements. Where there are no safety concerns, a phased approval revocation is implemented, specifying a 'wind-down' period to allow the using up of existing stocks by persons other than the approval holder.

Until April 2003 the wind-down period was normally set by PSD at 2 or 3 years, but it became apparent that the European Commission considers 18 months a reasonable period for the withdrawal or use of stocks in the supply chain.

Since 1 April 2003 phased revocations have been as follows.

- A **maximum** of six months for the approval holder to advertise, sell, supply and use the product.
- A further **maximum** of 12 months for sale, supply and use by persons other than the approval holder. This allows a total of 18 months from the date of revocation for stocks to pass through the supply chain and be used. Anyone may store the product during this 18-month period.

Shorter periods will continue to be applied where necessary.

The expiry dates shown in this *Guide* continue to be the final date of legal use.

Approval of Commodity Substances

Commodity substances are compounds that have a variety of alternative and often widespread non-pesticidal uses, but also have potential for use as a pesticide. For a commodity substance to be used as a pesticide, it requires approval under the Control of Pesticides Regulations. Approval is given only for the *use* of the substance; approval is not given for sale, storage, supply or advertisement. There is no approval holder, and no pesticide product label, but the approval and associated conditions of use are published in the Guide to Approved Pesticides on the PSD website. Twenty such substances have been approved for certain specified uses as laid down in the Guide to Approved Pesticides. Of these, ethylene, carbon dioxide, formaldehyde, methyl bromide, sodium chloride, sodium hypochlorite, strychnine hydrochloride and sulphuric acid are approved for agricultural or horticultural use, and are included in this *Guide*.

Off-label Extension of Use

Products may legally be used in a manner not covered by the printed label in several ways:

- in accordance with the 'Long-Term Arrangements for Extension of Use' (see below)
- in accordance with a specific off-label approval (SOLA). SOLAs are uses for which approval has been sought by individuals or organisations other than the manufacturer. The Notices of Approval are published by Defra and are widely available from ADAS or National Farmers' Union offices. Users of SOLAs must first obtain a copy of the relevant Notice of Approval and comply strictly with the conditions laid down therein
- in tank mixture with other approved pesticides in accordance with Consent C(i) made under FEPA. Full details of Consent C(i) are given in Annex A of the Guide to Approved Pesticides on the PSD website (currently www.pesticides.gov.uk/publications.asp?id=499), but there are two essential requirements for tank mixes. First, all the conditions of approval of all the components of a mixture must be complied with. Second, no person may mix or combine pesticides which are cholinesterase compounds unless allowed by the label of at least one of the pesticides in the mixture
- in conjunction with authorised adjuvants
- in reduced spray volume under certain conditions
- the use of certain herbicides on specified set-aside areas subject to restrictions which differ between Scotland and the rest of the UK
- by mutual recognition of a use fully approved in another Member State of the European Union and authorised by PSD.

Although approved, off-label uses are not endorsed by manufacturers and such treatments are made entirely at the risk of the user.

Long-Term Arrangements for Extension of Use

Since 1 January 1990, arrangements have been in place allowing many approved products to be used for additional minor uses, without the need for Specific Off-Label Approvals (SOLAs). The arrangements – the Long-Term Arrangements for Extension of Use (LTAEU) – permit growers to use pesticides on certain named minor crops provided an approval exists for use on specified major crops. Growers of minor crops are thus provided with access to a large number of important crop protection products. There is no approval documentation associated with these uses.

The scheme is founded on the principle that the risks to the operator, the consumer, wildlife and the environment arising from the extended (off-label) use are acceptable and no greater than those which arise from the approved uses of a product, which appear on its label. All off-label approvals, including those covered by these arrangements, undergo a full safety assessment by the PSD, in consultation with the Advisory Committee on Pesticide. There is, however, no test of efficacy – **use of these extensions of use, as with any off-label use, is made at the user's choosing, and the commercial risk is entirely theirs.**

Because the LTAEU are national arrangements which have no parallel in European Community (EC) legislation they are becoming increasingly difficult to sustain as they are not compatible with the Plant Protection Products Directive and procedures for setting maximum residue levels. After consultation, the Pesticides Safety Directorate concluded in November 2004 that the best and most cost-effective way of retaining these uses is by converting them to SOLAs. The LTAEU will not be revoked until the exercise is complete, which is anticipated to be the end of 2005. This exercise relates only to edible uses; ornamentals and other non-edible uses will be reviewed at a later date.

The following is a **summary** of the main points and uses of the current arrangements.

Specific restrictions for extension of use

Certain restrictions are necessary to ensure that the extension of use does not increase the risk to the operator, the consumer or the environment.

- The arrangements apply to label and SOLA recommendations, but ONLY for the use of products approved for use as Agricultural and/or Horticultural pesticides.
- Safety precautions and statutory conditions relating to use (shown in the statutory box on the label) MUST be observed. If extrapolation from a SOLA is to be used, all conditions specified on the SOLA Notice of Approval MUST be observed in addition to the safety precautions and statutory conditions on the specified product's label.
- Pesticides must only be used in the same situation (i.e. outdoor or protected) as that specified on the product label or SOLA Notice. Approvals for use on tomatoes, cucumbers, lettuces, chrysanthemums and mushrooms include protected crops unless otherwise stated. Use on other protected crops may only occur when the product label specifically allows use under protection on the crop on which the extrapolation is based. Similarly, pesticides approved only for use in protected situations must not be used outdoors. **If a label or SOLA Notice does not specify a situation, then extrapolation only to an outdoor use is permitted**.
- The application method must be as stated on the product label and in accordance with relevant Codes of Practice and COSHH requirements. Where application of a pesticide under these arrangements is planned to be made by hand-held equipment users MUST ensure that hand-held use is appropriate for the current label or SOLA Notice conditions.
- **Note:** unless otherwise stated, spray applications to protected crops include hand-held uses.

In cases where hand-held application is not appropriate to the use on which extrapolation is to be made, hand-held application may nevertheless be permitted provided certain conditions specified in the arrangements are met (consult the PSD website, currently www.pesticides.gov.uk/publications.asp?id=52). When planning to apply a pesticide under these arrangements by broadcast air-assisted sprayer, only those products with a specific label or SOLA recommendation for such use may be used. Any associated buffer zone restrictions must also be observed.

- Use in or near water (including coastal waters), or by aerial application, is not permitted under the arrangements.
- Rodenticides and other vertebrate control agents are not included, nor is use on land not intended for cropping.
- Pesticides classed as harmful, dangerous, extremely dangerous or high risk to bees must not be used off-label during flowering unless otherwise permitted. Applications must also not be made when flowering weeds are present or when bees are actively foraging. If the label or SOLA Notice specifies an aquatic buffer zone, users must conduct a Local Environment Risk Assessment for Pesticides (LERAP) assessment for the extension of use.
- The user of a pesticide under these arrangements must take all reasonable precautions to safeguard wildlife and the environment.

The arrangements apply as follows.

Non-edible crops and plants

Subject to the specific restrictions set out above:

- Pesticides approved for use on any growing crop may be used on commercial agricultural and horticultural holdings and forest nurseries on: (i) ornamental crops including hardy nursery stock, plants, bulbs, flowers and seed crops where neither the seed nor any part of the plant is to be consumed by humans or animals; (ii) forest nursery crops prior to final planting out..
- In addition, pesticides approved for use on any growing edible crop (except seed treatments) may be used on non-ornamental crops grown for seed subject to the same consumption restrictions as above (but not including seed crops of potatoes, cereals, oilseeds, peas, beans and other pulses).
- Pesticides (except seed treatments) approved for use on oilseed rape may be used on hemp grown for fibre and woad (*Isatis tinctoria* and *I. indigotica*).
- Pesticides (except seed treatments) approved for use on cereals, grass and forage maize may be used on commercial agricultural and horticultural holdings on *Miscanthus* spp. (Elephant grass), but not after the crop is 1 m high. The crops and its products must not be used for food or feed.
- Herbicides approved for use on oilseed rape may be used on sunflower, quinoa, fodder radish and sweet clover grown as game cover.
- Herbicides approved for use on cereals and forage maize may be used on canary grass, tanka millet, white millet and sorghum grown as game cover.

Farm forestry and rotational cropping

Subject to the specific restrictions set out above:

- Herbicides approved for use on cereals may be used in the first 5 years of establishment in farm forestry on land previously under arable cultivation or improved grassland, and reclaimed brownfield sites.
- Herbicides approved for use on cereals, oilseed rape, sugar beet, potatoes, peas and beans may be used in the first year of regrowth after cutting in coppices established on land previously under arable cultivation or improved grassland, or reclaimed brownfield sites.

Nursery fruit crops

Subject to the specific restrictions set out above:

- Pesticides approved for use on any crop for human or animal consumption may be used on commercial agricultural and horticultural holdings on nursery fruit trees, nursery grape vines prior to final planting out, bushes, canes and non-fruiting strawberry plants provided any fruit harvested within 12 months is destroyed. Applications must NOT be made if fruit is present.

Hops

Subject to the specific restrictions set out above:

- On commercial and horticultural holdings pesticides may be used on mature stock or mother plants kept specifically for propagation, hop propagules prior to final planting out, and first year 'nursery hops' in their final planting position but not taken to harvest in that year. Treated hops must not be harvested for human or animal consumption (including idling) within 12 months of treatment.

Crops used partly or wholly for consumption by humans or livestock

Subject to the specific restrictions set out above:

- Pesticides may be used on commercial agricultural and horticultural holdings on crops in the first column of Tables 1.2 and 1.3 below if they have been approved for use on the crop(s) opposite them in the second column. **However, it is the responsibility of the user to ensure that the proposed use does not result in any statutory UK MRL being exceeded. These extrapolations may NOT be used where the MRL for the crop in column 1 is lower than the MRL for the crop in column 2, or where an MRL has been established for the crop in column 1 but not for the crop in column 2.**

Table 1.2 Crops used partly or wholly for consumption by humans or livestock

Minor use	Crops on which use is approved	Additional special conditions
Arable crops		
Poppy (grown for oilseed), sesame	Sunflower	
Mustard, linseed, evening primrose, honesty	Oilseed rape	
Borage (grown for oilseed), canary flower e.g. *Echium vulgare, E. plantagineum* (grown for oilseed)	Oilseed rape	Seed treatments are not permitted
Rye, triticale	Barley	Treatments applied before first spikelet of inflorescence just visible
Rye, triticale	Wheat	
Grass seed crop	Grass for grazing or fodder	
Grass seed crop	Wheat, barley, oats, rye, triticale	Treated crops must not be grazed or cut for fodder until 90 days after treatment Seed treatments are not permitted. Use of chlormequat-containing products is not permitted
Lupins	Combining peas or field beans	

SECTION 5

Minor use	Crops on which use is approved	Additional special conditions
Fruit crops		
Almond, chestnut, hazelnut, walnut	Apple, cherry or plum	For herbicides used on the orchard floor ONLY
Quince, crab apple	Apple or pear	
Almond, chestnut, hazelnut, walnut	Any two of the following: almond, chestnut, hazelnut, walnut	
Nectarine, apricot	Peach	
Blackberry, dewberry, *Rubus* species (e.g. tayberry, loganberry)	Raspberry	
Whitecurrant, bilberry, cranberry, blueberry, other *Vaccinium* species	Blackcurrant or redcurrant	
Gooseberry	Blackcurrant	
Redcurrant	Blackcurrant	
Vegetable crops		
Parsley root	Carrot or radish	
Fodder beet, mangel	Sugar beet	
Horseradish	Carrot or radish	
Parsnip	Carrot	
Salsify	Carrot or celeriac	
Swede	Turnip	
Turnip	Swede	
Garlic, shallot	Bulb onion	
Leek	Salad onion	Herbicides only Latest timing up to two true leaves of the crop
Cayenne pepper	Pepper	
Aubergine	Tomato	
Squash, pumpkin, marrow, watermelon,	Melon	
Broccoli	Calabrese	
Calabrese	Broccoli	

Minor use	Crops on which use is approved	Additional special conditions
Roscoff cauliflower	Cauliflower	
Collards	Kale	
Chinese cabbage, pak choi, choi sum	Kale	
Lamb's lettuce, frisée/ frise, radicchio, cress, scarole	Lettuce	
Leaf herbs and edible flowers*	Lettuce or spinach or parsley or sage or mint or tarragon	
Leafy brassicas grown for baby leaf production	Lettuce	Seed treatments are not permitted Pre-crop emergence use not permitted
Spinach	Lettuce	Seed treatments are not permitted
Beet leaves, red chard, white chard, yellow chard	Spinach	
Edible podded peas (e.g. mange-tout, sugar snap)	Edible podded beans	
Runner beans	Dwarf French beans	
Broad beans	Vining pea	Applications must NOT be made during flowering or where bees are actively foraging Seed treatments are not permitted
Rhubarb, cardoon	Celery	
Edible fungi other than mushroom (e.g. oyster mushroom)	Mushroom	

* This extension of use applies to the following leaf herbs and edible flowers: *Agastache* spp., angelica, balm, basil, bay, borage, burnet (salad), caraway, catnip, camomile, chervil, chives, clary, coriander, curry plant, dill, dragonhead, fennel, fenugreek, feverfew, hyssop, land cress, lavender, lavandin, lovage, marjoram, marigold, mint, nasturtium, nettle, oregano, *Origanum heracleoticum*, parsley, rocket, rosemary, rue, sage, savory, sorrel, spike lavender, tarragon, thyme, verbena (lemon), woodruff, violet.

For applications in store on crops **partly or wholly for human or animal consumption**, the extensions of use are set out in Table 1.3.

Table 1.3 Applications in store on crops used partly or wholly for consumption by humans or animals

Minor use	Crops on which use is approved
Rye, barley, oats, buckwheat, millet, sorghum, triticale	Wheat
Dried peas	Dried beans
Dried beans	Dried peas
Mustard, sunflower, honesty, sesame, evening primrose, poppy (grown for oilseed), borage (grown for oilseed), canary flower e.g. *Echium vulgare/E. plantaginium* (grown for oilseed)	Oilseed rape

Clarifications

Under these arrangements the crops listed in Table 1.4 are considered synonymous or equivalent and, as such, uses on crops in column 1 can be read across to uses in column 2.

Table 1.4 Crops considered synonymous or equivalent

Column 1	Column 2 (equivalent)
Hazelnut	Cobnuts, filberts
French bean	Navy bean
Vining pea	Picking pea, shelling pea, non-edible podded pea
Linseed	Linola, flax
Wheat	Durum wheat

Using Crop Protection Chemicals

Use of Herbicides In or Near Water

Products in this *Guide* approved for use in or near water are listed in Table 1.5. Before use of any product in or near water, the appropriate water regulatory body (Environment Agency/Local Rivers Purification Authority; or in Scotland the Scottish Environment Protection Agency) must be consulted. Guidance and definitions of the situation covered by approved labels are given in the Defra publication *Guidelines for the Use of Herbicides on Weeds in or near Watercourses and Lakes*. Always read the label before use.

Table 1.5 Products approved for use in or near water

Chemical	Product
2,4-D	Dormone
dichlobenil	Casoron G, Luxan Dichlobenil Granules, Midstream GSR
glyphosate	Asteroid, Barclay Barbarian, Barclay Gallup 360, Barclay Gallup Amenity, Barclay Gallup Biograde 360, Barclay Gallup Biograde Amenity, Barclay Gallup Hi-Aktiv, Buggy SG, Clinic, Dow Agrosciences Glyphosate 360, Envision, Flame, Glyfos, Glyfos Gold, Glyfos ProActive, Glyphogan, Glyphosate 360, Greenaway Gly-490, Habitat, Kernel, Manifest, Nufosate, Romany, Roundup Amenity, Roundup Biactive, Roundup Pro Biactive, Roundup Pro-Green, Roundup ProVide, Spasor, Spasor Biactive, Tangent, Trustee Elite, Typhoon 360
maleic hydrazide	Regulox K
terbutryn	Clarosan

Use of Pesticides in Forestry

Table 1.6 Products in this *Guide* approved for use in forestry

Chemical	Product	Use
2,4-D	Dicotox Extra	Herbicide
2,4-D + dicamba + triclopyr	Broadsword, Greengard, Nu-Shot	Herbicide
4-indol-3-yl-butyric acid + 2-(1-naphthyl)acetic acid with dichlorophen	Synergol	Growth regulator/stimulator/sprout suppressant
alpha-cypermethrin	Alphaguard 100 EC	Insecticide
aluminium ammonium sulphate	Curb Crop Spray Powder, Liquid Curb Crop Spray	Animal deterrent/repellent
ammonium sulphamate	Amcide, Root-Out	Herbicide, Tree killer
asulam	Asulox, Greencrop Frond, I T Asulam, Inter Asulam, Spitfire	Herbicide
atrazine	Atrazol, Greencrop Amaize	Herbicide
carbosulfan	Marshal Soil Insecticide suSCon CR granules, Marshal Soil Insecticide suSCon CR Sachets	Insecticide
chlorpyrifos	Cyren, Dursban WG, Equity, Greencrop Pontoon, Lorsban T	Acaricide, Insecticide
copper hydroxide	Spin Out	Root acting plant growth regulator
cycloxydim	Greencrop Pomeroy, Greencrop Valentia, Laser, Marnoch Clodim, Standon Cycloxydim	Herbicide
dicamba	I T Dicamba	Herbicide
diflubenzuron	Dimilin Flo	Insecticide
diquat + paraquat	ASAP, Fernpath Pronto, PDQ	Desiccant, Herbicide
fluazifop-P-butyl	Fusilade Max	Herbicide
glufosinate-ammonium	Challenge, Harvest, Kaspar	Herbicide
isoxaben	Flexidor 125	Herbicide
metazachlor	Alpha Metazachlor 50 SC, Greencrop Monogram, Me2 Booty, Me2 Booty 2, Standon Metazachlor 500, Sultan 50 SC	Herbicide
paraquat	Agriguard Paraquat, Fernpath Graminite, Gramoxone 100	Herbicide

Chemical	Product	Use
propaquizafop	Bulldog, Cleancrop GYR, Falcon, Greencrop Satchmo, Landgold PQF 100, Raptor, Shogun, Standon Propaquizafop	Herbicide
propyzamide	Agriguard Propyzamide 50 WP, Agriguard Propyzamide Flo, Bulwark Flo, Kerb 50 W, Kerb 80 EDF, Kerb Granules, Kerb Pro Flo, Kerb Pro Granules, Menace 80 EDF, Propose 50 W	Herbicide
simazine	Alpha Simazine 50 SC, Gesatop, Sipcam Simazine Flowable	Herbicide
thiabendazole	Storite Clear Liquid	Fungicide
triclopyr	Garlon 4, Nomix Garlon 4, Timbrel	Herbicide
ziram	AAprotect	Animal deterrent/repellent

Pesticides Used as Seed Treatments

Information on the target pests for these products can be found in the relevant pesticide profile in Section 2.

Table 1.7 Products used as seed treatments (including treatments on seed potatoes)

Chemical	Product	Formulation	Crop(s)
aluminium ammonium sulphate	Guardsman STP Seed Dressing Powder	DS	Corms, Flower bulbs, Seeds
beta-cyfluthrin + imidacloprid	Chinook Blue	LS	Spring oilseed rape, Winter oilseed rape
	Chinook Colourless	LS	Winter oilseed rape
bitertanol + fuberidazole	Sibutol	FS	Spring oats, Spring rye, Spring wheat, Triticale, Winter oats, Winter rye, Winter wheat
bitertanol + fuberidazole + imidacloprid	Sibutol Secur	LS	Winter oats, Winter wheat
carboxin + thiram	Anchor	FS	Spring barley, Spring oats, Spring rye, Spring wheat, Triticale, Winter barley, Winter oats, Winter rye, Winter wheat
cymoxanil + fludioxonil + metalaxyl-M	Wakil XL	WS	Broad beans, Carrots, Combining peas, Parsnips, Poppies, Red beet, Spring field beans, Vining peas, Winter field beans
fludioxonil	Beret Gold	FS	Spring barley, Spring oats, Spring wheat, Winter barley, Winter oats, Winter wheat
fluquinconazole	Galmano	FS	Winter wheat
	Jockey F	FS	Winter barley, Winter wheat
fluquinconazole + prochloraz	Galmano Plus	FS	Winter wheat
	Jockey	FS	Winter barley, Winter wheat
flutolanil	Rhino	FS	Potatoes
fuberidazole + imidacloprid + triadimenol	Baytan Secur	FS	Winter barley, Winter oats, Winter wheat

Chemical	Product	Formulation	Crop(s)
fuberidazole + triadimenol	Baytan Flowable	FS	Spring barley, Spring oats, Spring rye, Spring wheat, Triticale, Winter barley, Winter oats, Winter rye, Winter wheat
guazatine	Panoctine	LS	Spring barley, Spring oats, Winter barley, Winter oats
	Ravine	LS	Spring barley, Spring oats, Spring wheat, Winter barley, Winter oats, Winter wheat
guazatine + imazalil	Panoctine Plus	LS	Spring barley, Spring oats, Winter barley, Winter oats
	Ravine Plus	LS	Spring barley, Spring oats, Winter barley, Winter oats
guazatine + triticonazole	Premis	FS	Spring wheat, Winter wheat
imazalil	Fungazil 100 SL	LS	Seed potatoes, Ware potatoes
imazalil + pencycuron	Monceren IM	DS	Potatoes
imazalil + thiabendazole	Extratect Flowable	FS	Potatoes, Seed potatoes
imazalil + triticonazole	Robust	FS	Spring barley, Winter barley
imidacloprid	Gaucho	WS	Broccoli, Brussels sprouts, Cabbages, Calabrese, Cauliflowers, Collards, Kale, Lettuce, Sugar beet
imidacloprid + tebuconazole + triazoxide	Raxil Secur	LS	Winter barley
iprodione	Rovral Liquid FS	FS	Flax, Poppies, Seed potatoes, Spring oilseed rape, Winter oilseed rape
pencycuron	Me2 Penny	DS	Potatoes
	Monceren DS	DS	Potatoes
	Standon Pencycuron DP	DS	Potatoes
prochloraz	Prelude 20LF	LS	Flax, Linseed
prochloraz + triticonazole	Kinto	FS	Winter barley, Winter wheat
silthiofam	Latitude	FS	Spring wheat, Winter barley, Winter wheat
tebuconazole + triazoxide	Raxil S	LS	Spring barley, Winter barley

Chemical	Product	Formulation	Crop(s)
thiabendazole	Hykeep	DS	Ware potatoes
	Storite Flowable	FS	Seed potatoes, Ware potatoes
thiabendazole + thiram	Hy-TL	FS	Broad beans, Bulb onions, Peas, Spring field beans, Winter field beans
thiram	Agrichem Flowable Thiram	LS	Broad beans, Cabbages, Carrots, Cauliflowers, Celery, Dwarf beans, Fodder beet, Grass seed, Leeks, Lettuce, Maize, Mangels, Onions, Parsley, Peas, Poppies, Radishes, Red beet, Runner beans, Salad onions, Spring field beans, Spring oilseed rape, Sugar beet, Turnips, Winter field beans, Winter oilseed rape
	Thiraflo	FS	Broad beans, Combining peas, Dwarf beans, Grass seed, Maize, Runner beans, Spring field beans, Spring oilseed rape, Vining peas, Winter field beans, Winter oilseed rape
	Thyram Plus	LS	Broad beans, Cabbages, Carrots, Cauliflowers, Dwarf beans, Grass seed, Leeks, Lettuce, Maize, Onions, Peas, Radishes, Runner beans, Salad onions, Spring field beans, Spring oilseed rape, Turnips, Winter field beans, Winter oilseed rape
tolclofos-methyl	Me2 Cindy	FS	Potatoes
	Rizolex	DS	Potatoes
	Rizolex Flowable	FS	Potatoes

Aerial Application of Pesticides

Only those products specifically approved for aerial application may be so applied, and they may only be applied to specific crops or for specified uses. A complete list of products approved for application from the air can be accessed in Guide to Pesticides on the website of the Pesticides Safety Directorate (PSD, www.pesticides.gov.uk). The list in Table 1.8 is taken mainly from this source, with updating from issues of *The Pesticides Monitor* and product labels.

It is emphasised that the list is for guidance only – reference must be made to the product labels for detailed conditions of use, which must be complied with. The list does not include those products which have been granted restricted aerial application approval limiting the area which may be treated.

Detailed rules are imposed on aerial application regarding prior notification of the Nature Conservancy, water authorities, bee keepers, Environmental Health Officers, neighbours, hospitals, schools, etc. and the conditions under which application may be made. The full conditions are available from Defra and must be consulted before any aerial application is made.

Table 1.8 Products approved for aerial application

Chemical	Product	Crops
2-chloroethylphosphonic acid	Agriguard Cerusite	Winter barley
	Cerone	Winter barley
asulam	Asulox	Amenity grass, Forest, Permanent pasture, Rough grazing
	Greencrop Frond	Amenity grass, Forest, Permanent pasture, Rough grazing
	I T Asulam	Amenity grass, Forest, Permanent pasture, Rough grazing
	Spitfire	Amenity grass, Forest, Permanent pasture, Rough grazing
benalaxyl + mancozeb	Galben M	Potatoes
	Intro Plus	Potatoes
	Tairel	Potatoes
chlormequat	Agriguard Chlormequat 720	Spring oats, Spring wheat, Winter oats, Winter wheat
	BASF 3C Chlormequat 720	Spring oats, Spring rye, Spring wheat, Triticale, Winter barley, Winter oats, Winter rye, Winter wheat
	Greencrop Carna	Spring oats, Spring wheat, Winter barley, Winter oats, Winter wheat
	Hive	Winter wheat
	Mandops Chlormequat 700	Spring oats, Spring wheat, Winter oats, Winter wheat

SECTION 5

Chemical	Product	Crops
	Mirquat	Spring oats, Spring wheat, Winter oats, Winter wheat
	Sigma PCT	Spring wheat, Winter barley, Winter wheat
chlormequat with choline chloride	Barclay Take 5	Spring oats, Spring wheat, Triticale, Winter barley, Winter oats, Winter rye, Winter wheat
	New 5C Cycocel	Spring oats, Spring rye, Spring wheat, Triticale, Winter barley, Winter oats, Winter rye, Winter wheat
	New 5C Quintacel	Spring oats, Spring rye, Spring wheat, Triticale, Winter barley, Winter oats, Winter rye, Winter wheat
chlorothalonil	Agriguard Chlorothalonil	Potatoes
	Bombardier FL	Potatoes
	Bravo 500	Potatoes
	Clortosip 500	Potatoes
	Flute	Potatoes
	Greencrop Orchid	Potatoes
	Repulse	Potatoes
	Sipcam Echo 75	Potatoes
chlorotoluron	Lentipur CL 500	Durum wheat, Triticale, Winter barley, Winter wheat
copper oxychloride	Cuprokylt	Potatoes
diflubenzuron	Dimilin Flo	Forest
dimethoate	BASF Dimethoate 40	Spring rye, Spring wheat, Sugar beet, Sugar beet seed crops, Triticale, Winter rye, Winter wheat
	Rogor L40	Spring rye, Spring wheat, Sugar beet, Triticale, Winter rye, Winter wheat
mancozeb	Dithane 945	Potatoes
	Dithane NT	Potatoes

Chemical	Product	Crops
	Dithane NT Dry Flowable	Potatoes
	Dithane Superflo	Potatoes
	Micene 80	Potatoes
	Micene DF	Potatoes
	Penncozeb WDG	Potatoes
	Quell Flo	Potatoes
metaldehyde	Brits	All edible crops, All non-edible crops, Natural surfaces not intended to bear vegetation
	Dixie 6	All edible crops, All non-edible crops, Natural surfaces not intended to bear vegetation
	Doff Horticultural Slug Killer Blue Mini Pellets	All edible crops, All non-edible crops, Cultivated land/soil
	Luxan Deal	All edible crops, All non-edible crops
pirimicarb	Agriguard Pirimicarb	Spring barley, Spring oats, Spring wheat, Winter barley, Winter oats, Winter wheat
	Aphox	Durum wheat, Spring barley, Spring oats, Spring rye, Spring wheat, Triticale, Winter barley, Winter oats, Winter rye, Winter wheat
	Greencrop Glenroe	Durum wheat, Spring barley, Spring oats, Spring rye, Spring wheat, Triticale, Winter barley, Winter oats, Winter rye, Winter wheat
	Phantom	Durum wheat, Spring barley, Spring oats, Spring rye, Spring wheat, Triticale, Winter barley, Winter oats, Winter rye, Winter wheat
tri-allate	Avadex Excel 15G	Combining peas, Fodder beet, Lucerne, Red beet, Red clover, Sainfoin, Spring barley, Spring field beans, Sugar beet, Triticale, Vetches, Vining peas, White clover, Winter barley, Winter field beans, Winter wheat

Resistance Management

Pest species are, by definition, adaptable organisms. The development of resistance to some crop protection chemicals is just one example of this adaptability. Repeated use of products with the same mode of action will clearly favour those individuals in the pest population able to tolerate the treatment. This leads to a situation where the tolerant (or resistant) individuals can dominate the population and the product is ineffective. In general, the more rapidly the pest species reproduces and the more mobile it is, the faster is the emergence of resistant populations, although some weeds seem able to evolve resistance more quickly than would be expected.

The speed of appearance of resistance depends on the mode of action of the crop protection chemicals, as well as the manner in which they are used. Resistance among insects and fungal diseases has been evident for much longer than weed resistance to herbicides, but examples in all three categories are now widespread and increasing.

This has created a need for agreement on the advice given for the use of crop protection chemicals in order to reduce the likelihood of the development of resistance and to avoid the loss of potentially valuable products in the chemical armoury.

In the UK, key independent research organisations, chemical manufacturers and other organisations have collaborated to share knowledge and expertise on resistance issues through three action groups. Participants include ADAS, the Pesticides Safety Directorate, universities, colleges and the Home Grown Cereals Authority. The groups have a common aim of monitoring resistance in UK and devising and publishing management strategies designed to combat it where it occurs. The groups are:

The Weed Resistance Action Group (WRAG), formed in 1989 (Secretary: Dr Stephen Moss, IACR-Rothamsted, Harpenden, Herts AL5 2JQ *Tel: 01582 7631330 ext. 2521*)

The Fungicide Resistance Action Group (FRAG), formed in 1995 (Secretary: Mr Oliver Macdonald, Pesticides Safety Directorate, Mallard House, King's Pool, 3 Peasholme Green, York YO1 2PX *Tel: 01904 455864*)

The Insecticide Resistance Action Group (IRAG), formed in 1997 (Secretary: Dr Ian Denholm, IACR-Rothamsted, Harpenden, Herts AL5 2JQ *Tel: 01582 763133 ext. 2324*)

The above groups publish detailed advice on resistance management relevant for each sector and, in some cases, specific to a pest problem. This information, together with further details about the function of each group, can be obtained from the Pesticides Safety Directorate web site: www.pesticides.gov.uk/rags.asp

The general guidelines for resistance management are similar for all three problem areas.

Preparation in advance

- Be aware of the factors that favour the development of resistance, such as repeated annual use of the same product, and assess the risk.
- Plan ahead and aim to integrate all possible means of control.
- Use cultural measures such as rotations, stubble hygiene, variety selection and, for fungicides, removal of primary inoculum sources, to reduce reliance on chemical control.
- Monitor crops regularly.
- Keep aware of local resistance problems.
- Monitor effectiveness of actions taken and take professional advice, especially in cases of unexplained poor control.

Using crop protection products

- Optimise product efficacy by using it as directed, at the right time, in good conditions.
- Treat pest problems early.
- Mix or alternate chemicals with different modes of action.

- Avoid repeated applications of very low doses.
- Keep accurate field records.

Label guidance depends on the appropriate strategy for the product. Most frequently it consists of a warning of the possibility of poor performance due to resistance, and a restriction on the number of treatments that should be applied in order to minimise the development of resistance. This information is summarised in the profiles in this *Guide*, but detailed guidance must always be obtained by reading the label itself before use.

International action committees

Resistance to crop protection products is an international problem. Agrochemical industry collaboration on a global scale is via three action committees whose aims are to support a coordinated industry approach to the management of resistance worldwide. In particular they produce lists of crop protection chemicals classified according to their mode of action. These lists and other information can be obtained from the respective websites:

Herbicide Resistance Action Committee (HRAC): www.plantprotection.org/hrac

Fungicide Resistance Action Committee (FRAC): www.frac.info

Insecticide Resistance Action Committee (IRAC): www.irac-online.org

Poisons and Poisoning

Chemicals Subject to the Poisons Law

Certain products in this book are subject to the provisions of the Poisons Act 1972, the Poisons List order 1982 and the Poisons Rules 1982 (copies of all these are obtainable from The Stationery Office). These rules include general and specific provisions for the storage, sale and supply of listed non-medicine poisons. Full details can be accessed in Annex C of the Guide to Approved Pesticides on the PSD website (currently www.pesticides.gov.uk/publications.asp?id=499). The nature of the formulation and the concentration of the active ingredient allow some products to be exempted from the rules (see below).

The chemicals approved for use in the UK and included in this book are specified under Parts I and II of the Poisons List as follows.

Part I Poisons

Sale restricted to registered retail pharmacists and to registered non-pharmacy businesses provided sales do not take place on retail premises:

aluminium phosphide
chloropicrin
magnesium phosphide
strychnine

Part II Poisons

Sale restricted to registered retail pharmacists and listed sellers registered with a local authority:

aldicarb
alphachloralose
formaldehyde
nicotine (b)
oxamyl (a)
paraquat (c)
sulphuric acid
zinc phosphide

(a) Granular formulations containing up to 12% w/w of this, or a combination of similarly flagged poisons, are exempt.
(b) Formulations containing not more than 7.5% of nicotine are exempt.
(c) Pellets containing not more than 5% paraquat ion are exempt.

Note:

1. The European Commission Ozone Depleting Substances Regulation (EC Regulation 2037/2000) banned the use of methyl bromide in the UK, with the exception of a few 'Critical Use Exemptions', which will be reviewed annually by the EC. Suppliers of products containing methyl bromide which had previously been listed in this Guide withdrew them in 2004.
2. All approvals for dichlorvos-containing products expired in April 2004.

Occupational Exposure Limits

A fundamental requirement of the COSHH Regulations is that exposure of employees to substances hazardous to health should be prevented or adequately controlled. Exposure by inhalation is usually the main hazard, and in order to measure the adequacy of control of exposure by this route, various substances have been assigned occupational exposure limits.

There are two types of occupational exposure limits defined under COSHH: Occupational Exposure Standards (OES) and Maximum Exposure Limits (MEL). The key difference is that an OES is set at a level at which there is no indication of risk to health; for an MEL a residual risk may exist and the level takes socio-economic factors into account. In practice, MELs have most often been allocated to carcinogens and other substances for which no threshold of effect can be identified, and for which there is no doubt about the seriousness of the effects of exposure.

OESs and MELs are set on the recommendation of the Advisory Committee on Toxic Substances. Full details are published by the Health and Safety Executive (HSE) in *Occupational Exposure Limits* 1995, EH 40/95 (ISBN 0 7176 0876 X).

As far as pesticides are concerned, OESs and MELs have been set for relatively few active ingredients. This is because pesticide products usually contain other substances in their formulation, including solvents, which may have their own OES/MEL. In practice, inhalation of solvent may be at least, or more, important than that of the active ingredient. These factors are taken into account in the approval process of the pesticide product under the Regulations. This indicates one of the reasons why a change of pesticide formulation usually necessitates a new approval assessment.

First Aid Measures

If pesticides are handled in accordance with the required safety precautions, as given on the container label, poisoning should not occur. It is difficult, however, to guard completely against the occasional accidental exposure. Thus if a person handling, or exposed to, pesticides becomes ill, it is a wise precaution to apply first aid measures appropriate to pesticide poisoning even though the cause of illness may eventually prove to have been quite different. An employer has a legal duty to make adequate first aid provision for employees. Regular pesticide users should consider appointing a trained first aider even if numbers of employees are not large, as there is a specific hazard.

The first essential in a case of suspected poisoning is for the person involved to stop work, to be moved away from any area of possible contamination, and for a doctor to be called at once. If no doctor is available the patient should be taken to hospital as quickly as possible. In either event it is most important that the name of the chemical being used should be recorded, and preferably the whole product label or leaflet should be shown to the doctor or hospital concerned.

Some pesticides which are unlikely to cause poisoning in normal use are extremely toxic if swallowed accidentally or deliberately. In such cases get the patient to hospital as quickly as possible, with all the information you have.

General measures

Measures appropriate in all cases of suspected poisoning include:

- remove any protective or other contaminated clothing (taking care to avoid personal contamination)
- wash any contaminated areas carefully with water or with soap and water if available
- in cases of eye contamination, flush with plenty of clean water for at least 15 minutes
- lay the patient down, keep at rest and under shelter; cover with one clean blanket or coat, and avoid overheating
- monitor level of consciousness, breathing and pulse rate
- if consciousness is lost, place the casualty in the recovery position (on his or her side with head down and tongue forward to prevent inhalation of vomit)
- if breathing ceases or weakens, commence mouth-to-mouth resuscitation: ensure that the mouth is clear of obstructions such as false teeth, that the breathing passages are clear, and that tight clothing around the neck, chest and waist has been loosened; if a poisonous chemical has been swallowed, it is essential that the first aider is protected by the use of a resuscitation device (several types are available on the market).

Specific measures

In case of poisoning with particular chemical groups, the following measures may be taken before transfer to hospital.

Organophosphorus and carbamate insecticides

Keep the patient at rest. The patient may suddenly stop breathing, so be ready to give artificial respiration.

Organochlorine compounds

If convulsions occur, do not interfere unless the patient is in danger of injury; if so any restraint must be gentle. When convulsions cease, place the casualty on his or her side with head down, tongue forward (recovery position).

Paraquat, diquat

Irrigate skin and eye splashes copiously with water. If any chemical has been swallowed, take to hospital for tests.

Cyanide (including sodium cyanide)

Send for medical aid. Remove casualty to fresh air, if necessary using breathing apparatus and protective clothing. Remove casualty's contaminated clothing. Gently brush solid particles from the skin, making sure you protect your own skin from contamination. Wash the skin and eyes copiously with water. Transfer casualty to nearest accident and emergency hospital by the quickest possible means together with the first aid cyanide antidote, if held on the premises.

Reporting pesticide poisoning

Any cases of poisoning by pesticides must be reported without delay to an HM Agricultural Inspector of the Health and Safety Executive. In addition, any cases of poisoning by substances named in Schedule 2 of The Reporting of Injuries, Diseases and Dangerous Occurrences Regulations 1985 must also be reported to HM Agricultural Inspectorate (this includes organophosphorus chemicals, mercury and some fumigants).

Cases of pesticide poisoning should also be reported to the manufacturer concerned.

Additional information

General advice on the safe use of pesticides is given in a range of Health and Safety Executive leaflets available from HSE Books (see Appendix 2).

A useful booklet, *Guidelines for Emergency Measures in Cases of Pesticide Poisoning*, is available from CropLife International (see Appendix 2).

The major agrochemical companies are able to provide authoritative medical advice about their own pesticide products. Detailed advice is also available to doctors from the National Poisons Information Service, New Cross Hospital, London SE14 5ER (020 7635 9191) and from regional centres (see Appendix 2).

Environmental Protection

Protection of Bees

Honey bees

Honey bees are a source of income for their owners, and important to farmers and growers as pollinators of their crops. It is irresponsible and unnecessary to use pesticides in such a way that may endanger them. Pesticides vary in their toxicity to bees, but those that present a special hazard are classed as 'harmful', 'dangerous' or 'extremely dangerous' to bees. Products so classified carry a specific warning in the precautions section of the label. These are indicated in this *Guide* in the **personal protective equipment/label precautions** section of the pesticide profile by the numbers E12c, E12d, E12e, E12f, depending on which phrase applies. The phrases are stated in full in Appendix 4.

In July 1996 a new classification was introduced that more accurately reflects the actual risk to honey bees when a product is used. Products assessed by PSD as posing such a risk are classified as 'High risk to bees'. Such products carry a new warning in the precautions section of the label, indicated in this *Guide* by the numbers E12a and E12b.

Product labels indicate the necessary environmental precautions to take, but where use of an insecticide on a flowering crop is contemplated, the British Beekeepers Association (see Appendix 2) has produced the following guidelines for growers:

- target insect pests with the most appropriate product
- choose a product that will cause minimal harm to beneficial species
- follow the manufacturer's instructions carefully
- inspect and monitor crops regularly
- avoid spraying crops in flower or where bees are actively foraging
- keep down flowering weeds
- spray late in the day, in still conditions
- avoid excessive spray volume and run-off
- adjust sprayer pressure to reduce production of fine droplets and drift
- give local beekeepers as much warning of your intention as possible.

Wild bees

Wild bees also play an important role. Bumblebees are useful pollinators of spring-flowering crops and fruit trees because they forage in cool, dull weather when honey bees are inactive. They play a particularly important part in pollinating field beans, red and white clover, lucerne and borage. Bumblebees nest and overwinter in field margins and woodland edges. Avoidance of direct or indirect spray contamination of these areas, in addition to the creation of hedgerows and field margins, and late cutting or grazing of meadows, all helps the survival of these valuable insects.

The Campaign Against Illegal Poisoning of Wildlife

The Campaign Against Illegal Poisoning of Wildlife, aimed at protecting some of Britain's rarest birds of prey and wildlife while also safeguarding domestic animals, was launched in March 1991 by the (then) Ministry of Agriculture, Fisheries and Food and the Department of the Environment, Transport and the Regions. The main objective is to deter those who may be considering using pesticides illegally.

The Campaign is supported by a range of organisations associated with animal welfare, nature preservation, field sports and game keeping including the Royal Society for the Protection of Birds, English Nature, the British Field Sports Society and the Game Conservancy Trust.

The three objectives are to:

- advise farmers, gamekeepers and other land managers on legal ways of controlling pests
- advise the public on how to report illegal poisoning incidents and to respect the need for legal alternatives
- investigate incidents and prosecute offenders.

A freephone number (0800 321 600) is available to make it easier for the public to report incidents, and numerous leaflets, posters, postcards, coasters and stickers have been created to publicise the existence of the Campaign.

The Campaign arose from the results of the Wildlife Incident Investigation Scheme for the investigation of possible cases of illegal poisoning. Under this scheme, all reported incidents are considered and thoroughly investigated where appropriate. Enforcement action is taken wherever sufficient evidence of an offence can be obtained, and numerous prosecutions have been made since the start of the Campaign.

Further information about the Campaign is available from:

Helena Cooke
Pesticides Safety Directorate
Room 317, Mallard House
3 Peasholme Green
York YO1 7PX
e-mail: helena.cooke@psd.defra.gsi.gov.uk
website: www.pesticides.gov.uk/environment.asp?id=504

Water Quality

The Food and Environment Protection Act 1985 (FEPA) places a special obligation on users of pesticides to 'safeguard the environment and in particular avoid the pollution of water'. Under the Water Resources Act 1991 it is an offence to pollute any controlled waters (watercourses or groundwater), either deliberately or accidentally. Protection of controlled waters from pollution is the responsibility of the Environment Agency.

Users of pesticides therefore have a duty to adopt responsible working practices and, unless they are applying herbicides in or near water, to prevent them getting into water. Guidance on how to achieve this is given in the Defra *Code of Good Agricultural Practice for the Protection of Water* (available from The Stationery Office; see Appendix 2).

The duty of care covers not only the way in which a pesticide is sprayed, but also its storage, preparation and disposal of surplus, sprayer washings and the container. Products in this *Guide* that are a major hazard to fish, other aquatic life or aquatic higher plants carry one of several specific label precautions in their profile, depending on the assessed hazard level.

Advice on any water pollution problems is available from the Environment Agency and the Crop Protection Association (see Appendix 2).

Protecting surface waters

Surface waters are particularly vulnerable to contamination. One of the best ways of preventing those pesticides that carry the greatest risk to aquatic wildlife from reaching surface waters is to prohibit their application within a boundary adjacent to the water. Such areas are known as no-spray, or buffer, zones. Certain products in this *Guide* are restricted in this way, and have a legally binding label precaution to make sure the potential exposure of aquatic organisms to pesticides that might harm them is minimised.

Before 1999 the protected zones were measured from the edge of the water. The distances were 2 m for hand-held or knapsack sprayers, 6 m for ground crop sprayers, and a variable distance (but often 18 m) for broadcast air-assisted applications, such as in orchards. The introduction of LERAPs (see below) has changed the method of measuring buffer zones.

'Surface water' includes lakes, ponds, reservoirs, streams, rivers and watercourses (natural or artificial). It also includes temporarily or seasonally dry ditches, which have the potential to carry water at different times of the year.

Buffer zone restrictions do not necessarily apply to all products containing the same active ingredient. Those in formulations that are not likely to contaminate surface water through spray drift do not pose the same risk to aquatic life and are not subject to the restrictions.

Local Environmental Risk Assessments for Pesticides

Local Environment Risk Assessments for Pesticides (LERAPs) were introduced in March 1999, and revised guidelines were issued in January 2002. They give users of most products currently subject to a buffer zone restriction the option of continuing to comply with the existing buffer zone restriction (using the new method of measurement), or carrying out a LERAP and possibly reducing the size of the buffer zone as a result. In either case, there is a new legal obligation for users to record his decision, including the results of the LERAP.

The scheme has changed the method of measuring the buffer zone. Previously the zone was measured from the edge of the water, but it is now the distance from the top of the bank of the watercourse to the edge of the spray area.

The LERAP provides a mechanism for taking into account other factors that may reduce the risk, such as dose reduction, the use of low drift nozzles, and whether the watercourse is dry or flowing. The previous arrangements applied the same restriction regardless of whether there was actually water present. Now there is a standard zone of 1 m from the top of a dry ditch bank.

Other factors to include in a LERAP that may allow a reduction in the buffer zone are:

- size of the watercourse: the wider it is, the greater the dilution factor and the lower the risk of serious pollution
- dose applied: the lower the dose, the less is the risk
- application equipment: sprayers and nozzles are star-rated according to their ability to reduce spray drift fallout - equipment offering the greatest reductions achieves the highest rating of three stars. The scheme was initially restricted to ground crop sprayers; new, more flexible rules introduced in February 2002 included broadcast air-assisted orchard and hop sprayers.

Other changes introduced in 2002 allow the reduction of a buffer zone if there is an appropriate living windbreak between the sprayed area and a watercourse.

Not all products that had a label buffer zone restriction are included in the LERAP scheme. The option to reduce the buffer zone does not apply to organophosphorus or synthetic pyrethroid insecticides. These groups are classified as Category A products. All other products that had a label buffer zone restriction are classified as Category B. The wording of the buffer zone label precautions has been amended for all products to take account of the new method of measurement, and whether or not the particular product qualifies for inclusion in the LERAP scheme.

Products in this *Guide* that are in Category A or B are identified in the **Environmental Safety** section of the profile. Updates to the list are published regularly by PSD and details can be obtained from www.pesticides.gov.uk/psd_databases.asp?id=325

The introduction of LERAPs is an important step forward because it demonstrates a willingness to reduce the impact of regulation on users of pesticides by allowing flexibility where local conditions make it safe to do so. This places a legal responsibility on users to ensure the risk assessment is done either personally, or by the spray operator or a professional consultant or advisor. It is compulsory to record the LERAP and make it available for inspection by enforcement authorities.

Groundwater regulations

New Groundwater Regulations were introduced in 1999 to complete the implementation in the UK of the EU Groundwater Directive (Protection of Groundwater Against Pollution Caused by Certain Dangerous Substances - 80/68/EEC). These Regulations help prevent the pollution of groundwater by controlling discharges or disposal of certain substances, including all pesticides.

Groundwater is defined under the Regulations as any water contained in the ground below the water table. Pesticides must not enter groundwater unless it is deemed by the appropriate Agency to be permanently unsuitable for other uses.

The Regulations make it a criminal offence to dispose of pesticides onto land without official authorisation from the Environment Agency (in England and Wales) or the Scottish Environmental Protection Agency. Normal use of a pesticide in accordance with product approval does not require authorisation. This includes spraying the washings and rinsings back on the crop provided that, in so doing, the maximum approved dose for that product on that crop is not exceeded.

The Agencies will review all authorisations within a 4-year interval. They may also grant authorisations for a limited period. The Agencies can serve notice at any time to modify the conditions of an authorisation where necessary to prevent pollution of groundwater.

In practice, the best advice to farmers and growers is to plan to use all diluted spray within the crop and to dispose of all washings via the same route, making sure they stay within the conditions of approval of the product.

SECTION 6
APPENDICES

Appendix 1
Suppliers of Pesticides and Adjuvants

Ag-Chem: Ag-Chem Direct Limited
P O Box 82
Thirsk
N. Yorks
YO7 4YY
Tel: (0797) 423 5105
Fax: (0797) 113 8857

Agrichem: Agrichem (International) Ltd
Industrial Estate
Station Road
Whittlesey
Cambs.
PE7 2EY
Tel: (01733) 204019
Fax: (01733) 204162
Email: info@agrichem.co.uk
Web: www.agrichem.co.uk

AgriGuard: AgriGuard Ltd
Unit 3
Tally House
Broomfield Business Park
Malahide
Co. Dublin
Ireland
Tel: (+353) 1 846 2044
Fax: (+353) 1 846 2489
Email: info@agriguard.ie
Web: www.agriguard.ie

Amega: Amega Sciences
Lanchester Way
Royal Oak Industrial Estate
Daventry
Northants.
NN11 5PH
Tel: (01327) 704444
Fax: (01327) 71154
Email: admin@amega-sciences.com
Web: www.amega-sciences.com

Antec Biosentry: Antec Biosentry
DuPont Animal Health Solutions
Windham Road
Chilton Industrial Estate
Sudbury
Suffolk
CO10 2XD
Tel: (01787) 377305
Fax: (01787) 310846
Email: biosecurity@antecint.com
Web: www.antecint.com

Aquaspersions: Aquaspersions Ltd
Beacon Hill Road
Halifax
W. Yorks.
HX3 6AQ
Tel: (01422) 386200
Fax: (01422) 386239
Email: info@aquaspersions.co.uk
Web: www.aquaspersions.com

B H & B: Battle Hayward & Bower Ltd
Victoria Chemical Works
Crofton Drive
Allenby Road Industrial Estate
Lincoln
LN3 4NP
Tel: (01522) 529206
Fax: (01522) 538960

Banks Cargill: Banks Cargill Agriculture Ltd
Fleet Road Industrial Estate
Holbeach
Spalding
Lincs.
PE12 8LY
Tel: (01406) 421405
Email: les_sykes@bankscargill.co.uk
Web: www.bankscargill.co.uk

Barclay: Barclay Chemicals Manufacturing Ltd
Damastown Way
Damastown Industrial Park
Mulhuddart
Dublin 15
Ireland
Tel: (+353) 1 822 4555
Fax: (+353) 1 822 4678
Email: info@barclay.ie
Web: www.barclay.ie

Barrier: Barrier BioTech Ltd
36/37 Haverscroft Industrial Estate
New Road
Attleborough
Norfolk
NR17 1YE
Tel: (01953) 456363
Fax: (01953) 455594
Email: sales@barrier-biotech.com
Web: www.barrier-biotech.com

BASF: BASF plc.
Agricultural Divison
PO Box 4, Earl Road
Cheadle Hulme
Cheshire
SK8 6QG
Tel: (0845) 602 2553
Fax: (0161) 485 2229
Web: www.agricentre.co.uk

Bayer CropScience:
Bayer CropScience Limited
Hauxton
Cambridge
CB2 5HU
Tel: (01223) 870312
Fax: (01223) 872142
Web: www.bayercropscience.co.uk

Bayer Environ.: Bayer Environmental Science
Durkan House
214-224 High Street
Waltham Cross
Herts.
EN8 7DP
Tel: (01992) 784270
Fax: (01992) 784276
Email: john.hall1@bayercropscience.com
Web: www.bayer-escience.co.uk

Belchim: Belchim Crop Protection Ltd
Suite 2, Unit 3
Phoenix Park
Eaton Socon
St Neots
Cambs.
PE19 8EP
Tel: (01480) 403333
Fax: (01480) 403444
Email: info@belchim.com
Web: www.belchim.com

Cardel: Cardel Agro SPRL
Avenue de Tervuren 270-272
11, Rue Pascal
B-1150 Brussels
Belgium
Tel: (+32) 2776 7652

Certis: Certis
1b Mills Way
Boscombe Down Business Park
Amesbury
Wilts.
SP4 7RX
Tel: (01980) 676500
Fax: (01980) 626555
Email: certis@certiseurope.co.uk
Web: www.certiseurope.co.uk

Cheminova: Cheminova Agro (UK) Ltd
Norfolk House
Gt. Chesterford Court
Gt. Chesterford
Saffron Walden
Essex
CB10 1PF
Tel: (01799) 530146
Fax: (01799) 530229

Chiltern: Chiltern Farm Chemicals Ltd
East Mellwaters
Stainmore
Bowes
Barnard Castle
Co. Durham
DL12 9RH
Tel: (01833) 628282
Fax: (01833) 628020
Web: www.chilternfarm.com

Ciba Specialty: Ciba Specialty Chemicals
Water Treatments Limited
P O Box 38
Low Moor
Bradford
W. Yorks.
BD12 0JZ
Tel: (01274) 417549
Fax: (01274) 417305
Email: soil.additives@cibasc.com
Web: www.cibasc.com

Coalite: Coalite Chemicals
PO Box 152
Buttermilk Lane
Bolsover
Chesterfield
Derbyshire
S44 6AZ
Tel: (01246) 826816
Fax: (01246) 240309
Email: sales@coalitechemicals.com
Web: www.coalitechemicals.com

Crompton: Crompton Europe Ltd
Kennet House
4 Langley Quay
Slough
Berks.
SL3 6EH
Tel: (01753) 603000
Fax: (01753) 603077

DAPT: DAPT Agrochemicals Limited
14 Monks Walk
Southfleet
Gravesend
Kent
DA13 9NZ
Tel: (01474) 834448
Fax: (01474) 834449
Email: rkjltd@supanet.com

Dax: Dax Products Ltd
18 Marlborough Road
Woodthorpe
Nottingham
NG5 4FG
Tel: (0115) 926 9996
Fax: (0115) 966 1173
Email: info@daxproducts.co.uk
Web: www.daxproducts.co.uk

De Sangosse: De Sangosse UK
PO Box 191
Romsey
Hants
SO51 5YP
Tel: (023) 8064 4567
Fax: (023) 8064 2121
Email: info.desangosse@btconnect.com
Web: www.nomoreslugs.com

Dewco-Lloyd: Dewco-Lloyd Ltd
Cyder House
Ixworth
Suffolk
IP31 2HT
Tel: (01359) 230555
Fax: (01359) 232553

Doff Portland: Doff Portland Ltd
Aerial Way
Hucknall
Nottingham
NG15 6DW
Tel: (0115) 963 2842
Fax: (0115) 963 8657
Email: info@doff.co.uk
Web: www.doff.co.uk

Dow: Dow AgroSciences
Latchmore Court
Brand Street
Hitchin
Herts.
SG5 1NH
Tel: (01462) 457272
Fax: (01462) 426605
Email: fhihotl@dow.com
Web: www.dowagro.com/uk

DuPont: DuPont (UK) Ltd
Agricultural Products Department
Wedgwood Way
Stevenage
Herts.
SG1 4QN
Tel: (01438) 734000
Fax: (01438) 734452

Elliott: Thomas Elliott Ltd
Bencewell Granary
Oakley Road
Bromley Common
Kent
BR2 8HG
Tel: (0208) 462 6622
Fax: (0208) 462 5599

Fargro: Fargro Ltd
Toddington Lane
Littlehampton
Sussex
BN17 7PP
Tel: (01903) 721591
Fax: (01903) 730737
Email: promos@fargro.co.uk
Web: www.fargro.co.uk

FCC: Farmers Crop Chemicals Ltd
P O Box 12379
Alcester
Warwicks.
B49 9AA
Tel: (01789) 774726
Fax: (01789) 774726
Email:
enquiries@farmerscropchemicals.com
Web: www.farmerscropchemicals.com

Fine: Fine Agrochemicals Ltd
Hill End House
Whittington
Worcester
WR5 2RQ
Tel: (01905) 361800
Fax: (01905) 361810
Email: enquire@fine-agrochemicals.com
Web: www.fine-agrochemicals.com

Ford Smith: Ford Smith & Co. Ltd
Lyndean Industrial Estate
Felixstowe Road
Abbey Wood
London
SE2 9SG
Tel: (020) 8310 8127
Fax: (020) 8310 9563
Email: fordsmithltd@aol.com

Gowan: Gowan International
Onen House
Onen
Tal-y-Coed
Monmouth
NP25 5EN
Tel: (01600) 780543
Fax: (01600) 780543

Greenaway: Greenaway Amenity Ltd
7 Browntoft Lane
Donington
Spalding
Lincs
PE11 4TQ
Tel: (01775) 821031
Fax: (01775) 821034
Email: greenawayamenity@aol.com
Web: www.greenawaycda.com

Greencrop: Greencrop Technology Ltd
Burren House
2 Cowbrook Court
Glossop
Derbyshire
SK13 8SL
Tel: (01457) 856001
Fax: (01457) 857137
Email: mail@greencrop-technology.co.uk
Web: www.greencrop-technology.co.uk

Headland: Headland Agrochemicals Ltd
Norfolk House
Gt. Chesterford Court
Gt. Chesterford
Saffron Walden
Essex
CB10 1PF
Tel: (01799) 530146
Fax: (01799) 530229
Email: enquiry@headlandgroup.com
Web: www.headland-ag.co.uk

Headland Amenity:
Headland Amenity Limited
1010 Cambourne Business Park
Cambourne
Cambs.
CB3 6DP
Tel: (01223) 597834
Fax: (01223) 598052
Email: info@headlandamenity.com
Web: www.headlandamenity.com

Helena: Helena Chemical Company
Cambridge House
Nottingham Road
Stapleford
Nottingham
NG9 8AB
Tel: (0115) 939 0202
Fax: (0115) 939 8031

I T Agro: I T Agro Ltd
805 Salisbury House
31 Finsbury Circus
London
EC2M 5SQ
Tel: (+45) 98 93 89 77
Fax: (+45) 98 93 80 01
Email: kiil@inter-trade.dk
Web: www.inter-trade.dk

Interagro: Interagro (UK) Ltd
Sworders Barn
Sworders Yard
North Street
Bishop's Stortford
Herts.
CM23 2LD
Tel: (01279) 501995
Fax: (01279) 501996
Email: info@interagro.co.uk
Web: www.interagro.co.uk

Interfarm: Interfarm (UK) Ltd
Kinghams's Place
36 Newgate Street
Doddington
Cambs.
PE15 0SR
Tel: (01354) 741414
Fax: (01354) 741004
Email: technical@interfarm.co.uk
Web: www.interfarm.co.uk

Intracrop: Intracrop
Byemoor Farm
Melmerby
Leyburn
N. Yorks.
DL8 4TW
Tel: (01969) 640655
Fax: (01969) 640633

Irish Drugs: Irish Drugs Ltd
Burnfoot
Lifford
Co. Donegal
Ireland
Tel: (+353) 74 9368104
Fax: (+353) 74 9368311
Email: idl@eircom.net

JohnsonDiversey: JohnsonDiversey Ltd
Weston Favell Centre
Northampton
NN3 8PD
Tel: (01604) 783505
Fax: (01604) 783506

K & S Fumigation:
K & S Fumigation Services Ltd
Shirley Farmhouse
Moor Lane
Woodchurch
Ashford
Kent
TN26 3SS
Tel: (01233) 758252
Fax: (01233) 758343
Email: k.s.treatments@talk21.com

Killgerm: Killgerm Chemicals Ltd
115 Wakefield Road
Flushdyke
Ossett
W. Yorks.
WF5 9AR
Tel: (01924) 268400
Fax: (01924) 264757
Email: info@killgerm.com
Web: www.killgerm.com

Koppert: Koppert (UK) Ltd
Homefield Road
Haverhill
Suffolk
CB9 8QP
Tel: (01440) 704488
Fax: (01440) 704487
Web: www.koppert.com

Landgold: Landgold & Co. Ltd
PO Box 829
Charles House
Charles Street
St. Helier
Jersey
JE4 0UE
Tel: (01534) 768446
Fax: (01534) 732843

Landseer: Landseer Ltd
Lodge Farm
Goat Hall Lane
Galleywood
Chelmsford
Essex
CM2 8PH
Tel: (01245) 357109
Fax: (01245) 494165

Loveland: Loveland Industries Ltd
Swaffham Bulbeck
Cambridge
CB5 0LU
Tel: (01223) 811215
Fax: (01223) 810020
Email: info@loveland.co.uk
Web: www.loveland.co.uk

Luxan: Luxan (UK) Ltd
Crown Business Park
Old Dalby
Leics.
LE14 3NQ
Tel: (01664) 850052
Fax: (01664) 820216
Email: enquiry@luxan.co.uk
Web: www.luxan.co.uk

Makhteshim: Makhteshim-Agan (UK) Ltd
Unit 16
Thatcham Business Village
Colthrop Way
Thatcham
Berks.
RG19 4LW
Tel: (01635) 860555
Fax: (01635) 861555
Email: admin@mauk.co.uk
Web: www.mauk.co.uk

Mandops: Mandops (UK) Ltd
36 Leigh Road
Eastleigh
Hants.
SO50 9DT
Tel: (023) 8064 1826
Fax: (023) 8062 9106
Email: enquiries@mandops.co.uk
Web: www.mandops.co.uk

Marnoch: Marnoch Ventures Ltd
Orchard House
Craighill Road
Omagh
BT79 7PD
Tel: (02808) 771604

Me2: Me2 Crop Protection Ltd
Tower House
Fishergate
York
YO10 4HA
Tel: (01904) 567331
Fax: (01904) 720094
Email: sales@me2cpl.co.uk
Web: www.me2cpl.co.uk

Microcide: Microcide Ltd
Shepherds Grove
Stanton
Bury St. Edmunds
Suffolk
IP31 2AR
Tel: (01359) 251077
Fax: (01359) 251545
Email: microcide@microcide co.uk
Web: www.microcide.co.uk

Mitchell Cotts: Mitchell Cotts Chemicals Ltd
PO Box 6
Steanard Lane
Mirfield
W. Yorks.
WF14 8QB
Tel: (01924) 493861
Fax: (01924) 490972

Monsanto: Monsanto (UK) Ltd
The Maris Centre
Hauxton Road
Trumpington
Cambridge
CB2 2LQ
Tel: (01223) 849200
Fax: (01223) 849414
Email: technical.helpline.uk@monsanto.com
Web: www.monsanto-ag.co.uk

Nickerson: Nickerson UK Ltd
Rothwell
Market Rasen
Lincs.
LN7 6DT
Tel: (01472) 371471
Fax: (01472) 371602
Email: enquiries@Nickerson.co.uk
Web: www.nickerson.co.uk

Nomix Enviro: Nomix Enviro Limited
Portland Building
Portland Street
Staple Hill
Bristol
BS16 4PS
Tel: (0117) 957 4574
Fax: (0117) 956 3461
Email: info@nomix.co.uk
Web: www.nomix.co.uk

Nufarm UK: Nufarm UK Ltd
Crabtree Manorway North
Belvedere
Kent
DA17 6BQ
Tel: (020) 8319 7222
Fax: (020) 8319 7280
Email: infouk@uk.nufarm.com
Web: www.ag.nufarm.co.uk

Pan Agriculture: Pan Agriculture Ltd
8 Cromwell Mews
Station Road
St Ives
Huntingdon
Cambs.
PE27 5HJ
Tel: (01480) 467790
Fax: (01480) 467041
Email: info@panagriculture.co.uk

Rentokil: Rentokil Initial plc
Felcourt
East Grinstead
W. Sussex
RH19 2JY
Tel: (01342) 833022
Fax: (01342) 326229
Web: www.ri-research.com

Rhizopon: Rhizopon UK Ltd
Croda Rosa
12 Bixley Road
Ipswich
Suffolk
IP3 8PL
Tel: (01473) 712666
Fax: (01473) 712666

Rigby Taylor: Rigby Taylor Ltd
Rigby Taylor House
Crown Lane
Horwich
Bolton
Lancs.
BL6 5HP
Tel: (01204) 677777
Fax: (01204) 677765
Email: info@rigbytaylor.com
Web: www.rigbytaylor.com

Roebuck Eyot: Roebuck Eyot Ltd
7a Hatfield Way
South Church Enterprise Park
Bishop Auckland
Co. Durham
DL14 6XF
Tel: (01388) 772233
Fax: (01388) 775233
Email: sales@roebuck-eyot.co.uk
Web: www.roebuck-eyot.co.uk

Scotts: The Scotts Company (UK) Ltd
Paper Mill Lane
Bramford
Ipswich
Suffolk
IP8 4BZ
Tel: (01473) 830492
Fax: (01473) 830386

Sinclair: William Sinclair Horticulture Ltd
Firth Road
Lincoln
LN6 7AH
Tel: (01522) 537561
Fax: (01522) 513609

Sipcam: Sipcam UK Ltd
3 The Barn
27 Kneesworth Street
Royston
Herts.
SG8 5AB
Tel: (01763) 212100
Fax: (01763) 212101

Sorex: Sorex Ltd
St Michael's Industrial Estate
Widnes
Cheshire
WA8 8TJ
Tel: (0151) 420 7151
Fax: (0151) 495 1163
Email: enquiries@sorex.com
Web: www.sorex.com

Sphere: Sphere Laboratories (London) Ltd
The Yews
Main Street
Chilton
Oxon.
OX11 0RZ
Tel: (01235) 831802
Fax: (01235) 833896

Standon: Standon Chemicals Ltd
48 Grosvenor Square
London
W1K 2HT
Tel: (020) 7493 8648
Fax: (020) 7493 4219

SumiAgro: SumiAgro (UK) Ltd
Merlin House
Falconry Court
Bakers Lane
Epping
Essex
CM16 5DQ
Tel: (01992) 563700
Fax: (01992) 563800
Email: sumiagro@sumiagro.co.uk
Web: www.sumiagro.co.uk

SumiAgro Amenity: SumiAgro Amenity, a division of SumiAgro (UK) Ltd
See SumiAgro (UK) Ltd

Syngenta:
Syngenta Crop Protection UK Limited
Whittlesford
Cambridge
CB2 4QT
Tel: (0800) 169 6058
Fax: (01223) 493700
Web: www.syngenta-crop.co.uk

Syngenta Bioline: Syngenta Bioline
Telstar Nursery
Holland Road
Little Clacton
Essex
CO16 9QG
Tel: (01255) 863200
Fax: (01255) 863206
Email: syngenta.bioline@syngenta.com
Web: www.syngenta-bioline.co.uk

Tomen: Tomen (UK) plc
Tomen House
13 Charles II Street
London
SW1Y 4QT
Tel: (020) 7321 6621
Fax: (020) 7321 6624
Email: andy_meads@ov.tomen.com
Web: www.arvesta.com

Tripart: Tripart Farm Chemicals Ltd
See AgriGuard Ltd

Truchem: Truchem Ltd
The Knoll
The Cross
Horsley
Stroud
Gloucestershire
GL6 0PR
Tel: (01453) 833293
Fax: (01453) 833293
Email: truchem@btopenworld.com

Unicrop: Universal Crop Protection Ltd
Park House
Maidenhead Road
Cookham
Berks.
SL6 9DS
Tel: (01628) 526083
Fax: (01628) 810457
Email: enquiries@unicrop.com

United Phosphorus: United Phosphorus Ltd
Chadwick House
Birchwood Park
Warrington
Cheshire
WA3 6AE
Tel: (01925) 819999
Fax: (01925) 817425
Email: chris@upluk6.demon.co.uk

Vitax: Vitax Ltd
Owen Street
Coalville
Leicester
LE67 3DE
Tel: (01530) 510060
Fax: (01530) 510299
Email: info@vitax.co.uk
Web: www.vitax.co.uk

Whyte Agrochem.: Whyte Agrochemicals Ltd
Denaby Lane Industrial Estate
Denaby Lane
Old Denaby
Doncaster
S. Yorks.
DN12 4LQ
Tel: (01709) 772200
Fax: (01709) 772201

Whyte Agrochemicals:
Whyte Agrochemicals Ltd
Marlborough House
298 Regents Park Road
Finchley
London
N3 2UA
Tel: (020) 8346 5946
Fax: (020) 8349 4589

Appendix 2
Useful Contacts

Agricultural Industries Confederation Ltd (AIC)
Confederation House
East of England Showground
Peterborough PE2 6XE
Tel: (01733) 385236
Fax: (01733) 385270
Web: www.agindustries.org.uk

BASIS Ltd
Bank Chambers
34 St John Street
Ashbourne
Derbyshire DE6 1GH
Tel: (01335) 343945/346138
Fax: (01335) 346488
Web: www.basis-reg.co.uk

British Beekeepers' Association
National Agricultural Centre
Stoneleigh
Kenilworth
Warwickshire CV8 2LZ
Tel: (024) 7669 6679
Fax: (024) 7669 0682

BCPC (British Crop Production Council)
7 Omni Business Centre
Omega Park
Alton
Hampshire GU34 2QD
Tel: (01420) 593200
Fax: (01420) 593209
Web: www.bcpc.org

BCPC Publications Sales
7 Omni Business Centre
Omega Park
Alton
Hampshire GU34 2QD
Tel: (01420) 593200
Fax: (01420) 593209
Web: www.bcpc.org/bookshop
e-mail: publications@bcpc.org

British Pest Control Association
1 Gleneagles House
Vernon Gate
South Street
Derby DE1 1UP
Tel: (01332) 294288
Fax: (01332) 295904
Web: www.bpca.org.uk

Crop Protection Association Ltd
4 Lincoln Court
Lincoln Road
Peterborough
Cambs. PE1 2RP
Tel: (01733) 349225
Fax: (01733) 562523
Web: www.cropprotection.org.uk

CropLife International
(previously the Global Crop Protection Federation)
Avenue Louise 143
B-1050 Brussels
Belgium
Tel: (+32) 2 542 0410
Fax: (+32) 2 542 0419
Web: www.croplife.org

Department of Agriculture and Rural Development (Northern Ireland)
Pesticides Section
Dundonald House
Upper Newtownards Road
Belfast BT4 3SB
Tel: (028) 9052 4704
Fax: (028) 9052 4266

Department of Environment, Food and Rural Affairs (Defra)
Nobel House
17 Smith Square
London SW1P 3JR
Tel: (020) 7238 6000
Fax: (020) 7238 6591
Web: www.defra.gov.uk

Environment Agency
Rio House
Waterside Drive
Aztec West
Almondsbury
Bristol BS12 4UD
Tel: (01454) 624400
Fax: (01454) 624409
Web: www.environment-agency.gov.uk

European Crop Protection Association (ECPA)
Avenue E van Nieuwenhuyse 6
B-1160 Brussels
Belgium
Tel: (+32) 2 663 1550
Fax: (+32) 2 663 1560
Web: www.ecpa.be

Farmers' Union of Wales
Llys Amaeth
Queen's Square
Aberystwyth
Dyfed SY23 2EA
Tel: (01970) 612755
Fax: (01970) 624369

Forestry Commission
231 Corstorphine Road
Edinburgh EH12 7AT
Tel: (0131) 334 0303
Fax: (0131) 334 3047
Web: www.forestry.gov.uk

Health and Safety Executive
Information Services
Room 318, Daniel House
Stanley Precinct
Bootle
Merseyside L20 3TW
Tel: (0151) 951 3191
Fax: (0151) 951 3467

Health and Safety Executive
Biocides & Pesticides Assessment Unit
Room 123, Magdalen House
Bootle
Merseyside L20 3QZ
Tel: (0151) 951 3535
Fax: (0151) 951 3317

Health and Safety Executive – Books
PO Box 1999
Sudbury
Suffolk CO10 2WA
Tel: (01787) 881165
Fax: (01787) 313995

Lantra
Lantra House
Stoneleigh Park
Nr Coventry
Warwickshire CV8 2LG
Tel: (024) 7669 6996
Fax: (024) 7669 6732
Web: www.lantra.co.uk

National Association of Agricultural Contractors (NAAC)
Samuelson House
Paxton Road
Orton Centre
Peterborough
Cambs. PE2 5LT
Tel: (01733) 362920
Fax: (01733) 362921

National Farmers' Union
Agriculture House
164 Shaftesbury Avenue
London WC2H 8HL
Tel: (020) 7331 7200
Fax: (020) 7331 7313
Web: www.nfu.org.uk

National Poisons Information Service
Guys' and St Thomas' Hospital Trust
London SE14 5ER
Tel: (020) 7635 9191

The Royal Hospitals
Belfast BT12 6BA
Tel: (028) 9024 0503

City Hospital NHS Trust
Birmingham B18 7QH
Tel: (0121) 507 5588/5589

Llandough Hospital
Penarth CF64 2XX
Tel: (029) 2070 9901

Royal Infirmary
Edinburgh
Tel: (0131) 536 2300

The General Infirmary
Leeds LS1 3EX
Tel: (0113) 243 0715

Royal Victoria Infirmary
Newcastle NE1 4LP
Tel: (0191) 232 5131

National Turfgrass Council
Hunter's Lodge
Dr Brown's Road
Minchinhampton
Gloucestershire GL6 9BT
Tel: (01453) 883588
Fax: (01453) 731449

NPTC
National Agricultural Centre
Stoneleigh
Kenilworth
Warwickshire CV8 2LG
Tel: (024) 7669 6553
Fax: (024) 7669 6128
Web: www.nptc.org.uk

Pesticides Safety Directorate
Mallard House
King's Pool
3 Peasholme Green
York YO1 2PX
Tel: (01904) 640500
Fax: (01904) 455733
Web: www.pesticides.gov.uk

Processors and Growers Research Organisation
The Research Station
Great North Road
Thornhaugh
Peterborough
Cambs. PE8 6HJ
Tel: (01780) 782585
Fax: (01780) 783993
Web: www.pgro.co.uk

Scottish Beekeepers' Association
North Trinity House
114 Trinity Road
Edinburgh EH5 3JZ
Tel: (0131) 552 5341

Scottish Environment Protection Agency (SEPA)
Erskine Court
The Castle Business Park
Stirling FK9 4TR
Tel: (01786) 457 700
Fax: (01786) 446 885
Web: www.sepa.org.uk

TSO (The Stationery Office)
Publications Centre
PO Box 276
London SW8 5DT
Tel: (020) 7873 9090 (orders)/
(020) 7873 0011 (enquiries)
Fax: (020) 7873 8200
Web: www.thestationeryoffice.com

Ulster Beekeepers' Association
57 Liberty Road
Carrickfergus
Co. Antrim BT38 9DJ
Tel: (01960) 362998

Welsh Beekeepers' Association
Trem y Clawdd
Fron Isaf
Chirk
Wrexham
Clwyd LL14 5AH
Tel/Fax: (01691) 773300

Appendix 3
Keys to Crop and Weed Growth Stages

Decimal Code for the Growth Stages of Cereals

Illustrations of these growth stages can be found in the reference indicated below and in some company product manuals.

0 Germination

- 00 Dryseed
- 03 Imbibition complete
- 05 Radicle emerged from caryopsis
- 07 Coleoptile emerged from caryopsis
- 09 Leaf at coleoptile tip

1 Seedling growth

- 10 First leaf through coleoptile
- 11 First leaf unfolded
- 12 2 leaves unfolded
- 13 3 leaves unfolded
- 14 4 leaves unfolded
- 15 5 leaves unfolded
- 16 6 leaves unfolded
- 17 7 leaves unfolded
- 18 8 leaves unfolded
- 19 9 or more leaves unfolded

2 Tillering

- 20 Main shoot only
- 21 Main shoot and 1 tiller
- 22 Main shoot and 2 tillers
- 23 Main shoot and 3 tillers
- 24 Main shoot and 4 tillers
- 25 Main shoot and 5 tillers
- 26 Main shoot and 6 tillers
- 27 Main shoot and 7 tillers
- 28 Main shoot and 8 tillers
- 29 Main shoot and 9 or more tillers

3 Stem elongation

- 30 ear at 1 cm
- 31 1st node detectable
- 32 2nd node detectable
- 33 3rd node detectable
- 34 4th node detectable
- 35 5th node detectable
- 36 6th node detectable
- 37 Flag leaf just visible
- 39 Flag leaf ligule/collar just visible

4 Booting

- 41 Flag leaf sheath extending
- 43 Boots just visibly swollen
- 45 Boots swollen
- 47 Flag leaf sheath opening
- 49 First awns visible

5 Inflorescence

- 51 First spikelet of inflorescence just visible
- 52 1/4 of inflorescence emerged
- 55 1/2 of inflorescence emerged
- 57 3/4 of inflorescence emerged
- 59 Emergence of inflorescence completed

6 Anthesis

- 60
- 61 } Beginning of anthesis
- 64
- 65 } Anthesis half way
- 68
- 69 } Anthesis complete

7 Milk development

- 71 Caryopsis watery ripe
- 73 Early milk
- 75 Medium milk
- 77 Late milk

8 Dough development

- 83 Early dough
- 85 Soft dough
- 87 Hard dough

9 Ripening

- 91 Caryopsis hard (difficult to divide by thumb-nail)
- 92 Caryopsis hard (can no longer be dented by thumb-nail)
- 93 Caryopsis loosening in daytime

(From Tottman, 1987. *Annals of Applied Biology*, **110**, 441–454)

Stages in Development of Oilseed Rape

Illustrations of these growth stages can be found in the reference indicated below and in some company product manuals.

0 Germination and emergence

1 Leaf production

1,0 Both cotyledons unfolded and green
1,1 First true leaf
1,2 Second true leaf
1,3 Third true leaf
1,4 Fourth true leaf
1,5 Fifth true leaf
1,10 About tenth true leaf
1,15 About fifteenth true leaf

2 Stem extension

2,0 No internodes ('rosette')
2,5 About five internodes

3 Flower bud development

3,0 Only leaf buds present
3,1 Flower buds present but enclosed by leaves
3,3 Flower buds visible from above ('green bud')
3,5 Flower buds raised above leaves
3,6 First flower stalks extending
3,7 First flower buds yellow ('yellow bud')

4 Flowering

4,0 First flower opened
4,1 10% all buds opened
4,3 30% all buds opened
4,5 50% all buds opened

5 Pod development

5,3 30% potential pods
5,5 50% potential pods
5,7 70% potential pods
5,9 All potential pods

6 Seed development

6,1 Seeds expanding
6,2 Most seeds translucent but full size
6,3 Most seeds green
6,4 Most seeds green-brown mottled
6,5 Most seeds brown
6,6 Most seeds dark brown
6,7 Most seeds black but soft
6,8 Most seeds black and hard
6,9 All seeds black and hard

7 Leaf senescence

8 Stem senescence

8,1 Most stem green
8,5 Half stem green
8,9 Little stem green

9 Pod senescence

9,1 Most pods green
9,5 Half pods green
9.9 Few pods green

(From Sylvester-Bradley, 1985. *Aspects of Applied Biology*, **10**, 395–400)

Stages in Development of Peas

Illustrations of these growth stages can be found in the reference indicated below and in some company product manuals.

0 Germination and emergence

- 000 Dry seed
- 001 Imbibed seed
- 002 Radicle apparent
- 003 Plumule and radicle apparent
- 004 Emergence

1 Vegetative stage

- 101 First node (leaf with one pair leaflets, no tendril)
- 102 Second node (leaf with one pair leaflets, simple tendril)
- 103 Third node (leaf with one pair leaflets, complex tendril)
- •
- •
- l0x X nodes (leaf with more than one pair leaflets, complex tendril)
- •
- •
- 10n Last recorded node

2 Reproductive stage (main stem)

- 201 Enclosed buds
- 202 Visible buds
- 203 First open flower
- 204 Pod set (small immature pod)
- 205 Flat pod
- 206 Pod swell (seeds small, immature)
- 207 Podfill
- 208 Pod green, wrinkled
- 209 Pod yellow, wrinkled (seeds rubbery)
- 210 Dry seed

3 Senescence stage

- 301 Desiccant application stage. Lower pods dry and brown, middle yellow, upper green. Overall moisture content of seed less than 45%
- 302 Pre-harvest stage. Lower and middle pods dry and brown, upper yellow. Overall moisture content of seed less than 30%
- 303 Dry harvest stage. All pods dry and brown, seed dry

(From Knott, 1987. *Annals of Applied Biology*, **111**, 233–244)

Stages in Development of Faba Beans

Illustrations of these growth stages can be found in the reference indicated below and in some company product manuals.

0 Germination and emergence

000 Dry seed
001 Imbibed seed
002 Radicle apparent
003 Plumule and radicle apparent
004 Emergence
005 First leaf unfolding
006 First leaf unfolded

1 Vegetative stage

101 First node
102 Second node
103 Third node
•
•
I0x X nodes
•
•
10n N, last reoorded node

2 Reproductive stage (main stem)

201 Flower buds visible
203 First open flowers
204 First pod set
205 Pods fully formed, green
207 Pod fill, pods green
209 Seed rubbery, pods pliable, turning black
210 Seed dry and hard, pods dry and black

3 Pod senescence

301 10% pods dry and black
•
•
305 50% pods dry and black
•
•
308 80% pods dry and black, some upper pods green
309 90% pods dry and black, most seed dry. Desiccation stage.
310 All pods dry and black, seed hard. Pre-harvest (glyphosate application stage)

4 Stem senescence

401 10% stem brown/black
•
•
405 50% stem brown/black
•
•
409 90% stem brown/black
410 All stems brown/black. All pods dry and black, seed hard.

(From Knott, 1990. *Annals of Applied Biology*, **116**, 391–404)

Stages in Development of Potato

Illustrations of these growth stages can be found in the reference indicated below and in some company product manuals.

0 Seed germination and seedling emergence

000 Dry seed
001 Imbibed seed
002 Radicle apparent
003 Elongation of hypocotyl
004 Seedling emergence
005 Cotyledons unfolded

1 Tuber dormancy

100 Innate dormancy (no sprout development under favourable conditions)
150 Enforced dormancy (sprout development inhibited by environmental conditions)

2 Tuber sprouting

200 Dormancy break, sprout development visible
21x Sprout with 1 node
22x Sprout with 2 nodes
•
•
29x Sprout with 9 nodes
21x(2) Second generation sprout with 1 node
22x(2) Second generation sprout with 2 nodes
•
•
29x(2) Second generation sprout with 9 nodes

Where x = 1, sprout <2 mm; 2, 2-5 mm; 3, 5-20 mm; 4, 20-30 mm; 5, 50-100 mm; 6, 100-150 mm long

3 Emergence and shoot expansion

300 Main stem emergence
301 Node 1
302 Node 2
•
•
319 Node 19
Second order branch
321 Node 1
•
•
Nth order branch
3N1 Node 1
•
•
3N9 Node 9

4 Flowering

Primary flower
400 No flowers
410 Appearance of flower bud
420 Flower unopen
430 Flower open
440 Flower closed
450 Berry swelling
460 Mature berry
Second order flowers
410(2) Appearance of flower bud
420(2) Flower unopen
430(2) Flower open
440(2) Flower closed
450(2) Berry swelling
460(2) Mature berry

5 Tuber development

500 No stolons
510 Stolon initials
520 Stolon elongation
530 Tuber initiation
540 Tuber bulking (>10 mm diam)
550 Skin set
560 Stolon development

6 Senescence

600 Onset of yellowing
650 Half leaves yellow
670 Yellowing of stems
690 Completely dead

(From Jefferies & Lawson, 1991. *Annals of Applied Biology*, **119**, 387–389)

Stages in Development of Linseed

Illustrations of these growth stages can be found in the reference indicated below and in some company product manuals.

0 Germination and emergence

- 00 Dry seed
- 01 Imbibed seed
- 02 Radicle apparent
- 04 Hypocotyl extending
- 05 Emergence
- 07 Cotyledon unfolding from seed case
- 09 Cotyledons unfolded and fully expanded

1 Vegetative stage (of main stem)

- 10 True leaves visible
- 12 First pair of true leaves fully expanded
- 13 Third pair of true leaves fully expanded
- 1*n* *n* leaf fully expanded

2 Basal branching

- 21 One branch
- 22 Two branches
- 23 Three branches
- 2*n* *n* branches

3 Flower bud development (on main stem)

- 31 Enclosed bud visible in leaf axils
- 33 Bud extending from axil
- 35 Corymb formed
- 37 Buds enclosed but petals visible
- 39 First flower open

4 Flowering (whole plant)

- 41 10% of flowers open
- 43 30% of flowers open
- 45 50% of flowers open
- 49 End of flowering

5 Capsule formation (whole plant)

- 51 10% of capsules formed
- 53 30% 0f capsules formed
- 55 50% of capsules formed
- 59 End of capsule formation

6 Capsule senescence (on most advanced plant)

- 61 Capsules expanding
- 63 Capsules green and full size
- 65 Capsules turning yellow
- 67 Capsules all yellow brown but soft
- 69 Capsules brown, dry and senesced

7 Stem senescence (whole plant)

- 71 Stems mostly green below panicle
- 73 Most stems 30% brown
- 75 Most stems 50% brown
- 77 Stems 75% brown
- 79 Stems completely brown

8 Stems rotting (retting)

- 81 Outer tissue rotting
- 85 Vascular tissue easily removed
- 89 Stems completely collapsed

9 Seed development (whole plant)

- 91 Seeds expanding
- 92 Seeds white but full size
- 93 Most seeds turning ivory yellow
- 94 Most seeds turning brown
- 95 All seeds brown and hard
- 98 Some seeds shed from capsule
- 99 Most seeds shed from capsule

(From Freer, 1991. *Aspects of Applied Biology*, **28**, 33–40)

SECTION 6

Stages in Development of Annual Grass Weeds

Illustrations of these growth stages can be found in the reference indicated below and in some company product manuals.

0 Germination and emergence

00 Dry seed
01 Start of imbibition
03 Imbibition complete
05 Radicle emerged from caryopsis
07 Coleoptile emerged from caryopsis
09 Leaf just at coleoptile tip

1 Seedling growth

10 First leaf through coleoptile
11 First leaf unfolded
12 2 leaves unfolded
13 3 leaves unfolded
14 4 leaves unfolded
15 5 leaves unfolded
16 6 leaves unfolded
17 7 leaves unfolded
18 8 leaves unfolded
19 9 or more leaves unfolded

2 Tillering

20 Main shoot only
21 Main shoot and 1 tiller
22 Main shoot and 2 tillers
23 Main shoot and 3 tillers
24 Main shoot and 4 tillers
25 Main shoot and 5 tillers
26 Main shoot and 6 tillers
27 Main shoot and 7 tillers
28 Main shoot and 8 tillers
29 Main shoot and 9 or more tillers

3 Stem elongation

31 First node detectable
32 2nd node detectable
33 3rd node detectable
34 4th node detectable
35 5th node detectable
36 6th node detectable
37 Flag leaf just visible
39 Flag leaf ligule just visible

4 Booting

41 Flag leaf sheath extending
43 Boots just visibly swollen
45 Boots swollen
47 Flag leaf sheath opening
49 First awns visible

5 Inflorescence emergence

51 First spikelet of inflorescence just visible
53 1/4 of inflorescence emerged
55 1/2 of inflorescence emerged
57 3/4 of inflorescence emerged
59 Emergence of inflorescence completed

6 Anthesis

61 Beginning of anthesis
65 Anthesis half-way
69 Anthesis complete

(From Lawson & Read, 1992. *Annals of Applied Biology*, **12**, 211–214)

Growth Stages of Annual Broad-leaved Weeds

Preferred Descriptive Phrases

Illustrations of these growth stages can be found in the reference indicated below and in some company product manuals.

- Pre-emergence
- Early cotyledons
- Expanded cotyledons
- One expanded true leaf
- Two expanded true leaves
- Four expanded true leaves
- Six expanded true leaves
- Plants up to 25 mm across/high
- Plants up to 50 mm across/high
- Plants up to 100 mm across/high
- Plants up to 150 mm across/high
- Plants up to 250 mm across/high
- Flower buds visible
- Plant flowering
- Plant senescent

(From Lutman & Tucker, 1987. *Annals of Applied Biology*, **110**, 683–687)

Appendix 4
Key to Hazard Classifications and Safety Precautions

Every product label contains information to warn users of the risks from using the product, together with precautions that must be followed in order to minimise the risks. A hazard classification (if any) and symbol is shown with associated risk phrases, followed by a series of safety precautions. These are represented in the pesticide profiles in Section 2 by code letters and numbers under the heading **Hazard classification and safety precautions**.

The codes are defined below, under the same sub-headings as they appear in the pesticide profiles.

Where a product label specifies the use of personal protective equipment (PPE), the requirements are listed under the sub-heading **Operator protection**, using letter codes to denote the protective items, according to the list below. Often PPE requirements are different for specified operations, e.g. handling the concentrate, cleaning equipment etc., but it is not possible to list them separately. The lists of PPE are therefore an indication of what the user may need to have available to use the product in different ways. **When making a COSHH assessment it is therefore essential that the product label is consulted for information on the particular use that is being assessed**.

Where the generalised wording includes a phrase such as '... for *xx* days', the specific requirement for each pesticide is shown in brackets after the code.

Hazard

H01 Very toxic
H02 Toxic
H03 Harmful
H04 Irritant
H05 Corrosive
H06 Extremely flammable
H07 Highly flammable
H08 Flammable
H09 Oxidising agent
H10 Explosive
H11 Dangerous for the environment

Risk phrases

R08 Contact with combustible material may cause fire
R09 Explosive when mixed with combustible material
R16 Explosive when mixed with oxidising substances
R20 Harmful by inhalation
R21 Harmful in contact with skin
R22a Harmful if swallowed
R22b May cause lung damage if swallowed
R23 Toxic by inhalation
R24 Toxic in contact with skin
R25 Toxic if swallowed
R26 Very toxic by inhalation
R27 Very toxic in contact with skin
R28 Very toxic if swallowed
R34 Causes burns
R35 Causes severe burns
R36 Irritating to eyes
R37 Irritating to respiratory system

R38	Irritating to skin
R39	Danger of very serious irreversible effects
R40	Limited evidence of a carcinogenic effect
R41	Risk of serious damage to eyes
R42	May cause sensitization by inhalation
R43	May cause sensitization by skin contact
R45	May cause cancer
R46	May cause heritable genetic damage
R48	Danger of serious damage to health by prolonged exposure
R50	Very toxic to aquatic organisms
R51	Toxic to aquatic organisms
R52	Harmful to aquatic organisms
R53a	May cause long-term adverse effects in the aquatic environment
R53b	Dangerous to aquatic organisms
R54	Toxic to flora
R55	Toxic to fauna
R56	Toxic to soil organisms
R57	Toxic to bees
R58	May cause long-term adverse effects in the environment
R60	May impair fertility
R61	May cause harm to the unborn child
R62	Possible risk of impaired fertility
R63	Possible risk of harm to the unborn child
R64	May cause harm to breast-fed babies
R66	Repeated exposure may cause skin dryness or cracking
R67	Vapours may cause drowsiness and dizziness
R68	Possible risk of irreversible effects

Operator protection

U01	To be used only by operators instructed or trained in the use of chemical/product/type of produce and familiar with the precautionary measures to be observed
U02a	Wash all protective clothing thoroughly after use, especially the inside of gloves
U02b	Avoid excessive contamination of coveralls and launder regularly
U02c	Remove and wash contaminated gloves immediately
U03	Wash splashes off gloves immediately
U04a	Take off immediately all contaminated clothing
U04b	Take off immediately all contaminated clothing and wash underlying skin. Wash clothes before re-use
U04c	Wash clothes before re-use
U05a	When using do not eat, drink or smoke
U05b	When using do not eat, drink, smoke or use naked lights
U06	Handle with care and mix only in a closed container
U07	Open the container only as directed
U08	Wash concentrate/dust from skin or eyes immediately
U09a	Wash any contamination/splashes/dust/powder/concentrate from skin or eyes immediately
U09b	Wash any contamination/splashes/dust/powder/concentrate from eyes immediately
U10	After contact with skin or eyes wash immediately with plenty of water
U11	In case of contact with eyes rinse immediately with plenty of water and seek medical advice
U12	In case of contact with skin rinse immediately with plenty of water and seek medical advice
U13	Avoid contact by mouth
U14	Avoid contact with skin
U15	Avoid contact with eyes
U16	Ensure adequate ventilation in confined spaces
U18	Extinguish all naked flames, including pilot lights, when applying the fumigant/dust/liquid/product
U19a	Do not breathe dust/fog/fumes/gas/smoke/spray mist/vapour. Avoid working in spray mist
U19b	Do not work in confined spaces or enter spaces in which high concentrations of vapour are present. Where this precaution cannot be observed distance breathing or self-

contained breathing apparatus must be worn, and the work should be done by trained operators

U20a Wash hands and exposed skin before eating, drinking or smoking and after work
U20b Wash hands and exposed skin before meals and after work
U20c Wash hands before meals and after work
U21 Before entering treated crops, cover exposed skin areas, particularly arms and legs
U22a Do not touch sachet with wet hands or gloves/Do not touch water soluble bag directly
U22b Protect sachets from rain or water
U23a Do not apply by knapsack sprayer/hand-held equipment
U23b Do not apply through hand held rotary atomisers
U24 Do not handle grain unnecessarily

Environmental protection

E02a Keep unprotected persons/animals out of treated/fumigation areas for at least xx hours/days
E02b Prevent access by livestock, pets and other non-target mammals and birds to buildings under fumigation and ventilation
E03 Label treated seed with the appropriate precautions, using the printed sacks, labels or bag tags supplied
E05a Do not apply directly to livestock/poultry
E05b Keep poultry out of treated areas for at least xx days/weeks
E06a Keep livestock out of treated areas for at least x days/weeks after treatment
E06b Dangerous to livestock. Keep all livestock out of treated areas/away from treated water for at least xx days/weeks. Bury or remove spillages
E06c Harmful to livestock. Keep all livestock out of treated areas/away from treated water for at least xx days/weeks. Bury or remove spillages
E07a Keep livestock out of treated areas for at least two weeks following treatment and until poisonous weeds, such as ragwort, have died down and become unpalatable
E07b Dangerous to livestock. Keep livestock out of treated areas/away from treated water for at least xx weeks and until foliage of any poisonous weeds, such as ragwort, has died and become unpalatable
E07c Harmful to livestock. Keep livestock out of treated areas/away from treated water for at least xx days/weeks and until foliage of any poisonous weeds such as ragwort has died and become unpalatable
E08 Do not feed treated straw or haulm to livestock within x days of spraying
E09 Do not use on crops if the straw is to be used as animal feed/bedding
E10a Dangerous to game, wild birds and animals
E10b Harmful to game, wild birds and animals
E11 Paraquat can be harmful to hares; spray stubbles early in the day
E12a High risk to bees. Do not apply to crops in flower or to those in which bees are actively foraging. Do not apply when flowering weeds are present
E12b High risk to bees. Do not apply to crops in flower, or to those in which bees are actively foraging, except as directed on [crop]. Do not apply when flowering weeds are present
E12c Extremely dangerous to bees. Do not apply to crops in flower or to those in which bees are actively foraging. Do not apply when flowering weeds are present
E12d Dangerous to bees. Do not apply to crops in flower or to those in which bees are actively foraging. Do not apply when flowering weeds are present
E12e Dangerous to bees. Do not apply to crops in flower, or to those in which bees are actively foraging, except as directed on [crop]. Do not apply when flowering weeds are present
E12f Harmful to bees. Do not apply to crops in flower or to those in which bees are actively foraging. Do not apply when flowering weeds are present
E12g Apply away from bees
E13a Extremely dangerous to fish or other aquatic life. Do not contaminate surface waters or ditches with chemical or used container
E13b Dangerous to fish or other aquatic life. Do not contaminate surface waters or ditches with chemical or used container
E13c Harmful to fish or other aquatic life. Do not contaminate surface waters or ditches with chemical or used container
E13d Apply away from fish
E14a Extremely dangerous to aquatic higher plants. Do not contaminate surface waters or ditches with chemical or used container

E14b Dangerous to aquatic higher plants. Do not contaminate surface waters or ditches with chemical or used container

E15a Do not contaminate surface waters or ditches with chemical or used container

E15b Do not contaminate water with product or its container. Do not clean application equipment near surface water. Avoid contamination via drains from farmyards or roads

E16a Do not allow direct spray from horizontal boom sprayers to fall within 5 m of the top of the bank of a static or flowing waterbody, unless a Local Environment Risk Assessment for Pesticides (LERAP) permits a narrower buffer zone, or within 1 m of the top of a ditch which is dry at the time of application. Aim spray away from water

E16b Do not allow direct spray from hand-held sprayers to fall within 1 m of the top of the bank of a static or flowing waterbody. Aim spray away from water

E16c Do not allow direct spray from horizontal boom sprayers to fall within 5 m of the top of the bank of a static or flowing waterbody, or within 1m of the top of a ditch which is dry at the time of application. Aim spray away from water. This product is not eligible for buffer zone reduction under the LERAP horizontal boom sprayers scheme.

E16d Do not allow direct spray from hand-held sprayers to fall within 1 m of the top of the bank of a static or flowing waterbody. Aim spray away from water. This product is not eligible for buffer zone reduction under the LERAP horizontal boom sprayers scheme scheme.

E16e Do not allow direct spray from horizontal boom sprayers to fall within 5 m of the top of the bank of a static or flowing water body or within 1 m from the top of any ditch which is dry at the time of application. Spray from hand held sprayers must not in any case be allowed to fall within 1 m of the top of the bank of a static or flowing water body. Always direct spray away from water. The LERAP scheme does not extend to adjuvants. This product is therefore not eligible for a reduced buffer zone under the LERAP scheme.

E16f Do not allow direct spray/granule applications from vehicle mounted/drawn hydraulic sprayers/applicators to fall within 6 m of surface waters or ditches/Do not allow direct spray/granule applications from hand-held sprayers/applicators to fall within 2 m of surface waters or ditches. Direct spray/applications away from water

E17a Do not allow direct spray from broadcast air-assisted sprayers to fall within xx m of surface waters or ditches. Direct spray away from water

E17b Do not allow direct spray from broadcast air-assisted sprayers to fall within xx m of the bank of a static or flowing waterbody, unless a Local Environmental Risk Assessment for Pesticides (LERAP) permits a narrower buffer zone, or within 5 m of the top of a ditch which is dry at the time of application. Aim spray away from water

E18 Do not spray from the air within 250 m horizontal distance of surface waters or ditches

E19a Do not dump surplus herbicide in water or ditch bottoms

E19b Do not empty into drains

E20 Prevent any surface run-off from entering storm drains

E21 Do not use treated water for irrigation purposes within xx days/weeks of treatment

E22a High risk to non-target insects or other arthropods. Do not spray within 6 m of the field boundary

E22b Risk to certain non-target insects or other arthropods. For advice on risk management and use in Integrated Pest Management (IPM) see directions for use

E22c Risk to non-target insects or other arthropods

E23 Avoid damage by drift onto susceptible crops or water courses

E31a Wash out container thoroughly and dispose of safely

E31b Wash out container thoroughly, empty washings into spray tank and dispose of safely

E31c Rinse container thoroughly by using an integrated pressure rinsing device or manually rinsing three times. Add washings to sprayer at time of filling and dispose of container safely

E32a Empty container completely and dispose of safely

E32b Empty container completely and dispose of it in the specified manner

E32c Treat used container as if it contained pesticide

E32d This material (and its container) must be disposed of in a safe way

E32e This material and its container must be disposed of as hazardous waste

E33 Return empty container as instructed by supplier

E34 Do not re-use container for any purpose/Do not re-use container for any other purpose

E35 Do not burn this container

E36 Do not rinse out the container

E37 Do not use with any pesticide which is to be applied in or near water

E38 Use appropriate containment to avoid environmental contamination

Consumer protection

C01 Do not use on food crops
C02 Do not harvest for human or animal consumption for at least xx days/weeks after last application
C04 Do not apply to surfaces on which food/feed is stored, prepared or eaten
C05 Remove/cover all foodstuffs before application
C06 Remove exposed milk before application
C07 Collect eggs before application
C08 Protect food preparing equipment and eating utensils from contamination during application
C09 Cover water storage tanks before application
C10 Protect exposed water/feed/milk machinery/milk containers from contamination
C11 Remove all pets/livestock/fish tanks before treatment/spraying
C12 Ventilate treated areas thoroughly when smoke has cleared/Ventilate treated rooms thoroughly before occupying

Storage and disposal

E01 Keep out of reach of children
E04 Keep away from food, drink and animal feeding-stuffs
E24 Store away from seeds, fertilizers, fungicides and insecticides
E25 Store well away from corms, bulbs, tubers and seeds
E26 Protect from frost
E27 Store away from heat
E28 Do not store near heat or open flame
E29a Store under cool, dry conditions
E29b Store in a safe, dry, frost-free place designated as an agrochemical store
E30a Keep in original container, tightly closed, in a safe place
E30b Keep in original container, tightly closed, in a safe place, under lock and key

Treated Seed

S01 Do not handle treated seed unnecessarily
S02 Do not use treated seed as food or feed
S03 Keep treated seed secure from people, domestic stock/pets and wildlife at all times during storage and use
S04a Bury or remove spillages
S04b Harmful to birds/game and wildlife. Treated seed should not be left on the soil surface. Bury or remove spillages
S04c Dangerous to birds/game and wildlife. Treated seed should not be left on the soil surface. Bury or remove spillages
S05 Do not reuse sacks or containers that have been used for treated seed for food or feed
S06a Wash hands and exposed skin before meals and after work
S06b Wash hands and exposed skin after cleaning and re-calibrating equipment
S07 Do not apply treated seed from the air
S08 Treated seed should not be broadcast

Vertebrate/Rodent control products

V01a Prevent access to baits/powder by children, birds and other animals, particularly cats, dogs, pigs and poultry
V01b Prevent access to bait/gel/dust by children, birds and non-target animals, particularly dogs, cats, pigs, poultry
V02 Do not prepare/use/lay baits/dust/spray where food/feed/water could become contaminated
V03a Remove all remains of bait, tracking powder or bait containers after use and burn or bury
V03b Remove all remains of bait and bait containers/exposed dust/after treatment (except where used in sewers) and dispose of safely (e.g. burn/bury). Do not dispose of in refuse sacks or on open rubbish tips.
V04a Search for and burn or bury all rodent bodies. Do not place in refuse bins or on rubbish tips
V04b Search for rodent bodies (except where used in sewers) and dispose of safely (e.g. burn/bury). Do not dispose of in refuse sacks or on open rubbish tips

V04c Dispose of safely any rodent bodies and remains of bait and bait containers that are recovered after treatment (e.g. burn/bury). Do not dispose of in refuse sacks or on open rubbish tips

V05 Use bait containers clearly marked POISON at all surface baiting points

Medical advice

M01 This product contains an anticholinesterase organophosphorus compound. DO NOT USE if under medical advice NOT to work with such compounds

M02 This product contains an anticholinesterase carbamate compound. DO NOT USE if under medical advice NOT to work with such compounds

M03 If you feel unwell, seek medical advice immediately (show the label where possible)

M04 In case of accident or if you feel unwell, seek medical advice immediately (show the label where possible)

M05a If swallowed, seek medical advice immediately and show this container or label

M05b If swallowed, do not induce vomiting: seek medical advice immediately and show this container or label

M05c If swallowed induce vomiting if not already occurring and take patient to hospital immediately

M06 This product contains an anticholinesterase carbamoyl triazole compound. DO NOT USE if under medical advice NOT to work with such compounds

Appendix 5
Key to Abbreviations and Acronyms

The abbreviations of formulation types in the following list are used in Section 2 (Pesticide Profiles) and are derived from the *Catalogue of Pesticide Formulation Types and International Coding System* (CropLife International Technical Monograph 2, 4th edn, April 1999)

1 Formulation Types

AB	Grain bait
AE	Aerosol generator
AL	Other liquids to be applied undiluted
AP	Any other powder
BB	Block bait
BR	Briquette
CB	Bait concentrate
CF	Capsule suspension for seed treatment
CG	Encapsulated granule (controlled release)
CL	Contact liquid or gel (for direct application)
CP	Contact powder (for direct application)
CR	Crystals
CS	Capsule suspension
DC	Dispersible concentrate
DP	Dustable powder
DS	Powder for dry seed treatment
EC	Emulsifiable concentrate
EG	Emulsifiable granule
ES	Emulsion for seed treatment
EW	Oil in water emulsion
FG	Fine granules
FP	Smoke cartridge
FS	Flowable concentrate for seed treatment
FT	Smoke tablet
FU	Smoke generator
FW	Smoke pellets
GA	Gas
GB	Granular bait
GE	Gas-generating product
GG	Macrogranules
GL	Emulsifiable gel
GP	Flo-dust (for pneumatic application)
GR	Granules
GS	Grease
GW	Water soluble gel
HN	Hot fogging concentrate
KK	Combi-pack (solid/liquid)
KL	Combi-pack (liquid/liquid)
KN	Cold-fogging concentrate
KP	Combi-pack (solid/solid)
LA	Lacquer
LI	Liquid, unspecified
LS	Solution for seed treatment
ME	Microemulsion
MG	Microgranules
OL	Oil miscible liquid
PA	Paste
PC	Gel or paste concentrate
PS	Seed coated with a pesticide
PT	Pellet
RB	Ready-to-use bait

RH	Ready-to-use spray in hand-operated sprayer
SA	Sand
SC	Suspension concentrate (= flowable)
SE	Suspo-emulsion
SG	Water soluble granules
SL	Soluble concentrate
SP	Water soluble powder
SS	Water soluble powder for seed treatment
ST	Water soluble tablet
SU	Ultra low-volume suspension
TB	Tablets
TC	Technical material
TP	Tracking powder
UL	Ultra low-volume liquid
VP	Vapour releasing product
WB	Water soluble bags
WG	Water dispersible granules
WP	Wettable powder
WS	Water dispersible powder for slurry treatment of seed
WT	Water dispersible tablet
XX	Other formulations
ZZ	Not Applicable

2 Other Abbreviations and Acronyms

ACP	Advisory Committee on Pesticides
ACTS	Advisory Committee on Toxic Substances
ADAS	Agricultural Development and Advisory Service
a.i.	active ingredient
AIC	Agriculture Industries Confederation
CD	Acontrolled droplet application
CPA	Crop Protection Association
cm	centimetre(s)
COPR	Control of Pesticides Regulations 1986
COSHH	Control of Substances Hazardous to Health Regulations
d	day(s)
Defra	Department for Environment Food and Rural Affairs
EA	Environment Agency
EBDC	ethylene-bis-dithiocarbamate fungicide
FEPA	Food and Environment Protection Act 1985
g	gram(s)
GCPF	Global Crop Protection Federation (now CropLife International)
GS	growth stage (unless in formulation column)
h	hour(s)
ha	hectare(s)
HBN	hydroxybenzonitrile herbicide
HI	harvest interval
HSE	Health and Safety Executive
ICM	integrated crop management
IPM	integrated pest management
kg	kilogram(s)
l	litre(s)
LERAP	Local Environmental Risk Assessments for Pesticides
m	metre(s)
MAFF	Ministry of Agriculture, Fisheries and Food (now Defra)
MBC	methyl benzimidazole carbamate fungicide
MEL	maximum exposure limit
min	minute(s)
mm	millimetre(s)
MRL	maximum residue level
mth	month(s)
NA	Notice of Approval
NFU	National Farmers' Union

OES	Occupational Exposure Standard
OLA	off-label approval
PPE	personal protective equipment
PPPR	Plant Protection Products Regulations
PSD	Pesticides Safety Directorate
SOLA	specific off-label approval
ULV	ultra-low volume
VI	Voluntary Initiative
w/v	weight/volume
w/w	weight/weight
wk	week(s)
yr	year(s)

Appendix 6
Definitions

The descriptions used in this *Guide* for the crops or situations in which products are approved for use are those used on the approved product labels. These are now standardised in a Crop Hierarchy published by the Pesticides Safety Directorate in which definitions are given. To assist users of this *Guide* the definitions of some of the terminology where misunderstandings can occur are reproduced below.

Rotational grass: Short-term grass crops grown on land that is likely to be growing different crops in future years (*e.g. short-term intensively managed leys for one to three years that may include clover*)

Permanent grassland: Grazed areas that are intended to be permanent in nature (*e.g. permanent pasture and moorland that can be grazed*).

Ornamental Plant Production: All ornamental plants that are grown for sale or are produced for replanting into their final growing position (*e.g. flowers, house plants, nursery stock, bulbs grown in containers or in the ground*).

Managed Amenity Turf: Areas of frequently mown, intensively managed, turf that is not intended to flower and set seed. It includes areas that may be for intensive public use (*e.g. all types of sports turf*).

Amenity Grassland: Areas of semi-natural or planted grassland subject to minimal management. It includes areas that may be accessed by the public (*e.g. railway and motorway embankments, airfields, and grassland nature reserves*). These areas may be managed for their botanical interest, and the relevant authority should be contacted before using pesticides in such locations.

Amenity Vegetation: Areas of semi-natural or ornamental vegetation, including trees, or bare soil around ornamental plants, or soil intended for ornamental planting. It includes areas to which the public have access. It does NOT include hedgerows around arable fields.

Natural surfaces not intended to bear vegetation: Areas of soil or natural outcroppings of rock that are not intended to bear vegetation, including areas such as sterile strips around fields. It may include areas to which the public have access. It does not include the land between rows of crops.

Hard surfaces: Man-made impermeable surfaces that are not intended to bear vegetation (*e.g. pavements, tennis courts, industrial areas, railway ballast*).

Permeable surfaces overlying soil: Any man-made permeable surface (excluding railway ballast) such as gravel that overlies soil and is not intended to bear vegetation

Green Cover on Land Temporarily Removed from Production: Includes fields covered by natural regeneration or by a planted green cover crop that will not be harvested (*e.g. green cover on setaside*). It does NOT include industrial crops.

Forest Nursery: Areas where young trees are raised outside for subsequent forest planting.

Forest: Groups of trees being grown in their final positions. Covers all woodland grown for whatever objective, including commercial timber production, amenity and recreation, conservation and landscaping, ancient traditional coppice and farm forestry, and trees from natural regeneration, colonisation or coppicing. Also includes restocking of established woodlands and new planting on both improved and unimproved land.

Farm forestry: Groups of trees established on arable land or improved grassland including those planted for short rotation coppicing. It includes mature hedgerows around arable fields.

Indoors (for rodenticide use): Situations where the bait is placed within a building or other enclosed structure, and where the target is living or feeding predominantly within that building or structure.

Herbs: Reference to Herbs or Protected Herbs when used in Section 2 may include any or all of the following. The particular label or SOLA Notice will indicate which species are included in the approval.

Agastache spp.
Angelica
Applemint
Balm
Basil
Bay
Borage (except when grown for oilseed)
Camomile
Caraway
Catnip
Chervil
Clary
Clary sage
Coriander
Curry plant
Dill
Dragonhead
English chamomile
Fennel
Fenugreek
Feverfew
French lavender
Gingermint
Hyssop
Korean mint
Land cress
Lavandin
Lavender
Lemon balm
Lemon peppermint
Lemon thyme
Lemon verbena
Lovage
Marigold
Marjoram
Mint
Mother of thyme
Nasturtium
Nettle
Oregano
Origanum heracleoticum
Parsley root
Peppermint
Pineapplemint
Rocket
Rosemary
Rue
Sage
Salad burnet
Savory
Sorrel
Spearmint
Spike lavender
Tarragon
Thyme
Thymus camphoratus
Violet
Winter savory
Woodruff

Appendix 7
References

The information given in *The UK Pesticide Guide* provides some of the answers needed to assess health risks, including the hazard classification and the level of operator protection required. However, the Guide cannot provide all the details needed for a complete hazard assessment, which must be based on the product label itself and, where necessary, the Health and Safety Data Sheet and other official literature.

Detailed guidance on how to comply with the Regulations is available from several sources.

Pesticides: Code of Practice

Code of Practice for the Safe Use of Pesticides on Farms and Holdings, 1998 (ISBN 0 11 242892 4)

Known as the 'Green Code', this Defra publication (PB3528), which promotes the safe use of pesticides, covers the requirements of the Control of Pesticides Regulations 1986 (COPR) and the Control of Substances Hazardous to Health Regulations 2002 (COSHH). The principal source of information for making a COSHH assessment is the approved product label. In most cases the label provides all the necessary information but in certain circumstances other sources must be consulted, and these are listed in the Code of Practice. **Note that this Code was being revised, and amalgamated with the 'Orange Code' (which covers pesticide use in amenity and industrial areas), when this edition went to press. The new code is due for publication in 2005. Details can be found on the PSD website (www.pesticides.gov.uk).**

Other Codes of Practice

Code of Practice for Suppliers of Pesticides to Agriculture, Horticulture and Forestry (the 'Yellow Code') (Defra Booklet PB 3529)

Code of Good Agricultural Practice for the Protection of Soil (Defra Booklet PB 0617)

Code of Good Agricultural Practice for the Protection of Water (Defra Booklet PB 0587)

Code of Good Agricultural Practice for the Protection of Air (Defra Booklet PB 0618)

Control of Substances Hazardous to Health Regulations 2002. Approved Code of Practice and Guidance (ISBN 0 7176 2534 6)

Approved Code of Practice for the Control of Substances Hazardous to Health in Fumigation Operations. Health and Safety Commission (ISBN 0 7176 1195 7)

Other Guidance and Practical Advice

HSE (by mail order from HSE Books) (See Appendix 2)

Recommendations for Training Users of Non-Agricultural Pesticides. Health and Safety Commission (ISBN 0 11 885548 4)

Defra (from The Stationery Office – see Appendix 2)

Local Environment Risk Assessments for Pesticides – A Practical Guide (PB4168)

Crop Protection Association (see Appendix 2)

Every Drop Counts: Keeping Water Clean

For the Benefit of Biodiversity – A Biodiversity Strategy and Action Plan for a Better Environment

Best Practice Guides. A range of leaflets giving guidance on best practice when dealing with pesticides before, during and after application.

British Crop Production Council (see Appendix 2)

The Pesticide Manual (13th edition) (ISBN 1 90139 613 4)

The e-Pesticide Manual PC CD-ROM (Version 3.1) (ISBN 1 90139 636 3)

The Manual of Biocontrol Agents (3rd edition of *The BioPesticide Manual*) (ISBN 1 90139 635 5)

IdentiPest PC CD-ROM (ISBN 1 90139 605 3)

Garden Detective PC CD-ROM (ISBN 1 90139 632 0)

Hand-held and Amenity Sprayers Handbook (ISBN 1 90139 603 7)

Boom and Fruit Sprayers Handbook (ISBN 1 90139 602 9)

Using Pesticides: A Complete Guide to Safe Effective Spraying (ISBN 1 90139 601 0)

Safety Equipment Handbook (ISBN 1 90139 606 1)

The Environment Agency (see Appendix 2)

The Prevention of Pollution by Pesticides (Leaflet PPG9)

SECTION 7
INDEX

Index of Proprietary Names of Products

The references are to entry numbers, not to pages. Adjuvant names are referred to as 'Adj' and are listed separately in Section 4

50/50 Liquid Mosskiller . . . 161
9363 . . . 313
AAprotect . . . 433
Abacus . . . *Adj*
AC 650 . . . 331
Acanto . . . 352
Acanto Prima . . . 136
Aconite 50 . . . 16
Actellic D . . . 354
Actellic Smoke Generator No. 20 . . . 354
Activator 90 . . . *Adj*
Acumen . . . 28
Adagio . . . 88
Adjust . . . 74
Admire . . . 273
Admix-P . . . *Adj*
Aducksbackside . . . 378
Afalon . . . 295
After . . . 374
Agate . . . 357
Ag-Chem Metribuzin . . . 324
Ag-Chem Prefect . . . 391
Agral . . . *Adj*
Agri-50E . . . 3
Agrichem DB Plus . . . 149
Agrichem Flowable Thiram . . . 414
Agricola Lenacil FL . . . 292
Agricola Lens . . . 293
Agricola Lens Plus . . . 202
Agricorn 500 II . . . 304
Agricorn D II . . . 137
Agriguard Bentazone 480 . . 27
Agriguard Cerusite . . . 81
Agriguard Chlormequat 720 . . . 74
Agriguard Chlorothalonil . . . 85
Agriguard Cymoxanil Plus . . 129
Agriguard Diquat . . . 185
Agriguard Epoxiconazole . . . 193
Agriguard Ethofumesate 200 . . . 201
Agriguard Ethofumesate Flo . . . 201
Agriguard Fluroxypyr . . . 245
Agriguard Lenacil . . . 292
Agriguard Metazachlor . . . 316
Agriguard Metribuzin . . . 324
Agriguard Paraquat . . . 340
Agriguard Pencycuron . . . 342
Agriguard Pendimethalin . . . 343
Agriguard Pirimicarb . . . 353
Agriguard Propyzamide 50 WP . . . 367
Agriguard Propyzamide Flo . . . 367
AgriGuard Pro-Turf . . . 55
Agritox 50 . . . 304
Agritox Dry . . . 304
Agroxone . . . 304
Aliette 80 WG . . . 255
Aligran . . . 283
Alistell . . . 148
Allure . . . 313
Ally . . . 325
Ally Express . . . 67
Alpha Atrazine 50 SC . . . 16
Alpha Briotril 24/16 . . . 49
Alpha Briotril Plus 19/19 . . . 49
Alpha Bromolin 225 EC . . . 46
Alpha Bromotril P . . . 46
Alpha Captan 80 WDG . . . 54
Alpha Captan 83 WP . . . 54
Alpha Chlorpyrifos 48 EC . . 99
Alpha Chlortoluron 500 . . . 92
Alpha IPU 500 . . . 283
Alpha Isoproturon 500 . . . 283
Alpha Linuron 50 SC . . . 295
Alpha Metamitron . . . 314
Alpha Metazachlor 50 SC . . 316
Alpha Prochloraz 40 EC . . . 355
Alpha Prometryne 50 WP . . 359
Alpha Propachlor 50 SC . . . 360
Alpha Protugan Plus . . . 285
Alpha Simazine 50 SC . . . 381
Alpha Trifluralin 48 EC . . . 424
Alphaguard 100 EC . . . 5
Alphamouse . . . 4
Alphathrin . . . 5
Alto Elite . . . 86
Alto Xtra . . . 133
Amaize . . . 16
Amber . . . *Adj*
Amcide . . . 14
Amega Pro TMF . . . 262
Amenitywise Iprodione Green . . . 281
Aminotriazole Technical . . . 12
Amistar . . . 19
Amistar Opti . . . 20
Amistar Pro . . . 22
Anchor . . . 64
Antec Durakil 1.5 SC . . . 5
Antec Durakil 6SC . . . 5
Aphox . . . 353
Apollo 50 SC . . . 109
Appeal . . . 313
Applaud . . . 53
Apres . . . 374
Aramo . . . 401
Ardent . . . 178
Arelon 500 . . . 283
Arizona . . . 296
Arma . . . *Adj*
Artist . . . 239
Asana . . . 195
ASAP . . . 186
Ashlade Carbendazim Flowable . . . 55
Ashlade CP . . . 72
Asteroid . . . 262
Asulox . . . 15
Atlacide Soluble Powder . . . 382
Atladox HI . . . 144
Atlantis WG . . . 279
Atlas Brown . . . 98
Atlas Somon . . . 385
Atol . . . 92
Atrazol . . . 16
Attract . . . 313
Attribut . . . 366
Aurora 50 WG . . . 65
Autumn Kite . . . 286
Avadex Excel 15G . . . 418
Axit GR . . . 288
Aztec . . . 419
Azural . . . 262
Bacara . . . 175
BackRow . . . *Adj*
Ballad . . . 99
Bandrift Plus . . . *Adj*
Bandu . . . 150
Banka . . . *Adj*
Banlene Super . . . 158
Banshee . . . 82
Barbarian . . . 262
Barclay Actol . . . *Adj*
Barclay Banshee . . . 82
Barclay Barbarian . . . 262
Barclay Busker . . . 395
Barclay Clinger . . . *Adj*
Barclay Eyetak 40 . . . 355
Barclay Gallup 360 . . . 262
Barclay Gallup Amenity . . . 262
Barclay Gallup Biograde 360 . . . 262
Barclay Gallup Biograde Amenity . . . 262

Barclay Gallup Hi-Aktiv 262
Barclay Hurler 245
Barclay Mutiny 46
Barclay Seismic 314
Barclay Take 5 78
Barleyquat B 74
Barramundi . . . . *Adj*
Barrier H 106
BAS 493F 195
Basagran SG 27
Basamid 146
BASF 3C Chlormequat 720 74
BASF Dimethoate 40 179
Basilex 415
Bastion T 246
Batallion 403
Baytan Flowable 259
Baytan Secur 258
Beam 389
Beetup 347
Bellmac Plus 305
Bellmac Straight 307
Bema 343
Beret Gold 237
Beta 347
Betanal Carrera 153
Betanal Expert 152
Betanal Flow 347
Betapal Concentrate 330
Bettaquat B 74
Better DF 68
Better Flowable 68
Bettix 70 WG 314
Bettix Flo 314
Bio Syl . . . . *Adj*
Bioduo . . . . *Adj*
Biofilm . . . . *Adj*
BioPower . . . . *Adj*
Biotite 380 49
Biplay PX 327
Bison 83 WG 283
BL 500 94
Blaster 117
Blazer 343
Blue Rat Bait 45
B-Nine 145
Bombardier FL 85
Bond . . . . *Adj*
Bonzi 338
Boomerang 185
Booty 316
Booty 2 316
Boulevard 107
Boxer 232
Brasson 360
Bravado 46
Bravo 500 85
Briotril 19/19 49
Briotril 24/16 49
Brits 313
Broadsword 140
Bromolin 225 EC 46
Bromotril P 46
Brown 98
Budburst 192
Buggy SG 262
Bulldog 362
Bullet 123
Bulwark Flo 367
Bumper 250 EC 364
Bumper P 356
Burex 430 SC 68
Butisan S 316
Butoxone 307
Buy 423
Byo-Flex . . . . *Adj*
Cabaret 131
Caddy 240 EC 131
Calibre 411
Calypso 409
Camber 159
Campbell's MCPA 50 304
Capitan 40 249
Capricorn 56
Capture 47
Caramba 318
Carbetamex 59
carbon dioxide 61
Carna 74
Casoron G 160
Casoron G4 160
C-Cure . . . . *Adj*
CDA Vanquish Biactive 262
Cedar 149
Celect . . . . *Adj*
Centium 360 CS 110
Ceres Platinum . . . . *Adj*
Cerone 81
Cerround . . . . *Adj*
Certis 2-Aminobutane 10
Certis Spraying Oil 346
Challenge 261
Challenge 60 261
Champion 111
Charisma 209
Cheetah Super 219
Chekker 9
Chess WG 368
Chevron 48 99
Chiltern Blues 313
Chiltern Hundreds 313
Chinook Blue 33
Chinook Colourless 33
Chipco Green 281
Chloropicrin Fumigant 84
Chlortoluron 500 92
Chrome 96
Chryzoplus Grey 0.8% 276
Chryzopon Rose 0.1% 276
Chryzosan White 0.6% 276
Chryzotek Beige 276
Chryzotop Green 276
Cindy 415
Circium II 304
Citation 70 324
Clarosan 404
Claymore 343
Cleancrop GYR 362
Cleancrop Phenmedipham 347
Cleanrun 3 156
Clenecorn Super 308
Clinic 262
Clortosip 500 85
Clovotox 308
Codacide Oil . . . . *Adj*
Colstar 224
Comet 369
Companion PCT12 . . . . *Adj*
Compass 282
Compitox Plus 308
Comrade 94
Consento 213
Conserve 386
Consul 251
Contact Plus . . . . *Adj*
Contest 5
Contrast 57
Copal 500 92
CoPilot 375
Corbel 223
Corniche 169
Cosavet DF 392
Covershield 198
Crawler 59
Cropsafe 5C Chlormequat . . 78
Croptex Bronze 345
Croptex Chrome 96
Croptex Fungex 118
Croptex Pewter 95
Croptex Steel 385
Crossfire 480 99
Crotale 347
Crystal 240
Cultar 338
Cuprokylt 120
Cuprokylt FL 120
Curb Crop Spray Powder . . 6
Curzate M68 129
Cyclade 76
Cymoxeb 129
Cyperguard 100 EC 130
Cyperkill 5 130
Cyren 99
Daconil Turf 85
Dacthal W-75 101
Dagger 271
Dairy Fly Spray 370
Danadim 179
Dancer FL 347
Dart 5
Dazide 145
DB Straight 147
Debut 425
Decimate 102
Decis 150
Decis Protech 150
Decoy Wetex 319

REFERENCES ARE TO ENTRY NUMBERS NOT PAGES

Defensor FL 55
Degesch Fumigation
Tablets 7
Degesch Plates 297
Delsene 50 Flo 55
Depitox 137
Deputy 152
Designer . . . . *Adj*
Devrinol 331
Dextrone X 340
Dexuron 190
DG90 16
Diamant 196
Dicotox Extra 137
Di-Farmon R 159
Dimilin Flo 172
Dioweed 50 137
Dipel DF 23
Dithane 945 300
Dithane NT 300
Dithane NT Dry Flowable . . 300
Dithane Superflo 300
Dithianon Flowable 187
Diurex 50SC 188
Diuron 80 WP 188
Dixie 6 313
Dockmaster 159
Doff Horticultural Slug Killer
Blue Mini Pellets 313
Doff Sodium Chlorate
Weedkiller 382
Dormone 137
Dosaflo 322
Dovetail 291
Dow Agrosciences
Glyphosate 360 262
Dow Shield 111
Doxstar 247
DP 911 PX 327
DP 911 WSB 327
Drat Rat Bait 83
Drill . . . . *Adj*
Du Pont Adjuvant . . . . *Adj*
DUK 118 410
Duplosan KV 308
Durakil 1.5SC 5
Durakil 6SC 5
Dursban WG 99
Dynamec 1
Eagle 8
Eclipse 194
Electis 75 WG 303
Elliott's Lawn Sand 230
Elliott's Mosskiller 230
Elvaron Multi 416
Embargo G 160
Emblem 46
Encore 284
Endorats 83
Endorats Premium Rat
Killer 45
Enforcer 161
Ensign 225
Envision 262
Epic 193
Epok 236
Equity 99
Erysto 374
Escar-Go 6 313
Esterol . . . . *Adj*
Estuary 343
Ethos 366
Ethosat 500 201
Ethrel C 81
ethylene 206
Euro 199
Evict 400
Exit Wetex 319
Exodus 412
Extratect Flowable 269
Eyetak 40 355
F238 191
Falcon 362
Fargro Chlormequat 74
Fazor 299
Felix . . . . *Adj*
Fernpath Dart 5
Fernpath Graminite 340
Fernpath Haptol 314
Fernpath Haptol Flo 314
Fernpath Hatchet 245
Fernpath Ipex 177
Fernpath Lenzo Flo 292
Fernpath Pronto 186
Fernpath Torate 111
Field Marshal 158
Fieldgard 283
Fiesta T 73
Filan 42
Filex 361
Finale 261
Finish PX 326
Flagon 400 EC 46
Flame 262
Flamenco 243
Flash 361
Flex 252
Flexidor 125 287
Flexity 323
Flomide 367
Flute 85
FMC Oil . . . . *Adj*
Foil 244
Folicur 395
Folio Gold 89
Force ST 400
formaldehyde 254
Fortress 374
Fortrol 122
Fosetyl-AL 80 WG 255
Foundation 159
Fox 34
Freeway 188
Frigate . . . . *Adj*
Frond 15
Frupica 309
Frupica SC 309
FS Liquid Derris 379
Fubol Gold WG 301
Fumite Pirimiphos Methyl
Smoke 354
Fungaflor Smoke 267
Fungazil 100 SL 267
Fungex 118
Fungo 161
Fury 10 EW 431
Fusilade Max 234
Fusonil Turf 85
Fyfanon 440 298
Galahad 203
Galben M 25
Galera 116
Galion . . . . *Adj*
Gallup 360 262
Gallup Amenity 262
Gallup Biograde 360 262
Gallup Biograde Amenity . . 262
Gallup Hi-Aktiv 262
Gallup Hi-Aktiv (A) 262
Galmano 243
Galmano Plus 244
Galore 194
Garlon 2 422
Garlon 4 422
Gaucho 273
Genie 25 249
Gesagard 359
Gesaprim 16
Gesatop 381
Gex 1664 . . . . *Adj*
GF 184 233
Gharda Bonanza 316
Glenroe 353
Globe 129
Gly-Flex . . . . *Adj*
Glyfos 262
Glyfos Gold 262
Glyfos ProActive 262
Glyphogan 262
Glyphosate 360 262
Goldbeet 314
Golmet 97
Goltix 90 314
Goltix Flowable 314
Goltix WG 314
Graduate 175
Graminite 340
Gramoxone 100 340
Grasp 417
Grazon 90 117
Green Gold . . . . *Adj*
Greenaway Gly-490 262
Greencrop Amaize 16
Greencrop Astra . . . . *Adj*
Greencrop Boomerang 185
Greencrop Boulevard 107
Greencrop Budburst 192
Greencrop Carna 74
Greencrop Champion 111

REFERENCES ARE TO ENTRY NUMBERS NOT PAGES

Greencrop Dolmen . . . . . . Adj
Greencrop Doonbeg WG . . 279
Greencrop Estuary 343
Greencrop Frond 15
Greencrop Galore 194
Greencrop Glenroe 353
Greencrop Gweedore 417
Greencrop Gypsy 262
Greencrop Malin 394
Greencrop Monogram 316
Greencrop Monsoon 225
Greencrop Orchid 85
Greencrop Pomeroy 125
Greencrop Pontoon 99
Greencrop Reaper 245
Greencrop Reaper 2 245
Greencrop Saffron FL 367
Greencrop Satchmo 362
Greencrop Solanum 235
Greencrop Storeclean 225 100
Greencrop Tabloid 395
Greencrop Tassle 46
Greencrop Triathlon 158
Greencrop Twinstar 356
Greencrop Tycoon 75
Greencrop Valentia 125
Greengard 140
Greenmaster Autumn 230
Greenmaster Extra 306
Greenmaster Mosskiller . . . 230
Greenor 114
Gremlin 70
Grey Squirrel Liquid Concentrate 430
Grip . . . . Adj
Gro-Stop Fog 94
Gro-Stop HN 94
Guardsman STP Seed Dressing Powder 6
Guide . . . . Adj
Guilder 82
Guru 88
Gweedore 417
Gypsy 262
Habitat 262
Hallmark 290
Hallmark with Zeon Technology 290
Halo 87
Haptol 314
Haptol Flo 314
Hardy 313
Harlequin 500 SC 285
Harmony M 326
Harvest 261
Hatchet 245
Hawk 108
Headland Cedar 149
Headland Copper 120
Headland Fortune . . . . Adj
Headland Guard 2000 . . . . Adj
Headland Guard Pro . . . . . . Adj
Headland Inflo XL . . . . Adj
Headland Intake . . . . Adj
Headland Link 164
Headland Polo 142
Headland Relay P 158
Headland Relay Turf 158
Headland Rhino . . . . Adj
Headland Saxon 159
Headland Spear 304
Headland Staff 137
Headland Sulphur 392
Headland Tolerate 92
Headland Transfer 158
Headland Trinity 158
Helmsman 60
Herbasan Flow 347
Herboxone 137
Heritage 19
High Load Mircam 159
Hiker 233
Hilite 262
Hive 74
Holster 139
Hurler 245
Huron 319
Hussar 278
Hyban-P 159
Hycamba Plus 158
HY-D Super 137
Hygrass-P 159
Hykeep 407
HY-MCPA 304
Hymec Triple 167
Hyprone-P 158
Hyspray . . . . Adj
Hysward-P 158
Hy-TL 408
I T Alpha-Cyper 5
I T Asulam 15
I T Cyanazine 122
I T Cyper 130
I T Dicamba 154
I T Fosetyl-AL 255
I T Glyphosate 262
I T Iprodione 281
Ibex 199
Ice 240
Impact Excel 87
Impetus 305
Impuls 30
Indar 5EW 215
Ingot 176
Inorganic Liquid Copper . . . 120
Insegar WG 220
Insignia 369
Inter Asulam 15
Inter Pendimethalin 343
Intercept 5GR 273
Intercept 70WG 273
Inter-Metribuzin WG 324
Intracrop Bla-Tex . . . . Adj
Intracrop Neotex . . . . Adj
Intracrop Predict . . . . Adj
Intracrop Questor . . . . Adj
Intracrop Retainer . . . . Adj
Intracrop Rigger . . . . Adj
Intracrop Saturn . . . . Adj
Intracrop Status . . . . Adj
Intracrop Super Rapeze MSO . . . . Adj
Intrepid 2 156
Intro Plus 25
Invader 181
Inzacur 85
Ipex 177
Isomec 308
Isotron 500 283
Javelin 177
Javelin Gold 177
Jenton 226
Jester 51
Jockey 244
Jockey F 243
Joules 85
Juggler 405
Jundi 100 EC 130
K & S Chlorofume 84
K2 74
Karamate Dry Flo Newtec . . 300
Karan 319
Karate Ready to Use Rat Bait 83
Karate Ready to Use Rodenticide Sachets . . . 83
Karathane Liquid 184
Karmex 188
Kaspar 261
Katalyst . . . . Adj
Katamaran 317
Kerb 50 W 367
Kerb 80 EDF 367
Kerb Flo 367
Kerb Granules 367
Kerb Pro Flo 367
Kerb Pro Granules 367
Kernel 262
Killgerm Py-Kill W 406
Killgerm ULV 400 370
Killgerm ULV 500 349
Kinetic . . . . Adj
Kinto 358
Klartan 394
Klerat 44
Klerat Wax Blocks 44
Klipper . . . . Adj
KME 197
KN 540 262
Kubist Flo 201
Kumulus DF 392
Landgold Amidosulfuron . . . 8
Landgold Bentazone SL . . . 27
Landgold Clodinafop 107
Landgold Cycloxydim 125
Landgold Deltaland 150
Landgold Deputy 152
Landgold Diquat 185

Landgold Epoxiconazole . . . 193
Landgold Fenpropimorph 750 . . . 223
Landgold Fluazinam . . . 235
Landgold Fluroxypyr . . . 245
Landgold Glyphosate 360 . . 262
Landgold Mecoprop-P 600 . . . 308
Landgold Metamitron . . . 314
Landgold Metazachlor 50 . . 316
Landgold Metsulfuron . . . 325
Landgold Pendimethalin 400 . . . 343
Landgold Pirimicarb 50 . . . 353
Landgold PQF 100 . . . 362
Landgold Strobilurin 250 . . . 19
Landgold Sulfosulfuron . . . 391
Landgold Tepraloxydim . . . 401
Landgold Tralkoxydim . . . 417
Landgold Tribenuron 75 . . . 421
Landmark . . . 197
Laser . . . 125
Latitude . . . 380
Lawn Sand . . . 230
Lawn Sand Plus . . . 162
Leaf-Koat . . . *Adj*
Lens . . . 293
Lentagran WP . . . 371
Lentipur CL 500 . . . 92
Lenzo Flo . . . 292
Lexone 70DF . . . 324
Lexus Millenium . . . 242
Leyclene . . . 48
LI-700 . . . *Adj*
Liberator . . . 174
Libsorb . . . *Adj*
Lincon 50 . . . 295
Link . . . 164
Linuron 500 . . . 295
Liquid Curb Crop Spray . . . 6
Liquid Mosskiller . . . 161
Logic . . . *Adj*
Logic Oil . . . *Adj*
Loncid . . . 111
Lonpar . . . 112
Lorsban T . . . 99
Lorsban WG . . . 99
Lupus . . . 319
Luxan 9363 . . . 313
Luxan 9363 Red . . . 313
Luxan Deal . . . 313
Luxan Dichlobenil Granules . . . 160
Luxan Gro-Stop Fog . . . 94
Luxan Gro-Stop HN . . . 94
Luxan Metaldehyde . . . 313
Luxan Micro-Sulphur . . . 392
Luxan Non-Ionic Wetter . . . *Adj*
Luxan Talunex . . . 7
Luxan Trigger . . . 313
Lyric . . . 249
Magician . . . 90
Magnate 100 SL . . . 267
Magnum . . . 70
Majestik . . . 332
Malin . . . 394
Mandolin Flow . . . 347
Mandops Chlormequat 700 . . . 74
Mangard . . . *Adj*
Manifest . . . 262
Manipulator . . . 74
Mantra . . . 195
Manzate 75 WG . . . 300
Marathon . . . 107
Maraud . . . 99
Marnoch Chlorothalonil . . . 85
Marnoch Clodim . . . 125
Marnoch Phorm . . . 223
Marquise . . . 314
Marshal Soil Insecticide suSCon CR granules . . . 62
Marshal Soil Insecticide suSCon CR Sachets . . . 62
Masai . . . 398
Mascot Systemic . . . 55
Mastiff . . . 195
Mavrik . . . 394
Mazide 25 . . . 299
Mazide Selective . . . 157
Me2 Aducksbackside . . . 378
Me2 After . . . 374
Me2 Azoxystrobin . . . 19
Me2 Booty . . . 316
Me2 Booty 2 . . . 316
Me2 Buy . . . 423
Me2 Cindy . . . 415
Me2 Cymoxeb . . . 129
Me2 Exodus . . . 412
Me2 KME . . . 197
Me2 New Aldee . . . 2
Me2 Penny . . . 342
Me2 Puddy . . . 193
Me2 Succotash . . . 42
Me2 Sylvester . . . 177
Me2 Tebuconazole . . . 395
Me2 Terpitz . . . 82
Medley . . . 209
Menace 80 EDF . . . 367
Menara . . . 133
Meothrin . . . 221
Merlin . . . 90
Metarex Amba . . . 313
Metarex Green . . . 313
Metarex RG . . . 313
Meteor . . . 79
Mextrol Biox . . . 49
Micene 80 . . . 300
Micene DF . . . 300
Microgran 2 . . . 230
Midstream GSR . . . 160
Mildothane Turf Liquid . . . 413
Mimas WG . . . 279
Minuet EW . . . 431
Mirage 40 EC . . . 355
Mircam . . . 159
Mircam Plus . . . 158
Mirquat . . . 74
Mitac HF . . . 11
Mithras 80 EDF . . . 367
Mixture B . . . *Adj*
MM 70 . . . 314
MM 70 Flo . . . 314
Mocap 10G . . . 205
Moddus . . . 426
Molotov . . . 313
Monceren DS . . . 342
Monceren IM . . . 268
Monitor . . . 391
Monogram . . . 316
Monsoon . . . 225
Mossicide . . . 161
MSS CIPC 5 G . . . 94
MSS CIPC 50 M . . . 94
MSS Diuron 500 FL . . . 188
MSS Diuron 80 WP . . . 188
MSS Sprout Nip . . . 94
Mutiny . . . 46
Mycotal . . . 428
Nectec Paste . . . 18
Nemathorin 10G . . . 256
Nemolt . . . 399
Neosorexa . . . 170
Neosorexa Ratpacks . . . 170
Nettle . . . *Adj*
New 5C Cycocel . . . 78
New 5C Quintacel . . . 78
New Aldee . . . 2
New Draza . . . 319
New Estermone . . . 138
Newman Cropspray 11E . . . *Adj*
Newman's T-80 . . . *Adj*
Newtron . . . 69
Nico Soap . . . 334
Nicotine 40% Shreds . . . 334
Nimrod . . . 52
Nion . . . *Adj*
Nocweed . . . 158
No-Fid . . . 334
Nomix Diuron 80 . . . 188
Nomix Diuron Flowable . . . 188
Nomix Garlon 4 . . . 422
Nomix Mosskiller . . . 161
Nortron Flo . . . 201
Novagib . . . 260
Novall . . . 317
Nu Film P . . . *Adj*
Nufarm Adjuvant Oil . . . *Adj*
Nufarm Cropoil . . . *Adj*
Nufarm Cropoil Gold . . . *Adj*
Nufosate . . . 262
Nu-Shot . . . 140
Nuturf Carbendazim . . . 55
Oberon . . . 387
Octave . . . 355
Oncol 10G . . . 26
Opera . . . 199
Opogard . . . 403
Opponent . . . 198

Optica 308
Option 126
Opus 193
Opus Team 194
Orchid 85
Orka 227
Output . . . . *Adj*
Outrun 158
Oxytril CM 49
Pan Ethephon 81
Pan Magician 90
Pan Oasis . . . . *Adj*
Pan Panorama . . . . *Adj*
Panache 40 355
Panacide M 161
Panacide TS 161
Panoctine 263
Panoctine Plus 264
Panther 177
paraffin oil 339
paraformaldehyde 254
Pastor 115
Pasturol Plus 158
PDQ 186
Pearl Micro 150
Penncozeb WDG 300
Pennine 50 359
Penny 342
Permasect C 130
Pewter 95
Phantom 353
Phase II . . . . *Adj*
Phostoxin 7
Pico Stomp 344
PicoMax 344
Picona 344
PicoPro 344
Pierce 158
Pirlid 111
Planet . . . . *Adj*
Plantvax 75 337
Platform S 66
Plenum WG 368
Plinth 343
Plover 171
Pluton 224
Podquat 80
Pointer 251
Polo 142
Pomeroy 125
Poncho Beta 32
Pontoon 99
Poraz 355
Posse 10G 62
Powertwin 204
PP Captan 80-WG 54
Precis 367
Prefect 391
Prelude 20LF 355
Premiere 288
Premis 265
Prima . . . . *Adj*
Primer 283
Priori Xtra 21
Profit Oil . . . . *Adj*
Prompt 159
Pronto 186
Proplant 361
Propose 50 W 367
Prospect 410
Prospero 355
Prostore 157 UL 36
Prostore 420 EC 36
Protugan 283
Protugan 80 WDG 283
Pro-Turf 55
Puddy 193
Pulsar 29
Punch C 57
Pursuit 8
Py-Kill W 406
Pyramin DF 68
Pyrinex 48EC 99
Python 324
Quantum 75 DF 421
Quaver Flo 367
Quell Flo 300
QuikSilver . . . . *Adj*
Radius 132
Radspor FL 192
Ramrod Flowable 360
Ranger . . . . *Adj*
Ranman Twinpack 124
Rapide . . . . *Adj*
Raptor 362
Ravine 263
Ravine Plus 264
Raxil S 397
Raxil Secur 274
Reaper 245
Reaper 2 245
Redlegor 149
Reflex T 253
Regel 313
Reglone 185
Regulox K 299
Relay P 158
Relay Turf 158
Reldan 22 100
Reliance 262
Renardine 72-2 40
Renovator 2 155
Repulse 85
Reward Oil . . . . *Adj*
Rhino 250
Rhizopon A Powder 275
Rhizopon A Tablets 275
Rhizopon AA Powder (0.5%) 276
Rhizopon AA Powder (1%) 276
Rhizopon AA Powder (2%) 276
Rhizopon AA Tablets 276
Rhizopon B Powder (0.1%) 329
Rhizopon B Powder (0.2%) 329
Rhizopon B Tablets 329
Rhythm 129
Rimidin 214
Ringer 55
Rivet 319
Rizolex 415
Rizolex Flowable 415
Robust 270
Rogor L40 179
Romany 262
Rombus 365
Ronilan FL 429
Ronstar 2G 335
Ronstar Liquid 335
Root-Out 14
Rouge 299
Roundup Amenity 262
Roundup Biactive 262
Roundup Gold 262
Roundup Max 262
Roundup Pro Biactive 262
Roundup Pro-Green 262
Roundup ProVide 262
Roundup Rail 262
Roundup Ultra ST 262
Rover 85
Rovral Flo 281
Rovral Green 281
Rovral Liquid FS 281
Rovral WP 281
Rowent 313
Roxam 75 WG 303
Royal MH 180 299
Rubigan 214
Runner 320
Ryda . . . . *Adj*
Safeguard 391
Saffron FL 367
Sage 389
Sahara 243
Sakarat Bromabait 45
Sakarat D (Whole Wheat) . . 170
Sakarat D Wax Bait 170
Sakarat Ready-to-Use (Cut Wheat Base) 430
Sakarat Ready-to-Use (Whole Wheat) 430
Sakarat X 430
Samson 333
Samurai 262
Sanction 25 249
Saracen . . . . *Adj*
SAS 90 . . . . *Adj*
Satchmo 362
Savona 210
Saxon 159
Scala 372
Sceptre 375
Scorpio 500 SC 85
Scorpion 262
Scotts Octave 355

Scout . . . . *Adj*
Sculptor 68
Seismic 314
Sencorex WG 324
Sentinel 2 360
Sequel 228
Seradix 1 276
Seradix 2 276
Seradix 3 276
Sewarin Extra 430
Sewarin P 430
Sewercide Cut Wheat Rat Bait 430
Sewercide Whole Wheat Rat Bait 430
Shield 111
Shirlan 235
SHL Granular Feed and Weed 166
SHL Granular Feed, Weed & Mosskiller 165
SHL Lawn Sand 230
SHL Lawn Sand Plus 162
SHL Turf Feed and Weed . . 166
SHL Turf Feed, Weed & Mosskiller 165
Shogun 362
Shortcut 426
Shotput 324
Sibutol 38
Sibutol Secur 39
Sigma PCT 74
Signal . . . . *Adj*
Signum 43
Siltex . . . . *Adj*
Silvacur 396
Silwet L-77 . . . . *Adj*
Sipcam C 50 126
Sipcam Echo 75 85
Sipcam Simazine Flowable 381
Sistan 51 315
Skater 314
SL 567A 312
Slippa . . . . *Adj*
SM99 . . . . *Adj*
SmartFresh 321
Snare 413
Snooker 282
sodium chloride 383
sodium hypochlorite 384
Solace 129
Solan 40 345
Solanum 235
Solar . . . . *Adj*
Solfa WG 392
Somon 385
Sonata 212
Sorex Fatal 103
Sorex Super Fly Spray 349
Sorex Warfarin 250 ppm Rat Bait 430
Sorex Warfarin 500 ppm Rat Bait 430
Sorex Warfarin Sewer Bait 430
Sorex Wasp Nest Destroyer 377
Sorexa CD Mouse Bait 104
Sorexa Checkatube 44
Sorexa Gel 170
Source II 299
Spannit 99
Spannit Granules 99
Spartan . . . . *Adj*
Spasor 262
Spasor Biactive 262
Spear 304
Spearhead 113
Speedway 2 186
Sphere 134
Spin Out 119
Spitfire 15
Spotlight 24 EC 65
Sprayfast . . . . *Adj*
Sprayfix . . . . *Adj*
Spraygard . . . . *Adj*
Spraying Oil 346
Spraymac . . . . *Adj*
Sprayprover . . . . *Adj*
Spread and Seal . . . . *Adj*
Squire 8
Stabilan 700 74
Stacato 262
Staff 137
Stalwart 334
Stamina . . . . *Adj*
Standon Azoxystrobin 19
Standon Bentazone S 27
Standon Chlorpyrifos 48 . . . 99
Standon Clodinafop 240 . . . 107
Standon Cycloxydim 125
Standon Diflufenican 625 . . 177
Standon Diflufenican-IPU . . 177
Standon Diquat SL 185
Standon Epoxiconazole 193
Standon Epoxifen 194
Standon Etridiazole 35 207
Standon Fenpropimorph 750 223
Standon FFA60 Plus 240
Standon Fluroxypyr 245
Standon Fluroxypyr 200 . . . 245
Standon Fosetyl-AL 80 WG 255
Standon Imazaquin 5C 79
Standon Iprodione 50 WP . . 281
Standon Kresoxim FM 225
Standon Kresoxim Super . . 195
Standon Kresoxim-Epoxiconazole 197
Standon Mepiquat Plus . . . 82
Standon Metazachlor 50 . . . 316
Standon Metazachlor 500 . . 316
Standon Metazachlor-Q . . . 317
Standon Mimas WG 279
Standon Pencycuron DP . . . 342
Standon Pirimicarb 50 353
Standon Propaquizafop 362
Standon Propyzamide 400 SC 367
Standon Propyzamide 50 . . 367
Standon Rimsulfuron 378
Standon Tebuconazole 395
Standon Tepraloxydim 401
Standon Tralkoxydim 417
Standon Vinclozolin 429
Starane 2 245
Starane Gold 233
Starane XL 233
Starion 35
Stealth 290
Steel 385
Stika . . . . *Adj*
Sting ECO 262
Stirrup 262
Stomp 400 SC 343
Storeclean 225 100
Storite Clear Liquid 407
Storite Excel 407
Storite Flowable 407
Strate 75
Stroby WG 289
Stronghold 77
strychnine 390
Succotash 42
Sulphur Flowable 392
sulphuric acid 393
Sultan 50 SC 316
Sumi-Alpha 200
Sunorg Pro 318
Super Fly Spray 349
Super Mosstox 161
Super Nova . . . . *Adj*
Supertox 30 143
suSCon Green Soil Insecticide 99
Sward . . . . *Adj*
Swift SC 423
Swing Gold 183
Swipe P 50
Sydex 143
Syford 137
Sylvester 177
Synergol 277
Sypex 75
Systhane 20EW 328
Systhane 6 W 328
T2 Green 158
Tabloid 395
Tachigaren 70 WP 266
Tairel 25
Take 5 78
Takron 68
Talat 218
Talon Rat & Mouse Bait (Cut Wheat) 44

Talon Rat & Mouse Bait (Whole Wheat) 44
Talstar 35
Tandril 48 424
Tangent 262
Tanos 127
Target SC 314
Tarot 378
Tasker 75 304
Tassle 46
Tattoo 302
Taylors Lawn Sand 230
Teldor 217
Telone II 163
Temik 10G 2
Tern 222
Terpal 82
Terpitz 82
Terrazole 35 WP 207
Thianosan DG 414
Thiovit Jet 392
Thiraflo 414
Thistlex 117
Thripstick 150
Thunder 204
Thyram Plus 414
Tigress Ultra 169
Timbrel 422
Titus 378
Toil . . . . *Adj*
Tolugan 700 92
Tolugan Extra 93
Tolurex 90 WDG 92
Tomahawk 245
Tomcat 2 45
Tomcat 2 Blox 45
Topas 341
Topik 107
Toppel 10 130
Torate 111
Torch 388
Tordon 101 144
Tordon 22K 350
Torero 203
Torpedo . . . . *Adj*
Torpedo II . . . . *Adj*
Torq 216
Totril 280
Touchdown 262
Touchdown Quattro 262
Touché 189
Transcend . . . . *Adj*
Treflan 424
Triathlon 158
Tribute 158
Tribute Plus 158
Triflurex 48 EC 424
Trifolex-Tra 305
Trik 13
Trimaran 424
Tripart Beta 347
Tripart Culmus 92
Tripart Defensor FL 55
Tripart Gladiator 2 68
Tripart Sentinel 2 360
Trireme 158
Tritox 158
Triumph 219
Trooper 240
Tropotox 307
Tropotox Plus 305
Trump 284
Trustee Elite 262
Try-Flex . . . . *Adj*
Tuberon 324
Tucana 369
Tumbleweed Pro-Active . . . 262
Turbair Super Flydown 370
Turf Tonic 230
Turfclear 55
Twin 204
Twinstar 356
Twist 423
Tycoon 75
Typhoon 360 262
Unicrop Flowable Diuron . . 188
Unicrop Thianosan DG 414
Unikat 75 WG 303
Unix 135
Upgrade 75
UPL Camppex 141
UPL Grassland Herbicide . . 158
UPL Linuron 45% Flowable 295
Uranus 296
Valentia 125
Validate . . . . *Adj*
Venus 392
Venzar Flowable 292
Vertalec 428
Veto F 396
Vi-Nil 231
Vitax Microgran 2 230
Vitax Turf Tonic 230
Vitesse 58
Vivid 369
Volcan 314
Volcan Combi 71
Voodoo 91
Vydate 10G 336
Wakil XL 128
Warefog 25 94
Warfarin 0.5% Concentrate 430
Warfarin Ready Mixed Bait 430
Wasp Nest Destroyer 377
Waterloo 185
Weedazol-TL 12
Wetcol 3 41
Wetta Plus . . . . *Adj*
Xanadu 189
XL-All Insecticide 334
XL-All Nicotine 95% 334
Zenon 388
ZP Rodent Pellets 432

THE UK PESTICIDE GUIDE 2005

RE-ORDERS

☐ Please send me ____ more copies of *The UK Pesticide Guide 2005* at £35 each

☐ Please send me —— copies of *The e-UK Pesticide Guide 2005* (PC CD-ROM, single-user version) at £49.95 + VAT (£58.69 incl. VAT)

Name ______________________ Position ______________________

Institution ______________________ Department ______________________

Address ______________________

City ______________ Region ______________ Postcode ______________

Country ______________

Tel ______________ Fax ______________ E-mail ______________

EU countries except UK – VAT No: ______________________

Pre-payment is required. **Note: add £8.00 for postage outside the UK**

☐ I enclose a cheque/draft for £________ payable to BCPE Ltd. Please send me a receipt.

☐ I wish to pay by credit card: ☐ Visa ☐ Mastercard ☐ Amex ☐ Switch

Please charge to my card £________ and send me a receipt. Name of issuing bank ______________

Card no. ☐☐☐☐ ☐☐☐☐ ☐☐☐☐ ☐☐☐☐

Expiry date ☐☐/☐☐ Security code ☐☐/☐☐ Switch cards only: Start date ☐☐/☐☐ Issue no. ☐

Signature ______________________ Date ______________

Name and address of cardholder if different from above: ______________________

____

BCPC

Please photocopy and return to:
BCPC Publications Sales, 7 Omni Business Centre, Omega Park, Alton, Hants GU34 2QD, UK
Tel: 01420 593 200, Fax: 01420 593 209, Email: publications@bcpc.org, Web: www.bcpc.org

FUTURE EDITIONS

☐ I wish to take out an annual order for ____ copies of each new edition of *The UK Pesticide Guide*

☐ Please send me advance price details for the 2006 edition of *The UK Pesticide Guide* when available

Order by phone: 01420 593 200 or online: www.bcpc.org/bookshop

Bookshop orders to:
Customer Services, CABI Publishing, CAB International,
Nosworthy Way, Wallingford, Oxfordshire OX10 8DE, UK
Tel: 01491 832111, Fax: 01491 829292,
Email: publishing@cabi.org, Web: www.cabi-publishing.org

Bulk discount:

100+ copies	30%
50–99	25%
10–49	15%
List price £35	

Thank you for your order